D1582306

APPLIED OPTIMIZATION
Formulation and Algorithms for Engineering Systems

The starting point in the formulation of any numerical problem is to take an intuitive idea about the problem in question and to translate it into precise mathematical language. This book provides step-by-step descriptions of how to formulate numerical problems so that they can be solved by existing software. It examines various types of numerical problems and develops techniques for solving them. A number of engineering case studies are used to illustrate in detail the formulation process. The case studies motivate the development of efficient algorithms that involve, in some cases, transformation of the problem from its initial formulation into a more tractable form.

Five general problem classes are considered: linear systems of equations, non-linear systems of equations, unconstrained optimization, equality-constrained optimization, and inequality-constrained optimization.

The book contains many worked examples and homework exercises and is suitable for students of engineering or operations research taking courses in optimization. Supplementary material including solutions, lecture slides, and appendices, are available online at www.cambridge.org/9780521855648.

ROSS BALDICK is a professor of electrical and computer engineering at The University of Texas at Austin. His current research involves optimization and economic theory applied to electric power system operations, and the public policy and technical issues associated with electric transmission under deregulation. He is an editor of *IEEE Transactions on Power Systems*.

620.001
1
BAL

APPLIED OPTIMIZATION

Formulation and Algorithms for Engineering Systems

ROSS BALDICK

Department of Electrical and Computer Engineering
The University of Texas at Austin

CAMBRIDGE
UNIVERSITY PRESS

10S2959
OS-3131
ELEC

CAMBRIDGE UNIVERSITY PRESS
Cambridge, New York, Melbourne, Madrid, Cape Town, Singapore, São Paulo

Cambridge University Press
The Edinburgh Building, Cambridge CB2 2RU, UK

Published in the United States of America by Cambridge University Press, New York

www.cambridge.org
Information on this title: www.cambridge.org/9780521855648

© Cambridge University Press 2006

This publication is in copyright. Subject to statutory exception
and to the provisions of relevant collective licensing agreements,
no reproduction of any part may take place without
the written permission of Cambridge University Press.

First published 2006

Printed in the United Kingdom at the University Press, Cambridge

A catalog record for this publication is available from the British Library

ISBN-13   978-0-521-85564-8 hardback
ISBN-10   0-521-85564-0 hardback

Cambridge University Press has no responsibility for the persistence or accuracy of URLs for
external or third-party internet websites referred to in this publication, and does not guarantee that
any content on such websites is, or will remain, accurate or appropriate.

To Ann

Contents

Contents

Appendices (downloadable from www.cambridge.org)

List of illustrations

Preface

There are many excellent books on optimization and it is important to justify the need for yet another one. The motivation for this book stems from my observations of the orientation of typical optimization texts used in optimization courses compared to the needs of students in our engineering program. Many optimization books and courses concentrate on the design of algorithms, with less attention to adapting a problem to make it amenable to solution by an existing algorithm. That is, many optimization books are about how to *design* optimization algorithms, about how to *write* optimization software, or about how to *apply* optimization software to an existing problem formulation.

While this book is about the solution of simultaneous equations and optimization problems, it is not primarily about how to design algorithms, write software, or solve existing problems. Instead, it is about how to *formulate* new problems so that they can be solved by existing software.

Given the fabulous panoply of well-written optimization software available today, the skill of formulating a problem so that it is solvable with standard software is the most widely applicable skill for most engineers in our graduate program. Increasingly, the "scarce resource" is the ability to formulate problems, rather than the hardware or optimization software itself. This book is primarily designed for people who have a simultaneous equations problem or an optimization problem in mind and who want to formulate it so that it is solvable with standard software.

Educators in various disciplines have recognized the need to focus on problem formulation. Business schools have been the most agile in shifting the emphasis of optimization and operations research courses from algorithms towards applications. This book is an attempt to make a similar shift that concentrates on engineering applications to illustrate the process of formulating engineering optimization problems and then solving them with existing general-purpose software. The book also discusses simultaneous equations problems, since many signal processing and other engineering applications involve the solution of equations.

This book began as the notes for a one semester elective course that I teach at The University of Texas at Austin. The course is taken by many of our incoming graduate students and is also accessible to advanced undergraduates. A typical class has approximately thirty to fifty students with a background in electrical or other engineering.

The slides that I use in teaching this course are available for downloading from www.cambridge.org/9780521855648. Over two hundred pages of worked solutions to selected exercises are available to instructors. The book can also be used as a reference for engineers who want to formulate an optimization problem and apply optimization software to it.

The level of development may appear somewhat formal to a typical first year engineering graduate student and certainly requires a mathematical background. *Calculus and Analytic Geometry* by Thomas and Finney [114], together with the material in the downloadable Appendix A of mathematical preliminaries, provide such background. Nevertheless, the definitions and proofs have been deliberately spelled out to make them accessible and provide motivation, even at the expense of stating the "obvious," particularly in the early chapters. In the interest of clarity and of elementary development, many of the theorems are not stated in their sharpest form, with more general versions cited in the references.

Engineering students often have considerable past experience with tools such as MATLAB [74], but little experience in using optimization packages. Use of the MATLAB Optimization Toolbox [17] in this book builds on typical student experience and avoids the considerable "start-up costs" of introducing students to a completely new software package.

I have benefited from extensive feedback from students at The University of Texas at Austin who have taken this course. Students and graders who have been particularly helpful with making suggestions, correcting mistakes, and helping me to improve the presentation include: Seung Jun Baek, Seyeong Choi, Jerome Froment-Curtil, Philippe Girolami, Sergey Gorinsky, Hyun-moo Kim, Aditya Lele, Caleb Lo, and Lin Xu. Of course, remaining errors are my own responsibility. If you come across any errors or have any comments about the material, please send me email, baldick@mail.utexas.edu, and I will endeavor to improve the presentation for the next edition. I would also like to hear about descriptions of novel problems that might fit into this framework.

In teaching this material, one of the most delightful comments I have received is "I finally understand the need for the formal development." If you are unconvinced at the beginning, I hope that by the end of the book you are a convert to precise mathematical descriptions.

Ross Baldick, Austin, Texas, November 2005.

1

Introduction

In this book, we are going to examine various types of numerical problems and develop techniques for solving them. We will gradually build up the necessary tools and use a number of case studies to:

(i) illustrate the process of **formulating** a problem, by which we mean translating from an intuitive idea of the problem by writing it down mathematically,

(ii) motivate and develop **algorithms** to solve problems, that is, descriptions of operations that can be implemented in software to take a problem specification and return a solution, and

(iii) illustrate how to match the formulation of the problem to the capabilities of available algorithms, involving, in some cases, **transformation** of the problem from its initial formulation.

We illustrate how to think about and describe problems to make them amenable to solution by optimization software. In brief: formulation of problems to facilitate their solution.

In the five parts of this book, we will consider five general problem classes. The first problem class is the solution of **linear systems of equations** (Part I). Solving linear systems of equations will turn out to be at the center of much of our subsequent work. For example, our second problem class, solving **non-linear systems of equations** (Part II), will be solved by an algorithm that requires repeated solution of linear systems of equations.

We will then go on to use these two algorithms as building blocks in algorithms for:

- **unconstrained optimization** (Part III), which will in turn form part of the algorithms for:

1

- **equality-constrained optimization** (Part IV), which will again form part of the algorithms for:
- **inequality-constrained optimization** (Part V).

These problem classes will be defined explicitly in Chapter 2.

In the rest of this chapter, we discuss a number of issues in preparation for the rest of the book. In Section 1.1 goals are discussed. Several course plans to achieve these goals are outlined in Section 1.2. In Section 1.3 we discuss model formulation and development. An overview of the organization of the book is then presented in Section 1.4, with pre-requisites described in Section 1.5.

1.1 Goals

The most important purpose of this book is to give you facility in taking your own problem and thinking about it in a way that allows you to:

(i) write down equations describing it, and
(ii) use optimization software to solve it.

Although we will outline the major ingredients of algorithms and write our own experimental software in order to understand some of the issues, we will leave the detailed development of production software to specialists. Furthermore, we will omit many details involved with practical implementation, most notably many of the effects of calculating using finite precision arithmetic. We will provide references to some of the details and if you are planning to develop production software you will need to consult these and many other references that are more oriented towards algorithm development. Outstanding books in this area include [6, 15, 45, 70, 84].

You can make use of general purpose tools such as GAMS [18], AMPL [38], Xpress-Mosel [49], LINDO [68], MATLAB [74], and Excel [75], and make use of the NEOS website (http://www-neos.mcs.anl.gov/) to solve small-scale to medium-scale problems. This book will provide the skills to formulate your own problems and you should then be able to use these or other software packages to solve at least small- to medium-scale versions of your own problems. There are also callable libraries such as CPLEX, IMSL, LAPACK, NAG, and Xpress-BCL that can be linked to user software for larger-scale problems [81].

Production software, such as is used in these tools, can be expected to differ from the algorithms we develop, at least in the details. The most extreme example of this is our concentration on a particular algorithmic approach called "interior point" algorithms for inequality-constrained optimization. In contrast, most currently available commercial software uses classical "active set" algorithms to solve these types of problems. Nevertheless, many of the fundamental considerations apply to all algorithms for a particular problem class. Our development will therefore help you to build intuition that is generally applicable.

Our reason for analyzing the algorithms is therefore not so much to be able to write software, but instead to understand under what circumstances we can expect numerical software to work and when we should be suspicious of results. Getting a useful answer in a reasonable time depends on understanding:

   (i) whether a problem is solvable or not solvable with a particular algorithm,

  (ii) how to formulate a problem in a way that *is* amenable to solution, and

 (iii) what computational resources are necessary to get a solution.

At the end of the book, you should be able to:

   (i) take a description of your problem,

  (ii) translate it into a mathematical formulation in a way that is most amenable to solution, (and intelligible to others in your field),

 (iii) evaluate if optimization techniques will be successful, (and estimate how much more computational resources a "large" problem will take compared to a "small" test problem),

 (iv) solve small- to medium-scale versions of the problem using commercial software, such as the MATLAB Optimization Toolbox or the software listed in [81], and

  (v) use the solution of the problem to calculate sensitivities to changes in problem specifications.

For many problems, such an "off-the-shelf" approach will be adequate for obtaining useful results. The fantastic performance of today's software and hardware compared to capabilities of even just a few years ago means that off-the-shelf software can be used to solve relatively large and sophisticated problems, so long as the formulation has been performed with care.

Even if your ultimate goal is the solution of a "large-scale" problem it is worthwhile to start with the formulation of a smaller or simpler problem that can be solved with standard software on a personal computer or workstation. Such a small start can reveal modeling flaws, identify the most important issues for more detailed treatment, and help point towards appropriate directions for larger-scale development.

In some cases, your problem may be so large or there will be application-specific issues that will necessitate some tailoring of the algorithm to your purposes. This book will give you the tools to understand the directions needed for the tailoring and includes some references to the literature. You should also expect to read more widely and deeply before developing your own optimization software for a "serious" problem. By the end of the book, you should have the skills to pursue such literature further.

However, before attempting to develop software for a difficult problem, you should "try out" several commercial software packages. GAMS [18] is a generic interface to a number of different optimization packages and can be very useful for finding the most suitable algorithm and implementation for a particular problem. The use of GAMS to solve various problems is described in [22].

The MATLAB Optimization Toolbox [17] also provides several optimization algorithms. If you already use MATLAB in your work then this book will provide you with the skills to interface your formulation to the Optimization Toolbox.

1.2 Course plans

In a one semester course that includes all five problem classes, it is generally possible to cover only a subset of the case studies and only sketch the non-linearly constrained versions of equality- and inequality-constrained optimization. Chapter 3, on problem transformations, can be initially assigned as reading and referenced as needed throughout the course for details of particular transformations.

Selection of a subset of the case studies allows some tailoring of the case studies to the audience; however, it is important to also expose students to case studies outside their immediate field of interest to cross-fertilize from field to field. Instructors could add case studies from other fields to supplement or replace the case studies in this book.

A two semester course could comfortably cover all problem classes and all the case studies and include both linearly and non-linearly constrained optimization. Alternatively, omitting the solution of linear and non-linear equations would allow for all the optimization material to be covered in one semester, given that students had already been exposed to the solution of linear and non-linear equations.

1.3 Model formulation and development

As mentioned in Section 1.1, the principal purpose of this book is in describing the process of model formulation in the context of the capability of algorithms. It is typical that a model of a problem that is faithful to our intuitive notion of the problem will yield a formulation that exceeds the capabilities of available algorithms. We must then balance computational tractability against fidelity to salient effects that we are trying to model. We may not be able to exactly solve the precise problem we have in mind; however, an optimization formulation can often improve on *ad hoc* designs.

A guiding principle in model development is **Occam's razor** [115]: "keep it simple." More precisely, the model should be no more complicated than is necessary to represent the important issues. A thorough understanding of the problem together with "engineering judgment" are central to identifying the important issues that must be represented and identifying the less central issues that can be neglected. Often, this process involves incremental model development as we revise our opinion of what can be neglected from the model.

The context of our treatment of problem formulation is as follows. We imagine a historical progression from a simplified analytic model of an engineering system to a more accurate, perhaps numerical, model. The analytic model may have been used to roughly evaluate systems, and even to guide design choices somewhat, but eventually more accurate numerical models became necessary to validate designs that involve more variables with more complicated interactions. Furthermore, as performance goals were raised, the need arose to systematically improve designs. With the analytic model no longer accurate enough or too complicated to analyze directly, we must embed the numerical model into an optimization framework [79, chapter 7].

We will see that *qualitative* understanding from an approximate analytic model combined with more accurate *computational* evaluations from a numerical model can be a powerful aid in formulating and solving such optimization problems. The qualitative understanding provided by the analytic model can help us to formulate the problem in a way that makes the computational issues tractable. The case study in Part V involving design of interconnects in integrated circuits particularly illustrates this interplay. (See Section 15.5.)

If an *accurate* analytic model exists, then qualitative analysis of the analytic model is directly applicable to the computations. Circuit theory case studies (in Sections 4.1, 6.1, and 6.2) illustrate this category of problems. These case studies also illustrate the development of progressively more complicated models.

However, accurate analytic models are not always convenient or tractable so that deliberate approximations can sometimes be judiciously applied to simplify the analysis. We will also illustrate this issue with some of the circuit theory case studies. In some applications, we do not even have a detailed analytic model of the underlying process and we must posit some generic model and identify the parameters of the model.

For some problems, it may also be difficult to obtain an appropriate analytic or numerical model to evaluate the performance criterion. Instead we may have to posit a proxy criterion to evaluate whether or not the decision is satisfactory and we may have to be content with avoiding bad decisions, rather than obtaining optimal decisions. There will typically also be some constraints that have to be satisfied. In this context, we say that we are seeking **satisficing** solutions, where the word satisficing is a contraction of "satisfying" and "sufficient" [109]. Three examples of such problems are:

- making routing decisions in a communications network, to be described in Section 15.2,
- developing a criterion to classify patterns into classes, to be described in Section 15.4, and
- choosing the widths of interconnects between latches and gates in integrated circuits, to be described in Section 15.5.

The progression in the book from formulation to algorithm to solution of the formulation should not be taken as the only way that applications and algorithms are developed: in practice, algorithm development suggests new possible applications and vice versa. The most dramatic example of the interplay between algorithm and application is the development of **linear programming**; that is, the optimization of **linear objectives** with respect to **affine constraints**. (We will define objectives and constraints formally in Chapter 2.) The first practical algorithm for this problem was developed by George Dantzig in the late 1940s. Linear programming has since become pervasive in many fields, opening up many applications that have in turn driven algorithm development.

Similarly, the algorithms we present have gone through many stages of refinement, iterated with reformulation of application problems. Our presentation glosses over many important details of algorithm design, but hopefully leaves intact most of the issues that are relevant to formulating problems so that they are susceptible to solution by optimization software. Of course, limited knowledge can be a dangerous thing!

In any new problem you encounter and try to formulate, you should expect that finding the best formulation will involve several iterations. At the outset it may not even be clear to you what sort of problem you have nor what is the best way to solve it. (As an example, one of the cases of solving linear systems of equations discussed in Section 5.8.1 will turn out to have a more natural interpretation and solution in the context of unconstrained optimization, discussed in Section 11.1.) The purpose of the case studies is to build up your expertise in problem formulation and in finding an algorithm to solve the problem.

For each case study, we will start with a physical description and move towards an algebraic description. We will emphasize the use of an explicit physical understanding of the system to develop each model rather than trying to fit a generic model to observed behavior. In other applications, a generic model may be necessary or desirable; however, in the context of explaining the process of problem formulation, applications with a well-defined physical model provide more concrete examples.

Some cases are more fleshed out than others, reflecting mostly my own personal understanding of the problem rather than any judgment as to the importance of the problem. Some of the case studies will be introduced with a specific small-scale example problem having only a few variables. Others will be introduced more generally.

You may already be very familiar with some of the models we describe in the case studies; however, you may not be so familiar with the *development* of the models. In working on your own application, the construction of a model will be of central importance. A little time spent understanding the modeling process for these familiar problems may prove valuable in your own application. As we describe each case, you should think about your own application and how to translate your understanding of it into a set of equations that models its behavior and its performance. The case studies that are drawn from outside of your field of interest may also provide some useful cross-fertilization of ideas.

You should not look on a case study as showing simply how to solve problems of that type, nor even showing how to formulate problems of that type. Instead, you should look at the case studies as examples of how generally to formulate problems. You should think about how to apply the ideas to your own problems.

Most of our applications are in electrical engineering, but the issues we present are applicable in a wide variety of areas. For formulation of problems in mechanical and structural engineering, see [117]. Although we touch on control theory problems, a much more detailed discussion of optimal control appears in, for example, [16]. Industrial control is discussed in [98]. A wide variety of problems is discussed in [6, chapter 1], [15, Part II], [22], [54], [84, chapter 1], and in [99, chapters 4 and 8].

1.4 Overview

In this section, we overview the organization of the book, describing notational and other conventions, functions and variables, case studies, algorithms, proofs, exercises, and a "road map" to the chapters.

1.4.1 Notational and other conventions

Notational conventions are introduced as needed but also collected together for easy reference in Appendix A, which is available for downloading from the website www.cambridge.org/9780521855648. Words with technical meanings are shown in **bold face** when they are first introduced and defined and, occasionally, when they are referred to again later in the book.

Many sections in this book are only one or two paragraphs long, with definitions and other sectional units divided-off so that they are easy to identify and locate and so that each idea can be read and digested. This has been done to:

- make it easy to locate specific topics, and
- make sure each topic is well "sign-posted" so that a reader always knows where the argument is going.

In each of the chapters describing algorithms, key issues are listed and summarized at the beginning of the chapter.

The order and development of topics and the terminology generally follows standard literature. However, there are several exceptions.

For example, the two concepts of "convexity" and "duality" are introduced before developing any algorithms. Convexity is introduced as a problem characteristic that helps with judging the quality of solutions from optimization algorithms. Duality is introduced as one of several ways to transform problems. The full significance of convexity and duality is detailed as the book progresses.

As examples of deviations from standard terminology:

- the "standard form" of linear programming is referred to by saying that a problem has "non-negativity constraints," and
- "quasi-convex" functions are referred to as having "convex level sets."

These deviations are made in the interest of simplifying the terminology to be more descriptive.

1.4.2 Functions and variables

The emphasis in this book will be on problems that are defined in terms of **smooth** functions of **continuous** variables, by which we mean functions that possess first and (usually) second derivatives across a continuous, as opposed to discrete, space of candidate solutions. The reasons for this concentration are three-fold.

(i) Although we will not cover the optimization of **non-smooth functions** in detail this book, optimization of smooth functions provides a basis for subsequent study of non-smooth functions [96, 100, 107]. It is worthwhile to begin, at least, with the simpler case of smooth functions. (Furthermore, we will consider some transformation techniques to deal with particular types of non-smooth functions. See Sections 3.1.3 and 3.1.4.)

(ii) For many engineering problems, we empirically estimate the functions involved using some combination of theory and imperfect observations. Even if the underlying functions have a complicated non-smooth structure, our measurement and estimation processes may not warrant a model with such detail [45, section 7.3]. The process of estimating such a function forms one of our case studies. (See Section 9.1.)

(iii) Design problems involving **discrete** variables can sometimes be approximately solved by "rounding-off" the solution to a related problem that neglects the discreteness of the variables [45, section 7.7]. (See the sizing of interconnects case study in Section 15.5 for an example.) In some cases, neglecting the discreteness yields a solution that happens to satisfy the discreteness condition automatically [84, chapter 8]. (See Section 16.3.6.1 for an example.) Even if rounding does not provide a good approximation to the answer, many problems involve a mixture of discrete and continuous variables, so that study of continuous optimization forms an important start for the overall problem.

However, you should always be cautious before assuming that a real world problem is as smooth and well-behaved as our analysis requires. You should be prepared to delve into the literature if your problem has features that are outside the scope of this book. For example, the journal *Optimization and Engineering* focuses on engineering applications of optimization.

Good starts for non-smooth problems are [100, 107]. Good starts for discrete problems are [22, 54, 67, 85, 92, 122]. We will make considerable use of a property called "convexity;" however, a variety of problems that do not have the property of convexity are covered in the series *Nonconvex Optimization and Its Applications* by Kluwer Academic Publishers. A good start for approximation algorithms for discrete problems is [52], while a good start for heuristics for discrete problems is [122, chapter 12].

For some problems, there may be multiple optima. References [87, 91] discuss the search for multiple optima. The *Journal of Global Optimization* focuses on such optimization problems.

1.4.3 Case studies

The most unique aspect of this book is the emphasis on case studies to illustrate the process of problem formulation. The problems are not simply introduced as a recitation of formulated problems. Instead, a relatively few case studies are motivated and formulated in detail to introduce each of the five problem classes. These fourteen case studies are chosen from a variety of sub-fields to enable readers to "get their bearings" for each problem class. The variety also enables readers to sample problems from other sub-fields and so foster cross-fertilization.

Many of the later case studies build on case studies that were treated in earlier chapters, demonstrating incremental model development. Figure 1.1 shows the

Problem class   Case studies

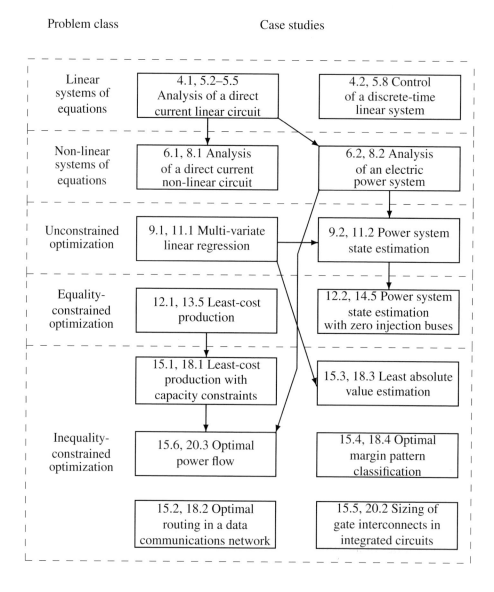

Fig. 1.1. Problem classes, case studies, and their dependencies. The dashed lines enclose the case studies for each problem class. Each case study is represented as a box. In each box, the first section number refers to where the case study is introduced. The subsequent section number or numbers refer to where the case study is solved. The arrows indicate dependencies between case studies.

dependency of the various case studies. The section number and title of each case study are shown in the figure. An arrow with its head pointing to a particular case study means that the case study at its tail is one of its pre-requisites. There are also case studies in most of the problem classes that do not require material from earlier chapters. For example, for the final problem class of inequality-constrained optimization, three out of the six case studies do not have any pre-requisite case studies.

1.4.4 Algorithms

Although our major purpose is the formulation of problems, it is not possible to formulate problems without some knowledge of how optimization software functions. It is not possible to completely eschew the details of optimization algorithms in the way that, for example, a manual describing word-processing software can eschew the details of typesetting. A book about optimization, particularly non-linear optimization, should discuss *both* formulation of problems and algorithms for their solution.

To this end, we will introduce key issues that have proved pivotal in algorithm development and problem formulation. Five of the most important are:

- **monotonicity** (introduced in Section 2.5.3),
- **convexity** (introduced in Section 2.6.3),
- **problem transformations** (introduced in Chapter 3),
- **symmetry** (introduced in Section 5.4), and
- **sparsity** (introduced in Section 5.5).

We will define these concepts carefully as they arise in the formulation of problems and in the development of algorithms and we will see how they can guide us in problem formulation to bridge the gap between:

- our intuitive notion of the problem to be solved, and
- the capabilities of algorithms,

to obtain a useful answer as quickly as possible.

We tease out the characteristics of our case studies, such as sparsity, to provide "hooks" to "attach" to an algorithm. That is, the formulation of problems is centered around identifying characteristics that will fit well with the operation of an optimization algorithm or software.

We will describe algorithms, but only generally, emphasizing the broad-brush issues for one or two particular example algorithms for each problem class, chosen because they:

- illustrate many of the general issues important to formulating and solving the problem, and

- provide a bridge to an algorithm for the next problem class.

The development of algorithms is therefore incremental. We owe a significant debt to the authors of [73] for this way of thinking about and ordering the development of these problems. We have also used a variety of other sources, particularly drawing from [11, 15, 22, 45, 54, 58, 59, 70, 84]. You are encouraged to delve into these books on optimization for further details and different perspectives.

Moreover, in moving from the solution of a linear to a non-linear problem, the linear case will be used as a base for developing the non-linear case. Results for problems with non-linearities will often be sketched as incremental extensions of the corresponding linear case.

We gloss over many of the details in the design of algorithms. This book will probably frustrate readers who have developed production optimization software and who know from first hand the importance of such details. Algorithm and software design requires intelligence, effort, and talent. But it is the very success of the developers of these algorithms and software that allows the rest of us to "take it easy."

For most problems, there are usually several algorithms available. However, we typically describe only one or two generic algorithms for each type of problem and try to focus on the issues that are commonly important for successfully solving each type of problem. Amongst software available today there are a host of detailed differences between algorithms that tailor them to particular problem characteristics. Successfully using optimization software in an application necessitates a careful matching of software capabilities to the problem. In practice, it is often crucial to utilize an algorithm that can take advantage of special problem structure. There are several very good guides that can aid in identifying suitable software, such as [81]. A brief introduction is given in [84, appendix C]. The reference manuals of optimization packages, such as [49, 74, 88, 120], also provide useful guidance.

1.4.5 Proofs

There are a number of theorems in the book. In long proofs it can be very easy to get bogged down in details and lose sight of the thread of the argument. Most of the longer proofs have been divided up into several parts to make them easier to follow, with a proof sketch in the body of the text listing the main ideas and the proof itself relegated to Appendix B, which is available for downloading from www.cambridge.org/9780521855648. In general, proofs are included in the body of the text only if they are particularly important for understanding why an algorithm works or why it must account for a particular issue. The emphasis is on issues that directly affect how one should formulate an optimization problem.

1.4.6 Exercises

There are a large number of exercises in the book. The exercises cover theory and solution of problems, both analytically and using software. Most of the case studies are solved in exercises.

Many of the exercises involving the use of software are designed explicitly around the MATLAB Optimization Toolbox. There are notes in the exercises about how to use the functions in the MATLAB Optimization Toolbox. From time to time, however, these functions are updated and some experimentation might be necessary if calling options change. If you use another software package instead of MATLAB then you may find that some of the numerical answers differ slightly from those produced by MATLAB.

The proofs of many of the earlier theorems are left to the exercises. It could be argued that many of these results, particularly in Chapter 3 on problem transformations, are rather too trivial to have been given the formality of a separate theorem. However, it is valuable to get started on easy proofs before introducing more sophisticated results. Part of the value of proving these results formally is in gaining familiarity with notation, encouraging rigorous thinking, and developing intuition about the ability of an algorithm to solve a particular type of problem.

1.4.7 Road map

In this section, we briefly summarize the organization of the chapters in the book. We will begin in Chapter 2 with a brief introduction to and a small example of each problem class. We then discuss broadly the nature of the algorithms we will develop and the quality of solutions. In Chapter 3, we introduce several ways to transform problems to change their characteristics, with a view to making them easier to solve. Chapters 2 and 3 serve only to *introduce* these and a variety of other concepts. We will try to suggest some of the reasons why these issues are important to problem formulation. However, their full significance will not be apparent until much later.

In subsequent chapters in each of the five parts of the book, we will revisit in detail the topics introduced in Chapters 2 and 3 in the context of the five problem classes. For each problem class, we will:

- **formulate** at least two case studies,
- develop an **algorithm** to solve problems in the problem class,

- develop **sensitivity analysis**,
- apply the algorithm to **solve** the corresponding small example problem from Chapter 2, and
- apply the algorithm to **solve** an instance of each case study.

Sometimes, solving the case study requires further transformation of the formulation of the case study using the transformations introduced in Chapter 3. For other case studies, the formulation fits well with the algorithm. As mentioned in Section 1.4.6, solution of most of the case studies is completed in the exercises.

1.5 Pre-requisites

In Appendix A, which is downloadable from www.cambridge.org/9780521855648, we present notational conventions and a number of results that we will use in the book. You should review Appendix A to ensure that you are familiar with all the topics. Many, but not all, of the results will be familiar to graduates of electrical engineering undergraduate degree programs. You should ensure that you understand these results before undertaking this book.

The book assumes familiarity with MATLAB and MATLAB M-files and that you have access to the MATLAB Optimization Toolbox.

Finally, we will also develop and prove a number of theoretical results. If you have not previously had experience with understanding, stating, and proving results rigorously, you should spend the time to prove several of the results quoted in Chapter 3 to prepare yourself for the more ambitious analysis in subsequent chapters.

2

Problems, algorithms, and solutions

In this chapter we will define the various types of problems that we will treat in the rest of the book and define various concepts that will help us to characterize the problems. In Section 2.1, we first define the notion of a **decision vector**. In Section 2.2 we define two problems involving **solution of simultaneous equations**. Then in Section 2.3 we describe three **optimization problems**.

For each problem, we will provide an elementary example, without any context, to illustrate the type of problem. The case studies in later chapters will provide more interesting problems and contexts. In this chapter we will concentrate on basic definitions without explicitly considering applications.

In later chapters, we will also develop algorithms to solve the problems, starting with the elementary example problems introduced here and then progressing to solution of the case studies. We will explicitly define what we mean by an **algorithm** in Section 2.4 in reference to two general schemata:

- **direct** algorithms, which, in principle, obtain the exact solution to the problem in a finite number of operations, and
- **iterative** algorithms, which generate a sequence of approximate solutions or "iterates" that, in principle, converge to the exact solution to the problem.

We will also consider some of the issues involved in ensuring that these algorithms provide useful solutions. In particular, in Section 2.5, we will discuss solutions of simultaneous equations problems, introducing the concepts of a **monotone function** and of **convex sets** and presenting conditions for uniqueness of solutions of simultaneous equations. In Section 2.6, we will discuss solutions of optimization problems, introducing the concept of **convex functions** and discussing the nature of **global** and **local** solutions to optimization problems and various other issues. In Section 2.7, we discuss **sensitivity analysis**; that is, the way in which solutions change with changes in the specification of a problem.

This chapter and the next are designed as overviews of the topics to be studied in detail in subsequent chapters. Several new concepts are introduced without much preparation. We will return to them in much greater detail as we progress in the book. The purpose of introducing them here is to motivate the discussion of problem formulation in the later chapters. You may want to skim this chapter and the next at first and then re-read them more carefully when the ideas are applied in later chapters.

2.1 Decision vector

All the problems that we consider will involve choices of the value of a **decision vector** from n-dimensional Euclidean space $\mathbb{R}^n$ [82, section 1-5] or from some subset $\mathbb{S}$ of $\mathbb{R}^n$, where:

- $\mathbb{R}$ is the set of real numbers, and
- $\mathbb{R}^n$ is the set of n-tuples of real numbers.

(We also call $\mathbb{R}^n$ the n-fold **Cartesian product** of $\mathbb{R}$ with itself [104, section 1.1]. See Definition A.4.)

We will usually denote the decision vector by x. Individual entries in the vector x are denoted by a subscript. There are more general types of decision vectors than the ones we consider, particularly in **optimal control** problems [16, 55, 63, 89], where x is considered to be an element of a more general set than $\mathbb{R}^n$.

Our goal will be to find a value of the decision vector x that satisfies some criterion. We will usually write $x^\star$ to denote a value of x that satisfies this criterion. If there are several values of x that satisfy the criterion then we sometimes distinguish them by writing $x^\star, x^{\star\star}, \ldots$.

2.2 Simultaneous equations

In this section, we will define simultaneous equations, both linear and non-linear.

2.2.1 Definition

Consider a vector function g that takes values of a decision vector in a **domain** $\mathbb{R}^n$ and returns values of the function that lie in a **range** $\mathbb{R}^m$. We write $g : \mathbb{R}^n \to \mathbb{R}^m$ to denote the domain and range of the function. Suppose we want to find a value $x^\star$ of the argument x that satisfies:

$$g(x) = \mathbf{0}, \tag{2.1}$$

where $\mathbf{0}$ is the m-vector of all zeros and is called the **zero vector**. To satisfy (2.1), each entry $g_\ell(x)$ of $g(x)$ must be equal to zero. That is we have m equations, for $\ell = 1, \ldots, m$, to be satisfied in all.

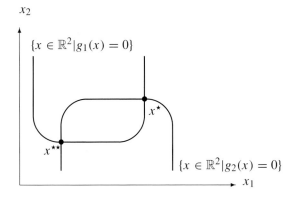

Fig. 2.1. An example of simultaneous equations and their solution.

A value, $x^\star$, that satisfies (2.1) is called a solution of the **simultaneous equations** $g(x) = 0$. (Even in the case that $m = 1$ we will refer to $g(x) = 0$ as simultaneous equations, although there is only one equation to solve in that case.) The set of *all* solutions is denoted by $\{x \in \mathbb{R}^n | g(x) = 0\}$. In words, this is the set of all vectors x in $\mathbb{R}^n$ such that $g(x)$ equals the zero vector. The vertical bar "|" can be interpreted as meaning "such that." (Some authors use a colon ":" instead of the vertical bar to denote "such that.")

For example, Figure 2.1 shows a case with a function $g : \mathbb{R}^2 \to \mathbb{R}^2$. There are two sets illustrated by the solid curves. One of the curves shows the points in the set $\{x \in \mathbb{R}^2 | g_1(x) = 0\}$, while the other shows the set $\{x \in \mathbb{R}^2 | g_2(x) = 0\}$. These two sets intersect at two points, $x^\star, x^{\star\star}$, illustrated as bullets •. The points $x^\star$ and $x^{\star\star}$ are the two solutions of the simultaneous equations $g(x) = 0$, so that $\{x \in \mathbb{R}^n | g(x) = 0\} = \{x^\star, x^{\star\star}\}$.

If there are no solutions to the equations then $\{x \in \mathbb{R}^n | g(x) = 0\} = \emptyset$, where $\emptyset$ is the empty set, and we say that the equations are **inconsistent**. For example, the sets illustrated in Figure 2.2 do not intersect, and so the equations are inconsistent.

Consider a linear combination of the entries of g. (See Definition A.54.) If some linear combination (with coefficients that are not all zero) yields a function that is identically zero then we say that the equations are **redundant**. For example, if two entries of g are the same then the equations are redundant. We will discuss redundant equations further in the context of ill-conditioned problems in Section 2.7.6.2.

2.2.2 Types of problems

To specify a particular instance of a simultaneous equations problem, we must specify g. We start with the simplest case for g and then generalize.

x_2

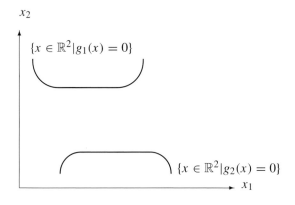

$\{x \in \mathbb{R}^2 | g_1(x) = 0\}$

$\{x \in \mathbb{R}^2 | g_2(x) = 0\}$

x_1

Fig. 2.2. Example of inconsistent simultaneous equations.

2.2.2.1 Linear simultaneous equations

Suppose that $g : \mathbb{R}^n \to \mathbb{R}^m$ in (2.1) is **affine**, that is, of the form:

$$\forall x \in \mathbb{R}^n, \, g(x) = Ax - b.$$

(See Definitions A.1 and A.19. The symbol $\forall$ means "for all" and is used to specify that a relation holds true for all possible values of the immediately following variable.) Then we have a set of **linear simultaneous equations**:

$$Ax - b = \mathbf{0}.$$

For brevity, the linear case is often written $Ax = b$ and b is called the **right-hand side vector**. The matrix A is called the **coefficient matrix**.

Examples For example, if:

$$A = \begin{bmatrix} 1 & 2 \\ 3 & 4 \end{bmatrix}, \, b = \begin{bmatrix} 1 \\ 1 \end{bmatrix}, \tag{2.2}$$

then it can be verified that:

$$x^\star = \begin{bmatrix} -1 \\ 1 \end{bmatrix}$$

is a solution. Figure 2.3 illustrates these simultaneous equations with the point $x^\star$ illustrated as a bullet •.

As another example, if:

$$A = \begin{bmatrix} 2 & 3 & 4 \\ 7 & 6 & 5 \\ 8 & 9 & 11 \end{bmatrix}, \, b = \begin{bmatrix} 9 \\ 18 \\ 28 \end{bmatrix}, \tag{2.3}$$

x_2

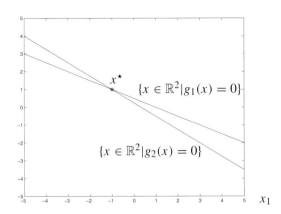

$\{x \in \mathbb{R}^2 | g_1(x) = 0\}$

$\{x \in \mathbb{R}^2 | g_2(x) = 0\}$

x_1

Fig. 2.3. Solution of linear simultaneous equations $g(x) = Ax - b = \mathbf{0}$ with A and b defined as in (2.2).

then it can be verified that:

$$x^\star = \begin{bmatrix} 1 \\ 1 \\ 1 \end{bmatrix}$$

is a solution. (We will return to this particular problem in Section 5.3.2.5.)

Number of solutions In both of the previous cases, we say "a" solution because we consider the possibility that there is more than one solution satisfying our criterion $Ax = b$. In other words, as in Figure 2.1, there may be several values that satisfy the equations. In fact, for each of the two particular example problems specified by A and b in (2.2) and (2.3), respectively, there is exactly one solution so that in each case the set of solutions $\{x \in \mathbb{R}^n | g(x) = \mathbf{0}\}$ is a **singleton set**, consisting of exactly one element. The solution can be obtained by **eliminating** one of the entries in x and solving for the other or others. (We will discuss elimination more generally as an example of a problem transformation in Section 3.2.2.)

Case studies In Part I, we will consider systematic ways to solve simultaneous linear equations of arbitrary size, if a solution exists. The development will begin with two case studies:

- nodal analysis of a direct current linear circuit (in Section 4.1), and
- control of a discrete-time linear system (in Section 4.2).

We will also consider the conditions under which there is exactly one solution, more than one solution, and no solutions.

x_2

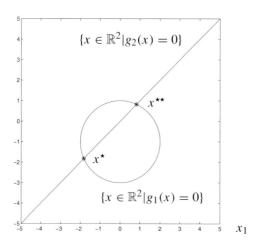

Fig. 2.4. Solution of non-linear simultaneous equations $g(x) = \mathbf{0}$ with g defined as in (2.5).

2.2.2.2 Non-linear simultaneous equations

The conditions for the existence of a unique solution to linear simultaneous equations will also prove relevant to solution of **non-linear simultaneous equations**; that is, the case of (2.1) where g is non-linear, which will be covered in Part II.

Examples For example, suppose that the function $g : \mathbb{R} \to \mathbb{R}$ is defined by:

$$\forall x \in \mathbb{R}, g(x) = (x)^2 + 2x - 3. \tag{2.4}$$

This is an example of a **quadratic function**. (See Definition A.20 for a general definition of a quadratic function.) The familiar "quadratic equation" shows that

$$x^\star = -3, \ x^{\star\star} = 1,$$

are the two solutions to $g(x) = 0$. That is, $\{x \in \mathbb{R} | g(x) = 0\} = \{-3, 1\}$.
  As another example, let $g : \mathbb{R}^2 \to \mathbb{R}^2$ be defined by:

$$\forall x \in \mathbb{R}^2, g(x) = \begin{bmatrix} (x_1)^2 + (x_2)^2 + 2x_2 - 3 \\ x_1 - x_2 \end{bmatrix}. \tag{2.5}$$

Figure 2.4 illustrates these simultaneous equations, with the two solutions illustrated with bullets • and labeled $x^\star$ and $x^{\star\star}$. (We will return to this particular problem in Exercise 7.2.)
  As a third example, let $g : \mathbb{R} \to \mathbb{R}$ be defined by:

$$\forall x \in \mathbb{R}, g(x) = (x - 2)^3 + 1. \tag{2.6}$$

By inspection, $x^\star = 1$ is the unique solution to $g(x) = 0$. (We will return to this particular problem in Section 7.4.1.)

Algorithms and number of solutions For larger non-linear problems, with more variables and more equations, there is usually no general formula to solve for the solutions directly. It is reasonable to expect that larger problems might also possess multiple solutions or no solutions under some circumstances. However, we will concentrate on case studies where we have reason to expect that there is a unique solution or, if there are multiple solutions, where we have a "preferred" solution in mind and we are seeking the preferred solution.

Case studies We will discuss these issues in detail in Part II, beginning with two case studies:

- nodal analysis of a non-linear direct current electric circuit (in Section 6.1), and
- analysis of an electric power system (in Section 6.2).

We will also consider the conditions under which there is exactly one solution, more than one solution, and no solutions.

2.2.2.3 Eigenvalue problems

One important class of non-linear equations that we will not treat in detail is the **eigenvalue** problem. (See Definition A.31.) Let $\mathbb{K}$ be the set of **complex numbers**. The n (not necessarily distinct) eigenvalues of a matrix $A \in \mathbb{R}^{n \times n}$ are given by the (possibly complex) solutions of the **characteristic equation** of A:

$$g(\lambda) = 0,$$

where $g : \mathbb{K} \to \mathbb{K}$ is the **characteristic polynomial**, defined by:

$$\forall \lambda \in \mathbb{K}, g(\lambda) = \det(A - \lambda \mathbf{I}),$$

where det is the determinant of the matrix. (See Definition A.10 of the determinant of a matrix.) The function g is an n-th degree polynomial in one complex variable. (See Definition A.22.) The characteristic equation $g(\lambda) = 0$ is one non-linear equation in one variable with n (possibly not all distinct) solutions.

The **eigenvectors** associated with an eigenvalue λ are the solutions of:

$$(A - \lambda \mathbf{I})x = \mathbf{0}.$$

(In general, there will be multiple solutions of this equation and if an eigenvalue λ is complex then each eigenvector x will be a complex n-vector.)

Example For example, for the matrix $A \in \mathbb{R}^{2 \times 2}$ defined by:

$$A = \begin{bmatrix} 2 & 1 \\ -5 & -4 \end{bmatrix},$$

we have that:

$$
\begin{aligned}
\forall \lambda \in \mathbb{K}, g(\lambda) &= \det(A - \lambda \mathbf{I}), \\
&= \det \begin{bmatrix} 2 - \lambda & 1 \\ -5 & -4 - \lambda \end{bmatrix}, \\
&= (2 - \lambda)(-4 - \lambda) - (1)(-5), \\
&= (\lambda)^2 + 2\lambda - 3.
\end{aligned}
$$

From the previous example, we already know that the two solutions to this equation are:

$$
\lambda^\star = -3, \lambda^{\star\star} = 1,
$$

so these are the eigenvalues of A. The eigenvectors associated with $\lambda^\star = -3$ are the vectors in the set:

$$
\{x \in \mathbb{R}^2 | (A + 3\mathbf{I})x = \mathbf{0}\}.
$$

The eigenvectors associated with $\lambda^{\star\star} = 1$ are the vectors in the set:

$$
\{x \in \mathbb{R}^2 | (A - \mathbf{I})x = \mathbf{0}\}.
$$

Discussion It turns out that there are special iterative algorithms to solve this sort of problem that are somewhat different in flavor to the algorithms we will describe for solving general linear and non-linear equations. Some algorithms for eigenvalue problems, for example, calculate simultaneously all the eigenvalues of A. The MATLAB function `eig` finds the eigenvalues and eigenvectors of a matrix. Further details about finding eigenvalues and eigenvectors and finding solutions of polynomial equations can be found in [30, 78, 90, 94, 121]. (The second reference describes how to solve *generalized* eigenvalue problems where the function g becomes:

$$
\forall \lambda \in \mathbb{K}, g(\lambda) = \det(A - \lambda B),
$$

with $A, B \in \mathbb{R}^{n \times n}$.) We will have occasion to calculate eigenvalues of some particular matrices in this book, but we will not discuss general algorithms for eigenvalue calculation.

2.3 Optimization

In Section 2.3.1 we define **optimization problems** and some associated concepts. We discuss various types of problems and provide some examples in Section 2.3.2. In Sections 2.3.3–2.3.6 we then discuss the existence of solutions to optimization problems and present some generalizations of optimization problems.

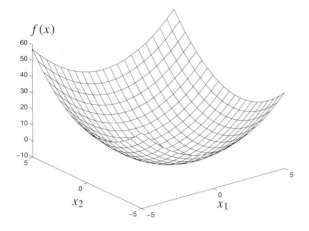

$f(x)$

Fig. 2.5. Graph of the example objective function defined in (2.7).

2.3.1 Definitions

To define an optimization problem, we must first define the concepts of an **objective** and a **feasible set**.

2.3.1.1 Objective

Consider a function $f : \mathbb{R}^n \to \mathbb{R}$ that denominates the "cost" or lack of desirability of solutions for a particular model or system. That is, $f(x)$ is the cost of using x as the solution. The function is called an **objective function** or an **objective**. The optimization problems we consider involve finding a value of x that minimizes the objective function $f(x)$.

Example An example of $f : \mathbb{R}^2 \to \mathbb{R}$ is given by:

$$\forall x \in \mathbb{R}^2, \ f(x) = (x_1)^2 + (x_2)^2 + 2x_2 - 3. \tag{2.7}$$

This function is graphed in Figure 2.5 for values of its argument in the range $-5 \le x_1 \le 5, -5 \le x_2 \le 5$. This objective function is **quadratic**, since the highest power of any entry of x is two; however, linear objectives and other non-linear objectives arise in various applications. There are also linear and constant terms in the function defined in (2.7). Quadratic functions will be our canonical examples of objective functions. (See Definition A.20.)

Discussion We have, so far, categorized objectives according to the highest power of any entry in the argument. We will categorize objectives in a different way in Section 2.6.3.4 once we have discussed optimization in more detail.

In some cases, a single objective does not capture the cost of decisions appropriately. For example, there may be two issues that are **incommensurable**; that is, that cannot be combined into a single measure or cost. In this case, we must use **multi-objective optimization** [76].

As another example, we may have several decision-makers who are each trying to optimize their own objective. If the objective of one decision-maker depends partly on the decision of another then the outcome of the decision-makers depends partly on the interaction between their decisions. In this case, we can seek the resulting **equilibrium** [119].

For the rest of this book, however, we will assume that a single objective function is suitable for characterizing the cost of decisions.

2.3.1.2 *Feasible set*

Our problem might involve restrictions on the choices of values of x. We can imagine a **feasible set** $\mathbb{S} \subseteq \mathbb{R}^n$ from which we must select a solution. For example, the feasible set for a particular application could be $\mathbb{S} = \{x \in \mathbb{R}^2 | -5 \leq x_1 \leq 5, -5 \leq x_2 \leq 5\}$. A point $x \in \mathbb{S}$ is called a **feasible point** for the problem.

2.3.1.3 *Problem*

A **minimization problem** is to minimize $f(x)$ over choices of x that lie in the feasible set $\mathbb{S}$. We define the following symbol:

$$\min_{x \in \mathbb{S}} f(x),$$

to mean the minimum *value* that $f(x)$ can take on over values of $x \in \mathbb{S}$, assuming that such a minimum exists. To be more precise, we make:

Definition 2.1 Let $\mathbb{S} \subseteq \mathbb{R}^n$, $f : \mathbb{S} \to \mathbb{R}$, and $f^\star \in \mathbb{R}$. Then by:

$$f^\star = \min_{x \in \mathbb{S}} f(x), \tag{2.8}$$

we mean that:

$$\exists x^\star \in \mathbb{S} \text{ such that: } (f^\star = f(x^\star)) \text{ and } ((x \in \mathbb{S}) \Rightarrow (f(x^\star) \leq f(x))). \tag{2.9}$$

$\square$

The symbol $\exists$ means "there exists." (See Definition A.1.) It means that there is at least one value of the immediately following variable that satisfies the subsequent conditions.

The set $\mathbb{S}$ is called the **feasible set**, the **constraint set**, or the **feasible region**. The value $f^\star$ is called the **minimum** of the problem $\min_{x \in \mathbb{S}} f(x)$, while $x^\star$ is called a **minimizer**. We say that the problem $\min_{x \in \mathbb{S}} f(x)$ **possesses** or **has a minimum** if there exists $f^\star \in \mathbb{R}$ satisfying (2.8). We also say that this $x^\star$ **achieves** the minimum.

In describing the process of seeking the minimum, we say that we are trying to find the *minimum* of $f(x)$ *over* $x \in \mathbb{S}$. The placement of the notation "$x \in \mathbb{S}$" as a subscript to the "min" provides a visual correspondence to the description of the minimum being "over" $x \in \mathbb{S}$.

We will often define f on a larger set than $\mathbb{S}$, for example on the whole of $\mathbb{R}^n$. For points $x \in (\mathbb{R}^n \setminus \mathbb{S})$, the value of $f(x)$ does not affect the minimum over $\mathbb{S}$.

If a problem possesses a minimum $f^\star$ then, by definition, there is a minimizer $x^\star$ satisfying (2.9). In general, there may be none, one, or more than one $x^\star$ that satisfies (2.9). If there is no $f^\star$ and $x^\star$ satisfying (2.9), then we say that the problem has no minimum or that it has no minimizer. This can occur, for example, if $\mathbb{S} = \emptyset$, where $\emptyset$ is the **empty set**. If $\mathbb{S} = \emptyset$, we say that the problem is **infeasible**. However, as we will see in Section 2.3.3, it is also possible for there to be no minimum even if the feasible set is non-empty.

2.3.1.4 Set of minimizers

If there exists one or more values of $x^\star$ satisfying (2.9), then we say that the problem has a minimizer or minimizers, respectively. The set of *all* the minimizers of $\min_{x \in \mathbb{S}} f(x)$ is denoted by:

$$\underset{x \in \mathbb{S}}{\mathrm{argmin}}\, f(x).$$

The word argmin abbreviates "the argument of the minimum;" that is, the value (or values) of the argument x that yields the minimum value of the objective. If the problem has no minimum (and, therefore, no minimizers) then we define:

$$\underset{x \in \mathbb{S}}{\mathrm{argmin}}\, f(x) = \emptyset.$$

If the problem has exactly one minimizer, $x^\star$, say, then $\mathrm{argmin}_{x \in \mathbb{S}} f(x) = \{x^\star\}$, a singleton set. To emphasize the role of $\mathbb{S}$, we also use the following notations:

$$\min_{x \in \mathbb{R}^n} \{f(x) | x \in \mathbb{S}\} \text{ and } \underset{x \in \mathbb{R}^n}{\mathrm{argmin}} \{f(x) | x \in \mathbb{S}\},$$

for the minimum and for the set of minimizers, respectively, of $\min_{x \in \mathbb{S}} f(x)$. In words, we are seeking the minimum and the set of minimizers of $f(x)$ over vectors x in $\mathbb{R}^n$ such that x is contained in $\mathbb{S}$. We will often use a more explicit notation if $\mathbb{S}$ is defined as the set of points satisfying a criterion. For example, if $f : \mathbb{R}^n \to \mathbb{R}$, $g : \mathbb{R}^n \to \mathbb{R}^m$, $h : \mathbb{R}^n \to \mathbb{R}^r$, and $\mathbb{S} = \{x \in \mathbb{R}^n | g(x) = \mathbf{0}, h(x) \leq \mathbf{0}\}$ then we will write $\min_{x \in \mathbb{R}^n} \{f(x) | g(x) = \mathbf{0}, h(x) \leq \mathbf{0}\}$ for $\min_{x \in \mathbb{S}} f(x)$.

Some authors use the word "minimum" for what we call the minimizer. The distinction that we make is important when there are multiple minimizers of a

$f(x)$

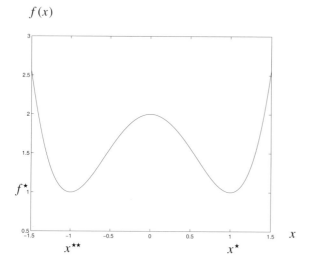

Fig. 2.6. Function having multiple unconstrained minimizers.

problem corresponding to the same minimum. To see an example of a problem with multiple minimizers, consider the function $f : \mathbb{R} \to \mathbb{R}$ defined by:

$$\forall x \in \mathbb{R}, \ f(x) = (x+1)^2(x-1)^2 + 1.$$

This function is illustrated in Figure 2.6. The unconstrained minimum of f is $f^\star = 1$. The corresponding minimizers are $x^\star = 1$ and $x^{\star\star} = -1$.

Exercise 2.3 provides a simple example of a minimization problem with $x \in \mathbb{R}$. In Exercise 2.3 and other simple problems, it is possible to determine the minimum and minimizer by inspection of f because of its particular form. In general, we will need to develop systematic approaches for functions having more than one or two variables.

2.3.1.5 Lower bound

In many applications, it can be useful to establish a standard or bound for evaluating the minimum. This is embodied in the following:

Definition 2.2 Let $\mathbb{S} \subseteq \mathbb{R}^n$, $f : \mathbb{S} \to \mathbb{R}$, and $\underline{f} \in \mathbb{R}$. If $\underline{f}$ satisfies:

$$\forall x \in \mathbb{S}, \ \underline{f} \le f(x),$$

then we say that $\underline{f}$ is a **lower bound** for the problem $\min_{x \in \mathbb{S}} f(x)$ or that the problem $\min_{x \in \mathbb{S}} f(x)$ is **bounded below** by $\underline{f}$. If $\mathbb{S} \ne \emptyset$ but no such $\underline{f}$ exists, then we say that the problem $\min_{x \in \mathbb{S}} f(x)$ is **unbounded below** (or unbounded if the "below" is clear from context.) □

Consider again $f : \mathbb{R}^2 \to \mathbb{R}$ defined in (2.7), which we repeat here:

$$\forall x \in \mathbb{R}^2, \ f(x) = (x_1)^2 + (x_2)^2 + 2x_2 - 3.$$

This function is illustrated in Figure 2.5. For the feasible set $\mathbb{S} = \mathbb{R}^2$, the value $\underline{f} = -10$ is a lower bound for the problem $\min_{x \in \mathbb{S}} f(x)$, as can be verified by inspection of Figure 2.5. A lower bound, such as $\underline{f} = -10$, provides a gauge for the quality of a candidate solution to a problem. If we find $x \in \mathbb{S}$ such that $f(x)$ is not significantly worse (that is, not significantly larger) than the lower bound $\underline{f}$ then we know that we have a good, if not exact, solution to the problem and that the lower bound is **strong**. That is, the lower bound is close to the minimum of the problem, if it exists. (On the other hand, if the lower bound is far below the objective value of the candidate solution then it is not a good gauge of the quality or otherwise of the candidate solution.)

As we will see in Section 2.3.3, it is possible for a problem to be bounded below even if it does not possess a minimum; however, if a problem possesses a minimum then it is bounded below by its minimum. Moreover, a problem that possesses a minimum is bounded below by every number that is less than or equal to its minimum. (See Exercise 2.5.)

2.3.1.6 Level and contour sets

To characterize the set of minimizers of the problem we make:

Definition 2.3 Let $\mathbb{S} \subseteq \mathbb{R}^n$, $f : \mathbb{S} \to \mathbb{R}$, and $\tilde{f} \in \mathbb{R}$. Then the **level set** at value $\tilde{f}$ of the function f is the set:

$$\mathbb{L}_f(\tilde{f}) = \{x \in \mathbb{S} | f(x) \leq \tilde{f}\}.$$

The **contour set** at value $\tilde{f}$ of the function f is the set:

$$\mathbb{C}_f(\tilde{f}) = \{x \in \mathbb{S} | f(x) = \tilde{f}\}.$$

For each possible function f, we can think of $\mathbb{L}_f$ and $\mathbb{C}_f$ themselves as *set-valued functions* from $\mathbb{R}$ to $(2)^{(\mathbb{R}^n)}$, where $(2)^{(\mathbb{R}^n)}$ denotes the **set of all subsets** of $\mathbb{R}^n$, sometimes called the **power set** of $\mathbb{R}^n$. □

Some authors use the phrase "sub-level sets" for what we call "level sets" [6, section 3.5.2][15, section 3.1.6]. The level and contour sets of a function are closely related: under mild conditions, the contour set at a particular value is the **boundary** (see Definition A.42) of the level set at the same value and we will often think of contour and level sets almost interchangeably. The level sets of a function will turn out to be more useful in analysis; however, the contour sets are easier to draw. We usually draw a family of contour sets for various values and collectively refer to the family as "the contour sets."

The contour sets of a function are often useful in visualizing functions that would otherwise be very difficult to sketch. This is because the contour set of $f : \mathbb{R}^n \to \mathbb{R}$ is a subset of $\mathbb{R}^n$ whereas the graph of f is a subset of $\mathbb{R}^n \times \mathbb{R}$ and therefore requires

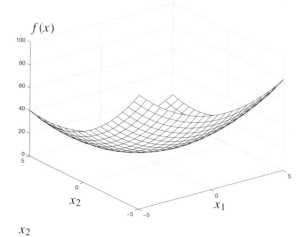

Fig. 2.7. The graph of the function defined in (2.10).

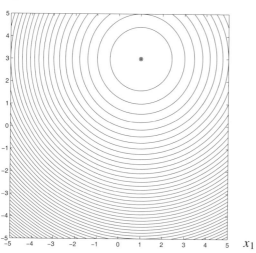

Fig. 2.8. The contour sets $\mathbb{C}_f(\tilde{f})$ of the function defined in (2.10) for values $\tilde{f} = 0, 2, 4, 6, \ldots$. The heights of the contours decrease towards the point $\begin{bmatrix} 1 \\ 3 \end{bmatrix}$, which is illustrated with a • and is the contour of height 0.

one more dimension to illustrate. The contour sets of a function $f : \mathbb{R}^2 \to \mathbb{R}$ can be drawn with the MATLAB function `contour`.

If $f^\star$ is the minimum of the problem $\min_{x \in \mathbb{S}} f(x)$, then $\mathbb{L}_f(f^\star) = \mathbb{C}_f(f^\star) = \operatorname{argmin}_{x \in \mathbb{S}} f(x)$. Consider the function $f : \mathbb{R}^2 \to \mathbb{R}$ defined by:

$$\forall x \in \mathbb{R}^2, \ f(x) = (x_1 - 1)^2 + (x_2 - 3)^2, \tag{2.10}$$

The graph of this function is shown in Figure 2.7 and requires three dimensions to represent.

The contour sets $\mathbb{C}_f(\tilde{f})$ of the function f defined in (2.10) are shown in Figure 2.8 for $\tilde{f} = 2, 4, 6, 8, \ldots$ over the region $\{x \in \mathbb{R}^2 | -5 \le x_1 \le 5, -5 \le x_2 \le 5\}$. The contour sets can be shown in a two-dimensional representation. It is helpful to show or explain the heights of the contours explicitly, or at least indicate how the heights of the contours vary on the diagram. For example, the caption of Figure 2.8 includes a description of how the contours vary.

In Section 2.3.2, we will use the function defined in (2.10) as an example objective to illustrate optimization problems.

2.3.2 Types of problems

In this section, we will categorize optimization problems by the type of feasible set. An optimization problem involving a specific form of the set $\mathbb{S}$ can be referred to by the form of $\mathbb{S}$. (In Section 2.6.3.4, we will discuss further categorization based on the properties of the objective.)

The three general forms of $\mathbb{S}$ that we will consider in detail in this book are:

- **unconstrained** optimization,
- **equality-constrained** optimization, and
- **inequality-constrained** optimization.

We describe these forms of $\mathbb{S}$ in detail in Sections 2.3.2.1–2.3.2.3. There are also more general forms of $\mathbb{S}$ that we will not discuss, particularly in the context of more general forms of decision vector. For example:

- **optimal control** problems [16, 55, 63, 89], where the decision vector may be a *function* and the constraints are defined in terms of properties of the function, such as requirements on its derivative, and
- **semi-definite programming** [15], where the decision "vector" is a matrix and the constraints are defined in terms of properties of the matrix.

2.3.2.1 Unconstrained optimization

If $\mathbb{S} = \mathbb{R}^n$, then the problem is said to be **unconstrained**, since there are no additional restrictions on x besides that it lies in $\mathbb{R}^n$. To specify an unconstrained problem, we must specify f.

Example For example, consider the objective $f : \mathbb{R}^2 \to \mathbb{R}$ defined in (2.10):

$$\forall x \in \mathbb{R}^2, \ f(x) = (x_1 - 1)^2 + (x_2 - 3)^2.$$

From Figure 2.8, which shows the contour sets of f, we can see that:

$$\min_{x \in \mathbb{R}^2} f(x) \ = \ f^\star = 0,$$

$$\operatorname*{argmin}_{x \in \mathbb{R}^2} f(x) \ = \ \left\{ \begin{bmatrix} 1 \\ 3 \end{bmatrix} \right\},$$

so that there is a minimum $f^\star = 0$ and a unique minimizer $x^\star = \begin{bmatrix} 1 \\ 3 \end{bmatrix}$ of this problem. (We will return to this particular problem in Section 10.1.1.)

As another example, suppose that we have linear simultaneous equations $Ax - b = 0$ that do not have a solution. We may try to seek a value of the decision vector that "most nearly" satisfies $Ax = b$ in the sense of minimizing a criterion. A natural criterion is to consider a **norm** $\|\bullet\|$ (see Definition A.28) and then seek x that minimizes $\|Ax - b\|$. The unconstrained minimization problem is then:

$$\min_{x \in \mathbb{R}^n} \|Ax - b\|. \tag{2.11}$$

Case studies Unconstrained problems will be investigated generally in Part III, beginning with two case studies:

- multi-variate linear regression (in Section 9.1), and
- power system state estimation (in Section 9.2).

Both of these case studies will involve a minimization problem that is similar to Problem (2.11).

2.3.2.2 *Equality-constrained optimization*

If $g : \mathbb{R}^n \to \mathbb{R}^m$ and $\mathbb{S} = \{x \in \mathbb{R}^n | g(x) = 0\}$, so that the feasible set is the set of values that satisfy the simultaneous equations $g(x) = 0$, then the problem is said to be **equality-constrained**. To emphasize the functional form of the equality constraints, we usually use the notation $\min_{x \in \mathbb{R}^n} \{f(x) | g(x) = 0\}$ for an equality-constrained problem $\min_{x \in \mathbb{R}^n} \{f(x) | x \in \mathbb{S}\}$, where $\mathbb{S} = \{x \in \mathbb{R}^n | g(x) = 0\}$.

Sub-types of equality-constrained problems To specify an equality-constrained problem, we must specify f and g. It is common to divide this type of problem further depending on the form of g.

Linearly constrained If g is affine then the problem is called **linearly constrained**. For example, consider the objective: $f : \mathbb{R}^2 \to \mathbb{R}$ defined in (2.10):

$$\forall x \in \mathbb{R}^2, f(x) = (x_1 - 1)^2 + (x_2 - 3)^2,$$

and include the constraint $g(x) = 0$ where we define the function $g : \mathbb{R}^2 \to \mathbb{R}$ by

$$\forall x \in \mathbb{R}^2, g(x) = x_1 - x_2. \tag{2.12}$$

The constraint $g(x) = 0$ requires that $x_1 = x_2$. Figure 2.9 repeats the contours of f for values $\tilde{f} = 2, 4, 6, \ldots$ from Figure 2.8 but adds a line that represents the set of feasible points under the equality constraint $g(x) = 0$.

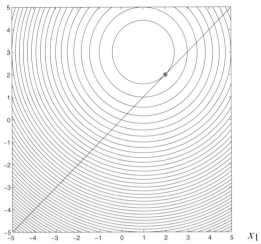

Fig. 2.9. The contour sets $\mathbb{C}_f(\tilde{f})$ of the function repeated from Figure 2.8 with feasible set from Problem (2.13) superimposed. The heights of the contours decrease towards the point $\begin{bmatrix} 1 \\ 3 \end{bmatrix}$.

The minimizer $x^\star = \begin{bmatrix} 2 \\ 2 \end{bmatrix}$ is illustrated with a •.

Consider the equality-constrained problem:

$$\min_{x \in \mathbb{R}^2}\{f(x)|g(x) = 0\} = \min_{x \in \mathbb{R}^2}\{f(x)|x_1 - x_2 = 0\}. \qquad (2.13)$$

The constraint restricts x to lie on the line where $x_1 = x_2$. We seek the smallest value of the objective that is consistent with this constraint. Inspection of Figure 2.9 shows that this occurs for $x^\star = \begin{bmatrix} 2 \\ 2 \end{bmatrix}$, which is the unique minimizer of Problem (2.13) and is illustrated with a • in Figure 2.9. The minimum of the problem is $f^\star = 2$. (We will return to this particular problem in Section 13.1.2.)

Non-linearly constrained If there is no restriction on g then the problem is called **non-linearly constrained**. For example, consider the same objective as previously, $f : \mathbb{R}^2 \to \mathbb{R}$ defined in (2.10):

$$\forall x \in \mathbb{R}^2, \ f(x) = (x_1 - 1)^2 + (x_2 - 3)^2.$$

However, let $g : \mathbb{R}^2 \to \mathbb{R}$ be defined by:

$$\forall x \in \mathbb{R}^2, \ g(x) = (x_1)^2 + (x_2)^2 + 2x_2 - 3.$$

The constraint $g(x) = 0$ requires that $(x_1)^2 + (x_2)^2 + 2x_2 - 3 = 0$. This is a circle of radius 2 and center $\begin{bmatrix} 0 \\ -1 \end{bmatrix}$. Figure 2.10 repeats Figure 2.8 but adds a circle that represents the set of feasible points for the equality constraint $g(x) = 0$.

Consider the equality-constrained problem:

$$\min_{x \in \mathbb{R}^2}\{f(x)|g(x) = 0\}. \qquad (2.14)$$

x_2

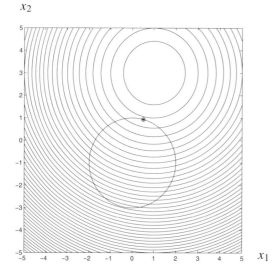

Fig. 2.10. Contour sets $\mathbb{C}_f(\tilde{f})$ of function repeated from Figure 2.8 with feasible set from Problem (2.14) superimposed. The heights of the contours decrease towards the point $\begin{bmatrix} 1 \\ 3 \end{bmatrix}$. The minimizer $x^\star \approx \begin{bmatrix} 0.5 \\ 0.9 \end{bmatrix}$ is illustrated as a •.

We seek the smallest value of the objective that is consistent with this constraint. Inspection of Figure 2.10 shows that this occurs for $x^\star \approx \begin{bmatrix} 0.5 \\ 0.9 \end{bmatrix}$, which is the unique minimizer of Problem (2.14) and is illustrated as a • in Figure 2.10. The minimum of the problem is $f^\star \approx 4.7$. (We will return to this particular problem in Section 14.2 and solve it more accurately.)

Case studies Equality-constrained problems will be investigated in Part IV, beginning with two case studies:

- least-cost production of a group of manufacturing facilities that must collectively meet a demand constraint (in Section 12.1), and
- power system state estimation with zero injection buses (in Section 12.2).

2.3.2.3 Inequality-constrained optimization

If $g : \mathbb{R}^n \to \mathbb{R}^m$, $h : \mathbb{R}^n \to \mathbb{R}^r$, and $\mathbb{S} = \{x \in \mathbb{R}^n | g(x) = \mathbf{0}, h(x) \leq \mathbf{0}\}$ so that the feasible points satisfy $g_\ell(x) = 0, \ell = 1, \ldots, m$, and $h_\ell(x) \leq 0, \ell = 1, \ldots, r$, then the problem is said to be **inequality-constrained**. Notice that, by our definition, an inequality-constrained problem can include equality constraints. Again, to emphasize the functional form of the equality and inequality constraints, we usually use the notation $\min_{x \in \mathbb{R}^n} \{f(x) | g(x) = \mathbf{0}, h(x) \leq \mathbf{0}\}$ for an inequality-constrained problem $\min_{x \in \mathbb{R}^n} \{f(x) | x \in \mathbb{S}\}$, where $\mathbb{S} = \{x \in \mathbb{R}^n | g(x) = \mathbf{0}, h(x) \leq \mathbf{0}\}$.

Sub-types of inequality-constrained optimization problems To specify a problem that is inequality-constrained, we must specify f, g, and h. It is common to divide this type of problem further depending on the form of objective and constraints. The following is an (overlapping) list of sub-types.

Non-negatively constrained If h is of the form:

$$\forall x, h(x) = -x,$$

so that the constraints are of the form $x \geq \mathbf{0}$ then the problem is **non-negatively constrained**.

Linear inequality constraints If h is affine then the problem is **linear inequality-constrained**.

Linear program If the objective is linear and g and h are affine then the problem is called a **linear program** or **linear optimization problem**. (The word "program" is a historical term and does not refer to the software used to solve the problem.)

For example, if $n = 2$ and the functions $f : \mathbb{R}^2 \to \mathbb{R}$, $g : \mathbb{R}^2 \to \mathbb{R}$, and $h : \mathbb{R}^2 \to \mathbb{R}^2$ are defined by:

$$
\begin{aligned}
\forall x \in \mathbb{R}^2, f(x) &= x_1 - x_2, \\
\forall x \in \mathbb{R}^2, g(x) &= x_1 + x_2 - 1, \\
\forall x \in \mathbb{R}^2, h(x) &= \begin{bmatrix} -x_1 \\ -x_2 \end{bmatrix},
\end{aligned}
$$

then the problem:

$$\min_{x \in \mathbb{R}^2}\{f(x)|g(x) = 0, h(x) \leq \mathbf{0}\} = \min_{x \in \mathbb{R}^2}\{x_1 - x_2|x_1 + x_2 - 1 = 0, x_1 \geq 0, x_2 \geq 0\},$$

$$(2.15)$$

is a linear program. This problem is linearly constrained and, in particular, the linear inequality constraints are non-negativity constraints. (We will return to this particular example problem in Section 16.1.1.2.) The contour sets of the objective and the feasible set are illustrated in Figure 2.11. The minimizer $x^{\star} = \begin{bmatrix} 0 \\ 1 \end{bmatrix}$ of Problem (2.15) is illustrated as a ●.

Linear programs are usually written so as to emphasize the linear and affine functions by showing the coefficient matrices explicitly as in, for example:

$$\min_{x \in \mathbb{R}^2}\{c^{\dagger}x|Ax = b, Cx \leq d\},$$

where $c \in \mathbb{R}^n$, $A \in \mathbb{R}^{m \times n}$, $b \in \mathbb{R}^m$, $C \in \mathbb{R}^{r \times n}$, $d \in \mathbb{R}^r$, and where $c^{\dagger}$ is the **transpose** of c. That is, $c^{\dagger}$ is a row vector with entries the same as the entries of

x_2

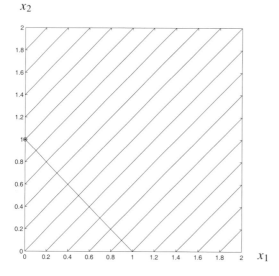

Fig. 2.11. Contour sets $\mathbb{C}_f(\tilde{f})$ of objective function and feasible set for Problem (2.15). The contour sets are the parallel lines. The feasible set is shown as the line joining the two points $\begin{bmatrix} 1 \\ 0 \end{bmatrix}$ and $\begin{bmatrix} 0 \\ 1 \end{bmatrix}$. The heights of the contours decrease to the left and up. The minimizer $x^\star = \begin{bmatrix} 0 \\ 1 \end{bmatrix}$ is illustrated as a •.

the column vector c. (See Definition A.6 in Appendix A.) In the case of Problem (2.15), the appropriate vectors and matrices are:

$$c = \begin{bmatrix} 1 \\ -1 \end{bmatrix}, \ A = \begin{bmatrix} 1 & 1 \end{bmatrix}, \ b = [1], \ C = \begin{bmatrix} -1 & 0 \\ 0 & -1 \end{bmatrix}, \ d = \begin{bmatrix} 0 \\ 0 \end{bmatrix}.$$

We can write this non-negatively constrained problem even more concisely as:

$$\min_{x \in \mathbb{R}^2} \{c^\dagger x \,|\, Ax = b, x \geq \mathbf{0}\}. \tag{2.16}$$

There is a rich body of literature on linear programming and there are special purpose algorithms to solve linear programming problems. The best known are:

- the **simplex algorithm** (and variants), and
- **interior point algorithms**.

Considerable effort has been spent on tailoring software for linear programming. The performance of such software on linear programming problems will generally be better than the performance on linear programming problems of software that can handle more general problems. In other words, if the problem at hand is linear, then it is sensible to use linear programming software to solve it.

Furthermore, the linear objective and affine constraint functions allow a number of simplifications and special cases that are not available in the general case. There are various special cases of linear programming problems that can be solved with very fast special-purpose algorithms. See, for example, [67, chapter 4] and [70, chapter 5] for linear programming problems involving **transportation** and **flows on networks**. Even some **integer** and **discrete optimization** problems turn out to have elegant and simple solutions if the objective is linear and the constraint functions are affine and have particular properties. We will not describe the special cases of linear programming in detail in this book, but they are covered in great detail in many books. For example, see [12][28][54, chapters 5–6][70, part I][84, part II]. We will not consider integer and discrete linear problems in this book, except briefly in Sections 15.5.2.1, 15.6.1.4, and 16.3.6.1. However, [12, 22, 46, 54, 67, 85, 92, 113, 122] cover many such problems.

Finally, it is sometimes possible to approximate a non-linear function in terms of piece-wise linear functions and subsidiary linear constraints. (See Section 3.1.4.3.) This allows linear programming software to be used on the problem. Because of the high performance of linear programming software, it is sometimes possible to solve the piece-wise linearized problem faster than by direct application of an algorithm for non-linear problems.

Standard format If g is affine and the inequality constraints are non-negativity constraints then the problem is said to be in the **standard format** [45, section 5.6.1][84, section 4.2]. Problem (2.16) is a linear program in standard format.

Quadratic program If f is quadratic and g and h are affine then the problem is called a **quadratic program** or a **quadratic optimization problem**. For example, consider the objective $f : \mathbb{R}^2 \to \mathbb{R}$ defined in (2.10):

$$\forall x \in \mathbb{R}^2, \ f(x) = (x_1 - 1)^2 + (x_2 - 3)^2,$$

the equality constraint function $g : \mathbb{R}^2 \to \mathbb{R}$ defined in (2.12):

$$\forall x \in \mathbb{R}^2, \ g(x) = x_1 - x_2,$$

and include the inequality constraint $h(x) \leq 0$ where the inequality constraint function $h : \mathbb{R}^2 \to \mathbb{R}$ is defined by:

$$\forall x \in \mathbb{R}^2, \ h(x) = 3 - x_2. \tag{2.17}$$

To satisfy $h(x) \leq 0$, we must have that $x_2 \geq 3$. The problem having (2.10) as objective as well as the equality constraint function defined in (2.12) and the inequality constraint function defined in (2.17) is:

$$\min_{x \in \mathbb{R}^2} \{ f(x) | g(x) = 0, h(x) \leq 0 \}. \tag{2.18}$$

The contour sets of the objective function and the feasible set are illustrated in Figure 2.12. By inspection of Figure 2.9 and comparison to Figure 2.12, we can

x_2

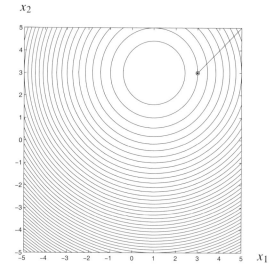

Fig. 2.12. The   contour
sets $\mathbb{C}_f(\tilde{f})$ of the objective
function and feasible set
for Problem (2.18).   The
heights of the contours
decrease towards the point
$\begin{bmatrix} 1 \\ 3 \end{bmatrix}$.   The feasible set
is the "half-line" starting
at the point $\begin{bmatrix} 3 \\ 3 \end{bmatrix}$.   The
minimizer $x^\star = \begin{bmatrix} 3 \\ 3 \end{bmatrix}$ is
illustrated with a •.

see that the minimum and minimizer of Problem (2.18) are:

$$\min_{x\in\mathbb{R}^2}\{f(x)|g(x) = 0, h(x) \leq 0\} = 4,$$

$$\operatorname*{argmin}_{x\in\mathbb{R}^2}\{f(x)|g(x) = 0, h(x) \leq 0\} = \left\{\begin{bmatrix} 3 \\ 3 \end{bmatrix}\right\} = \{x^\star\}.$$

The minimizer $x^\star = \begin{bmatrix} 3 \\ 3 \end{bmatrix}$ of Problem (2.18) is illustrated as a • in Figure 2.12.
(We will return to this example problem in Section 17.1.1.2.)

As with linear programs, quadratic programs are usually written so as to emphasize the quadratic and affine functions by showing the coefficient matrices explicitly as in, for example:

$$\min_{x\in\mathbb{R}^2}\left\{\frac{1}{2}x^\dagger Qx + c^\dagger x|Ax = b, Cx \leq d\right\},$$

where the objective has quadratic and linear terms but for brevity we have omitted
the constant term that was present in f since it does not affect the minimizer of
the problem. To represent Problem (2.18) in this way, the appropriate vectors and
matrices are:

$$Q = \begin{bmatrix} 2 & 0 \\ 0 & 2 \end{bmatrix}, c = \begin{bmatrix} -2 \\ -6 \end{bmatrix},$$
$$A = \begin{bmatrix} 1 & -1 \end{bmatrix}, b = [0], C = \begin{bmatrix} 0 & -1 \end{bmatrix}, d = [-3].$$

(Again, we emphasize that the function specified by Q and c differs from f in
Problem (2.18) by a constant but this does not affect the minimizer.)

As with linear programming, there has been considerable effort spent on developing software for this type of problem. If the problem at hand is quadratic, then it is sensible to use quadratic programming software to solve it.

Non-linear program If there are no restrictions on f, g, and h, then the problem is called a **non-linear program** or a **non-linear optimization problem**. Non-linear programs include, as special cases, both linear and quadratic programming.

For example, consider the objective $f : \mathbb{R}^3 \to \mathbb{R}$ defined by:

$$\forall x \in \mathbb{R}^3, \; f(x) = (x_1)^2 + 2(x_2)^2,$$

the equality constraint function $g : \mathbb{R}^3 \to \mathbb{R}^2$ defined by:

$$\forall x \in \mathbb{R}^3, \; g(x) = \begin{bmatrix} 2 - x_2 - \sin(x_3) \\ -x_1 + \sin(x_3) \end{bmatrix},$$

and the inequality constraint function $h : \mathbb{R}^3 \to \mathbb{R}$ defined by:

$$\forall x \in \mathbb{R}^3, \; h(x) = \sin(x_3) - 0.5.$$

The problem is:

$$\min_{x \in \mathbb{R}^3} \{ f(x) | g(x) = \mathbf{0}, h(x) \le 0 \}. \tag{2.19}$$

(We will return to this problem in Exercise 3.35 and Section 19.2.1.3.)

Convexity We will see in Section 2.6.3 that we can also classify problems on the basis of the notion of **convexity**, to be introduced in that section.

Satisfaction of constraints The inequality constraint $3 - x_2 \le 0$ is satisfied with equality at the solution of Problem (2.18). That is, $3 - x_2^\star = 0$. To describe this we make:

Definition 2.4 Let $h : \mathbb{R}^n \to \mathbb{R}^r$. An inequality constraint $h_\ell(x) \le 0$ is called a **binding constraint** or an **active constraint** at $x^\star$ if $h_\ell(x^\star) = 0$. It is called **non-binding** or **inactive** at $x^\star$ if $h_\ell(x^\star) < 0$. The set:

$$\mathbb{A}(x^\star) = \left\{ \ell \in \{1, \ldots, r\} \, | \, h_\ell(x^\star) = 0 \right\}$$

is called the **set of active constraints** or the **active set** for $h(x) \le \mathbf{0}$ at $x^\star$. $\square$

The active constraints at $x^\star$ are the inequality constraints that are satisfied with equality by $x^\star$. Note that the entries in the active set depend on the ordering of the constraints as specified by the entries in h. Typically, only some of the inequality constraints are satisfied with equality at a point $x^\star \in \mathbb{R}^n$ so that $\mathbb{A}(x^\star)$ is usually a strict subset of $\{1, \ldots, r\}$. (See Section A.1.2 in Appendix A for the definition of set relations such as "strict subset.")

x_2

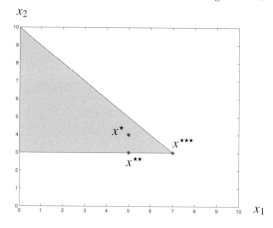

Fig. 2.13. Points $x^\star, x^{\star\star}$, and $x^{\star\star\star}$ that are feasible with respect to inequality constraints. The feasible set is the shaded triangular region for which $x_2 \geq 3$ and $x_1 + x_2 \leq 10$.

Naturally, if $x^\star$ is feasible for equality constraints $g(x) = \mathbf{0}$ then these equality constraints are also satisfied with equality by $x^\star$. However, by definition, we do not include equality constraints in the set of active constraints. We make the following definition.

Definition 2.5 Let $h : \mathbb{R}^n \to \mathbb{R}^r$. The point $x^\star$ is called **strictly feasible** for the inequality constraint $h_\ell(x) \leq 0$ if $h_\ell(x^\star) < 0$. The point $x^\star$ is called **strictly feasible** for the inequality constraints $h(x) \leq \mathbf{0}$ if $h(x^\star) < \mathbf{0}$. □

If $h : \mathbb{R}^n \to \mathbb{R}^r$ is continuous and satisfies certain other conditions then, by Exercise A.15 in Appendix A, the **boundary** (see Definition A.42) of $\mathbb{S} = \{x \in \mathbb{R}^n | h(x) \leq \mathbf{0}\}$ is the set $\{x \in \mathbb{R}^n | h(x) \leq \mathbf{0}$ and, for at least one ℓ, $h_\ell(x) = 0\}$ and its **interior** (see Definition A.40) is the set $\{x \in \mathbb{R}^n | h(x) < \mathbf{0}\}$; that is, the set of strictly feasible points for the inequality constraints is the interior of $\mathbb{S}$.

Example Consider the inequality constraint function $h : \mathbb{R}^2 \to \mathbb{R}^2$ defined by:

$$\forall x \in \mathbb{R}^2, h(x) = \begin{bmatrix} 3 - x_2 \\ x_1 + x_2 - 10 \end{bmatrix}.$$

Re-arranging the constraint $h(x) \leq \mathbf{0}$ we obtain that feasible points satisfy $x_2 \geq 3$ and $x_1 + x_2 \leq 10$. We consider the points:

$$x^\star = \begin{bmatrix} 5 \\ 4 \end{bmatrix}, x^{\star\star} = \begin{bmatrix} 5 \\ 3 \end{bmatrix}, x^{\star\star\star} = \begin{bmatrix} 7 \\ 3 \end{bmatrix}.$$

These points are all feasible with respect to the inequality constraints $h(x) \leq \mathbf{0}$ and are illustrated in Figure 2.13 as •.

We have the following.

• For the point $x^\star$, constraints $h_1(x) \leq 0$ and $h_2(x) \leq 0$ are non-binding so that the active set is $\mathbb{A}(x^\star) = \emptyset$. This point is in the interior of the set $\{x \in \mathbb{R}^2 | h(x) \leq \mathbf{0}\}$.

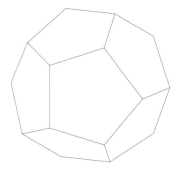

Fig. 2.14. Dodecahedron
in $\mathbb{R}^3$.

- For the point $x^{\star\star}$, the constraint $h_2(x) \leq 0$ is non-binding while the constraint $h_1(x) \leq 0$ is binding so that the active set is $\mathbb{A}(x^{\star\star}) = \{1\}$. This point is on the boundary of the set $\{x \in \mathbb{R}^2 | h(x) \leq \mathbf{0}\}$.
- For the point $x^{\star\star\star}$, constraints $h_1(x) \leq 0$ and $h_2(x) \leq 0$ are both binding so that the active set is $\mathbb{A}(x^{\star\star\star}) = \{1, 2\}$. This point is on the boundary of the set $\{x \in \mathbb{R}^2 | h(x) \leq \mathbf{0}\}$.

Example in higher dimension Consider Figure 2.14, which shows a **dodecahedron**, a twelve-sided solid, in $\mathbb{R}^3$. Six of the twelve sides are visible in Figure 2.14. The set of points on the surface and interior to the dodecahedron is defined by the intersection of 12 half-spaces. These half-spaces can be specified by a vector of 12 affine inequality constraints. That is, each inequality specifies one of the half-spaces. The boundary of each half-space in $\mathbb{R}^3$ is a plane, but for spaces $\mathbb{R}^n$ with $n > 3$, the boundary of each space is a **hyperplane**. (See Definition A.52.) The dodecahedron is an example of a set that can be described in the form $\mathbb{S} = \{x \in \mathbb{R}^3 | h(x) \leq \mathbf{0}\}$ with $h : \mathbb{R}^3 \to \mathbb{R}^{12}$ affine.

We consider various cases for a point in $\mathbb{S}$.

- $x^\star$ is in the interior of the dodecahedron. We have $h(x^\star) < \mathbf{0}$ and $\mathbb{A}(x^\star) = \emptyset$.
- $x^{\star\star}$ is on a **face** of the dodecahedron but not on an **edge** or **vertex**. That is, exactly one constraint ℓ is binding, $\mathbb{A}(x^{\star\star}) = \{\ell\}$, and $x^{\star\star}$ is on the boundary.
- $x^{\star\star\star}$ is on an **edge** but not a **vertex** of the dodecahedron. That is, exactly two constraints ℓ, ℓ' are binding, $\mathbb{A}(x^{\star\star\star}) = \{\ell, \ell'\}$, and $x^{\star\star\star}$ is on the boundary.
- $x^{\star\star\star\star}$ is on a **vertex** of the dodecahedron. That is, exactly three constraints ℓ, ℓ', and ℓ'' are binding, $\mathbb{A}(x^{\star\star\star\star}) = \{\ell, \ell', \ell''\}$, and $x^{\star\star\star\star}$ is on the boundary.

Discussion The importance of the notion of binding constraints is that it is typical for some of the inequality constraints to be binding at the optimum. Otherwise, we could omit the constraints and solve an unconstrained or equality-constrained problem. On the other hand, it is typical for not all the inequality constraints to be binding at the optimum. Otherwise, we could just consider them all to be equality constraints and solve an equality-constrained problem.

Representation of inequality constraints In the definition of inequality-constrained problems, we have considered only **single-sided inequalities** of the form $h(x) \leq 0$; however, an inequality of the form $\hat{h}(x) \geq 0$ can be re-expressed in the form $h(x) \leq 0$ by defining $h(x) = -\hat{h}(x), \forall x$. In practice, most optimization software can deal directly with both types of inequalities. Furthermore, most optimization software can deal directly with:

- **double-sided functional inequalities** such as $\underline{h} \leq h(x) \leq \overline{h}$, and
- **double-sided inequalities on variables** such as $\underline{x} \leq x \leq \overline{x}$,

Moreover, the software will usually take advantage of several simplifications that arise due to the double-sided inequalities. That is, if a problem is naturally formulated with a double-sided inequality, then the problem should be specified to the software as a double-sided inequality rather than as two single-sided inequalities. Furthermore, an inequality on a variable should always be specified to software as an inequality on the variable rather than as a special case of a functional inequality.

   For notational simplicity, we will usually restrict ourselves to inequalities of the form $h(x) \leq 0$, but recognize that problems may be easier to express in terms of the more comprehensive form $\underline{x} \leq x \leq \overline{x}, \underline{h} \leq h(x) \leq \overline{h}$. As remarked above, it is almost always worthwhile to take advantage of the more comprehensive form when the software has the capability.

Case studies Inequality-constrained problems will be investigated generally in Part V, beginning with six case studies:

- least-cost production with capacity constraints (in Section 15.1),
- optimal routing in a data communications network (in Section 15.2),
- least absolute value data fitting (in Section 15.3),
- optimal margin pattern classification (in Section 15.4),
- sizing of gate interconnects in integrated circuits (in Section 15.5), and
- optimal power flow (in Section 15.6).

2.3.2.4 Summary

For small example problems, whether they are unconstrained, equality-constrained, or inequality-constrained, inspection of a carefully drawn diagram can yield the minimum and minimizer. In general, a careful sketch can often be useful in helping to understand a problem. However, for larger problems where the dimension of x increases significantly past two, or the dimension of g or h increases, the geometry becomes more difficult to visualize and intuition becomes less reliable in predicting the solution. In the rest of this book, we will develop more abstract characterizations of the conditions for a point to be an optimizer of a problem. These characterizations will apply to problems where the dimensions of the decision vector and constraint functions are fixed, but arbitrary, finite values. This will lead to systematic procedures or "algorithms" for finding such points. We will discuss algorithms generally in Section 2.4.

2.3.3 Problems without minimum and the infimum

2.3.3.1 Analysis

We emphasize that it is easily possible for a problem to not possess a minimum. That is, we can easily construct a problem with no minimum. We first enumerate the possibilities for the existence of a minimum and the existence of lower bounds.

(i) $\min_{x \in \mathbb{S}} f(x)$ has a minimum, $f^\star$ say. Then every number $\underline{f}$ such that $\underline{f} \leq f^\star$ is a lower bound for the problem. Amongst all the lower bounds for the problem, $f^\star$ is the **greatest lower bound** for $\min_{x \in \mathbb{S}} f(x)$.

(ii) $\min_{x \in \mathbb{S}} f(x)$ does not have a minimum, but f is bounded below on $\mathbb{S}$. Then we can still consider the set of all lower bounds for the problem. It turns out that there is always a well-defined largest element of this set [6, section A.3][104, section 2.1], which we again call the greatest lower bound for $\min_{x \in \mathbb{S}} f(x)$.

(iii) $\min_{x \in \mathbb{S}} f(x)$ is unbounded below. Then, no real number is a lower bound.

(iv) $\mathbb{S} = \emptyset$, so that the problem is infeasible. Then we can adopt the convention that *every* real number satisfies the definition of the lower bound.

The following definition embodies the idea of the largest number that is a lower bound for the problem, but generalizes it to include the cases where the problem is unbounded or infeasible. To cover these additional cases we must consider "numbers" that can either be an element of $\mathbb{R}$ or equal to ∞ or to $-\infty$. Such a number is called an **extended real number** [104, section 2.3]. The numbers $-\infty$ and ∞ are defined to have the following property:

$$\forall x \in \mathbb{R}, \ -\infty < x < \infty.$$

(See Definition A.17.)

Definition 2.6 Let $\mathbb{S} \subseteq \mathbb{R}^n$, $f : \mathbb{S} \to \mathbb{R}$. Then, $\inf_{x \in \mathbb{S}} f(x)$, the **infimum** of the corresponding minimization problem, $\min_{x \in \mathbb{S}} f(x)$, is defined by:

$$\inf_{x \in \mathbb{S}} f(x) = \begin{cases} \text{the greatest lower bound} \\ \quad \text{for } \min_{x \in \mathbb{S}} f(x), & \text{if } \min_{x \in \mathbb{S}} f(x) \text{ is bounded below,} \\ -\infty, & \text{if } \min_{x \in \mathbb{S}} f(x) \text{ is unbounded below,} \\ \infty, & \text{if } \min_{x \in \mathbb{S}} f(x) \text{ is infeasible.} \end{cases}$$

By definition, the infimum is equal to the minimum of the corresponding minimization problem $\min_{x \in \mathbb{S}} f(x)$ if the minimum exists, but the infimum exists even if the problem has no minimum. To emphasize the role of $\mathbb{S}$, we also use the notation $\inf_{x \in \mathbb{R}^n} \{ f(x) | x \in \mathbb{S} \}$ and analogous notations for the infimum. $\square$

2.3.3.2 Examples

The following five examples illustrate problems without a minimum.

Unconstrained problem with unbounded objective For an example of an unconstrained problem that has no minimum, let $\mathbb{S} = \mathbb{R}$ and consider the linear objective $f : \mathbb{R} \to \mathbb{R}$ defined by:

$$\forall x \in \mathbb{R}, \ f(x) = x. \tag{2.20}$$

There is no $f^\star \in \mathbb{R}$ such that $\forall x \in \mathbb{R}$, $f^\star \leq f(x)$. The problem $\min_{x \in \mathbb{R}} f(x)$ is unbounded below. The infimum is $\inf_{x \in \mathbb{R}} f(x) = -\infty$.

Unconstrained problem with objective that is bounded below For an example of an unconstrained problem that has no minimum, but has an objective that is bounded below, let $\mathbb{S} = \mathbb{R}$ and consider the objective $f : \mathbb{R} \to \mathbb{R}$ defined by:

$$\forall x \in \mathbb{R}, \ f(x) = \exp(x).$$

In this case, $\underline{f} = 0$ is a lower bound for the problem $\min_{x \in \mathbb{R}} f(x)$, but there is no value of $x \in \mathbb{R}$ that achieves this bound. The problem has no minimum but the infimum is $\inf_{x \in \mathbb{R}} f(x) = 0$. The function exp is illustrated in Figure 2.15.

Strict inequalities Again consider the objective $f : \mathbb{R} \to \mathbb{R}$ defined in (2.20), but let the feasible set be

$$\mathbb{S} = \{ x \in \mathbb{R} | x > 0 \}.$$

The inequality defining the feasible set is a **strict inequality**. Figure 2.16 shows $f(x)$ versus x for $x \in \mathbb{S}$. The circle $\circ$ at $x = 0$ indicates that this point is not included in the graph but that points to the right of $x = 0$ and arbitrarily close to $x = 0$ are included in the graph.

Note that $\forall x \in \mathbb{S}$, $f(x) \geq 0$, so that the problem is bounded below by 0. (If the "below" is clear from context, we simply say that the problem is bounded by 0.)

$f(x)$

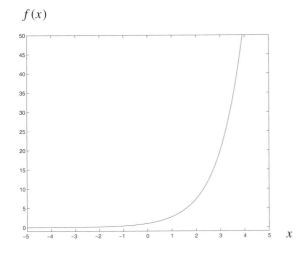

Fig. 2.15. The function exp is bounded below on the feasible set $\mathbb{R}$ but has no minimum.

$f(x)$

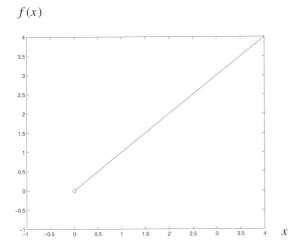

Fig. 2.16. Function that is bounded below on feasible set but where the problem has no minimum because the feasible set is defined by a strict inequality. The function is illustrated only on the feasible set. The circle o at $x = 0$, $f(x) = 0$ indicates that this point is not included in the graph but that points to the right of $x = 0$ and arbitrarily close to $x = 0$ are included in the graph.

However, there is no $x^\star \in \mathbb{S}$ such that $f(x^\star) = 0$. In this example, the problem is bounded, but does not have a minimum nor a minimizer. For this problem, the infimum is $\inf_{x \in \mathbb{R}} \{f(x) | x > 0\} = 0$.

Inconsistent constraints Consider any objective $f : \mathbb{R} \to \mathbb{R}$ and let

$$\mathbb{S} = \{x \in \mathbb{R} | g(x) = \mathbf{0}\},$$

where $g : \mathbb{R} \to \mathbb{R}^2$ is defined by:

$$\forall x \in \mathbb{R}, g(x) = \begin{bmatrix} x + 1 \\ x - 1 \end{bmatrix}.$$

$f(x)$

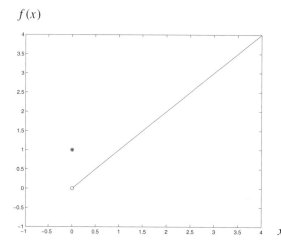

Fig. 2.17. Function (2.21) that is bounded below on feasible set but where the problem has no minimum because the function is discontinuous. The function is illustrated only on the feasible set. The bullet • at $x = 0$, $f(x) = 1$ indicates that this is the value of the function at $x = 0$.

Then there are no feasible solutions, since the equality constraints are **inconsistent** and so $\mathbb{S} = \emptyset$. In this example, there are no feasible values of x and therefore no minimum. The infimum is $\inf_{x \in \mathbb{R}} \{ f(x) | g(x) = \mathbf{0} \} = \infty$.

Discontinuous objective Finally, let

$$\mathbb{S} = \{ x \in \mathbb{R} | x \geq 0 \},$$

and define $f : \mathbb{S} \to \mathbb{R}$ by:

$$\forall x \in \mathbb{S},\ f(x) = \begin{cases} 1, & \text{if } x = 0, \\ x, & \text{if } x \neq 0. \end{cases} \tag{2.21}$$

The function f has a "jump" at $x = 0$ as illustrated in Figure 2.17. Such a function is called **not continuous** or **discontinuous**. (See Definition A.35.) The problem $\min_{x \in \mathbb{S}} f(x)$ is bounded below by zero, but there is again no minimum nor minimizer. The infimum is $\inf_{x \in \mathbb{R}} \{ f(x) | x \geq 0 \} = 0$.

2.3.3.3 Summary

In all five examples in Section 2.3.3.2, $\operatorname{argmin}_{x \in \mathbb{R}} f(x)$ is the empty set $\emptyset$. Careful formulation of a problem can sometimes avoid these issues. The first example can only occur if the constraint set $\mathbb{S}$ is unbounded and the objective is not bounded below. In practical problems, constraint sets are often bounded; however, this is not the case, by definition, in unconstrained problems. Practical unconstrained problems typically involve objectives that are bounded below. As the second example showed, having an objective that is bounded below is not sufficient to guarantee that a minimum exists. However, many unconstrained problems have objectives that grow rapidly as any entry in their argument grows large. For such functions there is usually a minimizer.

The third example illustrates the difficulties that can arise when *strict* inequalities are used to define a feasible region. Such inequalities typically define an **open set**. (See Definition A.41.) In contrast, it is often the case in engineering design problems that either:

- there is no physical interpretation of strict inequality constraints such as $x > 0$, or
- there is, practically speaking, no difference between a **strict** inequality constraint $x > 0$ and the **non-strict** inequality constraint $x \geq 0$.

In these cases we should always specify inequalities as non-strict inequalities. This choice is made explicitly in our definition of inequality-constrained optimization. Non-strict inequalities specified in terms of continuous functions define **closed** sets. (See Definition A.41 and Exercise A.14.)

The fourth example arises because there are no feasible points. This can be a sign of an incorrect problem formulation; however, in some applications, testing for the existence of a feasible point may be an important part of the decision process.

Finally, the fifth example illustrates possibilities when the objective is not continuous. For an engineering design problem, the presence of discontinuities in the model poses great difficulties if the location of the discontinuity depends on experimental data. We will usually consider continuous objectives.

2.3.4 Conditions for problems to possess a minimum and minimizer

We will usually restrict ourselves to continuous objective and constraint functions that are, furthermore, also once or twice partially differentiable. (See Definitions A.35 and A.36 for formal definitions of continuous and of partially differentiable.) Such functions are also called **smooth**. It is important to realize that the assumption of smoothness is different from assuming that we can explicitly calculate the derivative of a function. In particular, our algorithms will make use of the assumption that the functions are partially differentiable; however, in some cases we will develop subsidiary calculations to *numerically estimate* partial derivatives if they are not explicitly known analytically.

We are not going to consider optimization of **non-smooth** functions in detail. (However, we will introduce a particular class of non-smooth functions in Section 2.6.3.6 and then discuss in Sections 3.1.3 and 3.1.4 methods of transforming problems with this type of non-smooth objective into a problem with a smooth objective.) Non-smooth objectives are discussed in a variety of references, including [45, section 4.2][100].

In the following, we have a sufficient condition for an optimization problem with a continuous objective to possess a minimum.

Theorem 2.1 *Let $\mathbb{S} \subseteq \mathbb{R}^n$ be closed (see Definition A.41) and bounded (see Definition A.46) and let $f : \mathbb{S} \to \mathbb{R}$ be continuous (see Definition A.35.) Then the problem $\min_{x \in \mathbb{S}} f(x)$ possesses a minimum and minimizer.*

Proof See [104, propositions 2.17 and 9.10]. □

By choosing an objective function, $f : \mathbb{R}^n \to \mathbb{R}$, and constraint functions, $g : \mathbb{R}^n \to \mathbb{R}^m$ and $h : \mathbb{R}^n \to \mathbb{R}^r$, that are continuous and such that the feasible set is bounded, we can therefore ensure that a problem of the form $\min_{x \in \mathbb{R}^n} \{ f(x) | g(x) = 0, h(x) \le 0 \}$ possesses a minimum and minimizer.

2.3.5 Maximization problems and the supremum

We can also define a **maximization** problem by:

$$\max_{x \in \mathbb{S}} f(x) = -\min_{x \in \mathbb{S}} (-f(x)). \tag{2.22}$$

Maximization problems are natural when the objective is related to a measure of "profit" and we want to maximize the profit. We also define $\operatorname{argmax}_{x \in \mathbb{S}} f(x) = \operatorname{argmin}_{x \in \mathbb{S}} (-f(x))$. We speak of the maximum and maximizer of a maximization problem. Analogously with minimization problems, we define the **supremum** as follows:

Definition 2.7 Let $\mathbb{S} \subseteq \mathbb{R}^n$, $f : \mathbb{S} \to \mathbb{R}$. Then, $\sup_{x \in \mathbb{S}} f(x)$, the **supremum** of the corresponding maximization problem $\max_{x \in \mathbb{S}} f(x)$ is defined by:

$$
\sup_{x \in \mathbb{S}} f(x) =
\begin{cases}
\text{the \textbf{least upper bound}} & \\
\quad \text{for } \max_{x \in \mathbb{S}} f(x), & \text{if } \max_{x \in \mathbb{S}} f(x) \text{ is bounded above,} \\
\infty, & \text{if } \max_{x \in \mathbb{S}} f(x) \text{ is unbounded above,} \\
-\infty, & \text{if } \max_{x \in \mathbb{S}} f(x) \text{ is infeasible.}
\end{cases}
$$

The supremum is equal to the maximum of the corresponding maximization problem $\max_{x \in \mathbb{S}} f(x)$ if the maximum exists. □

We will usually discuss theoretical results in terms of minimization problems, but this does not limit our applications, since we can always re-write a maximization problem in terms of a minimization problem. Some optimization software is designed to minimize objectives, while other software is designed to maximize objectives, and some software can either maximize or minimize depending on user specification. To include both minimization and maximization, we can speak of the optimum and optimizer of an optimization problem.

2.3.6 Extended real functions

We can generalize Definitions 2.6 and 2.7 to include the possibility that f itself takes on the special values of ∞ or $-\infty$. Such a function is called an **extended real function** [104, section 2.3] (See Definition A.17.) First we must make the natural generalization of the definition of an optimization problem to allow objectives that are extended real functions:

Definition 2.8 Let $\mathbb{S} \subseteq \mathbb{R}^n$, $f : \mathbb{S} \to \mathbb{R} \cup \{-\infty, \infty\}$, and $f^\star \in \mathbb{R}$. Then by

$$f^\star = \min_{x \in \mathbb{S}} f(x),$$

we mean that:

$$\exists x^\star \in \mathbb{S} \text{ such that } f^\star = f(x^\star) \in \mathbb{R} \text{ and } (x \in \mathbb{S}) \Rightarrow (f(x^\star) \leq f(x)).$$

□

If no finite $f^\star$ exists satisfying the definition, then there is no minimum according to our definition. In particular, this can happen if every feasible $x \in \mathbb{S}$ has $f(x) = \infty$. In this case, we generalize Definition 2.6 of infimum by defining the infimum of such a problem to be ∞. Similarly, for a maximization problem for which the objective is always equal to $-\infty$ at feasible points, we also generalize Definition 2.7 of supremum by defining the supremum of the problem to be $-\infty$ in this case. These generalizations of Definitions 2.6 and 2.7 will be useful in Section 3.4 when we discuss some functions that are themselves defined in terms of the supremum or infimum of associated problems.

Naturally, objectives that arise directly from physical systems do not take on the values ∞ or $-\infty$; however, we will see that the concept of such an objective can be useful theoretically. Moreover, ∞ can be loosely thought of as a positive number that is much larger than the largest element of the range of the objective.

2.4 Algorithms

We distinguish two basic types of algorithms in this book:

- **direct**, to be described in Section 2.4.1, and
- **iterative**, to be described in Section 2.4.2.

We appeal to intuitive notions of computation to describe these types of algorithms. More formal definitions of some of these ideas can be found in [40].

2.4.1 Direct

By a **direct algorithm**, we mean that from the problem specification it is possible to write down a finite list of **operations** that calculates the solution of the problem. (Some authors use a similar term, **direct search**, to mean an iterative algorithm that does not make use of derivatives. The word "direct" is being used in two different senses in these two terms.)

Each operation must be a **basic operation** of our computer; that is, each operation requires an amount of computation time that is bounded by a constant. We will see that the algorithm we develop for the solution of linear equations in Chapter 5 is an example of a direct algorithm, given the reasonable assumption that manipulating any element of a vector or matrix, comparisons of numbers, and the arithmetic operations of addition, subtraction, multiplication, and division are all basic operations of our computer. We will usually characterize the computational effort to execute a direct algorithm in terms of the number of such operations required to execute the algorithm.

2.4.1.1 Discussion

Under the (usually unrealistic) assumptions that:

- all numbers in the problem specification are represented to **infinite precision**,
- all arithmetic operations are carried out to infinite precision, and
- the answers to each arithmetic operation are represented to infinite precision,

then the answer obtained from a direct algorithm would be exact. In practice, finite-precision representation and round-offs during calculations mean that the solutions are at best approximate. For example, when a large number and a small number are added together then the calculated answer may not be the true sum because some of the significant figures of the smaller number were discarded when the numbers were loaded into the arithmetic unit of the computer.

With modern computational systems and carefully designed algorithms, the errors are often very small; however, certain problem specifications are such that their solution is extremely sensitive to changes in the values of numbers in the specification. Such problems are called **ill-conditioned**. Finite-precision representations and calculations will often produce answers that are poor solutions to ill-conditioned problems, even if the values of numbers in the problem specification *were* exactly known. In practice, even if values were known exactly, they can only be represented to finite precision. More typically, the values are only known approximately and their representation introduces further error so that calculations will produce answers that may be very poor solutions to the exactly specified problem.

In summary, while direct algorithms in principle can be used to calculate the exact answer to a problem, there is in practice always some error. The error is typically small but can be large for ill-conditioned problems. Ill-conditioned problems will be discussed briefly in Section 2.7.6 and then in several other places throughout the book as they arise.

2.4.1.2 Applicability

As mentioned above, we will develop a direct algorithm for solving linear equations in Chapter 5. Special cases of some of our later problems will also be amenable to direct algorithms. However, some problems *cannot* be solved by direct algorithms.

For example, consider non-linear simultaneous equations. To see the issues involved, let us first restrict ourselves to the case $m = n = 1$ and consider $g : \mathbb{R} \to \mathbb{R}$ such that g is a **polynomial** in a single variable. (See Definition A.22.) If g is a quadratic polynomial, then we have already remarked in Section 2.2.2 that the "quadratic equation" provides a direct algorithm to find the solution or solutions to $g(x) = 0$, if taking a square root is a basic operation of our computer (in addition to the "standard" operations of addition, subtraction, multiplication, and division.) There are also direct algorithms to solve a non-linear equation $g(x) = 0$ involving a single cubic or a quartic function of one variable, so long as taking square and cube roots is a basic arithmetic operation of our computer.

However, for non-linear equations involving arbitrary fifth or higher degree polynomials, there is *provably* no direct algorithm available to find the solution, even allowing square and cube (and even higher) roots as basic operations [112, theorem 15.7]. Generally speaking, there are also no direct algorithms for solving arbitrary systems of non-linear equations $g(x) = \mathbf{0}$, $g : \mathbb{R}^n \to \mathbb{R}^m$ when m and n are both greater than one. (For some special cases, there are direct algorithms. See Exercise 2.9.)

To solve arbitrary systems of non-linear simultaneous equations (and optimization problems) we will have to develop alternative approaches, namely iterative algorithms, that yield solutions that are sufficiently accurate according to some appropriate criteria. (See Exercise 2.10.) We will introduce iterative algorithms in the next section.

2.4.2 Iterative

In an **iterative** algorithm, a sequence of intermediate values are generated successively by the algorithm. We begin with an **initial guess** of the solution and try to successively improve it. Each intermediate value is called an **iterate**. Under certain circumstances, the sequence of iterates "approaches" a solution of the problem, if a solution exists. Ideally, each successive iterate is a better approximation to the solution than the previous. Most of the algorithms in this book are iterative: in particular, we will develop iterative algorithms for each problem that cannot be solved by a direct algorithm.

In general, none of the iterates produced by an iterative algorithm will exactly solve the problem; however, we cannot keep iterating forever. As a practical matter, therefore, we must also include a **termination criterion** or **stopping criterion** in an iterative algorithm [45, section 8.2.3]. That is, we must specify conditions that, when satisfied, will cause the algorithm to terminate with a suitable approximation to the exact solution or with a report that no suitable approximation has been found. We will return to this topic as we develop specific iterative algorithms for each problem.

In Section 2.4.2.1, we define an **update recursion** to generate each successive iterate. Then, in Section 2.4.2.2, using the notion of a norm, we will define carefully the way in which successive iterates can be thought of as converging to a solution.

We will then define the rate of convergence to the solution in Section 2.4.2.3. We will use this information to characterize how many iterations are needed to find a solution of desired accuracy. The product of the number of iterations and the number of operations per iteration then characterizes the total number of operations required to find a solution of desired accuracy.

2.4.2.1 Recursion to define iterates

The iterative algorithms that we consider can be represented with the following general form of recursion:

$$x^{(\nu+1)} = x^{(\nu)} + \alpha^{(\nu)} \Delta x^{(\nu)}, \nu = 0, 1, 2, \ldots,$$

where (see Definition A.18):

- $x^{(0)}$ is the **initial guess** of the solution,
- ν is the iteration counter, $\nu = 0, 1, 2, 3, \ldots$,
- $x^{(\nu)}$ is the value of the iterate at the ν-th iteration,
- $\alpha^{(\nu)} \in \mathbb{R}_+$ is the **step-size** (where $\mathbb{R}_+$ is the set of non-negative real numbers) and, usually, $0 < \alpha^{(\nu)} \leq 1$,
- $\Delta x^{(\nu)} \in \mathbb{R}^n$ is the **step direction**, and
- the product $\alpha^{(\nu)} \Delta x^{(\nu)}$ is the **update** to add to the current iterate $x^{(\nu)}$ to obtain the new iterate $x^{(\nu+1)}$.

We call $\Delta x^{(\nu)}$ the step direction because it points in the direction that we step from $x^{(\nu)}$ to obtain the (hopefully) improved solution $x^{(\nu+1)}$. The step-size $\alpha^{(\nu)}$ is chosen

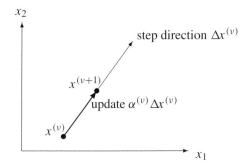

Fig. 2.18. Update of iterate in $\mathbb{R}^2$. The bullets • indicate the locations of the points $x^{(\nu)}$ and $x^{(\nu+1)}$, while the arrows ↗ indicate the magnitudes and directions of the vectors $\Delta x^{(\nu)}$ and $\alpha^{(\nu)} \Delta x^{(\nu)}$.

so that $x^{(\nu+1)}$ is a "sufficient" improvement over $x^{(\nu)}$ according to criteria we will develop. We often try to arrange that $\alpha^{(\nu)} = 1$ at many or most of the iterations.

The situation is illustrated in Figure 2.18 for $x \in \mathbb{R}^2$, where $x^{(\nu)}$ and the updated iterate $x^{(\nu+1)}$ are illustrated for a step-size of $\alpha^{(\nu)} = 0.5$. The vectors $\Delta x^{(\nu)}$ and $\alpha^{(\nu)} \Delta x^{(\nu)}$ are drawn as arrows ↗ with their tails at $x^{(\nu)}$ to show a step from $x^{(\nu)}$ towards the new iterate $x^{(\nu+1)}$.

Usually, the values of $\Delta x^{(\nu)}$ and $\alpha^{(\nu)}$ can each be thought of as the result of a direct algorithm for a problem that is specified by $x^{(\nu)}$; however, sometimes the calculation of $\alpha^{(\nu)}$ or $\Delta x^{(\nu)}$ is itself an iterative algorithm. If the calculation of $\Delta x^{(\nu)}$ and $\alpha^{(\nu)}$ is the result of a direct algorithm, then we can characterize the effort per iteration in terms of the number of basic operations required by that direct algorithm. Again, calculations involved in each iteration are usually subject to representation and round-off errors. These errors can be compounded over several iterations.

There are variations on this algorithm scheme that involve simultaneous computation of step direction and step-size. However, the separation into step direction and step-size allows us to introduce a basic approach to finding a step direction and then refine it with a step-size calculation.

2.4.2.2 Sequence of iterates and closeness to a solution

Iterative algorithms generate a sequence, $\{x^{(\nu)}\}_{\nu=0}^{\infty}$, of iterates. We hope that the successive elements in the sequence get closer to a solution $x^\star$ of our problem. To formalize this notion, we make:

Definition 2.9 Let $\|\bullet\|$ be a norm on $\mathbb{R}^n$. (See Definition A.28 for the definition of norm.) Let $\{x^{(\nu)}\}_{\nu=0}^{\infty}$ be a sequence of vectors in $\mathbb{R}^n$. Then, the sequence $\{x^{(\nu)}\}_{\nu=0}^{\infty}$ **converges** to a **limit** $x^\star$ if:

$$\forall \epsilon > 0, \exists N^\epsilon \in \mathbb{Z}_+ \text{ such that } (\nu \in \mathbb{Z}_+ \text{ and } \nu \geq N^\epsilon) \Rightarrow \left(\left\| x^{(\nu)} - x^\star \right\| \leq \epsilon \right).$$

The set $\mathbb{Z}_+$ is the set of non-negative integers. (See Definition A.3 in Appendix A.)

If the sequence $\{x^{(\nu)}\}_{\nu=0}^{\infty}$ converges to $x^{\star}$ then we write $\lim_{\nu\to\infty} x^{(\nu)} = x^{\star}$ or $\lim_{\nu\to\infty} x^{(\nu)} = x^{\star}$ and call $x^{\star}$ the **limit** of the sequence $\{x^{(\nu)}\}_{\nu=0}^{\infty}$. □

In the definition of convergence, we write N^{ϵ} to emphasize that N^{ϵ} will in general depend on ϵ. Typically, the smaller the value of ϵ, the larger the value of N^{ϵ} to satisfy the condition:

$$(\nu \in \mathbb{Z}_+ \text{ and } \nu \geq N^{\epsilon}) \Rightarrow \left(\left\|x^{(\nu)} - x^{\star}\right\| \leq \epsilon\right).$$

If a sequence $\{x^{(\nu)}\}_{\nu=0}^{\infty}$ converges to $x^{\star}$ then, for any "distance" ϵ, there must eventually be elements of the sequence that are close to $x^{\star}$ in the sense that their distance from $x^{\star}$, as measured by the norm, is less than ϵ. Moreover, after some iterate N^{ϵ}, all the subsequent iterates are within this distance ϵ of $x^{\star}$.

For example, consider the sequence $\{1/(\nu + 1)\}_{\nu=0}^{\infty}$. This sequence converges to $x^{\star} = \mathbf{0}$. (See Exercise 2.11.) Similarly, the sequences $\{(2)^{-\nu}\}_{\nu=0}^{\infty}$ and $\{(2)^{-((2)^{\nu})}\}_{\nu=0}^{\infty}$ also converge to $x^{\star} = \mathbf{0}$.

The definition of convergence is "theoretical" in that we cannot ever "experimentally" verify for all ϵ that there exists N^{ϵ} such that elements of the infinite sequence satisfy the requirements. Nevertheless, the notion of convergence will guide our search for a practical criterion for termination of iterations generated by an iterative algorithm. We will prove theoretical results of the form: "For a particular algorithm, if the specification of the problem satisfies certain properties then the sequence of iterates converges to the solution."

A further issue is that since we usually do not know the solution to the problem, we usually cannot use Definition 2.9 directly to describe convergence to the solution. We must usually rely on a combination of theoretical analysis and empirical observation of a finite number of the iterates to judge whether or not we are approaching a solution and to decide when to terminate calculations.

Sometimes, we cannot even prove that the sequence of iterates converges. In some cases, the best we can prove is that some **sub-sequence** (see Definition A.18) converges. A limit of a sub-sequence is called an **accumulation point** of the sequence.

2.4.2.3 Rate of convergence

Analysis The basic definition of convergence does not characterize *how fast* we are converging to the limit. Sometimes it is possible to theoretically characterize this rate. We make the following ([84, section 2.5]):

Definition 2.10 Let $\|\bullet\|$ be a norm. A sequence $\{x^{(\nu)}\}_{\nu=0}^{\infty}$ that converges to $x^{\star} \in \mathbb{R}^n$ is said to converge at **rate** $R \in \mathbb{R}_{++}$ (where $\mathbb{R}_{++}$ is the set of strictly positive real numbers)

$\left\| x^{(\nu)} - x^\star \right\|$

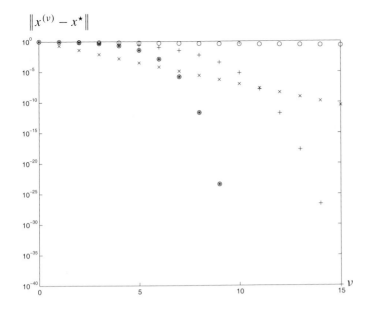

Fig. 2.19. Rates of convergence for several sequences, with: $R = 1$ and $C = 0.9$ shown as ○; $R = 1$ and $C = 0.2$ shown as ×; $R = 2$ and $C = 0.9$ shown as •; and $R = 1.5$ and $C = 0.9$ shown as +.

and with **rate constant** $C \in \mathbb{R}_{++}$ if:

$$\lim_{\nu \to \infty} \frac{\left\| x^{(\nu+1)} - x^\star \right\|}{\left\| x^{(\nu)} - x^\star \right\|^R} = C. \tag{2.23}$$

If (2.23) is satisfied for $R = 1$ and some value of C in the range $0 < C < 1$ then the rate is called **linear**. If (2.23) is satisfied for $R = 2$ and some C in the range $0 < C < \infty$ then the rate is called **quadratic**. If (2.23) is satisfied for some R in the range $1 < R < 2$ and some C in the range $0 < C < \infty$ then the rate is called **super-linear**. □

Sometimes, in Definition 2.10, the rate of convergence R is called the **asymptotic convergence rate** to emphasize that it may not apply for the first iterates, but only applies asymptotically as $\nu \to \infty$.

Under a linear rate of convergence, the iterates are getting closer to the limit $x^\star$ by a fixed factor C at each iteration, at least asymptotically. Notice that for a linear rate, if C is close to 1 then the progress towards the limit may be very slow. Roughly speaking, under quadratic convergence, if $C \approx 1$ then once the iterates become close enough to $x^\star$ then the number of correct digits in the iterate doubles with each iteration.

Example Figure 2.19 shows values of $\left\| x^{(\nu)} - x^\star \right\|$ versus ν for a linear rate of convergence (that is, $R = 1$) with:

- $C = 0.9$, shown as circles ○, and
- $C = 0.2$, shown as crosses ×.

The vertical scale in Figure 2.19 is logarithmic and so a "linear" rate of convergence appears as a line on this graph. Clearly, the progress towards the solution is slow if $C = 0.9$ but much faster if $C = 0.2$.

Figure 2.19 shows values of $\left\| x^{(\nu)} - x^\star \right\|$ versus ν for a quadratic rate of convergence with $C = 0.9$ as •. On the logarithmic scale of the graph, this curve appears "quadratic." Figure 2.19 also shows values of $\left\| x^{(\nu)} - x^\star \right\|$ versus ν for a super-linear rate of convergence with $R = 1.5$ and $C = 0.9$ as plus signs $+$.

Discussion For any given sequence there are generally several combinations of R and C satisfying (2.23). We can seek the largest value of R (together with some value of C in the range $0 < C < \infty$ for $R > 1$ or some value of C in the range $0 < C < 1$ for $R = 1$) that satisfies (2.23). It is customary to quote this largest value of R as *the* rate of convergence.

Qualitatively, the larger the value of R, the faster the iterates converge, at least asymptotically. However, as in Figure 2.19, an algorithm with a linear rate of convergence but with a small value of C may out-perform an algorithm with quadratic convergence, at least in the first few iterations. Moreover, these asymptotic rates of convergence may not even be a useful guide to the performance in the first few iterations if the asymptotic performance differs markedly from the performance in the first few iterations. Nevertheless, the convergence rate can be a guide to algorithm performance if the asymptotic rate applies at least roughly for the early iterations.

If we know the convergence rate as specified in (2.23), we can estimate how many iterations are required to reduce the initial error $\left\| x^{(0)} - x^\star \right\|$ by some specified factor. This is explored in Exercise 2.12. The total effort to reduce the initial error is the product of the effort per iteration and the number of iterations required. If we know the computational effort per iteration and the convergence rate then we can bound the total effort required to reduce the initial error to some desired final error.

2.5 Solutions of simultaneous equations

We now turn to the solution of simultaneous equations problems. In Section 2.5.1 we discuss the potential for multiple solutions of simultaneous equations while, in Sections 2.5.2 and 2.5.3, we consider conditions for uniqueness of solutions of simultaneous equations, first in the linear case and then the non-linear case.

2.5.1 Number of solutions

There may in general be none, one, or several solutions to systems of simultaneous equations. To demonstrate this, we first consider one linear equation $Ax = b$ in one variable so that $A, b \in \mathbb{R}$. The three possible cases are as follows:

$$0x = 0, \qquad \text{infinitely many solutions,}$$
$$0x = b, \qquad b \neq 0, \text{ no solutions,}$$
$$Ax = b, \qquad A \neq 0, \text{ one solution.}$$

The first two cases are trivial in one dimension, but have non-trivial generalizations in more than one variable. The generalization of the first case is the **null space**, to be discussed in Section 5.8.1.2. (See Definition A.50.) The second case is an example of inconsistent equations.

Non-linear simultaneous equations are more complicated. For example, for one quadratic equation $Q(x)^2 + Ax = b$ in one variable, so that $A, b, Q \in \mathbb{R}$, we have six cases:

$$0(x)^2 + 0x = 0, \qquad \text{infinitely many solutions,}$$
$$0(x)^2 + 0x = b, \qquad b \neq 0, \text{ no solutions,}$$
$$0(x)^2 + Ax = b, \qquad A \neq 0, \text{ one solution,}$$
$$Q(x)^2 + Ax = b, \qquad Q \neq 0, A^2 + 4Qb < 0, \text{ no (real) solutions,}$$
$$Q(x)^2 + Ax = b, \qquad Q \neq 0, A^2 + 4Qb = 0, \text{ one solution,}$$
$$Q(x)^2 + Ax = b, \qquad Q \neq 0, A^2 + 4Qb > 0, \text{ two solutions.}$$

The first three cases are repetitions of the corresponding cases for linear equations. The last three cases show the variety of possibilities with even a single equation. The situation becomes even more complicated for non-linear equations in more than one variable and for polynomials of higher degree and other functions. In general, for a polynomial of degree D in one variable, the equation $g(x) = 0$ can have up to D real solutions [112, theorem 2.8]. In the next two sections, we consider sufficient conditions for there to be no more than one solution for linear and non-linear simultaneous equations, respectively.

2.5.2 *Uniqueness of solution for linear equations*

Consider a matrix $A \in \mathbb{R}^{m \times n}$ and right-hand side $b \in \mathbb{R}^m$. If $m = n$ so that A has the same number of rows and columns then we say that the system of equations $Ax = b$ is a **square system** and that the matrix A is **square**. (See Definition A.7.) Necessary and sufficient conditions for there to be a unique solution to a square system of equations is that the coefficient matrix A be **non-singular**. (See Definition A.49.) We will consider non-square systems of linear simultaneous equations in Section 5.8.

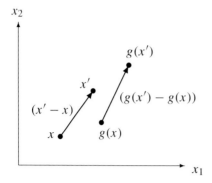

Fig. 2.20. Illustration of the definition of monotone. For all x and x' in $\mathbb{S}$, the vectors $(x' - x)$ and $(g(x') - g(x))$ point in directions that are within less than or equal to 90° of each other.

2.5.3 *Uniqueness of solution for non-linear equations*

In this section we will consider simultaneous equations where the number of equations equals the number of variables. We again refer to this as a **square system**. We will develop conditions for uniqueness of solution in terms of a property of the function specifying the equations.

2.5.3.1 *Monotone functions*

For a function $g : \mathbb{R} \to \mathbb{R}$ of one variable, there is the familiar notion of a monotonically increasing function. (See Definition A.24.) In the following definition, we generalize this to functions of several variables.

Definition 2.11 Let $\mathbb{S} \subseteq \mathbb{R}^n$ and let $g : \mathbb{S} \to \mathbb{R}^n$. We say that g is **monotone** on $\mathbb{S}$ if:

$$\forall x, x' \in \mathbb{S}, \left(g(x') - g(x)\right)^\dagger (x' - x) \geq 0. \tag{2.24}$$

We say that g is **strictly monotone** on $\mathbb{S}$ if:

$$\forall x, x' \in \mathbb{S}, (x \neq x') \Rightarrow \left(g(x') - g(x)\right)^\dagger (x' - x) > 0.$$

If g is monotone on $\mathbb{R}^n$ then we say that g is monotone. If g is strictly monotone on $\mathbb{R}^n$ then we say that g is strictly monotone. □

Geometrically, g is monotone on $\mathbb{S}$ if, for all pairs of vectors x and x' in $\mathbb{S}$, the vectors $(x' - x)$ and $(g(x') - g(x))$ point in directions that are within less than or equal to 90° of each other. This is the case for the vectors illustrated in Figure 2.20. The function g is strictly monotone if, for $x \neq x'$, the vectors $(x' - x)$ and $(g(x') - g(x))$ point in directions that are within less than 90° of each other.

Roughly speaking, monotone functions approximately preserve relative directions. For example, a monotone function could involve a translation of coordinates and a rotation by less than 90°.

Example Even if a function $\hat{g} : \mathbb{R}^n \to \mathbb{R}^n$ is not strictly monotone, we may find that by permuting the entries of $\hat{g}$ it is possible to create a strictly monotone function. For example, the function $\hat{g} : \mathbb{R}^2 \to \mathbb{R}^2$ defined by:

$$\forall x \in \mathbb{R}^2, \hat{g}(x) = \begin{bmatrix} x_2 \\ x_1 \end{bmatrix},$$

is not strictly monotone since:

$$
\begin{aligned}
\left(\hat{g}(x') - \hat{g}(x)\right)^\dagger (x' - x) &= 2(x_2' - x_2)(x_1' - x_1), \\
&< 0, \text{ if } x_2' > x_2 \text{ and } x_1' < x_1.
\end{aligned}
$$

However, the function $g : \mathbb{R}^2 \to \mathbb{R}^2$ obtained by swapping the entries of $\hat{g}$ is strictly monotone, since:

$$
\begin{aligned}
\left(g(x') - g(x)\right)^\dagger (x' - x) &= \|x' - x\|_2^2, \\
&> 0, \text{ for } x' \neq x.
\end{aligned}
$$

This example shows that the property of being strictly monotone depends, among other things, on the choice of the labeling of the entries of x and g.

Analysis The usefulness of the definition of a strictly monotone function lies in the following:

Theorem 2.2 *Let $\mathbb{S} \subseteq \mathbb{R}^n$ and $g : \mathbb{S} \to \mathbb{R}^n$ be strictly monotone on $\mathbb{S}$. Then there is at most one solution of the simultaneous equations $g(x) = \mathbf{0}$ that is an element of $\mathbb{S}$.*

Proof Suppose that there are two solutions $x^\star, x^{\star\star} \in \mathbb{S}$ with $x^\star \neq x^{\star\star}$. That is, $g(x^\star) = g(x^{\star\star}) = \mathbf{0}$. Consequently, $(g(x^\star) - g(x^{\star\star}))^\dagger (x^\star - x^{\star\star}) = 0$. But by definition of strictly monotone applied to $x^\star$ and $x^{\star\star}$, $(g(x^\star) - g(x^{\star\star}))^\dagger (x^\star - x^{\star\star}) > 0$. This is a contradiction. $\square$

Discussion The hypothesis in Theorem 2.2 that g be *strictly* monotone is important since the contradiction relies on the strict inequality in the definition of strict monotone. (See Exercise 2.16.)

Theorem 2.2 does not guarantee that there are any solutions to $g(x) = \mathbf{0}$. For example, consider the exponential function $\exp : \mathbb{R} \to \mathbb{R}$. This function is strictly monotone (see Exercise 2.18); however, there is no solution to $\exp(x) = 0$.

Moreover, it is possible for a function g to be not strictly monotone and yet there may be a unique solution or no solution to the equations $g(x) = \mathbf{0}$. For example, consider the function $g : \mathbb{R} \to \mathbb{R}$ defined by:

$$\forall x \in \mathbb{R}, g(x) = (x)^3 - x - 6.$$

$g(x)$

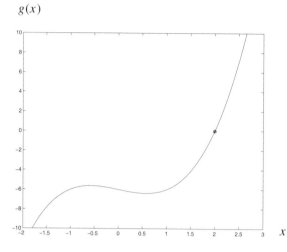

Fig. 2.21. Function g that is not strictly monotone but for which there is only one solution, $x^\star = 2$, to $g(x) = 0$. The solution is illustrated with the •.

This function is illustrated in Figure 2.21 and is not strictly monotone, yet there is only one solution to $g(x) = 0$, namely $x^\star = 2$.

Sometimes, we first define $g : \mathbb{R}^n \to \mathbb{R}^m$ and then consider its **restriction** (see Definition A.13) to a set $\mathbb{S} \subseteq \mathbb{R}^n$ on which we can show that g is strictly monotone. Theorem 2.2 makes no claim about the solutions of $g(x) = \mathbf{0}$ outside of the set $\mathbb{S}$. (See Exercise 2.19.)

Exercise 2.21 explores several systems of simultaneous equations defined in terms of monotonically increasing functions. As remarked above, monotonically increasing functions are a special case of monotone functions.

We can also consider more general conditions for there to be no more than one solution to simultaneous equations. For example, see Exercise 2.22.

2.5.3.2 Characterizing monotone and strictly monotone functions

The definition of monotone and strictly monotone functions is not convenient to apply directly because the condition in (2.24) involves checking over all possible pairs of points in the domain of the function. In this section, we present a theorem that provides a condition for a function to be strictly monotone. In practice, the condition is often easier to check than applying the definition of strictly monotone directly.

The test for strictly monotone involves a property of the matrix of partial derivatives of g. This matrix of partial derivatives is called the Jacobian and is introduced next. The notion of a positive definite matrix is then introduced. Then, we define the notion of a convex set. Finally, the condition for g to be strictly monotone is presented.

Jacobian Given a function $g : \mathbb{R}^n \to \mathbb{R}^m$ that is partially differentiable, we can consider a matrix-valued function consisting of these partial derivatives. This matrix-valued function is called the **Jacobian** and we will denote it by J. The entries of $J : \mathbb{R}^n \to \mathbb{R}^{m \times n}$ are defined by:

$$\forall k = 1, \ldots, n, \forall \ell = 1, \ldots, m, \; J_{\ell k} = \frac{\partial g_\ell}{\partial x_k}.$$

The ℓ-th row of J corresponds to an entry g_ℓ of g. The k-th column of J corresponds to an entry x_k of x. (Sometimes, we also use a special symbol, ∇, called the gradient, for the operation of taking the transpose of the Jacobian of a function. See Definition A.36.)

Positive definite and positive semi-definite A matrix $Q \in \mathbb{R}^{n \times n}$ is positive semi-definite if:

$$\forall x \in \mathbb{R}^n, x^\dagger Q x \geq 0.$$

The matrix is positive definite if:

$$\forall x \in \mathbb{R}^n, (x \neq \mathbf{0}) \Rightarrow (x^\dagger Q x > 0).$$

(See Definitions A.58 and A.59.)

Convex sets We introduce the notion of a convex set in:

Definition 2.12 Let $\mathbb{S} \subseteq \mathbb{R}^n$. We say that $\mathbb{S}$ is a **convex set** or that $\mathbb{S}$ is **convex** if:

$$\forall x, x' \in \mathbb{S}, \forall t \in [0, 1], (1 - t)x + tx' \in \mathbb{S}.$$

$\square$

That is, the line segment joining any two points in a convex set $\mathbb{S}$ is itself entirely contained in $\mathbb{S}$. Notice that $(1 - t)x + tx' = x + t(x' - x)$. The second expression is often useful in proving results.

Examples of convex and non-convex sets Convex sets are connected and do not have any indentations; that is, every point can be reached from any other point in the set via a line segment that stays within the set. Figure 2.22 shows four convex sets. In each of them, a pair of points is shown with a line segment joining the pair of points. Each line segment lies wholly within the set that contains its end-points. This applies for any pair of points in each convex set.

Four non-convex sets are illustrated in Figure 2.23. These sets have "indentations." In each set, there are many pairs of points such that a line segment drawn

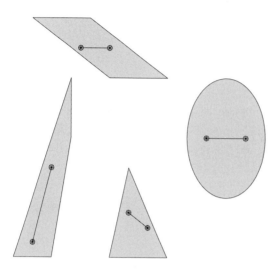

Fig. 2.22. Four examples of convex sets with pairs of points joined by line segments.

between the points will lie partly outside the set. Dis-connected sets are also non-convex. For example, if we re-interpret Figure 2.22 as showing a *single* set that is made up of four dis-connected pieces, then this single set is non-convex.

Conditions for strictly monotone The following theorem characterizes strictly monotone functions $g : \mathbb{S} \to \mathbb{R}^m$. The proof involves integrating the Jacobian of g along line segments between pairs of points in $\mathbb{S}$. To ensure that the integration is well defined, we require $\mathbb{S}$ to be a convex set.

Theorem 2.3 *Let $\mathbb{S} \subseteq \mathbb{R}^n$ be a convex set and $g : \mathbb{S} \to \mathbb{R}^n$. Suppose that g is partially differentiable with continuous partial derivatives on $\mathbb{S}$. Moreover, suppose that the Jacobian J is positive semi-definite throughout $\mathbb{S}$. Then g is monotone on $\mathbb{S}$. If J is positive definite throughout $\mathbb{S}$ then g is strictly monotone on $\mathbb{S}$.*

Proof Suppose that J is positive semi-definite throughout $\mathbb{S}$. Let $x, x' \in \mathbb{S}$. For $0 \leq t \leq 1$ we have that $(x + t[x' - x]) \in \mathbb{S}$ since $\mathbb{S}$ is a convex set. As t varies from 0 to 1, $(x + t[x' - x])$ traces out the line segment joining x and x'. Define $\phi : [0, 1] \to \mathbb{R}$ by:

$$\forall t \in [0, 1], \phi(t) = (x' - x)^{\dagger} g(x + t[x' - x]),$$
$$= g(x + t[x' - x])^{\dagger} (x' - x).$$

We have:

$$\phi(1) - \phi(0) = g(x')^{\dagger} (x' - x) - g(x)^{\dagger} (x' - x),$$
$$= (g(x') - g(x))^{\dagger} (x' - x),$$

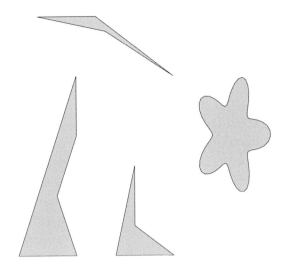

Fig. 2.23. Four examples of non-convex sets.

and so we must prove that $\phi(1) - \phi(0) \geq 0$. Notice that:

$$\frac{d\phi}{dt}(t) = (x' - x)^{\dagger} J(x + t[x' - x])(x' - x), \text{ by the chain rule [72, section 2.4],}$$

$$\geq 0, \text{ for } 0 \leq t \leq 1, \text{ since } J(x + t[x' - x]) \text{ is positive semi-definite.} \quad (2.25)$$

We have:

$$\phi(1) = \phi(0) + \int_{t=0}^{1} \frac{d\phi}{dt}(t)\,dt,$$

by the fundamental theorem of integral calculus applied to ϕ,

(see Theorem A.2 in Section A.4.4.1 of Appendix A),

$$\geq \phi(0), \text{ since the integrand is non-negative everywhere by (2.25),}$$

(see Theorem A.3 in Section A.4.4.2 of Appendix A).

This is the result we were trying to prove. A similar analysis applies for J positive definite, noting that the integrand is then strictly positive and continuous. □

In the proof of Theorem 2.3, we defined a subsidiary function ϕ and considered its derivative. We will take a similar approach in several other proofs.

2.6 Solutions of optimization problems

We now consider solutions to optimization problems. We distinguish **local** and **global** minima of optimization problems in Section 2.6.1 and **strict** and **non-strict** minimizers in Section 2.6.2. In Section 2.6.3 we present some conditions to guarantee that a local minimum is a global minimum.

2.6.1 Local and global minima

2.6.1.1 Definitions

Recall Problem (2.8) and its minimum $f^\star$:

$$f^\star = \min_{x \in \mathbb{S}} f(x).$$

Sometimes, we call $f^\star$ in Problem (2.8) the **global** minimum of the problem to emphasize that there is no $x \in \mathbb{S}$ that has a smaller value of $f(x)$. We can also define the weaker notion of a **local** minimum using a norm $\|\bullet\|$ to specify that we are confining our attention *locally* to a particular subset of the feasible set $\mathbb{S}$. Given $x^\star$, a norm $\|\bullet\|$, and a "distance" $\epsilon > 0$, the set of points x that satisfy $\|x - x^\star\| < \epsilon$ can be considered "close" or "local" to $x^\star$ in that they are all within a distance ϵ, as measured by the norm, from $x^\star$. We make:

Definition 2.13 Let $\|\bullet\|$ be a norm on $\mathbb{R}^n$, $\mathbb{S} \subseteq \mathbb{R}^n$, $x^\star \in \mathbb{S}$, and $f : \mathbb{S} \to \mathbb{R}$. We say that $x^\star$ is a **local minimizer** of the problem $\min_{x \in \mathbb{S}} f(x)$ if:

$$\exists \epsilon > 0 \text{ such that } \forall x \in \mathbb{S}, \left(\|x - x^\star\| < \epsilon\right) \Rightarrow (f(x^\star) \leq f(x)). \qquad (2.26)$$

The value $f^\star = f(x^\star)$ is called a **local minimum** of the problem. □

A local minimum may or may not be a global minimum but if a problem possesses a minimum then there is exactly one global minimum, by definition. The global minimum is also a local minimum. There can be multiple global minimizers, multiple local minimizers, and multiple local minima.

   If a point $\hat{x}$ is *not* a local minimizer, then there are feasible points arbitrarily close to $\hat{x}$ having a lower value of objective. Formally, $\hat{x}$ is not a local minimizer if:

$$\forall \epsilon > 0, \exists x^\epsilon \in \mathbb{S} \text{ such that } \left(\|\hat{x} - x^\epsilon\| < \epsilon\right) \text{ and } (f(\hat{x}) > f(x^\epsilon)). \qquad (2.27)$$

Notice that (2.27) is the negation of the statement in (2.26). It is very important to get the order of the universal and existential quantifiers, $\forall$ and $\exists$, correct in (2.27). (See Definition A.1.)

2.6.1.2 Examples

We illustrate local and global minima and minimizers (and points that are not minimizers) with three example problems. In each case, we must specify an objective and a feasible region. In the first two examples, the functions are defined on $\mathbb{R}$, while in the third example, the functions are defined on $\mathbb{R}^2$.

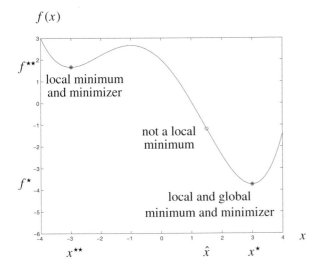

Fig. 2.24. Local minima, $f^\star$ and $f^{\star\star}$, with corresponding local minimizers $x^\star$ and $x^{\star\star}$, over a set $\mathbb{S}$. The point $x^\star$ is the global minimizer and $f^\star$ the global minimum over $\mathbb{S}$.

Multiple local minimizers over a convex set In Figure 2.24, the function $f : \mathbb{R} \to \mathbb{R}$ illustrated has two local minima, $f^\star$ and $f^{\star\star}$, with corresponding local minimizers at $x^\star = 3$, $x^{\star\star} = -3$ over the convex feasible region defined by $\mathbb{S} = \{x \in \mathbb{R} | -4 \le x \le 4\}$. The two local minima and minimizers are indicated with bullets •. Only one of the two minima is the global minimum, with corresponding global minimizer $x^\star = 3$.

In Figure 2.24, the point $\hat{x} = 1.5$, which is illustrated with a ∘, is one of many points that are *not* local minimizers. For any arbitrarily small but positive ϵ, we can find a point that is within a distance ϵ of $\hat{x} = 1.5$ and which has a smaller value of objective. In particular, if the norm is the absolute value norm, then the point $1.5 + (\epsilon/2)$ is within a distance ϵ of $\hat{x} = 1.5$ and has a lower value of the objective than the point $\hat{x} = 1.5$.

For example, Figure 2.25 shows the same function as illustrated in Figure 2.24. This figure illustrates that for $\epsilon = 1$ there is a point, namely $\hat{x} + (\epsilon/2) = 2$ and illustrated with a bullet • in Figure 2.25, that is within a distance ϵ of $\hat{x} = 1.5$ and which has lower value of objective than the point $\hat{x} = 1.5$. As another example, Figure 2.26 shows that for $\epsilon = 0.5$ there is a point, namely $\hat{x} + (\epsilon/2) = 1.75$ and illustrated with a bullet • in Figure 2.26, that is within a distance ϵ of $\hat{x} = 1.5$ and which has lower value of objective than the point $\hat{x} = 1.5$. The point $\hat{x} = 1.5$ is illustrated with a ∘ in both Figures 2.25 and 2.26. Moreover, for *any* value of $\epsilon > 0$ it is possible to find a point within a distance ϵ of $\hat{x} = 1.5$ that has a lower value of objective than the point $\hat{x} = 1.5$.

$f(x)$

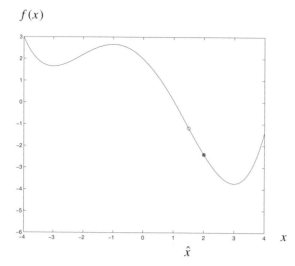

Fig. 2.25. A point $\hat{x}$ = 1.5, illustrated with a ○, that is not a local minimizer and another point, $\hat{x} + (\epsilon/2) = 2$, illustrated with a •, that is within a distance $\epsilon = 1$ of $\hat{x}$ and has a lower objective value.

$f(x)$

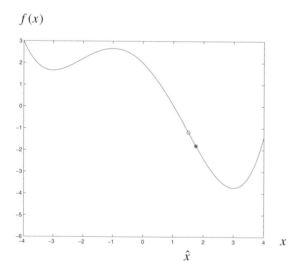

Fig. 2.26. A point $\hat{x}$ = 1.5, illustrated with a ○, that is not a local minimizer and another point, $\hat{x} + (\epsilon/2) = 1.75$, illustrated with a •, that is within a distance $\epsilon = 0.5$ of $\hat{x}$ and has a lower objective value.

Multiple local minimizers over a non-convex set Consider the non-convex set:

$$\mathbb{P} = \{x \in \mathbb{R} | -4 \leq x \leq 1 \text{ or } 2 \leq x \leq 4\}.$$

Figure 2.27 shows the restriction of the function in Figure 2.24 to $\mathbb{P}$. (See Definition A.13.) In this case, over the set $\mathbb{P}$ there are three local minima with corresponding local minimizers, $x^\star = 3$, $x^{\star\star} = -3$, and $x^{\star\star\star} = 1$, each indicated by a •. The additional local minimum compared to Figure 2.24 is due to the characteristics of the feasible set. Again, there is only one global minimum.

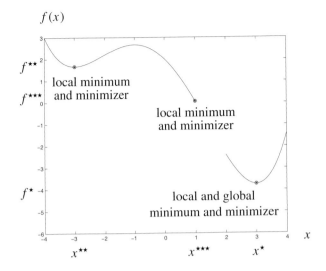

$f(x)$

$f^{\star\star}$

$f^{\star\star\star}$

local minimum
and minimizer

local minimum
and minimizer

$f^{\star}$

local and global
minimum and minimizer

$x^{\star\star}$ $x^{\star\star\star}$ $x^{\star}$

Fig. 2.27. The local and global minima and minimizers of a problem over a set $\mathbb{P} = \{x \in \mathbb{R} \mid -4 \leq x \leq 1 \text{ or } 2 \leq x \leq 4\}$.

Multiple local minimizers over a non-convex set in higher dimension As a third example, consider Figure 2.28, which shows the contour sets of the function $f : \mathbb{R}^2 \to \mathbb{R}$ defined in (2.10) and repeated from Figure 2.8. The feasible set is the shaded region on the contour plot in Figure 2.28. In this case there are two local minima. The local minimizers, $x^{\star} \approx \begin{bmatrix} 2.4 \\ -0.1 \end{bmatrix}$ and $x^{\star\star} \approx \begin{bmatrix} 0.8 \\ -0.7 \end{bmatrix}$ are again indicated with bullets. There is only one global minimum and one global minimizer $x^{\star} \approx \begin{bmatrix} 2.4 \\ -0.1 \end{bmatrix}$.

2.6.1.3 Discussion

The significance of the concept of a local minimum is that iterative algorithms involve generating a sequence of successively "better" points that provide successively better values of the objective or closer satisfaction of the constraints or both. That is, we will iteratively improve the solution by moving to a nearby better solution until we decide that no further improvement is possible. With an iterative improvement algorithm, we can usually only guarantee, at best, that we are moving towards a local minimum and minimizer.

For example, if a minimization algorithm is applied to the objective and feasible set in Figure 2.24 with the initial guess $x^{(0)} = -4$, then it is reasonable to expect that the algorithm will terminate with a point that is close to the local minimizer $x^{\star\star} = -3$, not the global minimizer $x^{\star} = 3$. (We say "reasonable to expect" because the progress of the algorithm will actually depend on implementation details.) If the initial guess $x^{(0)} = 0.5$ is chosen for the algorithm then it can be expected to terminate close to the global minimizer.

x_2

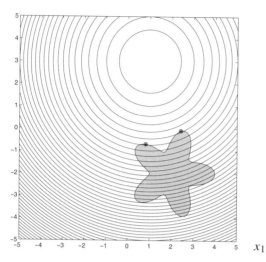

Fig. 2.28. Contour sets of the function defined in (2.10) with feasible set shaded. The two local minimizers are indicated by bullets. The heights of the contours decrease towards the point $\begin{bmatrix} 1 \\ 3 \end{bmatrix}$.

On the other hand, if $x^{(0)} = 0.5$ is chosen as the initial guess for the objective and feasible set in Figure 2.27, then the algorithm can be expected to terminate close to the local minimizer $x^{\star\star\star} = 1$ of that problem. For the problem illustrated in Figure 2.28, initial guesses such as $x^{(0)} = \begin{bmatrix} 1 \\ -2 \end{bmatrix}$ can be expected to result in the algorithm terminating close to the local minimizer $x^{\star\star}$ but not the global minimizer, while initial guesses such as $x^{(0)} = \begin{bmatrix} 2 \\ -1 \end{bmatrix}$ can be expected to result in the algorithm terminating close to the global minimizer $x^\star$.

This observation suggests that one approach to the issue of multiple local minima is to apply an algorithm several times to the problem with randomly chosen initial guesses. For general discussions of finding global optima where there are several local optima, see [91, 87]. The *Journal of Global Optimization* treats global optimization.

However, as we will see in Section 2.6.3, if a problem has a certain property, called "convexity," then a local minimizer is a global minimizer. In this case, under appropriate circumstances, we can guarantee that an iterative improvement algorithm moves closer to a global minimizer at each iteration. As we will see in Exercises 2.28 and 2.29, the problems illustrated in Figures 2.24–2.28 do not have the property of convexity.

Most of the theorems in this chapter and the next that are stated in terms of global optimality can be extended to results in terms of local optimality. Conversely, under

$f(x)$

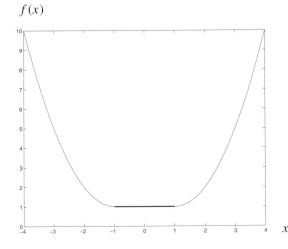

Fig. 2.29. A function with multiple global minimizers. The set of minimizers is indicated by a thick line.

assumptions of convexity, local optimality results can usually be extended to global optimality results.

2.6.2 Strict and non-strict minimizers

2.6.2.1 Definitions

There can be more than one minimizer even if the minimum is global. In Figure 2.29, the function $f : \mathbb{R} \to \mathbb{R}$ illustrated has a single local minimum over the feasible set $\mathbb{S} = \{x \in \mathbb{R} | -4 \leq x \leq 4\}$. This local minimum is also the global minimum. The global minimum is achieved by infinitely many (global) minimizers, namely, each point in the set $\{x \in \mathbb{R} | -1 \leq x \leq 1\}$. This set is indicated by a thick line in Figure 2.29. We sometimes call these **non-strict** minimizers to contrast with:

Definition 2.14 We say that $x^\star \in \mathbb{S}$ is a **strict global** minimizer of the problem $\min_{x \in \mathbb{S}} f(x)$ if:

$$\forall x \in \mathbb{S}, (x \neq x^\star) \Rightarrow (f(x^\star) < f(x)).$$

The value $f^\star = f(x^\star)$ is called a **strict global** minimum of the problem. □

We also define a local version:

Definition 2.15 We say that $x^\star \in \mathbb{S}$ is a **strict local** minimizer of the problem $\min_{x \in \mathbb{S}} f(x)$ if:

$$\exists \epsilon > 0 \text{ such that } \forall x \in \mathbb{S}, \left(0 < \|x - x^\star\| < \epsilon\right) \Rightarrow (f(x^\star) < f(x)).$$

The value $f^\star = f(x^\star)$ is called a **strict local** minimum of the problem. □

2.6.2.2 Examples

The two local minimizers, $x^\star = 3$ and $x^{\star\star} = -3$, in Figure 2.24 are strict local minimizers. One of them, $x^\star = 3$, is a also a strict global minimizer. All three local minimizers, $x^\star = 3$, $x^{\star\star} = -3$, $x^{\star\star\star} = 1$, in Figure 2.27 are strict local minimizers. One of them $x^\star = 3$, is also a strict global minimizer. The two local minimizers,

$x^\star \approx \begin{bmatrix} 2.4 \\ -0.1 \end{bmatrix}$ and $x^{\star\star} \approx \begin{bmatrix} 0.8 \\ -0.7 \end{bmatrix}$, in Figure 2.28 are strict local minimizers. One

of them, $x^\star \approx \begin{bmatrix} 2.4 \\ -0.1 \end{bmatrix}$, is also a strict global minimizer. The points in the set

$\{x \in \mathbb{R} | -1 \le x \le 1\}$ in Figure 2.29 are global minimizers but are neither strict local nor strict global minimizers.

2.6.3 Convex functions

We will introduce the powerful concept of convexity of a function and use it to derive conditions for when a local minimum is in fact a global minimum. We have already introduced the definition of convex sets in Section 2.5.3.2. The concept of convexity applies both to sets and to functions, but the definitions are different for each case. We will define convexity for functions in Section 2.6.3.1 and provide some examples in Section 2.6.3.2. We then present a theorem that connects convexity to uniqueness of solutions of optimization problems in Section 2.6.3.3 and discuss the significance in Section 2.6.3.4. In Section 2.6.3.5 we provide two ways to characterize convexity of functions that are often easier than verifying the definition directly. We then present the important special case of quadratic functions in Section 2.6.3.6. Much of the material is based on [70, section 6.4].

2.6.3.1 Definitions

The definition of convex function requires us to consider values of the function on a convex "test" set. This is embodied in the following definition.

Definition 2.16 Let $\mathbb{S} \subseteq \mathbb{R}^n$ be a convex set and let $f : \mathbb{S} \to \mathbb{R}$. Then, f is a **convex function** on $\mathbb{S}$ if:

$$\forall x, x' \in \mathbb{S}, \forall t \in [0, 1], \ f([1-t]x + tx') \le [1-t]f(x) + tf(x'). \tag{2.28}$$

If $f : \mathbb{R}^n \to \mathbb{R}$ is convex on $\mathbb{R}^n$ then we say that f is convex. A function $h : \mathbb{S} \to \mathbb{R}^r$ is convex on $\mathbb{S}$ if each of its components h_ℓ is convex on $\mathbb{S}$. If $h : \mathbb{R}^n \to \mathbb{R}^r$ is convex on $\mathbb{R}^n$ then we say that h is convex. The set $\mathbb{S}$ is called the **test set**.
Furthermore, f is a **strictly convex function** on $\mathbb{S}$ if:

$$\forall x, x' \in \mathbb{S}, (x \ne x') \Rightarrow \left(\forall t \in (0, 1), \ f([1-t]x + tx') < [1-t]f(x) + tf(x') \right).$$

If $f : \mathbb{R}^n \to \mathbb{R}$ is strictly convex on $\mathbb{R}^n$ then we say that f is strictly convex. A function

$f(x)$

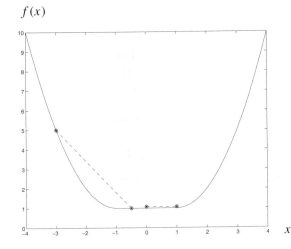

Fig. 2.30. Linear interpolation of a convex function between points never under-estimates the function. (For clarity, the line interpolating f between $x = 0$ and $x = 1$ is drawn slightly above the solid curve: it should be coincident with the solid curve.)

$h : \mathbb{S} \to \mathbb{R}^r$ is strictly convex on $\mathbb{S}$ if each of its components h_ℓ is strictly convex on $\mathbb{S}$. If $h : \mathbb{R}^n \to \mathbb{R}^r$ is strictly convex on $\mathbb{R}^n$ then we say that h is strictly convex. □

The condition in (2.28) means that linear interpolation of a convex function *between* any two points never under-estimates the function. Figure 2.30 repeats the function of Figure 2.29 but with two pairs of points indicated and a linear interpolation of f between each member of the pair shown as a dashed line. The linear interpolation of f between points is never below the function values.

In practice, we often define a function $f : \mathbb{R}^n \to \mathbb{R}$ and then want to consider its convexity on a convex set $\mathbb{S} \subseteq \mathbb{R}^n$. In this case, in Definition 2.16, we consider the restriction of f to $\mathbb{S}$. (See Definition A.13.) If the restriction $f : \mathbb{S} \to \mathbb{R}$ is convex on $\mathbb{S}$ then we say that f is convex on $\mathbb{S}$.

To test whether or not a function is convex we must also specify a convex test set $\mathbb{S}$. In the most straightforward case, the test set $\mathbb{S}$ is the whole of $\mathbb{R}^n$. If the test set is not specified, then we will assume that the test set is the whole of $\mathbb{R}^n$ and that the function is defined on the whole of $\mathbb{R}^n$; however, it is important to bear in mind that *without* a convex test set, it is meaningless to ask whether or not a function is convex. The identity $[1-t]f(x)+tf(x') = f(x)+t[f(x')-f(x)]$ is often useful in proving results.

Convex functions are partially differentiable "almost everywhere." That is, they are partially differentiable at all points except for a set of points of "measure" zero [100, theorem 25.5].

We can also define the notion of concavity of a function:

Definition 2.17 Let $\mathbb{S} \subseteq \mathbb{R}^n$ be a convex set and let $f : \mathbb{S} \to \mathbb{R}$. We say that f is a **concave function** on $\mathbb{S}$ if $(-f)$ is a convex function on $\mathbb{S}$. □

f

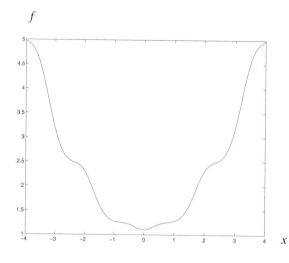

Fig. 2.31. A non-convex
function with convex level
sets.

2.6.3.2 Examples

A linear or affine function is convex and concave on any convex set. (See Exercise 2.33.) The convexity and concavity of affine functions has important implications for linear programming problems.

The function $f : \mathbb{R} \to \mathbb{R}$ shown in Figure 2.24 is not convex on the convex set $\mathbb{S} = \{x \in \mathbb{R} | -4 \le x \le 4\}$. The function $f : \mathbb{R} \to \mathbb{R}$ shown in Figures 2.29 and 2.30 is convex on $\mathbb{S} = \{x \in \mathbb{R} | -4 \le x \le 4\}$ but not strictly convex on this set. The function shown in Figure 2.5 is strictly convex on $\mathbb{R}^2$.

Qualitatively, convex functions are "bowl-shaped" [70, section 6.4] and have level sets that are convex sets as specified in:

Definition 2.18 Let $\mathbb{S} \subseteq \mathbb{R}^n$ and $f : \mathbb{S} \to \mathbb{R}$. Then the function f has **convex level sets** on $\mathbb{S}$ if for all $\tilde{f} \in \mathbb{R}$ we have that $\mathbb{L}_f(\tilde{f})$ is a convex set. If $f : \mathbb{R}^n \to \mathbb{R}$ has convex level sets on $\mathbb{R}^n$ then we say that f has convex level sets. ◻

(Some authors use the term **quasi-convex** for a function with convex level sets.)

A convex function has convex level sets. (See Exercise 2.34.) However, the converse is not true. That is, a function with convex level sets need not itself be a convex function. For example, Figure 2.31 shows a non-convex function having convex level sets.

The convexity of the level sets of a convex function is illustrated in Figure 2.32 for the function $f : \mathbb{R}^2 \to \mathbb{R}$ defined by:

$$\forall x \in \mathbb{R}^2, \ f(x) = (x_1 - 1)^2 + (x_2 - 3)^2 - 1.8(x_1 - 1)(x_2 - 3). \qquad (2.29)$$

In Exercise 2.48, this function is shown to be convex. The contour sets of this function are elliptical as illustrated in Figure 2.32 because the function is quadratic. The

x_2

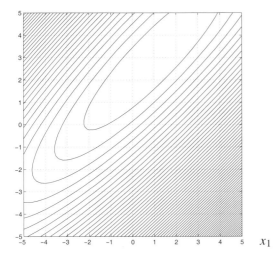

Fig. 2.32. Contour sets $\mathbb{C}_f(\tilde{f})$ of the function defined in (2.29). The heights of the contours decrease towards the point $\begin{bmatrix} 1 \\ 3 \end{bmatrix}$.

level sets of this function are filled ellipses. Filled ellipses are convex sets. Non-quadratic convex functions have level sets that are convex but which are typically not elliptical.

Figure 2.7 shows another example of a convex function. Figure 2.8 illustrates the contour sets of this function. It has circular contour sets. A circle is a special case of an ellipse.

The function shown in Figure 2.27 is defined on a non-convex set $\mathbb{P} = \{x \in \mathbb{R} | -4 \le x \le 1 \text{ or } 2 \le x \le 4\}$. The feasible set shown in Figure 2.28 is also non-convex. To discuss convexity of the functions in these figures, we must restrict consideration to a convex subset of the domain of definition as discussed in Exercise 2.29.

2.6.3.3 Relationship to optimization problems

Theorem 2.4 *Let $\mathbb{S} \subseteq \mathbb{R}^n$ be a convex set and $f : \mathbb{S} \to \mathbb{R}$. Then:*

(i) *If f is convex on $\mathbb{S}$ then it has at most one local minimum over $\mathbb{S}$.*

(ii) *If f is convex on $\mathbb{S}$ and has a local minimum over $\mathbb{S}$ then the local minimum is the global minimum.*

(iii) *If f is strictly convex on $\mathbb{S}$ then it has at most one minimizer over $\mathbb{S}$.*

Proof We prove all three items by contradiction.

(i) For the sake of a contradiction, suppose that f is convex, yet that it has two local minima over $\mathbb{S}$; that is, there are two distinct values $f^\star \in \mathbb{R}$ and $f^{\star\star} \in \mathbb{R}$, say, with $f^\star \ne f^{\star\star}$ that each satisfy Definition 2.13.

$f(x)$

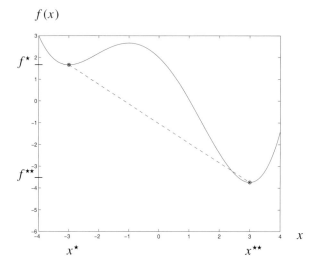

Fig. 2.33. Multiple minima and minimizers in proof of Theorem 2.4, Item (i).

For concreteness, suppose that $f^\star > f^{\star\star}$ and let $x^\star \in \mathbb{S}$ and $x^{\star\star} \in \mathbb{S}$ be any two local minimizers associated with $f^\star$ and $f^{\star\star}$, respectively. The situation is illustrated in Figure 2.33. The solid line shows $f(x)$ as a function of x while the dashed line shows the linear interpolation of f between $x^\star$ and $x^{\star\star}$.

We are going to show that $x^\star$ satisfies the condition (2.27) for $x^\star$ not to be a local minimizer, which we repeat here for reference:

$$\forall \epsilon > 0, \exists x^\epsilon \in \mathbb{S} \text{ such that } \left(\left\|x^\star - x^\epsilon\right\| < \epsilon\right) \text{ and } \left(f(x^\star) > f(x^\epsilon)\right).$$

We have:

$$
\begin{aligned}
\forall t \in [0, 1], \; f(x^\star + t[x^{\star\star} - x^\star]) \;\; &\leq \;\; f(x^\star) + t[f(x^{\star\star}) - f(x^\star)], \\
&\qquad \text{by convexity of } f, \\
&= \;\; f^\star + t[f^{\star\star} - f^\star], \\
&\qquad \text{by definition of } f^\star \text{ and } f^{\star\star}, \\
&< \;\; f^\star, \text{ for } 0 < t \leq 1, \text{ since } f^\star > f^{\star\star}, \\
&= \;\; f(x^\star). \qquad\qquad\qquad\qquad\qquad (2.30)
\end{aligned}
$$

For $0 \leq t \leq 1$, we have $x^\star + t(x^{\star\star} - x^\star) \in \mathbb{S}$ since $\mathbb{S}$ is convex. But this means that there are feasible points arbitrarily close to $x^\star$ that have a lower objective value. In particular, given any norm $\|\bullet\|$ and any number $\epsilon > 0$, we can define $x^\epsilon = x^\star + t(x^{\star\star} - x^\star)$ where t is specified by:

$$t = \min\left\{1, \frac{\epsilon}{2\left\|x^{\star\star} - x^\star\right\|}\right\}.$$

Note that $x^\epsilon \in \mathbb{S}$ since $0 \le t \le 1$ and that x^ϵ satisfies:

$$
\begin{aligned}
\left\| x^\star - x^\epsilon \right\| &= \left\| x^\star - [x^\star + t(x^{\star\star} - x^\star)] \right\|, \text{ by definition of } x^\epsilon, \\
&= \left\| -t(x^{\star\star} - x^\star) \right\|, \\
&= |t| \times \left\| x^{\star\star} - x^\star \right\|, \text{ by Property (iv) of norms,} \\
&\le \frac{\epsilon}{2 \left\| x^{\star\star} - x^\star \right\|} \left\| x^{\star\star} - x^\star \right\|, \text{ by definition of } t, \\
&= \epsilon/2, \\
&< \epsilon.
\end{aligned}
$$

Furthermore $0 < t \le 1$ by construction, so by (2.30):

$$f(x^\star) > f(x^\epsilon).$$

That is, $x^\star$ satisfies (2.27) and is therefore not a local minimizer of f, which is a contradiction. As suggested by the "hump" in f at $x \approx -1$, the situation illustrated in Figure 2.33 is inconsistent with the assumption that f is convex. We conclude that f has at most one local minimum.

(ii) ([11, proposition B.10]) Suppose that the local minimum is $f^\star \in \mathbb{R}$ with corresponding local minimizer $x^\star \in \mathbb{S}$. Suppose that it is not a global minimum and minimizer. That is, there exists $x^{\star\star} \in \mathbb{S}$ such that $f^{\star\star} = f(x^{\star\star}) < f(x^\star)$. Then the same argument as in Item (i) shows that $f^\star$ is not a local minimum.

(iii) Suppose that f is strictly convex, yet that it has two local minimizers, $x^\star \ne x^{\star\star}$, say. Since f is convex, then by Item (i), both minimizers correspond to the unique minimum, say $f^\star$, of f over $\mathbb{S}$. We have:

$$
\begin{aligned}
\forall t \in (0, 1), f(x^\star + t[x^{\star\star} - x^\star]) &< f(x^\star) + t[f(x^{\star\star}) - f(x^\star)], \\
&\quad \text{by strict convexity of } f, \\
&= f^\star + t[f^\star - f^\star], \text{ by definition of } f^\star, \\
&= f^\star,
\end{aligned}
$$

which means that neither $x^\star$ nor $x^{\star\star}$ were local minimizers of f, since feasible points of the form $x^\star + t(x^{\star\star} - x^\star)$ have a lower objective value for all $t \in (0, 1)$. That is, by a similar argument to that in Item (i), we can construct a feasible x^ϵ that is within a distance ϵ of $x^\star$ having a smaller value of objective than $x^\star$. ☐

We combine the notions of convexity of a set and of a function in the following definition.

Definition 2.19 If $\mathbb{S} \subseteq \mathbb{R}^n$ is a convex set and $f : \mathbb{R}^n \to \mathbb{R}$ is convex on $\mathbb{S}$, then $\min_{x \in \mathbb{S}} f(x)$ is called a **convex optimization problem** or a **convex problem**. ☐

2.6.3.4 Discussion

Local versus global minimizers Theorem 2.4 shows that a convex problem has at most one local minimum. Moreover, if we find a local minimum for a convex problem, it is in fact the global minimum. If we have a strictly convex function and find a local minimizer, then the minimizer is unique. We emphasize: we will see that the nature of our iterative algorithms is that we can only guarantee, at best, that the sequence of iterates converges to a local optimizer. If the problem is convex, however, the local optimizer is global. Convexity is therefore a very important property that we should seek in problem formulation.

Choice of step directions Convexity enables us to relate the two goals of:

(i) moving from the current iterate in a direction that decreases the objective while still maintaining feasibility, and

(ii) moving from the current iterate towards the minimizer of the problem.

The second goal is important; however, since we do not know the value of the minimizer, we must usually be content to choose step directions that satisfy the first goal. If we have a convex problem, then these goals are not inconsistent. For example, suppose that the current iterate is $x^{(\nu)}$, while $x^\star$ is a minimizer of the convex function f over the convex feasible set $\mathbb{S}$. The step direction $\Delta x^{(\nu)} = x^\star - x^{(\nu)}$ points in a direction from $x^{(\nu)}$ that satisfies both goals as Exercise 2.35, Parts (i) and (ii) show.

In contrast, with a non-convex problem, moving in the direction of the minimizer might yield a point that is not feasible or increase the objective as Exercise 2.35, Parts (iii) and (iv) show.

Convex problems and generalizations Exercise 2.36, Part (iii) shows that a minimization problem:

$$\min_{x \in \mathbb{R}^n} \{ f(x) | g(x) = \mathbf{0}, h(x) \leq \mathbf{0} \}$$

that has:

- a convex objective $f : \mathbb{R}^n \to \mathbb{R}$,
- an affine equality constraint function $g : \mathbb{R}^n \to \mathbb{R}^m$, with $\forall x \in \mathbb{R}^n, g(x) = Ax - b$, and
- a convex inequality constraint function $h : \mathbb{R}^n \to \mathbb{R}^r$,

is a convex problem. Our choice of a minimization problem and form of the inequality constraints was chosen specifically so that convex objective, convex inequality constraint function, and affine equality constraint function yield a convex problem.

Exercise 2.36, Part (iii) shows that it is possible to have a convex problem that involves non-convex constraint functions; however, the analysis is much simpler with convex functions. In Section 2.6.3.5, we will investigate several tests for convexity of a function.

If a problem is not convex, then it is still possible to apply the various algorithms we will describe. It is, however, not in general possible to guarantee global or even local optimality for the results of applying the algorithms. As mentioned in Section 2.6.1.3, a typical approach to this situation is to re-start an iterative algorithm from various randomly chosen points and check if the solution from each starting point is consistent.

Furthermore, there are various conditions that are weaker than the ones we present that still guarantee that there is a unique local optimum to the problem. The most straightforward of these is to allow the objective function to have convex level sets. (See Definition 2.18.) A function with convex level sets has only one local minimum over a convex set. For example, Figure 2.31 shows a non-convex function that has convex level sets on $\mathbb{S} = \{x \in \mathbb{R} | -4 \leq x \leq 4\}$. It has a single local minimum over any convex region contained in $\mathbb{S}$. In particular, it has a single local minimum over the feasible set $\mathbb{S} = \{x \in \mathbb{R} | -4 \leq x \leq 4\}$.

Analysis under these and other generalizations of convex functions appears in [6, chapter 3][15, chapter 3 and appendix B][105]. Exercises 2.39 and 2.40 provide the flavor of some of these generalizations.

Most of the objectives that we will consider in this book are partially differentiable with continuous partial derivatives as well as being convex. We will briefly discuss convex functions having points of non-differentiability in Section 3.1.3 and Section 3.1.4. Further discussion of such convex, non-differentiable functions appears in [6, chapter 3][11, section 6.3][106].

Maximizing a convex function Although we will mostly be concerned with minimizing a function, here we will briefly consider maximizing a convex function. We first make:

Definition 2.20 Let $\mathbb{S} \subseteq \mathbb{R}^n$ and $x \in \mathbb{S}$. We say that x is an **extreme point** of $\mathbb{S}$ if:

$$\forall x', x'' \in \mathbb{S}, \left((x' \neq x) \text{ and } (x'' \neq x)\right) \Rightarrow \left(x \neq \frac{1}{2}(x' + x'')\right).$$

☐

That is, x is an extreme point of $\mathbb{S}$ if it cannot be expressed as the "average" of two other points in $\mathbb{S}$. Moreover, x is not an extreme point if it can be expressed as the average of two other points. For a set defined in terms of affine equalities and inequalities, the extreme points are its vertices. For example, in Figure 2.22, there are three polygons. The extreme points of each polygon are its vertices. The interior and edges of the polygon (not including its vertices) are not extreme points. As another example, the extreme points of the dodecahedron in Figure 2.14 are its vertices.

On the other hand, for a set such as a disk or a filled ellipse, all of the points on its boundary are extreme points. In Figure 2.22, the extreme points of the filled ellipse are the points on the ellipse.

We have the following:

Theorem 2.5 *Let* $\mathbb{S} \subseteq \mathbb{R}^n$ *be a convex set and* $f : \mathbb{S} \to \mathbb{R}$ *be convex on* $\mathbb{S}$*. Consider the maximization problem:*

$$\max_{x \in \mathbb{S}} f(x),$$

Suppose this problem possesses a maximum. Then there is a maximizer of this problem that is an extreme point of $\mathbb{S}$*.*

Proof See Exercise 2.44. □

In principle, we can maximize a convex objective over a convex set by searching over all the extreme points of the feasible set. There may be a very large number of extreme points of a set and this approach is not practical in general. However, for affine objectives and affine constraints (and some other cases), this approach leads to a practical method of optimization: the **simplex method** of linear programming. We will discuss the simplex method in Chapter 16.

2.6.3.5 *Characterizing convex functions*

The definition of convexity of a function, Definition 2.16, is not very convenient to apply in practice since the condition in (2.28) involves testing every pair of points in a set. This is analogous to the difficulty with characterizing monotone functions using the condition in (2.24). As with testing for monotone functions, there are alternative characterizations that are easier to apply than the definition. We have the following theorems relating convexity to the first and second derivatives of the function.

First derivative We will first describe a test for convexity that involves the first derivative of the function.

Theorem 2.6 *Let* $\mathbb{S} \subseteq \mathbb{R}^n$ *be a convex set and suppose that* $f : \mathbb{S} \to \mathbb{R}$ *is partially differentiable with continuous partial derivatives on* $\mathbb{S}$*. Then* f *is convex on* $\mathbb{S}$ *if and only if:*

$$\forall x, x' \in \mathbb{S}, \ f(x) \geq f(x') + \nabla f(x')^{\dagger}(x - x'). \tag{2.31}$$

$f(x), \phi(x)$

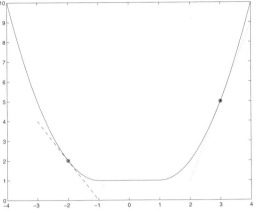

Fig. 2.34. First-order Taylor approximation about $x = -2$ (shown dashed) and about $x = 3$ (shown dotted) of a convex function (shown solid).

Proof See Appendix B. □

The function $\phi : \mathbb{R}^n \to \mathbb{R}$ on the right-hand side of (2.31) defined by:

$$\forall x \in \mathbb{R}^n, \phi(x) = f(x') + \nabla f(x')^\dagger (x - x'),$$

is called the **first-order Taylor approximation** of the function f, linearized about x'. The expression $\nabla f(x')^\dagger \Delta x$ is called the **directional derivative** of f at x' in the direction Δx. (See Definition A.37.) The function ϕ captures the "first-order behavior" of f nearby to x' in the direction $\Delta x = x - x'$.

The inequality in (2.31) shows that the first-order Taylor approximation of a convex function never over-estimates the function as illustrated in Figure 2.34. We will return to the topic of first-order Taylor approximations and the related topic of Taylor's theorem in Section 7.1.2.

The lower bound on f provided by the inequality in (2.31) should be compared to the upper bound on f in the definition of convexity as illustrated in Figure 2.30. Combining the upper bound on a convex function using the definition of convexity with the lower bound using the first-order Taylor approximation allows us to "sandwich" the values of a function between two affine functions. This is illustrated in Figure 2.35. (See Exercise 2.47.)

Second derivative We will also describe a test for convexity involving positive semi-definiteness of the matrix of second derivatives, which is called the **Hessian** and is denoted $\nabla^2 f$ or $\nabla^2_{xx} f$. (See Definition A.38.)

In the particular case that $f : \mathbb{R} \to \mathbb{R}$ then $\nabla^2 f$ is a 1×1 matrix-valued function. A 1×1 matrix is positive semi-definite if its entry is non-negative and it is positive definite if its entry is strictly positive. For $f : \mathbb{R}^2 \to \mathbb{R}$, $\nabla^2 f$ is a symmetric 2×2

$f(x), \phi(x)$

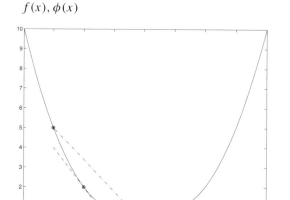

Fig. 2.35. Sandwiching of convex function between two affine functions. The first-order Taylor approximation about $x = -2$ (shown dashed) is a lower bound to the function. The linear interpolation of f between $x = -3$ and $x = -0.5$ (shown dash-dotted) is an upper bound to the function on the interval $\{x \in \mathbb{R}| -3 \le x \le -0.5\}$.

matrix-valued function. A symmetric 2×2 matrix is positive definite if its diagonal entries and its determinant are strictly positive. (See Exercises 2.45 and 2.46.)

In general, a positive semi-definite $n \times n$ matrix may have both positive and negative entries so that it may not be obvious that a given matrix is positive definite or positive semi-definite. That is, non-negativity of the entries of a matrix is neither sufficient nor necessary for positive semi-definiteness of the matrix. (See Exercise 2.49.)

In Section 5.4.6, we will see how to systematically identify positive definite and positive semi-definite matrices. Here we will be satisfied with making the connection between the positive semi-definiteness of $\nabla^2 f$ and convexity of f in:

Theorem 2.7 *Let $\mathbb{S} \subseteq \mathbb{R}^n$ be convex and suppose that $f : \mathbb{S} \to \mathbb{R}$ is twice partially differentiable with continuous second partial derivatives on $\mathbb{S}$. Suppose that $\nabla^2 f$ is positive semi-definite throughout $\mathbb{S}$. (See Definition A.59.) Then f is convex on $\mathbb{S}$. If $\nabla^2 f$ is positive definite throughout $\mathbb{S}$ then f is strictly convex throughout $\mathbb{S}$.*

Proof See Appendix B. □

Exercise 2.50 explores the case when the second derivative matrix is not positive semi-definite. This yields a result concerning conditions for a function to not be convex.

Note that if $\nabla^2 f$ is positive definite then by applying Theorems 2.2 and 2.3 to ∇f we find that the simultaneous equations $\nabla f(x) = \mathbf{0}$ have at most one solution. We will see in Part III that necessary conditions for $x^\star$ to be an unconstrained minimizer of f are that $\nabla f(x^\star) = \mathbf{0}$. We will apply this in Section 10.1.4 to identify conditions for a unique minimizer of an unconstrained optimization problem.

2.6.3.6 Further examples of convex functions

In this section, we present some more examples of convex functions.

Quadratic functions Consider a quadratic function $f : \mathbb{R}^n \to \mathbb{R}$ defined by:

$$\forall x \in \mathbb{R}^n, \ f(x) = \frac{1}{2}x^\dagger Qx + c^\dagger x, \tag{2.32}$$

where $Q \in \mathbb{R}^{n \times n}$ and $c \in \mathbb{R}^n$ are constants and Q is symmetric. (See Definitions A.20 and A.21.) The Hessian of this function is Q, which is constant and independent of x. (See Exercise A.10.)

Consider a quadratic function $f : \mathbb{R}^n \to \mathbb{R}$ defined as in (2.32). If the matrix Q is positive definite then the contour sets of f are elliptical, while if Q is positive semi-definite and not positive definite, then the contour sets are "cylindrical." In both of these cases, f is convex as shown in Exercise 2.49, Parts (i)–(iii).

If Q is not positive semi-definite, then the contour sets are "hyperbolic." The corresponding function f is not convex. (See Exercise 2.49, Part (iv).)

Piece-wise functions and point-wise maxima Sometimes a function is defined by building up its definition over several regions. That is, we think of the function as being made of pieces that fit together. In general, piece-wise functions may or may not be convex even if the underlying pieces are convex on their respective regions. For example, consider the function $f : \mathbb{R} \to \mathbb{R}$ defined by:

$$\forall x \in \mathbb{R}, \ f(x) = \begin{cases} (x+5)^2, & \text{if } x \leq 0, \\ (x-5)^2, & \text{if } x > 0. \end{cases}$$

This function is defined in terms of two pieces: one of the pieces defines the values of $f(x)$ for the region consisting of negative values of x. The other piece defines the values of $f(x)$ for the region consisting of positive values of x. On each piece, the function defining the piece is a convex quadratic function. This function is illustrated in Figure 2.36 and it is clearly not convex. In summary, functions that are defined piece-wise in terms of convex functions are not necessarily convex.

It turns out, however, that if a function can be interpreted as the **point-wise maximum** of underlying convex functions then it is convex. This situation arises in problems where the objective itself is defined as the result of a subsidiary optimization problem. For example, suppose that $f_\ell : \mathbb{R}^n \to \mathbb{R}$ for $\ell = 1, \dots, r$ and define $f : \mathbb{R}^n \to \mathbb{R}$ by:

$$\forall x \in \mathbb{R}^n, \ f(x) = \max_{\ell=1,\dots,r} f_\ell(x). \tag{2.33}$$

That is, f is the **point-wise maximum** of the individual functions f_ℓ. Such objectives arise when there are a number of issues that must be considered simultaneously and we are concerned about the worst-case of all the issues. If each of the

$f(x)$

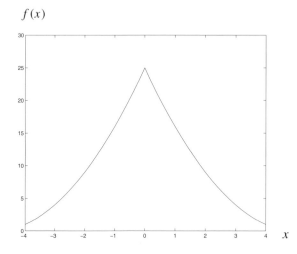

Fig. 2.36. Example of a piece-wise quadratic non-convex function.

functions f_ℓ are convex then the point-wise maximum of them is also convex. (See Exercise 2.52.)

Figure 2.37 shows the two functions f_1, $f_2 : \mathbb{R} \to \mathbb{R}$ defined by:

$$\forall x \in \mathbb{R}, \ f_1(x) \ = \ (x+5)^2,$$
$$\forall x \in \mathbb{R}, \ f_2(x) \ = \ (x-5)^2.$$

Each of these functions is convex.

Figure 2.38 shows the function $f : \mathbb{R} \to \mathbb{R}$ defined by:

$$\forall x \in \mathbb{R}, \ f(x) \ = \ \max\{f_1(x), f_2(x)\},$$
$$= \ \max\{(x+5)^2, (x-5)^2\}.$$

We can also interpret this function as being defined as a piece-wise quadratic function:

$$\forall x \in \mathbb{R}, \ f(x) = \begin{cases} (x+5)^2, & \text{if } x \geq 0, \\ (x-5)^2, & \text{if } x < 0. \end{cases}$$

By Exercise 2.52, this function is convex.

2.7 Sensitivity and large change analysis

2.7.1 Motivation

In many cases, the solution of a particular set of simultaneous equations or a particular optimization problem forms only a part of a larger design process. For example, consider a factory owner faced with the problem of operating equipment in the factory to optimize some criterion, such as maximizing profits. The solution of the maximum operating profit problem will suggest optimal settings of the

$f_1(x), f_2(x)$

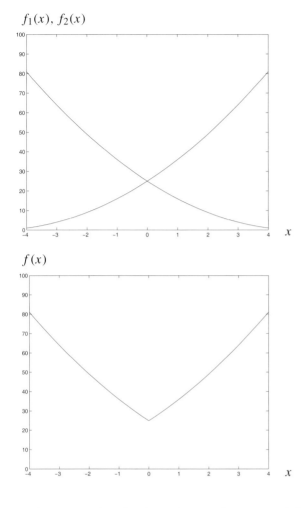

Fig. 2.37. Functions used to define point-wise maximum.

$f(x)$

Fig. 2.38. Example of a piece-wise quadratic convex function.

equipment and optimal purchases of raw materials to maximize operating profits, given the configuration of equipment in the factory.

The owner may also be interested in purchasing more pieces of equipment or otherwise modifying the configuration of the factory in a way that is not directly captured by the decision vector of the optimal operation problem. The owner will want to calculate the potential operating profits of the new configuration given that the *new* configuration is also to be operated optimally. The optimal operation of the new configuration can be expected to differ from the optimal operation of the old configuration. Furthermore, the owner will want to be able to estimate the change in operating profits from the current configuration of the factory to the new configuration to assess whether or not the new purchase is justified by the higher operating profits. The owner may also be interested in assessing by how much the operating profits change if the cost of raw materials changes.

A straightforward way to calculate:

- the optimal settings of the new configuration, or
- the change in operating profit due to a change in raw materials costs,

is to formulate and solve the operational problem corresponding to the new configuration or new costs. Exercise 2.54 gives an example of this kind of analysis, which involves explicitly solving a "base-case" corresponding to the old configuration and a "change-case" corresponding to the new configuration.

2.7.2 Parameterization

Let us represent the change in the problem by supposing that the problem is *parameterized* by a vector $\chi \in \mathbb{R}^s$. The vector of parameters χ represents the parts of the problem that are being changed. For example, it might represent a specification of the configuration of the factory in the example above.

In the case of linear equations, instead of a fixed coefficient matrix and right-hand side vector, we assume that we have matrix-valued and vector-valued *functions* $A : \mathbb{R}^s \rightarrow \mathbb{R}^{m \times n}$ and $b : \mathbb{R}^s \rightarrow \mathbb{R}^m$ and that we want to solve the linear simultaneous equations:

$$A(\chi)x = b(\chi).$$

We might have particular values of the parameters, for example $\chi = 0$, in mind for the base-case and we solve the base-case equations $A(0)x = b(0)$ for a base-case solution $x^\star$. We might then want to solve the equations for another value of χ and we consider solving $A(\chi)x = b(\chi)$ for the change-case solution.

More generally, we can imagine a function $g : \mathbb{R}^n \times \mathbb{R}^s \rightarrow \mathbb{R}^m$, where the second (vector) argument is the parameter $\chi \in \mathbb{R}^s$. We will represent the dependence on a parameter notationally by putting a semi-colon between the first and second vector argument of the function. That is, we write $g(x; \chi)$ for g evaluated at x, given the value χ of the parameter. If we want to consider the function for a particular value of χ, we write $g(\bullet; \chi)$.

In the case of non-linear simultaneous equations, we consider solving:

$$g(x; \chi) = 0,$$

for x, given a particular value of the parameter χ. We typically have a particular set of base-case conditions in mind, for example specified by $\chi = 0$, and we are also interested in solving the simultaneous equations for change-case conditions.

Even more generally, we can imagine functions $f : \mathbb{R}^n \times \mathbb{R}^s \rightarrow \mathbb{R}$, $g : \mathbb{R}^n \times \mathbb{R}^s \rightarrow \mathbb{R}^m$, and $h : \mathbb{R}^n \times \mathbb{R}^s \rightarrow \mathbb{R}^r$. In each case, χ is a parameter in the evaluation of the function and we again put a semi-colon between the first and second argument of the function. In the case of an optimization problem, we consider solving:

$$\min_{x \in \mathbb{R}^n} \{f(x; \chi) | g(x; \chi) = \mathbf{0}, h(x; \chi) \leq \mathbf{0}\}.$$

Again, we typically have a set of base-case conditions corresponding to $\chi = \mathbf{0}$ and also want to solve the problem for non-zero values of χ.

In Exercise 2.54, for example, $\chi = \begin{bmatrix} \gamma \\ \eta \end{bmatrix} \in \mathbb{R}^2$ and γ and η are parameters in the definition of the constraint functions in an inequality-constrained problem. The particular value $\chi = \mathbf{0}$ specifies the base-case. We find both the base-case and change-case solutions in Exercise 2.54.

2.7.3 Sensitivity

It is typical that many new configurations must be analyzed. Moreover, each new configuration may represent only a small change to the existing configuration. In these circumstances, it would be very convenient to be able to use the solution to the existing problem and apply **sensitivity analysis**; that is, calculate the partial derivatives of the minimum and minimizer with respect to the entries of χ, evaluated at the base-case solution corresponding to $\chi = \mathbf{0}$, and estimate the change in the solution based on the partial derivatives. Abusing notation, we will consider $f^\star$ and $x^\star$ to be *functions* of χ and write $\dfrac{\partial f^\star}{\partial \chi}$ and $\dfrac{\partial x^\star}{\partial \chi}$ for the sensitivities of the minimum and minimizer with respect to χ. We will generally only evaluate these sensitivities for $\chi = \mathbf{0}$.

In Exercise 2.54, Parts (ii) and (iii), we calculate these sensitivities by explicitly solving the change-case problem as a function of the parameters and differentiating the results. In general, we would prefer not to have to solve the change-case explicitly in order to calculate the derivatives. We will develop techniques to calculate sensitivities directly from the solution of the base-case problem without having to explicitly re-solve the problem [34, chapter 1].

2.7.4 Large changes

We also may need to estimate the change in the solution due to a **large change** in the value of χ in the problem, where by "large change" we mean a change that is so large that analysis based on the derivatives is or may be inaccurate. Again, we would prefer not to have to solve the change-case explicitly as in Exercise 2.54, Part (i). We will also develop techniques to be able to calculate the effect of large changes in linear simultaneous equations using the solution of the base-case problem.

2.7.5 Examples

In this section we consider examples of sensitivity analysis for each of the five problem classes.

2.7.5.1 Linear simultaneous equations

Consider the linear simultaneous equations defined in (2.2) in Section 2.2.2 but suppose that the coefficient matrix and right-hand side were parameterized so that $A : \mathbb{R} \to \mathbb{R}^{2 \times 2}$ and $b : \mathbb{R} \to \mathbb{R}^2$ were specified by:

$$\forall \chi \in \mathbb{R}, \; A(\chi) = \begin{bmatrix} 1 & 2 + \chi \\ 3 & 4 \end{bmatrix}, b(\chi) = \begin{bmatrix} 1 \\ 1 + \chi \end{bmatrix}. \tag{2.34}$$

We may be interested in the sensitivity of the solution $x^\star$ of $A(\chi)x = b(\chi)$ to χ, evaluated at $\chi = 0$. We will return to this example in Section 5.6.1.3.

2.7.5.2 Non-linear simultaneous equations

Consider the non-linear equations defined in (2.6) in Section 2.2.2.2 but suppose that the equations were parameterized so that $g : \mathbb{R} \times \mathbb{R} \to \mathbb{R}$ was specified by:

$$\forall x \in \mathbb{R}, \forall \chi \in \mathbb{R}, \; g(x; \chi) = (x - 2 - \sin \chi)^3 + 1.$$

We may be interested in the sensitivity of the solution $x^\star$ of $g(x; \chi) = 0$ to χ, evaluated at $\chi = 0$. We will return to this example in Section 7.5.1.2.

2.7.5.3 Unconstrained minimization

Consider the unconstrained minimization problem with the objective f defined in (2.10) in Section 2.3.1 but suppose that the objective was parameterized so that $f : \mathbb{R}^2 \times \mathbb{R} \to \mathbb{R}$ was specified by:

$$\forall x \in \mathbb{R}^2, \forall \chi \in \mathbb{R}, \; f(x; \chi) = (x_1 - \exp(\chi))^2 + (x_2 - 3 \exp(\chi))^2 + 5\chi.$$

We may be interested in the sensitivity of the minimum and minimizer of the problem $\min_{x \in \mathbb{R}^2} f(x; \chi)$ to χ, evaluated at $\chi = 0$. We will return to this example in Section 10.3.2.

2.7.5.4 Equality-constrained minimization

Consider the equality-constrained problem defined in (2.13) in Section 2.3.2:

$$\min_{x \in \mathbb{R}^2} \{ f(x) | Ax = b \},$$

where the objective $f : \mathbb{R}^2 \to \mathbb{R}$ and the coefficient matrix $A \in \mathbb{R}^{1 \times 2}$ are defined by:

$$\begin{aligned} \forall x \in \mathbb{R}^2, \; f(x) &= (x_1 - 1)^2 + (x_2 - 3)^2, \\ A &= \begin{bmatrix} 1 & -1 \end{bmatrix}. \end{aligned}$$

However, suppose that the right-hand side was parameterized so that $b : \mathbb{R} \to \mathbb{R}$ was defined by:

$$\forall \chi \in \mathbb{R}, b(\chi) = [-\chi].$$

We may be interested in the sensitivity of the minimum and minimizer of the problem $\min_{x \in \mathbb{R}^2} \{f(x) | Ax = b(\chi)\}$ to χ, evaluated at $\chi = 0$. We will return to this example in Section 13.4.4.

2.7.5.5 Inequality-constrained minimization

Consider the inequality-constrained Problem (2.18) in Section 2.3.2:

$$\min_{x \in \mathbb{R}^2} \{f(x) | g(x) = 0, h(x) \leq 0\},$$

where the objective $f : \mathbb{R}^2 \to \mathbb{R}$ and equality constraint function $g : \mathbb{R}^2 \to \mathbb{R}$ are defined by:

$$\begin{aligned} \forall x \in \mathbb{R}^2, f(x) &= (x_1 - 1)^2 + (x_2 - 3)^2, \\ \forall x \in \mathbb{R}^2, g(x) &= x_1 - x_2. \end{aligned}$$

However, suppose that the inequality constraint function was parameterized so that $h : \mathbb{R}^2 \times \mathbb{R} \to \mathbb{R}$ was defined by:

$$\forall x \in \mathbb{R}^2, \forall \chi \in \mathbb{R}, h(x; \chi) = 3 - x_2 - \chi.$$

We may be interested in the sensitivity of the minimum and minimizer of the problem $\min_{x \in \mathbb{R}^2} \{f(x) | g(x) = 0, h(x; \chi) \leq 0\}$ to χ, evaluated at $\chi = 0$. We will return to this example in Section 17.4.3.

2.7.6 Ill-conditioned problems

2.7.6.1 Motivation

For many problems, a small change in the specification of the problem leads to a correspondingly small change in the minimum and minimizer. That is, the sensitivity of the minimizer and minimum to a change in the problem specification is small. In contrast, we make:

Definition 2.21 A problem is said to be **ill-conditioned** if a relatively small change in the problem specification leads to a relatively large change in the solution. □

We will interpret "large change" in two ways. In particular, a large change in the solution could be:

• *qualitative*, if a change in the problem specification changed the nature of the solution or affected the feasibility of an optimization problem, or

- *quantitative,* if, for some parameter χ in the problem specification, the sensitivities $\dfrac{\partial f^\star}{\partial \chi}$ or $\dfrac{\partial x^\star}{\partial \chi}$ were large.

Ill-conditioned problems are often a sign of an impractical formulation because numerical values in a problem formulation are inevitably subject to error.

For example, suppose that a measured quantity q in a problem specification is subject to as small relative error of, say, $\epsilon \ll 1$. That is, the error in the measured quantity is ϵq. However, suppose that such an error leads to a relative change in the minimum $f^\star$ or minimizer $x^\star$ that is much larger than $\epsilon f^\star$ or $\epsilon x^\star$, respectively. In this case, we should be very cautious about using the results from this model because errors in the data will produce large errors in the results. In a practical implementation of the calculation that involves finite precision arithmetic, representation and round-off errors can introduce significant discrepancies between the computed and the exact solution of an ill-conditioned problem.

2.7.6.2 Simultaneous equations example

Consider simultaneous equations that are redundant. For example, suppose that two entries, g_1 and g_2, of $g : \mathbb{R}^n \to \mathbb{R}^m$ are the same. Suppose that $x^\star$ is a solution of $g(x) = \mathbf{0}$, so that $g_1(x^\star) = g_2(x^\star) = 0$. Now suppose that g_1 changes to $\tilde{g}_1 = g_1 + \epsilon$. Note that for all $\epsilon \neq 0$, the equations $g(x) = \mathbf{0}$ are inconsistent. That is, an arbitrarily small change in the problem specification results in a large qualitative change in the solution: the problem changes from having a solution to having no solution. That is, redundant simultaneous equations are ill-conditioned. For this reason, we will generally try to avoid redundant equations in the formulation of simultaneous equations problems and avoid redundancy in formulating equality constraints in optimization problems.

2.7.6.3 Optimization example

As another example, suppose that we wish to minimize a convex function and consider the problem of finding a step direction that points towards the minimizer of the problem based on "local" derivative information about the function at a particular iterate $x^{(\nu)}$. This problem is important in unconstrained optimization because we usually cannot expect to know a step direction such as $\Delta x^{(\nu)} = x^\star - x^{(\nu)}$ that would bring us directly to the minimizer since it requires knowledge of the minimizer. Usually, we must rely on local information at the point $x^{(\nu)}$. It will turn out that the direction *perpendicular* to the surface of the contour set at a point is particularly easy to find. As we will see in Section 10.2.1, this direction is the negative of the gradient of f evaluated at the point. (See Definition A.36.)

In the case that the contour sets of f are circular, as illustrated in Figure 2.8, the direction perpendicular to the surface of the contour set points directly towards the

x_2

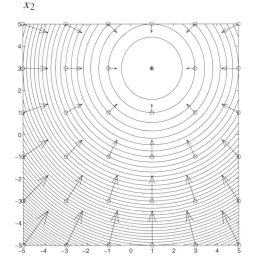

x_1

Fig. 2.39. Directions perpendicular to contour sets.

unconstrained minimizer of f. Figure 2.39 repeats Figure 2.8, but superimposed on the contour plot is a collection of points denoted by $\circ$. At each of these points there is the tail of an arrow. Each arrow points in the direction given by the negative of the gradient of f, which happens to be towards the minimizer of this function, $x^\star = \begin{bmatrix} 1 \\ 3 \end{bmatrix}$, which is shown as the $\bullet$. In this particular case, the gradient of f is sufficient to provide a direction that points towards the minimizer.

However, as illustrated in Figure 2.32, in the case that the contour sets are elliptical but not circular, movement perpendicular to the contour set will not point directly towards the minimizer. Figure 2.40 repeats Figure 2.32 and again the arrows show directions that are perpendicular to contour sets; that is, proportional to the negative of the gradient of f. In this case, the arrows still point in a direction that reduces the objective. However, the arrows do not point directly towards the minimizer, which is shown as the $\bullet$. That is, in this case, the negative of the gradient of f does not provide a direction that points directly towards the minimizer.

Unfortunately, if the contour sets are highly eccentric then the problem of finding the direction that points towards the minimizer becomes ill-conditioned. To understand why this is the case, imagine that the function changes slightly, so that its minimizer is at $x^{\star\star} = \begin{bmatrix} 2 \\ 4 \end{bmatrix}$ instead of $x^\star = \begin{bmatrix} 1 \\ 3 \end{bmatrix}$.

A contour plot of the changed function is shown in Figure 2.41. The ellipses are shifted up by one unit and to the right by one unit. The arrows in Figure 2.41 again show directions that are proportional to the negative of the gradient of the changed

Problems, algorithms, and solutions

x_2

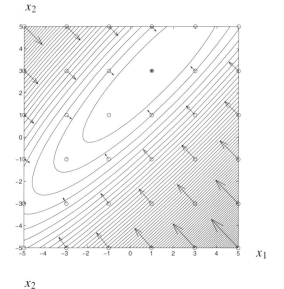

Fig. 2.40. Directions perpendicular to contour sets.

x_2

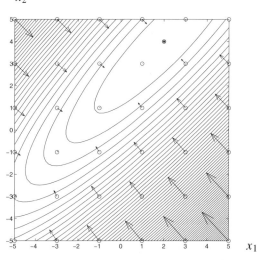

Fig. 2.41. Directions perpendicular to contour sets
for changed function.

function. The arrows in Figure 2.41 are in essentially the same direction as those
shown in Figure 2.40.

In summary, the change in minimizer has had negligible effect on the negative
of the gradient of the function. Conversely, a small change in the negative of the
gradient of f corresponds to a large change in the direction that points towards
the minimizer. Consider the problem of finding a direction that points towards the
minimizer of f using the information provided by the gradient of f. This problem
is ill-conditioned.

2.7.6.4 Discussion

In both examples, small changes in the problem led to large changes in the solution, in either a qualitative or quantitative sense. We will consider ill-conditioning in several contexts throughout the book.

2.8 Summary

In this chapter we have defined two main classes of problems:

   (i) solution of simultaneous equations, and
  (ii) optimization problems,

illustrating particular types of problems with elementary examples. We defined **direct** and **iterative** algorithms and characterized:

- conditions for uniqueness of solution of simultaneous equations using the notion of a monotone function,
- local and global and strict and non-strict minima and minimizers of optimization problems using the notion of convexity, and
- conditions for uniqueness of a local minimum and minimizer.

We also discussed sensitivity analysis and ill-conditioned problems.
  In subsequent chapters we will:

- discuss transformations of problems,
- describe case studies that illustrate the various types of problems in detail,
- present algorithms for each case study that take advantage of various problem characteristics,
- apply problem transformations where necessary, and
- analyze the solutions of the case studies and their sensitivity to changes in problem specification.

Exercises

Problems

2.1 Find the set of solutions to the linear simultaneous equations $Ax = b$, with $A \in \mathbb{R}^{1 \times 1}$, $x \in \mathbb{R}^1$, and $b \in \mathbb{R}^1$, and where:

   (i) $A = [1], b = [1]$,
  (ii) $A = [0], b = [1]$,
 (iii) $A = [0], b = [0]$.

2.2 Prove that $x^\star = -3, 1$ are the *only* solutions to $g(x) = 0$, where $g : \mathbb{R} \to \mathbb{R}$ was defined in (2.4). (Hint: You must prove that any other value cannot satisfy the equation. A sketch of $g(x)$ versus x may be useful to suggest an approach.)

2.3 Let $f : \mathbb{R} \to \mathbb{R}$ be defined by:

$$\forall x \in \mathbb{R}, \ f(x) = (x - 2)^2 + 1,$$

and let $\mathbb{S} = \mathbb{R}$.

(i) Find $\min_{x \in \mathbb{S}} f(x)$,
(ii) Find $\mathrm{argmin}_{x \in \mathbb{S}} f(x)$.

2.4 Show that any number $\underline{f} \in \mathbb{R}$ such that $\underline{f} \leq 1$ is a lower bound for the problem $\min_{x \in \mathbb{S}} f(x)$, where $f : \mathbb{R} \to \mathbb{R}$ is the function defined in Exercise 2.3.

2.5 Let $\mathbb{S} \subseteq \mathbb{R}^n$ and $f : \mathbb{S} \to \mathbb{R}$ and suppose that $f^\star$ is the minimum of $\min_{x \in \mathbb{S}} f(x)$. Also suppose that $\underline{f} \in \mathbb{R}$ satisfies $\underline{f} \leq f^\star$. Show that $\underline{f}$ is a lower bound for $\min_{x \in \mathbb{S}} f(x)$.

2.6 Let $h : \mathbb{R}^2 \to \mathbb{R}^2$ be defined by:

$$\forall x \in \mathbb{R}^2, \ h(x) = -x,$$

(that is, $\forall x \in \mathbb{R}^2$, $h_1(x) = -x_1$, $h_2(x) = -x_2$) and consider the constraints $h(x) \leq 0$. For each of the points:

$$x^\star = \begin{bmatrix} 0 \\ 0 \end{bmatrix}, x^{\star\star} = \begin{bmatrix} 0 \\ 1 \end{bmatrix}, x^{\star\star\star} = \begin{bmatrix} 1 \\ 1 \end{bmatrix},$$

answer the following:

(i) Is $h_1(x) \leq 0$ active for the point?
(ii) Is $h_2(x) \leq 0$ active for the point?
(iii) What is the active set for the point?
(iv) Is the point strictly feasible for the constraint $h_1(x) \leq 0$?
(v) Is the point strictly feasible for the constraint $h_2(x) \leq 0$?
(vi) Is the point strictly feasible for the constraints $h(x) \leq 0$?
(vii) Is the point on the boundary of $\{x \in \mathbb{R}^2 | h(x) \leq 0\}$?

Arrange your answer as a table with a column for each of the points $x^\star, x^{\star\star}$, and $x^{\star\star\star}$ and seven rows for Parts (i)–(vii).

2.7 Consider the function $f : \mathbb{R}^2 \to \mathbb{R}$ defined in (2.7):

$$\forall x \in \mathbb{R}^2, \ f(x) = (x_1)^2 + (x_2)^2 + 2x_2 - 3.$$

(i) Sketch $\mathbb{C}_f(\tilde{f})$ for $\tilde{f} = 0, 1, 2, 3$.
(ii) Sketch on the same graph the set of points satisfying $g(x) = 0$ where $g : \mathbb{R} \to \mathbb{R}$ is defined by:

$$\forall x \in \mathbb{R}^2, \ g(x) = x_1 + 2x_2 - 3.$$

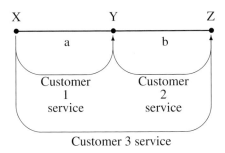

Fig. 2.42. Data communications network.

(iii) Use your sketch to find the minimum and minimizer of $\min_{x \in \mathbb{R}^2} \{f(x) | g(x) = 0\}$.

2.8 Suppose that you own a data communications network that has just three customers: customers $k = 1, 2, 3$. The network consists of two backbone "links:" link a that joins point X to point Y and link b that joins point Y to point Z, respectively. Links a and b have capacities c_a and c_b, respectively, that represent the maximum bandwidth that the links can carry. The arrangement is shown in Figure 2.42.

The three customers desire the following services.

- Customer 1 desires service from point X to point Y, requiring bandwidth on link a.
- Customer 2 desires service from point Y to point Z, requiring bandwidth on link b.
- Customer 3 desires service from point X to point Z, requiring bandwidth on both link a and link b.

Utilization of bandwidth on a link is additive. That is, the total load on a link is equal to the sum of the loads of the customers on that link. For each link, the total load on the link must be less than or equal to the link capacity for the load to be feasible.

You would like to allocate bandwidth to the customers in a systematic way. Somehow, the customers can communicate their "willingness-to-pay" or "utility" for services. That is, each customer k can provide a function $f_k : \mathbb{R}_+ \to \mathbb{R}$ that represents how much customer k values any particular desired (non-negative) level of service. Let us assume that these desires are **commensurable**; that is, they can legitimately be added together to determine overall utility for all the customers. (See Section 2.3.1.1.) We would like to maximize the overall value of all customers' use of the network.

Cast this problem into an optimization problem of the form:

$$\max_{x \in \mathbb{R}^n} \{f(x) | Ax = b, Cx \le d\},$$

where $f : \mathbb{R}^n \to \mathbb{R}$ is a function, A and C are matrices, and b and d are vectors. Make sure that the feasible set does not contain any points for which the functions f_k are not defined. Specify the following explicitly.

(i) The variables in the formulation and their definitions. (Hint: Use $n = 3$ variables to represent the service delivered to the three customers.)

(ii) The objective. (You can specify this function in terms of functions already defined.)

(iii) The matrix and vector specifying the equality constraints, if any.

(iv) The matrix and vector specifying the inequality constraints, if any.

Algorithms

2.9 Consider the non-linear simultaneous equations:

$$a(x)^2 + b(y)^2 = c,$$
$$A(x)^2 + B(y)^2 = C,$$

where $x, y \in \mathbb{R}$ are unknowns and $a, b, c, A, B, C \in \mathbb{R}$ are fixed parameters that satisfy:

$$aB - Ab \neq 0.$$

Consider the transformation $X = (x)^2$ and $Y = (y)^2$.

(i) Write down the solutions of:

$$aX + bY = c,$$
$$AX + BY = C.$$

(ii) What are the conditions on a, b, c, A, B, C to yield real solutions in x, y?

(iii) Given that the conditions in Part (ii) hold, find all the solutions of the non-linear simultaneous equations.

2.10 Show that there is no direct algorithm that, given an arbitrary differentiable function $f : \mathbb{R}^n \to \mathbb{R}$, can find the minimum and all minimizers of the unconstrained minimization problem $\min_{x \in \mathbb{R}^n} f(x)$. (Hint: Use the result from Section 2.4.1.2 that there is no direct algorithm that can solve simultaneous non-linear equations in arbitrary fifth or higher degree polynomials [112, theorem 15.7].)

2.11 Suppose that the sequence of iterates generated by a particular algorithm satisfied:

$$\forall \nu \in \mathbb{Z}_+, x^{(\nu)} = \frac{1}{\nu + 1}.$$

Apply Definition 2.9 to show that the sequence of iterates $\{x^{(\nu)}\}_{\nu=0}^{\infty}$ converges to $x^\star = 0$ using the norm $\|\bullet\|$ given by the absolute value $| \bullet |$. (Hint: *Do not* assume that $\lim_{\nu \to \infty} 1/(\nu + 1) = 0$. This is the result that you are being asked to prove.)

2.12 Suppose that our initial guess is in the right "ballpark." That is, we know that the initial error satisfies:

$$\left\| x^{(0)} - x^\star \right\| \leq \overline{\rho},$$

where $\overline{\rho}$ is a relatively large number. Furthermore, assume that our algorithm generates a sequence of iterates that converge to $x^\star$ satisfying not just the definition of the asymptotic rate in (2.23) in Definition 2.10, but also satisfying:

$$\forall \nu \in \mathbb{Z}_+, \frac{\left\| x^{(\nu+1)} - x^\star \right\|}{\left\| x^{(\nu)} - x^\star \right\|^R} \leq C.$$

For each of the following specifications of R and C, calculate a bound on the number of iterations N required to reduce the "final error" $\left\| x^{(N)} - x^\star \right\|$ to be less than some desired tolerance $\epsilon \overline{\rho}$, where $\epsilon < 1$ is the desired reduction in the error.

(i) $R = 1$ and $0 < C < 1$, so that convergence is linear,

(ii) $R = 2$ and $0 < C < \infty$, so that convergence is quadratic,

(iii) $1 < R < 2$ and $0 < C < \infty$, so that convergence is super-linear.

Your answer will be in terms of ϵ, C, and (possibly) $\bar{\rho}$.

2.13 For each of the following sequences and each of the following values of R, calculate the following limit (or state that the limit diverges to ∞):

$$\lim_{\nu \to \infty} \frac{\left\| x^{(\nu+1)} - x^\star \right\|}{\left\| x^{(\nu)} - x^\star \right\|^R}.$$

Use the following values of R: (a) $R = 0$, (b) $R = \frac{1}{2}$, (c) $R = 1$, and (d) $R = 2$. Consider the following sequences:

(i) $\forall \nu \in \mathbb{Z}_+, x^{(\nu)} = 1/(\nu + 1)$,

(ii) $\forall \nu \in \mathbb{Z}_+, x^{(\nu)} = (2)^{-\nu}$,

(iii) $\forall \nu \in \mathbb{Z}_+, x^{(\nu)} = (2)^{-((2)^\nu)}$.

That is, you will have to calculate or bound twelve limits. Show working for at least three of the limits you calculate. Use the bounds you calculate to find the rate of convergence of the sequence. That is, find the largest value of R such that the limit is finite. Also, specify if the convergence is linear, quadratic, or super-linear, or not, according to Definition 2.10. Arrange your answer as a table with three rows corresponding to the different sequences and four columns corresponding to the different values of R, with a final column to specify the rate of convergence. (For each sequence and value of R, use the absolute value as the norm. You can assume that the limit of each sequence $\{x^{(\nu)}\}_{\nu=0}^\infty$ is $x^\star = 0$.)

2.14 Let $\|\bullet\|$ be a norm on $\mathbb{R}^n$ and suppose that $\mathbb{S} \subseteq \mathbb{R}^n$ is convex and bounded. Suppose that $f : \mathbb{S} \to \mathbb{R}$ is twice partially differentiable with continuous second partial derivatives. (See Definition A.36.) Suppose that there exist bounds $\bar{\kappa}, \underline{\kappa} \in \mathbb{R}_{++}$ such that:

$$\forall x' \in \mathbb{S}, \left\| \nabla f(x') \right\| \leq \bar{\kappa},$$

$$\forall x \in \mathbb{R}^n, \forall x' \in \mathbb{S}, x^\dagger \nabla^2 f(x') x \geq \underline{\kappa} \|x\|^2 .$$

Suppose that the sequence $\{x^{(\nu)}\}_{\nu=0}^\infty$ converges to a point $x^\star \in \mathbb{S}$ such that $\nabla f(x^\star) = 0$ and with a rate $R' \in \mathbb{R}_+$ and rate constant C'. Calculate a rate, R, of convergence and an upper bound, C, on the rate constant for the sequence $\{f(x^{(\nu)})\}_{\nu=0}^\infty$. Note that the sequence $\{f(x^{(\nu)})\}_{\nu=0}^\infty$ converges to $f(x^\star)$. You can assume that $f(x^{(\nu)}) \geq f(x^\star), \forall \nu$. Use the absolute value as norm for the sequence $\{f(x^{(\nu)})\}_{\nu=0}^\infty$. (Hint: Let $x \in \mathbb{S}$ and define $\phi : [0, 1] \to \mathbb{R}$ by:

$$\forall t \in [0, 1], \phi(t) = f(x^\star + t(x - x^\star)).$$

Use the fundamental theorem of calculus, Theorem A.2 in Section A.4.4.1 in Appendix A, as in the proof of Theorem 2.7.)

Solutions of simultaneous equations

2.15 Consider a function $f : \mathbb{R} \to \mathbb{R}$. Show that f is monotonically increasing (see Definition A.24) if and only if it is monotone. (See Definition 2.11.)

2.16 Give an example of a monotone function $g : \mathbb{R}^n \to \mathbb{R}^n$ such that the simultaneous equations $g(x) = 0$ have multiple solutions.

2.17 Let $\mathbb{S} \subseteq \mathbb{R}^n$ and let $g : \mathbb{R}^n \to \mathbb{S}$ be strictly monotone and onto $\mathbb{S}$. Show that the inverse function $g^{-1} : \mathbb{S} \to \mathbb{R}^n$ of g exists. (See Definition A.27 for the definition of inverse function.)

2.18 In this exercise we consider the exponential and logarithmic functions.

(i) Show that the exponential function $\exp : \mathbb{R} \to \mathbb{R}$ is strictly monotone.
(ii) Show that the logarithmic function $\ln : \mathbb{R}_{++} \to \mathbb{R}$ is strictly monotone.

2.19 Consider the function $g : \mathbb{R} \to \mathbb{R}$ defined by:

$$\forall x \in \mathbb{R}, g(x) = (x)^2 - 1.$$

(i) Show that g is strictly monotone on the set $\mathbb{S} = \mathbb{R}_+$. (Note that $0 \in \mathbb{S}$.)
(ii) What are the solutions of $g(x) = 0$ on $\mathbb{R}$?
(iii) How many of these solutions are elements of $\mathbb{S}$?

2.20 Let $g : \mathbb{R}^n \to \mathbb{R}^n$ be affine and of the form:

$$\forall x \in \mathbb{R}^n, g(x) = Ax - b,$$

where $A \in \mathbb{R}^{n \times n}$ and $b \in \mathbb{R}^n$. Show that if there is more than one solution to the linear simultaneous equations $Ax = 0$ then g is not strictly monotone.

2.21 In this exercise we consider the number of solutions of simultaneous equations.

(i) Consider a function $g : \mathbb{R} \to \mathbb{R}$ having the following form:

$$\forall x \in \mathbb{R}, g(x) = \sum_{\ell=1}^{m} g_\ell(x),$$

where each $g_\ell : \mathbb{R} \to \mathbb{R}$ is a strictly monotonically increasing function. Consider solving the equation $g(x) = 0$. Prove that there is no more than one solution.
(ii) Consider the function g in Part (i) but suppose that the functions $g_\ell : \mathbb{R} \to \mathbb{R}$ are monotonically increasing, but not necessarily strictly monotonically increasing. Is there still only one solution to $g(x) = 0$? (Prove or give a counter-example.)
(iii) Now consider a two variable function $g : \mathbb{R}^2 \to \mathbb{R}$ of the form:

$$\forall x \in \mathbb{R}^2, g(x) = g_1(x_1) + g_2(x_2),$$

where $x = \begin{bmatrix} x_1 \\ x_2 \end{bmatrix}$ and where $g_1 : \mathbb{R} \to \mathbb{R}$ and $g_2 : \mathbb{R} \to \mathbb{R}$. Again suppose that the g_ℓ are strictly monotonically increasing. Is there still no more than one possible solution to $g(x) = 0$? (Prove or give a counterexample.)

2.22 Let $\mathbb{S} \subseteq \mathbb{R}^n$ and $g : \mathbb{S} \to \mathbb{R}^n$ be strictly monotone on $\mathbb{S}$. Moreover, suppose that $h : \mathbb{R}^n \to \mathbb{R}^n$ specifies a transformation satisfying:

$$\forall y \in \mathbb{R}^n, ((h(y) = \mathbf{0}) \Leftrightarrow (y = \mathbf{0})).$$

Prove that there is at most one solution of the simultaneous equations $h(g(x)) = \mathbf{0}$ that is an element of $\mathbb{S}$.

2.23 Show that $\mathbb{R}^n$ is a convex set. (You can assume that $\mathbb{R}$ is closed under multiplication and addition.)

2.24 In this exercise we consider convexity of sets.

(i) Show that the open ball of radius ρ and center $x^{(0)}$, defined by

$$\left\{ x \in \mathbb{R}^n \,\middle|\, \left\| x - x^{(0)} \right\| < \rho \right\},$$

(see Definition A.44) is a convex set.
(ii) Show that the closed ball of radius ρ and center $x^{(0)}$, defined by

$$\left\{ x \in \mathbb{R}^n \,\middle|\, \left\| x - x^{(0)} \right\| \leq \rho \right\},$$

(see Definition A.43) is a convex set.

2.25 Let $\mathbb{S}', \mathbb{S}'' \subseteq \mathbb{R}^n$. Show that if $\mathbb{S}'$ and $\mathbb{S}''$ are convex sets then $\mathbb{S} = \mathbb{S}' \cap \mathbb{S}''$ is a convex set.

2.26 Let $\mathbb{S} \subseteq \mathbb{R}^n$ be a convex set and $g : \mathbb{S} \to \mathbb{R}^n$. Suppose that g is partially differentiable with Jacobian J. Prove that if $\frac{1}{2}(J + J^\dagger)$ is positive definite throughout $\mathbb{S}$ then g is strictly monotone. (Hint: Use the result of Exercise A.1 in Appendix A.)

2.27 Show the following.

(i) If $A \in \mathbb{R}^{m \times n}$ then $Q = A^\dagger A$ is positive semi-definite. (Hint: Consider $\|Ax\|_2^2$ using the properties of the norm as described in Definition A.28.)
(ii) If Q' and Q'' are positive semi-definite then $Q = Q' + Q''$ is also positive semi-definite.
(iii) If Q' is positive semi-definite and Q'' is positive definite then $Q = Q' + Q''$ is positive definite.
(iv) If $A \in \mathbb{R}^{n \times n}$ is non-singular, then $Q = A^\dagger A$ is positive definite. (Hint: Use the result from Section 2.5.2 that necessary and sufficient conditions for there to be a unique solution to a square system of equations is that the coefficient matrix A be non-singular. Then consider $\|Ax\|_2^2$.)
(v) If $A \in \mathbb{R}^{m \times n}$ has linearly independent columns (see Definition A.55), then $Q = A^\dagger A$ is positive definite.
(vi) If $A \in \mathbb{R}^{m \times n}$ has linearly independent columns and $R \in \mathbb{R}^{m \times m}$ is positive definite, then $Q = A^\dagger R A$ is positive definite.

Solutions of optimization problems

2.28 In this exercise we investigate convexity of sets.

(i) Show that the feasible set $\mathbb{P} = \{x \in \mathbb{R} | -4 \leq x \leq 1 \text{ or } 2 \leq x \leq 4\}$ in Figure 2.27 is not a convex set. (A sketch will suffice.)

(ii) Show that the feasible set in Figure 2.28 is not a convex set. (A sketch will suffice.)

2.29 In this exercise we will consider the function $f : \mathbb{P} \to \mathbb{R}$ illustrated in Figure 2.27 where $\mathbb{P} = \{x \in \mathbb{R} | -4 \leq x \leq 1 \text{ or } 2 \leq x \leq 4\}$. To analyze convexity of f, we will restrict consideration to test sets that are convex subsets of $\mathbb{P}$.

(i) Show that $\mathbb{S}_1 = \{x \in \mathbb{R} | -4 \leq x \leq 1\}$ is convex.

(ii) Show that f is not convex on $\mathbb{S}_1 = \{x \in \mathbb{R} | -4 \leq x \leq 1\}$.

(iii) Show that $\mathbb{S}_2 = \{x \in \mathbb{R} | 2 \leq x \leq 4\}$ is convex.

(iv) Is f convex on $\mathbb{S}_2 = \{x \in \mathbb{R} | 2 \leq x \leq 4\}$? (Inspection of the graph will suffice for an answer.)

2.30 Let $\mathbb{S}$ be a convex set. Show that if $f_1, f_2 : \mathbb{S} \to \mathbb{R}$ are convex on $\mathbb{S}$ and $a, b \in \mathbb{R}_+$ then $f = af_1 + bf_2$ is convex on $\mathbb{S}$.

2.31 Let $\mathbb{S} \subseteq \bar{\mathbb{S}} \subseteq \mathbb{R}^n$ with $\mathbb{S}$ and $\bar{\mathbb{S}}$ both convex sets. Suppose that $f : \bar{\mathbb{S}} \to \mathbb{R}$ is convex on $\bar{\mathbb{S}}$. Show that f is convex on $\mathbb{S}$.

2.32 Suppose that $f : \mathbb{R} \to \mathbb{R}$ is concave on $\mathbb{S} = \mathbb{R}_+$ and that $f(0) \geq 0$ with f continuous at $x = 0$. Let $\mathbb{P} = \mathbb{R}_{++}$. Show that the function $\phi : \mathbb{P} \to \mathbb{R}$ defined by:

$$\forall x \in \mathbb{P}, \phi(x) = \frac{f(x)}{x},$$

is non-increasing on $\mathbb{P}$.

2.33 Show that an affine function defined on $\mathbb{R}^n$ is convex on any convex subset of $\mathbb{R}^n$.

2.34 Suppose that $\mathbb{S}$ is convex and that $f : \mathbb{S} \to \mathbb{R}$ is convex on $\mathbb{S}$. Let $\tilde{f} \in \mathbb{R}$. Show that the level set $\mathbb{L}_f(\tilde{f})$ at value $\tilde{f}$ of f is a convex set.

2.35 Let $\mathbb{S} \subseteq \mathbb{R}^n$ be convex and let $f : \mathbb{R}^n \to \mathbb{R}$ be convex on $\mathbb{S}$. Suppose that $x^{(\nu)} \in \mathbb{S}$ and that $x^\star$ is a minimizer of f over $\mathbb{S}$. Let $\Delta x^{(\nu)} = x^\star - x^{(\nu)}$.

(i) Show that:

$$\forall \alpha \in [0, 1], x^{(\nu)} + \alpha \Delta x^{(\nu)} \in \mathbb{S}.$$

(ii) Show that if $x^{(\nu)}$ is not a local minimizer, then:

$$(0 < \alpha \leq 1) \Rightarrow (f(x^{(\nu)} + \alpha \Delta x^{(\nu)}) < f(x^{(\nu)})).$$

(iii) Show by an example that the result in Part (i) is false if we allow $\mathbb{S}$ to be non-convex.

(iv) Show by an example that the result in Part (ii) is false if we allow f to be non-convex. That is, let $\mathbb{S}$ be convex, but suppose that $f : \mathbb{R}^n \to \mathbb{R}$ is not convex on $\mathbb{S}$. Show that the result in Part (ii) is then not necessarily true.

2.36 In this exercise we investigate convexity of optimization problems.

(i) Show that if $g : \mathbb{R}^n \to \mathbb{R}^m$ is affine and $h : \mathbb{R}^n \to \mathbb{R}^r$ is convex then $\mathbb{S} = \{x \in \mathbb{R}^n | g(x) = \mathbf{0}, h(x) \leq \mathbf{0}\}$ is a convex set. (Hint: To prove convexity of the set $\{x \in \mathbb{R}^n | h(x) \leq \mathbf{0}\}$ apply the result of Exercise 2.34 with $\tilde{f} = 0$. To prove convexity of $\{x \in \mathbb{R}^n | g(x) = \mathbf{0}\}$, note that $\{x \in \mathbb{R}^n | g(x) = \mathbf{0}\} = \{x \in \mathbb{R}^n | g(x) \leq \mathbf{0}\} \cap \{x \in \mathbb{R}^n | - g(x) \leq \mathbf{0}\}$. Then use Exercise 2.25 to prove convexity of $\mathbb{S}$.)

(ii) Show that if $f : \mathbb{R}^n \to \mathbb{R}$ is convex, $g : \mathbb{R}^n \to \mathbb{R}^m$ is affine, and $h : \mathbb{R}^n \to \mathbb{R}^r$ is convex then $\min_{x \in \mathbb{S}} f(x)$ is a convex optimization problem, where $\mathbb{S} = \{x \in \mathbb{R}^n | g(x) = \mathbf{0}, h(x) \leq \mathbf{0}\}$.

(iii) Find $g : \mathbb{R}^n \to \mathbb{R}^m$ and $h : \mathbb{R}^n \to \mathbb{R}^r$, with g and h partially differentiable with continuous partial derivatives, such that $\mathbb{S} = \{x \in \mathbb{R}^n | g(x) = \mathbf{0}, h(x) \leq \mathbf{0}\}$ is a convex set, but such that g is not affine and at least one component of h is not convex. That is, give a specific example of functions g and h such that if f is convex then $\min_{x \in \mathbb{S}} f(x)$ is a convex optimization problem, but g is not affine and h is not convex.

2.37 Give an example of functions $f : \mathbb{R} \to \mathbb{R}$ and $h : \mathbb{R} \to \mathbb{R}^2$ such that:

- the constraint $h_2(x) \leq 0$ is not binding at the minimizer of the constrained problem $\min_{x \in \mathbb{R}} \{f(x) | h(x) \leq \mathbf{0}\}$, but
- the constraint $h_2(x) \leq 0$ is violated by the minimizer of the unconstrained problem $\min_{x \in \mathbb{R}} f(x)$.

(Hint: Consider an inequality constraint function such that $h_1(x) \leq 0$ is binding at the minimizer of the constrained problem.)

2.38 Give an example of a function $f : \mathbb{R} \to \mathbb{R}$ such that:

- the constraint $x \geq 0$ is not binding in the problem $\min_{x \in \mathbb{R}} \{f(x) | x \geq 0\}$, but the solution of this problem is different to
- the solution of the unconstrained problem $\min_{x \in \mathbb{R}} f(x)$.

(Hint: Consider a non-convex function with a local minimizer at $x = 1$, but a global minimizer at $x = -1$.)

2.39 Prove the following generalization of Item (i) of Theorem 2.4. Let $\mathbb{S} \subseteq \mathbb{R}^n$ be convex and let $f : \mathbb{S} \to \mathbb{R}$. Prove that if f has convex level sets on $\mathbb{S}$ then it has at most one strict local minimum over $\mathbb{S}$. (Hint: You only need to change one equation and one sentence of the proof. Specify the changes.)

2.40 Suppose that $\mathbb{S} \subseteq \mathbb{R}^n$ is convex, $f : \mathbb{S} \to \mathbb{R}$ is convex on $\mathbb{S}$, and $g : \mathbb{S} \to \mathbb{R}_{++}$ is concave and strictly positive on $\mathbb{S}$. Show that the function $\phi : \mathbb{S} \to \mathbb{R}$ defined by:

$$\forall x \in \mathbb{S}, \phi(x) = \frac{f(x)}{g(x)},$$

has convex level sets on $\mathbb{S}$.

2.41 Let $S \subseteq \mathbb{R}^n$ be convex and let $f : S \to \mathbb{R}$ be a convex function on S. In this exercise we will explore **Jensen's inequality**.

  (i) Prove Jensen's inequality: $\forall N \in \mathbb{Z}_+, \forall x^{(1)}, \ldots, x^{(N)} \in S, \forall t^{(1)}, \ldots, t^{(N)} \in [0, 1]$
  such that $\sum_{v=1}^{N} t^{(v)} = 1$, we have that $f\left(\sum_{v=1}^{N} t^{(v)} x^{(v)}\right) \leq \sum_{v=1}^{N} t^{(v)} f(x^{(v)})$.
  (Hint: Prove by induction on N.)
  (ii) Show that the function $f : \mathbb{R}_{++} \to \mathbb{R}$ defined by:

$$\forall x \in \mathbb{R}_{++}, \ f(x) = -\ln(x),$$

     is convex on $S = \mathbb{R}_{++}$. (The set $\mathbb{R}_{++}$ is the set of strictly positive real numbers. See Definition A.3.)
  (iii) Use the previous two results to prove the arithmetic mean-geometric mean inequality:

$$\forall N \in \mathbb{Z}_+, \forall x^{(1)}, \ldots, x^{(N)} \in \mathbb{R}_{++}, \ \frac{1}{N}\sum_{v=1}^{N} x^{(v)} \geq \left(\prod_{v=1}^{N} x^{(v)}\right)^{1/N}.$$

2.42 Let $S = \{x \in \mathbb{R} | a \leq x \leq b\}$, where $a, b \in \mathbb{R}$ and $a < b$. What are the extreme points of S? Prove your answer by applying Definition 2.20.

2.43 In this exercise we consider extreme points of sets.

  (i) Find a set that is convex but has no extreme points.
  (ii) What are the extreme points of a sector of a disk?

2.44 Prove Theorem 2.5 in the case of a strictly convex function. (Hint: Suppose that every maximizer is not an extreme point.)

2.45 Let $Q \in \mathbb{R}^{1 \times 1}$. That is, $Q = [Q_{11}]$.

  (i) Show that Q is symmetric.
  (ii) Show that Q is positive semi-definite if and only if $Q_{11} \geq 0$.
  (iii) Show that Q is positive definite if and only if $Q_{11} > 0$.

2.46 Let $Q \in \mathbb{R}^{2 \times 2}$ be symmetric. That is, $Q = \begin{bmatrix} Q_{11} & Q_{12} \\ Q_{12} & Q_{22} \end{bmatrix}$.

  (i) Show that Q is positive semi-definite if and only if $Q_{11} \geq 0, Q_{22} \geq 0$, and $\det(Q) = Q_{11}Q_{22} - (Q_{12})^2 \geq 0$.
  (ii) Show that Q is positive definite if and only if $Q_{11} > 0, Q_{22} > 0$, and $\det(Q) = Q_{11}Q_{22} - (Q_{12})^2 > 0$.

(Hint: For the $\Rightarrow$ direction, consider particular choices of x, such as $x = \mathbf{I}_1$. For the $\Leftarrow$ direction, write out $x^\dagger Q x$ explicitly and first consider it to be a quadratic function of x_1. Consider the conditions on the coefficients of this quadratic function for it to be positive. Note that these conditions are themselves a requirement on a quadratic function of x_2 for it to be positive. Consider these conditions as well.)

2.47 Consider the function $f : \mathbb{R} \to \mathbb{R}$ defined by:

$$\forall x \in \mathbb{R}, \ f(x) = \frac{1}{2}(x)^2.$$

(i) Show that this function is convex. (Hint: Use the result of Exercise 2.45.)

(ii) Use Theorem 2.6 to find an affine function that provides a lower bound to f. That is, find an affine function $\underline{f} : \mathbb{R} \to \mathbb{R}$ such that:

$$\forall x \in \mathbb{R}, \ \underline{f}(x) \leq f(x).$$

(iii) Use the definition of convexity to find an affine function that provides an upper bound to f in the range $[1, 3]$. That is, find an affine function $\overline{f} : \mathbb{R} \to \mathbb{R}$ such that:

$$\forall x \in [1, 3], \ \overline{f}(x) \geq f(x).$$

2.48 Consider the function $f : \mathbb{R}^2 \to \mathbb{R}$ defined in (2.29):

$$\forall x \in \mathbb{R}^2, \ f(x) = (x_1 - 1)^2 + (x_2 - 3)^2 - 1.8(x_1 - 1)(x_2 - 3).$$

Show that the function is convex. (Hint: Evaluate $\nabla^2 f$ and use Theorem 2.7 and Exercise 2.27.)

2.49 Consider the quadratic function $f : \mathbb{R}^3 \to \mathbb{R}$ defined by:

$$\forall x \in \mathbb{R}^3, \ f(x) = \frac{1}{2}x^{\dagger}Qx,$$

where:

(i) $Q = \begin{bmatrix} 1 & 0 & 0 \\ 0 & 1 & 0 \\ 0 & 0 & 1 \end{bmatrix}$.

    (a) Show that Q is positive definite,

    (b) Sketch $\mathbb{C}_f(2)$.

(ii) $Q = \begin{bmatrix} 1 & -0.2 & 0 \\ -0.2 & 1 & 0 \\ 0 & 0 & 1 \end{bmatrix}$.

    (a) Show that Q is positive definite,

    (b) Sketch $\mathbb{C}_f(2)$.

(iii) $Q = \begin{bmatrix} 1 & 0 & 0 \\ 0 & 1 & 0 \\ 0 & 0 & 0 \end{bmatrix}$.

    (a) Show that Q is positive semi-definite,

    (b) Sketch $\mathbb{C}_f(2)$.

(iv) $Q = \begin{bmatrix} 1 & 0 & 0 \\ 0 & -1 & 0 \\ 0 & 0 & 0 \end{bmatrix}$.

    (a) Show that Q is not positive semi-definite,

    (b) Sketch $\mathbb{C}_f(2)$.

(v) $Q = \begin{bmatrix} 1 & 2 & 0 \\ 2 & 1 & 0 \\ 0 & 0 & 0 \end{bmatrix}$.

   (a) Show that Q is not positive semi-definite,
   (b) Sketch $\mathbb{C}_f(2)$.

(Hint: Exercise 2.27 may be helpful. The MATLAB function `contour` can only draw the contour sets of functions $f : \mathbb{R}^2 \to \mathbb{R}$. To use MATLAB, you must use the definition of $\mathbb{C}_f(2)$ and the MATLAB functions `sphere` and `mesh`.)

2.50 Let $f : \mathbb{R}^n \to \mathbb{R}$ be twice partially differentiable with continuous second partial derivatives. Suppose that at a point $\hat{x}$ the matrix $\nabla^2 f(\hat{x})$ is not positive semi-definite. (See Definition A.59.) Show that the function f is not convex on $\mathbb{R}^n$.

2.51 Consider a quadratic function $f : \mathbb{R} \to \mathbb{R}$ defined by:

$$\forall x \in \mathbb{R}, \ f(x) = \frac{1}{2}Q(x)^2 + cx,$$

where $Q \in \mathbb{R}, c \in \mathbb{R}$, and let $a, b \in \mathbb{R}$ with $a < b$.

   (i) Suppose that $Q \in \mathbb{R}_+$ so that f is convex. Find the maximum of the problem:

$$\max_{x \in \mathbb{R}} \{f(x) | a \le x \le b\}.$$

   (ii) Now suppose that you do not know whether f is convex. Find the maximum of the same problem:

$$\max_{x \in \mathbb{R}} \{f(x) | a \le x \le b\}.$$

   (Hint: use Theorem 2.5 and Exercise 2.42.)
   (iii) Again suppose that you do not know whether f is convex. Find the minimum:

$$\min_{x \in \mathbb{R}} \{f(x) | a \le x \le b\}.$$

2.52 Let $f_\ell : \mathbb{R}^n \to \mathbb{R}, \ell = 1, \ldots, r$, each be convex and define their point-wise maximum $f : \mathbb{R}^n \to \mathbb{R}$ as in (2.33):

$$\forall x \in \mathbb{R}^n, \ f(x) = \max_{\ell=1,\ldots,r} f_\ell(x).$$

Prove that f is convex.

2.53 A colleague of yours has formulated an optimization problem that requires a particular variable, x_1, to be either 0 or 1. To represent the "integrality" of x_1, he adds the following constraint to the formulation:

$$g_1(x) = 0,$$

where $g_1 : \mathbb{R}^n \to \mathbb{R}$ is defined by:

$$\forall x \in \mathbb{R}^n, g_1(x) = x_1(x_1 - 1).$$

(i) Show that requiring $g_1(x) = 0$ is equivalent to requiring that $x_1 = 0$ or 1. You must show that:

$$\forall x \in \mathbb{R}^n, (g_1(x) = 0) \Leftrightarrow (x_1 = 0 \text{ or } 1).$$

(ii) Is the function $g_1 : \mathbb{R}^n \to \mathbb{R}$ convex on $\mathbb{R}^n$? Either prove or disprove.

(iii) Is the set $\mathbb{S} = \{x \in \mathbb{R}^n | g_1(x) \leq 0\}$ convex? Either prove or disprove. (Note that $\mathbb{S}$ is a subset of $\mathbb{R}^n$, not of $\mathbb{R}$.)

(iv) Is the set $\mathbb{P} = \{x \in \mathbb{R}^n | g_1(x) = 0\}$ convex? (Note that $\mathbb{P}$ is a subset of $\mathbb{R}^n$, not of $\mathbb{R}$.)

(v) Would you expect that software designed for convex optimization could guarantee to find the global optimum of a problem that included $g_1(x) = 0$ as one of the constraints? Give your reasons.

Sensitivity and large change analysis

2.54 Recall Problem (2.18):

$$\min_{x \in \mathbb{R}^2} \{ f(x) | g(x) = 0, h(x) \leq 0 \},$$

with:

- objective $f : \mathbb{R}^2 \to \mathbb{R}$ defined in (2.10):

$$\forall x \in \mathbb{R}^2, f(x) = (x_1 - 1)^2 + (x_2 - 3)^2,$$

- equality constraint function $g : \mathbb{R}^2 \to \mathbb{R}$ defined in (2.12):

$$\forall x \in \mathbb{R}^2, g(x) = x_1 - x_2,$$

and
- inequality constraint function $h : \mathbb{R}^2 \to \mathbb{R}$ defined in (2.17):

$$\forall x \in \mathbb{R}^2, h(x) = 3 - x_2.$$

This problem was solved by inspection of Figure 2.12 and had solution:

$$\min_{x \in \mathbb{R}^2} \{ f(x) | g(x) = 0, h(x) \leq 0 \} = 4,$$

$$\arg \min_{x \in \mathbb{R}^2} \{ f(x) | g(x) = 0, h(x) \leq 0 \} = \left\{ \begin{bmatrix} 3 \\ 3 \end{bmatrix} \right\} = \{x^\star\}.$$

We will call this the base-case problem.

Now consider a parameter $\chi = \begin{bmatrix} \gamma \\ \eta \end{bmatrix} \in \mathbb{R}^2$ and a corresponding change-case problem $\min_{x \in \mathbb{R}^2} \{ f(x) | g(x; \chi) = 0, h(x; \chi) \leq 0 \}$ where:

- $f : \mathbb{R}^2 \to \mathbb{R}$ is the same as in the base-case problem,
- the equality constraint function $g : \mathbb{R}^2 \times \mathbb{R}^2 \to \mathbb{R}$ is a parameterized version of the function g defined in (2.12):

$$\forall x \in \mathbb{R}^2, g(x; \chi) = x_1 - x_2 + \gamma,$$

and

- the inequality constraint function $h : \mathbb{R}^2 \times \mathbb{R} \to \mathbb{R}$ is a parameterized version of the function h defined in (2.17):

$$\forall x \in \mathbb{R}^2, h(x; \chi) = 3 - x_2 + \eta.$$

Note that $g(\bullet; 0)$ is the equality constraint function in the base-case problem and $h(\bullet; 0)$ is the inequality constraint function in the base-case problem.

(i) Solve the change-case problem in terms of the values of γ and η for values of γ and η that are close to zero. (Hint: Make a sketch similar to Figure 2.9. The sketch and a discussion will suffice for the answer.)

(ii) Find the partial derivative of the minimizer and minimum with respect to γ. That is, calculate:

(a) the partial derivative of the minimizer with respect to γ,
(b) the partial derivative of the minimum with respect to γ.

(iii) Find the partial derivative of the minimizer and minimum with respect to η. That is, calculate:

(a) the partial derivative of the minimizer with respect to η,
(b) the partial derivative of the minimum with respect to η.

2.55 Consider the communications network of Exercise 2.8. Suppose that you were considering expanding the capacity of the network and knew the costs of increasing the capacity of each link. Also suppose that you could estimate the change in the value to the customers due to a change in the capacity of the link. Suggest how to decide on whether or not to expand the capacity of the network.

3

Transformation of problems

In this chapter we consider ways to **transform** problems. Such transformations are critical in matching a problem to the characteristics of an algorithm. Sometimes the transformation is done implicitly by formulating the problem in a particular, perhaps non-obvious, way. In several of the examples in this book, however, we will first formulate the problem in what might be considered a "natural" way and then look for ways to transform the problem to allow an algorithm to be effective. We will see that problem transformation is one of the key elements in matching a problem to an effective algorithm.

For example, we can think of transforming:

(i) the variables or equations of a system of simultaneous equations, or

(ii) the objective, variables, or constraints of an optimization problem,

to create a new problem. Typically, to be useful, the numbers of variables and constraints (or equations) in such a transformed problem should be not significantly larger than the numbers of variables and constraints (or equations) in the original problem. We could then consider the original and transformed problems to be "equivalent" if:

(i) given a solution of the original simultaneous equations it was easy to calculate a solution of the transformed simultaneous equations and vice versa, or

(ii) given the optimum and an optimizer of the original optimization problem it was easy to calculate the optimum and an optimizer of the transformed optimization problem and vice versa.

More formal notions of problem equivalence and of "easy" can be found in [40].

The relationship between maximization and minimization in (2.22) is a trivial example of transformation of an optimization problem to create an equivalent problem. We will consider other pairs of equivalent problems in this chapter and also consider some transformations that create "almost" equivalent problems that can provide useful guidance or insight into the solution of the original problem by providing an approximate solution to the original problem.

Careful transformation of problems can significantly simplify a problem or even make an otherwise intractable problem tractable. For example, in some cases we can transform an equality-constrained problem into a related inequality-constrained problem, or vice versa. We can then check, either empirically or based on theoretical understanding, to see which form is easier to solve. Sometimes, one formulation is much easier to solve than the other. In other cases, the existence of a transformation to a particular form shows that the original problem formulation can be solved relatively easily. Transformations can help us to match a problem formulation to the capabilities of an algorithm. On the other hand, a careless transformation—or a careless formulation—can render an otherwise simple problem extremely difficult to solve.

In some cases, we may be able to directly apply a theorem to one formulation but not to the other, and so gain insight into the solution. For example, we have indicated that iterative algorithms can only be expected to find local solutions of problems. In Section 2.6.3 we showed that, for a convex problem, a local minimum is also a global minimum. Therefore, if a problem can be transformed to a convex problem in a way that does not qualitatively change the local minima, then we can guarantee that a local minimum of the original problem is the global minimum.

In practice, we will find that it is not always possible to find transformations that guarantee convexity. It may nevertheless be possible to find transformations, or to formulate a problem, so that it is "approximately" convex. Although we may not be able to prove that a local minimum is the global minimum, we may be able show that the global minimum is not significantly better than any local minimum.

Even if no suitable transformation is available, knowledge that the problem *is* non-convex alerts us to expect several local optima or to look for weaker conditions that still guarantee that there is a unique local minimum. If the possibility of multiple local minima cannot be eliminated, then various search techniques can be applied to try to investigate several local optima and obtain the best. We will not treat such problems in this book; however, there is a wide literature on global optimization in such cases [91]. (See also the series *Nonconvex Optimization and Its Applications* by Kluwer Academic Publishers, including [113].)

We will introduce several transformations, including:

- transformations of the objective in Section 3.1;
- transformations of the variables in Section 3.2;
- transformations of the constraints in Section 3.3; and
- transformation of the problem involving a notion called "duality" in Section 3.4.

We summarize all the transformations in this chapter for easy reference so that we become aware of the possibilities of problem transformation early in the book; however, at this stage it will be unclear how to best use the transformations, particularly in relation to duality. The presentation here may even seem overly formal. However, in later chapters, we will further develop the transformations and see that they are crucial in matching a problem formulation to the capabilities of available algorithms. You may want to skim this chapter at first and then refer back to it as we confront particular case studies. Various other transformations are discussed in [6][15][22, chapter 13][28, chapter 6][41][42][43][70, section 3.10].

3.1 Objective

We consider transformations of the objective of an optimization problem. The four basic techniques for transforming the objective that we will discuss are:

(i) **monotonically increasing transformations,**
(ii) **adding terms,**
(iii) **moving the objective into the constraints,** and
(iv) **approximating the objective.**

3.1.1 Monotonically increasing transformations

First we present a very straightforward theorem relating the optimizers of the original and transformed problems under a **monotonically increasing** or **monotonically decreasing** transformation of the objective. (See Definition A.24 for definition of monotonically increasing and decreasing functions. The definition is a special case of Definition 2.11 of a monotone function. See Exercise 2.15.) The basic idea is that a monotonically increasing transformation of the objective preserves ordering of objective values so that the original and transformed problem have the same optimizers.

Theorem 3.1 *Let* $\mathbb{S} \subseteq \mathbb{R}^n$, *let* $f : \mathbb{R}^n \to \mathbb{R}$, *and let* $\eta^\nearrow : \mathbb{R} \to \mathbb{R}$ *be strictly monotonically increasing on* $\mathbb{R}$. *Define* $\phi : \mathbb{R}^n \to \mathbb{R}$ *by:*

$$\forall x \in \mathbb{R}^n, \phi(x) = \eta^\nearrow(f(x)).$$

Consider the problems $\min_{x \in \mathbb{S}} \phi(x)$ *and* $\min_{x \in \mathbb{S}} f(x)$. *Then:*

(i) $\min_{x \in \mathbb{S}} f(x)$ *has a minimum if and only if* $\min_{x \in \mathbb{S}} \phi(x)$ *has a minimum.*
(ii) *If either one of the problems in Item* (i) *possesses a minimum (and consequently,*

by Item (i), *each one possesses a minimum), then:*

$$\eta^{\nearrow}\left(\min_{x\in\mathbb{S}} f(x)\right) = \min_{x\in\mathbb{S}} \phi(x),$$

$$\operatorname*{argmin}_{x\in\mathbb{S}} f(x) = \operatorname*{argmin}_{x\in\mathbb{S}} \phi(x).$$

Proof See Exercise 3.1. □

Similar results apply for strictly monotonically decreasing transformations of the objective.

Two transformations of objective that will prove particularly useful in our case studies involve the exponential and logarithmic functions, which are strictly monotonically increasing by Exercises 2.15 and 2.18. According to Theorem 3.1, a problem with f as objective will have the same set of minimizers as a problem with $\exp(f(\bullet))$ or $\ln(f(\bullet))$ as objective. (In the latter case, we must require $f > 0$, that is, $f : \mathbb{R}^n \to \mathbb{R}_{++}$, in order for $\ln(f(\bullet))$ to be well-defined. The set $\mathbb{R}_{++}$ is the set of strictly positive real numbers. See Definition A.3.) Because f, $\exp(f(\bullet))$, and $\ln(f(\bullet))$ each have different properties, we can use whichever objective turns out to be easier to handle with the software at hand. The squared function provides another example of a monotonically increasing transformation for a function $f : \mathbb{R}^n \to \mathbb{R}_+$.

3.1.2 Adding terms

Another approach to transforming the objective involves adding terms to the objective. We will consider adding terms that depend on the constraint function with a view to incorporating the constraints into the objective so that either:

- we do not have to consider the constraints explicitly, or
- the constraints are easier to deal with.

The basic idea is that there is a *tension* between minimizing the objective and satisfying the constraints. By incorporating terms into the objective that worsen the objective at points that do not satisfy the constraints, or at points which are close to not satisfying the constraints, we can reduce the tension.

The two basic approaches involve:

- adding a **penalty function** that makes the objective large for values of the decision vector that violate the constraints, to be discussed in Section 3.1.2.1, and
- adding a **barrier function** that erects a barrier to violating the constraints, to be introduced in Section 3.1.2.2.

3.1.2.1 Penalty function

In the following, recall that $\mathbb{R}_+$ is the set of non-negative real numbers. (See Definition A.3.)

Theorem 3.2 *Let* $\mathbb{S} \subseteq \mathbb{R}^n$ *and* $f : \mathbb{R}^n \to \mathbb{R}$. *Consider the optimization problem* $\min_{x \in \mathbb{S}} f(x)$. *Let* $f_p : \mathbb{R}^n \to \mathbb{R}_+$ *be such that* $(x \in \mathbb{S}) \Rightarrow (f_p(x) = 0)$ *and let* $\Pi \in \mathbb{R}_+$. *Then:*

(i) $\min_{x \in \mathbb{S}} f(x)$ *has a minimum if and only if* $\min_{x \in \mathbb{S}}(f(x) + \Pi f_p(x))$ *has a minimum.*

(ii) *If either one of the problems in Item* (i) *possesses a minimum (and consequently, by Item* (i), *each one possesses a minimum), then:*

$$\min_{x \in \mathbb{S}} f(x) = \min_{x \in \mathbb{S}}(f(x) + \Pi f_p(x)),$$
$$\operatorname*{argmin}_{x \in \mathbb{S}} f(x) = \operatorname*{argmin}_{x \in \mathbb{S}}(f(x) + \Pi f_p(x)).$$

Proof See Exercise 3.7. □

The function f_p in Theorem 3.2 is called a **penalty function**, the parameter Π is called a **penalty coefficient**, and the function $f + \Pi f_p$ is called the **penalized objective**.

Discontinuous penalty function

Example To see how Theorem 3.2 might be useful, consider the objective $f : \mathbb{R} \to \mathbb{R}$ defined by:

$$\forall x \in \mathbb{R}, \ f(x) = x, \tag{3.1}$$

and the feasible set $\mathbb{S} = \{x \in \mathbb{R} | 1 \le x \le 3\}$. The problem:

$$\min_{x \in \mathbb{R}}\{f(x) | 1 \le x \le 3\},$$

has minimum $f^\star = 1$ and minimizer $x^\star = 1$.

Let $\Pi = 1$ and consider the penalty function $f_p : \mathbb{R} \to \mathbb{R}$ defined by:

$$\forall x \in \mathbb{R}, \ f_p = \begin{cases} 0, & \text{if } 1 \le x \le 3, \\ 10, & \text{otherwise.} \end{cases} \tag{3.2}$$

The function f_p satisfies the conditions for Theorem 3.2 and is illustrated in Figure 3.1. The circles, o, in Figure 3.1 indicate that the function f_p has a point of discontinuity as x approaches 1 from below or approaches 3 from above. The functions f and $f + \Pi f_p$ are illustrated in Figure 3.2. The circles again indicate points of discontinuity.

The penalized objective function $f + \Pi f_p$ includes information about the feasible

$f_\mathrm{p}(x)$

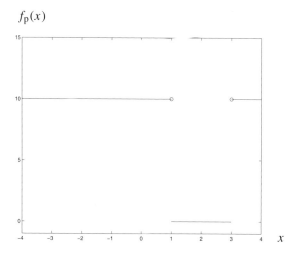

Fig. 3.1. The   penalty function $f_\mathrm{p}(x)$ versus x. In this figure and the next, the circles o indicate that the illustrated function has a point of discontinuity as x approaches 1 from below or 3 from above.

$f(x),\ f(x) + \Pi f_\mathrm{p}(x)$

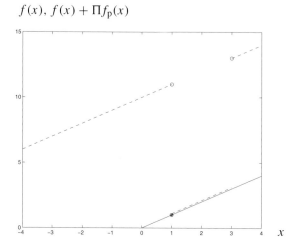

Fig. 3.2. The   objective function $f(x)$ versus x (shown solid) and the penalized objective function $f(x) + \Pi f_\mathrm{p}(x)$ versus x (shown dashed). (For clarity, for $1 \leq x \leq 3$ the penalized objective function is drawn slightly above the solid curve: it should be coincident with the solid curve.) One local minimizer of $f + \Pi f_\mathrm{p}$ in the region $\{x \in \mathbb{R}|-4 \leq x \leq 4\}$ is indicated by the bullet •.

set. The point $x^\star = 1$ is an unconstrained local minimizer of $f + \Pi f_\mathrm{p}$ in the region $\{x \in \mathbb{R}|-4 \leq x \leq 4\}$ and is indicated in Figure 3.2 by a bullet •. This point is also the minimizer of the constrained problem $\min_{x \in \mathbb{S}} f(x)$. That is, the penalty function allows us to consider the effect of the constraints by considering the penalized objective only.

Discussion The drawback of the penalty function f_p defined in (3.2) is that the penalized objective function $f + \Pi f_\mathrm{p}$ is not continuous because of the form of f_p. Moreover, local information at a feasible point in the interior $\underline{\mathbb{S}} = \{x \in \mathbb{R}|1 < x < 3\}$ of $\mathbb{S}$ does not inform about the boundary of the feasible region. It is still necessary to consider the feasible region explicitly and, from a practical perspective, it turns out that the discontinuity in the penalized objective imposes other difficulties [70, section 12.8].

We have indicated that we would like to consider continuous and differentiable objective functions. We will not treat non-differentiable penalty functions, such as (3.2), further in this book. However, further details can be found in [6, section 9.3][11, section 4.3.1] and an approach to particular types of non-differentiable functions will be explored in Section 3.1.3. In Section 3.1.2.2, we will see that some of the disadvantages of discontinuous penalty functions can also be avoided by solving a sequence of problems using a "barrier function."

Continuous penalty function

Analysis To avoid discontinuous functions, we will specialize Theorem 3.2 to a continuous penalty function that has important applications in equality-constrained and inequality-constrained optimization. We will consider the case of equality constraints in the following corollary to Theorem 3.2:

Corollary 3.3 *Suppose that* $f : \mathbb{R}^n \to \mathbb{R}, g : \mathbb{R}^n \to \mathbb{R}^m, \Pi \in \mathbb{R}_+,$ *and that* $\|\bullet\|$ *is a norm on* $\mathbb{R}^m$. *Consider the problems* $\min_{x \in \mathbb{R}^n} \{f(x) | g(x) = \mathbf{0}\}$ *and* $\min_{x \in \mathbb{R}^n} \{f(x) + \Pi \|g(x)\|^2 | g(x) = \mathbf{0}\}$. *Then:*

(i) *the problem* $\min_{x \in \mathbb{R}^n} \{f(x) | g(x) = \mathbf{0}\}$ *has a minimum if and only if the problem* $\min_{x \in \mathbb{R}^n} \{f(x) + \Pi \|g(x)\|^2 | g(x) = \mathbf{0}\}$ *has a minimum.*

(ii) *If either one of the problems in Item* (i) *possesses a minimum (and consequently, by Item* (i)*, each one possesses a minimum), then*

$$\min_{x \in \mathbb{R}^n} \{f(x) | g(x) = \mathbf{0}\} = \min_{x \in \mathbb{R}^n} \{f(x) + \Pi \|g(x)\|^2 | g(x) = \mathbf{0}\},$$

$$\arg\min_{x \in \mathbb{R}^n} \{f(x) | g(x) = \mathbf{0}\} = \arg\min_{x \in \mathbb{R}^n} \{f(x) + \Pi \|g(x)\|^2 | g(x) = \mathbf{0}\}.$$

Proof In the hypothesis of Theorem 3.2, let $\mathbb{S} = \{x \in \mathbb{R}^n | g(x) = \mathbf{0}\}$ and define $f_\mathrm{p} : \mathbb{R}^n \to \mathbb{R}_+$ by $f_\mathrm{p}(\bullet) = \|g(\bullet)\|^2$. Then:

$$(x \in \mathbb{S}) \quad \Leftrightarrow \quad (g(x) = \mathbf{0}),$$
$$\Rightarrow \quad (f_\mathrm{p}(x) = 0),$$

so that the hypothesis and therefore the conclusion of Theorem 3.2 holds. $\square$

The importance of Corollary 3.3 is that by adding the term $\Pi \|g(x)\|^2$ to the objective, we penalize deviations of the variable x outside the feasible set without changing the objective for feasible points and without destroying the continuity of the objective. By choosing the L_2 norm $\|\bullet\|_2$ (see Section A.3.1) we can also preserve partial differentiability of the objective if g is partially differentiable. (See Exercise 3.8.)

x_2

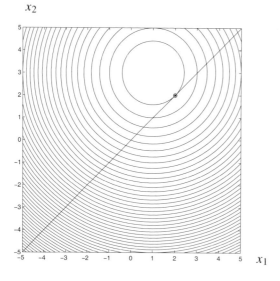

Fig. 3.3. The contour sets $\mathbb{C}_f(\tilde{f})$ of the objective function and the feasible set from Problem (2.13). The heights of the contours decrease towards the point $\begin{bmatrix} 1 \\ 3 \end{bmatrix}$.

x_2

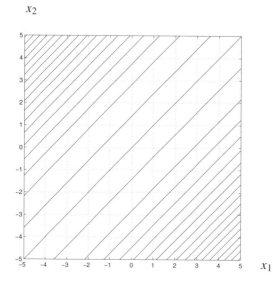

Fig. 3.4. The contour sets $\mathbb{C}_{(g)^2}(\tilde{f})$ of $(g(\bullet))^2$. The heights of the contours decrease towards the line $x_1 = x_2$.

Example Consider Figure 3.3, which repeats Figure 2.9. The figure shows the contour sets of the objective $f : \mathbb{R}^2 \to \mathbb{R}$ of Problem (2.13):

$$\forall x \in \mathbb{R}^2, \ f(x) = (x_1 - 1)^2 + (x_2 - 3)^2,$$

and a line that represents the feasible set $\{x \in \mathbb{R}^2 | g(x) = 0\}$, where $g : \mathbb{R}^2 \to \mathbb{R}$ is defined by:

$$\forall x \in \mathbb{R}^2, \ g(x) = x_1 - x_2.$$

As discussed in Section 2.2.2, the minimizer of $\min_{x \in \mathbb{R}^2}\{f(x)|g(x) = 0\}$ is $x^\star = \begin{bmatrix} 2 \\ 2 \end{bmatrix}$.

Consider using the L_2 norm $\|\bullet\|_2$, so that $\|g(\bullet)\|_2^2 = (g(\bullet))^2$. Figure 3.4 shows the contour sets of $(g(\bullet))^2$. The contours decrease towards the line representing the feasible set. The contours are parallel since g is affine.

Figure 3.5 shows the contour sets of the corresponding penalized objective $f + \Pi(g)^2$ for $\Pi = 1$, and again shows the line representing the feasible set. (The contours are spaced differently to those in Figure 3.3.) Adding the penalty to the objective makes infeasible points less "attractive" and does not change the objective values on the feasible set. Under certain circumstances, this can make it easier to find the minimum and minimizer of the problem. The unconstrained minimizer of $f + \Pi(g)^2$ for $\Pi = 1$ is $\begin{bmatrix} 5/3 \\ 7/3 \end{bmatrix}$, which is closer to the minimizer of the equality-constrained problem than is the unconstrained minimizer of f.

Larger values of the penalty coefficient Π, such as $\Pi = 10$ as shown in Figure 3.6, make infeasible points even less attractive. The unconstrained minimizer of $f + \Pi(g)^2$ for $\Pi = 10$ is very close to $\begin{bmatrix} 2 \\ 2 \end{bmatrix}$, which is the minimizer of the equality-constrained problem. (The contours in Figure 3.6 are spaced differently to those in both Figures 3.3 and 3.5.)

Sequence of problems If the formulation requires that the constraints must be met exactly then the continuous and differentiable penalty functions we are considering cannot enforce the constraints exactly for finite values of Π. In principle, a sequence of *unconstrained* problems can be solved with values of Π increasing without bound. Under certain conditions, the sequence of solutions of these unconstrained problems approaches a solution of the constrained problem as $\Pi \to \infty$. (See Exercise 3.10.) However, because we must eventually terminate at a finite value of Π, this approach is not directly applicable when we have constraints that must be satisfied to obtain a valid solution.

Soft constraints Sometimes, however, constraints in a formulation actually reflect "target" values rather than constraints that absolutely must be met. Such targets are sometimes called **soft constraints** to distinguish them from the constraints that we have considered previously, which might be called **hard constraints** in this context. Unconstrained optimization of a penalized objective can be a very effective means to approximately satisfy soft constraints. For example, in Figure 3.6, the unconstrained minimizer of $f + \Pi(g)^2$ is close to being on the feasible set and is nearby to the point $\begin{bmatrix} 2 \\ 2 \end{bmatrix}$, which is the solution to the constrained problem.

Transformation of problems

x_2

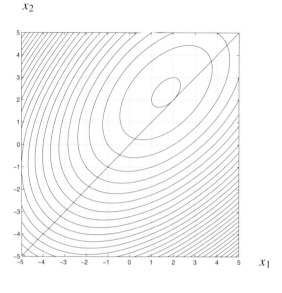

Fig. 3.5. The contour sets $\mathbb{C}_{f+1(g)^2}(\tilde{f})$ of the penalized objective function and the feasible set from Problem (2.13). The heights of the contours decrease towards the point $\begin{bmatrix} 5/3 \\ 7/3 \end{bmatrix}$.

x_2

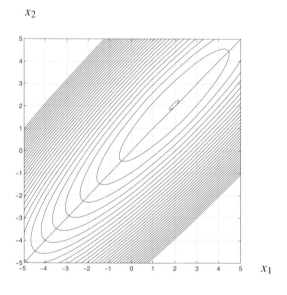

Fig. 3.6. The contour sets $\mathbb{C}_{f+10(g)^2}(\tilde{f})$ of the penalized objective function and the feasible set from Problem (2.13). The heights of the contours decrease towards a point that is near to $\begin{bmatrix} 2 \\ 2 \end{bmatrix}$.

If the constraint violation was still not acceptable for the value $\Pi = 10$, then the value of Π could be increased and the unconstrained problem re-solved. This process could be repeated until the constraint was satisfied to within an acceptable tolerance.

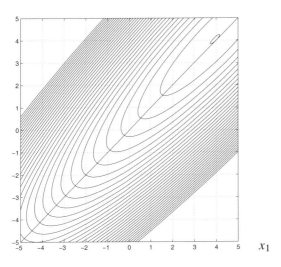

Fig. 3.7. The contour sets from Figure 3.6 shifted up and to the right. The feasible set from Problem (2.13) is also shown. The heights of the contours decrease towards a point that is near to $\begin{bmatrix} 4 \\ 4 \end{bmatrix}$.

Ill-conditioning For very tight tolerances, the required value of Π will be large. Unfortunately, as Π becomes large the unconstrained problem itself becomes difficult to solve. This is because the contour sets of the penalized objective become increasingly elliptical and eccentric as Π increases, even if the contour sets of the original objective were circular as in Figure 3.3. This is evident in the progression from Figure 3.3 to Figure 3.5 to Figure 3.6 [6, section 9.2].

As we discussed in Section 2.7.6.3, for a function with roughly circular contour sets, it is easy to find a direction that points towards the minimizer, while for a function with highly eccentric elliptical contour sets this can be more problematic: the problem of finding such a direction is ill-conditioned. As in Section 2.7.6.3, imagine that the objective function changes slightly, with the center of the ellipses in Figure 3.6 moving to the right and up. In particular, Figure 3.7 shows the case where the center of the ellipses are shifted up by two units and to the right by two units. As in the example in Section 2.7.6.3, the effect on local appearance of the levels sets at a point such as $x = \begin{bmatrix} 0 \\ -5 \end{bmatrix}$ is only small. If $x^{(\nu)} = \begin{bmatrix} 0 \\ -5 \end{bmatrix}$ were the current iterate, for example, then it would be difficult to accurately determine the direction of the minimizer of the penalized objective from local information at this point. There is a trade-off between satisfaction of the constraints and the difficulty of minimizing the penalized objective.

Inequality constraints It is also possible to generalize the penalty function approach to the inequality-constrained case. For example, if $h : \mathbb{R}^n \to \mathbb{R}^r$ then a penalty function for the constraint $h(x) \leq 0$ is $f_p : \mathbb{R}^n \to \mathbb{R}$ defined by:

$$\forall x \in \mathbb{R}^n, \ f_p(x) = \sum_{\ell=1}^{r} \max\{0, (h_\ell(x))^2\}.$$

See [6, chapter 9] and [70, section 13.5] for details and further development of penalty methods. However, we will take another approach in Part V to inequality-constrained problems, involving a "barrier function," to be introduced in Section 3.1.2.2.

Summary Large values of the penalty coefficient Π yield an ill-conditioned problem, while for finite values of Π we must still consider the constraints explicitly if we want them to be satisfied exactly. Nevertheless, modest values of the penalty coefficient can significantly enhance the performance of algorithms that explicitly treat the constraints, since the penalized objective works in concert with the explicit treatment of the constraints, reducing the tension between minimizing the objective and satisfying the constraints. Moreover, the penalty function can serve to make the penalized objective convex, which guarantees that a local minimum is a global minimum. (See Exercise 3.11.) We will see that convexity of the objective can also facilitate the process of *finding* a locally optimal point.

3.1.2.2 Barrier function

An alternative to the penalty method for inequality-constrained problems involves adding a function that grows large as we *approach* the boundary of the feasible region from the interior. Consider again the feasible set $\mathbb{S} = \{x \in \mathbb{R} | 1 \leq x \leq 3\}$ and its interior, $\underline{\mathbb{S}} = \{x \in \mathbb{R} | 1 < x < 3\}$. Figure 3.8 shows a **barrier function** $f_b : \underline{\mathbb{S}} \to \mathbb{R}$ that is designed to penalize values of x that are close to the boundary of the feasible region. (The barrier function f_b is not defined for values of x that are outside the feasible region or on its boundary.)

The barrier function grows very rapidly as x approaches 1 from above or approaches 3 from below. Moreover, derivative information at points near to the boundary provides information about proximity to the boundary. This allows us to avoid considering the inequality constraints explicitly.

Consider again the objective function $f : \mathbb{R} \to \mathbb{R}$ defined in (3.1) and illustrated in Figure 3.2. Figure 3.9 shows $f(x)$ for $0 \leq x \leq 4$ together with $f(x) + f_b(x)$ for values of x that are in the interior of the feasible set $\{x \in \mathbb{R} | 1 \leq x \leq 3\}$. A local minimizer of $f + f_b$ is illustrated with a •. This point is near to the minimizer of the original constrained problem $\min_{x \in \mathbb{R}}\{f(x) | x \in \mathbb{S}\}$.

$f_b(x)$

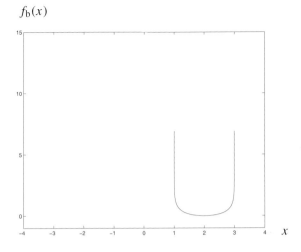

Fig. 3.8. The barrier function $f_b(x)$ versus x on the
interior of the feasible set.

$f(x), f(x) + f_b(x)$

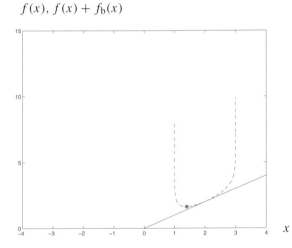

Fig. 3.9. The objective
function $f(x)$ versus x
(shown solid) and the
objective plus barrier function $f(x) + f_b(x)$ versus
x on the interior of the feasible set (shown dashed).
The local minimizer of
the objective plus barrier
function is indicated by the
bullet •.

The barrier function avoids the issue of discontinuity that we observed in Figure 3.2. However, the modified objective does not satisfy Theorem 3.2 because
the added term is non-zero on the feasible set. It turns out that this issue can be
dealt with by solving a *sequence* of problems where the added term is gradually
reduced towards zero. As the added term becomes smaller, the modified objective
becomes "almost" the same as the original objective except for values of x that are
very close to the boundary of the region defined by the inequality constraints. Under certain conditions, by repeatedly reducing the size of the barrier and re-solving
the resulting problem, we obtain a sequence of solutions that approach a solution
of the original problem; however, we still have to consider ill-conditioning in the
problems. We will investigate this approach in Part V.

$f(x)$

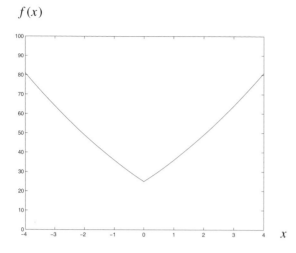

Fig. 3.10. Function that is defined as point-wise maximum.

3.1.3 Moving the objective into the constraints

The discussion in this section is based on [45, section 4.2.3]. Recall the discussion of point-wise maximum functions from Section 2.6.3.6. Such objectives arise when there are a number of issues that must be considered simultaneously and we are concerned about the worst-case of all the issues. (See Exercise 3.15.)

It is typical for such objectives to be non-differentiable. For example, Figure 3.10 repeats Figure 2.38, and shows the function $f : \mathbb{R} \to \mathbb{R}$ defined by:

$$\forall x \in \mathbb{R}, \ f(x) = \max\{(x+5)^2, (x-5)^2\}.$$

This function is not differentiable at the point $x = 0$. It is differentiable at every other point; however, the unconstrained minimizer of the function is at this point of non-differentiability, so methods that use the derivative of the function or assume that the function is differentiable will fail precisely at the minimizer.

There are techniques for treating non-differentiable objectives. See, for example, [11, 51, 97]. However, we will take the approach of removing the non-differentiabilities by transforming the problem. We shift the effects of the non-differentiability into the constraints, using the following theorem [28, section 6.5]:

Theorem 3.4 *Let* $\mathbb{S} \subseteq \mathbb{R}^n$ *and let* $f_\ell : \mathbb{R}^n \to \mathbb{R}$ *for* $\ell = 1, \ldots, r$. *Define* $f : \mathbb{R}^n \to \mathbb{R}$ *by:*

$$\forall x \in \mathbb{R}^n, \ f(x) = \max_{\ell=1,\ldots,r} f_\ell(x).$$

Consider the problems $\min_{x \in \mathbb{S}} f(x)$ *and*

$$\min_{x \in \mathbb{S}, z \in \mathbb{R}} \{z \mid f_\ell(x) - z \leq 0, \forall \ell = 1, \ldots, r\}. \tag{3.3}$$

Then:

(i) $\min_{x \in \mathbb{S}} f(x)$ *has a minimum if and only if* $\min_{x \in \mathbb{S}, z \in \mathbb{R}} \{z | f_\ell(x) - z \le 0, \forall \ell = 1, \dots, r\}$ *has a minimum.*

(ii) *If either one of the problems in Item* (i) *possesses a minimum (and consequently, by Item* (i)*, each one possesses a minimum), then*

$$\min_{x \in \mathbb{S}} f(x) \quad = \quad \min_{x \in \mathbb{S}, z \in \mathbb{R}} \{z | f_\ell(x) - z \le 0, \forall \ell = 1, \dots, r\},$$

$$\operatorname*{argmin}_{x \in \mathbb{S}} f(x) \quad = \quad \left\{ x \in \mathbb{R}^n \; \middle\| \begin{bmatrix} x \\ z \end{bmatrix} \in \arg \min_{x \in \mathbb{S}, z \in \mathbb{R}} \left\{ z \; \middle| \; \begin{array}{l} f_\ell(x) - z \le 0, \\ \forall \ell = 1, \dots, r \end{array} \right\} \right\}.$$

Proof See Exercise 3.12. $\square$

In Theorem 3.4, the transformed Problem (3.3) has objective $\phi : \mathbb{R}^n \times \mathbb{R} \to \mathbb{R}$ defined by:

$$\forall x \in \mathbb{R}^n, \forall z \in \mathbb{R}, \phi(x, z) = z.$$

This objective function is convex. Therefore, if $\mathbb{S}$ is convex and the functions f_ℓ are convex then the transformed Problem (3.3) is a convex optimization problem. (See Exercise 3.14.) This provides the key to the usefulness of Theorem 3.4.

To understand the transformation, consider Figure 3.11. This figure repeats Figure 2.37 and shows the functions f_1 and f_2 that were point-wise maximized to form the objective shown in Figure 3.10. Figure 3.12 re-interprets Figure 3.11 in terms of Problem (3.3). It shows the contour sets of the objective, which are horizontal lines of constant value of z. Problem (3.3) tries to find the minimum feasible value of z; that is, it seeks the "lowest" feasible line. Suppose that $\mathbb{S} = \{x \in \mathbb{R} | -4 \le x \le 4\}$. Then the two curves shown in Figure 3.12 determine the feasible region of Problem (3.3). The inequality constraints of Problem (3.3) require that $f_\ell(x) - z \le 0$. That is, at each value of x, feasible values of z are those that are greater than the value of all the $f_\ell(x)$, for $\ell = 1, \dots, r$. In the case of Figure 3.12, the feasible points are the ones that lie at or "above" the two curved lines.

Minimizing z forces us to find the "lowest" value of z that lies at or above each of the curved lines. The minimum is $z^\star = f^\star = 25$. This is the minimum of the original function f that was defined as a point-wise maximum. The minimizer is $x^\star = 0$.

3.1.4 Approximating the objective

In the previous sections, we have described transformations such that the transformed problem, in principle, yields an exact solution to the original problem. In some cases the equality between original and transformed problem is obtained only in a limiting process. In this section, we consider deliberately approximating the objective to make its characteristics more suitable for solution. Again, equality

$f_1(x),\ f_2(x)$

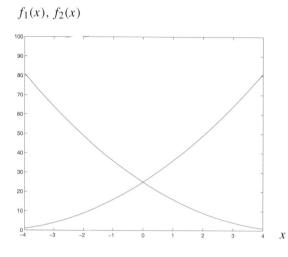

Fig. 3.11. Functions used to define point-wise maximum, repeated from Figure 2.37.

z

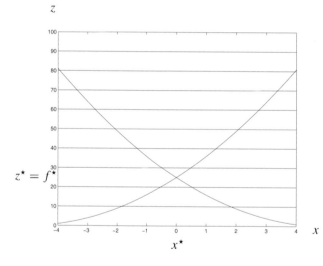

$z^\star = f^\star$

$x^\star$

Fig. 3.12. Feasible regions and contour sets of objective for transformed problem. The feasible region is the set of points $\begin{bmatrix} x \\ z \end{bmatrix}$ that lies "above" both of the curves. The contour sets of the objective decrease towards $z = 0$.

between the original and approximating problem may only occur in some limit so that in practice we may solve the problem repeatedly until we have a sufficiently accurate approximation. The four basic techniques we will discuss are:

(i) **linear approximation**,

(ii) **quadratic approximation**,

(iii) **piece-wise linearization**, and

(iv) **smoothing**.

The first three techniques are used, for example:

- when software designed for a linear objective is to be applied to a problem with non-linear objective, or
- when software designed for a quadratic objective is to be applied to a problem with a non-quadratic objective.

Linear approximation can also be applied to constraint functions so that software designed for linear constraints can be applied to non-linearly constrained problems.

Typically, a *sequence* of approximate problems is then solved. The approximation to the objective (and constraints) is updated for each iteration. Particularly during the early iterations, it may not be worth solving the approximate problem to high accuracy. We may instead choose to only solve the linear or quadratic approximating problem "roughly."

The fourth technique, smoothing, is useful when an objective is not differentiable. By carefully smoothing an objective, a partially differentiable objective can be created that differs only slightly from the exact objective. Again, a sequence of approximate problems can be solved if necessary to provide sufficient accuracy.

3.1.4.1 Linear approximation

We first consider linearizing an objective about a current estimate $x^{(\nu)}$. (As noted, non-linear constraints can also be linearized.) A linear programming algorithm is then used to solve for the optimal $x^{(\nu+1)}$ that minimizes the linearized objective while satisfying the (linearized) constraints. Usually, an extra set of constraints of the form:

$$\forall k = 1, \ldots, n, |x_k^{(\nu+1)} - x_k^{(\nu)}| \leq \overline{\Delta x_k},$$

where $\overline{\Delta x_k} \in \mathbb{R}_{++}$ is the maximum allowed step, is included in the approximating problem to keep $x^{(\nu+1)}$ close enough to $x^{(\nu)}$ so that the linear approximation is still approximately valid at $x^{(\nu+1)}$. The objective and constraints are then re linearized about $x^{(\nu+1)}$, and so on. If the successive iterates stop changing significantly, then the process is considered to have reached an optimum. This approach is called **successive linear programming**. The drawback of this approach is that if the objective and constraints are significantly non-linear, then many iterations will be required to obtain a useful solution. Furthermore, linearization ignores interdependencies between variables due to non-zero mixed second partial derivatives.

3.1.4.2 Quadratic approximation

Instead of a linear approximation, a quadratic approximation can be made to the objective at each iteration ν. (Additional terms involving the constraints are also incorporated into the objective [44, section 1].) This approach is called **successive**

quadratic programming and allows better approximation of functions with non-zero second partial derivatives [44].

3.1.4.3 Piece-wise linearization

An alternative to successive linearization and successive quadratic approximation is to approximate a non-linear function f by defining a piece-wise linear function that matches f at "break-points." To make the approximation for convex f, we define subsidiary variables that partition the decision space. For example, for a function $f : [0, 1] \to \mathbb{R}$ we might:

- define subsidiary variables $\xi_1, \ldots, \xi_5$,

- include constraints:

$$ x = \sum_{j=1}^{5} \xi_j, $$

$$ 0 \le \xi_j \le 0.2, $$

- define parameters:

$$ d = f(0), $$

$$ c_j = \frac{1}{0.2} [f(0.2 \times j) - f(0.2 \times (j-1))], \, j = 1, \ldots, 5, $$

and

- replace the objective f by the piece-wise linearized objective $\phi : \mathbb{R}^5 \to \mathbb{R}$ defined by:

$$ \forall \xi \in \mathbb{R}^5, \phi(\xi) = c^\dagger \xi + d. $$

The variables ξ are included in the problem together with the constraints. The variables ξ define "break-points" at $x = 0, 0.2, 0.4, 0.6, 0.8, 1.0$. The piece-wise linearized objective ϕ matches f exactly at the break-points and linearly interpolates f between the break-points. Since f is convex then the piece-wise linearized objective ϕ never under-estimates f in between the break-points. (For non-convex f, a more complicated approach involving discrete variables is necessary.)

For example, consider the quadratic function $f : [0, 1] \to \mathbb{R}$ defined by:

$$ \forall x \in [0, 1], f(x) = (x)^2. $$

$f(x), \phi(\xi)$

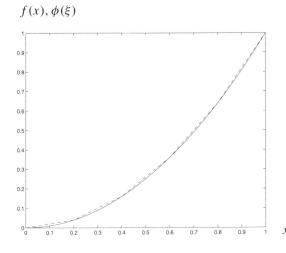

Fig. 3.13. Piece-wise linearization (shown dashed) of a function (shown solid).

This function is illustrated in Figure 3.13 as a solid line together with its piece-wise linearization, which is illustrated as a dashed line. In this case:

$$
\begin{aligned}
d &= f(0), \\
&= 0, \\
c_j &= \frac{1}{0.2} \left(f(0.2 \times j) - f(0.2 \times (j-1)) \right), \\
&= (0.4 \times j) - 0.2.
\end{aligned}
$$

To illustrate the approximation, the function ϕ has been evaluated at the values of ξ that correspond to each value of x shown in Figure 3.13. Even with only five pieces, the quadratic function is well approximated by the piece-wise linearization.

To piece-wise linearize f in an optimization problem, we use ϕ as the objective instead of f, augment the decision vector to include ξ, and include the constraints that link ξ and x. Similarly, non-linear constraints can also be piece-wise linearized.

Piece-wise linearization can be applied to both quadratic and more general non-linear functions. For non-linear functions, if the function is **additively separable**; that is, if it can be expressed as the sum of functions each of which depend only on one of the elements of x, then piece-wise linearization will involve an increase in the number of variables by a factor equal to the number of break-points. (See Definition A.23 and see [6, section 11.3] for a discussion of this case.) However, if the function is non-separable, then many extra variables will have to be defined to represent the function. The MATLAB functions `interp1` and `interp2` perform interpolation, including piece-wise linearization, of functions with one and two arguments, respectively.

A generalization of this idea is to partition the decision space into regions and to linearly approximate functions on each region. The leads to the **finite element method** [27, 77, 108].

3.1.4.4 Smoothing

In Section 3.1.3, we showed how to treat a non-differentiable objective defined in terms of the maximum of other functions. We transformed the problem into a constrained problem where all the functions were differentiable.

In some cases, we prefer not to create extra constraints and may be willing to accept some inaccuracy in the solution in return for avoiding the extra constraints. To do this, we **smooth** the objective at the points of non-differentiability [5, section III].

For example, consider the absolute value function $|\bullet| : \mathbb{R} \to \mathbb{R}_+$ defined by:

$$\forall x \in \mathbb{R}, |x| = \begin{cases} x, & \text{if } x \geq 0, \\ -x, & \text{if } x < 0. \end{cases}$$

This function is continuous but not differentiable. Consider the function $\phi :$ $\mathbb{R} \to \mathbb{R}$ defined by:

$$\forall x \in \mathbb{R}, \phi(x) = \sqrt{(|x|^2 + \epsilon)}. \tag{3.4}$$

We call ϕ a **smoothed version** of $|\bullet|$. It can be verified that, for all $\epsilon > 0$, the function ϕ is differentiable. (See Exercise 3.17.) Moreover, the error between ϕ and $|\bullet|$ decreases with decreasing ϵ. Consequently, the smoothed function can be used as an approximation to $|\bullet|$, with a controllable approximation error determined by the choice of ϵ. Figure 3.14 shows the absolute value function together with two smoothed versions, for $\epsilon = 0.1$ (shown dashed) and $\epsilon = 0.01$ (shown dotted). Typically, as ϵ is reduced, problems defined in terms of ϕ become more ill-conditioned, so there is a compromise between fidelity to the original function and solvability.

3.2 Variables

Theorems 3.1 and 3.2 in Section 3.1 show that by solving a problem with a transformed *objective*, we can also solve the original problem. We will now show that transformation of *variables* can also be useful. The two basic techniques that we will discuss are:

(i) **scaling**, and
(ii) **onto transformations**.

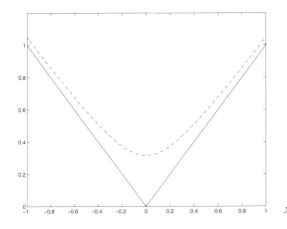

Fig. 3.14. Smoothed version for $\epsilon = 0.1$ (shown dashed) and $\epsilon = 0.01$ (shown dotted) of absolute value function (shown solid).

3.2.1 Scaling

This simplest way to transform variables is to "scale" them; that is, change their units. This can be useful when the variables in the initial formulation of the problem have widely differing magnitudes, since, as a practical matter, optimization software often makes the implicit assumption that the variables have similar magnitudes at the optimum [45, section 7.5].

For example, suppose that we have two variables x_1 and x_2 in our formulation. Suppose that typical values for x_1 are between -0.01 and 0.01, while typical values for x_2 are between -10^4 and 10^4. Furthermore, let us suppose that, based on our understanding of the application, a "significant error" in x_1 is around 10^{-3}, while a significant error in x_2 is around 10^3. In particular, consider the objective $f : \mathbb{R}^2 \to \mathbb{R}$ defined by:

$$\forall x \in \mathbb{R}^2, \ f(x) = (1000x_1)^2 + (x_2/1000)^2. \tag{3.5}$$

The contour sets of this function are shown in Figure 3.15. The axes have different scales in this figure: if the axes were shown at the same scale, the contour sets would be extremely elliptical.

If we want to obtain an solution that yields an objective that is within one unit of the minimum then we need to obtain a value of x such that (approximately):

$$|x_1^\star - x_1| \ < \ 0.001,$$
$$|x_2^\star - x_2| \ < \ 1000,$$

where $x^\star = 0$ is the minimizer. That is, errors in the different elements in x have differing effects on the objective.

In assessing closeness to the actual solution and in implementing stopping cri-

x_2

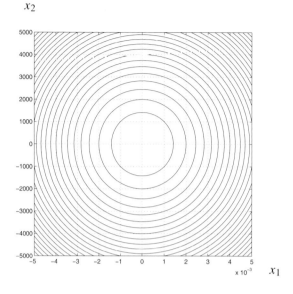

Fig. 3.15. Contour sets of
function f defined in (3.5).

ξ_2

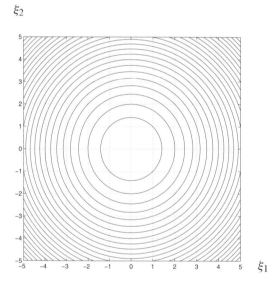

Fig. 3.16. Contour sets of
function ϕ defined in (3.6)
with scaled variables.

teria, we will make use of norms that combine the discrepancies or errors of all
elements in x. While we can define a **weighted norm** that weights the importance of individual elements of x differently, it is typical for software to use an
unweighted norm such as $\|\bullet\|_2$ or $\|\bullet\|_\infty$. If we apply an algorithm that uses un-
weighted norms to the formulation as described, then errors in x_2 will be accorded
too much importance relative to errors in x_1.

To appropriately weight the importance of errors in x_1 and x_2, suppose that we define scaled variables $\xi \in \mathbb{R}^2$ by:

$$\xi_1 = 1000x_1,$$
$$\xi_2 = x_2/1000.$$

Consider the objective $\phi : \mathbb{R}^2 \to \mathbb{R}$ defined by:

$$\forall \xi \in \mathbb{R}^2, \phi(\xi) = (\xi_1)^2 + (\xi_2)^2. \tag{3.6}$$

We observe that with the scaling of variables, the two functions f and ϕ represent the same underlying function. Moreover, errors in ξ_1 have the same relative effect on the objective as errors in ξ_2, and the contour sets of ϕ are circular as shown in Figure 3.16 with axes shown at the same scale. As we discussed in Section 2.6.3.4, for a function with roughly circular contour sets, it is easy to find a direction that points towards the minimizer.

In general, we should scale variables so that, at the solution, the values of each element of the design vector are of the same order of magnitude and so that an acceptable error for each element is roughly the same. It might also be necessary to add a constant offset to some of the variables to achieve this goal if the range of typical values is not symmetric about zero. Sometimes, software will perform scaling automatically.

3.2.2 Onto transformations

3.2.2.1 Analysis

Transformations that are more general in nature than scaling can change the structure of the problem. The basic idea is that we can re-write the problem in terms of new variables so long as "exploring" over the new variables also "covers" the whole of the original feasible set $\mathbb{S}$. This idea is embodied in the definition of an **onto** function. (See Definition A.25.)

Theorem 3.5 *Let* $\mathbb{S} \subseteq \mathbb{R}^n$, $\mathbb{P} \subseteq \mathbb{R}^{n'}$, $f : \mathbb{S} \to \mathbb{R}$, *let* $\tau : \mathbb{P} \to \mathbb{S}$ *be* **onto** $\mathbb{S}$, *and define* $\phi : \mathbb{P} \to \mathbb{R}$ *by:*

$$\forall \xi \in \mathbb{P}, \phi(\xi) = f(\tau(\xi)).$$

Consider the problems: $\min_{\xi \in \mathbb{P}} \phi(\xi)$ *and* $\min_{x \in \mathbb{S}} f(x)$. *Then:*

(i) *the problem* $\min_{x \in \mathbb{S}} f(x)$ *has a minimum if and only if* $\min_{\xi \in \mathbb{P}} \phi(\xi)$ *has a minimum.*

(ii) *If either one of the problems in Item* (i) *possesses a minimum (and consequently,*

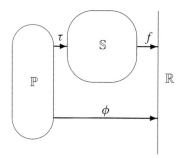

Fig. 3.17. Sets and transformations in Theorem 3.5.

by Item (i), *each one possesses a minimum), then*

$$\min_{\xi \in \mathbb{P}} \phi(\xi) \;=\; \min_{x \in \mathbb{S}} f(x),$$

$$\operatorname*{argmin}_{x \in \mathbb{S}} f(x) \;=\; \left\{ \tau(\xi) \,\middle|\, \xi \in \operatorname*{argmin}_{\xi \in \mathbb{P}} \phi(\xi) \right\}.$$

Proof   See Exercise 3.18. □

Figure 3.17 illustrates the relationship between the sets and transformations de-scribed in Theorem 3.5. The function ϕ is the composition of the onto function τ with f.

We will see in Part IV that it is sometimes possible to find an onto function such that $\mathbb{P} = \mathbb{R}^{n'}$ and so the transformed problem $\min_{\xi \in \mathbb{P}} \phi(\xi)$ is an *unconstrained* problem $\min_{\xi \in \mathbb{R}^{n'}} \phi(\xi)$. Even when it is not possible to find a transformation that yields an unconstrained problem, it is often possible to choose a transformation that simplifies the form of the constraints or objective.

For example, Exercise 3.19 considers rectangular to polar coordinate transfor-mation. In Exercise 3.19, the feasible set in rectangular coordinates is non-convex; however, in polar coordinates the feasible set is an interval on the real line, which is convex. Exercise 3.19 shows that a non-linear transformation can be used to create a convex feasible set.

To apply Theorem 3.5, it is sometimes easiest to first define a function $\tau : \mathbb{R}^{n'} \to \mathbb{R}^n$ that is onto $\mathbb{R}^n$ and then define $\mathbb{P} \subseteq \mathbb{R}^{n'}$ by:

$$\mathbb{P} = \{\xi \in \mathbb{R}^{n'} | \tau(\xi) \in \mathbb{S}\}.$$

Then we consider the **restriction** of τ to $\mathbb{P}$. (See Definition A.13.) The restriction of τ to $\mathbb{P}$ then satisfies the conditions of Theorem 3.5. (See Exercise 3.20.) That is, we now think of τ as a function $\tau : \mathbb{P} \to \mathbb{S}$. We can then apply the theorem.

Exercise 3.22 shows that linear transformation of variables preserves convexity of a function. However, linear transformations of variables can change the shape of the contour sets of a problem as shown in Exercise 3.23. As we discussed in

Section 2.6.3.4 and mentioned in Section 3.2.1, for a function with roughly circular contour sets, it is easy to find a direction that points towards the minimizer. As Exercise 3.24 suggests, it is possible to transform the variables of a function with elliptical level sets into one with circular level sets using a linear transformation of the variables. Nevertheless, objectives with highly elliptical contour sets can still prove problematic because finding the linear transformation itself requires knowledge of the second derivative of the function and, moreover, the evaluation of the linear transformation is an ill-conditioned problem if the level sets are highly eccentric.

3.2.2.2 Elimination of variables

An important special case of Theorem 3.5 occurs when we eliminate variables. We first present an elementary theorem involving elimination of variables for simultaneous equations and then a corollary of Theorem 3.5 for optimization problems.

Simultaneous equations

Analysis We can sometimes eliminate variables when solving simultaneous equations using:

Theorem 3.6 *Let* $g : \mathbb{R}^n \to \mathbb{R}^m$, $n' \leq n$ *and collect the last* n' *entries of* x *together*

into a vector $\xi = \begin{bmatrix} x_{n-n'+1} \\ \vdots \\ x_n \end{bmatrix} \in \mathbb{R}^{n'}$. *Suppose that functions* $\omega_\ell : \mathbb{R}^{n'} \to \mathbb{R}$ *for*

$\ell = 1, \ldots, (n - n')$, *can be found that satisfy:*

$$\forall \begin{bmatrix} x_1 \\ \vdots \\ x_{n-n'} \\ \xi \end{bmatrix} \in \{x \in \mathbb{R}^n | g(x) = \mathbf{0}\}, \forall \ell = 1, \ldots, (n - n'), x_\ell = \omega_\ell(\xi).$$

Collect the functions ω_ℓ, $\ell = 1, \ldots, (n - n')$, *into a vector function* $\omega : \mathbb{R}^{n'} \to \mathbb{R}^{n-n'}$. *Then, for* $x \in \{x \in \mathbb{R}^n | g(x) = \mathbf{0}\}$, *the vector function* $\omega : \mathbb{R}^{n'} \to \mathbb{R}^{n-n'}$ *expresses:*

• *the sub-vector of* x *consisting of the first* $(n - n')$ *components of* x,
• *in terms of the sub-vector* ξ *of* x *consisting of the last* n' *components of* x.

Suppose that $\xi^\star \in \mathbb{R}^{n'}$ *solves* $g\left(\begin{bmatrix} \omega(\xi) \\ \xi \end{bmatrix}\right) = \mathbf{0}$. *(Note that these equations involve*

only ξ.) *Then* $x^\star = \begin{bmatrix} \omega(\xi^\star) \\ \xi^\star \end{bmatrix}$ *satisfies* $g(x) = \mathbf{0}$.

Conversely, suppose that $x^\star \in \mathbb{R}^n$ *satisfies* $g(x^\star) = \mathbf{0}$. *Let* $\xi^\star \in \mathbb{R}^{n'}$ *be the sub-vector*

of $x^\star$ *consisting of its last* n' *components. Then* $\xi^\star$ *solves* $g\left(\begin{bmatrix} \omega(\xi) \\ \xi \end{bmatrix}\right) = \mathbf{0}$.

Proof See Exercise 3.18. □

Discussion In Theorem 3.6, we write the entries of g in terms of the vector ξ and the function ω by replacing x_ℓ, $\ell = 1, \ldots, (n - n')$ by $\omega_\ell(\xi)$, $\ell = 1, \ldots, (n - n')$, respectively. This eliminates x_ℓ, $\ell = 1, \ldots, (n - n')$. The functions ω typically involve re-arranging some of the entries of g. In this case, we can delete the corresponding entries of g when solving $g\left(\begin{bmatrix} \omega(\xi) \\ \xi \end{bmatrix}\right) = \mathbf{0}$ since these entries are satisfied identically by $x = \begin{bmatrix} \omega(\xi) \\ \xi \end{bmatrix}$. The variables ξ are called the **independent variables**, while the variables x_ℓ, $\ell = 1, \ldots, (n - n')$, are called the **dependent variables**.

Example Consider $g : \mathbb{R}^2 \to \mathbb{R}^2$ defined by:

$$\forall x \in \mathbb{R}^2, g(x) = \begin{bmatrix} x_1 - x_2 \\ (x_2)^2 - x_2 \end{bmatrix}.$$

The first entry of g can be re-arranged as $x_1 = \omega_1(x_2)$, where $\xi = x_2$ and $\omega_1 : \mathbb{R} \to \mathbb{R}$ is defined by:

$$\forall x_2 \in \mathbb{R}, \omega_1(x_2) = x_2.$$

We can delete the first entry g_1 from the equations to be solved since it is satisfied identically by $x = \begin{bmatrix} \omega_1(\xi) \\ \xi \end{bmatrix}$. We need only solve the smaller system consisting of the one equation $g_2\left(\begin{bmatrix} \omega_1(\xi) \\ \xi \end{bmatrix}\right) = 0$. That is, we must solve $(\xi)^2 - \xi = 0$, which has solutions $\xi^\star = 0$, $\xi^{\star\star} = 1$. The corresponding solutions of $g(x) = \mathbf{0}$ are $x^\star = \mathbf{0}$ and $x^\star = \mathbf{1}$.

Optimization

Analysis Consider the following:

Corollary 3.7 *Let* $\mathbb{S} \subseteq \mathbb{R}^n$, $f : \mathbb{R}^n \to \mathbb{R}$, *and* $n' \leq n$ *and collect the last* n' *entries of* x *together into a vector* $\xi = \begin{bmatrix} x_{n-n'+1} \\ \vdots \\ x_n \end{bmatrix} \in \mathbb{R}^{n'}$. *Consider the special case of the optimization problem* $\min_{x \in \mathbb{S}} f(x)$ *such that functions* $\omega_\ell : \mathbb{R}^{n'} \to \mathbb{R}$ *for* $\ell =$

$1, \ldots, (n - n')$, can be found that satisfy:

$$\forall \begin{bmatrix} x_1 \\ \vdots \\ x_{n-n'} \\ \xi \end{bmatrix} \in \mathbb{S}, \forall \ell = 1, \ldots, (n - n'), x_\ell = \omega_\ell(\xi).$$

(Typically, these functions correspond to $(n - n')$ of the equality constraints in the definition of $\mathbb{S}$. The condition means that these equality constraints can be re-arranged to express each of the first $n - n'$ entries of the decision vector in terms of the last n' entries.) Collect the functions ω_ℓ, $\ell = 1, \ldots, (n - n')$, into a vector function $\omega : \mathbb{R}^{n'} \to \mathbb{R}^{n-n'}$. Let $\mathbb{P} \subseteq \mathbb{R}^{n'}$ be the projection of $\mathbb{S}$ onto the last n' components of $\mathbb{R}^n$. (See Definition A.47.) Define $\phi : \mathbb{R}^{n'} \to \mathbb{R}$ by:

$$\forall \xi \in \mathbb{R}^{n'}, \phi(\xi) = f\left(\begin{bmatrix} \omega(\xi) \\ \xi \end{bmatrix} \right).$$

Consider the problems: $\min_{\xi \in \mathbb{P}} \phi(\xi)$ *and* $\min_{x \in \mathbb{S}} f(x)$. *Then:*

(i) *the problem* $\min_{x \in \mathbb{S}} f(x)$ *has a minimum if and only if* $\min_{\xi \in \mathbb{P}} \phi(\xi)$ *has a minimum.*

(ii) *If either one of the problems in Item* (i) *possesses a minimum (and consequently, by Item* (i), *each one possesses a minimum), then:*

$$\min_{x \in \mathbb{S}} f(x) = \min_{\xi \in \mathbb{P}} \phi(\xi),$$

$$\operatorname*{argmin}_{x \in \mathbb{S}} f(x) = \left\{ \begin{bmatrix} \omega(\xi) \\ \xi \end{bmatrix} \in \mathbb{R}^n \,\middle|\, \xi \in \operatorname*{argmin}_{\xi \in \mathbb{P}} \{\phi(\xi)\} \right\}.$$

Proof See Exercise 3.18. □

The function ϕ is called the **reduced function** [84, section 14.2].

Example Consider the linear equality-constrained problem from Section 2.3.2.2, with objective defined in (2.10) and equality constraint function defined in (2.12). That is, consider the problem:

$$\min_{x \in \mathbb{R}^2} \{(x_1 - 1)^2 + (x_2 - 3)^2 | x_1 - x_2 = 0\}. \tag{3.7}$$

As in the previous example, the equality constraint in this problem can be re-arranged as $x_1 = \omega_1(x_2)$, where $\xi = x_2$ and $\omega_1 : \mathbb{R} \to \mathbb{R}$ is defined by:

$$\forall x_2 \in \mathbb{R}, \omega_1(x_2) = x_2.$$

Moreover, the projection of $\mathbb{S} = \{x \in \mathbb{R}^2 | x_1 - x_2 = 0\}$ onto the last component of $\mathbb{R}^2$ is $\mathbb{P} = \mathbb{R}$. To see this, consider Figure 3.18, which repeats Figure 2.8. The feasible set is shown by the line. For each $x_2 \in \mathbb{R}$ there is a corresponding point

x_2

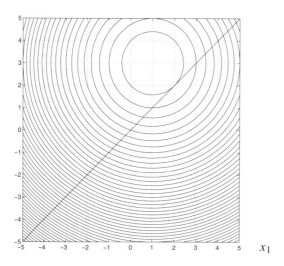

Fig. 3.18. The contour sets $\mathbb{C}_f(\tilde{f})$ of the function defined in (2.10) for values $\tilde{f} = 2, 4, 6, \ldots$ with feasible set superimposed. The heights of the contours decrease towards the point $\begin{bmatrix} 1 \\ 3 \end{bmatrix}$.

$\phi(x_2)$

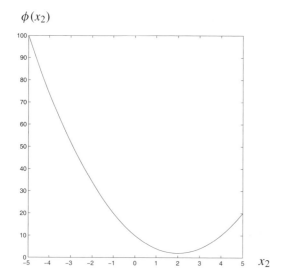

Fig. 3.19. Transformed objective function ϕ.

$x \in \mathbb{S}$, namely $\begin{bmatrix} x_2 \\ x_2 \end{bmatrix}$. Therefore, the projection of $\mathbb{S}$ onto the last component of $\mathbb{R}^2$ is the whole of $\mathbb{R}$.

Using the equality constraint, we can eliminate x_1 from the objective function by noticing that:

$$
\begin{aligned}
(x \in \mathbb{S}) \Rightarrow (x_1 - 1)^2 + (x_2 - 3)^2 &= (\omega_1(x_2) - 1)^2 + (x_2 - 3)^2, \\
&= (x_2 - 1)^2 + (x_2 - 3)^2, \\
&= 2(x_2)^2 - 8x_2 + 10.
\end{aligned}
$$

In the context of Corollary 3.7, we can define the transformed objective $\phi : \mathbb{R} \to \mathbb{R}$ by:

$$
\begin{aligned}
\forall x_2 \in \mathbb{R}, \phi(x_2) &= f\left(\begin{bmatrix} \omega(\xi) \\ \xi \end{bmatrix}\right), \\
&= f\left(\begin{bmatrix} x_2 \\ x_2 \end{bmatrix}\right), \\
&= 2(x_2)^2 - 8x_2 + 10.
\end{aligned}
$$

This objective is shown in Figure 3.19. We can visualize it by imagining the contours in Figure 3.18 along the feasible set and then consider the contours from the vantage of the x_2-axis. That is, Figure 3.19 shows the shape of the objective when we look from the x_2-axis in Figure 3.18 but see only the values of the objective for points that are in the feasible set.

By Corollary 3.7, Problem (3.7) is equivalent to:

$$
\min_{x_2 \in \mathbb{R}}\{2(x_2)^2 - 8x_2 + 10\}.
$$

In this case, the equality constraint has been completely eliminated from the formulation, leaving an unconstrained problem that can be solved for a minimizer $x_2^\star$. Inspection of Figure 3.19 yields $x_2^\star = 2$. The corresponding optimal value of $x_1^\star$ can be found by substituting according to $x_1^\star = \omega_1(x_2^\star)$. That is, $x_1^\star = 2$.

Discussion We will use elimination of variables in several places throughout the book beginning in Section 5.2. Moreover, it is possible to generalize the idea of elimination of variables to the case where ω is not known explicitly but can only be found *implicitly*. See [70, section A.6][72, section 4.4] and Section A.7.3 of Appendix A for a discussion of this generalization, known as the **implicit function theorem**. We will use the implicit function theorem in Section 7.5.1.1 and elsewhere to facilitate sensitivity analysis.

3.3 Constraints

Transformation of constraints is another technique that will prove useful. The five basic techniques we will discuss are:

(i)   **scaling** and **pre-conditioning**,
(ii) **slack variables**,
(iii) **changing the functional form**,
(iv) **altering the feasible region**, and
(v)   **hierarchical decomposition**.

There are a number of other transformations that are possible, particularly in the context of linear constraints. See, for example, [84, section 7.5].

3.3.1 Scaling and pre-conditioning

The most basic way to transform constraints is to **scale** them, as we discussed for variables. This can have a significant effect on the progress of the optimization algorithm if the units of the constraints are widely varying. For linear constraints, a generalization of the idea of scaling, called **pre-conditioning**, involves multiplying both the coefficient matrix and the right-hand side vector on the left by a suitably chosen matrix M that:

- does not change the set of points satisfying the constraints, but
- makes it easier to find points satisfying the constraints.

We will investigate pre-conditioning further in Section 5.7.2.

Scaling can also be useful for non-linear constraints. Similar to the discussion in Section 3.2.1, evaluation of whether or not a set of constraints are satisfied "closely enough" will usually involve testing a norm of the constraint function. As in the discussion in Section 3.2.1, it is sensible to scale the entries of the constraint function so that a "significant" violation of any constraint from the perspective of the application involves roughly the same numerical value for each of the entries of the scaled constraint function. Pre-conditioning can also be applied to non-linear constraints.

3.3.2 Slack variables

In this section, we consider a transformation that alters inequality constraints into a combination of equality constraints and non-negativity constraints through the use of **slack variables**, which account for the amount by which an inequality constraint is strictly satisfied. This is specified in the following:

Theorem 3.8 *Let* $f : \mathbb{R}^n \to \mathbb{R}, g : \mathbb{R}^n \to \mathbb{R}^m, h : \mathbb{R}^n \to \mathbb{R}^r$. *Consider the problems:*

$$\min_{x \in \mathbb{R}^n} \{ f(x) | g(x) = \mathbf{0}, h(x) \le \mathbf{0} \}, \tag{3.8}$$

$$\min_{x \in \mathbb{R}^n, w \in \mathbb{R}^r} \{ f(x) | g(x) = \mathbf{0}, h(x) + w = \mathbf{0}, w \ge \mathbf{0} \}. \tag{3.9}$$

We have that:

(i) *Problem (3.8) has a minimum if and only if Problem (3.9) has a minimum.*

(ii) *If either one of the problems in Item* (i) *possesses a minimum (and consequently, by Item* (i)*, each one possesses a minimum), then the minima are equal. Moreover, to each minimizer $x^\star$ of Problem (3.8) there corresponds a minimizer $\begin{bmatrix} x^\star \\ w^\star \end{bmatrix}$ of Problem (3.9) and vice versa.*

Proof See Exercise 3.30.

□

Theorem 3.8 will be used in Sections 17.3.1 and 19.4.1 to solve problems of the form of Problem (3.8) using algorithms for solving problems with non-negativity constraints of the form of Problem (3.9).

3.3.3 Changing the functional form

More generally than in the previous two sections, we can consider transforming the form of the functions specifying the feasible set without altering the feasible set. The usefulness of this observation is that the precise formulation of the functions can significantly affect the tractability of the problem. We will investigate this in Parts IV and V.

As an example of changing the functional form, Exercise 3.32 shows that by re-arranging the functional form of the constraints it is possible to transform some non-convex functions into convex functions. In particular, in Exercise 3.32, it is possible to transform the constraint function to be affine.

As another example, a monotonically increasing transformation of an equality or inequality constraint function (together with the corresponding transformation of its right-hand side) does not change the feasible region, but may transform the function into being convex. This is embodied in:

Theorem 3.9 *Let $f : \mathbb{R}^n \to \mathbb{R}$, $g : \mathbb{R}^n \to \mathbb{R}^m$, $b \in \mathbb{R}^m$, $h : \mathbb{R}^n \to \mathbb{R}^r$, and $d \in \mathbb{R}^r$. Let $\tau_\ell^\nearrow : \mathbb{R} \to \mathbb{R}$, $\ell = 1, \ldots, m$, and $\sigma_\ell^\nearrow : \mathbb{R} \to \mathbb{R}$, $\ell = 1, \ldots, r$, each be strictly monotonically increasing and continuous on $\mathbb{R}$. Define $\gamma : \mathbb{R}^n \to \mathbb{R}^m$, $\beta \in \mathbb{R}^m$, $\eta : \mathbb{R}^n \to \mathbb{R}^r$, and $\delta \in \mathbb{R}^r$ by:*

$$
\begin{aligned}
\forall \ell = 1, \ldots, m, \forall x \in \mathbb{R}^n, \gamma_\ell(x) &= \tau_\ell^\nearrow(g_\ell(x)), \\
\forall \ell = 1, \ldots, m, \beta_\ell &= \tau_\ell^\nearrow(b_\ell), \\
\forall \ell = 1, \ldots, r, \forall x \in \mathbb{R}^n, \eta_\ell(x) &= \sigma_\ell^\nearrow(h_\ell(x)), \\
\forall \ell = 1, \ldots, r, \delta_\ell &= \sigma_\ell^\nearrow(d_\ell).
\end{aligned}
$$

Consider the problems:

$$\min_{x\in\mathbb{R}^n}\{f(x)|g(x)=b, h(x)\le d\} \ and \ \min_{x\in\mathbb{R}^n}\{f(x)|\gamma(x)=\beta, \eta(x)\le\delta\}.$$

The second problem is obtained from the first by transforming corresponding functions and entries of each constraint. Then:

(i) *the problem* $\min_{x\in\mathbb{R}^n}\{f(x)|g(x)=b, h(x)\le d\}$ *has a minimum if and only if the problem* $\min_{x\in\mathbb{R}^n}\{f(x)|\gamma(x)=\beta, \eta(x)\le\delta\}$ *has a minimum.*

(ii) *If either one of the problems in Item* (i) *possesses a minimum (and consequently, by Item* (i)*, each one possesses a minimum), then the minima are equal and they have the same minimizers.*

Proof See Exercise 3.31. □

See Exercise 3.33 for an application of this theorem involving **posynomial functions** [6, section 11.5][15, section 4.5.3], as defined in:

Definition 3.1 Let $A \in \mathbb{R}^{m\times n}$ and $B \in \mathbb{R}^m_{++}$ and define $f : \mathbb{R}^n_{++} \to \mathbb{R}$ by:

$$\forall x \in \mathbb{R}^n_{++}, \ f(x) = \sum_{\ell=1}^{m} B_\ell(x_1)^{A_{\ell 1}}(x_2)^{A_{\ell 2}}\cdots(x_n)^{A_{\ell n}}.$$

The function f is called a **posynomial function**. If $m = 1$ then f is called a **monomial function**. □

In Exercise 3.33, a non-convex problem involving posynomial functions is transformed into a convex problem. (We will use this transformation in Section 20.2 in the sizing of interconnects in integrated circuits case study.)

3.3.4 Altering the feasible region

Even more generally, it can sometimes also be useful to alter the feasible region as the following theorem shows:

Theorem 3.10 *Let* $\underline{\mathbb{S}} \subseteq \mathbb{S} \subseteq \overline{\mathbb{S}} \subseteq \mathbb{R}^n$, $f : \mathbb{R}^n \to \mathbb{R}$ *and consider the problems:*

$$\min_{x\in\underline{\mathbb{S}}} f(x), \ \min_{x\in\mathbb{S}} f(x), \ \min_{x\in\overline{\mathbb{S}}} f(x),$$

and assume that they all have minima and minimizers. Then:

(i) $\min_{x\in\underline{\mathbb{S}}} f(x) \ge \min_{x\in\mathbb{S}} f(x) \ge \min_{x\in\overline{\mathbb{S}}} f(x).$

(ii) *If* $x^\star \in \mathrm{argmin}_{x\in\overline{\mathbb{S}}} f(x)$ *and* $x^\star \in \mathbb{S}$ *then* $\min_{x\in\mathbb{S}} f(x) = \min_{x\in\overline{\mathbb{S}}} f(x)$ *and, furthermore,* $\mathrm{argmin}_{x\in\mathbb{S}} f(x) = \left(\mathrm{argmin}_{x\in\overline{\mathbb{S}}} f(x)\right) \cap \mathbb{S}.$

(iii) *If* $x^\star \in \mathrm{argmin}_{x\in\mathbb{S}} f(x)$ *and* $x^\star \in \underline{\mathbb{S}}$ *then* $\min_{x\in\mathbb{S}} f(x) = \min_{x\in\underline{\mathbb{S}}} f(x)$ *and, furthermore,* $\mathrm{argmin}_{x\in\underline{\mathbb{S}}} f(x) = (\mathrm{argmin}_{x\in\mathbb{S}} f(x)) \cap \underline{\mathbb{S}}.$

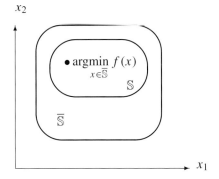

x_2

$\bullet \underset{x \in \overline{\mathbb{S}}}{\operatorname{argmin}} f(x)$

$\mathbb{S}$

$\overline{\mathbb{S}}$

x_1

Fig. 3.20. Illustration of relaxing the feasible set.

Proof See Exercise 3.34. □

We will discuss the implications of Theorem 3.10 in the following sections.

3.3.4.1 Enlarging or relaxing the feasible set

Item (ii) of Theorem 3.10 suggests a way to solve some problems: consider an alternative problem with the same objective but a "larger" feasible set $\overline{\mathbb{S}}$. The problem $\min_{x \in \overline{\mathbb{S}}} f(x)$ is called a **relaxation** of or a **relaxed** version of the original problem $\min_{x \in \mathbb{S}} f(x)$. If the minimizer of the relaxed problem $\min_{x \in \overline{\mathbb{S}}} f(x)$ happens to lie in $\mathbb{S}$, then a minimizer of the original problem $\min_{x \in \mathbb{S}} f(x)$ has been found. The situation is illustrated in Figure 3.20, where the minimizer of f over $\overline{\mathbb{S}}$ happens to be an element of $\mathbb{S}$ and consequently is also a minimizer of f over $\mathbb{S}$.

Surprisingly, it is sometimes easier to optimize over a larger set than a smaller set, if the larger set has a more suitable structure. For example, Exercises 3.36 and 3.37 show cases where, respectively:

(i) $\overline{\mathbb{S}}$ is convex while $\mathbb{S}$ is not, and
(ii) $\overline{\mathbb{S}}$ involves temporarily ignoring some of the constraints, yielding an easier problem.

As another example, in Section 15.6.4.1 we will show that although the feasible set $\mathbb{S}$ for the optimal power flow case study is not convex, a suitable relaxation $\overline{\mathbb{S}}$ of the feasible set is convex. In Section 15.6.4.1, we consider the circumstances where the minimizer of the relaxed problem is also an element of $\mathbb{S}$.

If the solution to the relaxed problem *is not* contained in the feasible set of the original problem, then at least one of the relaxed constraints was violated. We can add one or more of these constraints to form a revised problem. (A violated constraint might not be binding at the solution of the original problem, however. See Exercises 2.37 and 2.38.) This revised problem is "intermediate" between the original and relaxed problem. We can try to solve the revised problem in the hope

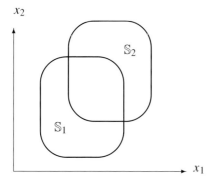

x_2

$\mathbb{S}_2$

$\mathbb{S}_1$

x_1

Fig. 3.21. Illustration of divide and conquer.

of finding a solution that satisfies all the constraints. This leads to **cutting plane** approaches [11, section 6.3.2][12, section 6.5][41][70, section 13.6].

3.3.4.2 Constricting the feasible set

Item (iii) in Theorem 3.10 simply formalizes a way to use *a priori* knowledge to narrow a search: if an optimizer is known to lie in a subset $\underline{\mathbb{S}}$ of $\mathbb{S}$ then we can confine our search to that subset. This can be useful if it is easier to search over $\underline{\mathbb{S}}$ than over $\mathbb{S}$.

3.3.4.3 Divide and conquer

We can generalize the idea of constricting the feasible set to develop a **divide and conquer** approach. For example, suppose that $\mathbb{S}_1 \subseteq \mathbb{S}$, $\mathbb{S}_2 \subseteq \mathbb{S}$, and $\mathbb{S}_1 \cup \mathbb{S}_2 = \mathbb{S}$. If the minimizer of the problem over $\mathbb{S}$ exists, then it must be contained in either $\mathbb{S}_1$ or $\mathbb{S}_2$ (or both). We solve both $\min_{x \in \mathbb{S}_1} f(x)$ and $\min_{x \in \mathbb{S}_2} f(x)$ and check for the smaller minimum and corresponding minimizer. This yields the minimum and minimizer of $\min_{x \in \mathbb{S}} f(x)$. (See Exercise 3.39.)

This approach can be very effective if $\mathbb{S}_1$ and $\mathbb{S}_2$ are chosen so that minimizing over $\mathbb{S}_1$ and minimizing over $\mathbb{S}_2$ is easier than minimizing over $\mathbb{S}$ itself. For example, if $\mathbb{S}$ is not itself convex, it may be the case that $\mathbb{S}$ can be represented as the union of convex subsets. If the objective is convex on each of the convex subsets then the individual minimization problems are convex. The situation is illustrated in Figure 3.21, where the two sets $\mathbb{S}_1$ and $\mathbb{S}_2$ illustrated are convex and their union is the non-convex feasible set $\mathbb{S} = \mathbb{S}_1 \cup \mathbb{S}_2$.

As another example, suppose that the feasible set is the subset of $\mathbb{R}^2$ consisting of the points that are within 1 unit of the point $x = \mathbf{0}$ or within 1 unit of the point $x = \mathbf{1}$. This set is not convex, but is clearly the union of two convex sets.

A third example involves discrete optimization. Suppose that a particular entry x_1 of $x \in \mathbb{R}^n$ is required to be either 0 or 1. Then, we can partition the feasible set $\mathbb{S}$ into $\mathbb{S}_1 = \mathbb{S} \cap \{x \in \mathbb{R}^n | x_1 = 0\}$ and $\mathbb{S}_2 = \mathbb{S} \cap \{x \in \mathbb{R}^n | x_1 = 1\}$. Optimizing

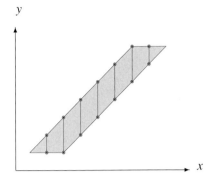

Fig. 3.22. Illustration of hierarchical decomposition.

over $\mathbb{S}_1$ and over $\mathbb{S}_2$ and further partitioning of the feasible set for other discrete variables leads to **branch and bound** techniques [113, 122].

3.3.5 Hierarchical decomposition

Sometimes problems can be decomposed into a hierarchy of levels in such a way that we can think of solving an **outer** or **master problem**, the objective of which is the solution to an **inner problem**. For example, consider a feasible set $\mathbb{S} \subseteq \mathbb{R}^{n+s}$ such that:

$$\mathbb{S} = \left\{ \begin{bmatrix} x \\ y \end{bmatrix} \in \mathbb{R}^{n+s} \,\middle|\, x \in \mathbb{S}_1, y \in \mathbb{S}_2(x) \right\},$$

where $\mathbb{S}_1 \subseteq \mathbb{R}^n$ and $\mathbb{S}_2 : \mathbb{S}_1 \to (2)^{(\mathbb{R}^s)}$ is a set-valued function. That is, for each element x in $\mathbb{S}_1$, there is a *set* of elements $\mathbb{S}_2(x) \subset \mathbb{R}^s$ such that for any $y \in \mathbb{S}_2(x)$, the vector $\begin{bmatrix} x \\ y \end{bmatrix}$ is an element of $\mathbb{S}$. We can think of $\mathbb{S}$ as the union of "slices," where there is one slice $\mathbb{S}_2(x)$ for each x in $\mathbb{S}_1$. This is illustrated in Figure 3.22, where several such slices are shown. If $\mathbb{S}_2(x)$ is independent of x then we write $\mathbb{S}_2$ for $\mathbb{S}_2(x)$. In this case, $\mathbb{S}$ is the **Cartesian product** $\mathbb{S} = \mathbb{S}_1 \times \mathbb{S}_2$.

We have the following ([41, theorem 1][106, section 8.2]):

Theorem 3.11 *Suppose that $\mathbb{S} \subseteq \mathbb{R}^{n+s}$ is of the form:*

$$\mathbb{S} = \left\{ \begin{bmatrix} x \\ y \end{bmatrix} \in \mathbb{R}^{n+s} \,\middle|\, x \in \mathbb{S}_1, y \in \mathbb{S}_2(x) \right\},$$

with $\mathbb{S}_1 \subseteq \mathbb{R}^n$ and, for each $x \in \mathbb{S}_1$, $\mathbb{S}(x) \subseteq \mathbb{R}^s$. Let $f : \mathbb{S} \to \mathbb{R}$ and assume that, for each $x \in \mathbb{S}_1$, the minimization problem $\min_{y \in \mathbb{S}_2(x)} f\left(\begin{bmatrix} x \\ y \end{bmatrix}\right)$ has a minimum. Consider the problems:

$$\min_{\begin{bmatrix} x \\ y \end{bmatrix} \in \mathbb{S}} f\left(\begin{bmatrix} x \\ y \end{bmatrix}\right) \quad and \quad \min_{x \in \mathbb{S}_1} \left\{ \min_{y \in \mathbb{S}_2(x)} f\left(\begin{bmatrix} x \\ y \end{bmatrix}\right) \right\}.$$

Then:

(i) $\min\limits_{\left[\begin{smallmatrix} x \\ y \end{smallmatrix}\right] \in \mathbb{S}} f\left(\left[\begin{smallmatrix} x \\ y \end{smallmatrix}\right]\right)$ *has a minimum if and only if* $\min\limits_{x \in \mathbb{S}_1}\left\{ \min\limits_{y \in \mathbb{S}_2(x)} f\left(\left[\begin{smallmatrix} x \\ y \end{smallmatrix}\right]\right)\right\}$ *has a minimum.*

(ii) *If either one of the problems in Item* (i) *possesses a minimum (and consequently, by Item* (i), *each one possesses a minimum), then:*

$$\min_{\left[\begin{smallmatrix} x \\ y \end{smallmatrix}\right] \in \mathbb{S}} f\left(\left[\begin{matrix} x \\ y \end{matrix}\right]\right) = \min_{x \in \mathbb{S}_1}\left\{ \min_{y \in \mathbb{S}_2(x)} f\left(\left[\begin{matrix} x \\ y \end{matrix}\right]\right)\right\},$$

$$\arg\min_{\left[\begin{smallmatrix} x \\ y \end{smallmatrix}\right] \in \mathbb{S}} f\left(\left[\begin{matrix} x \\ y \end{matrix}\right]\right) = \left\{\left[\begin{matrix} \hat{x} \\ \hat{y} \end{matrix}\right] \in \mathbb{R}^{n+s} \,\middle|\, \begin{matrix} \hat{x} \in \arg\min\limits_{x \in \mathbb{S}_1}\left\{ \min\limits_{y \in \mathbb{S}_2(x)} f\left(\left[\begin{matrix} x \\ y \end{matrix}\right]\right)\right\}, \\[2ex] \hat{y} \in \arg\min\limits_{y \in \mathbb{S}_2(\hat{x})} f\left(\left[\begin{matrix} \hat{x} \\ y \end{matrix}\right]\right) \end{matrix}\right\}.$$

Proof See Exercise 3.40. □

In Theorem 3.11, $\mathbb{S}_1$ is the projection of $\mathbb{S}$ onto the first n components of $\mathbb{R}^{n+s}$. (See Definition A.47.) For this reason, hierarchical decomposition is sometimes referred to as **projection** [41, 42]. Theorem 3.11 allows us to hold some of the decision vector *constant* temporarily while we optimize over the rest of the decision vector. That is, we keep $x \in \mathbb{S}_1$ constant *temporarily* or think of it as a parameter while we optimize the inner problem over $y \in \mathbb{S}_2(x)$. If we can solve for the solution of the inner problem as a function of x, or can approximate its dependence on x, then we can use this functional dependence in the outer problem.

For example, consider the feasible set:

$$\mathbb{S} = \left\{\left[\begin{matrix} x \\ y \end{matrix}\right] \in \mathbb{R}^2 \,\middle|\, (x)^2 + (y)^2 = 1\right\},$$

which is the set of points on the unit circle in the plane. We can re-write this set in the form:

$$\mathbb{S} = \left\{\left[\begin{matrix} x \\ y \end{matrix}\right] \,\middle|\, -1 \leq x \leq 1, y \in \left\{\sqrt{1-(x)^2}, -\sqrt{1-(x)^2}\right\}\right\},$$

where $\mathbb{S}_1 = \{x \in \mathbb{R} \mid -1 \leq x \leq 1\}$ is the projection of $\mathbb{S}$ onto the first component of $\mathbb{R}^2$. In this particular case, for each $x \in \mathbb{S}_1$, the inner minimization problem in Theorem 3.11 involves finding the minimum over a set with just two elements, namely $\mathbb{S}_2(x) = \{\sqrt{1-(x)^2}, -\sqrt{1-(x)^2}\}$. Even if the objective is non-convex, and despite the fact that $\mathbb{S}_2(x)$ is not a convex set, it may be easy to perform this minimization.

If $\mathbb{S}$ is convex and f is a convex function on $\mathbb{S}$ then both the inner problem and the outer problem are convex. (See Exercise 3.41.)

Hierarchical decomposition is also useful when holding $x \in \mathbb{S}_1$ constant yields an inner problem with a particular structure that is easy to solve or for which a convenient approximate solution is possible. This leads to **Bender's decomposition** [41, 42, 43].

Hierarchical decomposition is often used implicitly in an *ad hoc* basis, such as when outer level decision variables are implicitly held constant while inner level decision variables are optimized. See Section 15.5.1.5 for a discussion of this in the context of integrated circuit design. More details about hierarchical decomposition can be found in [106, chapters 8, 12, and 13].

3.4 Duality

In this section we briefly describe a transformation that is rather more radical and less straightforward than the transformations we have described so far. Its full explanation and justification will await a more thorough theoretical development in later chapters in Parts IV and V. We will only give the briefest outline of the issues involved and introduce it here only to complete our list of problem transformations.

Taking the **dual** of a problem is a process whereby a new problem is defined where the role of the variables and the constraints is either partially or completely exchanged. The variables and constraints have a significant influence on the computational effort to solve a problem, so that "dualizing" a problem may make it significantly easier to solve. The original problem is called the **primal** problem in this context.

Let $f : \mathbb{R}^n \to \mathbb{R}$, $g : \mathbb{R}^n \to \mathbb{R}^m$, and $h : \mathbb{R}^n \to \mathbb{R}^r$ and consider the problem:

$$\min_{x \in \mathbb{R}^n} \{f(x) | g(x) = \mathbf{0}, h(x) \leq \mathbf{0}\}. \tag{3.10}$$

In Sections 3.4.1 and 3.4.2, we define two functions associated with f, g, and h. In Section 3.4.3, we then consider the relationship between these functions and minimizing f, providing an example and further discussion in Sections 3.4.4 and 3.4.5, respectively.

3.4.1 Lagrangian

We make:

Definition 3.2 Consider the function $\mathcal{L} : \mathbb{R}^n \times \mathbb{R}^m \times \mathbb{R}^r \to \mathbb{R}$ defined by:

$$\forall x \in \mathbb{R}^n, \forall \lambda \in \mathbb{R}^m, \forall \mu \in \mathbb{R}^r, \mathcal{L}(x, \lambda, \mu) = f(x) + \lambda^\dagger g(x) + \mu^\dagger h(x). \tag{3.11}$$

The function $\mathcal{L}$ is called the **Lagrangian** and the variables λ and μ are called the **dual variables**. If there are no equality constraints then $\mathcal{L} : \mathbb{R}^n \times \mathbb{R}^r \to \mathbb{R}$ is defined by omitting the term $\lambda^\dagger g(x)$ from the definition, while if there are no inequality constraints then $\mathcal{L} : \mathbb{R}^n \times \mathbb{R}^m \to \mathbb{R}$ is defined by omitting the term $\mu^\dagger h(x)$ from the definition. $\square$

Sometimes, the symbol for the dual variables is introduced when the problem is defined by writing it in parenthesis after the constraint, as in the following:

$$\min_{x \in \mathbb{R}^n} f(x) \text{ such that } g(x) = \mathbf{0}, \qquad (\lambda).$$

3.4.2 Dual function

Associated with the Lagrangian, we make:

Definition 3.3 Consider the function $\mathcal{D} : \mathbb{R}^m \times \mathbb{R}^r \to \mathbb{R} \cup \{-\infty\}$ defined by:

$$\forall \begin{bmatrix} \lambda \\ \mu \end{bmatrix} \in \mathbb{R}^{m+r}, \mathcal{D}(\lambda, \mu) = \inf_{x \in \mathbb{R}^n} \mathcal{L}(x, \lambda, \mu). \tag{3.12}$$

The function $\mathcal{D}$ is called the **dual function.** It is an extended real function. (See Definition A.17 and see Exercise 3.43 to confirm that the specification of the range of $\mathcal{D}$ is appropriate.) If there are no equality constraints or there are no inequality constraints, respectively, then the dual function $\mathcal{D} : \mathbb{R}^r \to \mathbb{R} \cup \{-\infty\}$ or $\mathcal{D} : \mathbb{R}^m \to \mathbb{R} \cup \{-\infty\}$ is defined in terms of the corresponding Lagrangian.

The set of points on which the dual function takes on real values is called the **effective domain** $\mathbb{E}$ of the dual function. That is,

$$\mathbb{E} = \left\{ \begin{bmatrix} \lambda \\ \mu \end{bmatrix} \in \mathbb{R}^{m+r} \middle| \mathcal{D}(\lambda, \mu) > -\infty \right\},$$

and the restriction of $\mathcal{D}$ to $\mathbb{E}$ is a real-valued function $\mathcal{D} : \mathbb{E} \to \mathbb{R}$. $\square$

Recall Definition 2.17 of a concave function. The usefulness of the dual function stems in part from the following:

Theorem 3.12 *Let* $f : \mathbb{R}^n \to \mathbb{R}$, $g : \mathbb{R}^n \to \mathbb{R}^m$, *and* $h : \mathbb{R}^n \to \mathbb{R}^r$. *Consider the corresponding Lagrangian defined in (3.11), the dual function defined in (3.12), and the effective domain* $\mathbb{E}$ *of the dual function. The effective domain* $\mathbb{E}$ *of the dual function is a convex set. The dual function is concave on* $\mathbb{E}$.

Proof See Exercise 3.44 and [6, theorem 6.3.1][11, proposition 5.1.2]. $\square$

The convexity of the effective domain and the concavity of the dual function on the effective domain does not depend on any property of the objective nor of the constraint functions.

3.4.3 Dual problem

We have the following:

Theorem 3.13 *Let* $f : \mathbb{R}^n \to \mathbb{R}$, $g : \mathbb{R}^n \to \mathbb{R}^m$, *and* $h : \mathbb{R}^n \to \mathbb{R}^r$. *Let* $\lambda \in \mathbb{R}^m$ *and* $\mu \in \mathbb{R}^r_+$ *and suppose that* $\hat{x} \in \{x \in \mathbb{R}^n | g(x) = \mathbf{0}, h(x) \leq \mathbf{0}\}$. *That is,* $\hat{x}$ *is feasible for Problem (3.10). Then:*

$$f(\hat{x}) \geq \mathcal{D}(\lambda, \mu), \tag{3.13}$$

where $\mathcal{D} : \mathbb{R}^m \times \mathbb{R}^r \to \mathbb{R} \cup \{-\infty\}$ *is the dual function defined in (3.12).*

Proof ([6, theorem 6.2.1].) By definition of $\mathcal{D}$,

$$
\begin{aligned}
\mathcal{D}(\lambda, \mu) &= \inf_{x \in \mathbb{R}^n} \mathcal{L}(x, \lambda, \mu), \\
&= \inf_{x \in \mathbb{R}^n} \{f(x) + \lambda^\dagger g(x) + \mu^\dagger h(x)\}, \text{ by definition of } \mathcal{L}, \\
&\leq f(\hat{x}) + \lambda^\dagger g(\hat{x}) + \mu^\dagger h(\hat{x}), \text{ by definition of inf}, \\
&\leq f(\hat{x}),
\end{aligned}
$$

since $g(\hat{x}) = \mathbf{0}$, $h(\hat{x}) \leq \mathbf{0}$, and $\mu \geq \mathbf{0}$. $\square$

Theorem 3.13 enables us to gauge whether we are close to a minimum of Problem (3.10): for any $\lambda \in \mathbb{R}^m$ and $\mu \in \mathbb{R}^r_+$, we know that the minimum of Problem (3.10) is no smaller than $\mathcal{D}(\lambda, \mu)$. The bound in (3.13) will be incorporated into stopping criteria for iterative algorithms for equality-constrained problems in Sections 13.3.1.4 and 14.3.2 and for inequality-constrained problems in Sections 16.4.6.4, 17.3.1.4, and 19.4.1.4 [15, section 5.5.1].

We also have:

Corollary 3.14 *([6, corollaries 1–4 of theorem 6.2.1]) Let* $f : \mathbb{R}^n \to \mathbb{R}$, $g : \mathbb{R}^n \to \mathbb{R}^m$, *and* $h : \mathbb{R}^n \to \mathbb{R}^r$. *Then:*

$$
\begin{aligned}
\inf_{x \in \mathbb{R}^n} \{f(x) | g(x) = \mathbf{0}, h(x) \leq \mathbf{0}\} &\geq \sup_{\left[\begin{smallmatrix}\lambda\\\mu\end{smallmatrix}\right] \in \mathbb{R}^{m+r}} \{\mathcal{D}(\lambda, \mu) | \mu \geq \mathbf{0}\}, \\
&= \sup_{\left[\begin{smallmatrix}\lambda\\\mu\end{smallmatrix}\right] \in \mathbb{E}} \{\mathcal{D}(\lambda, \mu) | \mu \geq \mathbf{0}\},
\end{aligned}
$$

where $\mathbb{E}$ *is the effective domain of* $\mathcal{D}$. *Moreover, if Problem (3.10) has a minimum then:*

$$\min_{x \in \mathbb{R}^n} \{f(x) | g(x) = \mathbf{0}, h(x) \leq \mathbf{0}\} \geq \sup_{\left[\begin{smallmatrix}\lambda\\\mu\end{smallmatrix}\right] \in \mathbb{E}} \{\mathcal{D}(\lambda, \mu) | \mu \geq \mathbf{0}\}. \tag{3.14}$$

If Problem (3.10) is unbounded below then:

$$\forall \lambda \in \mathbb{R}^m, \forall \mu \in \mathbb{R}^r_+, \mathcal{D}(\lambda, \mu) = -\infty,$$

so that $\mathbb{E} = \emptyset$. *If* $\sup_{\left[\begin{smallmatrix}\lambda\\\mu\end{smallmatrix}\right]\in\mathbb{E}} \{\mathcal{D}(\lambda, \mu)|\mu \geq \mathbf{0}\}$ *is unbounded above then Problem (3.10) is infeasible.*

Proof   See Exercise 3.45. □

This result is called **weak duality** and the right-hand side of (3.14) is called the **dual problem**. If $\mathbb{E} = \emptyset$ we say that the dual problem is infeasible.

The inequalities in (3.13) and (3.14) can be strict, in which case the difference between the left- and right-hand sides is called the **duality gap**. If the left- and right-hand sides are the same, we say that there is no duality gap or that the duality gap is zero.

Evaluating the right-hand side of (3.14) requires:

- evaluating how the infimum of the **inner problem** $\inf_{x\in\mathbb{R}^n} \mathcal{L}(x, \lambda, \mu)$ in the definition of $\mathcal{D}$ depends on λ and μ, and
- finding the supremum of the **outer problem** $\sup_{\left[\begin{smallmatrix}\lambda\\\mu\end{smallmatrix}\right]\in\mathbb{E}} \{\mathcal{D}(\lambda, \mu)|\mu \geq \mathbf{0}\}$.

Exercise 3.46 shows that it is possible for there to be a duality gap. Moreover, the calculation of sup and inf is not very practical from a numerical point of view. However, in some circumstances, the inequality in (3.14) can be replaced by equality and the sup and inf can be replaced by max and min, so that the right-hand side of (3.14) equals the minimum of the primal Problem (3.10) and the dual problem on the right-hand side of (3.14) becomes:

$$\max_{\left[\begin{smallmatrix}\lambda\\\mu\end{smallmatrix}\right]\in\mathbb{E}} \{\mathcal{D}(\lambda, \mu)|\mu \geq \mathbf{0}\} = \max_{\left[\begin{smallmatrix}\lambda\\\mu\end{smallmatrix}\right]\in\mathbb{E}} \left\{ \min_{x\in\mathbb{R}^n}\{f(x) + \lambda^\dagger g(x) + \mu^\dagger h(x)\} \middle| \mu \geq \mathbf{0} \right\},$$

(3.15)

having an inner minimization problem embedded in an outer maximization problem. By Theorem 3.12, $\mathcal{D}$ is concave on $\mathbb{E}$, so that, by Theorem 2.4, it has at most one local maximum. Consequently, Problem (3.15) is amenable to solution by the optimization techniques we describe, so that if:

- the dual problem has maximum equal to the minimum of the primal, and
- the minimizer of the inner problem in the definition of the dual function sheds light on the minimizer of the primal problem,

then the dual formulation provides a useful transformation. The requirements for these conditions to hold depend on the convexity of the primal problem and on other technical conditions on the functions, which we will discuss in detail in Parts IV and V. In the next section, we will consider an example where such conditions happen to hold.

3.4.4 Example

Consider the problem $\min_{x \in \mathbb{R}} \{f(x) | g(x) = 0\}$ where $f : \mathbb{R} \to \mathbb{R}$ and where $g : \mathbb{R} \to \mathbb{R}$ are defined by:

$$\forall x \in \mathbb{R}, f(x) = (x)^2,$$
$$\forall x \in \mathbb{R}, g(x) = 3 - x.$$

Since there are no inequality constraints, we will omit the argument μ of $\mathcal{L}$ and of $\mathcal{D}$. We consider the dual function $\mathcal{D} : \mathbb{R} \to \mathbb{R} \cup \{-\infty\}$, defined by:

$$\begin{aligned}
\forall \lambda \in \mathbb{R}, \mathcal{D}(\lambda) &= \inf_{x \in \mathbb{R}} \mathcal{L}(x, \lambda), \\
&= \inf_{x \in \mathbb{R}} \{(x)^2 + \lambda(3 - x)\}, \\
&= \inf_{x \in \mathbb{R}} \left\{ \left(x - \frac{\lambda}{2}\right)^2 + 3\lambda - \frac{(\lambda)^2}{4} \right\}, \quad \text{on completing the square,} \\
&= 3\lambda - \frac{(\lambda)^2}{4},
\end{aligned}$$

since the minimum value of $(x - \lambda/2)^2$ occurs for $x = \lambda/2$. Moreover, $\mathbb{E} = \mathbb{R}$.

Since $\mathcal{D}$ is quadratic and strictly concave, the dual problem has a maximum and:

$$\begin{aligned}
\max_{\lambda \in \mathbb{E}} \{\mathcal{D}(\lambda)\} &= \max_{\lambda \in \mathbb{R}} \left\{ 3\lambda - \frac{(\lambda)^2}{4} \right\}, \\
&= \max_{\lambda \in \mathbb{R}} \left\{ -\left(\frac{\lambda}{2} - 3\right)^2 + 9 \right\}, \\
&= 9,
\end{aligned}$$

with maximizer $\lambda^\star = 6$. The value of the minimizer of $\mathcal{L}(\bullet, \lambda^\star)$ is $x^\star = \lambda^\star/2 = 3$, which is the minimizer of the equality-constrained problem. We have solved the primal equality-constrained problem by solving the dual problem.

3.4.5 Discussion

The combination of outer and inner optimization problems in duality is superficially similar to the situation discussed in Section 3.3.5 for hierarchical decomposition; however, in duality the outer problem seeks to maximize and the inner problem seeks to minimize, whereas in hierarchical decomposition *both* the inner and outer problems are minimization problems. Hierarchical decomposition and duality are therefore different approaches to transforming a problem. (In the context of duality, hierarchical decomposition is sometimes referred to as **primal decomposition** to emphasize this distinction [106, section 8.2].)

For each equality constraint $g_\ell(x) = 0$ in the primal problem we have created a new variable λ_ℓ in the dual problem. For each inequality constraint $h_\ell(x) \leq 0$ in the primal problem we have created a new variable μ_ℓ and a new constraint $\mu_\ell \geq 0$ in the dual problem. In some circumstances, such as the example in Section 3.4.4:

- the minimization over $x \in \mathbb{R}^n$ in the inner problem in (3.15) can be performed analytically or particularly easily numerically, or
- each entry x_k can be eliminated,

making the inner problem easy to solve.

The primal and dual problems are qualitatively different. The dual problem may be easier to solve than the primal. If so and if the primal and dual have equal optima for a particular problem then duality can provide a useful transformation. For example, if the primal problem has many variables but only a few constraints, then the dual problem will involve maximization of a concave function over just a few variables. We will see an example of such a problem in the least-cost production case study in Section 12.1. In some cases, the inner problem is very easy to solve so that the dual problem is, overall, much easier to solve than the primal. For example, if the objective is additively separable then we will see that the inner problem can be **decomposed** into smaller sub-problems.

Even if the problem at hand does not satisfy the conditions for equality of the primal and dual, it is still possible to try to solve the dual problem. In general, the solution of the dual problem will only be a lower bound on the solution of the primal, as illustrated in Exercise 3.46; however, this may still be useful in finding a solution of the primal problem or in providing information about the primal problem such as a bound on the minimum. We will investigate these issues in Parts IV and V. We will also see in Parts IV and V that the Lagrangian and the dual problem provide information for sensitivity analysis.

3.5 Summary

In this chapter we introduced various ways to transform problems. These transformations involved:

- the objective,
- the variables,
- the constraints, and
- duality.

We are now ready to begin discussing the case studies and the algorithms. The material in this chapter has been very technical in places and you may want to return to it as we proceed.

Exercises

Objective

3.1 Prove Theorem 3.1.

3.2 Let $f : \mathbb{R} \to \mathbb{R}$ be defined by:

$$\forall x \in \mathbb{R}, \ f(x) = (x)^2 + 1.$$

(i) Find the minimum and minimizer of $\min_{x \in \mathbb{R}} f(x)$. Explain your answer. (Hint: You do not need to perform any calculations to solve this problem.)

(ii) Find the minimum and minimizer of $\min_{x \in \mathbb{R}} \ln(f(x))$. Explain you answer. (Hint: You should invoke a theorem to solve this problem.)

3.3 Repeat Exercise 2.7 but change the objective to $\phi = \exp(f(\bullet))$ instead of f and in the first part, sketch $\mathbb{C}_\phi(\tilde\phi)$ for $\tilde\phi = 1, \exp(1), \exp(2), \exp(3)$. That is, perform the following.

(i) Sketch $\mathbb{C}_\phi(\tilde\phi)$ for $\tilde\phi = 1, \exp(1), \exp(2), \exp(3)$.

(ii) Sketch on the same graph the set of points satisfying $g(x) = 0$ where:

$$\forall x \in \mathbb{R}^2, \ g(x) = x_1 + 2x_2 - 3.$$

(iii) Find $\min_{x \in \mathbb{R}^2}\{\phi(x)|x_1 + 2x_2 - 3 = 0\}$ and $\operatorname{argmin}_{x \in \mathbb{R}^2}\{\phi(x)|x_1 + 2x_2 - 3 = 0\}$.

3.4 This exercise considers convexity under a transformation of the objective.

(i) Show that $f : \mathbb{R} \to \mathbb{R}$ defined by $\forall x \in \mathbb{R}, \ f(x) = -\exp(-\frac{1}{2}(x)^2)$ is not convex. A sketch and explanation will suffice.

(ii) Define $\phi : \mathbb{R} \to \mathbb{R}$ by $\forall x \in \mathbb{R}, \ \phi(x) = -\ln(-f(x))$ and prove that ϕ is convex.

3.5 Suppose that $\mathbb{S} \subseteq \mathbb{R}^n$ is convex and that $f : \mathbb{S} \to \mathbb{R}$ has convex level sets on $\mathbb{S}$ and that $\eta^\nearrow : \mathbb{R} \to \mathbb{R}$ is monotonically increasing. Define $\phi : \mathbb{S} \to \mathbb{R}$ by $\forall x \in \mathbb{S}, \ \phi(x) = \eta^\nearrow(f(x))$. Prove that ϕ has convex level sets on $\mathbb{S}$.

3.6 (Exercise 3.12 of [6].) Suppose that $\mathbb{S} \subseteq \mathbb{R}^n$ is convex, $f : \mathbb{S} \to \mathbb{R}$ is convex on $\mathbb{S}$, and $\eta^\nearrow : \mathbb{R} \to \mathbb{R}$ is monotonically increasing and convex. Define $\phi : \mathbb{S} \to \mathbb{R}$ by $\forall x \in \mathbb{S}, \ \phi(x) = \eta^\nearrow(f(x))$. Prove that ϕ is convex on $\mathbb{S}$.

3.7 Prove Theorem 3.2.

3.8 Suppose that: $g : \mathbb{R}^n \to \mathbb{R}^m$ is partially differentiable and $\Pi \in \mathbb{R}$. Calculate the gradient of $\Pi \|g(\bullet)\|_2^2$.

3.9 Repeat Exercise 2.7 with the objective $f + (g)^2$ instead of f where $f : \mathbb{R}^2 \to \mathbb{R}$ and $g : \mathbb{R}^2 \to \mathbb{R}$ are as defined in Exercise 2.7:

$$
\begin{aligned}
\forall x \in \mathbb{R}^2, \ f(x) &= (x_1)^2 + (x_2)^2 + 2x_2 - 3, \\
\forall x \in \mathbb{R}^2, \ g(x) &= x_1 + 2x_2 - 3.
\end{aligned}
$$

(i) Sketch $\mathbb{C}_{f+(g)^2}(\tilde{f})$ for $\tilde{f} = 0, 1, 2, 3$.
(ii) Sketch on the same graph the set of points satisfying $g(x) = 0$.
(iii) Find $\min_{x \in \mathbb{R}^2}\{f(x) + (g(x))^2 | x_1 + 2x_2 - 3 = 0\}$ and $\mathrm{argmin}_{x \in \mathbb{R}^2}\{f(x) + (g(x))^2 | x_1 + 2x_2 - 3 = 0\}$.

3.10 Suppose that $f : \mathbb{R}^n \to \mathbb{R}$ and $g : \mathbb{R}^n \to \mathbb{R}^m$ are partially differentiable with continuous partial derivatives. Let $\|\bullet\|$ be a norm on $\mathbb{R}^m$.

Suppose that the problem $\min_{x \in \mathbb{R}^n}\{f(x) | g(x) = 0\}$ has a minimizer. Furthermore, for each $\nu \in \mathbb{R}_+$, suppose that the problem $\min_{x \in \mathbb{R}^n}\{f(x) + \nu \|g(x)\|^2\}$ has a minimizer and let $x^{(\nu)} \in \mathrm{argmin}_{x \in \mathbb{R}^n}\{f(x) + \nu \|g(x)\|^2\}$. Moreover, suppose that the sequence $\{x^{(\nu)}\}_{\nu=0}^{\infty}$ converges to $x^\star$.

Prove that $x^\star$ is a minimizer of $\min_{x \in \mathbb{R}^n}\{f(x) | g(x) = 0\}$.

3.11 Define $f : \mathbb{R}^2 \to \mathbb{R}$ and $g : \mathbb{R}^2 \to \mathbb{R}$ by:

$$
\begin{aligned}
\forall x \in \mathbb{R}^2, \ f(x) &= -2(x_1 - x_2)^2 + (x_1 + x_2)^2, \\
\forall x \in \mathbb{R}^2, \ g(x) &= x_1 - x_2,
\end{aligned}
$$

and consider the problem $\min_{x \in \mathbb{R}^2}\{f(x) | g(x) = 0\}$.

(i) Show that f is not convex. (A sketch will suffice.)
(ii) Show that the penalized objective $f + \Pi(g)^2$ is convex for $\Pi = 2$.
(iii) Find all local minima and minimizers of the problem.
(iv) Show that the penalized objective $f + \Pi(g)^2$ is strictly convex for $\Pi = 3$.

3.12 Prove Theorem 3.4.

3.13 Let $f_\ell : \mathbb{R}^n \to \mathbb{R}$, $\ell = 1, \ldots, r$, be convex. Prove that the function $f : \mathbb{R}^n \to \mathbb{R}$ is convex, where f is defined by:

$$
\forall x \in \mathbb{R}^n, \ f(x) = \max_{\ell=1,\ldots,r} f_\ell(x).
$$

3.14 Let $\mathbb{S} \subseteq \mathbb{R}^n$ be a convex set and let $f_\ell : \mathbb{R}^n \to \mathbb{R}$, $\ell = 1, \ldots, r$, be convex on $\mathbb{S}$. Consider Problem (3.3):

$$
\min_{x \in \mathbb{S}, z \in \mathbb{R}} \{z | f_\ell(x) - z \le 0, \forall \ell = 1, \ldots, r\}.
$$

Prove that Problem (3.3) is a convex problem. (Hint: This problem has:

- decision vector $\begin{bmatrix} x \\ z \end{bmatrix} \in \mathbb{R}^n \times \mathbb{R}$,

- objective $\phi : \mathbb{R}^n \times \mathbb{R} \to \mathbb{R}$ defined by:

$$\forall x \in \mathbb{R}^n, \forall z \in \mathbb{R}, \phi(x, z) = z,$$

and
- feasible set:

$$(\mathbb{S} \times \mathbb{R}) \cap \left\{ \begin{bmatrix} x \\ z \end{bmatrix} \in \mathbb{R}^{n+1} \,\middle|\, h(x, z) \leq 0 \right\},$$

where $h : \mathbb{R}^n \times \mathbb{R} \to \mathbb{R}^r$ is defined by:

$$\forall \ell = 1, \ldots, r, \forall \begin{bmatrix} x \\ z \end{bmatrix} \in \mathbb{R}^n \times \mathbb{R}, h_\ell(x, z) = f_\ell(x) - z.$$

Show that:

(i) $\mathbb{S} \times \mathbb{R}$ is a convex set;
(ii) ϕ is a convex function on $\mathbb{S} \times \mathbb{R}$; and
(iii) for each $\ell = 1, \ldots, r$, h_ℓ is a convex function on $\mathbb{S} \times \mathbb{R}$,

and then use previous results.)

3.15 Give an example where $f_1, f_2 : \mathbb{R} \to \mathbb{R}$ are each differentiable, but $f : \mathbb{R} \to \mathbb{R}$ defined by:

$$\forall x \in \mathbb{R}, f(x) = \max\{f_1(x), f_2(x)\},$$

is not differentiable. (Hint: consider f to be the absolute value function and express it as the maximum of two other elementary functions.)

3.16 Suppose that $h : \mathbb{R}^n \to \mathbb{R}^r$ and that we want to decide if $\{x \in \mathbb{R}^n | h(x) \leq 0\}$ is a non-empty set. Suggest how to answer this question with an optimization algorithm that can treat inequality constraints but which must be furnished with an initial guess that is strictly feasible with respect to the inequalities. (Hint: Consider the problem:

$$\min_{x \in \mathbb{R}^n, z \in \mathbb{R}} \{z | h_\ell(x) - z \leq 0, \forall \ell = 1, \ldots, r\}.$$

Pick any initial value $x^{(0)}$. What should $z^{(0)}$ be so that $\begin{bmatrix} x^{(0)} \\ z^{(0)} \end{bmatrix}$ is strictly feasible for the inequality constraints?)

3.17 Consider the absolute value function $|\bullet| : \mathbb{R} \to \mathbb{R}_+$ and its smoothed version $\phi : \mathbb{R} \to \mathbb{R}_+$ as defined in (3.4), which we repeat here:

$$\forall x \in \mathbb{R}, \phi(x) = \sqrt{(|x|^2 + \epsilon)}.$$

(i) Show that $|\bullet|$ is not differentiable at $x = 0$. (Hint, use the definition of derivative and consider the limits from the left and from the right of $x = 0$ in the definition.)
(ii) Show that ϕ is differentiable for $\epsilon > 0$. (Hint: the only point that poses difficulty is at $x = 0$.)
(iii) Calculate the second derivative of ϕ.
(iv) What happens to the second derivative of ϕ as $\epsilon \to 0$.

Variables

3.18 In this exercise we consider onto functions.

  (i) Prove Theorem 3.5, Item (i).
  (ii) Prove Theorem 3.5, Item (ii).
  (iii) Prove Theorem 3.6.
  (iv) Prove Corollary 3.7.

3.19 This exercise considers the simplest case of **rectangular** to **polar coordinate** transformation. Consider the feasible set $S = \{x \in \mathbb{R}^2 | (x_1)^2 + (x_2)^2 = 1\}$. Let $\mathbb{P} = \{\xi \in \mathbb{R} | 0 \leq \xi < 2\pi\}$ and define $\tau : \mathbb{P} \to S$ by:

$$\forall \xi \in \mathbb{P}, \tau(\xi) = \begin{bmatrix} \cos \xi \\ \sin \xi \end{bmatrix}.$$

  (i) Show that S is not convex.
  (ii) Show that τ is onto S.
  (iii) Show that $\mathbb{P}$ is convex.

3.20 Suppose that $\tau : \mathbb{R}^{n'} \to \mathbb{R}^n$ is onto $\mathbb{R}^n$. Let $S \subseteq \mathbb{R}^n$ and define $\mathbb{P} = \{\xi \in \mathbb{R}^{n'} | \tau(\xi) \in S\}$. Show that the restriction of τ to $\mathbb{P}$ is onto S.

3.21 Let $x \in \mathbb{R}^n$ and consider the scaling of variables defined by:

$$\forall \ell = 1, \ldots, n, \, x_\ell = \kappa_\ell \xi_\ell,$$

where $\kappa_\ell \in \mathbb{R}, \ell = 1, \ldots, n$.

  (i) Show that this transformation can be expressed in the form $\tau : \mathbb{R}^n \to \mathbb{R}^n$, where:

$$\forall \xi \in \mathbb{R}^n, \tau(\xi) = K\xi,$$

  where K is a square matrix with all entries zero except along its diagonal. The ℓ-th diagonal entry of K is κ_ℓ.
  (ii) Show that if $\kappa_\ell \neq 0$ for each $\ell = 1, \ldots, n$ then τ is onto $\mathbb{R}^n$.

3.22 This exercise considers convexity under a linear transformation of the variables. Let $A \in \mathbb{R}^{n \times m}$ and let $f : \mathbb{R}^n \to \mathbb{R}$ be convex.

  (i) Show that the function $\phi : \mathbb{R}^m \to \mathbb{R}$ defined by $\forall \xi \in \mathbb{R}^m, \phi(\xi) = f(A\xi)$ is convex.
  (ii) Let $\mathbb{P} \subseteq \mathbb{R}^m$ be convex and define $S = \{A\xi \in \mathbb{R}^n | \xi \in \mathbb{P}\}$. Show that S is convex.

3.23 Consider the functions $f : \mathbb{R}^2 \to \mathbb{R}$ and $g : \mathbb{R}^2 \to \mathbb{R}$ as defined in Exercise 2.7.

$$\begin{aligned} \forall x \in \mathbb{R}^2, f(x) &= (x_1)^2 + (x_2)^2 + 2x_2 - 3, \\ \forall x \in \mathbb{R}^2, g(x) &= x_1 + 2x_2 - 3. \end{aligned}$$

Also consider the function $\tau : \mathbb{R}^2 \to \mathbb{R}^2$ defined by:

$$\forall \xi \in \mathbb{R}^2, x = \tau(\xi) = \begin{bmatrix} \xi_1 - \xi_2 \\ \xi_2 \end{bmatrix},$$

and define the following functions:

- $\phi : \mathbb{R}^2 \to \mathbb{R}, \forall \xi \in \mathbb{R}^2, \phi(\xi) = f(\tau(\xi))$,
- $\gamma : \mathbb{R}^2 \to \mathbb{R}, \forall \xi \in \mathbb{R}^2, \gamma(\xi) = g(\tau(\xi))$.

Let $\mathbb{S} = \{x \in \mathbb{R}^2 | g(x) = 0\}$ and define $\mathbb{P} = \{\xi \in \mathbb{R}^2 | \tau(\xi) \in \mathbb{S}\}$. Consider the restriction of τ to $\mathbb{P}$.

(i) Show that $\tau : \mathbb{P} \to \mathbb{S}$ is onto $\mathbb{S}$.
Now repeat Exercise 2.7 for the functions ϕ and γ. That is, perform the following.
(ii) Sketch $\mathbb{C}_\phi(\tilde{\phi})$ for $\tilde{\phi} = 0, 1, 2, 3$.
(iii) Sketch the set of points satisfying $\gamma(\xi) = 0$.
(iv) Find $\min_{\xi \in \mathbb{R}^2} \{\phi(\xi) | \gamma(\xi) = 0\}$ and $\operatorname{argmin}_{\xi \in \mathbb{R}^2} \{\phi(\xi) | \gamma(\xi) = 0\}$.

3.24 Recall the function $f : \mathbb{R}^2 \to \mathbb{R}$ defined in (2.29):

$$\forall x \in \mathbb{R}^2, \ f(x) = (x_1 - 1)^2 + (x_2 - 3)^2 - 1.8(x_1 - 1)(x_2 - 3).$$

Define the function $\phi : \mathbb{R}^2 \to \mathbb{R}$ by:

$$\forall \xi \in \mathbb{R}^2, \phi(\xi) = f(A\xi),$$

where $A \in \mathbb{R}^{2 \times 2}$ is given by:

$$A = \begin{bmatrix} 1 & 0 \\ 0.9 & 0.4359 \end{bmatrix}.$$

Show that the contour sets of ϕ are circular.

3.25 In this exercise we consider elimination of variables.

(i) Consider the problem:

$$\min_{x \in \mathbb{R}^2} \{(x_1)^2 + (x_2 - 3)^2 | x_1 - 2x_2 = 0\}.$$

Use Corollary 3.7 to solve this problem.
(ii) Consider the problem:

$$\min_{x \in \mathbb{R}^2} \{(x_1)^2 + (x_2 - 3)^2 | x_1 - 2x_2 = 0, x_2 \le 0\}.$$

Use Corollary 3.7 to solve this problem. (Hint: What is the projection of the feasible set onto the last component of $\mathbb{R}^2$?)

3.26 Consider a rectangle with sides x_1 and x_2. We collect the variables x_1 and x_2 together into a vector $x \in \mathbb{R}^2$ and consider three associated functions:

- $f : \mathbb{R}^2 \to \mathbb{R}$ defined by:

$$\forall x \in \mathbb{R}^2, \ f(x) = x_1 x_2,$$

the area of the rectangle,
- $\phi : \mathbb{R}^2 \to \mathbb{R}$ defined by:

$$\forall x \in \mathbb{R}^2, \ \phi(x) = \sqrt{(x_1)^2 + (x_2)^2},$$

the length of the diagonal of the rectangle, and
- $\varphi : \mathbb{R}^2 \to \mathbb{R}$ defined by:

$$\forall x \in \mathbb{R}^2, \ \varphi(x) = x_1 + x_2,$$

the sums of the lengths of the sides of the rectangle.

(i) Let $\tilde{\varphi} \in \mathbb{R}$ and consider the problem:

$$\max_{x \in \mathbb{R}^2} \{ f(x) | \varphi(x) = \tilde{\varphi} \}.$$

    (a) Show that f is not concave.
    (b) Use Corollary 3.7 to eliminate x_1.
    (c) Show that the resulting transformed problem has a concave objective.
    (d) Solve the transformed problem.

(ii) Let $\tilde{\varphi} \in \mathbb{R}$ and consider the problem:

$$\min_{x \in \mathbb{R}^2} \{ \phi(x) | \varphi(x) = \tilde{\varphi} \}.$$

    (a) Show that ϕ is not convex.
    (b) Use Corollary 3.7 to eliminate x_1.
    (c) Show that the resulting transformed problem has a convex objective.
    (d) Solve the transformed problem.

(iii) Let $\tilde{f} \in \mathbb{R}$ and consider the problem:

$$\max_{x \in \mathbb{R}^2} \{ \varphi(x) | f(x) = \tilde{f} \}.$$

    (a) Show that the feasible set is not convex.
    (b) Use Corollary 3.7 to eliminate x_1.
    (c) Show that the resulting transformed problem has a concave objective.
    (d) Solve the transformed problem.

(iv) Let $\tilde{f} \in \mathbb{R}$ and consider the problem:

$$\max_{x \in \mathbb{R}^2} \{ \phi(x) | f(x) = \tilde{f} \}.$$

    (a) Use Corollary 3.7 to eliminate x_1.
    (b) Show that the resulting transformed problem has a concave objective.
    (c) Solve the transformed problem.

(v) Let $\tilde{\phi} \in \mathbb{R}$ and consider the problem:

$$\max_{x \in \mathbb{R}^2} \{ \varphi(x) | \phi(x) = \tilde{\phi} \}.$$

(a) Show that the feasible set is not convex.
(b) Use Corollary 3.7 to eliminate x_1.
(c) Show that the resulting transformed problem has a concave objective.
(d) Solve the transformed problem.

(vi) Let $\tilde{\phi} \in \mathbb{R}$ and consider the problem:

$$\max_{x \in \mathbb{R}^2}\{f(x)|\phi(x) = \tilde{\phi}\}.$$

(a) Use Corollary 3.7 to eliminate x_1.
(b) Show that the resulting transformed problem has a concave objective.
(c) Solve the transformed problem.

3.27 Consider $A \in \mathbb{R}^{2 \times 2}$ and $b \in \mathbb{R}^2$ defined in (2.2):

$$A = \begin{bmatrix} 1 & 2 \\ 3 & 4 \end{bmatrix}, b = \begin{bmatrix} 1 \\ 1 \end{bmatrix}.$$

(i) Use the equation specified by the first row of A to find a function $\omega_1 : \mathbb{R} \to \mathbb{R}$ that allows x_1 to be expressed in terms of x_2.
(ii) Use ω_1 to eliminate x_1 in the second row of $Ax = b$.
(iii) Find $x_2^\star$ that satisfies the second row with x_1 eliminated.
(iv) Find $x^\star$ that satisfies the complete system $Ax = b$.

3.28 In this exercise, we consider posynomial functions and monomial functions [6, section 11.5][15, section 4.5.3]. (See Definition 3.1.) Let $A \in \mathbb{R}^{m \times n}$ and $B \in \mathbb{R}_{++}^m$ and define the **posynomial function** $f : \mathbb{R}_{++}^n \to \mathbb{R}$ by:

$$\forall x \in \mathbb{R}_{++}^n, f(x) = \sum_{\ell=1}^{m} B_\ell (x_1)^{A_{\ell 1}} (x_2)^{A_{\ell 2}} \cdots (x_n)^{A_{\ell n}}.$$

If $m = 1$ then f is a **monomial function**.
  Also consider the onto function $\tau : \mathbb{R}^n \to \mathbb{R}_{++}^n$ defined by:

$$\forall \xi \in \mathbb{R}^n, \tau(\xi) = \begin{bmatrix} \exp(\xi_1) \\ \vdots \\ \exp(\xi_n) \end{bmatrix}.$$

If $x = \tau(\xi)$ then $\xi_k = \ln(x_k), k = 1, \ldots, n$.

(i) Show that posynomial functions are not in general convex on $\mathbb{R}_{++}^n$. (Hint: Consider the function $f : \mathbb{R}_{++}^2 \to \mathbb{R}$ defined by:

$$\forall x \in \mathbb{R}_{++}^2, f(x) = x_1 x_2.$$

Consider $x = \begin{bmatrix} 1 \\ 3 \end{bmatrix} \in \mathbb{R}_{++}^2, x' = \begin{bmatrix} 3 \\ 1 \end{bmatrix} \in \mathbb{R}_{++}^2$, and $t = 0.5$.)

(ii) Show that posynomial functions do not in general have convex level sets. (A sketch and an explanation will suffice.)

(iii) Let $f : \mathbb{R}^n_{++} \to \mathbb{R}$ be posynomial and $g : \mathbb{R}^n_{++} \to \mathbb{R}$ be monomial. Show that $h = f/g$ is posynomial.

(iv) Let $\sigma_\ell \in \mathbb{R}, \ell = 1, \ldots, m$, and let A_ℓ be the ℓ-th row of A, $\ell = 1, \ldots, m$. Show that:

$$\left(\sum_{\ell=1}^m \sigma_\ell [A_\ell]^\dagger A_\ell\right)\left(\sum_{k=1}^m \sigma_k\right) - \left(\sum_{\ell=1}^m \sigma_\ell [A_\ell]^\dagger\right)\left(\sum_{k=1}^m \sigma_k A_k\right)$$

$$= \sum_{\ell=1}^m \sum_{k=1}^{\ell-1} \sigma_\ell \sigma_k [A_\ell - A_k]^\dagger [A_\ell - A_k].$$

(v) Show that $\phi : \mathbb{R}^n \to \mathbb{R}$ defined by:

$$\forall \xi \in \mathbb{R}^n, \phi(\xi) = \ln(f(\tau(\xi))),$$

is a convex function. (Hint: Define $b \in \mathbb{R}^m$ by $b_\ell = \ln(B_\ell), \ell = 1, \ldots, m$, and note that:

$$\forall \xi \in \mathbb{R}^n, \phi(\xi) = \ln\left(\sum_{\ell=1}^m \exp([A_\ell]^\dagger \xi + b_\ell)\right),$$

where A_ℓ is the ℓ-th row of A, $\ell = 1, \ldots, m$. Now apply Theorem 2.7 using the result from Part (iv).)

(vi) Suppose that f is monomial. What can you say about $\phi(\bullet)$ defined in Part (v)?

(vii) Show that $\varphi : \mathbb{R}^n \to \mathbb{R}$ defined by:

$$\forall \xi \in \mathbb{R}^n, \varphi(\xi) = f(\tau(\xi)),$$

has convex level sets.

Constraints

3.29 Consider $A \in \mathbb{R}^{2 \times 2}$ and $b \in \mathbb{R}^2$ defined in (2.2):

$$A = \begin{bmatrix} 1 & 2 \\ 3 & 4 \end{bmatrix}, b = \begin{bmatrix} 1 \\ 1 \end{bmatrix}.$$

Multiply both A and b on the left by $M \in \mathbb{R}^{2 \times 2}$ to form the pre-conditioned system:

$$MAx = Mb,$$

where

$$M = \begin{bmatrix} -2 & 1 \\ \frac{3}{2} & -\frac{1}{2} \end{bmatrix}.$$

(i) Find the solution of the pre-conditioned system $MAx = Mb$.

(ii) Compare it to the solution of $Ax = b$.

(iii) Would *any* choice of M be acceptable for facilitating a search for solution to $Ax = b$? (For example, would $M = \mathbf{0}$ be suitable?)

3.30 Prove Theorem 3.8. (Hint: First suppose that Problem (3.9) has a minimum and let $\begin{bmatrix} x^\star \\ w^\star \end{bmatrix}$ be any minimizer of Problem (3.9). Conversely, suppose that $x^\star$ is a minimizer of Problem (3.8).)

3.31 Prove Theorem 3.9.

3.32 Consider the function $h : (\mathbb{R} \times \mathbb{R}_{++}) \to \mathbb{R}^3$ defined by:

$$\forall x \in (\mathbb{R} \times \mathbb{R}_{++}), h(x) = \begin{bmatrix} \frac{x_1}{x_2} - 1 \\ -x_1 \\ -x_2 + 1 \end{bmatrix},$$

and consider the feasible region $\mathbb{S} = \{x \in (\mathbb{R} \times \mathbb{R}_{++}) | h(x) \le 0\}$.

(i) Show that $\mathbb{S}$ is a convex set.

(ii) Show that h_1 is not a convex function on $\mathbb{S}$. (Hint: Use Theorem 2.6 and try x' such that $x'_1 = 0$.)

(iii) Show that the set $\mathbb{S}$ is the same as the set $\hat{\mathbb{S}} = \{x \in \mathbb{R}^2 | \hat{h}(x) \le 0\}$, where $\hat{h} : \mathbb{R}^2 \to \mathbb{R}^3$ is defined by:

$$\forall x \in \mathbb{R}^2, \hat{h}(x) = \begin{bmatrix} x_1 - x_2 \\ -x_1 \\ -x_2 + 1 \end{bmatrix}.$$

(iv) Show that $\hat{h}$ is a convex function on $\mathbb{R}^2$.

Pay particular attention to the difference between a convex *set* and a convex *function*. You should review the definitions of both before answering this question.

3.33 Let $f : \mathbb{R}^n_{++} \to \mathbb{R}$ and $h_\ell : \mathbb{R}^n_{++} \to \mathbb{R}, \ell = 1, \ldots, r$, be posynomial functions and let $g_\ell : \mathbb{R}^n_{++} \to \mathbb{R}, \ell = 1, \ldots, m$, be monomial functions. (See Definition 3.1.) Let $\underline{x}, \overline{x} \in \mathbb{R}^n_{++}$ with $\underline{x} \le \overline{x}, \underline{g}, \overline{g} \in \mathbb{R}^m_{++}$ with $\underline{g} \le \overline{g}$, and $\overline{h} \in \mathbb{R}^r_{++}$. Collect the functions g_ℓ together into the vector function $g : \mathbb{R}^n_{++} \to \mathbb{R}^m$, collect the functions h_ℓ together into the vector function $h : \mathbb{R}^n_{++} \to \mathbb{R}^r$, and consider the **posynomial program** ([6, section 4.5][15, section 4.5]):

$$\min_{x \in \mathbb{R}^n_{++}} \{f(x) | \underline{g} \le g(x) \le \overline{g}, h(x) \le \overline{h}, \underline{x} \le x \le \overline{x}\}.$$

By Exercise 3.28, Parts (i) and (ii), the objective and constraints are neither convex functions nor do they have convex level sets.

Use the transformation in Exercise 3.28 and Theorems 3.1, 3.5, and 3.9 to transform this problem into a convex optimization problem. (Hint: Use Exercise 3.28, Part (vi) to identify the functional form of the transformed version of g. You must show that the transformed version of g is linear. Make sure that the representation of the transformed version of the constraints $\underline{x} \le x \le \overline{x}$ is as simple as possible.)

3.34 Prove Theorem 3.10.

3.35 Recall the example non-linear program, Problem (2.19), from Section 2.3.2.3:

$$\min_{x \in \mathbb{R}^3} \{f(x) | g(x) = 0, h(x) \le 0\},$$

where $f : \mathbb{R}^3 \to \mathbb{R}$, $g : \mathbb{R}^3 \to \mathbb{R}^2$, and $h : \mathbb{R}^3 \to \mathbb{R}$ are defined by:

$$\forall x \in \mathbb{R}^3, \; f(x) = (x_1)^2 + 2(x_2)^2,$$
$$\forall x \in \mathbb{R}^3, \; g(x) = \begin{bmatrix} 2 - x_2 - \sin(x_3) \\ -x_1 + \sin(x_3) \end{bmatrix},$$
$$\forall x \in \mathbb{R}^3, \; h(x) = \sin(x_3) - 0.5.$$

(i) Suppose that the problem has a minimizer. Show that it also has a minimizer $x^\star \in \mathbb{S} = \{x \in \mathbb{R}^3 | g(x) = \mathbf{0}, h(x) \le 0, -\frac{\pi}{2} \le x_3 \le \frac{\pi}{2}\}$.

(ii) Define $\tau : \mathbb{R}^2 \times [-1, 1] \to \mathbb{R}^3$ by:

$$\forall \xi \in (\mathbb{R}^2 \times [-1, 1]), \; \tau(\xi) = \begin{bmatrix} \xi_1 \\ \xi_2 \\ \sin^{-1}(\xi_3) \end{bmatrix}.$$

Define $\mathbb{P} = \{\xi \in (\mathbb{R}^2 \times [-1, 1]) | \tau(\xi) \in \mathbb{S}\}$. Show that the restriction of τ to $\mathbb{P}$ is onto $\mathbb{S}$.

(iii) Use the previous parts to solve the problem.

3.36 Let $f : \mathbb{R}^2 \to \mathbb{R}$ be defined by:

$$\forall x \in \mathbb{R}^2, \; f(x) = x_1.$$

Consider the problem:

$$\min_{x \in \mathbb{R}^2} \{f(x) | x \in \mathbb{S}\},$$

where $\mathbb{S} = \{x \in \mathbb{R}^2 | x_1 \le 0, (x_1)^2 + (x_2)^2 = 1\}$ and also consider the relaxed problem:

$$\min_{x \in \mathbb{R}^2} \{f(x) | x \in \overline{\mathbb{S}}\},$$

where $\overline{\mathbb{S}} = \{x \in \mathbb{R}^2 | x_1 \le 0, (x_1)^2 + (x_2)^2 \le 1\}$.

(i) Show that the objective of both problems is convex on $\mathbb{R}^2$.

(ii) Show that $\mathbb{S}$ is not convex. A sketch and an explanation will suffice.

(iii) Show that $\overline{\mathbb{S}}$ is convex. A sketch and an explanation will suffice.

(iv) Find a local minimum and minimizer of the relaxed problem $\min_{x \in \mathbb{R}^2} \{f(x) | x \in \overline{\mathbb{S}}\}$. A graphical argument showing the contour sets of f and showing the feasible set is sufficient.

(v) Show that the local minimum and minimizer you have found are, respectively, the global minimum and minimizer of the relaxed problem $\min_{x \in \mathbb{R}^2} \{f(x) | x \in \overline{\mathbb{S}}\}$. (Hint: apply Theorem 2.4.)

(vi) Show that the global minimum of the relaxed problem $\min_{x \in \mathbb{R}^2} \{f(x) | x \in \overline{\mathbb{S}}\}$ is also the global minimum of $\min_{x \in \mathbb{R}^2} \{f(x) | x \in \mathbb{S}\}$. (Hint: apply Theorem 3.10.)

3.37 Consider the inequality-constrained problem $\min_{x \in \mathbb{S}} f(x)$ where:

$$\mathbb{S} = \{x \in \mathbb{R}^2 | g(x) = 0, x_1 - 10 \le 0\},$$

and where $f : \mathbb{R}^2 \to \mathbb{R}$ and $g : \mathbb{R}^2 \to \mathbb{R}$ were defined in Exercise 2.7. That is:

$$\forall x \in \mathbb{R}^2, f(x) = (x_1)^2 + (x_2)^2 + 2x_2 - 3,$$
$$\forall x \in \mathbb{R}^2, g(x) = x_1 + 2x_2 - 3.$$

Re-interpret the equality-constrained problem $\min_{x \in \mathbb{R}^2} \{f(x) | g(x) = 0\}$, which was solved in Exercise 2.7, Part (iii), as a relaxation of the inequality-constrained problem $\min_{x \in S} f(x)$. Use Theorem 3.10 to solve the inequality-constrained problem.

3.38 Suppose that you own a timber plantation and mill. Because of the plantation's proximity to a nuclear fuel waste repository, the raw logs that we harvest come from mutated trees that have elliptical cross-sections. If we measure out from the "center" of the cross-section of the tree then the edge of the cross-section can be described by the equation:

$$\frac{1}{8}(x_1)^2 + \frac{1}{2}(x_2)^2 = 1,$$

where $\begin{bmatrix} x_1 \\ x_2 \end{bmatrix}$ are the coordinates of the edge.

The lumber processing equipment can cut a single rectangular cross-sectioned piece of sawn wood out of the elliptical log. The pre-1950s technology equipment that we own turns the rest of the log into sawdust. Sawdust has value equal to 0.1 times the value of sawn wood per unit cross-sectional area.

We would like to choose the optimum dimensions of the rectangular cross-section to maximize the sum of:

- the value of the sawn wood, and
- the value of the sawdust,

that we obtain from the log.

(i) Formulate the inequality-constrained problem to maximize the total value of sawn wood and sawdust we obtain from the elliptical log. Note the following.

- To operate, the processing equipment must be given the coordinates of the top-right corner of the final product. The equipment cuts a rectangular piece of wood that is symmetric about the center.
- To cut a "feasible" cross-section from the elliptical log, the top-right corner must lie inside or on the elliptical edge of the log.
- The coordinates must be non-negative.

The variables in the problem are the coordinates of the edge that is to be given to the processing equipment. That is, the variables are $x = \begin{bmatrix} x_1 \\ x_2 \end{bmatrix}$. Cast the problem as a maximization problem of the form:

$$\max_{x \in \mathbb{R}^2} \{f(x) | g(x) = \mathbf{0}, h(x) \leq \mathbf{0}\}.$$

In particular, specify the following.

- Define the objective function explicitly.
- Define any entries in the equality constraint function (or state that there are no equality constraints).
- Define any entries in the inequality constraint function (or state that there are no inequality constraints).

(ii) Re-formulate the problem as a minimization problem in the most straightforward way. Make sure you define anything that has not already been defined.
(iii) Sketch the contour sets of the objective of the problem defined in Part (ii).
(iv) Consider the equality-constrained problem formed by changing the inequality constraints that limit the top right corner to being inside the log into equality constraints. Show that the equality-constrained problem and the problem in Part (ii) have the same solution. (A sketch will suffice.)
(v) What is the minimum of the problem in Part (ii). What are the values of the coordinates to be specified to the processing equipment?

3.39 Use divide and conquer to prove that $x^\star = 3$ is the global minimizer of the function shown in Figure 2.27 over the set $\mathbb{P} = \{x \in \mathbb{R} | -4 \le x \le 1 \text{ or } 2 \le x \le 4\}$.

3.40 Prove Theorem 3.11.

3.41 Suppose that $\mathbb{S} \subseteq \mathbb{R}^{n+s}$ is convex and of the form:

$$\mathbb{S} = \left\{ \begin{bmatrix} x \\ y \end{bmatrix} \in \mathbb{R}^n \times \mathbb{R}^s \,\middle|\, x \in \mathbb{S}_1, y \in \mathbb{S}_2(x) \right\},$$

with $\mathbb{S}_1 \subseteq \mathbb{R}^n$ and, for each $x \in \mathbb{S}_1$, $\mathbb{S}_2(x) \subseteq \mathbb{R}^s$. Without loss of generality, assume that $\forall x \in \mathbb{S}_1, \mathbb{S}_2(x) \ne \emptyset$. (Otherwise, re-define $\mathbb{S}_1$ by omitting any elements x such that $\mathbb{S}_2(x) = \emptyset$.) Let $f : \mathbb{S} \to \mathbb{R}$ be convex on $\mathbb{S}$.

(i) Let $x \in \mathbb{S}_1$. Show that $\mathbb{S}_2(x)$ is convex.
(ii) Let $x \in \mathbb{S}_1$ and define $\varphi : \mathbb{S}_2(x) \to \mathbb{R}$ by:

$$\forall y \in \mathbb{S}_2(x), \varphi(y) = f\left(\begin{bmatrix} x \\ y \end{bmatrix}\right).$$

Show that φ is convex on $\mathbb{S}_2(x)$.
(iii) Show that $\mathbb{S}_1$ is convex.
(iv) Suppose that for each $x \in \mathbb{S}_1$ the minimization problem $\min_{y \in \mathbb{S}_2(x)} f\left(\begin{bmatrix} x \\ y \end{bmatrix}\right)$ has a minimum. Consider the function $\phi : \mathbb{S}_1 \to \mathbb{R}$ defined by:

$$\forall x \in \mathbb{S}_1, \phi(x) = \min_{y \in \mathbb{S}_2(x)} f\left(\begin{bmatrix} x \\ y \end{bmatrix}\right).$$

Show that ϕ is convex on $\mathbb{S}_1$. (Hint: Consider $x, x' \in \mathbb{S}_1$ and corresponding minimizers $y^\star \in \text{argmin}_{y \in \mathbb{S}_2(x)} f\left(\begin{bmatrix} x \\ y \end{bmatrix}\right)$ and $y'^\star \in \text{argmin}_{y \in \mathbb{S}_2(x')} f\left(\begin{bmatrix} x' \\ y \end{bmatrix}\right)$.)

3.42 In this exercise we apply hierarchical decomposition to solve a non-convex problem. Consider the function $f : \mathbb{R}^3 \to \mathbb{R}$ defined by:

$$\forall x \in \mathbb{R}, \forall y \in \mathbb{R}^2, f\left(\begin{bmatrix} x \\ y \end{bmatrix}\right) = |y_2 + x(y_1 - y_2)|.$$

(i) Show that f is not convex.

(ii) Show that f is not concave.

(iii) Show that:

$$\max_{x\in\mathbb{R},\, y\in\mathbb{R}^2}\left\{f\left(\begin{bmatrix} x \\ y \end{bmatrix}\right)\Big|\, 0 \le x \le 1, 0 \le y_1 \le \bar{\rho}, 0 \le y_2 \le \bar{\rho}\right\} = \bar{\rho}.$$

(Hint: Use Theorem 3.11 to decompose the problem hierarchically. First consider the optimizing values of y for given fixed x. Then optimize over values of x. Note that $y_2 + x(y_1 - y_2) = y_2(1 - x) + y_1 x$.)

Duality

3.43 Let $f : \mathbb{R}^n \to \mathbb{R}$, $g : \mathbb{R}^n \to \mathbb{R}^m$, and $h : \mathbb{R}^n \to \mathbb{R}^r$. Consider the associated Lagrangian $\mathcal{L} : \mathbb{R}^n \times \mathbb{R}^m \times \mathbb{R}^r \to \mathbb{R}$ defined by:

$$\forall x \in \mathbb{R}^n, \forall \lambda \in \mathbb{R}^m, \forall \mu \in \mathbb{R}^r, \mathcal{L}(x, \lambda, \mu) = f(x) + \lambda^\dagger g(x) + \mu^\dagger h(x),$$

and the corresponding dual function $\mathcal{D} : \mathbb{R}^m \times \mathbb{R}^r \to \mathbb{R} \cup \{-\infty, +\infty\}$ defined in (3.12):

$$\forall \begin{bmatrix} \lambda \\ \mu \end{bmatrix} \in \mathbb{R}^{m+r}, \mathcal{D}(\lambda, \mu) = \inf_{x\in\mathbb{R}^n} \mathcal{L}(x, \lambda, \mu).$$

Using the definition of inf, show that $\mathcal{D}$ is, in fact, an extended real function $\mathcal{D} : \mathbb{R}^m \times \mathbb{R}^r \to \mathbb{R} \cup \{-\infty\}$. That is, you must show that $\mathcal{D}$ never takes on the value $+\infty$.

3.44 Prove Theorem 3.12. (Hint: Let $\begin{bmatrix} \lambda \\ \mu \end{bmatrix}, \begin{bmatrix} \lambda' \\ \mu' \end{bmatrix} \in \mathbb{R}^{m+r}$, let $t \in [0, 1]$, and note that:

$$\forall x \in \mathbb{R}^n, \mathcal{L}(x, [1-t]\lambda + t\lambda', [1-t]\mu + t\mu') = [1-t]\mathcal{L}(x, \lambda, \mu) + t\mathcal{L}(x, \lambda', \mu').$$

Consider the definitions of $\mathcal{D}([1-t]\lambda + t\lambda', [1-t]\mu + t\mu')$, $\mathcal{D}(\lambda, \mu)$, and $\mathcal{D}(\lambda', \mu')$.)

3.45 Prove Corollary 3.14 using Theorem 3.13.

3.46 In this exercise we consider the left- and right-hand sides of (3.14) for the case where the feasible set of the primal problem is $\mathbb{S} = \{x \in \mathbb{R}^n | g(x) = 0\}$. That is, we only have equality constraints and we can neglect the dual variables μ corresponding to the inequality constraints.

(i) Consider the primal problem $\min_{x\in\mathbb{R}^2}\{f(x)|g(x) = 0\}$ where the functions $f : \mathbb{R}^2 \to \mathbb{R}$ and $g : \mathbb{R}^2 \to \mathbb{R}$ were defined in Exercise 3.11. That is:

$$\begin{aligned} \forall x \in \mathbb{R}^2, f(x) &= -2(x_1 - x_2)^2 + (x_1 + x_2)^2, \\ \forall x \in \mathbb{R}^2, g(x) &= x_1 - x_2, \end{aligned}$$

Evaluate the left- and right-hand sides of (3.14) for this f and g. That is, evaluate $\min_{x\in\mathbb{R}^2}\{f(x)|g(x) = 0\}$ and $\sup_{\lambda\in\mathbb{E}} \mathcal{D}(\lambda)$. Be careful that you actually find an infimum of the inner problem. Is there a duality gap?

(ii) Repeat the previous part but re-define f to be:

$$\forall x \in \mathbb{R}^2, f(x) = (x_1 + x_2)^2.$$

(iii) Repeat the previous part but re-define f to be:

$$\forall x \in \mathbb{R}^2, \ f(x) = (x_1 + x_2)^2 + (x_1 - x_2)^2.$$

Part I

Linear simultaneous equations

4

Case studies of linear simultaneous equations

In this chapter, we will develop two case studies of problems that involve the solution of linear simultaneous equations:

(i) the solution of Kirchhoff's laws in a simple electrical circuit (Section 4.1), and

(ii) the search for a set of inputs to a "discrete-time linear system" that will bring the system to a desired state (Section 4.2).

The first case study will be developed in some detail, while the second will be much more briefly described. As we proceed, we will introduce notation to help us express ideas concisely and precisely. We will try to describe the choices that are made in formulating a model. You may already be very familiar with these models; however, the *reasoning* we present here may help you to pose and answer questions that arise in *formulating* your own models. We emphasize that the formulation of a new problem may involve many stages and refinements and that our presentation hides some of this process.

4.1 Analysis of a direct current linear circuit

4.1.1 Motivation

In designing a large integrated circuit to be fabricated or even a large circuit consisting of discrete components, it is very important to be able to calculate the behavior of the circuit without going to the expense of actually building a prototype. Furthermore, because of manufacturing tolerances, manufactured component values may differ from nominal. The values may also drift over the lifetime of the circuit. Both manufacturing tolerances and drift of component values can be interpreted as changes in the values of components from their nominal values in the base-case circuit. It is useful to calculate the effect of such changes from the nominal values on the circuit behavior.

Electrical circuits are well characterized by a set of equations called **Kirchhoff's laws**. This case study assumes some familiarity with basic circuit analysis and, in particular, Kirchhoff's laws applied to resistive circuits. However, as with all of our case studies, the main issue is not the formulation of this particular case study, but rather the *process* of formulating case studies.

We will discuss the problem of finding a solution to Kirchhoff's laws and also how this solution changes with changes in component values. There are a variety of texts on circuit theory that cover this material such as [21, 95]. Our development follows [95].

After formulating the problem in Section 4.1.2, we will consider changes to it in Section 4.1.3 and explore some of its characteristics in Section 4.1.4.

4.1.2 Formulation

This material is based on an extended example in [95, section 1.3 and following]. Consider a circuit consisting of interconnected **resistors** and **current sources** as shown in Figure 4.1. The current sources are shown as circles with the value of the current source and a direction for positive conventional current flow shown inside the circle. The current sources are all direct current (DC). The resistors are shown as rectangles with a value of the **resistance** inside. The particular configuration shown in Figure 4.1 is called a **ladder circuit** because it resembles a ladder on its side; however, the arrangement of the interconnected resistors can, in principle, be arbitrary.

Collectively, the current sources and resistors are called the **components** in the circuit or the **branches** of the circuit. The lines joining the components together represent wires that are assumed to be **ideal conductors**. Each set of joined wires is called a **node** and is labeled uniquely with a number. We say that the branches **link** or are **incident to** to the nodes. A "node" in our circuit is not just a single point in the diagram, but includes all the wires associated with each node label. A collection of nodes joined by branches is called a **graph**. A circuit can be thought of as a special type of graph where the branches are components.

We assume that the resistances and current sources have known "nominal" values; however, we are also interested in the case where these values change from nominal. That is, we want to:

- calculate all the electrical quantities associated with the circuit, and
- characterize how these quantities change if the circuit changes.

4.1.2.1 Variables of interest

One of the most basic issues in formulating a problem is to identify and distinguish:

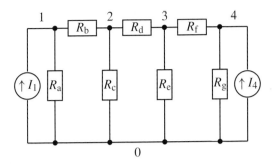

Fig. 4.1. A ladder circuit consisting of resistors and current sources.
Source: This figure is based on [95, figure 1.4].

- the variables of interest from
- the variables that are of less importance.

It is important to realize that for virtually any physical system there are a very large number of issues that could potentially be modeled. To develop a tractable model we will have to neglect most of the possible issues and concentrate on just a small number of them. This choice requires judgment and care and is one aspect of **Occam's razor** [115]. As suggested in Section 1.3, the model should be no more complicated than is necessary to represent the important issues, where "important" depends on our perspective.

In typical circuits, we seek values of:

- the voltages across the resistors and current sources, and
- the currents through the resistors.

We usually neglect most other quantities and assume that they do not affect the voltages and currents. For example, in a resistive circuit, we can usually neglect the effect of variation in temperature on the values of the resistances and on the values of the voltages and currents. We have already done this implicitly in our description of the circuit by specifying the current sources and resistors without reference to the temperature. It is worthwhile, however, to be explicit about these decisions.

In general, we can choose to neglect a quantity if we assess that neglecting it will not significantly impair our ability to calculate the values of the variables of interest. This assessment may have to be modified as we test and refine the model. For example, if we find that the *power* dissipated in the resistors in the circuit is large, then the assumption of negligible variation of temperature may prove to be false. In that case we would explore a more refined model that included:

- the dependence of the value of resistance on temperature, and
- the variation of temperature with power dissipated.

In this case study, we will initially represent voltage and current assuming that all other quantities, such as temperature, stay constant. As indicated above, we will then check to see if this assumption is consistent with the calculated power dissipation in each resistor and its power rating. We will also consider the sensitivity to temperature variation.

Given our interest in the currents and voltages, let us consider the relationship *between* them. By definition of a resistor, the voltage across any resistor is equal to the product of:

- the current through it, and
- its resistance.

That is, the **terminal characteristics** of the resistor are specified by its resistance. So, if we can find the currents through all the resistors then we can calculate the voltage across each resistor. On the other hand, if we can find the voltages across all the resistors then we can calculate the current through each resistor.

It is common when formulating a model to concentrate on particular variables as the ones of most interest or on the ones that are easiest to deal with. Other variables in the system that are of interest can sometimes be considered **dependent** in that they are easy to calculate once the variables of most interest are known. We can **eliminate** them from the formulation by substituting for them using their known dependence on the **independent variables**. (See Section 3.2.2.2.) The values of the dependent variables can be found once the independent variables have been evaluated or specified.

In systems such as circuits that are associated with graphs, we have the choice between concentrating on:

- nodal quantities, in this case the **nodal voltages**, as the independent variables, or
- branch quantities, in this the case **branch currents**, as the independent variables.

Knowing one set of variables enables us to calculate the other, recognizing that branch voltages can be expressed in terms of nodal voltages. The relationships between branch voltages and branch currents are called the **branch constitutive relations** and represent the terminal characteristics [95, appendix A.4]. For example, for a resistor, the branch constitutive relation says that branch voltage is proportional to branch current, with constant of proportionality given by the resistance. We could either:

- use the nodal voltages as the independent variables and calulate the current flowing through each resistor in terms of the nodal voltages, or

- use the branch currents as the independent variables and calculate the branch voltages in terms of the branch currents.

In some problems, the choice of the independent and dependent variables may be dictated by the relationship between cause and effect in the physical system. However, the choice of independent and dependent variables is somewhat flexible for the circuit case study. For example, we can also imagine concentrating on some mixture of nodal voltages and currents, the knowledge of which allows us to calculate all the other quantities. For circuit analysis, this is called **modified nodal analysis** [95, section 2.2]. Modified nodal analysis is useful when there is a mixture of voltage and current sources in the circuit. (See Exercise 4.1.) In general, the choice of the variables of interest will depend on several issues such as our point of view and the ease of calculation using a particular set of variables.

Usually, we consider circuits that:

- are **connected**; that is, where any node can be reached from any other node by traversing branches, and
- have at least one **loop**.

If a circuit is not connected, we can decompose it into its connected components for analysis. If a circuit has no loops, then no current can flow, so the analysis is trivial.

In the case of a connected circuit with at least one loop, one issue to consider in choosing variables of interest is that there are always at least as many branches as nodes. (See Exercise 4.2.) A nodal based description will have less variables than the branch based description. In the absence of other considerations, we will concentrate on the nodal voltages, realizing that the branch currents can be calculated once the voltages are known. In other applications, it might be more useful to concentrate on the currents or on a particular combination of currents and voltages.

4.1.2.2 Kirchhoff's voltage law

Kirchhoff's voltage law expresses the fact that the voltage around any loop is zero [21, chapter 1][95, chapter 1]. This means that we can single out one of the nodes and call it the datum or ground node. We can think of it as having zero voltage. We can pick any node in the circuit as datum, but we usually choose a node that has many branches connected to it. (We will discuss reasons for this choice in Section 4.1.4.3 and Exercise 4.6.) We label the datum node as node 0.

For any other node k in the circuit, we can measure the voltage from k to the datum node by summing the voltages across branches in any path from the datum node to node k. By Kirchhoff's voltage law, this sum is independent of the path taken from the datum node to node k. For example, consider a loop in Figure 4.1

that includes R_a, R_b, R_d, and R_e. By Kirchhoff's voltage law, the sum of the voltages across R_a, R_b, and R_d in the direction from the datum node to node 1 to node 2 to node 3 must be equal and opposite to the voltage across R_e in the direction from the datum node to node 3. Similarly, along any other path joining the datum node to node 3, the sum of the voltages across the branches is the same.

That is, Kirchhoff's voltage law enables the definition of the notion of a nodal voltage. We will write x_k for this nodal voltage, and often refer to it as *the* voltage at node k, although it is more properly described as the voltage between node k and the datum node. Naturally, $x_0 = 0$, since this is the voltage between the datum node and itself.

Kirchhoff's voltage law also enables us to express the voltages across branches as the differences between corresponding nodal voltages. For example, the voltage across R_a in Figure 4.1 is $x_1 - x_0 = x_1$.

Kirchhoff's voltage law is an example of a **conservation law**. That is, it is a property associated with position such that when its values are traced over a closed path, the value at the completion of the path is the same as the value at the beginning of the path. Many physical properties satisfy conservation laws. For example, the gravitational potential energy of an object is conserved under movement along a closed path. Consequently, we can define gravitational potential energy as a function of position alone.

4.1.2.3 Branch constitutive relations

As indicated above, the branch constitutive relation for each resistor in the circuit expresses the linear relationship between resistor current and voltage. The other type of component in our ladder circuit is a current source. For this component, the branch constitutive relation specifies that the branch current is constant. (For a voltage source, the branch constitutive relation specifies that the branch voltage is constant. See Exercise 4.1.)

4.1.2.4 Kirchhoff's current law

Having defined our variables of interest, we now use **Kirchhoff's current law** [21, chapter 1][95, chapter 1] and the branch constitutive relations to write down a series of equations to describe the circuit. Kirchhoff's current law expresses conservation of charge when current is flowing in a circuit. There is one equation for each node, expressing the fact that the net current flowing out of each node is zero. For example, consider node 1 in Figure 4.1. The net current flowing from node 1 into the components incident to node 1 is:

$$\frac{x_1 - x_0}{R_a} + \frac{x_1 - x_2}{R_b} - I_1,$$

where we have used the branch constitutive relations to express branch currents in terms of branch voltages and, in turn, used Kirchhoff's voltage law to express the branch voltages as the difference between appropriate nodal voltages.

By Kirchhoff's current law, the net current flowing from node 1 must be zero. Equating the net current to zero, noting that $x_0 = 0$, and re-arranging, we obtain:

$$\left(\frac{1}{R_a} + \frac{1}{R_b} \right) x_1 + \left(-\frac{1}{R_b} \right) x_2 = I_1. \tag{4.1}$$

Similarly, we can write down the equations for the other nodes:

$$\left(-\frac{1}{R_b} \right) x_1 + \left(\frac{1}{R_b} + \frac{1}{R_c} + \frac{1}{R_d} \right) x_2 + \left(-\frac{1}{R_d} \right) x_3 = 0, \tag{4.2}$$

$$\left(-\frac{1}{R_d} \right) x_2 + \left(\frac{1}{R_d} + \frac{1}{R_e} + \frac{1}{R_f} \right) x_3 + \left(-\frac{1}{R_f} \right) x_4 = 0, \tag{4.3}$$

$$\left(-\frac{1}{R_f} \right) x_3 + \left(\frac{1}{R_f} + \frac{1}{R_g} \right) x_4 = I_4. \tag{4.4}$$

There is also an equation for the datum node, but it is **redundant**. (See Exercise 4.3.)

Kirchhoff's current law is an example of a **flow balance constraint**. In a system where material is moving and cannot be created or destroyed, the sum of the flows at any point must be zero. Analogous constraints apply to many physical systems, particularly models of **transport** of materials.

4.1.2.5 *Nodal admittance matrix and voltage and current vector*

Define a matrix $A \in \mathbb{R}^{4 \times 4}$ and vectors $x \in \mathbb{R}^4$ and $b \in \mathbb{R}^4$ as follows:

$$A = \begin{bmatrix} A_{11} & A_{12} & A_{13} & A_{14} \\ A_{21} & A_{22} & A_{23} & A_{24} \\ A_{31} & A_{32} & A_{33} & A_{34} \\ A_{41} & A_{42} & A_{43} & A_{44} \end{bmatrix},$$

$$= \begin{bmatrix} \frac{1}{R_a} + \frac{1}{R_b} & -\frac{1}{R_b} & 0 & 0 \\ -\frac{1}{R_b} & \frac{1}{R_b} + \frac{1}{R_c} + \frac{1}{R_d} & -\frac{1}{R_d} & 0 \\ 0 & -\frac{1}{R_d} & \frac{1}{R_d} + \frac{1}{R_e} + \frac{1}{R_f} & -\frac{1}{R_f} \\ 0 & 0 & -\frac{1}{R_f} & \frac{1}{R_f} + \frac{1}{R_g} \end{bmatrix}, \tag{4.5}$$

$$x = \begin{bmatrix} x_1 \\ x_2 \\ x_3 \\ x_4 \end{bmatrix}, \tag{4.6}$$

$$b = \begin{bmatrix} b_1 \\ b_2 \\ b_3 \\ b_4 \end{bmatrix},$$

$$= \begin{bmatrix} I_1 \\ 0 \\ 0 \\ I_4 \end{bmatrix}. \tag{4.7}$$

The matrix A is called the **nodal admittance matrix** because it is made up of **admittances** in the circuit; that is, inverses of **impedances**. The matrix A has entries that correspond to the coefficients in (4.1)–(4.4). (We have a DC system, so that the admittances are **conductances** and do not have any reactive or capacitive components. In Section 6.2, we will re-examine the nodal admittance matrix in the case of an alternating current (AC) circuit.)

Although we have defined a variable x_0, this variable is not included in our definition of the vector x. In general, we can choose to include or not include particular variables in the definition of a vector. The vector b is the vector of current injections from current sources in the circuit, while x is the vector of voltages. Both b and x exclude the datum node. At nodes 2 and 3 there are no current sources, so the corresponding entries in the vector b are zero. We will relate A, x, and b in Section 4.1.2.6.

4.1.2.6 Linear equations

We will express Kirchhoff's equations as a set of linear simultaneous equations involving A, x, and b. Recall that the product of the matrix A and vector x is:

$$Ax = \begin{bmatrix} \sum_{k=1}^{4} A_{1k}x_k \\ \sum_{k=1}^{4} A_{2k}x_k \\ \sum_{k=1}^{4} A_{3k}x_k \\ \sum_{k=1}^{4} A_{4k}x_k \end{bmatrix}.$$

If we write $Ax = b$, we reproduce the nodal equations (4.1)–(4.4) for our system. (See Exercise 4.4.) So, if we could solve the equation $Ax = b$, which is a system of linear simultaneous equations, then we could calculate the voltages in the system and then calculate the branch currents using the branch constitutive relations. We call A the **coefficient matrix** of the linear system $Ax = b$, while b is called the **right-hand side**. We will discuss how to solve this general type of equation in Chapter 5 and also investigate how to take advantage of particular characteristics of equations to speed up calculations.

4.1.3 Changes

As well as solving the base-case circuit specified by the linear system $Ax = b$, we might also want to consider the solution when the circuit changes. That is, we want to solve a "change-case" circuit. In our ladder circuit, there are two types of circuit components that can change:

- either the currents from the current sources vary, corresponding to a change in the right-hand side of the linear system, b, or
- the resistances vary, corresponding to a change in the coefficient matrix of the system, A.

We would be interested in the effect of changes to the resistances, for example, to gauge how tight a tolerance we need to specify on the resistors to ensure correct operation of the circuit. Furthermore, we might want to consider the changes due to the change of a parameter such as temperature.

For each type of component or parameter change, we can consider two related notions of change:

(i) infinitesimal changes in component or parameter values, providing a **sensitivity analysis**, and
(ii) **large changes** in component values or parameters.

We will consider these in the next two sections. One way to calculate the effects of such changes is to take differences between a base-case and a change-case. However, we will avoid such an approach by taking advantage of the base-case solution for the nominal component or parameter values to yield information for the change-case.

We can also consider the addition of a new component or of a new node into the circuit. We observe that we can consider these types of changes as special cases of changing the resistance.

4.1.3.1 Sensitivity

We define **sensitivity analysis**, sometimes called **small-signal sensitivity analysis** in the context of circuit theory, to mean the calculation of a partial derivative of the solution $x^\star$, or a function of the solution, with respect to some parameter. For example, we might want to calculate the partial derivative of the solution for a particular voltage, say x_2, with respect to the value of:

- a current source, say I_4, to obtain $\dfrac{\partial x_2^\star}{\partial I_4}$,

- a resistor, say R_b, to obtain $\dfrac{\partial x_2^\star}{\partial R_b}$, or

- the temperature, T, to obtain $\dfrac{\partial x_2^\star}{\partial T}$.

We will evaluate the derivatives at the base-case or nominal circuit values. The calculation of $\dfrac{\partial x_2^\star}{\partial T}$ will require an explicit model of how the values of components in the circuit vary with temperature. The calculation will make use of the chain rule to express the total derivative of $x_2^\star$ with respect to T in terms of partial derivatives of $x_2^\star$ with respect to component values and the derivatives of the component values with respect to T.

As another example of a sensitivity analysis, we may also want to consider the sensitivity of a **performance criterion** or **objective** function to variations in parameters. For example, consider the function $f : \mathbb{R}^4 \to \mathbb{R}$ defined by:

$$\forall x \in \mathbb{R}^4, \ f(x) = (x_1)^2 + 2(x_2)^2 + 3(x_3)^2 + 4(x_4)^2. \tag{4.8}$$

We may be interested in the sensitivity of $f(x^\star)$ to various parameters. For example, we might want to calculate the derivative with respect to the value of:

- a current source, say I_4, to obtain $\dfrac{\partial [f(x^\star)]}{\partial I_4}$,

- a resistor, say R_b, to obtain $\dfrac{\partial [f(x^\star)]}{\partial R_b}$, or

- the temperature, T, to obtain $\dfrac{\partial [f(x^\star)]}{\partial T}$.

Again, we will evaluate the derivatives at the base-case or nominal circuit values.

4.1.3.2 Large changes

We will discuss, in turn, large changes in the current source and in the resistance values.

Change in current source   Consider a variation in the current injection b_ℓ at node ℓ by an amount Δb_ℓ. In the case that b_ℓ in the original circuit is zero, a change in b_ℓ means that we have added a *new* current source into the circuit between node ℓ and the datum node. For example, since b_2 and b_3 are zero in the circuit shown in Figure 4.1, a change in either of these currents means that we have added a new current source into the circuit. This is illustrated in Figure 4.2 for a current source added at node $\ell = 2$. Let us define Δb to be the vector that is zero everywhere except in the ℓ-th entry, where it equals Δb_ℓ. Then the new circuit must satisfy $Ax' = b + \Delta b$, where x' is the new value of voltages in the circuit. Simultaneous changes in several current injections are represented by a vector Δb with several non-zero entries. For example, if a current source is added between nodes 2 and 3 then both Δb_2 and Δb_3 will be non-zero (and opposite in sign).

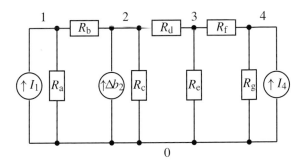

Fig. 4.2. The ladder circuit of Figure 4.1 with a change, Δb_ℓ, in the current injected at node $\ell = 2$.

Change in resistance Consider a variation in the resistance of the resistor joining nodes ℓ and k. This will change the equations in Kirchhoff's laws that pertain to current injections at nodes ℓ and k. We will calculate the change to our system of linear equations.

First consider the situation before the change and suppose that there was a resistor of resistance $R_{\ell k}$ between nodes ℓ and k. Kirchhoff's current law for node ℓ, for example, includes the terms: $(x_\ell/R_{\ell k}) - (x_k/R_{\ell k})$ representing the current flow from node ℓ to node k along $R_{\ell k}$. Corresponding to these terms in the current law, the expression for $A_{\ell\ell}$ includes the term $1/R_{\ell k}$, while the expression for $A_{\ell k}$ includes the term $-1/R_{\ell k}$. Similarly, the expression for $A_{k\ell}$ includes the term $-1/R_{\ell k}$, while the expression for A_{kk} includes the term $1/R_{\ell k}$.

After a change in the resistance, these terms in the admittance matrix are changed to plus or minus the reciprocal of the new resistance between ℓ and k. The change is equal to plus or minus the change in the reciprocal of the resistance between nodes ℓ and k, which is the change in the value of the conductance between these nodes. Let us write $\Delta G_{\ell k}$ for the change in the conductance of the resistor between ℓ and k. Figure 4.3 shows the circuit of Figure 4.1, but with the resistors re-labeled with their conductances and with a change in the conductance of the resistor between nodes $\ell = 2$ and $k = 3$. The change in A is ΔA, where ΔA has zeros everywhere except in the $\ell\ell$-th, ℓk-th, $k\ell$-th, and kk-th entries. For these entries:

- $\Delta A_{\ell\ell} = \Delta A_{kk} = \Delta G_{\ell k}$, and
- $\Delta A_{\ell k} = \Delta A_{k\ell} = -\Delta G_{\ell k}$.

The new circuit must satisfy $(A + \Delta A)x' = b$, where x' is again the new value of the voltages.

If a resistance between node ℓ and the datum node changes, then ΔA is zero except that $\Delta A_{\ell\ell} = \Delta G_{\ell 0}$, where $\Delta G_{\ell 0}$ is the corresponding change in the conductance. Simultaneous changes in the conductances of several components can be handled with an appropriate matrix ΔA having more non-zero entries.

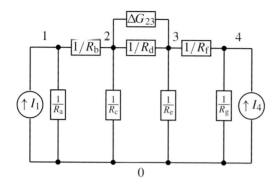

Fig. 4.3. The ladder circuit of Figure 4.1 with resistors re-labeled with their conductances and with a change in the conductance between nodes $\ell = 2$ and $k = 3$. Note that the convention for labeling the resistors has changed compared to the previous figures.

4.1.4 Problem characteristics

In this section, we will try to tease out some of the characteristics of the problem. The discussion will seem a little unmotivated until we develop algorithms explicitly in Chapter 5. In thinking about your own problem formulation, you might want to consider this discussion to be an example of probing the model to understand it better. Sometimes, a little undirected exploration can be useful in better understanding a model or finding a suitable method of attack. (Of course, for this description, we have chosen characteristics here that we know are relevant, while truly undirected exploration will usually turn up at least some unimportant features.) Exploration can also be useful in finding "bugs" in the model; that is, errors and unintended consequences of a particular way of looking at things or of a particular approximation.

4.1.4.1 Numbers of variables and equations

Each row of A represents one of the linear equations (4.1)–(4.4) and there is one such equation corresponding to every non-datum node. Each entry of x represents a variable, namely, the voltage at a non-datum node. We are trying to find the value of the variables in the vector x. Since A is a square matrix (having the same number of rows as columns) we have the same number of equations as variables. We call $Ax = b$ a **square system of linear simultaneous equations**.

4.1.4.2 Solvability

From physical principles, since each node in the ladder circuit has a resistor joining that node to the datum node then corresponding to *any* choice of current injections b by current sources, there will be a unique valid set of voltages. That is, $Ax = b$ is solvable for any given b and there is only one solution. This is true because of the arrangement of the resistors and current sources in our ladder circuit, but is not true for every possible circuit consisting of current sources and resistors [95, section 2.6]. (See Exercise 4.5.)

4.1.4.3 Admittance matrix

Let us try to describe a way to write down the admittance matrix A for a general circuit. There are two types of entries in A:

- diagonal entries $A_{\ell\ell}, \ell = 1, \ldots, n$, and
- off-diagonal entries $A_{\ell k}, \ell \neq k, \ell = 1, \ldots, n, k = 1, \ldots, n$.

The diagonal entries $A_{\ell\ell}$ relate the current b_ℓ injected by the current source at node ℓ to the voltage x_ℓ at node ℓ, while an off-diagonal entry $A_{\ell k}$ relates the current b_ℓ injected by the current source at node ℓ to the voltage x_k at node k. Generalizing from (4.1)–(4.4), the ℓk-th entry of A can be written down as follows (see Exercise 4.6):

$$\forall \ell = 1, \ldots, n, \forall k = 1, \ldots, n, A_{\ell k} =$$

$$\begin{cases} \text{sum of the conductances} \\ \quad\text{connected to node } \ell, & \text{if } \ell = k, \\ \text{minus the conductance joining } \ell \text{ and } k, & \text{if } \ell \neq k \text{ and there is a resistor} \\ & \quad\text{between } \ell \text{ and } k, \\ \quad\quad\quad 0, & \text{if } \ell \neq k \text{ and there is no resistor} \\ & \quad\text{between } \ell \text{ and } k. \end{cases}$$

$$(4.9)$$

We have progressed from a particular example circuit to an expression for the admittance matrix corresponding to an *arbitrary* circuit. It is often easier to proceed in two such steps:

(i) develop a particular example, and
(ii) generalize,

than it is to formulate the general case directly. (It is often worthwhile to then check the general case by applying it to the particular example. See Exercise 4.6.)
We now analyze the general form of A, considering the four issues of:

- symmetry,
- sparsity,
- diagonal dominance, and
- changes in the admittance matrix.

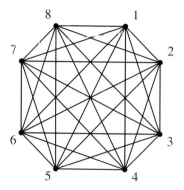

Fig. 4.4. A graph with $n = 8$ nodes and all $n(n - 1)/2 = 28$ possible branches. For clarity in this graph, each node is represented by a bullet •, while each branch is represented by a line. This is a different convention to that used in Figures 4.1–4.3.

Symmetry Our first observation about the general expression for entries of A is that A is **symmetric**. That is, the ℓk-th entry is equal to the $k\ell$-th entry. Symmetry in models can often help us to avoid some of the computational effort. Symmetry appears in many ways in many problems. (Ironically, a certain type of symmetry underlies the non-existence of direct algorithms for *non-linear* simultaneous equations. See [69].)

Sparsity If there is no component joining directly between a particular pair of nodes, then there is a zero entry in the corresponding entries of A. (We emphasized this in (4.9) by distinguishing the two cases of a resistor and no resistor between ℓ and k, although "no resistor" can be considered equivalent to a resistor having zero conductance between ℓ and k.)

Let us think about whether it is "typical" or "atypical" for an arbitrary pair of nodes to be joined directly by a resistor. In a circuit with n nodes, there are $n(n - 1)/2$ possible resistive branches in the circuit. (See Exercise 4.7.) For example, Figure 4.4 shows a graph with $n = 8$ nodes and $n(n - 1)/2 = 28$ branches. (In Figure 4.4, we adopt the convention that a node is drawn as a single •, while a branch is shown as a line segment. This convention is different to that used in the circuits shown in Figures 4.1–4.3, but is more standard in graph theory.)

While it is possible for an eight node circuit to have 28 branches, this is not typical. In typical circuits each node, except for the datum node, has relatively few branches connected to it. A typical number is three or four branches connected to each non-datum node, so that in total there are typically less than $2n$ branches. Therefore, for n large, the total number of branches is typically far less than the maximum possible number of $n(n - 1)/2$ branches.

In our example circuit in Figure 4.1, each node, except for the datum node, has two or three branches incident to it. However, $n = 5$, so that the maximum possible number of branches, $n(n - 1)/2 = 10$, is not much larger than 7, the actual number of branches. But consider if the ladder circuit was extended to, say, $n = 100$. We would then have approximately 200 branches, while the maximum possible number of branches in an $n = 100$ node circuit is $n(n - 1)/2 = 4450$, which is much larger than 200. In a large ladder circuit, the A matrix would be mostly zeros. We call a matrix **sparse** if most of its entries are zero.

The choice of datum node is arbitrary from the perspective of the circuit equations. However, by choosing as datum node the node with the most branches incident to it, we will minimize the number of non-zeros in the admittance matrix. (See Exercise 4.6.) This will turn out to reduce the computational effort involved in solving $Ax = b$.

In constructing a general algorithm for solving linear equations, we will be particularly concerned about the growth in computational effort as n increases. The characteristics of A for n relatively large is therefore of interest to us. Unfortunately, when we draw a picture of an example, such as in Figure 4.1, we will almost always have to draw a case for small n, which can be misleading in that the sparsity of typical, large circuits is not evident. We will see that sparsity in large-scale models allows considerable savings of computational effort.

Diagonal dominance Since each node in the ladder circuit is connected to the datum node by a resistor, each diagonal entry $A_{\ell\ell}$ in A is greater than the sum of the absolute values of the other entries in the ℓ-th column of A. Similarly, $A_{\ell\ell}$ is greater than the sum of the absolute values of the entries in the ℓ-th row of A. Such a matrix is called **strictly diagonally dominant**. (A matrix where each diagonal entry is greater than or equal to the sum of the absolute values of the other entries in its column and greater than or equal to the sum of the absolute values of the other entries in its row is called **diagonally dominant**. See Definition A.9.)

Changes in the admittance matrix When A is changed by the change of a resistor between nodes ℓ and k, the change ΔA is equal to $\Delta G_{\ell k}$, the change in conductance, times a matrix that has ones in the $\ell\ell$-th and kk-th places, minus ones in the ℓk-th and $k\ell$-th places and zeros elsewhere. That is:

$$
\Delta A = \Delta G_{\ell k}
\begin{bmatrix}
 & \overbrace{}^{\ell\text{-th column}} & & \overbrace{}^{k\text{-th column}} & & \\
 & 1 & & -1 & & \} \ \ell\text{-th row} \\
 & & & & & \\
 & -1 & & 1 & & \} \ k\text{-th row}
\end{bmatrix},
\qquad (4.10)
$$

where all entries in the matrix are zero except for the four non-zero entries that are explicitly shown.

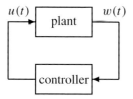

Fig. 4.5. A feedback con-
trol system applied to a
plant.

For a resistor between node ℓ and the datum node, ΔA is equal to $\Delta G_{\ell 0}$, the change in conductance, times a matrix that has ones in the $\ell\ell$-th place and zeros elsewhere. That is:

$$\Delta A = \Delta G_{\ell 0} \begin{bmatrix} & \overset{\ell\text{-th column}}{\frown} & \\ & 1 & \\ & & \} \ \ell\text{-th row} \end{bmatrix}, \tag{4.11}$$

where again all entries in the matrix are zero except for the one non-zero entry that is explicitly shown.

If we apply this argument to building up an admittance matrix "from scratch," we can think of A as the sum of terms, each one of which is of the form of (4.10) or (4.11). This is explored in Exercise 4.8, where it is shown that the admittance matrix is positive definite. (Positive definiteness of the admittance matrix can also be established from the fact that a resistive circuit dissipates energy. See [39] for a sketch of the proof.)

4.2 Control of a discrete-time linear system

4.2.1 Motivation

This case study assumes some familiarity with control theory and, in particular, discrete time linear systems. This material is covered in greater detail in many books on control and linear systems, such as [55, 89].

In a classical feedback control system, we use the outputs of plant sensors as the inputs to integrators and other elements, which in turn control the plant actuators. The integrators and other elements form an analog feedback control system, one purpose of which is to regulate plant behavior despite, say, random fluctuations in plant variables such as temperature and material inputs. The situation is illustrated in Figure 4.5, where the outputs of the plant are labeled $w(t)$ and the inputs are labeled $u(t)$.

In a digital control system, the function of the feedback control system is per-formed by a microprocessor or digital controller. Unlike an analog controller, which accepts continuous time inputs, the digital controller receives samples of plant sensor outputs that are sampled at discrete intervals of time. Furthermore, the controls to the plant actuators are usually adjusted only at discrete intervals of time.

In this case study, we are going to investigate the conditions under which we can shift the **state** of the system to a desired final value by adjusting the inputs over a sequence of time intervals. That is, we are going to consider the **open loop** control of the system. In a **feedback controller**, as illustrated in Figure 4.5, we use the output (or the state) of the system to decide on the controls. Furthermore, in **optimal control** we recognize the costs of certain control actions and states. These issues pose somewhat different problems to the one we investigate in this case study; however, several of the issues turn out to be similar. See Exercise 12.5 and [11, 16, 55, 63, 64, 89] for further details.

We formulate the problem in Section 4.2.2, consider changes in Section 4.2.3, and explore the problem characteristics in Section 4.2.4.

4.2.2 Formulation

4.2.2.1 Variables

In a typical plant, there will be some variables of interest that we are trying to control or manipulate. As with our DC circuit case study in Section 4.1, there will usually be a much larger number of variables that we could potentially study. In a given application we must identify a subset of salient issues of interest.

To represent the behavior of the plant over time, we will typically find that we must consider a somewhat larger group of variables than just the variables of interest, in order to be able to *predict* the behavior of the variables of interest over time. In particular, we define the **state** [89, section 5-1] of the system to be the smallest set of variables, w say, such that:

- w includes all the variables of interest as a sub-vector, and
- knowledge of the value of w for any particular time $t = t_0$, together with the knowledge of the values of the input $u(t)$ to the plant for $t_0 \leq t \leq t_1$ completely specifies the value of w for any time $t_0 \leq t \leq t_1$.

The usual convention in control theory is to use x for the state and for x to be a vector of length n; however, we will need to use the variable x to stand for something else and so we use w for the state. In Figure 4.5 we have implicitly assumed that all of the state can be measured directly as outputs of the plant; however, in some applications this is not the case and they must be estimated.

To illustrate the choice of variables of interest, consider the temperature of the system. In a chemical reactor, we might be very interested in the temperature because it strongly affects the yield of chemicals. We would model temperature explicitly as one of the state variables in such a system. In contrast, in an electromechanical actuator we may not be so interested in the temperature, except to verify that it does not exceed a limit. We might not explicitly model the temperature of the actuator. In summary, the choice of the variables of interest depends on our interests and judgment.

However, we may find that the specification of the state requires us to include more variables than just the variables of particular interest. For example, in the chemical reactor, we may not be directly interested in the pressure, say, but the pressure may be necessary to include in our state in order to be able to predict the future temperature.

The definition of state embodies the idea that as time elapses, the value of the state w changes and so we must think of the state as being a function of time, so that, for example, $w : \mathbb{R} \to \mathbb{R}^m$. In our application, we are considering a digital controller that samples the state of the system and updates the controls every, say, T units of time. Therefore, we concentrate on the values of the state at these sampling instants. We can write $w(kT)$ for the value of w at the k-th sampling instant.

We write u for the vector of plant actuator or process inputs. For simplicity, we assume that u is actually only one-dimensional; however, in general we can consider *multi*-input systems [19]. Again, since we are considering a digital controller that updates the controls every T units of time, we will assume that the input stays constant between the k-th and $(k + 1)$-th sampling instant so that $u(t) = u(kT), kT \leq t < (k + 1)T$. We will write $u(kT)$ for the value of the actuator input set by the controller for the period between the k-th and $(k + 1)$-th sampling instant.

4.2.2.2 Behavior of system

By the definition of state, the value of the state at time kT and the value of the input for period k determines value of the state at time $(k + 1)T$. That is, for each k, there exists a function $\phi^{(k)} : \mathbb{R}^m \times \mathbb{R} \to \mathbb{R}^m$ describing the behavior of the plant such that for any value of the state $w(kT)$ at time kT and any value of the input $u(kT)$ applied between time kT and time $(k + 1)T$, we can calculate the value of the state at time $(k + 1)T$ by:

$$\forall k \in \mathbb{Z}, \; w([k + 1]T) = \phi^{(k)}(w(kT), u(kT)).$$

If, for each k, the function $\phi^{(k)}$ is linear of the form:

$$\forall k \in \mathbb{Z}, \forall w \in \mathbb{R}^m, \forall u \in \mathbb{R}, \; \phi^{(k)}(w, u) = G^{(k)}w + h^{(k)}u,$$

with $G^{(k)} \in \mathbb{R}^{m \times m}$ and $h^{(k)} \in \mathbb{R}^m$, then we say that the system is **linear**.

If the functions $\phi^{(k)}$ are actually independent of k, so that $\phi : \mathbb{R}^m \times \mathbb{R} \to \mathbb{R}^m$ can be found satisfying:

$$\forall k \in \mathbb{Z}, \, w([k+1]T) = \phi(w(kT), u(kT)),$$

then we say that the system is **time-invariant**.

If the functions $\phi^{(k)}$ are linear and independent of k and so of the form:

$$\forall k \in \mathbb{Z}, \forall w \in \mathbb{R}^m, \forall u \in \mathbb{R}, \, \phi^{(k)}(w, u) = Gw + hu,$$

for $G \in \mathbb{R}^{m \times m}$ and $h \in \mathbb{R}^m$, then the system is **linear time-invariant**. A typical situation where this relationship occurs is when the plant can be described by a linear differential equation with time-invariant coefficients. (See Exercise 4.9.) Even when the behavior of the system is more complicated, linear approximations are often made to gauge qualitative behavior.

In summary, linear time-invariant systems behave according to the **difference equation** [55, section 2.3.3]:

$$\forall k \in \mathbb{Z}, \, w([k+1]T) = Gw(kT) + hu(kT). \tag{4.12}$$

The matrix G is called the **state transition matrix**. We will restrict ourselves to linear time-invariant systems for the rest of the case study.

4.2.2.3 *Changing the state of the system*

Let us suppose that at time $kT = 0$ the plant is in state $w(0) \in \mathbb{R}^m$ and that we would like it instead to be in some other desired final state $w^{\text{desired}} \in \mathbb{R}^m$. Our only way to affect the state is by adjusting the input variable $u(kT)$, perhaps over several successive time-steps. So, our problem is to pick values of the inputs to bring the state to the desired final value after, say, n time-steps. In other words, our goal is to have $w(nT) = w^{\text{desired}}$.

If we allow ourselves to choose values for $u(kT)$ for $k = 0, 1, \ldots (n-1)$, then we will find that:

$w(nT)$

$$
\begin{aligned}
=\ & Gw([n-1]T) + hu([n-1]T), \\
& \quad \text{on substituting for } w(nT) \text{ from (4.12) for } k = n-1, \\
=\ & (G)^2 w([n-2]T) + Ghu([n-2]T) + hu([n-1]T), \\
& \quad \text{on substituting for } w([n-1]T) \text{ from (4.12) for } k = n-2, \\
=\ & (G)^3 w([n-3]T) + (G)^2 hu([n-3]T) + Ghu([n-2]T) + hu([n-1]T), \\
& \quad \text{continuing,}
\end{aligned}
$$

$$
\vdots \qquad \vdots
$$

$$
=\ (G)^n w(0) + \sum_{k=0}^{n-1} (G)^{n-1-k} hu(kT),
$$

where $(G)^k$ means the k-fold product of G with itself. Let us write:

- A for the $m \times n$ matrix with k-th column equal to $(G)^{n-1-k}h$, (where k ranges from 0 to $(n-1)$), so that:

$$
A = \begin{bmatrix} (G)^{n-1}h & (G)^{n-2}h & \cdots & Gh & h \end{bmatrix},
$$

- x for the n-vector with k-th entry $u(kT)$, (where k again ranges from 0 to $(n-1)$), so that:

$$
x = \begin{bmatrix} u(0T) \\ u(1T) \\ \vdots \\ u([n-2]T) \\ u([n-1]T) \end{bmatrix},
$$

and
- $b = w^{\text{desired}} - (G)^n w(0)$.

Then, we have that:

$$
Ax = \begin{bmatrix} (G)^{n-1}h & (G)^{n-2}h & \cdots & Gh & h \end{bmatrix} \begin{bmatrix} u(0T) \\ u(1T) \\ \vdots \\ u([n-2]T) \\ u([n-1]T) \end{bmatrix},
$$

$$
= \sum_{k=0}^{n-1} (G)^{n-1-k} hu(kT).
$$

Therefore, $w(nT)$ will be equal to w^{desired} if:

$$w^{\text{desired}} = (G)^n w(0) + Ax.$$

That is, $w(nT)$ will be equal to w^{desired} if:

$$Ax = b. \tag{4.13}$$

The right-hand side of (4.13) is a constant vector (for a given n, $w(0)$, and w^{desired}). That is, we have a set of linear simultaneous equations in x with coefficient matrix A and right-hand side b.

4.2.2.4 Example

Suppose that $n = 2$, $m = 2$, and:

$$h = \begin{bmatrix} 0 \\ 1 \end{bmatrix},$$

$$G = \begin{bmatrix} 0 & 1 \\ 1 & 1 \end{bmatrix},$$

$$w(0) = \begin{bmatrix} 1 \\ 3 \end{bmatrix},$$

$$w^{\text{desired}} = \begin{bmatrix} 3 \\ 7 \end{bmatrix}.$$

Then:

$$\begin{aligned} A &= \begin{bmatrix} Gh & h \end{bmatrix}, \\ &= \begin{bmatrix} 1 & 0 \\ 1 & 1 \end{bmatrix}, \\ b &= w^{\text{desired}} - (G)^n w(0), \\ &= \begin{bmatrix} -1 \\ 0 \end{bmatrix}. \end{aligned}$$

Solving for x, we obtain:

$$\begin{aligned} \begin{bmatrix} u(0T) \\ u(1T) \end{bmatrix} &= x, \\ &= \begin{bmatrix} -1 \\ 1 \end{bmatrix}. \end{aligned}$$

For this particular example, it is also possible to find a control $u(0T)$ that achieved the desired final state in one time-step; that is, for $n = 1$. In particular, there is a solution $u(0T) \in \mathbb{R}$ to:

$$hu(0T) = w^{\text{desired}} - Gw(0),$$

namely $u(0T) = 3$. However, it will typically require more than one time-step to achieve a desired state. (See Exercise 4.10.)

4.2.2.5 Labeling of vector and matrix entries

The entries of x and the columns of A are labeled from 0 to $(n - 1)$. This contrasts with the labeling of the variables in the case study in Section 4.1 where the entries were labeled from 1 to n. In general, we can label the entries of a vector in any way we choose. For example, in some applications it may be convenient to use labels that are not consecutive. Moreover, we call x an n-vector if it has n entries, independent of the way in which the entries are labeled. For notational simplicity, most of the theoretical development in this book will assume that entries are labeled consecutively from 1. The changes in algorithms to accommodate alternative labelings, such as used in this discrete-time linear system case study, are straightforward and we will not mention them explicitly.

4.2.3 Changes

In this section we will consider changes in the initial state, the desired state, and in the system.

4.2.3.1 Initial and desired state

If w^{desired} or $w(0)$ change, then the right-hand side b in the linear equation (4.13) will also change correspondingly. We would like to be able to solve for the required values of actuator settings for a variety of possible values of w^{desired} and $w(0)$.

4.2.3.2 System

If the behavior of the plant changes, then the state transition matrix G and therefore the coefficient matrix A in (4.13) will change. This can occur as plant components drift over time or because of manufacturing tolerances. We can think of A as differing from a correct plant model by ΔA. We may be interested in calculating the *error* between $w(nT)$ and w^{desired} that results from using the incorrect model of the system. (See Exercise 5.50.)

4.2.4 Problem characteristics

As in our previous example, let us try to discern some characteristics of the linear equation.

4.2.4.1 Numbers of variables and equations

Unlike the previous case study in Section 4.1, the number of variables does not necessarily equal the number of equations. The number of variables is n, which is equal to the number of entries in x, but the number of equations is equal to m, which is the number of entries in b.

4.2.4.2 Solvability

It is not always the case that (4.13) is solvable. Solvability will depend on G, h, $w^{desired}$, $w(0)$, and on n. For some G, h, $w^{desired}$, and $w(0)$, it is impossible to achieve the desired state, whatever the value of n. For other choices, there is a minimum value of n, below which it is impossible, and at or above which it is possible to solve (4.13). (See Exercise 4.10.)

4.2.4.3 Coefficient matrix

In general, the coefficient matrix is not symmetric and has different numbers of rows and columns. The coefficient matrix is not necessarily diagonally dominant nor is it necessarily sparse, unless the physical system displays these characteristics. For example, if the plant consists of several independent and uncoupled sub-systems then A will be sparse, but if the plant state variables are tightly linked to each other then A will not be sparse.

Exercises

Analysis of a direct current linear circuit

4.1 Suppose that the circuit of Figure 4.1 is modified by adding a voltage source of value V_{23} between nodes 2 and 3, with a positive value of V_{23} meaning that the voltage at node 2 is positive with respect to the voltage at node 3. This voltage source is in parallel with the resistor between nodes 2 and 3 as illustrated in Figure 4.6. Define a matrix A' and vectors x' and b' so that the solution of the system $A'x' = b'$ yields the solution to the modified circuit. This formulation is called **modified nodal analysis**. (Hint: The vector x' no longer represents only the unknown voltages but now also includes an unknown current. The vector b' no longer represents only known currents but now also includes a known voltage. The vector x' will be of length 5. Its first four entries will be the entries of x, while its last entry will be the current, i_{23}, flowing into the voltage source at node 2 and flowing out of the voltage source at node 3. The vector b' will also be of length 5 and its first four entries will be the entries of b. The last entry will be V_{23}.)

4.2 Show that there are at least as many branches as nodes in a connected circuit with at least one loop.

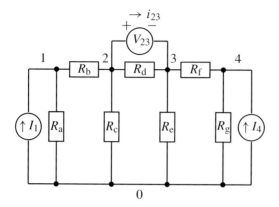

Fig. 4.6. Circuit for Exercise 4.1. The ladder circuit of Figure 4.1 with a voltage source V_{23} added between nodes 2 and 3.

4.3 Write down the equation for the datum node for the circuit of Figure 4.1. Show that it is redundant. That is, show that the equation for the datum node is a linear combination of the equations for the other nodes. (See Definition A.54.)

4.4 Show that the equation $Ax = b$ with $A, x,$ and b defined in (4.5)–(4.7) reproduces (4.1)–(4.4).

4.5 Give an example of a circuit with current sources for which Kirchhoff's current law cannot be satisfied.

4.6 In this exercise we verify the form of the admittance matrix A for the circuit.
  (i) Show that (4.9) is the correct definition of A to satisfy $Ax = b$ for an arbitrary circuit consisting of resistors and current sources.
  (ii) Show that (4.5) is the special case of (4.9) for to the circuit of Figure 4.1.
  (iii) Show that by choosing the datum node to be the node with the most branches incident to it, we can minimize the number of non-zeros in the admittance matrix.

4.7 Show that there are at most $n(n-1)/2$ branches in an n node circuit.

4.8 In this exercise we verify the form of the change in the admittance matrix ΔA.
  (i) Show that ΔA defined in (4.10) is of the form $\Delta A = \gamma w w^{\dagger}$ for some $\gamma \in \mathbb{R}$, $w \in \mathbb{R}^n$.
  (ii) Show that ΔA defined in (4.11) is of the form $\Delta A = \gamma w w^{\dagger}$ for some $\gamma \in \mathbb{R}$, $w \in \mathbb{R}^n$.
  (iii) Suppose that there are n non-datum nodes and $r \geq n$ resistors in the circuit. Suppose that the circuit is connected and that current can flow in it and that each pair of nodes is joined directly by at most one resistor. (That is, there are no resistors in parallel.) Show that we can write A in the form $A = WGW^{\dagger}$, where:
   - $W \in \mathbb{R}^{n \times r}$ has n linearly independent rows, and
   - $G \in \mathbb{R}^{r \times r}$ is a diagonal matrix with positive diagonal entries.
  (iv) Show that A is positive definite. (Hint: Use Exercise 2.27, Part (vi).)

Control of a discrete-time linear system

4.9 Consider a continuous time linear system described by the differential equation:

$$\forall t, \frac{dw}{dt}(t) = \hat{A}w(t) + \hat{b}u(t),$$

where $\hat{A} \in \mathbb{R}^{m \times m}$ and $\hat{b} \in \mathbb{R}^{m}$. This describes a linear time-invariant system in continuous time [55, 89]. The solution to this differential equation on an interval $[t_0, t_1]$, given that $w(t_0)$ is known and $u(\tau)$ is specified for $t_0 \leq \tau < t_1$, is:

$$w(t_1) = \exp(\hat{A}(t_1 - t_0))w(t_0) + \exp(\hat{A}t_1) \int_{t_0}^{t_1} \exp(-\hat{A}\tau)\hat{b}u(\tau) \, d\tau,$$

where $\exp(\hat{A}t) \in \mathbb{R}^{m \times m}$ is the **matrix exponential** [89, section 5-6]. Show that the behavior of this system can be described in the form (4.12) when the control input is constant for $kT \leq t < (k + 1)T$. (Hint: Set $t_0 = kT$ and $t_1 = (k + 1)T$. Make a substitution of variables in the integral on the right-hand side noting that for any $t, t' \in \mathbb{R}$, $\exp(\hat{A}(t + t')) = \exp(\hat{A}t) \exp(\hat{A}t')$.)

4.10 Choose values of G, h, w^{desired}, and $w(0)$ for which (4.13) has:

   (i) no solution, no matter what value of n;
   (ii) a solution for $n = 2$ but no solution for $n = 1$.

There is no need to prove that your values of G, h, w^{desired}, and $w(0)$ correspond to differential equations describing a physical plant.

5

Algorithms for linear simultaneous equations

In Chapter 4, we introduced two case studies that involved solution of linear equations. In the case of the direct current linear circuit case study described in Section 4.1, we could envision having to solve for a large circuit. We will encounter linear equations throughout the book in the development of algorithms and will find that we need to repeatedly solve potentially very large systems. Therefore, in this chapter we will consider generally how to solve large systems of the form:

$$Ax = b. \tag{5.1}$$

The matrix A is called the **coefficient matrix**, while b is called the **right-hand side vector**.

In Section 5.1 we will discuss solution of (5.1) by **inversion** of the coefficient matrix using **Cramér's rule** [55, appendix A.7]. Because of the computational burden and other problems with using Cramér's rule, we seek alternatives to inverting A that nevertheless allow us to solve $Ax = b$. **Gaussian elimination** is one such technique.

Gaussian elimination is a computationally efficient approach to solving $Ax = b$ in that the computational effort to perform Gaussian elimination is bounded by a polynomial in n, in fact by a cubic polynomial in n. This means that, for large n, Gaussian elimination takes far less computational effort than Cramér's rule. (See [40] for a more precise account of the notion of computational effort: here we will appeal to an intuitive notion. For example, as discussed in Section 2.4, we will assume that the arithmetic operations of addition, subtraction, multiplication, and division are basic operations requiring a constant amount of computational effort per operation.)

Before discussing Gaussian elimination in more detail, we will first consider in Section 5.2 the two special cases of (5.1) where the coefficient matrix is:

- **upper triangular** (that is, all the entries below the diagonal are zero), and

186

• **lower triangular** (that is, all the entries above the diagonal are zero).

For example, the matrix $U \in \mathbb{R}^{3 \times 3}$ defined by:

$$U = \begin{bmatrix} 2 & 3 & 4 \\ 0 & -\frac{9}{2} & -9 \\ 0 & 0 & 1 \end{bmatrix}, \qquad (5.2)$$

is upper triangular and the matrix $L \in \mathbb{R}^{3 \times 3}$ defined by:

$$L = \begin{bmatrix} 1 & 0 & 0 \\ \frac{7}{2} & 1 & 0 \\ 4 & \frac{2}{3} & 1 \end{bmatrix}, \qquad (5.3)$$

is lower triangular.

Linear simultaneous equations involving upper or lower triangular matrices are called **triangular systems**. Solution of triangular systems will also prove useful in the case of more general systems of linear simultaneous equations.

In Section 5.3 we will develop a method for solving our circuit problem and related problems. In particular, we will assume that A is $n \times n$ with rows and columns labeled consecutively from 1 to n and that there is exactly one solution to the simultaneous equations. The method we develop will be similar to Gaussian elimination and involves factorizing A into the product of a lower triangular matrix L and an upper triangular matrix U. We will then use the development from Section 5.2 to solve the simultaneous equations $Ax = b$ re-written in terms of the triangular factors of A. The calculations described in Sections 5.2 and 5.3 will constitute a direct algorithm for solving linear simultaneous equations and will be applied to the circuit case study from Section 4.1.

Because **symmetry** arises repeatedly in the systems we solve, we will pay special attention to symmetric systems, describing a variant of LU factorization for symmetric systems in Section 5.4. In Section 5.5 we will discuss how to exploit the **sparsity** of equations.

In Section 5.6, we will discuss both **sensitivity analysis** and also how to re-solve a system after a **large change** in either the right-hand side or the coefficient matrix. In Section 5.7 we will discuss the related issue of **ill-conditioning**, introducing an alternative factorization to the LU factorization, called the QR factorization, that can be useful in ill-conditioned problems.

In Section 5.8 we will discuss **non-square** systems, including the discrete-time linear system case study from Section 4.2. Finally, in Section 5.9, we will briefly discuss **iterative** solution methods for linear equations.

In the rest of the book, we will find that the solution of linear equations is embedded in algorithms for solving non-linear equations and algorithms for optimization.

The choice of linear equation solver used in the software implementing these algorithms may be beyond our control. In this case, we should think of LU factorization as representative of linear equation solving techniques; however, an alternative factorization or an iterative method may actually be used in a particular application. Sometimes we will have direct control over the linear solution method and the description of LU factorization and variations will then provide information to aid in making sensible decisions. In all cases, understanding of LU factorization and alternatives will help in the formulation of problems; however, we will gloss over many details and you may need to delve more deeply into these techniques for special purpose applications.

The key issues discussed in this chapter are:

- solution of **triangular systems** and **factorization of matrices**,
- **computational effort** and particular features of problems, such as **symmetry** and **sparsity** that can reduce the necessary computational effort,
- **sensitivity analysis** and **ill-conditioning**, and
- solution of **non-square systems**.

5.1 Inversion of coefficient matrix

Suppose that A is **invertible** with inverse A^{-1}. (See Definition A.49.) Let $x = A^{-1}b$. Then:

$$
\begin{aligned}
Ax &= AA^{-1}b, \\
&= \mathbf{I}b, \text{ by definition of inverse,} \\
&= b, \text{ by definition of } \mathbf{I}.
\end{aligned}
$$

That is, knowledge of the inverse allows us to calculate the solution of the linear equations explicitly. **Cramér's rule** [55, appendix A.7] shows how to calculate A^{-1} in terms of the **determinant** of A and the determinants of sub-matrices of A. (See Definition A.10.) In particular, Cramér's rule says that if the determinant of A is not equal to zero then A is invertible and the $k\ell$-th entry of A^{-1} is given by:

- $(-1)^{\ell+k}$ times
- the determinant of the matrix obtained from A by deleting its ℓ-th row and k-th column, divided by
- the determinant of A.

The most straightforward algorithm for calculating the determinant involves a recursive calculation requiring computational effort of on the order of $n!$ arithmetic operations for an $n \times n$ matrix. For a 2×2 or a 3×3 matrix, this calculation is fast. However, if n is large then Cramér's rule is impractical because the calculation of determinants is too computationally intensive. In the rest of this chapter we describe alternative approaches that require much less effort for large n.

Nevertheless, Cramér's rule can be extremely useful for:

- proving properties of matrices, as we will see, for example, in Exercise 5.8,
- inverting small matrices, since Cramér's rule allows the inverse to be written down explicitly, and
- inverting specific types of matrices. (See Exercise 5.1.)

For example, for $A \in \mathbb{R}^{2 \times 2}$, if $\det(A) = A_{11}A_{22} - A_{12}A_{21} \neq 0$ then A is invertible and by Cramér's rule:

$$A^{-1} = \frac{1}{A_{11}A_{22} - A_{12}A_{21}} \begin{bmatrix} A_{22} & -A_{12} \\ -A_{21} & A_{11} \end{bmatrix}. \tag{5.4}$$

5.2 Solution of triangular systems

A set of linear simultaneous equations where the coefficient matrix is lower or upper triangular is called a **triangular system**. We will show that:

- a triangular system with a lower triangular coefficient matrix can be solved using a direct algorithm called **forwards substitution**, while
- a triangular system with an upper triangular matrix can be solved using a direct algorithm called **backwards substitution**.

Both algorithms involve eliminating variables by re-arranging equations, as introduced in Section 3.2.2.2. In Sections 5.2.1–5.2.3, we will discuss forwards and backwards substitution and the computational effort. The development is based on [70, appendix C].

5.2.1 Forwards substitution

5.2.1.1 Analysis

Consider an $n \times n$ matrix L that is lower triangular. That is, $L_{\ell k} = 0, \forall \ell < k$, assuming that the rows and columns of L are labeled consecutively from 1 through n. Suppose that we want to find $y^\star \in \mathbb{R}^n$ satisfying $Ly = b$. (We will see in Section 5.3.1 the reason for choosing y as the decision variable instead of x.) This is a special case of our general Problem (5.1) of solving linear simultaneous equations.

In this special case, we have that:

$$
\begin{aligned}
b_1 &= L_{11}y_1, \\
b_2 &= L_{21}y_1 + L_{22}y_2, \\
b_3 &= L_{31}y_1 + L_{32}y_2 + L_{33}y_3, \\
&\vdots \qquad \vdots \\
b_n &= L_{n1}y_1 + L_{n2}y_2 + L_{n3}y_3 + \cdots + L_{nn}y_n.
\end{aligned}
$$

Assume that the diagonal entries of L are non-zero. The first row can be re-arranged to give:

$$
y_1 = \frac{b_1}{L_{11}}.
$$

We have re-arranged the first row to allow elimination of y_1 by expressing it as a function of the rest of the entries in y, as first discussed in Section 3.2.2.2. In fact, since the function on the right-hand side of this expression is independent of the rest of the entries in y, this means that we can directly evaluate y_1. Re-arranging the second row then yields:

$$
y_2 = \frac{b_2 - L_{21}y_1}{L_{22}},
$$

which can be calculated once y_1 is known. Similarly, once $y_1, \ldots, y_{\ell-1}$ are known, y_ℓ can be calculated as:

$$
y_\ell = \frac{b_\ell - \sum_{k=1}^{\ell-1} L_{\ell k} y_k}{L_{\ell\ell}}. \tag{5.5}
$$

This process is called **forwards substitution** since it starts at the first entry of the vector y and works forward towards the last entry of y. We can eliminate each y_ℓ in turn by expressing it in terms of $y_1, \ldots, y_{\ell-1}$.

5.2.1.2 Example

Consider the lower triangular matrix defined in (5.3):

$$
L = \begin{bmatrix} 1 & 0 & 0 \\ \frac{7}{2} & 1 & 0 \\ 4 & \frac{2}{3} & 1 \end{bmatrix},
$$

and the vector:

$$
b = \begin{bmatrix} 9 \\ 18 \\ 28 \end{bmatrix}.
$$

Performing forwards substitution on $Ly = b$, we obtain:

$$y^\star = \begin{bmatrix} 9 \\ -\frac{27}{2} \\ 1 \end{bmatrix}.$$

5.2.2 Backwards substitution

5.2.2.1 Analysis

Now consider an upper triangular matrix U. That is, $U_{\ell k} = 0, \forall \ell > k$, again assuming that the rows and columns of U are labeled consecutively from 1 through n. Suppose that $y \in \mathbb{R}^n$ is given and we want to solve $Ux = y$. (We will see in Section 5.3.1 the reason for using y as the symbol for the right-hand side vector instead of b.) This is another special case of our general problem of solving linear simultaneous equations. We have that:

$$U_{11}x_1 + \cdots + U_{1,n-2}x_{n-2} + U_{1,n-1}x_{n-1} + U_{1,n}x_n = y_1,$$
$$\vdots \quad \vdots$$
$$U_{n-2,n-2}x_{n-2} + U_{n-2,n-1}x_{n-1} + U_{n-2,n}x_n = y_{n-2},$$
$$U_{n-1,n-1}x_{n-1} + U_{n-1,n}x_n = y_{n-1},$$
$$U_{n,n}x_n = y_n.$$

Again assume that the diagonal entries of U are non-zero. In this case, we can re-arrange the *last* equation to give:

$$x_n = \frac{y_n}{U_{n,n}}.$$

Then we can substitute into the second last equation and obtain:

$$x_{n-1} = \frac{y_{n-1} - U_{n-1,n}x_n}{U_{n-1,n-1}}.$$

Once $x_n, \ldots, x_{\ell+1}$ are known, the entry x_ℓ can be calculated as:

$$x_\ell = \frac{y_\ell - \sum_{k=\ell+1}^{n} U_{\ell k}x_k}{U_{\ell\ell}}.$$

Calculation of x in this way is called **backwards substitution** because it starts at the last entry of x and works backwards. Again, we can eliminate each x_ℓ in turn by expressing it in terms of $x_{\ell+1}, \ldots, x_n$.

5.2.2.2 Example

Consider the upper triangular matrix defined in (5.2):

$$U = \begin{bmatrix} 2 & 3 & 4 \\ 0 & -\frac{9}{2} & -9 \\ 0 & 0 & 1 \end{bmatrix},$$

and the vector:

$$y = \begin{bmatrix} 9 \\ -\frac{27}{2} \\ 1 \end{bmatrix}.$$

Performing backwards substitution on $Ux = y$, we obtain:

$$x^{\star} = \begin{bmatrix} 1 \\ 1 \\ 1 \end{bmatrix},$$

which is the solution to $Ux = y$.

5.2.3 Computational effort

5.2.3.1 Forwards substitution

Forwards substitution calculates y_ℓ, $\ell = 1, \ldots, n$. Calculation of y_1 requires a division. Calculation of each y_ℓ for $\ell = 2, \ldots, n$ requires: $(\ell - 1)$ multiplications, $(\ell - 2)$ additions, a subtraction, and a division. In total, this is:

$$\sum_{\ell=2}^{n} (\ell - 1) = \frac{1}{2}(n - 1)n \text{ multiplications,}$$

$$\sum_{\ell=2}^{n} (\ell - 2) = \frac{1}{2}(n - 2)(n - 1) \text{ additions,}$$

$$(n - 1) \text{ subtractions, and}$$

$$n \text{ divisions.}$$

5.2.3.2 Backwards substitution

Backwards substitution calculates x_ℓ, $\ell = n, \ldots, 1$. Calculation of x_n requires a division. Calculation of each x_ℓ for $\ell = (n - 1), \ldots, 1$ requires: $(n - \ell)$ multiplications, $(n - \ell - 1)$ additions, a subtraction, and a division. In total, this is:

$$\sum_{\ell=1}^{n-1}(n - \ell) \;=\; \frac{1}{2}(n - 1)n \text{ multiplications,}$$

$$\sum_{\ell=1}^{n-1}(n - \ell - 1) \;=\; \frac{1}{2}(n - 2)(n - 1) \text{ additions,}$$

$$(n - 1) \text{ subtractions, and}$$

$$n \text{ divisions.}$$

5.2.3.3 Overall

The overall effort for forwards and backwards substitution is therefore on the order of the *square* of the number of variables. Given that accessing entries of vectors and matrices and that addition, subtraction, multiplication, and division are basic operations of our computer then we have described direct algorithms for solving upper triangular systems and for solving lower triangular systems.

5.3 Solution of square, non-singular systems

In this section, we will develop a method to solve linear simultaneous equations where the coefficient matrix is square and non-singular. We will first show in Section 5.3.1 that *if* a matrix A had been factorized into an appropriate lower triangular matrix L and upper triangular matrix U, so that $A = LU$, then it would be possible to solve the system $Ax = b$. The triangular factors allow the simultaneous equations $Ax = b$ to be solved in terms of two triangular systems involving L and U.

Then we will show in Section 5.3.2 that it is possible to factorize a square, non-singular matrix A into appropriate L and U. In Section 5.3.3 we will analyze the computational effort. Then in Section 5.3.4 we will discuss some variations on the basic LU factorization.

5.3.1 Combining forwards and backwards substitution

Suppose that we can factorize $A \in \mathbb{R}^{n \times n}$ into LU, with L lower triangular and U upper triangular. Then, we have:

$$
\begin{aligned}
b \;&=\; Ax, \text{ the equation we want to solve,}\\
&=\; LUx, \text{ since } A = LU,\\
&=\; L(Ux),\\
&=\; Ly,
\end{aligned}
$$

where $y = Ux$. So, if L has non-zero diagonal entries then we can solve for y in the equation $b = Ly$ using the algorithm developed in Section 5.2.1. If U has non-zero diagonal entries then we can solve for x in the equation $y = Ux$ using the algorithm developed in Section 5.2.2. We obtain a solution x to the system $b = Ax$. We conclude that if we can factorize a matrix A into LU with L lower triangular and U upper triangular and with both L and U having non-zero diagonal entries, then we can solve the system $b = Ax$. We have transformed the problem of solving $Ax = b$ into the solution of three successive problems:

(i) factorization of A into LU,

(ii) forwards substitution to solve $Ly = b$, and

(iii) backwards substitution to solve $Ux = y$.

If A is singular then we cannot factorize A into LU with L and U having non-zero diagonal entries. To see this, note that if we *could* factorize a singular matrix into LU with non-zero diagonals, then we could solve the linear equation $Ax = b$ for any b and hence invert A, which is a contradiction. (See Exercise 5.2.) In the following sections we will consider non-singular A.

5.3.2 LU factorization

In the following discussion we will specify a series of "stages" to implement the algorithm. To index the stages, we will add a superscript in parentheses, so that $M^{(j)}$ represents a matrix that is defined in the j-th stage of the algorithm. This is the same notation used for the iterates in an iterative algorithm as introduced in Section 2.4.2.1; however, here the calculation will terminate in a finite number of stages (in fact, in $(n-1)$ stages) with the exact answer (under the assumption that we could calculate in infinite precision arithmetic.) That is, we describe a direct algorithm for factorizing the matrix A. (Some authors refer to this algorithm as an iterative algorithm since it involves a number of stages and, indeed, we will number the matrices calculated in the stages in a similar fashion to the labeling we use in iterative algorithms. However, we reserve the word iterative for algorithms that generate a sequence of approximate solutions that converge to the exact answer. We have used the word "stages" to distinguish the matrices $M^{(j)}$ from the iterates in an iterative algorithm.)

To factorize A, we will multiply it on the left by the non-singular matrices $M^{(1)}, M^{(2)}, \ldots, M^{(n-1)}$ such that the matrix $U = M^{(n-1)}M^{(n-2)} \cdots M^{(1)}A$ is upper triangular. At each stage, the product:

$$A^{(j+1)} = M^{(j)}M^{(j-1)} \cdots M^{(1)}A,$$

will become successively "closer" to being upper triangular. In particular, each successive product will have one more column below the diagonal "zeroed."

There are many choices of the $M^{(j)}$, $j = 1, \ldots, n - 1$, that will zero out successive columns below the diagonal. We will choose the $M^{(j)}$, $j = 1, \ldots, n - 1$, to have two additional properties:

   (i) each $M^{(j)}$ will be lower triangular (and therefore, by Exercise 5.1, have a lower triangular inverse), and

  (ii) $[M^{(j)}]^{-1}$, the inverse of $M^{(j)}$, will be easy to compute.

We let:

$$L = [M^{(n-1)} \cdots M^{(1)}]^{-1}.$$

Recalling that the inverse of a product of matrices is equal to the product of the inverses in reverse order, we have:

$$
\begin{aligned}
L &= [M^{(n-1)} \cdots M^{(1)}]^{-1}, \\
  &= [M^{(1)}]^{-1} \cdots [M^{(n-1)}]^{-1}.
\end{aligned}
$$

That is, L is the product of $(n - 1)$ lower triangular matrices and, by Exercise 5.3, L is also lower triangular. Finally,

$$
\begin{aligned}
LU &= [M^{(n-1)} \cdots M^{(1)}]^{-1} M^{(n-1)} \cdots M^{(1)} A, \text{ by definition,} \\
   &= A,
\end{aligned}
$$

so that A has been factorized into LU. Using the terminology of Section 3.3.1, we could say that we have pre-conditioned the system $Ax = b$ by successively multiplying by the pre-conditioning matrices $M^{(1)}$, $M^{(2)}$, $\ldots$, $M^{(n-1)}$. We will return to this interpretation in Section 5.7.3.

5.3.2.1 First stage

Pivoting In the first stage of the algorithm, we let:

$$
M^{(1)} = \begin{bmatrix}
1 & 0 & \cdots & \cdots & 0 \\
-L_{21} & 1 & \ddots & & \vdots \\
-L_{31} & 0 & 1 & \ddots & \vdots \\
\vdots & \vdots & \ddots & \ddots & 0 \\
-L_{n1} & 0 & \cdots & 0 & 1
\end{bmatrix}, \tag{5.6}
$$

where $L_{\ell 1} = A_{\ell 1}/A_{11}, \ell = 2, \ldots, n$. Now define $A^{(2)} = M^{(1)}A$. By direct calculation:

$$A^{(2)} = \begin{bmatrix} A_{11} & A_{12} & \cdots & A_{1n} \\ 0 & A_{22}^{(2)} & \cdots & A_{2n}^{(2)} \\ \vdots & \vdots & & \vdots \\ 0 & A_{n2}^{(2)} & \cdots & A_{nn}^{(2)} \end{bmatrix}, \tag{5.7}$$

where, for example, $A_{22}^{(2)} = A_{22} - L_{21}A_{12}$ and generally:

$$A_{\ell k}^{(2)} = A_{\ell k} - L_{\ell 1}A_{1k}, 1 < \ell, k \leq n.$$

The first column of $A^{(2)}$ below the diagonal consists of zeros because:

$$\begin{aligned} A_{\ell 1}^{(2)} &= A_{\ell 1} - L_{\ell 1}A_{11}, \text{ for } 1 < \ell \leq n, \\ &= A_{\ell 1} - \frac{A_{\ell 1}}{A_{11}}A_{11}, \\ &= A_{\ell 1} - A_{\ell 1}, \\ &= 0. \end{aligned}$$

We have zeroed the entries in the first column of A below its first entry. We say that the entries have been **annihilated**. The operation of zeroing a column is called **pivoting** and we say that we have **pivoted** on the entry A_{11} or that we have used A_{11} as **pivot**. In subsequent stages, we will proceed to zero the entries below the diagonal in the other columns of A, but first note that:

$$\begin{bmatrix} 1 & 0 & \cdots & & 0 \\ -L_{21} & 1 & \ddots & & \vdots \\ -L_{31} & 0 & 1 & \ddots & \vdots \\ \vdots & \vdots & \ddots & \ddots & 0 \\ -L_{n1} & 0 & \cdots & 0 & 1 \end{bmatrix} \begin{bmatrix} 1 & 0 & \cdots & & 0 \\ L_{21} & 1 & \ddots & & \vdots \\ L_{31} & 0 & 1 & \ddots & \vdots \\ \vdots & \vdots & \ddots & \ddots & 0 \\ L_{n1} & 0 & \cdots & 0 & 1 \end{bmatrix} = \mathbf{I},$$

so that:

$$[M^{(1)}]^{-1} = \begin{bmatrix} 1 & 0 & \cdots & & 0 \\ L_{21} & 1 & \ddots & & \vdots \\ L_{31} & 0 & 1 & \ddots & \vdots \\ \vdots & \vdots & \ddots & \ddots & 0 \\ L_{n1} & 0 & \cdots & 0 & 1 \end{bmatrix}.$$

Small or zero pivot Unfortunately, our construction will fail if $A_{11} = 0$. Moreover, even if A_{11} is non-zero, if A_{11} is small in magnitude compared to $A_{\ell 1}$ then $L_{\ell 1} = A_{\ell 1}/A_{11}$ will be large in magnitude. For moderate to large values of A_{1k} this will mean that the product $L_{\ell 1} A_{1k}$ can be large compared to $A_{\ell k}$. In this case, when the difference $A_{\ell k} - L_{\ell 1} A_{1k}$ is calculated, the calculated result will have an error, due to round-off error, that is large compared to $A_{\ell k}$. That is, the calculated value $A_{\ell k}^{(2,\text{calc})}$ differs from the exact value by $A_{\ell k}^{(2,\text{error})}$ so that:

$$A_{\ell k}^{(2,\text{calc})} = A_{\ell k} - L_{\ell 1} A_{1k} + A_{\ell k}^{(2,\text{error})}. \tag{5.8}$$

Error analysis Let us try to gauge the effect of the error $A_{\ell k}^{(2,\text{error})}$ on the solution. We can think of the effect of this error as being the same as if we had changed the value of $A_{\ell k}$ in the original matrix by $A_{\ell k}^{(2,\text{error})}$ and then performed the calculation using infinite precision arithmetic. That is, we can re-arrange (5.8) to:

$$A_{\ell k}^{(2,\text{calc})} = (A_{\ell k} + A_{\ell k}^{(2,\text{error})}) - L_{\ell 1} A_{1k},$$

where we now imagine the error as being in the original entry of A and that the calculations are performed to infinite precision.

We will see in Section 5.7 that, for certain matrices A, even a relatively small change in the value of an entry of A will cause a relatively large change in the solution of the equation $Ax = b$. That is, according to Definition 2.21, the problem of solving the linear equations is **ill-conditioned** for such matrices. Consequently, for such matrices, using small pivots will lead to calculated solutions that differ significantly from the exact solution. We will discuss this further in Section 5.7.3, but for now we will just bear in mind that small pivots are undesirable. In the next few paragraphs, we will consider ways to avoid zero or small pivots by re-ordering equations and variables.

Permuting rows and columns If $A_{11} = 0$ (or if A_{11} is small in magnitude), but $A_{\ell k} \neq 0$ for some ℓ and k then we can re-order the rows and columns and pivot on $A_{\ell k}$ instead. This simply corresponds to permuting the equations and permuting the variables so that:

- equation ℓ is re-numbered to be equation 1, and
- variable k is re-numbered to be variable 1.

This approach is called **full pivoting** [45, section 2.2.5]. Re-ordering the equations or variables does not alter the solution of the equations, except that we must keep track of the re-ordering of the variables to reconstruct the solution in the original ordering of the variables. So long as we keep track of the permutations, we can reconstruct the solution in the original order of the variables.

With full pivoting, it is usual to seek the largest entry $A_{\ell k}$ in the matrix to use as the pivot. Full pivoting is aimed at creating entries in L that have small magnitudes. This minimizes the effect of round-off errors in the calculation of $A_{\ell k}^{(2)}$, as will be discussed in Section 5.7.3.

Partial pivoting An alternative to full pivoting is to permute only, say, the rows. This is called **partial pivoting** [45, section 2.2.5]. With partial pivoting, the largest entry below the diagonal is usually chosen as the pivot. The permutation of the rows can be represented by multiplying A on the left by a permutation matrix $P \in \mathbb{R}^{n \times n}$; that is, a matrix with exactly one 1 in each row and each column and zeros elsewhere. For example, for $A \in \mathbb{R}^{2 \times 2}$, the matrix:

$$ P = \begin{bmatrix} 0 & 1 \\ 1 & 0 \end{bmatrix}, $$

swaps the rows. (See Exercise 5.4.)

Because of the connection to a permutation matrix, pivoting on rows is sometimes called PLU factorization. The MATLAB function lu performs partial pivoting [74] and returns matrices P, L, and U such that $PA = LU$.

Diagonal pivoting Another alternative to full pivoting is possible if $A_{\ell k} \neq 0$ for some $\ell = k$. In this case we can re-order the equations and the variables of our system with the same permutation applied to both equations and variables. In other words, we pivot on $A_{\ell \ell}$ instead of A_{11} by re-ordering so that:

- equation ℓ is re-numbered to be equation 1, and
- variable ℓ is re-numbered to be variable 1.

This is called **diagonal pivoting**.

In the case that A is symmetric, then diagonal pivoting preserves the symmetry of the system in that the bottom right-hand $(n - 1) \times (n - 1)$ sub-matrix of $A^{(2)}$ is symmetric, as we will see in Section 5.4. This will turn out to have computational advantages for symmetric systems in that we can take advantage of symmetry to perform less calculations. On the other hand, unless the matrix is diagonally dominant, we generally will not be able to use the largest pivots if we restrict ourselves to diagonal pivoting.

Summary In the first stage of the algorithm, to calculate $A^{(2)}$ using A_{11} as pivot, we:

- copy the first row of A into $A^{(2)}$;
- zero the entries in the first column of $A^{(2)}$ below the diagonal; and

- explicitly calculate the entries $A_{\ell k}^{(2)}$ for $1 < \ell \leq n, 1 < k \leq n$ using $A_{\ell k}^{(2)} = A_{\ell k} - L_{\ell 1} A_{1k}$.

We call A_{11} the **standard pivot**. If $A_{11} = 0$ then the rows and/or columns of A can be re-ordered to replace A_{11} with a non-zero entry. Similarly, if A_{11} is a very small number, then we may also want to re-order the rows and/or columns of A to replace A_{11} with a larger entry to minimize the effects of round-off errors in calculations.

5.3.2.2 Second stage

Pivoting In the second stage of the algorithm, we now choose $M^{(2)}$ to zero the second column of $A^{(2)}$ below the diagonal. Let:

$$M^{(2)} = \begin{bmatrix} 1 & 0 & \cdots & & 0 \\ 0 & 1 & \ddots & & \\ & -L_{32} & 1 & & \vdots \\ \vdots & -L_{42} & 0 & \ddots & \\ & \vdots & \vdots & \ddots & \ddots & 0 \\ 0 & -L_{n2} & 0 & \cdots & 0 & 1 \end{bmatrix}, \tag{5.9}$$

where $L_{\ell 2} = A_{\ell 2}^{(2)}/A_{22}^{(2)}, \ell = 3, \ldots, n$. Now let $A^{(3)} = M^{(2)} M^{(1)} A = M^{(2)} A^{(2)}$. Then,

$$A^{(3)} = \begin{bmatrix} A_{11} & A_{12} & A_{13} & \cdots & A_{1n} \\ 0 & A_{22}^{(2)} & A_{23}^{(2)} & \cdots & A_{2n}^{(2)} \\ 0 & 0 & A_{33}^{(3)} & \cdots & A_{3n}^{(3)} \\ \vdots & \vdots & \vdots & & \vdots \\ 0 & 0 & A_{n3}^{(3)} & \cdots & A_{nn}^{(3)} \end{bmatrix}, \tag{5.10}$$

so that we have zeroed the second column below the diagonal. The other entries of $A^{(3)}$ are given by:

$$A_{\ell k}^{(3)} = A_{\ell k}^{(2)} - L_{\ell 2} A_{2k}^{(2)}, 2 < \ell, k \leq n.$$

We have:

$$[M^{(2)}]^{-1} = \begin{bmatrix} 1 & 0 & \cdots & & 0 \\ 0 & 1 & \ddots & & \\ & L_{32} & 1 & & \vdots \\ \vdots & L_{42} & 0 & 1 & \ddots \\ & \vdots & \vdots & \ddots & \ddots & 0 \\ 0 & L_{n2} & 0 & \cdots & 0 & 1 \end{bmatrix}.$$

Error analysis As in Section 5.3.2.1, we can interpret round-off errors in the calculation of $A^{(3)}$ in terms of a perturbation introduced into $A^{(2)}$, which we can, in turn, interpret in terms of a perturbation in the original matrix A.

Permuting rows and columns The construction may again fail if $A^{(2)}_{22} = 0$. Again, full, partial, or diagonal pivoting can be used if there is a suitable non-zero pivot $A^{(2)}_{\ell k}$ for some $2 \leq \ell \leq n$ and $2 \leq k \leq n$.

Summary At the second stage of the algorithm, to calculate $A^{(3)}$ using $A^{(2)}_{22}$ as pivot, we:

- copy the first two rows of $A^{(2)}$ into $A^{(3)}$;
- zero the entries in the first two columns of $A^{(3)}$ below the diagonal; and
- explicitly calculate the entries $A^{(3)}_{\ell k}$ for $2 < \ell \leq n, 2 < k \leq n$ using $A^{(3)}_{\ell k} = A^{(2)}_{\ell k} - L_{\ell 2} A^{(2)}_{2k}$.

We call $A^{(2)}_{22}$ the standard pivot. Again, if $A^{(2)}_{22} = 0$ or if it is small in magnitude, then the rows and/or columns of $A^{(2)}$ can be re-ordered to place another entry in the 22 place.

5.3.2.3 Subsequent stages

Pivot We continue in this way using either the standard pivot $A^{(j)}_{jj}$ at each stage j, or using some other entry of the matrix as pivot, annihilating each successive column under the diagonal. In particular, for the standard pivot $A^{(j)}_{jj}$ at stage j we have that:

$$\forall \ell > j, L_{\ell j} = A^{(j)}_{\ell j}/A^{(j)}_{jj},$$
$$\forall \ell > j, \forall k > j, A^{(j+1)}_{\ell k} = A^{(j)}_{\ell k} - L_{\ell j} A^{(j)}_{jk}. \tag{5.11}$$

Error analysis At each stage, errors in $A^{(j+1)}$ can be interpreted in terms of a perturbation in $A^{(j)}$, which can be interpreted in terms of a perturbation in the original matrix A.

Summary At stage j of the algorithm, to calculate $A^{(j+1)}$ using $A^{(j)}_{jj}$ as pivot, we:

- copy the first j rows of $A^{(j)}$ into $A^{(j+1)}$;
- zero the entries in the first j columns of $A^{(j+1)}$ below the diagonal; and
- explicitly calculate the entries $A^{(j+1)}_{\ell k}$ for $j < \ell \leq n, j < k \leq n$ using $A^{(j+1)}_{\ell k} = A^{(j)}_{\ell k} - L_{\ell j} A^{(j)}_{jk}$.

We call $A^{(j)}_{jj}$ the standard pivot. Again, if $A^{(j)}_{jj} = 0$ or if it is small in magnitude, then the rows and/or columns of $A^{(j)}$ can be re-ordered to place a suitable non-zero entry $A^{(j)}_{\ell k}$, where $j \leq \ell \leq n$ and $j \leq k \leq n$, in the jj place.

5.3.2.4 Last stage

After the $(n-1)$-th stage of the algorithm:

$$A^{(n)} = M^{(n-1)}M^{(n-2)}\cdots M^{(1)}A = \begin{bmatrix} A_{11} & A_{12} & A_{13} & \cdots & A_{1n} \\ 0 & A_{22}^{(2)} & A_{23}^{(2)} & \cdots & A_{2n}^{(2)} \\ 0 & 0 & A_{33}^{(3)} & \cdots & A_{3n}^{(3)} \\ \vdots & \vdots & \ddots & \ddots & \vdots \\ 0 & 0 & \cdots & 0 & A_{nn}^{(n)} \end{bmatrix}.$$

We let $U = A^{(n)}$. The diagonal entries of the upper triangular matrix U are the pivots. By direct calculation:

$$L = [M^{(n-1)}\cdots M^{(1)}]^{-1} = \begin{bmatrix} 1 & 0 & \cdots & 0 \\ L_{21} & 1 & \ddots & \vdots \\ \vdots & \ddots & \ddots & 0 \\ L_{n1} & \cdots & L_{n,n-1} & 1 \end{bmatrix},$$

with L lower triangular and $LU = A$. Factorization of A into L and U does not involve b, so that once we have factorized A we can solve $b = Ax$ for *any* b using forwards and backwards substitution as discussed in Section 5.3.1. (This contrasts with the usual description of Gaussian elimination, where b is intimately involved in the elimination.)

5.3.2.5 Example

To illustrate this process, let us perform the LU factorization for the example matrix $A = \begin{bmatrix} 2 & 3 & 4 \\ 7 & 6 & 5 \\ 8 & 9 & 11 \end{bmatrix}$ and solve the system $Ax = b$, where $b = \begin{bmatrix} 9 \\ 18 \\ 28 \end{bmatrix}$, which we first met in Section 2.2.2.

In the first stage, we have:

$$M^{(1)} = \begin{bmatrix} 1 & 0 & 0 \\ -\frac{7}{2} & 1 & 0 \\ -4 & 0 & 1 \end{bmatrix}, \quad A^{(2)} = M^{(1)}A = \begin{bmatrix} 2 & 3 & 4 \\ 0 & -\frac{9}{2} & -9 \\ 0 & -3 & -5 \end{bmatrix}.$$

Then, in the second stage:

$$M^{(2)} = \begin{bmatrix} 1 & 0 & 0 \\ 0 & 1 & 0 \\ 0 & -\frac{2}{3} & 1 \end{bmatrix}, \quad U = A^{(3)} = \begin{bmatrix} 2 & 3 & 4 \\ 0 & -\frac{9}{2} & -9 \\ 0 & 0 & 1 \end{bmatrix}, \quad L = \begin{bmatrix} 1 & 0 & 0 \\ \frac{7}{2} & 1 & 0 \\ 4 & \frac{2}{3} & 1 \end{bmatrix}.$$

We have already performed forwards and backwards substitution in Sections 5.2.1

and 5.2.2, respectively, for these L and U factors and the vector $b = \begin{bmatrix} 9 \\ 18 \\ 28 \end{bmatrix}$ and

obtained the solution $x^\star = \begin{bmatrix} 1 \\ 1 \\ 1 \end{bmatrix}$.

5.3.2.6 Singular matrices

Exercises 5.6 and 5.7 show that factorization can sometimes be performed on singular matrices. However, if factorization fails then (under the assumption of infinite precision arithmetic) the matrix is singular. (See Exercises 5.8 and 5.9.)

5.3.3 Computational effort

Let us estimate the amount of computational effort involved in performing the LU factorization. We divided the LU factorization into $(n-1)$ stages. At the j-th stage, we calculate:

- the $(n-j)$ entries of L that are in its j-th column and lying below the diagonal, and
- the $(n-j)^2$ values of $A^{(j+1)}$ that are in the lower right of the matrix.

The entries for L and $A^{(j+1)}$ are shown in (5.6) and (5.7) for stage 1 and are shown in (5.9) and (5.10) for stage 2. The $(n-j)$ entries calculated for L each require one division, while calculation of the lower right $(n-j)^2$ entries of $A^{(j+1)}$ each require one multiplication and one subtraction. The total effort therefore is:

$$\sum_{j=1}^{n-1}(n-j) = \frac{1}{2}n(n-1) \text{ divisions,}$$

$$\sum_{j=1}^{n-1}(n-j)^2 = \frac{1}{6}(2n-1)n(n-1) \text{ multiplications, and}$$

$$\sum_{j=1}^{n-1}(n-j)^2 = \frac{1}{6}(2n-1)n(n-1) \text{ subtractions.}$$

The overall effort for LU factorization is therefore on the order of the *cube* of the number of variables. To solve the system $Ax = b$ requires LU factorization followed by forwards and backwards substitution. The overall effort to solve $Ax = b$ is therefore on the order of $(n)^3$.

5.3.4 Variations

We present some variations on the basic theme of LU factorization in the following sections.

5.3.4.1 Factorization in place

To implement the LU factorization algorithm, we can start with a copy of A and apply the pivot operations directly to update the entries in the copy of A, thereby transforming it into the LU factors. The entries of the lower triangle of L can be entered into the lower triangle of A as they are calculated, while the entries of the diagonal and upper triangle of U can be entered into the diagonal and upper triangle of A as they are calculated. This procedure gives correct results because no information is lost in "re-using" the memory locations that contained the copy of A.

We do not need to store the diagonal entries of L explicitly, since they are all ones. Similarly, we do not need to store the entries of the upper triangle of L nor the lower triangle of U, since they are all zeros. That is, all the essential entries of L and U will fit in the memory locations that originally contained A. This process is called factorization in place and we will return to it in the discussion of sparsity in Section 5.5.3. (See Exercise 5.12.)

5.3.4.2 Diagonal entries of L and U

In the presentation in Section 5.3.2, L has ones on its diagonal, while the entries on the diagonal of U were the pivots. This allows us to avoid the n divisions described in Section 5.2.1 for forwards substitution. It is possible instead to factorize A into two matrices L' and U' so that U' has ones on its diagonal, while the entries on the diagonal of L' are the pivots. This allows us to instead avoid the n divisions described in Section 5.2.2 for backwards substitution.

To see that this alternative factorization is possible, define D to be a diagonal matrix with diagonal entries equal to the corresponding diagonal entries of U. Let $L' = LD$ and $U' = D^{-1}U$. Then $A = L'U'$ with L' lower diagonal and U' upper diagonal and U' has ones on its diagonal.

5.3.4.3 LDU factorization

Suppose we have factorized A into LU' with L lower triangular with ones on its diagonal and U' upper triangular. Define a diagonal matrix D to have diagonal entries equal to the diagonal entries of U' and then define $U = D^{-1}U'$. We now have a factorization of A into LDU, where D is a diagonal matrix and both L and U have ones on the diagonal. The entries in the diagonal of D are the pivots. We will explore this factorization further for the case of symmetric A in Section 5.4.

5.4 Symmetric coefficient matrix

If A is symmetric, we can save approximately half the work in factorization, so long as we only use diagonal pivots. Symmetric systems often arise in circuit applications, as we have seen, and also occur in optimization applications. Sometimes, a system that appears at first to be not symmetric can be made symmetric by **scaling** the rows or columns or re-arranging the rows or columns. (See Exercise 5.13.)

5.4.1 LU factorization

Let us see the modifications for *LU* factorization of a symmetric A. Consider the first stage of the construction in Section 5.3.2.1, where the entries in $A^{(2)}$ are calculated, and consider the sub-matrix of $A^{(2)}$ formed by deleting its first row and column. We have:

Lemma 5.1 *Suppose that A is symmetric and diagonal pivoting was used in the first stage of factorization to re-order rows and columns. Then:*

    (i) *the first row of A is equal to A_{11} times the transpose of the first column of L, (that is, the entries in the first column of L arranged into a row), and*

    (ii) *the sub-matrix of $A^{(2)}$ formed by deleting its first row and column is symmetric.*

Proof The proof involves calculation of the entries $A^{(2)}$. See Appendix B for details. □

Because the sub-matrix of $A^{(2)}$ is symmetric, it is only necessary to calculate and store its diagonal and upper triangle. (Techniques to store only part of a matrix will be introduced in Section 5.5.) This saves nearly half the effort of the first stage. Similar savings are possible at each subsequent stage. We have:

Lemma 5.2 *Let $2 \le j \le (n-1)$ and consider the matrix $A^{(j)}$ formed at the $(j-1)$-th stage of the factorization. Suppose that the sub-matrix of $A^{(j)}$ obtained by deleting its first $(j-1)$ rows and $(j-1)$ columns is symmetric. Assume that diagonal pivoting is used at the j-th stage of factorization. Consider the matrix $A^{(j+1)}$ formed at the j-th stage of factorization. Then:*

    (i) *the j-th row of $A^{(j+1)}$ is equal to $A^{(j)}_{jj}$ times the transpose of the j-th column of L, (that is, the entries in the j-th column of L arranged into a row), and*

    (ii) *the sub-matrix of $A^{(j+1)}$ formed by deleting its first j rows and j columns is also symmetric.*

Proof The proof is analogous to that of Lemma 5.1. See Appendix B for details. □

Corollary 5.3 *Suppose that A is symmetric and that diagonal pivoting is used at each stage of the factorization. Then for each j, $1 \leq j \leq n-1$, the sub-matrix of $A^{(j)}$ formed by deleting its first $(j-1)$ rows and $(j-1)$ columns is symmetric. Moreover, at the end of the factorization, for each ℓ, the ℓ-th row of U is equal to $U_{\ell\ell}$ times the transpose of the ℓ-th column of L.*

Proof By induction. Lemma 5.1 proves the result for $j = 1$. Lemma 5.2 then proves the induction step. □

5.4.2 Example

Let us perform LU factorization for the symmetric matrix $A = \begin{bmatrix} 2 & 3 & 4 \\ 3 & 5 & 7 \\ 4 & 7 & 13 \end{bmatrix}$.

5.4.2.1 First stage

We have $M^{(1)} = \begin{bmatrix} 1 & 0 & 0 \\ -\frac{3}{2} & 1 & 0 \\ -2 & 0 & 1 \end{bmatrix}$. The 2×2 sub-matrix obtained from $A^{(2)}$ by deleting the first row and column of A is symmetric. We need only calculate the three entries in the diagonal and upper triangle of this sub-matrix to determine all four entries. Then $A^{(2)} = M^{(1)}A = \begin{bmatrix} 2 & 3 & 4 \\ 0 & \frac{1}{2} & 1 \\ 0 & 1 & 5 \end{bmatrix}$, where $A^{(2)}_{32} = A^{(2)}_{23}$ so that we need only calculate $A^{(2)}_{32}$ to obtain the values of both entries.

5.4.2.2 Second stage

We now have $M^{(2)} = \begin{bmatrix} 1 & 0 & 0 \\ 0 & 1 & 0 \\ 0 & -2 & 1 \end{bmatrix}$. Then $U = A^{(3)} = M^{(2)}M^{(1)}A = \begin{bmatrix} 2 & 3 & 4 \\ 0 & \frac{1}{2} & 1 \\ 0 & 0 & 3 \end{bmatrix}$.

5.4.2.3 Last stage

Finally, $L = [M^{(1)}]^{-1}[M^{(2)}]^{-1} = \begin{bmatrix} 1 & 0 & 0 \\ \frac{3}{2} & 1 & 0 \\ 2 & 2 & 1 \end{bmatrix}$.

5.4.3 Computational savings

The savings in computational effort for the small example in Section 5.4.2 is modest. In general, for a matrix of size $n \times n$, the diagonal and upper triangle has $n(n+1)/2$ entries, while the full matrix has $(n)^2$ entries. Calculating only the diagonal and upper triangle in the first stage of factorization saves a fraction of the work equal to:

$$\frac{(n)^2 - n(n+1)/2}{(n)^2} = \frac{(n)^2/2 - n/2}{(n)^2},$$

$$= \frac{n-1}{2n},$$

$$\approx \frac{1}{2}, \text{ for } n \text{ large.}$$

Similarly, at each successive stage approximately half the work is saved, so that overall approximately half the computational effort is required compared to factorizing a non-symmetric matrix. (See Exercise 5.14.)

5.4.4 $LDL^\dagger$ and Cholesky factorization

Consider a symmetric matrix factorized into LU using diagonal pivots at each stage of the factorization. Let D be a diagonal matrix with entries equal to the diagonal of U. Then, by Corollary 5.3, $U = DL^\dagger$, where $L^\dagger$ is the *transpose* of L obtained by the swapping the rows and columns of L. That is, $A = LDL^\dagger$ with the diagonal entries of D being the pivots. We can save storage by factorizing A as $LDL^\dagger$ and storing only L and D since then we do not have to store U explicitly. We will make use of this advantage in Section 5.5.1.

If the entries of D are all positive, then a related factorization involves setting $R = D^{\frac{1}{2}}L^\dagger$, where the matrix $D^{\frac{1}{2}}$ is diagonal with each diagonal entry equal to the positive square root of the corresponding entry of D. Then $A = R^\dagger R$ is called the **Cholesky factorization** of A [45, section 2.2.5.2]. The matrix R is upper triangular.

5.4.5 Discussion of diagonal pivoting

Diagonal pivoting is not always possible for non-singular A and we may be forced to seek an off-diagonal pivot. (See Exercise 5.17.) Even if there are non-zero diagonals, if the diagonal entries are very small, then using them as pivots can prove problematic. We will discuss this further in Sections 5.5.3.4 and 5.7.3.

In our circuit case study from Section 4.1, the admittance matrix is strictly diagonally dominant and so the diagonal entries are relatively large compared to the off-diagonal. Diagonal pivoting is consequently adequate for the particular problem in our case study as we will see in Exercise 5.22. In other circuit formulations and more generally in other applications, this may not be the case and off-diagonal pivoting becomes necessary. (See, for example, [95, section 3.4] for a discussion in the context of circuit simulation and see [45, section 2.2.5] for a general discussion.)

5.4.6 Positive definite A

We will often be interested in factorizing symmetric matrices that are **positive definite** (see Definition A.58) or **positive semi-definite** (see Definition A.59.) A positive definite matrix is non-singular. (See Exercise 5.20.) If A is symmetric and can be factorized into $LDL^\dagger$, then we have the following test for positive definiteness.

Lemma 5.4 *Suppose that $A \in \mathbb{R}^{n \times n}$ is symmetric and can be factorized as $A = LDL^\dagger$, with $D \in \mathbb{R}^{n \times n}$ diagonal and L lower triangular with ones on the diagonal. Then A is positive definite if and only if all the diagonal entries of D are strictly positive.*

Proof

$\Rightarrow$ We first prove that A being positive definite implies that the diagonal entries of D are strictly positive. To prove this, we prove the contra-positive. So, suppose that there is at least one diagonal entry, $D_{\ell\ell}$, say, of D that is non-positive. We will exhibit $x \neq \mathbf{0}$ such that $x^\dagger A x \leq 0$. To find such a x, solve the equation $L^\dagger x = \mathbf{I}_\ell$ for x. (This is possible since L is lower triangular and has ones on its diagonal. We just perform backwards substitution on $L^\dagger$.) Notice that $x \neq \mathbf{0}$, for else $\mathbf{I}_\ell = L^\dagger x = L^\dagger \mathbf{0} = \mathbf{0}$, which is a contradiction. Furthermore,

$$
\begin{aligned}
x^\dagger A x &= x^\dagger L D L^\dagger x, \text{ by assumption on } A, \\
&= \mathbf{I}_\ell{}^\dagger D \mathbf{I}_\ell, \text{ by definition of } x, \\
&= D_{\ell\ell}, \text{ on direct calculation}, \\
&\leq 0, \text{ by supposition.}
\end{aligned}
$$

Therefore, A is not positive definite.

$\Leftarrow$ We now prove that the diagonal entries of D being positive implies that A is positive definite. So, suppose that all the diagonal entries of D are strictly positive. Define the matrix $D^{\frac{1}{2}}$ to be diagonal with each diagonal entry $D^{\frac{1}{2}}_{\ell\ell}$ equal to the square root of the corresponding diagonal entry of D. That is, $D^{\frac{2}{2}}_{\ell\ell} = \sqrt{D_{\ell\ell}}, \forall \ell$. Let $x \neq \mathbf{0}$ be given and define $y = D^{\frac{1}{2}} L^\dagger x$. We first claim that $y \neq \mathbf{0}$.

For suppose the contrary. That is, suppose that $y = \mathbf{0}$. Then, $\left[D^{\frac{1}{2}} \right]^{-1} y = \mathbf{0}$. (Notice that the diagonal entries of $D^{\frac{1}{2}}$ are all strictly positive, so that $D^{\frac{1}{2}}$ is invertible.) But then $\mathbf{0} = \left[D^{\frac{1}{2}} \right]^{-1} y = L^\dagger x$. Solving $L^\dagger x = \mathbf{0}$ by backwards substitution we obtain $x = \mathbf{0}$, a contradiction. Therefore, $y \neq \mathbf{0}$.

Second, we observe that:

$$\begin{aligned}
x^\dagger A x &= x^\dagger L D L^\dagger x, \text{ by assumption on } A, \\
&= x^\dagger L D^{\frac{1}{2}} D^{\frac{1}{2}} L^\dagger x, \text{ by definition of } D^{\frac{1}{2}}, \\
&= y^\dagger y, \text{ by definition of } y, \\
&= \|y\|_2^2, \text{ by definition of } \|\bullet\|_2, \\
&> 0,
\end{aligned}$$

since the length of a non-zero vector is strictly positive by Property (ii) of norms. That is, A is positive definite. □

We now use this lemma to prove the following result about symmetric positive definite matrices.

Theorem 5.5 *If A is symmetric and positive definite, then:*

(i) *A is invertible,*

(ii) *A is factorizable as $LDL^\dagger$, with D diagonal having strictly positive diagonal entries and L lower triangular with ones on the diagonal, and*

(iii) *A^{-1} is also symmetric and positive definite.*

Proof The proof is divided into three parts:

(i) A is invertible,

(ii) A is factorizable as $LDL^\dagger$, and

(iii) A^{-1} is symmetric and positive definite.

See Appendix B for details. □

Exercises 4.8, 5.21, and 5.22 show why the circuit case study in Section 4.1 has a unique solution: the coefficient matrix is guaranteed to be non-singular. In fact, it is symmetric and positive definite. Moreover, we can take advantage of the symmetry of the coefficient matrix and guarantee that the diagonal pivots will always be satisfactory. Exercise 5.22 shows that diagonal pivoting is equivalent to full pivoting for a symmetric strictly diagonally dominant matrix.

5.4.7 Indefinite coefficient matrix

In some applications, the coefficient matrix may be of the form:

$$\mathcal{A} = \begin{bmatrix} A & B \\ B^\dagger & C \end{bmatrix},$$

consisting of four **blocks**; that is sub-matrices, A, B, $B^\dagger$, and C. Suppose that A is a square symmetric matrix that is positive semi-definite or positive definite and that C is a square symmetric matrix that is negative semi-definite or negative definite. (See Definitions A.58 and A.59.) The coefficient matrix $\mathcal{A}$ is **indefinite**; that is, it

is neither positive semi-definite nor negative semi-definite. For example, consider
the matrix:

$$\mathcal{A} = \begin{bmatrix} 1 & 0 \\ 0 & -1 \end{bmatrix},$$

where: $A = [1]$ is positive definite; $B = [0]$; and $C = [-1]$ is negative definite.
The matrix $\mathcal{A}$ is indefinite since the top left-hand block, $A = [1]$, is positive definite
and the bottom right-hand block, $C = [-1]$, is negative definite. This particular
matrix is, however, non-singular.

For a non-singular indefinite matrix there are special purpose factorization techniques to factorize this matrix that take advantage of the known partitioning of the
matrix into positive definite and negative definite parts [37, 116]. We will need to
factorize such matrices as part of the solution of our later case studies and we will
assume that semi-definite factorization algorithms are available to us; however, we
will not treat the details of factorizing such matrices.

5.5 Sparsity techniques

Using the LU or $LDL^\dagger$ factorization as described in Sections 5.2–5.4 is not effective in solving very large systems because the computational effort of factorization
grows as the cube of the number of variables and the effort for forwards and backwards substitution grows as the square of the number of variables. As described
in Section 4.1.4 for our circuit case study, the non-zero entries in the admittance
matrix occur only:

- on the diagonal, and
- at those off-diagonal entries corresponding to resistors,

so that the admittance matrix is a **sparse matrix**. We may also have right-hand
side vectors b that only have a few non-zero entries. In the circuit case study, for
example, this occurs if there are only a small number of current sources in the
circuit.

In the case of a sparse matrix, we can arrange calculations in the LU or $LDL^\dagger$
factorization to reduce the computational effort considerably. We can also take
advantage of the sparsity of the right-hand side vector to reduce calculations.

Sparsity appears in many problems and techniques for treating sparse matrices
are highly developed. We will give an introduction that discusses only the sparsity
techniques that are directly relevant to solving the types of linear systems that we
will need to solve in the circuit case study and in subsequent chapters. The general
issue to bear in mind is that if a problem has a sparse structure then it can be
very worthwhile to exploit the sparsity, particularly for a large-scale problem. In

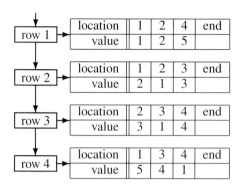

Fig. 5.1. Sparse matrix storage by rows of the matrix defined in (5.12).

our circuit problem, the pattern of non-zeros is somewhat random; however, if the non-zeros form a regular pattern in the matrix, then this can also lead to further advantages.

In Section 5.5.1 we discuss storage of sparse matrices and vectors. In Section 5.5.2 we discuss forwards and backwards substitution of sparse L and U. Then in Section 5.5.3, we discuss factorization of sparse matrices. In Section 5.5.4 we will then mention some other special types of sparse matrices including matrices that have a regular pattern of non-zeros.

The development is based on [125]. Other references on sparsity include [2, 95].

5.5.1 Sparse storage

5.5.1.1 Sparse matrices

Consider a circuit with 1000 nodes, which would in principle require a 1000×1000 matrix to represent the admittance matrix. Even if there is enough main memory to store this amount of data as a "dense" matrix, the disk traffic to access the 1,000,000 entries in the matrix will take significant time.

The key to sparse storage is to avoid storing the whole matrix, which includes both the non-zero and the zero entries, and instead store only the *values* and the indices of the *locations* of the non-zero entries in the matrix. It is usually convenient to break the matrix up into its rows or its columns. To be concrete, we will break up the matrix into a **linked list** of its rows as shown in Figure 5.1 [48, section 8.2][60]. Each downward-pointing arrow in Figure 5.1 represents a pointer in the linked list of rows.

Each row of the matrix can itself be stored as a list of pairs of numbers: the pairs are the locations and values of successive non-zero entries. This is illustrated in

Figure 5.1 for the matrix:

$$A = \begin{bmatrix} 1 & 2 & 0 & 5 \\ 2 & 1 & 3 & 0 \\ 0 & 3 & 1 & 4 \\ 5 & 0 & 4 & 1 \end{bmatrix}. \tag{5.12}$$

Each horizontal arrow in Figure 5.1 is a pointer to the list of pairs of locations and values for the corresponding row of the matrix. (These rows can themselves be stored as a linked list; however, this is not shown explicitly in Figure 5.1.) Row 1 has the values 1, 2, and 5 in, respectively, the 1-st, 2-nd, and 4-th locations in the row. We store the locations and values as *pairs*. We indicate the end of the list of pairs with a special symbol "end." This is called row-wise storage. The alternative is column-wise storage.

In the case of a symmetric matrix, we need only store the diagonal and the upper or (equivalently) the lower triangle. It may be useful to have both row-wise access and column-wise access to a matrix. We do this by maintaining a list of rows and a list of columns. In the case of a symmetric matrix, we need only keep one list, since the row-wise and the column-wise lists are the same.

5.5.1.2 Sparse vectors

Similarly, **sparse vectors** can be stored as a list of pairs of numbers representing the locations and values of the non-zero entries of the vector. For example, consider the change in the circuit case study of Section 4.1 illustrated in Figure 4.2, which is repeated for reference in Figure 5.2. In this circuit, the current injected at node 2 changes by Δb_2. Suppose that the value of the change in the current source was $\Delta b_2 = 1$. Then, we could define a vector $\Delta b \in \mathbb{R}^4$ that represents the changes at all nodes as specified by:

$$\Delta b = \begin{bmatrix} 0 \\ 1 \\ 0 \\ 0 \end{bmatrix}. \tag{5.13}$$

This vector can be stored sparsely as the pair of lists of locations and values shown in Figure 5.3.

5.5.1.3 Implementation

The illustration we have shown is applicable to a programming language with pointers and records [48, section 8.2][60]. A linked list of records can be easily modified by changing the pointers. However, there are other possible ways to represent sparse matrices. For example, in a language such as FORTRAN, the list can be implemented as a pair of arrays, but this has less flexibility than the use

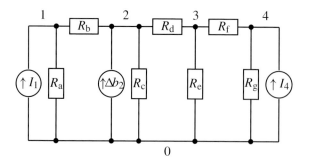

Fig. 5.2. The ladder circuit of Figure 4.2, showing a change, Δb_ℓ, in the current injected at node $\ell = 2$.

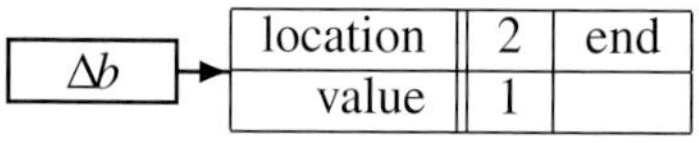

	location	2	end
Δb	value	1	

Fig. 5.3. Sparse storage of the vector defined in (5.13).

of a linked list. See [2] for details. MATLAB and other packages can define and treat sparsely stored matrices and vectors. Many of the MATLAB sparse matrix commands begin with the two letters sp [74].

5.5.2 Forwards and backwards substitution

5.5.2.1 Sparse matrices

Forwards and backwards substitution on matrices L and U that are stored sparsely is straightforward, since we need only access the entries in L and U that are non-zero. We first consider forward substitution to solve $Ly = b$. Suppose we store L as a list of rows, then for each substitution to solve for y_ℓ in (5.5), we use the list for row ℓ of L. We re-write (5.5) as:

$$
\begin{aligned}
y_\ell &= \frac{1}{L_{\ell\ell}} \left(b_\ell - \sum_{k=1}^{\ell-1} L_{\ell k} y_k \right), \\
&= b_\ell - \sum_{\substack{k < \ell \\ L_{\ell k} \neq 0 \\ y_k \neq 0}} L_{\ell k} y_k,
\end{aligned}
\tag{5.14}
$$

observing that, for each ℓ:

- $L_{\ell\ell} = 1$, and
- to calculate the sum in (5.14) we need only consider values of k for which both $L_{\ell k} \neq 0$ and $y_k \neq 0$.

To calculate y_ℓ, we first initialize $y_\ell = b_\ell$. For each non-zero entry $L_{\ell k}, k < \ell$, in row ℓ, if $y_k \neq 0$, we calculate $L_{\ell k} y_k$ and subtract it from the current value of

y_ℓ. This calculation can be performed as we "traverse" down the linked lists that contain row ℓ of L and contain y, noting that these lists specify only the values satisfying $L_{\ell k} \neq 0$ and $y_k \neq 0$.

A similar procedure applies to backwards substitution, except that we will have to explicitly divide by $U_{\ell\ell}$. The total effort for forwards and backwards substitution depends only on the number of non-zero entries in L and U. This may be only a very small fraction of the total number of entries in L and U.

5.5.2.2 Sparse vectors

If b has relatively few non-zero entries, that is, it is a **sparse vector**, then we can further reduce the computational effort in forwards substitution by only performing (5.14) for the entries y_ℓ for which there is a non-zero value of b_ℓ or non-zero values of $L_{\ell k} y_k$ for $k < \ell$. For example, in forwards substitution, if $b = \mathbf{I}_n$ then we only need perform (5.14) for $\ell = n$ since $y_\ell = 0$ for $\ell < n$. A similar savings is possible for backwards substitution to solve $Ux = y$ for sparse y. This is particularly useful when we are solving a change-case with a changed right-hand side if the change in the right-hand side Δb has only a few non-zero entries. (See Exercise 5.25.)

5.5.3 Factorization

In the above discussion of forwards and backwards substitution, we tacitly assumed that L and U would be sparse matrices. In this section, we discuss factorization of a sparse matrix A in such a way as to make the factors sparse.

5.5.3.1 Fill-ins

The matrix:

$$A = \begin{bmatrix} 1 & 2 & 0 & 5 \\ 2 & 1 & 3 & 0 \\ 0 & 3 & 1 & 4 \\ 5 & 0 & 4 & 1 \end{bmatrix}$$

has two zero entries above and two zero entries below the diagonal. However, its L factor has only one zero entry below the diagonal. (See Exercise 5.14.) We can represent the zeros and non-zeros of A with the following diagram:

where the circles ○ represent the *non-zero* entries and the blanks represent the zeros. (This typographical convention is standard but a little confusing: notice that the circles represent *non-zeros,* not zeros.) Similarly, we can represent the zeros and non-zeros of L with:

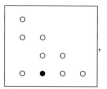

where ○ and ● both represent non-zeros, with the circles ○ corresponding to entries that were non-zero in A, while the bullet ● corresponds to an entry that was zero in A. We refer to the latter entries, indicated by bullets ●, as **fill-ins** because they correspond to a non-zero entry in L that was created at a position of a zero in A.

   If we represent A as a sparse matrix and try to factorize it in place as discussed in Section 5.3.2, we will not have allocated storage for the fill-ins. So, to factorize a matrix stored sparsely, we must first copy it into a sparse matrix with the extra storage for the fill-ins. The fill-ins are initialized to zero, but will become non-zero as factorization proceeds. There are several MATLAB functions available to allocate storage for fill-ins.

5.5.3.2 Choosing pivots to minimize fill-ins

If a fill-in is created at stage j of the factorization then that entry will have to be annihilated at a later stage of factorization, requiring calculations. Furthermore, fill-ins increase the number of non-zero entries in the L and U factors, which increases the computational effort required for forwards and backwards substitutions. By re-ordering the rows and columns of the matrix A, that is, by choosing a different order for the pivots, we will find that we can affect the number of fill-ins created as calculations proceed. In this section, we seek an ordering of the rows and columns of the matrix that minimizes the number of fill-ins during factorization. For simplicity, we will restrict ourselves to selecting from diagonal pivots, so that our discussion applies to $LDL^\dagger$ factorization of symmetric matrices; however, it is straightforward to generalize this discussion to the use of non-diagonal pivots.

Heuristic criteria A fill-in created at one stage of the factorization can cause further fill-ins to be created at later stages, so it is in general very difficult to find the optimal ordering to minimize the total number of fill-ins created during the complete factorization. Finding the optimal ordering could easily take more computational effort than factorization itself. There are several heuristics available to approximately minimize the number of fill-ins created. (See, for example, [2, 71].)

   If we ignore the effect of fill-ins created at one stage on the number of fill-ins created at subsequent stages then it is reasonable to choose the pivot at stage j so as to minimize the number of fill-ins created at stage j, ignoring the effect of this decision on the number of fill-ins created at later stages. This will often produce good, though potentially sub-optimal, orderings.

In choosing pivots, we must only consider non-zero entries as candidates. If our matrix is strictly diagonally dominant, for example, then Exercises 5.21 and 5.22 show that all of the diagonal entries will always be positive at each stage of fac-torization. However, if we choose the pivot at stage j to minimize the number of fill-ins created at stage j then we will typically not choose the pivot of largest magnitude. We may decide that the savings in computational effort by choosing the entry that minimizes the number of fill-ins outweighs the risk that we will choose a relatively small positive pivot from the diagonal. Section 5.5.3.4 discusses the case where we might encounter zero or small non-zero pivots.

Number of fill-ins with standard pivot To apply the heuristic, we must calculate the number of fill-ins created at stage j as a function of the entry chosen as pivot. To simplify the discussion, we will first calculate the number of fill-ins if we do not re-order rows and columns; that is, we will first assume that we are using the standard pivot, entry $A_{jj}^{(j)}$, at stage j. We will then consider the number of fill-ins when other pivots are chosen at stage j. For simplicity, we assume that A is symmetric, although this is not central to the development. We have the following upper bound on the number of fill-ins.

Lemma 5.6 *Let $N(j)$ be the number of fill-ins created at stage j of factorization using $A_{jj}^{(j)}$ as pivot. Then $N(j) \leq \overline{N}(j)$, where:*

$$\overline{N}(j) = [(\text{the number of non-zero entries in the } j\text{-th row of } A^{(j)}) \text{ minus } 1]^2.$$

Proof See Appendix B. □

The upper bound $\overline{N}(j)$ is very easy to evaluate and represents the worst possible case of creation of fill-ins where every non-zero entry in the j-th row of A creates a fill-in for every one of the non-zero elements in the j-th column of A below the diagonal that must be explicitly annihilated. That is, it ignores the entries that are already non-zero in $A^{(j)}$. (See the proof of Lemma 5.6 in Appendix B for details.) The calculation of $\overline{N}(j)$ can be facilitated by explicitly keeping track of the number of non-zeros in each row and column of the sparse storage.

Number of fill-ins with other pivots We similarly define $N(\ell)$ to be the number of fill-ins created at stage j if we pivot on the entry $A_{\ell\ell}^{(j)}$ at stage j instead of pivoting on the entry $A_{jj}^{(j)}$. Again, we can approximate $N(\ell)$ by ignoring the entries that are already non-zero in $A^{(j)}$ to obtain an upper bound $\overline{N}(\ell)$. Following a similar argument to Lemma 5.6, we find that $\overline{N}(\ell)$ is equal to the square of one less than the number of non-zero entries in the ℓ-th row of $A^{(j)}$. Again, this bound is very easy to evaluate.

Application of heuristic To minimize the number of fill-ins at stage j, we will choose to pivot on the entry $A_{\ell\ell}^{(j)}$ that minimizes $\overline{N}(\ell)$, which will also approximately minimize $N(\ell)$. That is, we pick the row ℓ of $A^{(j)}$, where $j \leq \ell \leq n$, that has the least number of non-zero entries. We break ties arbitrarily or can apply a more detailed heuristic to distinguish the ties. This scheme is intuitively reasonable and produces relatively good results, but there are others as discussed in [2]. Several algorithms for re-ordering the columns and rows of a matrix based on various heuristics are implemented in MATLAB functions. See, for example, the MATLAB functions `symmmd` and `symrcm`.

5.5.3.3 Computational effort

The computational effort for factorization is difficult to calculate exactly because it depends on the number of fill-ins as the factorization proceeds. However, the effort for factorization is typically much less than cubic in the number of variables and, in practice, sometimes grows only slightly faster than *linearly* in the number of variables. (See Exercise 5.26.)

The most important implication is that the solution time depends strongly on the number of non-zero entries in the A matrix and the computational effort typically grows less quickly than with the cube of the size of the system. A very large, but sparse, system can be faster to solve than a small dense system having more non-zeros than the sparse system.

5.5.3.4 Other criteria for pivot selection

In Section 5.5.3.2, the choice of pivot in performing LU factorization has been to minimize the number of fill-ins created. However, we cannot pivot on a zero entry and, moreover, if we pivot on a small but non-zero entry then the coefficients in the corresponding column of the L matrix will be large, since the entries in the column of L are inversely proportional to the pivot.

In calculating the entry $A_{\ell k}^{(j+1)}$ in $A^{(j+1)}$ at the j-th stage of the factorization algorithm, we subtract $L_{\ell j} A_{jk}^{(j)}$ from $A_{\ell k}^{(j)}$. If $L_{\ell j} A_{jk}^{(j)}$ is much larger than $A_{\ell k}^{(j)}$ then we may expect large round-off errors in the subtraction as discussed in Section 5.3.2.1. We can again interpret the large round-off error as having the same result as an infinite precision calculation done with a perturbed value of $A_{\ell k}^{(j)}$. That is, we will solve a system of equations that is different to the original system and therefore obtain an answer that is different from the exact solution. We will discuss this issue in more depth in Section 5.7.3. Here, we again observe that we should try to avoid small pivots to minimize errors.

One way to avoid small pivots is to choose the pivot based on the pivot magnitude as well as on the number of fill-ins created. As we have indicated above, this presents difficulties for our *LU* factorization algorithm for sparse matrices, because we would like to know the order of the pivots ahead of time so that we can create the appropriate fill-ins in the linked list representation of the matrix. However, we cannot easily estimate the size of the pivots before we perform the factorization, except in the case of special types of matrices such as diagonally dominant matrices.

Therefore, including magnitude as a criterion in pivot selection will force us to *dynamically* allocate storage for fill-ins as factorization proceeds. For this reason, some implementations of sparse *LU* factorization do not consider pivot magnitude. See [95, chapter 7] for a further discussion of this issue as it relates to circuit analysis and see [45, section 2.2.5] for a general discussion. We will return to this issue in Section 5.7.3.

5.5.4 Special types of sparse matrices

5.5.4.1 Banded matrices and matrices with regular structure

In some applications, the pattern of non-zeros in the matrix is regular. This occurs, for example, when the coefficient matrix represents some coupling between adjacent cells in a regular structure such as a tessellation of a plane or division of space into cubes. In such cases, it is not necessary to keep explicit track of the position of the non-zeros when storing the coefficient matrix. We can avoid explicit storage of the location information and only maintain a list of the values.

For example, a **banded matrix** has zeros everywhere except on the diagonal and on entries that are close to the diagonal. A **tri-diagonal matrix** is a banded matrix that has non-zero entries only on the diagonal and adjacent to the diagonal. There are special factorization algorithms that have been developed for these types of matrices. (See Exercise 5.28.)

There are also other types of matrices, such as Toeplitz, Hankel, Hessenberg, and "arrowhead" that have regular patterns that can facilitate factorization. (See Exercise 5.29.)

5.5.4.2 Block pivoting and sparsity

In the discussion so far, we have performed pivoting on individual entries. However, consider the matrix:

$$\mathcal{A} = \begin{bmatrix} A & B \\ C & D \end{bmatrix},$$

that consists of four blocks, A, B, C, and D. Assume that A and D are square. If A is invertible, then we can multiply $\mathcal{A}$ on the left by:

$$\mathcal{M}^{(1)} = \begin{bmatrix} \mathbf{I} & \mathbf{0} \\ -CA^{-1} & \mathbf{I} \end{bmatrix},$$

to obtain:

$$\begin{aligned} \mathcal{A}^{(2)} &= \begin{bmatrix} \mathbf{I} & \mathbf{0} \\ -CA^{-1} & \mathbf{I} \end{bmatrix} \begin{bmatrix} A & B \\ C & D \end{bmatrix}, \\ &= \begin{bmatrix} A & B \\ \mathbf{0} & D - CA^{-1}B \end{bmatrix}, \end{aligned} \quad (5.15)$$

where we have zeroed the columns beneath A. The pre-conditioning matrix $\mathcal{M}^{(1)}$ was obtained by "pretending" that the matrix:

$$\mathcal{A} = \begin{bmatrix} A & B \\ C & D \end{bmatrix},$$

was a 2×2 matrix and pivoting on the block A.

The first block column of $\mathcal{L}$ in the block $\mathcal{LU}$ factorization of $\mathcal{A} = \begin{bmatrix} A & B \\ C & D \end{bmatrix}$ is given by $\begin{bmatrix} \mathbf{I} \\ CA^{-1} \end{bmatrix}$ and the first block row of $\mathcal{U}$ is given by $\begin{bmatrix} A & B \end{bmatrix}$. We say that we have **pivoted on the block** A. In general, the factorization can continue with the remaining blocks, but for this matrix the block factorization is complete. We have that:

$$\begin{aligned} \mathcal{A} &= \mathcal{LU}, \\ &= \begin{bmatrix} \mathbf{I} & \mathbf{0} \\ CA^{-1} & \mathbf{I} \end{bmatrix} \begin{bmatrix} A & B \\ \mathbf{0} & D - CA^{-1}B \end{bmatrix}. \end{aligned}$$

If A is small in size, such as a 2×2 matrix, then we can explicitly calculate the inverse according to (5.4). Block pivoting can sometimes be used to avoid numerical problems caused by small pivots if the inverse of the block is calculated explicitly. (See Exercises 5.30 and 5.31.)

If the sparsity pattern of the system is such that non-zero entries occur in blocks, then it can be more efficient to store the matrix as a sparse collection of blocks. For example, in some applications such as the case study to be described in Section 6.2, we will need to treat **complex numbers**. As we will discuss in Section 6.2.2.1, complex numbers can be stored as pairs of numbers corresponding to either:

- the **real** and **imaginary** part of the complex number, or
- the **magnitude** and **angle** of the complex number.

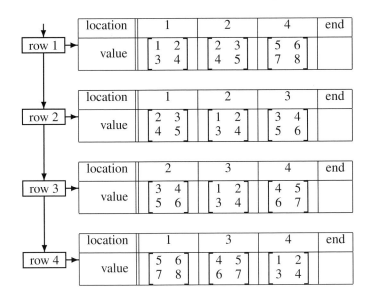

Fig. 5.4. An example of storage by block rows for a block matrix.

We can represent complex equations as pairs of real and imaginary parts. The "entries" of the coefficient matrix for the equations will therefore consist of 2×2 blocks and block pivoting can be used. For example, Figure 5.4 shows the storage of such a matrix. As in Figure 5.1, the matrix is stored by rows; however, in this case these rows are actually "block rows" and the entries in each block row are 2×2 blocks.

5.6 Changes

5.6.1 Sensitivity

5.6.1.1 Analysis

Let us generalize our coefficient matrix and right-hand side vector to be matrix- and vector-valued *functions*, respectively, of a parameter $\chi \in \mathbb{R}^s$. That is, we now consider $A : \mathbb{R}^s \to \mathbb{R}^{n \times n}$ and $b : \mathbb{R}^s \to \mathbb{R}^n$ to be matrix- and vector-valued functions of χ, respectively. We suppose that we have solved the linear equation for a particular base-case value of χ, say $\chi = \mathbf{0}$, for which $A(\mathbf{0}) \in \mathbb{R}^{n \times n}$ was non-singular, and found $x = x^{\star\star} \in \mathbb{R}^n$ that satisfied $A(\mathbf{0})x = b(\mathbf{0})$. We want to consider the sensitivity of the solution to χ for perturbations around the base-case value $\chi = \mathbf{0}$.

Theorem 5.7 *Suppose that $A : \mathbb{R}^s \to \mathbb{R}^{n \times n}$ and $b : \mathbb{R}^s \to \mathbb{R}^n$ are partially differentiable with continuous partial derivatives and that $A(\mathbf{0})$ is non-singular. Then, there exists a function $x^\star : \mathbb{R}^s \to \mathbb{R}^n$ such that:*

- *for χ in a neighborhood of $\mathbf{0}$, (see Definition A.45), the function $x^\star$ satisfies the linear simultaneous equations $A(\chi)x^\star(\chi) = b(\chi)$, and*
- *the function $x^\star$ is partially differentiable in the neighborhood with partial derivative with respect to χ_j at $\chi = \mathbf{0}$ given by:*

$$\frac{\partial x^\star}{\partial \chi_j}(\mathbf{0}) = [A(\mathbf{0})]^{-1}\left[\frac{\partial b}{\partial \chi_j}(\mathbf{0}) - \frac{\partial A}{\partial \chi_j}(\mathbf{0})x^{\star\star}\right], \qquad (5.16)$$

where $x^{\star\star} \in \mathbb{R}^n$ satisfies the base-case linear simultaneous equations $A(\mathbf{0})x^{\star\star} = b(\mathbf{0})$.

Proof The matrix $A(\chi)$ is invertible for all χ in a neighborhood of $\mathbf{0}$ by Exercise 5.32. Consequently, there is a well-defined solution of $A(\chi)x = b(\chi)$ for all χ in this neighborhood and for each such χ we can define the value of $x^\star(\chi)$ to be this solution. That is, for all χ within a neighborhood of $\mathbf{0}$ we have that $A(\chi)x^\star(\chi) = b(\chi)$. (Since $A(\mathbf{0})$ is non-singular, the solution is unique and we have that $x^\star(\mathbf{0}) = x^{\star\star}$.)
That is, $x^\star(\chi) = [A(\chi)]^{-1}b(\chi)$ for all χ in this neighborhood. By Exercise 5.32, the inverse $[A(\chi)]^{-1}$ is partially differentiable with respect to χ_j in the neighborhood. Moreover, the partial derivative is continuous. Therefore, $x^\star(\chi)$, being the product of partially differentiable functions with continuous partial derivatives, is also partially differentiable with respect to χ_j in the neighborhood.
Totally differentiating $A(\chi)x^\star(\chi) = b(\chi)$ with respect to χ_j, evaluating at $\chi = \mathbf{0}$, and re-arranging yields (5.16). $\square$

If we assume that the function $x^\star : \mathbb{R}^s \to \mathbb{R}^n$ defined in Theorem 5.7 is well-defined and partially differentiable then (5.16) can be calculated directly by totally differentiating the equation $A(\chi)x^\star(\chi) = b(\chi)$ with respect to χ_j, evaluating the terms at $\chi = \mathbf{0}$, and re-arranging. However, the difficulty in the proof is to prove that the function $x^\star$ exists and is actually partially differentiable. We can also use the **implicit function theorem** [72, section 4.4], Theorem A.9 in Section A.7.3 of Appendix A, to prove that the function $x^\star : \mathbb{R}^s \to \mathbb{R}^n$ exists and is partially differentiable.

5.6.1.2 Discussion

If we have already factorized the base-case coefficient matrix $A(\mathbf{0})$ then (5.16) shows that the sensitivity of $x^\star$ with respect to variation in χ_j can be calculated with one additional forwards and backwards substitution using the right-hand side $\frac{\partial b}{\partial \chi_j}(\mathbf{0}) - \frac{\partial A}{\partial \chi_j}(\mathbf{0})x^{\star\star}$. We have therefore achieved our goal of calculating the sensitivity with much less effort than would be required for explicitly solving for a change-case.

Finding the partial derivative of $x^\star$ with respect to all entries of $\chi \in \mathbb{R}^s$ requires s forwards and backwards substitutions. Each forwards and backwards substitution provides a sensitivity $\frac{\partial x^\star}{\partial \chi_j}(\mathbf{0})$.

Since the base-case solution $x^{\star\star}$ in Theorem 5.7 is equal to $x^{\star}(0)$, we will from now on abuse notation somewhat and usually write $x^{\star}$ for the base-case solution and *also* for the function that represents the dependence of the solution on χ. That is, depending on context:

- $x^{\star}$ will sometimes stand for the particular base-case optimal value $x^{\star\star} \in \mathbb{R}^n$, and
- $x^{\star}$ will sometimes stand for the function $x^{\star} : \mathbb{R}^s \to \mathbb{R}^n$ satisfying $A(\chi)x^{\star}(\chi) = b(\chi)$.

Since we are usually only interested in the base-case solution for $\chi = 0$ and its sensitivity evaluated at $\chi = 0$, this will not be ambiguous.

We will use Theorem 5.7 in two ways, one straightforward and one more subtle. The first way leads to **direct sensitivity analysis**, while the second leads to **adjoint sensitivity analysis**. Both approaches follow the discussion in [95, chapter 9].

5.6.1.3 Direct sensitivity analysis

As mentioned above, a single forwards and backwards substitution is required to calculate the sensitivity of $x^{\star}$ to an entry χ_j of $\chi \in \mathbb{R}^s$. Sensitivity analysis based on direct application of Theorem 5.7 is called **direct sensitivity analysis**.

Example Recall the example of sensitivity analysis for linear equations (2.34) introduced in Section 2.7.5.1, where $A : \mathbb{R} \to \mathbb{R}^{2\times 2}$ and $b : \mathbb{R} \to \mathbb{R}^2$ were specified by:

$$\forall \chi \in \mathbb{R}, A(\chi) = \begin{bmatrix} 1 & 2 + \chi \\ 3 & 4 \end{bmatrix}, b(\chi) = \begin{bmatrix} 1 \\ 1 + \chi \end{bmatrix}.$$

The base-case solution of these equations is:

$$x^{\star}(0) = \begin{bmatrix} -1 \\ 1 \end{bmatrix}.$$

Moreover:

$$\frac{\partial A}{\partial \chi}(\chi) = \frac{\partial A}{\partial \chi}(0),$$

$$= \begin{bmatrix} 0 & 1 \\ 0 & 0 \end{bmatrix},$$

$$\frac{\partial b}{\partial \chi}(\chi) = \frac{\partial b}{\partial \chi}(0),$$

$$= \begin{bmatrix} 0 \\ 1 \end{bmatrix},$$

$$\frac{\partial b}{\partial \chi}(0) - \frac{\partial A}{\partial \chi}(0)x^{\star}(0) = \begin{bmatrix} -1 \\ 1 \end{bmatrix}.$$

By Theorem 5.7, the sensitivity of the solution to χ is:

$$\frac{\partial x^\star}{\partial \chi}(0) = [A(0)]^{-1}\left[\frac{\partial b}{\partial \chi}(0) - \frac{\partial A}{\partial \chi}(0)x^\star\right].$$

Performing forwards and backwards substitution, we obtain:

$$\frac{\partial x^\star}{\partial \chi}(0) = \begin{bmatrix} 3 \\ -2 \end{bmatrix}.$$

Circuit case study Exercise 5.33 is an example of a direct sensitivity analysis for the ladder circuit case study. The linear equations that are solved in Exercise 5.33 correspond to circuits having the same components as the base-case circuit but with different current vectors. Instead of the current vector b as in the base-case, a circuit is solved with a "current vector" that is equal to $\frac{\partial b}{\partial \chi_j}(0) - \frac{\partial A}{\partial \chi_j}(0)x^\star(0)$. Each such circuit is called a **direct sensitivity circuit**. According to (5.16), the solution of this circuit yields the sensitivity $\frac{\partial x^\star}{\partial \chi_j}(0)$. (The phrase "current vector" is in quotes since the units of this expression is not the same as the units of current. Moreover, the solution has the units of voltage divided by the units of χ_j.)

For example, for the ladder circuit in Figure 4.3, which is repeated in Figure 5.5, there is an additional conductance of ΔG_{23} between nodes 2 and 3. The sensitivity of the solution of this circuit with respect to $\Delta G_{23} = \chi$, evaluated at $\Delta G_{23} = \chi = 0$, is given by the solution of a circuit with "current injections" (actually having units of voltage) equal to:

$$\frac{\partial b}{\partial \Delta G_{23}}(0) - \frac{\partial A}{\partial \Delta G_{23}}(0)x^\star = \mathbf{0} - \begin{bmatrix} 0 & 0 & 0 & 0 \\ 0 & 1 & -1 & 0 \\ 0 & -1 & 1 & 0 \\ 0 & 0 & 0 & 0 \end{bmatrix}x^\star,$$

where:

- $x^\star$ is the base-case solution,
- the current injections do not depend on ΔG_{23} so that $\frac{\partial b}{\partial \Delta G_{23}}(0) = \mathbf{0}$, and
- the dependence of the admittance matrix A on ΔG_{23} was discussed in Section 4.1.3.2.

The solution of the circuit is a vector of "voltages" (actually having units of voltage divided by impedance) that represent the sensitivities with respect to ΔG_{23}.

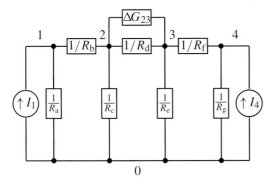

Fig. 5.5. The ladder cir-
cuit of Figure 4.3 that has a
change in the conductance
between nodes $\ell = 2$ and
$k = 3$.

5.6.1.4 Adjoint sensitivity

In this section we will suppose that there is an **objective function** $f : \mathbb{R}^n \to \mathbb{R}$
that provides the value or payoff of the solution x. We define $f^\star : \mathbb{R}^s \to \mathbb{R}$ by:

$$\forall \chi \in \mathbb{R}^s, \; f^\star(\chi) = f(x^\star(\chi)).$$

We are interested in calculating the partial derivatives of $f^\star$, again assuming that
we have a base-case solution $x^\star(\mathbf{0})$ corresponding to the parameter value $\chi = \mathbf{0}$
and also assuming that f is differentiable.

In this case, we seek:

$$
\begin{aligned}
\frac{\partial f^\star}{\partial \chi_j}(\mathbf{0}) &= \frac{d[f(x^\star(\chi))]}{d\chi_j}(\mathbf{0}), \\[2ex]
&= \frac{\partial f}{\partial x}(x^\star(\mathbf{0}))\frac{\partial x^\star}{\partial \chi_j}(\mathbf{0}), \text{ by the chain rule [72, section 2.4]}, \\[2ex]
&= \frac{\partial f}{\partial x}(x^\star(\mathbf{0}))[A(\mathbf{0})]^{-1}\left[\frac{\partial b}{\partial \chi_j}(\mathbf{0}) - \frac{\partial A}{\partial \chi_j}(\mathbf{0})x^\star(\mathbf{0})\right], \quad (5.17)
\end{aligned}
$$

by (5.16). Let us define $\xi \in \mathbb{R}^n$ to be the solution of:

$$[A(\mathbf{0})]^\dagger \xi = \nabla f(x^\star(\mathbf{0})). \qquad (5.18)$$

Solving for ξ in (5.18) and taking the transpose of the result yields:

$$\xi^\dagger = \frac{\partial f}{\partial x}(x^\star(\mathbf{0}))[A(\mathbf{0})]^{-1}.$$

Substituting into (5.17), we obtain:

$$\frac{\partial f^\star}{\partial \chi_j}(\mathbf{0}) = \xi^\dagger \left[\frac{\partial b}{\partial \chi_j}(\mathbf{0}) - \frac{\partial A}{\partial \chi_j}(\mathbf{0})x^\star(\mathbf{0})\right].$$

Calculation of the vector ξ in (5.18) requires the solution of a linear equation
with coefficient matrix $[A(\mathbf{0})]^\dagger$. If the base-case matrix $A(\mathbf{0})$ has been factorized

into $A(0) = LU$ then $[A(0)]^\dagger = [LU]^\dagger = U^\dagger L^\dagger$ by Exercise 5.18 and so the factors from the base-case can be re-used to obtain ξ. Forwards substitution is used to solve $U^\dagger \zeta = \nabla f(x^\star)$ and then backwards substitution is used to solve $L^\dagger \xi = \zeta$. The roles of L and U are swapped compared to solving an equation such as $Ax = b$. The calculation is called an **adjoint sensitivity analysis**. After ξ has been calculated with one forwards and backwards substitution, sensitivities of $f^\star$ with respect to all entries of χ can be evaluated.

Example Consider the previous example and the objective function $f : \mathbb{R}^2 \to \mathbb{R}$ defined in (2.7) from Section 2.3.1:

$$\forall x \in \mathbb{R}^2, \ f(x) = (x_1)^2 + (x_2)^2 + 2x_2 - 3.$$

We have that:

$$\forall x \in \mathbb{R}^2, \nabla f(x) = \begin{bmatrix} 2x_1 \\ 2x_2 + 2 \end{bmatrix},$$

$$\nabla f(x^\star) = \begin{bmatrix} -2 \\ 4 \end{bmatrix}.$$

In this case, (5.18) becomes:

$$\begin{bmatrix} 1 & 3 \\ 2 & 4 \end{bmatrix} \xi = \begin{bmatrix} -2 \\ 4 \end{bmatrix},$$

which has solution $\xi = \begin{bmatrix} 10 \\ -4 \end{bmatrix}$, so that:

$$\frac{\partial f^\star}{\partial \chi}(0) = \xi^\dagger \left[\frac{\partial b}{\partial \chi}(0) - \frac{\partial A}{\partial \chi}(0)x^\star(0) \right],$$

$$= [10 \ -4] \begin{bmatrix} -1 \\ 1 \end{bmatrix},$$

$$= -14.$$

Circuit case study Exercise 5.34 is an example of an adjoint sensitivity analysis for the DC circuit case study of Section 4.1. The linear equation that is solved in Exercise 5.34 corresponds to a circuit that has entries in its admittance matrix that are the *transpose* of those in the base-case and has entries in its "current vector" that are defined in terms of the sensitivity of the objective function. The circuit is called the **adjoint sensitivity circuit**. One solution of the adjoint sensitivity circuit suffices for sensitivities of an objective function with respect to all parameters of interest.

In the case of resistive circuits with current sources, the admittance matrix is symmetric, so that the resistors in the adjoint circuit are the same as those in the base-case circuit. For some circuits, however, the admittance matrix is not symmetric and the adjoint circuit has components that are different from those in the base-case circuit.

5.6.2 Large changes

In this section, we discuss large changes in the right-hand side and large changes in the coefficient matrix of a linear system.

5.6.2.1 Right-hand side

In the discussion so far we have developed techniques that enable us to solve the linear equations $Ax = b$ for a given square, non-singular A and an *arbitrary b*. That is, we can easily accommodate large changes in b, either by:

(i) re-solving the equations with the new value of b, or

(ii) solving for the change Δx in x to match the change Δb in b.

In the second case, we assume that we have already have a solution $x^\star$ that satisfies $Ax^\star = b$ and now we want to find Δx that satisfies $A(x^\star + \Delta x) = b + \Delta b$, where Δb is the change in the right-hand side. We must solve $A\Delta x = \Delta b$. The computational effort using forwards and backwards substitution as described in Section 5.2 is on the order of $(n)^2$. However, we saw a case in Section 5.5.2.2 where the effort is much smaller than $(n)^2$ if Δb has only a few non-zero entries. (See Exercise 5.25.)

To summarize, the solution of the system $Ax = b$ is a linear function of the right-hand side vector b. That is, if a sensitivity analysis is carried out with respect to parameters that are all entries of b then the sensitivity and large change analysis yield the same result.

5.6.2.2 Coefficient matrix

Suppose that we have factorized a symmetric matrix A into $LDL^\dagger$, but that we want to now solve the changed equation $(A + \Delta A)x = b$. Since the factorization of A involved considerable effort, we may wish to avoid the effort of factorizing $A + \Delta A$ by making use of the factors of A. We will see that this is practical for certain types of changes ΔA for which the effort to *update* the factorization of A is less than the effort to factorize $A + \Delta A$. The development follows [45, section 2.2.5.7].

Assuming that A has been factorized as $LDL^{\dagger}$, we have:

$$\begin{aligned} A + \Delta A &= LDL^{\dagger} + \Delta A, \\ &= LDL^{\dagger} + LL^{-1}\Delta A[L^{-1}]^{\dagger}L^{\dagger}, \text{ since } LL^{-1} = \mathbf{I}, \\ &= L(DL^{\dagger} + L^{-1}\Delta A[L^{-1}]^{\dagger}L^{\dagger}), \end{aligned}$$

collecting the common factor on the left,

$$= L(D + L^{-1}\Delta A[L^{-1}]^{\dagger})L^{\dagger},$$

collecting the common factor on the right.

Now suppose that we can factorize the matrix $D + L^{-1}\Delta A[L^{-1}]^{\dagger}$ into $\hat{L}\hat{D}\hat{L}^{\dagger}$ with $\hat{L}$ lower triangular with ones on its diagonal and $\hat{D}$ diagonal. Then, we would have:

$$\begin{aligned} A + \Delta A &= L(D + L^{-1}\Delta A[L^{-1}]^{\dagger})L^{\dagger}, \\ &= L\hat{L}\hat{D}\hat{L}^{\dagger}L^{\dagger}, \\ &= \tilde{L}\hat{D}\tilde{L}^{\dagger}, \end{aligned}$$

where $\tilde{L} = L\hat{L}$. By Exercise 5.3, the product of two lower triangular matrices is lower triangular, so that $\tilde{L} = L\hat{L}$ is lower triangular and, moreover, has ones on its diagonal. That is, $A + \Delta A = \tilde{L}\hat{D}\tilde{L}^{\dagger}$ and we have factorized the changed matrix.

The practicality of this approach depends on being able to factorize the matrix $D + L^{-1}\Delta A[L^{-1}]^{\dagger}$ using less effort than it takes to factorize $A + \Delta A$. This is not true if ΔA is an arbitrary change, but is true for some restricted forms of ΔA that are nevertheless extremely useful in applications.

For example, suppose that:

- $\gamma, \delta \in \mathbb{R}$ with γ and δ non-zero, and
- $w, u \in \mathbb{R}^n$ with w and u linearly independent (see Definition A.55).

Then, the particular forms:

- $\Delta A = \gamma ww^{\dagger} \in \mathbb{R}^{n \times n}$, which is called a **symmetric rank one update**, since the matrix is of rank one (see Definition A.57), and
- $\Delta A = \gamma ww^{\dagger} + \delta uu^{\dagger} \in \mathbb{R}^{n \times n}$, which is called a **symmetric rank two update**, since the matrix is of rank two (see Definition A.57),

allow for convenient calculations of $\tilde{L}$ and $\hat{D}$ from L, D, and γ, δ, w, u. The computational effort involved is on the order of $(n)^2$, which is considerably less than the effort involved in factorizing $A + \Delta A$ directly. The forwards and backwards substitution to solve for the new value of x also requires effort on the order of $(n)^2$. The details for $\Delta A = \gamma ww^{\dagger}$, the symmetric rank one update, are presented in [45, section 2.2.5.7].

Exercise 4.8 showed that changes in resistance in circuit problems involve a change in the admittance matrix that is a symmetric rank one update. Therefore, the effect of changes in resistance can be calculated using this technique. It is possible to repeat this process if there are several successive changes to the matrix. However, it is prudent to occasionally factorize the resulting matrix directly to avoid accumulation of round-off errors. There are also various special considerations if the changed matrix is nearly equal to a singular matrix. As with small-signal sensitivity analysis, for the circuit case study there are interpretations of the changed solution in terms of circuits [95, chapter 8].

5.6.3 New variables and equations

We can also consider augmenting a system of equations by adding a new variable and a new equation. Methods to make use of the existing factorization in this case are described in [45, section 2.2.5.7]. In the context of circuit analysis, this allows us to add a node to the circuit. See [95, chapter 8] for details.

5.7 Ill-conditioning

In describing algorithms, we have mostly assumed that all numbers are represented to infinite precision and that all calculations are performed to infinite precision. In practice, we can often assume that numbers are represented with enough precision and that calculations are performed to enough precision so that the answers are "almost exact" and they do not "underflow" or "overflow" the representation of the numbers in the computer. However, this is not always a valid assumption when issues of **numerical conditioning** prove critical to understand the nature of errors in the calculations [45, section 2.1][121, chapter 4]. We mentioned this issue in Sections 5.3.2.1 and 5.5.3.4 in reference to pivot selection. In Section 5.7.1 we will discuss issues related to numerical conditioning in more detail.

In Section 5.7.2 we will discuss the ideas of scaling and pre-conditioning and then relate them to numerical conditioning. In Section 5.7.3, we will discuss the factorization of ill-conditioned matrices. We will describe the issue for LU and $LDL^{\dagger}$ factorization and then introduce another factorization, QR, that can be helpful in solving systems of equations that involve ill-conditioned matrices.

5.7.1 Numerical conditioning and condition number

5.7.1.1 Discussion

In Section 5.3 we showed that if A is non-singular then we can factorize it, while if it is singular then at some stage we will find that there are no non-zero pivots. We avoided discussion of the issue of when a coefficient matrix is "nearly" singular in the sense that a small perturbation of the matrix (for example, due to representation or round-off error) would make it singular. This issue is related to the problem of small pivots mentioned in Sections 5.3.2.1 and 5.5.3.4.

5.7.1.2 Example

For example, consider:

$$A = \begin{bmatrix} 1 & \delta \\ 1 & 0 \end{bmatrix}. \tag{5.19}$$

If $\delta \neq 0$, then under the assumption of infinite precision arithmetic, we could reliably factorize A and solve $Ax = b$ exactly for the solution $x^\star$.

However, if δ is small in magnitude, then A is "nearly" singular in that perturbing δ to make it equal to zero would make A singular. In this case, we will see that small relative errors in the specification of A or b (or in the calculations to factorize A or to perform forwards or backwards substitution) lead to large relative errors in the value of the solution $x^\star$. That is, the problem of solving $Ax = b$ given the A defined in (5.19) is ill-conditioned according to Definition 2.21.

We demonstrate that the problem of solving $Ax = b$, with A specified as in (5.19), is ill-conditioned. First note that by Cramér's rule,

$$A^{-1} = \begin{bmatrix} 0 & 1 \\ 1/\delta & -1/\delta \end{bmatrix}.$$

Also, given $b = \begin{bmatrix} b_1 \\ b_2 \end{bmatrix}$, then:

$$x^\star = A^{-1}b = \begin{bmatrix} b_2 \\ (b_1/\delta) - (b_2/\delta) \end{bmatrix}. \tag{5.20}$$

As a concrete example, let $b = \begin{bmatrix} 1 \\ 1 \end{bmatrix}$ so that $\|b\|_2 = \sqrt{2}$. According to (5.20), the solution to $Ax = b$ is $x^\star = \begin{bmatrix} 1 \\ 0 \end{bmatrix}$, so that $\|x^\star\|_2 = 1$. We consider, in turn, changes to b and to A assuming infinite precision arithmetic.

Right-hand side Consider the "perturbed" system $Ax = b + \Delta b$, with $\Delta b = \begin{bmatrix} \chi \\ 0 \end{bmatrix}$, so that $\|\Delta b\|_2 = |\chi|$. The solution $x^\star + \Delta x^\star$ to this system satisfies $\Delta x^\star = \begin{bmatrix} 0 \\ \chi/\delta \end{bmatrix}$, so that $\|\Delta x^\star\|_2 = |\chi/\delta|$. The relative change in the norm of the solution is:

$$\frac{\|\Delta x^\star\|_2}{\|x^\star\|_2} = |\chi/\delta|,$$

while the relative change in the norm of the right-hand side was:

$$\frac{\|\Delta b\|_2}{\|b\|_2} = \frac{|\chi|}{\sqrt{2}}.$$

Combining these observations, the relative change in the norm of the solution is on the order of $|1/\delta|$ times the relative change in the norm of the right-hand side. That is, the norm of the sensitivity of the solution to χ is on the order of $|1/\delta|$. For small $|\delta|$, this is a large sensitivity to χ. That is, by Definition 2.21, the problem of solving $Ax = b$ is ill-conditioned because a relatively small change in the problem specification leads to a relatively large change in the solution.

Coefficient matrix Now consider the perturbed system $(A + \Delta A)x = b$, with $\Delta A = \begin{bmatrix} \chi & 0 \\ 0 & 0 \end{bmatrix}$, so that $\|\Delta A\|_2 = |\chi|$, where $\|\bullet\|_2$ is the induced matrix L_2 norm. (See Definition A.30 and Exercise 5.35.) Also $\|A\|_2 \approx \sqrt{2}$. The solution $x^\star + \Delta x^\star$ to this system satisfies $x^\star + \Delta x^\star = \begin{bmatrix} 1 \\ \chi/\delta \end{bmatrix}$, $\Delta x^\star = \begin{bmatrix} 0 \\ \chi/\delta \end{bmatrix}$, so that $\|\Delta x^\star\|_2 = |\chi/\delta|$. The relative change in the norm of the solution is:

$$\frac{\|\Delta x^\star\|_2}{\|x^\star + \Delta x^\star\|_2} \approx |\chi/\delta|,$$

if $|\chi/\delta|$ is small. The relative change in the norm of the coefficient matrix is:

$$\frac{\|\Delta A\|_2}{\|A\|_2} \approx \frac{|\chi|}{\sqrt{2}}.$$

Again, the relative change in the solution is on the order of $|1/\delta|$ times the relative change in the coefficient matrix. That is, the sensitivity to χ is large and, by Definition 2.21, the problem of solving $Ax = b$ is again ill-conditioned because a relatively small change in the problem specification leads to a relatively large change in the solution.

5.7.1.3 Analysis

As well as the sensitivity of the solution to changes in the exact value of A and b, evaluation of (5.20) may be significantly affected by the representation and round-off error in the computational system being used. For example, if the exact value of δ is 1.3×10^{-20}, but the closest number that can be represented in the computer is 1×10^{-20}, then there will be a large error in the computed solution compared to a value calculated with infinite precision representation and arithmetic. This issue is exacerbated when there are many calculations that must be performed to obtain the solution as, for example, in the calculations required for factorization and forwards and backwards substitution. As each calculation is performed, its solution is rounded off to the nearest number in the finite precision representation used by the computer. Errors accumulate as calculations proceed [45, section 2.1]. We will return to this issue in Section 5.7.3.

The degree of ill-conditioning is characterized by a measure known as the **condition number** of the matrix [45, section 2.2.4.3].

Definition 5.1 Let $\|\bullet\|$ stand for vector and matrix norms on $\mathbb{R}^n$ and $\mathbb{R}^{n \times n}$ that are compatible. (See Definition A.32.) For example, the matrix norm $\|\bullet\|$ could be the matrix norm induced by the vector norm. (See Definition A.30.) Suppose that $A \in \mathbb{R}^{n \times n}$ is non-singular. Then the **condition number** of A is defined by $\|A\| \|A^{-1}\|$. If $A \in \mathbb{R}^{n \times n}$ is singular then the condition number is defined to be ∞. $\square$

The condition number lies in the range from one to infinity. A matrix with a "large" condition number is said to be "ill-conditioned," while a matrix with a "small" condition number is said to be "well-conditioned." The MATLAB function cond evaluates the condition number using the L_2 norm. The significance of the condition number is contained in the following theorem ([45, section 2.2.4.3]).

Theorem 5.8 *Let $\|\bullet\|$ stand for vector and matrix norms on $\mathbb{R}^n$ and $\mathbb{R}^{n \times n}$ that are compatible. (See Appendix A.3.2.) Suppose that $A \in \mathbb{R}^{n \times n}$ is non-singular and $b \in \mathbb{R}^n$. We consider the relation between solutions of the system $Ax = b$ and solutions of the perturbed systems $Ax = b + \Delta b$ and $(A + \Delta A)x = b$. We have the following bounds.*

(i) *Consider the perturbed system $Ax = b + \Delta b$. The solution $x^\star + \Delta x^\star$ to this perturbed system satisfies:*

$$\frac{\|\Delta x^\star\|}{\|x^\star\|} \leq \|A\| \|A^{-1}\| \frac{\|\Delta b\|}{\|b\|},$$

where $x^\star$ is the solution to $Ax = b$. That is, the relative change in the solution is bounded by the product of the condition number and the relative change in the right-hand side.

(ii) *Consider the perturbed system $(A + \Delta A)x = b$. The solution $x^\star + \Delta x^\star$ to this system satisfies:*

$$\frac{\|\Delta x^\star\|}{\|x^\star + \Delta x^\star\|} \leq \|A\| \|A^{-1}\| \frac{\|\Delta A\|}{\|A\|},$$

where $x^\star$ is the solution to $Ax = b$. That is, the relative change in the solution is bounded by the product of the condition number and the relative change in the coefficient matrix.

Proof   See [45, section 2.2.4.3] and Exercise 5.36. $\square$

Theorem 5.8 suggests that changes in A or b may result in changes in the solution that are "amplified" by as much as the condition number of A. Note that the condition number is closely related to the sensitivity of the solution to changes in A and b as determined by the sensitivity analysis in Theorem 5.7. In particular, if the condition number is large then the sensitivity as calculated in Theorem 5.7 will also be large.

Consider the matrix A defined in (5.19) and suppose that δ is small. If the induced matrix norm $\|\bullet\|_2$ is chosen, then:

- $\|A\|_2 \approx \sqrt{2}$, and
- $\left\|A^{-1}\right\|_2 \approx |1/\delta|$,

so that the condition number is proportional to $|1/\delta|$. According to Theorem 5.8, relatively small changes in either the right-hand side b or the coefficient matrix A of the system $Ax = b$ can potentially produce large relative changes in the solution with the amplification proportional to $|1/\delta|$. Moreover, applying Theorem 5.7, we obtain that the norm of the sensitivity to χ is $|1/\delta|$. These observations are consistent with the above calculations for the matrix defined in (5.19) since the changes in A and b were indeed amplified by $|1/\delta|$ in the solution.

5.7.2 Scaling and pre-conditioning

Scaling can sometimes be used effectively to reduce the condition number of a matrix. For example, consider the matrix:

$$A = \begin{bmatrix} \delta & 0 \\ 1 & 1 \end{bmatrix}, \tag{5.21}$$

which has inverse:

$$A^{-1} = \begin{bmatrix} 1/\delta & 0 \\ -1/\delta & 1 \end{bmatrix},$$

so that the solution to $Ax = b$ is given by:

$$x^\star = A^{-1}b = \begin{bmatrix} b_1/\delta \\ -(b_1/\delta) + b_2 \end{bmatrix}. \tag{5.22}$$

If the $\|\bullet\|_2$ induced matrix norm is again used, then for small δ we have that:

- $\|A\|_2 \approx 1$,
- $\left\|A^{-1}\right\|_2 \approx \left|\sqrt{2}/\delta\right|$,

so that the condition number is again proportional to $|1/\delta|$. By scaling the first equation of $Ax = b$ by multiplying it by $1/\delta$ we obtain the new system:

$$\begin{bmatrix} 1 & 0 \\ 1 & 1 \end{bmatrix} x = \begin{bmatrix} b_1/\delta \\ b_2 \end{bmatrix}, \tag{5.23}$$

and the coefficient matrix now has a condition number that is a small constant that is independent of δ. Of course, we still face the issue that the solution in (5.22) is very dependent on the value of δ; however, the condition number of the coefficient matrix has improved. We will return to the implications of this observation in Section 5.7.3 and Exercise 5.38.

We can generalize this idea by multiplying the equations through by a matrix M. We called this **pre-conditioning** in Section 3.3.1. The reason for this name is that multiplying the system can change the condition number of the coefficient matrix. The ideal would be to multiply the system $Ax = b$ by the inverse A^{-1} because then we would have the system $\mathbf{I}x = A^{-1}b$, and $\mathbf{I}$ is trivial to factorize (and has condition number 1 using the $\|\bullet\|_2$ norm).

In practice, if an approximate inverse for A can be found then this can be used to pre-condition the system. A common choice is to pre-condition by a diagonal matrix. For example, the diagonal matrix that has entries equal to the inverse of the diagonal entries of A can be used. Diagonal pre-conditioning was used for the matrix in (5.21) to produce the coefficient matrix in (5.23) and simply involves scaling the equations.

It is important to realize that pre-conditioning will not remove the sensitivity of the solution to changes in the originally specified coefficient matrix A and vector b. However, the condition number of the resulting coefficient matrix can be improved. Scaling of variables can also be applied to improve the condition number of the matrix. (See Exercise 5.39.) Both approaches can facilitate factorization by avoiding the factorization of an ill-conditioned matrix, which *exacerbates* the ill-conditioning of the originally specified problem. We will discuss factorization of ill-conditioned matrices in the next section, using the notion of pre-conditioning to help to explain how the steps in factorization exacerbate ill-conditioning.

5.7.3 Matrix factorization

5.7.3.1 LU factorizing ill-conditioned systems

We observed in Sections 5.3.2.1 and 5.5.3.4 that if we use small pivots then it is possible that large numerical errors will be introduced into our calculations. We showed that these errors would yield the same result as if we had calculated to infinite precision but had started with a perturbed matrix. The analysis of ill-conditioning has shown that for an ill-conditioned system, such a perturbation will result in a large change in the solution of the perturbed system compared to the exact system. Unfortunately, many of the candidate pivots for an ill-conditioned matrix may be small as shown in Exercise 5.38.

As Exercise 5.38 suggests, if we LU factorize the coefficient matrix of an ill-conditioned system, we may find that some or all of the pivots are small. In the presence of finite precision representation of numbers and round-off errors during computation, large errors can arise in the entries of L and U. We have noted that pivot schemes can seek the largest available pivot to try to avoid this issue; however, if the system is sparse and a fixed ordering of the pivots is chosen in advance to minimize the number of pivots, then it is not always possible to guarantee that the pivots will be acceptably large. (See Exercise 5.40.)

Even if the largest available pivots are selected at each step of factorization, some of the pivots will still be small for an ill-conditioned matrix. Representation and round-off errors will mean that the calculated factors differ significantly from the exact factors. Solving the system using forwards and backwards substitution will produce results that do not even approximately solve the original system $Ax = b$.

To gain further insights into the accumulation of errors throughout a factorization, we will interpret the steps involved in factorization as "pre-conditioning" as discussed in Section 5.7.2. If we encounter small pivots in LU factorization then there will be some large entries in the matrices $M^{(j)}$ and $[M^{(j)}]^{-1}$. (See Section 5.3.2.) Because of this, we can make an already ill-conditioned matrix worse through the LU factorization. In particular, the condition number of the lower triangular matrix:

$$L = [M^{(n-1)} \ldots M^{(1)}]^{-1},$$

is:

$$\|L\| \, \|L^{-1}\| = \left\| [M^{(n-1)} \ldots M^{(1)}]^{-1} \right\| \, \left\| M^{(n-1)} \ldots M^{(1)} \right\|,$$

which by Lemma A.1 is bounded by:

$$\left\| [M^{(n-1)}]^{-1} \right\| \ldots \left\| [M^{(1)}]^{-1} \right\| \, \left\| M^{(n-1)} \right\| \ldots \left\| M^{(1)} \right\| =$$

$$\left\| M^{(n-1)} \right\| \, \left\| [M^{(n-1)}]^{-1} \right\| \ldots \left\| M^{(1)} \right\| \, \left\| [M^{(1)}]^{-1} \right\|,$$

which is the product of the condition numbers of the matrices $M^{(n-1)}, \ldots, M^{(1)}$. Since forward substitution solves the system $Ly = b$, if any of the $M^{(j)}$ are ill-conditioned then the solution $y^\star$ to $Ly = b$ will depend sensitively on b and on round-off errors in the calculation of the $M^{(j)}$.

Similarly, the condition number of the upper triangular matrix:

$$U = M^{(1)} \ldots M^{(n-1)} A,$$

is given by:

$$\|U\| \, \|U^{-1}\| = \left\| M^{(1)} \ldots M^{(n-1)} A \right\| \, \left\| [M^{(1)} \ldots M^{(n-1)} A]^{-1} \right\|,$$

which by Lemma A.1 is bounded by:

$$\left\| M^{(n-1)} \right\| \ldots \left\| M^{(1)} \right\| \, \|A\| \, \left\| [M^{(n-1)}]^{-1} \right\| \ldots \left\| [M^{(1)}]^{-1} \right\| \, \|A^{-1}\| =$$

$$\left\| M^{(n-1)} \right\| \, \left\| [M^{(n-1)}]^{-1} \right\| \ldots \left\| M^{(1)} \right\| \, \left\| [M^{(1)}]^{-1} \right\| \, \|A\| \, \|A^{-1}\|,$$

which is the product of:

- the condition numbers of $M^{(n-1)}, \ldots, M^{(1)}$, and
- the condition number of A.

Since backward substitution solves the system $Ux = y^\star$, if any of the $M^{(j)}$ are ill-conditioned or if A is ill-conditioned then the solution $x^\star$ to $Ux = y^\star$ will depend sensitively on $y^\star$ and on round-off errors in the calculation of the $M^{(j)}$.

Unfortunately, both $M^{(j)}$ and $[M^{(j)}]^{-1}$ have entries that are proportional to the inverse of the pivot used at the j-th stage and their norms will both be correspondingly large. (See Exercise 5.41.) The bound on the condition numbers of L and U will grow large if we do not or cannot avoid small pivots. Even if A were itself well-conditioned, if we choose small pivots we would find that the matrices L and U are ill-conditioned. That is, in factorizing we can transform a well-conditioned problem into the solution of two ill-conditioned problems.

Our suggestions for pivoting included choosing the largest available pivot; however, it is still generally the case that the condition number of U can be much larger than that of A. Similarly, the condition number of L can be much larger than 1. In summary, the effect of pre-conditioning by $M^{(1)}, \ldots, M^{(n-1)}$ is to increase the condition number of the resulting system, which exacerbates the ill-conditioning of A. If A is ill-conditioned then the resulting systems $Ly = b$ and $Ux = y$ can be extremely ill-conditioned. (For detailed analysis of the "growth" of the condition number, see [45, section 2.2.5.1].)

5.7.3.2 $LDL^\dagger$ for positive definite A

The key issue in LU factorization is that the pre-conditioning matrices $M^{(j)}$ worsen the condition number of the resulting coefficient matrix U compared to the condition number of A and worsen the condition number of L compared to the condition number of $\mathbf{I}$. It turns out that under some circumstances the worsening of the condition number is relatively mild. For example, in the case of a strictly diagonally dominant matrix such as in our circuit case study of Section 4.1.1, the largest pivots are on the diagonal and diagonal pivoting will keep the condition number of the system as low as possible. This favorable circumstance also occurs for $LDL^\dagger$ factorization of any symmetric positive definite matrix. This property of symmetric positive definite matrices allowed us to concentrate on pivoting to preserve sparsity rather than worrying about the size of pivots for factorization of symmetric positive definite systems. See [45, section 2.2.5.2] for further details.

5.7.3.3 QR

If the matrix is not symmetric and positive definite, then we have indicated that LU factorization can lead to unacceptable performance because the ill-conditioning of the original matrix is effectively worsened by the construction of the LU factors. The basic problem is that the matrices $M^{(j)}$ are themselves ill-conditioned if small pivots are selected. An alternative factorization involves multiplying by a sequence of matrices $M^{(j)}$ for which $\left\|M^{(j)}\right\|_2 = \left\|[M^{(j)}]^{-1}\right\|_2 = 1$ so that the condition number of L is one and the condition number of U is the same as the condition number of A.

One particular choice of the pre-conditioning matrices $M^{(j)}$ yields a product, Q, that has the property of being **unitary**. That is, $Q^\dagger Q = \mathbf{I}$. Each column of a unitary matrix Q is "perpendicular" to every other column and each column has $\|\bullet\|_2$ norm equal to 1.

(We have previously used the symbol Q to stand for the quadratic coefficient matrix of a quadratic function. For example, see Section 2.5.3.2. In the context of factorization, however, we will follow the literature and also use the symbol Q for the factor of A.)

The resulting factorization is called the QR factorization and can be applied to an $m \times n$ matrix A with $m \geq n$ and having linearly independent columns to produce a factorization $A = QR$ where $Q \in \mathbb{R}^{m \times m}$ is unitary and $R \in \mathbb{R}^{m \times n}$ is upper triangular. That is, $R_{\ell k} = 0$ for $\ell > k$.

(We have previously used the symbol R to stand for the rate of convergence of an iterative algorithm and for electrical resistance. See Definition 2.10 and Section 4.1. In the context of factorization, we will follow the literature and also use the symbol R for the factor of the matrix.)

The condition number of Q is 1 using the $\|\bullet\|_2$ norm. The system $Qy = b$ has a coefficient matrix Q that is well-conditioned. The system $Rx = y$ has a coefficient matrix with condition number that is equal to the condition number of A. That is, unlike LU factorization, QR factorization does not exacerbate the ill-conditioning of the original matrix A.

As with LU factorization, we may consider full or partial pivoting, in which case the product QR is equal to a re-ordered version of the columns or rows of A. If the matrix A has some linearly dependent columns, and so has rank $n' < n$, say, then we may have to re-order the columns to avoid zero pivots. In this case, we will obtain a factorization $AP = QR$ where:

- $P \in \mathbb{R}^{n \times n}$ is a permutation matrix,
- $Q \in \mathbb{R}^{m \times m}$ is unitary, and
- $R \in \mathbb{R}^{m \times n}$ is upper triangular, but will have zero entries in rows $n' + 1$ through m.

We will not discuss the calculations involved in QR factorization in detail, but many software packages, such as MATLAB and LAPACK, can perform QR factorization. For example, the MATLAB command for QR factorization is qr [74]. The MATLAB command qrinsert takes a QR factorization of a matrix and finds the QR factors of the matrix with an additional column augmented to it.

The main drawbacks of QR factorization are that:

- it takes more computational effort than LU factorization, and
- the matrix Q will not usually be sparse even if A is sparse. (See Exercise 5.42.)

5.8 Non-square systems

In this section we will consider the solution of non-square systems. We will first consider the case of more variables than equations in Section 5.8.1 and then the case of more equations than variables in Section 5.8.2. Finally, we will unify the two cases using the notion of the **pseudo-inverse** in Section 5.8.3.

5.8.1 More variables than equations

Consider the system $Ax = b$ where $A \in \mathbb{R}^{m \times n}$, $b \in \mathbb{R}^m$, and $m < n$.

5.8.1.1 Inconsistent equations

Recall that a system of equations is called **inconsistent** if there is no solution. The system is inconsistent if some of the rows of $A \in \mathbb{R}^{m \times n}$ can be expressed as a linear combination of the other rows while the corresponding entries of b do not have the same linear relation. (See Exercise 5.43.)

Faced with an inconsistent system, we may still like to find a "solution" that satisfies the constraints as nearly as possible in some sense. This problem will turn out to be an optimization problem and we will treat it in Section 11.1.

5.8.1.2 Consistent equations and the null space

If we assume that the m rows of A are **linearly independent**; that is, no row can be expressed as a combination of the rest of the rows (see Definition A.55), then the system is **consistent**. In fact, there will be multiple solutions of $Ax = b$. (We will return to the case of consistent systems with linearly dependent rows in Section 5.8.2.2.)

If the m rows of A are linearly independent then there is an $m \times m$ sub-matrix of A with linearly independent columns. We first consider the case where these linearly independent columns happen to be the first m columns of A. Later, we will consider the more general case where we do not know which m columns are linearly independent by using the QR factorization that was introduced in Section 5.7.3.3.

First m columns linearly independent Suppose we are seeking solutions to $Ax = b$. Let $n' = n - m$ and partition A into $\begin{bmatrix} A^{\|} & A^{\perp} \end{bmatrix}$ where $A^{\|} \in \mathbb{R}^{m \times m}$ and $A^{\perp} \in \mathbb{R}^{m \times n'}$. That is, $A^{\|}$ is the sub-matrix of A consisting of the first m columns of A. Similarly, partition x into $\begin{bmatrix} \omega \\ \xi \end{bmatrix}$ where $\omega \in \mathbb{R}^m$ and $\xi \in \mathbb{R}^{n'}$.

Suppose that $A^{\|}$ has linearly independent columns, so that $A^{\|}$ is non-singular (or, more generally, suppose that we can permute the entries of x and the corresponding columns of A so that this is true.) Then:

$$
\begin{aligned}
Ax = b &\Leftrightarrow \begin{bmatrix} A^{\|} & A^{\perp} \end{bmatrix} \begin{bmatrix} \omega \\ \xi \end{bmatrix} = b, \text{ by definition of } \begin{bmatrix} A^{\|} & A^{\perp} \end{bmatrix} \text{ and } \begin{bmatrix} \omega \\ \xi \end{bmatrix}, \\
&\Leftrightarrow A^{\|}\omega + A^{\perp}\xi = b, \\
&\Leftrightarrow A^{\|}\omega = b - A^{\perp}\xi, \\
&\Leftrightarrow \omega = [A^{\|}]^{-1}(b - A^{\perp}\xi),
\end{aligned}
$$

since $A^{\|}$ is non-singular.

Let $\hat{\xi} = 0$ and $\hat{\omega} = [A^{\|}]^{-1}b$. Then define:

$$
\begin{aligned}
\hat{x} &= \begin{bmatrix} \hat{\omega} \\ \hat{\xi} \end{bmatrix}, \\
&= \begin{bmatrix} [A^{\|}]^{-1}b \\ 0 \end{bmatrix}.
\end{aligned}
$$

The vector $\hat{x}$ is one **particular solution** to $Ax = b$.

In practice, to calculate the particular solution $\hat{x}$ we would, for example:

- factorize $A^{\|} = L^{\|}U^{\|}$ where $L^{\|}, U^{\|} \in \mathbb{R}^{m \times m}$ are lower and upper triangular, respectively, and
- perform forwards and backwards substitution to solve $A^{\|}\omega = b$.

The set of all solutions to $Ax = b$ is given by $\{\hat{x} + \Delta x \in \mathbb{R}^n | A\Delta x = 0\}$. (See Exercise 5.45.) The set:

$$
\mathcal{N}(A) = \{\Delta x \in \mathbb{R}^n | A\Delta x = 0\},
$$

is called the **null space** of A. (See Definition A.50.) With this definition, we can express the set of all solutions to $Ax = b$ as:

$$
\{x \in \mathbb{R}^n | Ax = b\} = \{\hat{x} + \Delta x \in \mathbb{R}^n | \Delta x \in \mathcal{N}(A)\}.
$$

To characterize the null space $\mathcal{N}(A)$, partition Δx into $\begin{bmatrix} \Delta\omega \\ \Delta\xi \end{bmatrix}$, where $\Delta\omega \in \mathbb{R}^m$

and $\Delta\xi \in \mathbb{R}^{n'}$. We have:

$$A\Delta x = 0 \quad \Leftrightarrow \quad \begin{bmatrix} A^{\|} & A^{\perp} \end{bmatrix} \begin{bmatrix} \Delta\omega \\ \Delta\xi \end{bmatrix} = 0,$$

$$\Leftrightarrow \quad A^{\|}\Delta\omega + A^{\perp}\Delta\xi = 0,$$

$$\Leftrightarrow \quad A^{\|}\Delta\omega = -A^{\perp}\Delta\xi,$$

$$\Leftrightarrow \quad \Delta\omega = -[A^{\|}]^{-1}A^{\perp}\Delta\xi.$$

That is,

$$\begin{aligned} \mathcal{N}(A) &= \{\Delta x \in \mathbb{R}^n | A\Delta x = 0\}, \\ &= \left\{ \begin{bmatrix} -[A^{\|}]^{-1}A^{\perp}\Delta\xi \\ \Delta\xi \end{bmatrix} \middle| \Delta\xi \in \mathbb{R}^{n'} \right\}, \\ &= \{Z\Delta\xi | \Delta\xi \in \mathbb{R}^{n'}\}, \end{aligned}$$

where:

$$Z = \begin{bmatrix} -[A^{\|}]^{-1}A^{\perp} \\ I \end{bmatrix}.$$

We say that the columns of Z form a **basis** for the null space of A. That is, any element of the null space of A can be written as a linear combination of the columns of Z. (See Definition A.56.) Any element of the null space corresponding to a value of $\Delta\xi$ could be evaluated by factorizing $A^{\|}$ and solving $A^{\|}\Delta\omega = -A^{\perp}\Delta\xi$ by forwards and backwards substitution.

In summary, every solution of $Ax = b$ is of the form:

$$x = \begin{bmatrix} \hat{\omega} \\ 0 \end{bmatrix} + \begin{bmatrix} \Delta\omega \\ \Delta\xi \end{bmatrix},$$

where:

- $\hat{x} = \begin{bmatrix} \hat{\omega} \\ 0 \end{bmatrix}$ is a particular solution of $Ax = b$, and

- $\begin{bmatrix} \Delta\omega \\ \Delta\xi \end{bmatrix} \in \mathcal{N}(A)$.

The situation is illustrated in Figure 5.6. The null space is shown as a dashed line. One typical point $\begin{bmatrix} \Delta\omega \\ \Delta\xi \end{bmatrix} \in \mathcal{N}(A)$ is shown as a $\circ$. The particular solution $\hat{x} = \begin{bmatrix} \hat{\omega} \\ 0 \end{bmatrix}$ is also shown as a $\circ$.

The solid line represents the set of points in the set:

$$\{x \in \mathbb{R}^n | Ax = b\} = \left\{ \begin{bmatrix} \omega \\ \xi \end{bmatrix} \in \mathbb{R}^n \middle| A \begin{bmatrix} \omega \\ \xi \end{bmatrix} = b \right\}.$$

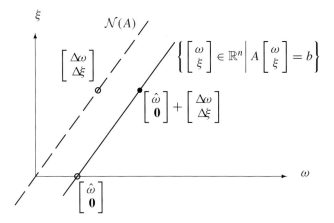

Fig. 5.6. Solution of linear equations. The solid line represents the set of points satisfying the linear equations. The null space of the coefficient matrix is shown as the dashed line.

Each point in this set can be represented as the sum of $\hat{x} = \begin{bmatrix} \hat{\omega} \\ 0 \end{bmatrix}$ and a point in the null space. The point $\begin{bmatrix} \hat{\omega} \\ 0 \end{bmatrix} + \begin{bmatrix} \Delta\omega \\ \Delta\xi \end{bmatrix} \in \{x \in \mathbb{R}^n | Ax = b\}$ is illustrated with a bullet •. For each choice of $\Delta\xi \in \mathbb{R}^{n'}$ there is a value $\Delta\omega \in \mathbb{R}^m$ such that:

$$\begin{bmatrix} \hat{\omega} \\ 0 \end{bmatrix} + \begin{bmatrix} \Delta\omega \\ \Delta\xi \end{bmatrix} \in \left\{ \begin{bmatrix} \omega \\ \xi \end{bmatrix} \in \mathbb{R}^n \middle| A \begin{bmatrix} \omega \\ \xi \end{bmatrix} = b \right\}.$$

Linearly independent columns unknown The above construction required that the first m columns of A were linearly independent, or that we knew how to re-order the columns of A so that the first m columns were linearly independent. We develop a systematic approach for the more typical case where we do not know which columns are linearly independent [84, section 3.3.4]. First, an analogous factorization to the QR factorization can be used to write $PA = LQ$, where now:

- $P \in \mathbb{R}^{m \times m}$ is a permutation matrix,
- $L \in \mathbb{R}^{m \times n}$ is lower triangular, with its first m' columns linearly independent and its last $n' = n - m'$ columns zero, and
- $Q \in \mathbb{R}^{n \times n}$ is unitary.

(For example, performing a QR factorization on $A^\dagger$ will yield the transpose of the appropriate factors. See Exercise 5.46.)

Partition L into $\begin{bmatrix} L^\| & 0 \end{bmatrix}$ where $L^\| \in \mathbb{R}^{m \times m'}$ is lower triangular with its m' columns linearly independent. (If A has m linearly independent columns then $m' = m$.) Let $n' = n - m'$ and $y = Qx$ and partition $y \in \mathbb{R}^n$ into $y = \begin{bmatrix} \omega \\ \xi \end{bmatrix}$ where

$\omega \in \mathbb{R}^{m'}$ and $\xi \in \mathbb{R}^{n'}$. Then:

$$
\begin{aligned}
Ax = b \quad &\Leftrightarrow \quad PAx = Pb, \text{ since } P \text{ is non-singular,} \\
&\Leftrightarrow \quad LQx = Pb, \text{ by definition of } LQ, \\
&\Leftrightarrow \quad Ly = Pb \text{ and } y = Qx, \\
&\Leftrightarrow \quad \begin{bmatrix} L^{\parallel} & \mathbf{0} \end{bmatrix} \begin{bmatrix} \omega \\ \xi \end{bmatrix} = Pb \text{ and } \begin{bmatrix} \omega \\ \xi \end{bmatrix} = Qx, \\
&\Leftarrow \quad L^{\parallel}\omega = Pb \text{ and } y = \begin{bmatrix} \omega \\ \mathbf{0} \end{bmatrix} = Qx.
\end{aligned}
$$

If $m' = m$ then similar arguments to before show that $\hat{y} = \begin{bmatrix} [L^{\parallel}]^{-1} Pb \\ \mathbf{0} \end{bmatrix}$ satisfies $Ly = Pb$ and that $\hat{x} = Q^{-1}\hat{y} = Q^{\dagger}\hat{y}$ satisfies $Ax = b$. That is, $\hat{x}$ is a particular solution to the linear equations. Moreover, the null space of A is:

$$
\begin{aligned}
\mathcal{N}(A) \quad &= \quad \{\Delta x \in \mathbb{R}^n \mid A\Delta x = \mathbf{0}\}, \\
&= \quad \left\{ Q^{\dagger} \begin{bmatrix} \mathbf{0} \\ \Delta\xi \end{bmatrix} \middle| \Delta\xi \in \mathbb{R}^{n'} \right\}, \\
&= \quad \{Z\Delta\xi \mid \Delta\xi \in \mathbb{R}^{n'}\},
\end{aligned}
$$

where Z is the last n' columns of $Q^{\dagger}$. That is, the last n' columns of $Q^{\dagger}$ form a **basis** for the null space of A. Moreover, the first m' columns of $Q^{\dagger}$ form a basis for the **range space** of $A^{\dagger}$. (See Definition A.50 and Exercise 5.47.) The set of all solutions to $Ax = b$ is given by $\{\hat{x} + Z\Delta\xi \mid \Delta\xi \in \mathbb{R}^{n'}\}$, where $\hat{x}$ is any particular solution.

5.8.2 *More equations than variables*

Consider the system $Ax = b$ where $A \in \mathbb{R}^{m \times n}$, $b \in \mathbb{R}^m$, and $m > n$.

5.8.2.1 *Inconsistent equations*

We saw in Exercise 5.5 that if A is square and singular, then there are values of b for which the system is inconsistent. **Inconsistent equations** typically also occur if A is non-square with more equations than variables. We saw an inconsistent system of simultaneous equations with more equations than variables in Exercise 4.10 where controlling the input of the discrete time linear system during the 0-th period only was insufficient to be able to move the two variables representing the state of the system into an arbitrary state: we had two or more equations and only one variable. We will investigate this type of problem in Section 11.1.

5.8.2.2 Consistent equations

Consider a coefficient matrix $A \in \mathbb{R}^{m \times n}$ with $m > n$. Let $\hat{x} \in \mathbb{R}^n$ be an arbitrary vector and define $b = A\hat{x}$. Then, obviously, the linear simultaneous equations $Ax = b$ have at least one solution, namely $x = \hat{x}$. If A has rank n then this is the only solution; however, if A has rank less than n then there will be multiple solutions. We can use a generalization of the QR factorization, called the **complete orthogonal factorization**, to find the null space in the case that there are multiple solutions. (See [45, section 2.2.5.3].)

In summary, for some values of b, namely those in the range space of A, the system $Ax = b$ will have one or more solutions, even if there are more equations than variables. For such b, we say that the equations are **consistent**. In this case, we can identify a subset of no more than n of the equations such that each of the m equations can be expressed as a linear combination of the equations in the subset. (See Definition A.55.) There are redundant equations.

For example, consider the equation for the current injected at the datum node in our circuit example described in Section 4.1. If we include this equation in the set of linear simultaneous equations, then we have $m = n + 1$ equations, so that there are more equations than variables. The n nodal equations for the non-datum nodes form a linearly independent subset. Each of the m equations can be expressed as a linear combination of the n nodal equations for the non-datum nodes. (See Exercise 4.3.) There is a redundant equation.

5.8.3 The pseudo-inverse

The preceding discussion can be unified by defining the notion of the **pseudo-inverse**, which is defined to be the (unique) matrix $A^+ \in \mathbb{R}^{n \times m}$ such that the vector $x = A^+ b$ is the vector having the minimum value of norm $\|x\|_2$ over all vectors that minimize $\|Ax - b\|_2$ [45, section 2.2.5.6][89, appendix 8]. If A has rank n then $A^+ = [A^\dagger A]^{-1} A^\dagger$; however, this is generally a poor method to calculate A^+. To see this, note that if A is square then the condition number of $A^\dagger A$ is the square of the condition number of A. It is better to calculate A^+ from the QR factorization or the **complete orthogonal factorization** of A. (See [45, section 2.2.25.6] for further details and also see Section 11.1.)

5.9 Iterative methods

The algorithms we have presented so far in this chapter for solving linear equations are direct. Very large, but sparse, systems can be solved effectively with these methods. However, if the coefficient matrix is extremely large or is dense, then the factorization approaches become too time consuming. It may not even be possible to store the matrix conveniently.

An alternative approach involves an **iterative algorithm**, which calculates a sequence $\{x^{(\nu)}\}_{\nu=0}^{\infty}$ of approximations to the solution of $Ax = b$. At iteration ν of an iterative algorithm for solving the linear equations, the coefficient matrix is multiplied into a vector. Either implicitly or explicitly, this matrix–vector multiplication is used to evaluate the difference between $Ax^{(\nu)}$ and b and the difference is then used to obtain the update $\Delta x^{(\nu)}$ to calculate the new value of the iterate $x^{(\nu+1)}$. For a dense matrix, the matrix–vector multiplication requires $(n)^2$ operations, so if an iterative algorithm can be terminated with a useful approximate answer in much less than n iterations, then the total effort will be smaller than for factorization of a dense matrix. Moreover, if the matrix is sparse or has a special structure such as:

- a **Toeplitz matrix** where $A_{\ell k}$ depends only on $(\ell - k)$, or

- a **Hankel matrix** where $A_{\ell k}$ depends only on $(\ell + k)$,

then it may be possible to perform a matrix–vector multiplication in far less than $(n)^2$ operations. In some cases, it may be possible to perform a matrix–vector multiplication without even explicitly storing the non-zero entries of the matrix.

Iterative approaches are typically used for linear systems with large and dense coefficient matrices. We will not explore iterative algorithms for linear systems in this book; however, one commonly used iterative algorithm is called the **conjugate gradient method**. Description of this algorithm (and the reason for its name) and of other iterative algorithms is contained in [29, 45, 58]. Pre-conditioning is often used in conjunction with iterative techniques for the solution of linear equations [70, section 9.7].

5.10 Summary

In this chapter we have described LU factorization (and its variants) and forwards and backward substitution as an efficient approach to solving systems of linear equations, paying particular attention to symmetric systems. We considered the selection of pivots and discussed the solution of perturbed systems, sparse methods, and the issue of ill-conditioning. We briefly discussed the solution of non-square systems and iterative techniques. In later chapters we will need to solve large linear systems repeatedly so that the algorithms developed in this chapter will be incorporated into all subsequent algorithms.

Exercises

Solution of triangular systems

5.1 Let $L \in \mathbb{R}^{n \times n}$ be lower triangular and let $U \in \mathbb{R}^{n \times n}$ be upper triangular.

(i) Show that $\det(L)$ is equal to the product of the diagonal entries.
(ii) Suppose that the diagonal entries of L are all non-zero. Use Cramér's rule to show that L^{-1} is lower triangular.
(iii) Suppose that the diagonal entries of U are all non-zero. Use Cramér's rule to show that U^{-1} is upper triangular.

Solution of square, non-singular systems

5.2 Let $A \in \mathbb{R}^{n \times n}$. Suppose that for every $b \in \mathbb{R}^n$ there is a solution to $Ax = b$. Show that A^{-1} exists. (Hint: consider, in turn, the solutions of $Ax = b$ for $b = \mathbf{I}_1, \mathbf{I}_2, \ldots, \mathbf{I}_n$, where we recall that $\mathbf{I}_k$ is the n-vector with zeros everywhere except at the k-th entry, which is equal to 1.)

5.3 Prove that the product of any number of lower triangular matrices is lower triangular. (Hint: Prove by mathematical induction.)

5.4 Consider the 2×2 system $Ax = b$ where $A = \begin{bmatrix} 0 & 1 \\ 1 & 0 \end{bmatrix}$.

(i) Show that the matrix A is non-singular.
(ii) Show that A cannot be factorized into the product of an upper triangular and a lower triangular matrix. (Hint: Suppose that A could be factorized into LU with L lower triangular and U upper triangular. Multiply out the entries of L and U and derive a contradiction.)
(iii) Show that by re-arranging the order of the equations we can factorize A into upper and a lower triangular factors. Specify the factors. (Hint: $\mathbf{I}$ is both upper triangular and lower triangular.)

5.5 Factorize $A = \begin{bmatrix} 2 & 3 & 4 & 5 \\ 9 & 8 & 7 & 6 \\ 10 & 11 & 12 & 13 \\ 24 & 22 & 20 & 18 \end{bmatrix}$. Use the standard pivot at each stage. Show each of three stages required. That is, show $M^{(1)}$, $A^{(2)}$, $M^{(2)}$, $A^{(3)}$, $M^{(3)}$, $U = A^{(4)}$, and $L = [M^{(3)} M^{(2)} M^{(1)}]^{-1}$. (Hint: You will encounter a zero pivot but it will not prevent you from factorizing the matrix. What should $M^{(3)}$ be so that it is non-singular and makes $L = [M^{(3)} M^{(2)} M^{(1)}]^{-1}$ lower triangular?)

5.6 Consider the matrix $A = \begin{bmatrix} 2 & 3 & 4 \\ 7 & 6 & 5 \\ 8 & 9 & 10 \end{bmatrix}$.

(i) Factorize this matrix into L and U factors, using the standard pivot at each stage.

(ii) Is A invertible?

(iii) Is it possible to solve $Ax = b$ for all b?

(iv) Is it possible to solve $Ax = b$ for some particular values of b?

(v) Solve the system $Ax = b$, where $b = \begin{bmatrix} 9 \\ 18 \\ 27 \end{bmatrix}$,

(vi) Try factorizing this matrix using the MATLAB function `lu`.

5.7 Consider the matrix $A = \begin{bmatrix} 2 & 3 & 4 & 4 \\ 7 & 6 & 5 & 5 \\ 8 & 9 & 10 & 10 \\ 1 & 1 & 1 & 2 \end{bmatrix}$.

(i) Apply the first two stages of the LU factorization algorithm using the standard pivot at each stage and show that $A_{33}^{(3)} = 0$.

(ii) Factorize the matrix A' obtained from A by swapping the third and fourth rows and swapping the third and fourth columns.

(iii) Is A invertible?

(iv) Is it possible to solve $Ax = b$ for all b?

(v) Is it possible to solve $Ax = b$ for some particular values of b?

(vi) Try factorizing this matrix using the MATLAB function `lu`. Why does the answer differ from your previous factorization?

5.8 Let $A \in \mathbb{R}^{n \times n}$ and consider LU factorization of A. Suppose at the second stage of factorization we had $A_{\ell k}^{(2)} = 0$, $\forall \ell, k$ such that $\ell \geq 2$ and $k \geq 2$, so that there is no choice for a pivot, either diagonal or off-diagonal. Show that A is singular. (Assume that all numbers are represented to infinite precision and all calculations are carried out to infinite precision.) (Hint: Try using Cramér's rule to invert A. Recall that:

- by Cramér's rule, the inverse of a matrix exists if and only if its determinant is non-zero,
- the determinant of the product of two matrices equals the product of the determinants, and
- by Exercise 5.1, the determinant of an upper or lower triangular matrix equals the product of its diagonal entries.)

5.9 Let $A \in \mathbb{R}^{n \times n}$ and consider LU factorization of A. Suppose that partial pivoting (on rows, say) fails to find any non-zero pivots at a particular stage of factorization. Show that the matrix is singular. (Assume that all numbers are represented to infinite precision and all calculations are carried out to infinite precision.) (Hint: Suppose that partial pivoting fails to find a non-zero pivot at the j-th stage. Consider the matrix $A^{(j)}$ and the hint from Exercise 5.8.)

5.10 Recall the circuit of Section 4.1 shown in Figure 4.1 and repeated in Figure 5.7 for reference.

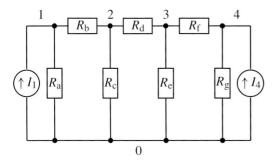

Fig. 5.7. Ladder circuit consisting of resistors and current sources repeated from Figure 4.1.
Source: This figure is based on [95, figure 1.4].

(i) Factorize the nodal admittance matrix for the circuit and solve the base-case circuit equation $Ax = b$ to obtain the base-case solution x^*. Assume that all resistors have value 1 and that the two current sources have value 1.
(ii) What is the power dissipation in each resistor?
(iii) Suppose that each resistor was 0.5 watt in rating, meaning that the resistor will have resistance that is close to its nominal resistance and that its temperature will not rise too high so long as the dissipation is below 0.5 watt. Will each resistor be within its ratings?
(iv) How about if each resistor was 0.25 watt in rating?

5.11 Factorize the matrix from Exercise 4.1 and solve the circuit for all unknown voltages and currents. Assume that all resistors have value 1, all current sources have value 1, and all voltage sources have value 1.

5.12 Write a MATLAB M-file to perform factorization in place of a matrix A.

Symmetric coefficient matrix

5.13 Consider the matrix $A = \begin{bmatrix} 1 & 2 & 3 \\ 6 & 16 & 20 \\ 2 & 5 & 8 \end{bmatrix}$. Show that A can be transformed into a symmetric matrix by multiplying the second row by 0.5 and then swapping the second and third rows.

5.14 Factorize $A = \begin{bmatrix} 1 & 2 & 0 & 5 \\ 2 & 1 & 3 & 0 \\ 0 & 3 & 1 & 4 \\ 5 & 0 & 4 & 1 \end{bmatrix}$ into LU factors. Make use of symmetry to minimize your calculations.

5.15 Recall the MATLAB M-file developed in Exercise 5.12.

(i) Extend the MATLAB M-file to be able to factorize matrices and solve linear systems by forwards and backwards substitution.
(ii) Modify the M-file to reduce the computational effort for symmetric matrices.

(iii) Use the M-file to factorize A from Exercise 5.14.

(iv) Solve the system $Ax = \begin{bmatrix} 1 \\ 2 \\ 3 \\ 4 \end{bmatrix}$.

5.16 Find the $LDL^\dagger$ factorization of the matrix in Exercise 5.14.

5.17 In this exercise we consider diagonal pivoting.

(i) Consider the matrix $A = \begin{bmatrix} 0 & 1 \\ 1 & 0 \end{bmatrix}$. Show that A cannot be factorized into $LDL^\dagger$ with L lower triangular with ones on its diagonal and D diagonal. (Hint: See Exercise 5.4.)

(ii) Show that A is not positive definite. (See Definition A.58.)

5.18 Show the following.

(i) If $L \in \mathbb{R}^{n \times n}$ is invertible then $[L^\dagger]^{-1} = [L^{-1}]^\dagger$. (Hint: Take the transpose of $LL^{-1} = \mathbf{I}$.)

(ii) If $A \in \mathbb{R}^{n \times n}$ and $A = LU$ then $A^\dagger = U^\dagger L^\dagger$.

5.19 Suppose that $A \in \mathbb{R}^{n \times n}$ is symmetric and positive definite.

(i) Consider a diagonal entry A_{pp} of A and a number $\Delta A_{pp} \in \mathbb{R}$. Let us define the new matrix A' from A by adding ΔA_{pp} to the pp-th entry of A. That is:

$$
A'_{\ell k} = \begin{cases} A_{pp} + \Delta A_{pp}, & \text{if } \ell = k = p, \\ A_{\ell k}, & \text{otherwise.} \end{cases}
$$

Show that there exists $\epsilon \in \mathbb{R}_{++}$ such that if ΔA_{pp} satisfies $0 \leq |\Delta A_{pp}| < \epsilon$ then A' is positive definite. (Hint: Note that:

- addition, multiplication, and subtraction are continuous functions,
- division is continuous at any point where the divisor is non-zero, and
- composition of continuous functions is continuous.)

(ii) Now consider an off-diagonal entry $A_{pq} = A_{qp}$ and a number $\Delta A_{pq} \in \mathbb{R}$. Let:

$$
A'_{\ell k} = \begin{cases} A_{pq} + \Delta A_{pq}, & \text{if } \ell = p \text{ and } k = q, \\ A_{qp} + \Delta A_{pq}, & \text{if } \ell = q \text{ and } k = p, \\ A_{\ell k}, & \text{otherwise.} \end{cases}
$$

Note that A' is symmetric. Show that there exists $\epsilon \in \mathbb{R}_{++}$ such that if ΔA_{pq} satisfies $0 \leq |\Delta A_{pq}| < \epsilon$ then A' is positive definite.

(iii) Again consider an off-diagonal entry $A_{pq} = A_{qp}$ and a number $\Delta A_{pq} \in \mathbb{R}$. Let:

$$
A'_{\ell k} = \begin{cases} A_{pq} + \Delta A_{pq}, & \text{if } \ell = p \text{ and } k = q, \\ A_{\ell k}, & \text{otherwise.} \end{cases}
$$

Note that A' is not symmetric. Show that there exists $\epsilon \in \mathbb{R}_{++}$ such that if ΔA_{pq} satisfies $0 \leq |\Delta A_{pq}| < \epsilon$ then A' is positive definite. (Hint: Use Exercise A.1.)

(iv) Define the following function $\pi : \mathbb{R}^{n \times n} \to \mathbb{R}$ by:

$$\forall A \in \mathbb{R}^{n \times n}, \pi(A) = \begin{cases} 1, & \text{if } A \text{ is positive definite,} \\ 0, & \text{if } A \text{ is not positive definite.} \end{cases}$$

Consider the norm $\| \bullet \|$ on $\mathbb{R}^{n \times n}$ given by $\| A \| = \max_{\ell, k} |A_{\ell k}|$. Show that, under the defined norm, the function π is continuous at a "point" A such that $\pi(A) = 1$. That is, show that if $\pi(A) = 1$ then there exists $\epsilon \in \mathbb{R}_{++}$ such that if $\| \Delta A \| < \epsilon$ then $\pi(A + \Delta A) = 1$. This shows that perturbing a symmetric positive definite matrix by a small enough perturbation does not destroy its positive definiteness.

(v) Is π continuous at a point A such that $\pi(A) = 0$? (Hint: consider $A = [0] \in \mathbb{R}^{1 \times 1}$.)

5.20 Let $A \in \mathbb{R}^{n \times n}$ be symmetric and positive definite. Show that A is non-singular. (Hint: Prove the contra-positive, using the property of a singular matrix presented in Theorem A.5.)

5.21 Let $A \in \mathbb{R}^{n \times n}$ be symmetric and strictly diagonally dominant. Show that A is positive definite.

5.22 Let $A \in \mathbb{R}^{n \times n}$ be symmetric and strictly diagonally dominant. Consider the factorization of A into $LDL^{\dagger}$.

(i) Show that at each stage of factorization *any* diagonal pivot can be used, since they are all strictly positive.

(ii) Show that at each stage of the factorization, the largest element in the remaining matrix is on the diagonal.

Sparsity techniques

5.23 In this exercise we use MATLAB to investigate the speed-up in factorization that is possible by taking advantage of sparsity.

(i) Create a 100×100, random, symmetric matrix that has only 0.1% of its entries non-zero using the following MATLAB command:

```
R100 = sprandsym(100, 0.001);
```
This matrix is stored sparsely by MATLAB. Find the elapsed time to LU factorize the sparse matrix R100 by issuing the command:

```
tic; [l100, u100] = lu(R100); toc
```
Make sure all the command is on one line, otherwise the time you take to type will be included in the elapsed time. If you are using MATLAB in a time-shared environment, you should repeat the factorization several times and record the shortest time, since the elapsed time depends, in part, on how many other users there are on the system.

(ii) Create a full matrix (that is a square array) with the same non-zeros as R100 by issuing the command:

```
F100 = full(R100);
```
Now factorize this matrix and find the elapsed time with:

```
tic; [L100, U100] = lu(F100); toc
```

Again, you should repeat the factorization several times and record the shortest time.

(iii) Repeat Parts (i) and (ii) for a 500×500 matrix. (Again, repeat each factorization several times.)

(iv) Repeat Parts (i) and (ii) for a 1000×1000 matrix. (Again, repeat each factorization several times.)

(v) Explain your results.

5.24 In this exercise we develop MATLAB M-files for sparse backwards substitution.

(i) Write a MATLAB M-file to perform backwards substitution for a sparsely stored matrix U. Assume that the location and values of the non-zero entries of the $n \times n$ upper triangular matrix U have been stored row-wise in lists, `location[ℓ]` and `value[ℓ]`, respectively. For each row ℓ, `location[ℓ][k]` is the location of the k-th non-zero entry in row ℓ of U and `value[ℓ][k]` is its value. For example, for each row, `location[ℓ][1] = ℓ`, the location of the diagonal entry, which is the first non-zero entry in row ℓ of the upper triangular matrix U. Moreover, `value[ℓ][1]` is the value of the diagonal entry of U in row ℓ.

Assume that the results of the forwards substitution are in an array `y` and use an array `x` to store the results of the backwards substitution. If there are K non-zero entries in row ℓ, then assume that the end of the row is indicated by the value of `location[ℓ][K+1]` being zero. For example, if there are three non-zero entries in row `row`, then `location[row][4] = 0`.

(ii) Sparsely store the values of entries from the factor U calculated in Exercise 5.14 and then solve $Uy = \begin{bmatrix} 1 \\ 2 \\ 3 \\ 4 \end{bmatrix}$.

5.25 Consider the change in the circuit of Section 4.1 shown in Figure 5.2. Assume that all resistors have value 1 and that the two current sources in the base-case circuit have value 1.

(i) Assume that $\Delta b_2 = 1$ and solve the equation $A \Delta x = \Delta b$ using forwards and backwards substitution. Use the factorization from Exercise 5.10 and take advantage of sparsity to minimize the number of calculations.

(ii) Use the answer from Exercise 5.10 to write down the solution of $Ax' = b + \Delta b$.

5.26 Suppose that there were no fill-ins added in the course of factorization. (This is unrealistic, but we have developed techniques to minimize the number of fill-ins and in practice the number of fill-ins may be relatively small.) Calculate the order of the computational effort for factorization in terms of the size of the system, n, and the **density**, d, of non-zeros in the matrix, where the density of non-zeros is defined to be the ratio of the number of non-zero entries in the matrix divided by $(n)^2$. Assume that:

- each diagonal entry is non-zero and no zero pivots are encountered,
- the matrix is symmetric,
- the off-diagonal entries are spread evenly throughout the matrix so that in each row or column there are approximately dn non-zeros in each row, and

- $d \ll 1$.

5.27 In many applications, the density of a matrix as defined in Exercise 5.26 decreases with the size of the system. For example, in circuit problems the number of branches incident to non-datum nodes is typically less than three or four and so the number of non-zero entries in the matrix is less than three or four times n. Suppose that we are considering a problem such that the number of non-zero entries in the matrix is κn, where κ is a positive constant. Use the results of Exercise 5.26 to estimate the computational effort for factorization. (Hint: The density is given by $d = \kappa / n$.)

5.28 Show that the factors of a tri-diagonal matrix are **bi-diagonal**; that is, they have non-zeros only on the diagonal and either adjacent to and above the diagonal or adjacent to and below the diagonal.

5.29 Consider an **arrowhead matrix** $A \in \mathbb{R}^{n \times n}$, which has non-zeros only on the diagonal, the last column, and the last row, so that:

$$
\begin{bmatrix}
A_{11} & 0 & \cdots & & 0 & A_{1n} \\
0 & \ddots & \ddots & & \vdots & \vdots \\
\vdots & \ddots & \ddots & & 0 & \vdots \\
0 & \cdots & 0 & A_{n-1,n-1} & A_{n-1,n} \\
A_{n1} & \cdots & \cdots & A_{n,n-1} & A_{nn}
\end{bmatrix}.
$$

  (i) Perform LU factorization on A. Do not re-order rows nor columns. Assume that all pivots are non-zero.
  (ii) Comment on the sparsity of the factors obtain in Part (i).
  (iii) Now re-order rows and columns so that the last row becomes the first row and the last column becomes the first column. Comment on the sparsity of the factors obtained with this matrix ordering.

5.30 Let:

$$
A = \begin{bmatrix} A_{11} & A_{12} \\ A_{21} & A_{22} \end{bmatrix} \in \mathbb{R}^{2 \times 2},
$$

$$
B = \begin{bmatrix} B_1 \\ B_2 \end{bmatrix} \in \mathbb{R}^{2 \times 1},
$$

$$
C = \begin{bmatrix} C_1 & C_2 \end{bmatrix} \in \mathbb{R}^{1 \times 2},
$$

$$
D \in \mathbb{R}.
$$

  (i) Perform block pivoting on $\begin{bmatrix} A & B \\ C & D \end{bmatrix}$. That is, explicitly calculate $-CA^{-1}$ and $D - CA^{-1}B$.
  (ii) Suppose that $A_{11} = A_{22} = 0$. Calculate $-CA^{-1}$ and $D - CA^{-1}B$.

5.31 In this exercise we use block LU factorization to calculate the inverse of a block matrix.

(i) Calculate the inverse of $\begin{bmatrix} A & B \\ 0 & D - CA^{-1}B \end{bmatrix}$, where A, B, C, D are matrices, with A and D square. Assume that $D - CA^{-1}B$ is invertible.

(ii) Evaluate the inverse of $\begin{bmatrix} A & B \\ C & D \end{bmatrix}$ using the previous part together with the block LU factorization of $\begin{bmatrix} A & B \\ C & D \end{bmatrix}$ shown in (5.15):

$$\begin{bmatrix} \mathbf{I} & \mathbf{0} \\ -CA^{-1} & \mathbf{I} \end{bmatrix} \begin{bmatrix} A & B \\ C & D \end{bmatrix} = \begin{bmatrix} A & B \\ 0 & D - CA^{-1}B \end{bmatrix}.$$

Changes

5.32 Suppose that $A \in \mathbb{R}^{n \times n}$ is non-singular.

(i) Consider a diagonal entry A_{pp} of A and a number $\Delta A_{pp} \in \mathbb{R}$. Let us define the new matrix A' from A by adding ΔA_{pp} to the pp-th entry of A. That is:

$$A'_{\ell k} = \begin{cases} A_{pp} + \Delta A_{pp}, & \text{if } \ell = k = p, \\ A_{\ell k}, & \text{otherwise.} \end{cases}$$

Show that there exists $\epsilon \in \mathbb{R}_{++}$ such that if ΔA_{pp} satisfies $0 \leq |\Delta A_{pp}| < \epsilon$ then A' is invertible and moreover its inverse is a differentiable function of ΔA_{pp} with a continuous derivative. (Hint: Note that Cramér's rule allows us to express the inverse of A' in terms of the elements of A and ΔA_{pp} and that:

- addition, multiplication, and subtraction are differentiable functions with continuous derivatives,
- division is differentiable with a continuous derivative at any point where the divisor is non-zero, and
- composition of differentiable functions with continuous derivatives produces a differentiable function with continuous derivative.)

(ii) Now consider an off-diagonal entry A_{pq} and a number $\Delta A_{pq} \in \mathbb{R}$. Let:

$$A'_{\ell k} = \begin{cases} A_{pq} + \Delta A_{pq}, & \text{if } \ell = p \text{ and } k = q, \\ A_{\ell k}, & \text{otherwise.} \end{cases}$$

Show that there exists $\epsilon \in \mathbb{R}_{++}$ such that if ΔA_{pq} satisfies $0 \leq |\Delta A_{pq}| < \epsilon$ then A' is invertible and moreover its inverse is a differentiable function of ΔA_{pq} with a continuous derivative.

(iii) Show that if the matrix-valued function $A : \mathbb{R}^s \to \mathbb{R}^{n \times n}$ is partially differentiable with continuous partial derivatives and $A(\mathbf{0})$ is non-singular then its inverse exists for all χ in a neighborhood of $\mathbf{0}$ and the inverse is a partially differentiable function of χ with continuous partial derivatives in this neighborhood.

5.33 Again consider the circuit of Section 4.1 shown in Figures 4.1 and 5.7 that was solved for the base-case values of components in Exercise 5.10.

(i) Calculate the sensitivity of all entries in the solution to variations in I_4.
(ii) Calculate the sensitivity of all entries in the solution to variations in R_b.
(iii) Suppose that all resistors increased linearly with temperature T, with a variation of 1% in resistance for each Celsius degree change in temperature. Suppose that the solution from Exercise 5.10 was for a base-case temperature of 27°C. Calculate the sensitivity of all entries in the solution to variations in temperature T about the base-case temperature.

5.34 Again consider the circuit of Section 4.1 shown in Figures 4.1 and 5.7 that was solved for the base-case values of components in Exercise 5.10 to obtain the base-case solution $x^\star$. Consider the objective $f : \mathbb{R}^4 \to \mathbb{R}$ defined in (4.8). That is:

$$\forall x \in \mathbb{R}^4, \ f(x) = (x_1)^2 + 2(x_2)^2 + 3(x_3)^2 + 4(x_4)^2.$$

(i) Calculate $\dfrac{\partial f}{\partial x}(x^\star)$.
(ii) Solve (5.18) to obtain ξ corresponding to this objective function for the base-case circuit. That is, solve $[A(0)]^\dagger \xi = \nabla f(x^\star(0))$.
(iii) Calculate the sensitivity of the objective to variations in I_4.
(iv) Calculate the sensitivity of the objective to variations in R_b.
(v) Suppose that all resistors increased linearly with temperature T, with a variation of 1% in resistance for each Celsius degree change in temperature. Suppose that the solution from Exercise 5.10 was for a base-case temperature of 27°C. Calculate the sensitivity of the objective to variations in temperature T about the base-case temperature.

Ill-conditioning

5.35 Let:

$$\Delta A = \begin{bmatrix} \chi & 0 \\ 0 & 0 \end{bmatrix}.$$

Show that $\|\Delta A\|_2 = \chi$, where $\|\bullet\|_2$ is the induced matrix L_2 norm. (See Section A.3.2.)

5.36 Prove Theorem 5.8. (Hint: Write down the equations satisfied by $x^\star$ and $\Delta x^\star$ and then use the definition of compatible norms to bound $\|\Delta x^\star\|$ in terms of $\|x^\star\|$ and $\|x^\star + \Delta x^\star\|$.)

5.37 In this exercise we calculate the condition number of two matrices.

(i) Let $A = \begin{bmatrix} \delta & 1 \\ 1 & 1 \end{bmatrix}$. Calculate $\|A\|_2$, $\|A^{-1}\|_2$, and the condition number of A. (Recall that $\|A\|_2$ for a symmetric matrix is equal to the largest of the absolute values of the eigenvalues of A. (See Section A.3.2.)) Assume that $\delta \neq 1$.

(ii) Let $A = \begin{bmatrix} 1+\delta & 1 \\ 1 & 1 \end{bmatrix}$. Calculate $\|A\|_2$, $\|A^{-1}\|_2$, and the condition number of A.

Assume that $\delta \neq 0$.

5.38 Factorize:

(i) the matrix $A = \begin{bmatrix} 1 & \delta \\ 1 & 0 \end{bmatrix}$ in (5.19),

(ii) the matrix $A = \begin{bmatrix} \delta & 0 \\ 1 & 1 \end{bmatrix}$ in (5.21),

(iii) the matrix $A = \begin{bmatrix} 1 & 0 \\ 1 & 1 \end{bmatrix}$ in (5.23), and

(iv) the matrix $A = \begin{bmatrix} \delta & 1 \\ 1 & 1 \end{bmatrix}$ from Exercise 5.37,

(v) the matrix $A = \begin{bmatrix} 1+\delta & 1 \\ 1 & 1 \end{bmatrix}$ from Exercise 5.37,

into LU factors. Do not re-order rows and columns and do not scale rows nor columns. In each case comment on the size of the pivots if $|\delta| \ll 1$. Also comment on the sensitivity of the entries in the factors to:

- changes in δ, and
- round-off errors during factorization.

5.39 Show that by a suitable scaling of the variables, the linear system specified by the matrix in (5.19),

$$A = \begin{bmatrix} 1 & \delta \\ 1 & 0 \end{bmatrix},$$

can be transformed into a system with a coefficient matrix having a small condition number.

5.40 Consider the following matrix:

$$A = \begin{bmatrix} 0.1 & 2 & 1 & 0 \\ 2 & 10 & 2 & 0 \\ 1 & 2 & 0.1 & 0 \\ 0 & 0 & 0 & 0.0001 \end{bmatrix}.$$

(i) Use diagonal pivoting to re-order the rows and columns of the matrix to minimize the number of fill-ins at the first stage of factorization.

(ii) Use diagonal pivoting to re-order the rows and columns of the matrix to maximize the size of the pivot at the first stage of factorization.

(iii) Discuss the implications of the results of the previous parts.

5.41 Show that for any matrix $A \in \mathbb{R}^{m \times n}$, $\|A\|_2$ is at least as large as the absolute value of each of the elements, where $\|\bullet\|_2$ is the induced matrix L_2 norm. (See Section A.3.2.)

5.42 In this exercise, we use MATLAB to perform QR factorization.

(i) Factorize the matrix $A = \begin{bmatrix} 2 & 3 & 4 & 4 \\ 7 & 6 & 5 & 5 \\ 8 & 9 & 10 & 10 \\ 1 & 1 & 1 & 2 \end{bmatrix}$ from Exercise 5.7 using the MATLAB command qr. Comment on the sparsity (or otherwise) of the factors.

(ii) Factorize the matrix $A = \begin{bmatrix} 1 & 2 & 0 & 5 \\ 2 & 1 & 3 & 0 \\ 0 & 3 & 1 & 4 \\ 5 & 0 & 4 & 1 \end{bmatrix}$ from Exercise 5.14 using the MATLAB command qr. Comment on the sparsity (or otherwise) of the factors.

Non-square systems

5.43 Give an example of a system with more variables than equations that is inconsistent.

5.44 Suppose that $A \in \mathbb{R}^{m \times n}$, with $m < n$ is of the form $A = \begin{bmatrix} I & A^\perp \end{bmatrix}$. Let $b \in \mathbb{R}^m$ and find the set of solutions to $Ax = b$.

5.45 Let $A \in \mathbb{R}^{m \times n}$, $b \in \mathbb{R}^m$ and suppose that $\hat{x} \in \mathbb{R}^n$ satisfies $A\hat{x} = b$. Show that the set of all solutions to $Ax = b$, (that is, the set $\mathbb{S} = \{x \in \mathbb{R}^n | Ax = b\}$) is the same as the set $\mathbb{P} = \{\hat{x} + \Delta x | \Delta x \in \mathcal{N}(A)\}$, where $\mathcal{N}(A)$ is the null space of A. (Hint: First let $x \in \mathbb{S}$ and show that $x \in \mathbb{P}$. Then let $x \in \mathbb{P}$ and show that $x \in \mathbb{S}$.)

5.46 Let:

$$A = \begin{bmatrix} 1 & 0 & 2 & 0 & 1 \\ 0 & 3 & 0 & 1 & 1 \end{bmatrix}, b = \begin{bmatrix} 1 \\ 1 \end{bmatrix}.$$

(i) Consider the matrix $A^\|$ that consists of the first two columns of A. Find the inverse of $A^\|$.
(ii) Find a particular solution $\hat{x}$ to the equations $Ax = b$.
(iii) Find a matrix Z with columns that form a basis for the null space of A. Evaluate the entries in Z explicitly.
(iv) Explicitly characterize all solutions of the equations $Ax = b$.
(v) Use the MATLAB function qr to find the LQ factorization of A. (Hint: Perform QR factorization of $A^\dagger$.)

5.47 Let $A \in \mathbb{R}^{m \times n}$ with $m < n$ and suppose that A has been factorized as $PA = LQ$, with:

- $P \in \mathbb{R}^{m \times m}$ is a permutation matrix,
- $L \in \mathbb{R}^{m \times n}$ is lower triangular, with its first m' columns linearly independent and its last $n' = n - m'$ columns zero, and
- $Q \in \mathbb{R}^{n \times n}$ is unitary.

(i) Show that the null space of A is given by:

$$\mathcal{N}(A) \;=\; \{\Delta x \in \mathbb{R}^n \,|\, A\Delta x = \mathbf{0}\},$$

$$=\; \left\{ Q^\dagger \begin{bmatrix} \mathbf{0} \\ \Delta \xi \end{bmatrix} \,\middle|\, \Delta \xi \in \mathbb{R}^{n'} \right\},$$

$$=\; \{Z\Delta\xi \,|\, \Delta\xi \in \mathbb{R}^{n'}\},$$

where Z is the last n' columns of $Q^\dagger$.

(ii) Show that the range space of $A^\dagger$ is given by:

$$\mathcal{R}(A^\dagger) \;=\; \{A^\dagger \lambda \,|\, \lambda \in \mathbb{R}^m\},$$

$$=\; \left\{ Q^\dagger \begin{bmatrix} \Delta\omega \\ \mathbf{0} \end{bmatrix} \,\middle|\, \Delta\omega \in \mathbb{R}^{m'} \right\},$$

$$=\; \{T\Delta\omega \,|\, \Delta\omega \in \mathbb{R}^{m'}\},$$

where T is the first m' columns of $Q^\dagger$.

(iii) Prove Theorem A.4 from Appendix A. That is, prove that:

$$\forall x \in \mathbb{R}^n, \exists \lambda \in \mathbb{R}^m, \exists z \in \mathbb{R}^n \text{ with } Az = \mathbf{0} \text{ such that } x = z + A^\dagger \lambda.$$

(Hint: Use the previous parts. Note that, since Q is unitary then, for $x \in \mathbb{R}^n$, $x =$
$\mathbf{I}x = Q^\dagger Q x = \begin{bmatrix} T & Z \end{bmatrix} \begin{bmatrix} T^\dagger \\ Z^\dagger \end{bmatrix} x$. Let $z = ZZ^\dagger x \in \mathbb{R}^n$ and $\Delta\omega = T^\dagger x \in \mathbb{R}^{m'}$.)

5.48 Let $A \in \mathbb{R}^{m \times n}$ and let $Z \in \mathbb{R}^{n \times n'}$ be a matrix with columns that form a basis for the null space of A. Show that $AZ = \mathbf{0}$. (Hint: Note that the ℓk-th element of AZ is $\mathbf{I}_\ell^\dagger AZ\mathbf{I}_k$.)

5.49 Let $A : \mathbb{R}^s \to \mathbb{R}^{m \times n}$ be a matrix-valued function. Suppose that $A(\mathbf{0})$ has linearly independent rows. Show that there exists $Z : \mathbb{R}^s \to \mathbb{R}^{n \times n'}$ that is partially differentiable with continuous partial derivatives such that for χ in a neighborhood of $\chi = \mathbf{0}$ the matrix $Z(\chi)$ has columns that form a basis for the null space of $A(\chi)$. (Hint: Permute the columns of $A(\mathbf{0})$, if necessary, so that the first m columns of $A(\mathbf{0})$ are linearly independent. As in Section 5.8.1.2, partition A into $\begin{bmatrix} A^\| & A^\perp \end{bmatrix}$ where $A^\| : \mathbb{R}^s \to \mathbb{R}^{m \times m}$ and $A^\perp : \mathbb{R}^s \to \mathbb{R}^{m \times n'}$. Use Part (iii) of Exercise 5.32 to show that the inverse of $A^\|$ exists for all χ in a neighborhood of $\mathbf{0}$ and that the inverse is a partially differentiable function of χ with continuous partial derivatives in this neighborhood.)

5.50 Let $m = 2$. Recall the discrete-time linear system case study from Section 4.2. Suppose that:

$$G = \begin{bmatrix} 1 & 1 \\ 0 & 1 \end{bmatrix}, h = \begin{bmatrix} 0 \\ 1 \end{bmatrix}, w(0) = \begin{bmatrix} 0 \\ 0 \end{bmatrix}, w^{\text{desired}} = \begin{bmatrix} 1 \\ 1 \end{bmatrix}.$$

Recall that we defined:

- A to be the $m \times n$ matrix with k-th column equal to $(G)^{n-1-k}h$, (where k ranges from 0 to $(n-1)$),
- x to be the n-vector with k-th component $u(kT)$, (where k again ranges from 0 to $(n-1)$), and

- $b = w^{\text{desired}} - (G)^n w(0)$.

For this specification of A and b:

(i) Solve $Ax = b$ for $n = 2$.

(ii) Now suppose that A changes to $A + \Delta A$ where $\Delta A = \begin{bmatrix} 0.1 & 0 \\ 0 & 0 \end{bmatrix}$. Assume that you apply the input calculated in Part (i). Find the actual value of $w(2T)$ and the error between $w(2T)$ and w^{desired}.

(iii) Find any solution of $Ax = b$ for $n = 3$.

5.51 Suppose that in my pocket I have some pennies (worth 1c each), nickels (worth 5c each), dimes (worth 10c each), and quarters (worth 25c each.) I do not know how many of each type of coin that I have, but I do know the following:

- the value of the pennies and quarters is 47c in total,
- the value of the nickels and the quarters is $1.00 in total, and
- the value of the dimes and quarters is $2.00 in total.

I want to know how many pennies, nickels, dimes, and quarters I have in my pocket using the above information.

(i) Cast this problem into a linear system of equations of the form $Ax = b$. Specify explicitly:

- The variables in the formulation and their definitions.
- The coefficient matrix A.
- The right-hand side vector b.

(ii) Find a matrix Z with columns that form a basis for the null space of A.

(iii) Find a particular solution $\hat{x}$ of $Ax = b$.

(iv) Find the set of all $x \in \mathbb{R}^n$ that satisfy $Ax = b$.

(v) Some of the solutions to the linear equations from the previous part are not physically feasible. What extra conditions on x would guarantee that the solution is physically feasible? Express your solution in a similar way to the previous part.

Part II

Non-linear simultaneous equations

6

Case studies of non-linear simultaneous equations

In this chapter, we will develop two case studies of problems that involve the solution of non-linear simultaneous equations. Both case studies will involve circuits, but the non-linearities will arise in different ways. The case studies are:

(i) the solution of Kirchhoff's laws in a non-linear direct current (DC) circuit (Section 6.1), and

(ii) the solution of Kirchhoff's laws in a linear alternating current (AC) circuit where the variables of interest are not currents and voltages but instead are power (and "reactive power") injections (Section 6.2).

The first case study will draw on the development from the linear circuit study described in Section 4.1 and we will not repeat in detail those issues that were already presented in Section 4.1. The second case study will be discussed in considerable detail.

The progression from the case study of Section 4.1 to the case studies of Sections 6.1 and 6.2 are examples of model extension and development. By developing the model incrementally, we can treat a few issues at a time without being overwhelmed by trying to analyze all the issues at once. In developing your own models, you might also build them step-wise, rather than all-at-once. Even an unrealistically simple *initial* model can provide valuable insights into more realistic cases. For example, most DC circuits in practice include at least some non-linear components. Nevertheless, formulation of the linear DC circuit case study in Section 4.1 has provided most of what we need to formulate a non-linear DC circuit case study.

6.1 Analysis of a non-linear direct current circuit

6.1.1 Motivation

In the case study in Section 4.1 we analyzed a direct current (DC) circuit where all the components were *linear*; that is, the voltage and current in each resistor are related by a linear function. In contrast, in this case study, some of the components will be *non-linear*. As indicated in Section 4.1.1, in both linear and non-linear circuits it is economically important to be able to predict:

- the behavior of the circuit without actually building a prototype, and
- the effect of changes in component values on the circuit behavior.

This case study assumes some familiarity with non-linear circuits [95, chapter 10].

6.1.2 Formulation

This material is based on an example in [95, section 10.2]. We will first discuss the modeling of non-linear devices and then write down Kirchhoff's laws for a circuit consisting of current sources, resistors, and diodes.

6.1.2.1 Device models

Terminal characteristics If we are interested in the detailed internal behavior of electronic components then we must use a model that represents these effects. In circuit applications, however, we are often only interested in the **terminal charac-teristics** of the device. We can often avoid considerable internal detail if we are content to only model the terminal behavior of our components.

   We have already restricted ourselves to terminal characteristics in our linear cir-cuit model in Section 4.1, where we asserted that we knew the resistance of the resistors; that is, that we knew that the ratio of terminal voltage to current was a known constant or, equivalently, that the current is a linear function of voltage. One parameter, the resistance, is sufficient to represent the terminal characteristics of each resistor, at least if the temperature is constant. We were not interested in a detailed internal model of the resistor.

Non-linear devices Terminal characteristics of non-linear devices can also be mod-eled, but the models will typically require more than one parameter. For example, consider a **diode**, which is a two-terminal element that allows current to flow in one direction, the **forward direction**, but limits the current in the other direction, the **reverse direction**, to a very small value. For voltages applied across the diode in the forward direction, the current is a rapidly growing function of the voltage. That is, the current flows with only a relatively small voltage drop across the diode. For

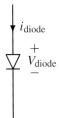

i_{diode}

$+$

V_{diode}

$-$

Fig. 6.1. Symbol for diode together with voltage and current conventions.

voltages across the diode in the opposite direction, the current flow is small. Semi-conductor diodes can be approximately modeled with a function $i_{\text{diode}} : \mathbb{R} \to \mathbb{R}$ of the form ([95, section 1.5]):

$$\forall V_{\text{diode}} \in \mathbb{R}, i_{\text{diode}}(V_{\text{diode}}) = I_{\text{sat}} \left[\exp \left(\frac{q V_{\text{diode}}}{\eta \mathcal{K} T} \right) - 1 \right], \qquad (6.1)$$

where:

- i_{diode} is the function representing current through the diode, with positive values corresponding to currents flowing in the forward direction,
- V_{diode} is the voltage across the diode, with positive values corresponding to voltages that cause current to flow in the forward direction,
- I_{sat} is the reverse saturation current of the diode,
- q is the charge on the electron,
- η is the "non-ideality factor" of the diode,
- $\mathcal{K}$ is Boltzmann's constant, (we use the non-standard symbol $\mathcal{K}$ to distinguish Boltzmann's constant from k used as an index for nodes), and
- T is the thermodynamic temperature of the diode.

Figure 6.1 shows the symbol for a diode together with the conventions for positive values of voltage and current.

While q and $\mathcal{K}$ are fundamental physical constants,

- the parameters I_{sat} and η depend on the design of the diode, and
- the temperature T depends on the history of power dissipation in the diode, the thermal characteristics of the diode package, and the operating environment of the diode.

That is, there are three parameters necessary to specify the function i_{diode}. Figure 6.2 illustrates i_{diode} defined in (6.1) for $q/(\eta \mathcal{K} T) = 40 \text{V}^{-1}$ and $I_{\text{sat}} = 10^{-6} \text{A}$.

As in the case of the linear circuit, we will assume that T is a given constant in this case study. A typical choice would be that T is equal to the ambient temperature. However, this assumption and similar ones should be analyzed for their

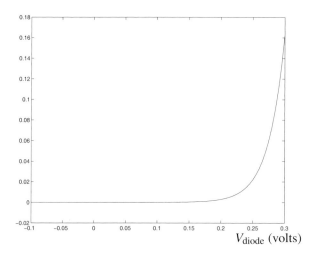

V_{diode} (volts)

Fig. 6.2. Current to voltage relationship for diode.

reasonableness based on the results of the case study calculations. For example, if
the power dissipated by the diode, $i_{\text{diode}}(V_{\text{diode}}) \times V_{\text{diode}}$, is nearly as large as the
rated maximum power dissipation of the diode then the assumed temperature T
may be in error because it may deviate significantly from ambient. Similar observations were made for resistors in Section 4.1.2.1 and Exercise 5.10.

Models of terminal characteristics are also available for other non-linear devices
such as **bipolar junction transistors** and **metal-oxide semiconductor field-effect
transistors (MOSFETs)** [56, chapter 4][95, section 10.6]. Since these devices
have three terminals, we generally require more than one equation to characterize their terminal behavior. The salient point is that by concentrating on terminal
behavior we can obtain a simplified model compared to a model of the detailed
internal behavior. This is particularly important when we try to analyze large systems; however, the terminal model by definition "hides" the internal behavior of
the component, so that the simplified model does not necessarily characterize everything about the system. Moreover, the models will typically only represent the
terminal behavior approximately. Nevertheless, in some applications, such as the
modelling of MOSFETs that are very small physically, the terminal models can
themselves become very complicated.

Choice of terminal model As we discussed in Section 4.1.2.1, we should model
components in our system so that the aspects of the system of greatest interest
are included. We may choose to suppress other aspects of the system to keep the
overall model as simple as possible; however, in doing so we must keep aware of

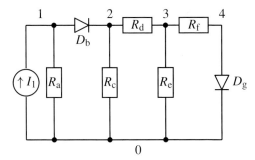

Fig. 6.3. A simple non-linear circuit.

the deficiencies and inaccuracies in the overall model. For example, (6.1) does not include the effects of **junction capacitance** [95, section 10.8] so that it is, strictly speaking, only valid for direct current (DC) circuits. This is adequate for our DC calculation, but the diode model must be modified for alternating current (AC) analysis. Again, Occam's razor is important in selecting a terminal model.

6.1.2.2 *Kirchhoff's current law*

Consider a circuit consisting of interconnected resistors, current sources, and non-linear elements such as diodes. An example circuit consisting of resistors, a current source, and two diodes is shown in Figure 6.3.

As in Section 4.1, we will write down Kirchhoff's current law for the nodes in this circuit. Again, we label the nodes 0 through 4 and we let x_k be the voltage between node k and the datum node 0. We assume that the resistances and currents sources have known values and assume that the current through diode D_b is a known function, $i_b(V_b)$ of the voltage V_b across it and that, similarly, the current through diode D_g is $i_g(V_g)$, where V_g is the voltage across diode D_g. The functional forms of i_b and i_g are assumed to be given by (6.1), with known values of:

- reverse saturation currents I_{satb} and I_{satg}, respectively, and
- non-ideality factors, η_b and η_g, respectively,

for diodes D_b and D_g and a known common operating temperature T for both diodes.

Noting that the voltage across diode D_b is $V_b = x_1 - x_2$, while the voltage across diode D_g is $V_g = x_4$, we then have, by Kirchhoff's current law applied to nodes 1, 2, 3, and 4:

$$\left(\frac{1}{R_a}\right) x_1 + i_b(x_1 - x_2) - I_1 \;=\; 0, \tag{6.2}$$

$$-i_{\rm b}(x_1 - x_2) + \left(\frac{1}{R_{\rm c}} + \frac{1}{R_{\rm d}}\right) x_2 + \left(-\frac{1}{R_{\rm d}}\right) x_3 = 0, \tag{6.3}$$

$$\left(-\frac{1}{R_{\rm d}}\right) x_2 + \left(\frac{1}{R_{\rm d}} + \frac{1}{R_{\rm e}} + \frac{1}{R_{\rm f}}\right) x_3 + \left(-\frac{1}{R_{\rm f}}\right) x_4 = 0, \tag{6.4}$$

$$\left(-\frac{1}{R_{\rm f}}\right) x_3 + \left(\frac{1}{R_{\rm f}}\right) x_4 + i_{\rm g}(x_4) = 0. \tag{6.5}$$

Note that, for example, $i_{\rm b}(x_1 - x_2)$ means the function $i_{\rm b} : \mathbb{R} \to \mathbb{R}$ evaluated at $x_1 - x_2$, whereas $\left(\frac{1}{R_{\rm a}}\right) x_1$ is the product of the conductance $1/R_{\rm a}$ and the voltage at node 1.

As in the direct current linear circuit case study in Section 4.1, the equation for the datum node is **redundant**. (See Exercise 6.1.)

In general, for multi-terminal devices, the functions specifying the non-linear relations will have two or more arguments to represent the voltages at all the terminals. See [95, section 10.6] for a discussion of three-terminal elements.

6.1.2.3 Non-linear equations

Unlike in the case of linear equations, we cannot represent the relationship between voltage and current in (6.2)–(6.5) by a matrix of real numbers. We must develop a more general representation. Define the **vector function** $g : \mathbb{R}^4 \to \mathbb{R}^4$ by:

$$\forall x \in \mathbb{R}^4, g(x) = \begin{bmatrix} \left(\frac{1}{R_{\rm a}}\right) x_1 + i_{\rm b}(x_1 - x_2) - I_1 \\ -i_{\rm b}(x_1 - x_2) + \left(\frac{1}{R_{\rm c}} + \frac{1}{R_{\rm d}}\right) x_2 + \left(-\frac{1}{R_{\rm d}}\right) x_3 \\ \left(-\frac{1}{R_{\rm d}}\right) x_2 + \left(\frac{1}{R_{\rm d}} + \frac{1}{R_{\rm e}} + \frac{1}{R_{\rm f}}\right) x_3 + \left(-\frac{1}{R_{\rm f}}\right) x_4 \\ \left(-\frac{1}{R_{\rm f}}\right) x_3 + \left(\frac{1}{R_{\rm f}}\right) x_4 + i_{\rm g}(x_4) \end{bmatrix}. \tag{6.6}$$

If we write $g(x) = \mathbf{0}$ then we have reproduced (6.2)–(6.5). These are a set of non-linear simultaneous equations. (Note that g is a vector function while g refers to the diode.)

Our convention in defining g is to move any constant terms in the equations, such as the current I_1, into the definition of g. For example, to represent linear equations $Ax = b$ in this way we would define $g : \mathbb{R}^n \to \mathbb{R}^m$ by:

$$\forall x \in \mathbb{R}^n, g(x) = Ax - b.$$

6.1.3 Circuit changes

Changes in the values of resistors, current sources, or diode parameters will change the functional form of corresponding entries in g. For example, if a resistor or a diode between nodes ℓ and k changes then the functional form of g_ℓ and g_k will change. If a resistor, current source, or diode between node ℓ and the datum node changes then the functional form of g_ℓ will change. In the case of changes in the resistance of the resistors and the current of the current sources, the changes were described in Section 4.1.3: a change can be represented as the addition of a new conductance or new current source into the circuit. Changes in the diode could be due to changes in I_{sat}, η, or T, for example, and would change the functional relationship between the diode current and diode voltage.

6.1.4 Problem characteristics

6.1.4.1 Numbers of variables and equations

As in the linear circuit, we have the same number of variables as equations.

6.1.4.2 Number of solutions

Consider the diode model defined in (6.1) and illustrated in Figure 6.2. The current to voltage characteristic is strictly monotonically increasing so that increasing voltage corresponds to increasing current.

Exercise 2.21, Part (i) showed that in a one variable equation involving the sum of strictly monotonically increasing functions, there can be at most one solution. In our more complicated resistor–diode circuit, which requires several equations and variables to represent, we could observe from our physical experience that there is only one solution. The uniqueness of solution is, in part, because the diode current to voltage relationship is strictly monotonically increasing.

Our physical experience is based on a limited number of observations and does not constitute a general proof. Exercise 6.2 considers the general problem of solving a circuit with current sources, resistors, and diodes and shows that strict monotonicity of component model functions is sufficient to guarantee that there is at most one solution for the circuit. Exercise 6.2 illustrates the process of backing up an empirical observation with a theoretical analysis. It also illustrates the generalization of a result developed for linear simultaneous equations into a result that applies for non-linear simultaneous equations.

Not every two-terminal electronic component has a strictly monotonically increasing terminal model. For example, a **tunnel diode** has a characteristic that is not strictly monotonically increasing. Figure 6.4 shows a typical tunnel diode characteristic [13]. (Note that the axes are scaled differently in Figure 6.4 compared to Figure 6.2.) The characteristic of the tunnel diode shown in Figure 6.4 is neither monotonically increasing nor monotonically decreasing.

Circuits that include devices such as tunnel diodes can have multiple solutions. (In fact, the non-monotonic characteristics of a tunnel diode together with an energy storage element can be used to make an oscillator.) Moreover, three terminal

$I_{\text{tunnel diode}}(V_{\text{tunnel diode}})$ (amps)

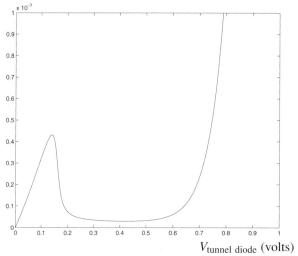

Fig. 6.4. Current to voltage relationship for tunnel
diode.

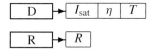

Fig. 6.5. Storage of pa-
rameters for diode and resistor as linked lists.

devices such as bipolar junction transistors and MOSFETs typically have models
that are not strictly monotone [95, section 10.6]. Indeed, by design, circuits such
as **flip-flops**, which include bipolar junction transistors and MOSFETS, have two
stable solutions [56, chapter 8].

6.1.4.3 Sparsity

The vector function g is "sparse" in the sense that a typical entry of g depends only
on a few entries of x. Although we cannot store the representation of a non-linear
function as a sparse matrix, we can still store the parameters necessary to specify
the functions in a sparse structure.

   For example, we could store the specification in a linked list of records. Each
record would include a flag, R or D, to specify whether the component was a resistor or a diode. If the component is a resistor, then one value would be required
to specify the resistance. If the component is a diode then three values would be
needed to specify the parameters I_{sat}, η, and T. The circuit simulation package,
SPICE [95], for example, allows circuits to be specified in such a manner. Figure 6.5 shows such a representation, with the parameters for a diode and for a
resistor stored as linked lists.

6.1.4.4 *Non-existence of direct algorithms*

Because of the presence of the non-linear diode elements, there is in general no direct algorithm for solving an arbitrary circuit consisting of current sources, resistors, and diodes.

6.2 Analysis of an electric power system

6.2.1 *Motivation*

An electric power system is a physically large electrical circuit consisting of voltage sources (generators) and loads interconnected by transmission and distribution lines. It therefore satisfies Kirchhoff's laws. In addition, there is a variety of subsidiary **protection equipment** that monitors for undesirable conditions such as:

- overload conditions on generators and on the transmission and distribution lines,
- under- or over-voltage conditions on loads, and
- short-circuits due to lightning strikes on transmission lines.

In each case, if the undesirable condition is detected and is severe enough then the affected element is disconnected from the rest of the system using **circuit breakers**, which are switches capable of interrupting large currents. For example, suppose that a lightning strike causes a short-circuit to occur between lines. The severe current that then flows will be detected and the line will be disconnected from the rest of the system by the protection equipment through the action of circuit breakers at either end of the line.

It is important to be able to predict the power flows on lines and the voltage magnitudes at loads in advance of actual operations so that overloads and under- or over-voltage conditions can be anticipated and pre-emptive action taken. Typically, an estimate is made of the loads at some particular time of the day, such as when loads are anticipated to be greatest. Then the line flows and voltages due to these loads are calculated and checked against limits.

If a lightning strike results in a transmission line being disconnected from the system, then the remaining transmission system usually has enough connectivity so that all loads are still supplied. However, the power flows on the transmission lines will change due to this **outage** and the flows in the "outaged system" will usually be heavier than before the outage. Again, it is important to calculate the power flows to see if overloads will result in the outaged system.

This case study presents the calculation of the power flows on the transmission lines and the voltages at various nodes to provide the information for an analysis of transmission line overloads and of the satisfaction of voltage limits. This calculation is called a **power flow study**. The calculation of flows in the system after a transmission line outage is called a **contingency study**. This case study assumes some familiarity with electric power systems [8][79]. The case study also assumes some familiarity with complex numbers and their representation [8, chapter 2][79, chapter 2].

6.2.2 Formulation

The material in this section is based on [8, chapter 10][79, chapter 3], but is considerably simplified. We consider the components in the generation–transmission system. As with the circuit case study in Section 4.1, we can think of a transmission network as a set of nodes with branches joining certain pairs of them. In power systems, it is customary to use the word **bus** to refer to a node. The branches include transmission lines, generators, loads, and also various other components that we will omit for simplicity.

6.2.2.1 Variables

Phasors and reference angle In all modern power systems, generators produce AC voltages and currents at their terminals that are maintained close to a nominal frequency of either 50 Hz or 60 Hz, depending on the country. We can use complex numbers, called **phasors**, to represent the magnitude and angle of the AC voltages and currents at a fixed frequency. The magnitude of the complex number represents the root-mean-square magnitude of the voltage, while the angle of the complex number represents the angle displacement between the sinusoidal voltage or current and a reference sinusoid. The specification of the root-mean-square magnitude and angle displacement completely determines the sinusoid, given a fixed reference sinusoid.

The angles of the voltages and currents in the system would all change if we changed the angle of our reference sinusoid, but this would have no effect on the physical system. In fact, we can arbitrarily assign the angle at one of the buses to be zero and measure all the other angles with respect to this angle. We call this bus the **reference bus**. It is customary to choose the reference bus to be a bus that has a generator and to assign a **reference angle** of $0°$ to the reference bus. We will assume that the reference bus is numbered to be bus 1. Given the arbitrary reference angle, there is a one-to-one correspondence between the value of a complex phasor and the sinusoidally varying quantity that it represents.

In other problems, it is also common to find that one of the variables can be arbitrarily assigned. Some thought should be given to this choice because a careful choice can often simplify the problem. For example, in our earlier circuit case studies, we picked out a node to be the datum node and assigned its voltage to be zero. A logical choice for the datum node in the earlier circuit case studies was the node with the most branches incident to it, since choosing this node as datum node yields the sparsest admittance matrix. The choice of datum node and of the reference bus for the power flow case study will be discussed in more detail in Sections 6.2.2.4 and 6.2.2.6, respectively.

Representation of complex numbers Although some computers support complex arithmetic operations, for our application there will turn out to be advantages to using pairs of real numbers to represent complex numbers. To represent a complex number $V \in \mathbb{K}$ with real numbers requires two real numbers, either:

- the **magnitude** $|V|$ and the **angle** $\angle V$, so that $V = |V| \exp(\angle V \sqrt{-1})$, or
- the **real** $\Re\{V\}$ and **imaginary** $\Im\{V\}$ parts, so that $V = \Re\{V\} + \Im\{V\} \sqrt{-1}$.

(We will use the symbol $\sqrt{-1}$ to avoid confusion with the symbol i that we have used for current. Since we will occasionally use the symbol j as a counter and also use j to index entries of vectors, we will also avoid the electrical engineering convention of using j to stand for $\sqrt{-1}$.)

We indicated that we need to compare voltage magnitudes to limits. Therefore, it seems sensible to represent the voltages as magnitudes and phases. Magnitude and angle will also turn out to be a convenient representation for currents. In other applications, representation with real and imaginary parts might turn out to be more useful and we will see that power is conveniently represented by real and imaginary parts.

Scaling and "per unit" The voltage limits in a typical system require that the voltage magnitude be maintained within about 5% of the "nominal" value. We use the solution of the power flow to check whether the voltages are within limits. If the voltages are not within limits, then some action must be taken to avoid the situation. The voltage magnitudes in the solution of the equations will occasionally be outside the 5% tolerance; however, typically, the voltages will not fall very far outside the 5% tolerance unless the system is in a very poor operating state.

There are voltage transformers throughout a typical power system. This means that the nominal voltage magnitude varies considerably throughout the system, in fact, by several orders of magnitude. We remarked in Section 3.2.1 that it can be advantageous to scale all the variables so that they have roughly equal magnitude at the solution of the problem. If we normalize each voltage by the corresponding nominal value, then all the normalized voltage magnitudes will be around 1. For example, one part of a system may have a nominal operating voltage of 110 kV, while another part of the system may have a nominal operating voltage of 765 kV. By scaling all the voltages by their nominal values, then an actual value of 121 kV in the 110 kV part of the system would be represented by a scaled value of:

$$\frac{121 \text{ kV}}{110 \text{ kV}} = 1.1,$$

while an actual value of 688.5 kV in the 765 kV part of the system would be represented by a scaled value of:

$$\frac{688.5 \text{ kV}}{765 \text{ kV}} = 0.9.$$

This scaling makes it easy to determine if the actual voltages are within 5% of the nominal voltage. Moreover, in assessing the accuracy of a solution, an error of 0.01, say, in the scaled values represents the same relative error in voltage magnitudes anywhere in the system. This allows us to use an unweighted norm to assess error in the voltage values.

It is also customary to scale the power quantities in terms of **base units** for the system. For example, a base power of 10 MW might be chosen and all power quantities will be normalized by 10 MW to give a **per unit** quantity. The effect of this scaling is that most magnitudes of power in the system will fall into the range of approximately 0.1 to 100.

It also turns out that the transmission line limits require that the angles of the voltages between ends of lines in the system be typically between $-45°$ and $45°$ or $-\pi/4$ to $\pi/4$ radians. (See Exercise 6.4.)

6.2.2.2 Symmetry

Three-phase circuits Generation–transmission systems are usually operated as balanced **three-phase systems**, with three equal-magnitude but 120° out-of-phase sinusoidal voltages imposed on triplets of virtually identical components. This arrangement is illustrated in Figure 6.6, which shows a three-phase generator connected through a three-phase transmission line to a three-phase load. The dashed lines separate the generator, transmission line, and load models. The three voltages generated by the three-phase generator are called the three **phases** and are conventionally labeled a, b, and c. This arrangement has various advantages over a **single-phase system** [8, chapter 2].

The three circles on the left-hand side of Figure 6.6 represent the three **stator** windings of the generator. A moving magnetic field induces voltages in the three stator windings that are 120° apart. The wires interconnecting the windings are drawn so that they are roughly 120° apart on the diagram, symbolizing the relationship between the angles of the generated voltages in each phase. The point where all three generator windings are connected together is called the **star point** and is denoted by an n′.

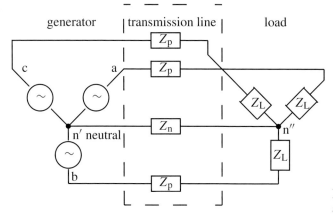

Fig. 6.6. An example balanced three-phase system.

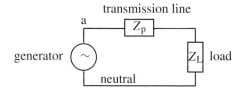

Fig. 6.7. Per-phase equivalent circuit for the three-phase circuit in Figure 6.6.

Similarly, the transmission line in the middle of the figure consists of three identical "phase" wires each of **impedance** $Z_p = R_p + X_p\sqrt{-1}$ together with a **neutral wire** of impedance $Z_n = R_n + X_n\sqrt{-1}$. (We will qualify the modeling of the transmission line further in Section 6.2.2.3.) The load on the right-hand side consists of three identical impedances $Z_L = R_L + X_L\sqrt{-1}$. The symbols R_p, R_n, and R_L represent the transmission resistance, the neutral resistance, and the load resistance, respectively. The symbols X_p, X_n, and X_L represent the transmission reactance, the neutral reactance, and the load reactance, respectively.

The load impedances are connected together at the load star point, denoted by the n''. The neutral wire joins the star points of the generator and load. (Many transmission lines do not have a neutral wire. This can be represented in our model as an infinite neutral impedance.)

Per-phase equivalent Under certain assumptions (see [8, chapter 2]), the behavior of a balanced three-phase circuit can be completely determined from the behavior of a **per-phase equivalent circuit**, which has components that are derived from and in many cases are identical to the components of one of the three phases. To see this, note that the three voltages in the generator result in currents in the a, b, and c phases that are identical in magnitude and 120° apart in angle. The sum of these currents is zero. (See Exercise 6.3.) But this means that the current flowing in the neutral impedance Z_n in Figure 6.6 is equal to zero so that nodes n' and n''

are at the same voltage. (This is true even if there is no neutral wire; that is, even if the neutral impedance is infinite.) This means that the currents in each phase can be calculated by considering each phase separately and all neutral points shorted together. This yields a per-phase equivalent circuit as shown in Figure 6.7. In this per-phase equivalent circuit, the other phases have been decoupled. (See [8, section 2.5] for more details.)

Conventionally, the a-phase quantities are used in the per-phase equivalent. Figure 6.7 shows the a-phase equivalent circuit of the three-phase circuit of Figure 6.6. In this case, the a-phase equivalent consists of a single generator winding, a single transmission line, a single load impedance, and a neutral wire that has zero impedance.

Model transformation The determination of the behavior of a three-phase system through the analysis of a related per-phase equivalent is an example of **model transformation**. We have not discussed the transformation in detail in this case because it is very specific to the power flow problem; however, the general principle is to try to find aspects of the system that allow us to simplify the model. In this case, it is the symmetry of the balanced three-phase system that allows the simplification. Even if the loads and other components in the system are not exactly symmetric, the per-phase equivalent can still give a good approximation to the behavior of the three-phase circuit.

Model transformation is application-specific and usually *ad hoc*. It is often worth doing if it significantly lessens the computational requirements for analyzing the model. In the present problem, we cut down the calculations by a factor of three by only considering one phase. The other phase quantities can be calculated by adding or subtracting 120° from the angles of the a-phase quantities. (There are also other more subtle advantages of solving the per-phase equivalent. See [8, chapter 12] for details.)

In other problems, an investigation of the symmetry may also lead to a simplification in the representation. Particularly if there are a large number of symmetrically organized components in the system, symmetry can provide profound insights into the model. If the system is only approximately symmetric, the analysis based on the assumption of symmetry will not yield exact solutions but can still be useful as a verification of the results of a more detailed model that represents the asymmetries. For example, there are typically slight asymmetries between phases in a power system. However, the per-phase equivalent provides a good approximation to behavior. When there are significant asymmetries, the per-phase equivalent is still useful as part of the calculation of the asymmetric circuit. (See [8, chapter 12].)

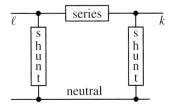

Fig. 6.8. Equivalent π circuit of per-phase equivalent of transmission line.

6.2.2.3 Transmission lines

Since transmission lines are physically extended objects, the **wave equations** are necessary to describe their internal behavior [8, chapter 4]. That is, the boxes used in Figures 6.6 and 6.7 should be construed as representing **distributed parameter circuits**, not just impedances as suggested in Section 6.2.2.2. We can represent the distributed parameter per-phase equivalents of the transmission lines with a π-**equivalent circuit**, which describes the **terminal characteristics** of the line for nominal-frequency voltages and currents in terms of lumped parameter elements [8, section 4.4]. A π-equivalent circuit for the per-phase equivalent of a transmission line joining two buses ℓ and k is shown in Figure 6.8. There are two **shunt components** connected from the terminals to neutral and a **series component** bridging the terminals. Each component has an impedance (or, equivalently, an **admittance**) determined by the characteristics of the line. Sometimes, the shunt components have values such that their effect on the circuit is negligible. In this case, we can neglect them as was done implicitly in Figures 6.6 and 6.7.

The π-equivalent circuit is another example of simplifying a model of a component by only considering its terminal behavior. Naturally, if we want to find out about the current and voltage along the line, we must model the internal behavior explicitly. For transmission lines operating in steady state and excited at a single frequency, it is possible to infer these internal currents and voltages from the terminal current and voltage phasors, so that the π-equivalent does not omit any essential information about the transmission line. The π-equivalent circuit exactly represents all the information needed to reconstruct the internal behavior of the transmission lines, given that the line is in steady state excitement at a single frequency.

In other cases, terminal models do not always provide enough information to infer all the internal operating behavior of the system. The choice of terminal model will depend on the application.

6.2.2.4 Bus admittance matrix and power flow equations

Consider the per-phase equivalent of a three-bus, three-line transmission system as illustrated in Figure 6.9. The buses 1, 2, and 3 are shown to be interconnected by

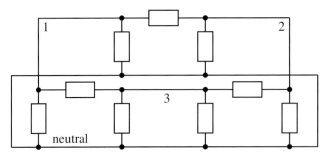

Fig. 6.9. Per-phase equivalent circuit model for the
three-bus, three-line system.

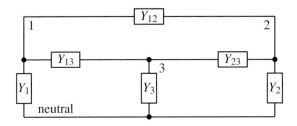

Fig. 6.10. The per-phase
equivalent circuit model
for the three-bus, three-
line system with parallel
components combined.

the π-equivalent models of the lines. Generators and loads are omitted from this circuit but will be re-introduced in Section 6.2.2.5. For each bus $\ell = 1, 2, 3$, the pair of shunt π elements joining bus ℓ to neutral can be combined together to form a single shunt element. This yields a circuit with:

- one element corresponding to each of the buses $\ell = 1, 2, 3$, joining bus ℓ to neutral, and
- one element corresponding to each line,

as illustrated in Figure 6.10. Let us write:

- Y_ℓ for the admittance of the element joining bus ℓ to neutral, so that Y_ℓ is the sum of the shunt admittances of the π-equivalent models of the lines incident to ℓ, and
- $Y_{\ell k}$ for the admittance of the series element corresponding to a line joining buses ℓ and k,

as illustrated in Figure 6.10. (We are abusing notation here by using Y with single and double subscripts to represent different parts of the circuit.)

The series element is most easily characterized in terms of its impedance. For a series impedance $Z_{\ell k} = R_{\ell k} + X_{\ell k}\sqrt{-1}$ between buses ℓ and k, the admittance $Y_{\ell k}$

is given by:

$$Y_{\ell k} = \frac{1}{Z_{\ell k}},$$

$$= \frac{1}{R_{\ell k} + X_{\ell k}\sqrt{-1}},$$

$$= \frac{1}{R_{\ell k} + X_{\ell k}\sqrt{-1}} \times \frac{R_{\ell k} - X_{\ell k}\sqrt{-1}}{R_{\ell k} - X_{\ell k}\sqrt{-1}},$$

$$= \frac{R_{\ell k} - X_{\ell k}\sqrt{-1}}{(R_{\ell k})^2 + (X_{\ell k})^2}. \tag{6.7}$$

Typically, $R_{\ell k}, X_{\ell k} > 0$ for a line. Therefore, $Y_{\ell k}$ typically has positive real part and negative imaginary part. Typically, a shunt admittance Y_ℓ has positive real and positive imaginary parts.

Let I_ℓ be the complex phasor current injected at bus ℓ into the network. As in Section 4.1, we define the voltage at any given bus in terms of the voltage between that bus and a datum node. In particular, we let V_ℓ be the phasor voltage at bus ℓ with respect to neutral, so that the datum node for the circuit is the neutral point. Since every shunt element is connected to the neutral point, choosing the neutral point as the datum node will yield the sparsest admittance matrix.

As in Section 4.1, we can collect the currents and voltages into vectors, which we denote by I and V, respectively. For each bus ℓ:

- V_ℓ is the phasor voltage at bus ℓ, and
- I_ℓ is the phasor current flowing from any generator or load at bus ℓ into the network at bus ℓ (including the current flowing into any shunt elements at bus ℓ).

As in the previous circuit case studies, we can again obtain a relationship of the form $AV = I$ between current and voltage, where:

$$\forall \ell, k, A_{\ell k} = \begin{cases} Y_\ell + \sum_{j \in \mathbb{J}(\ell)} Y_{\ell j}, & \text{if } \ell = k, \\ -Y_{\ell k}, & \text{if } k \in \mathbb{J}(\ell) \text{ or } \ell \in \mathbb{J}(k), \\ 0, & \text{otherwise,} \end{cases} \tag{6.8}$$

where $\mathbb{J}(\ell)$ is the set of buses joined directly by a transmission line to bus ℓ. The **bus admittance matrix** A has essentially the same characteristics as in the DC circuit, except that it is now a complex matrix.

6.2.2.5 Generators and loads

When electricity is bought and sold, prices are usually set for the power and energy, not for the voltage or current. Because this **real power** is the variable of most economic interest in power systems, we will model generators and loads as constant sources or sinks of real power located at buses in the system. However, real power does not completely describe the interaction between generators or loads and the system. We also have to characterize the injected **reactive power**. (Recall that reactive power is a measure of the energy that is moved into and out of the inductors and capacitors in the system during each cycle. See [8, chapter 2] for more details.)

We can combine the real and reactive powers into the **complex power**, which is the sum of:

- the real power, and
- the reactive power times $\sqrt{-1}$.

That is, we have used the real and imaginary parts of a complex number to represent the real and reactive power. The usefulness of this representation is that, for example, the complex power S_ℓ injected at bus ℓ into the network is given by:

$$S_\ell = V_\ell I_\ell^*,$$

where the superscript $*$ indicates **complex conjugate**. (This notation should be carefully distinguished from the similar looking notation for optimum value, indicated by superscript $\star$.) The current I_ℓ equals the sum of:

- the current flowing into the shunt element Y_ℓ, and
- the sum of the currents flowing into each line connecting ℓ to a bus $k \in \mathbb{J}(\ell)$ through admittance $Y_{\ell k}$.

We can substitute for the currents to obtain:

$$
\begin{aligned}
S_\ell &= V_\ell \left[A_{\ell\ell} V_\ell + \sum_{k \in \mathbb{J}(\ell)} A_{\ell k} V_k \right]^* , \\
&= |V_\ell|^2 A_{\ell\ell}^* + \sum_{k \in \mathbb{J}(\ell)} A_{\ell k}^* V_\ell V_k^* ,
\end{aligned}
\tag{6.9}
$$

This equation encompasses various cases:

- if a generator at a bus ℓ injects real power into the network, then I_ℓ will be in phase with V_ℓ such that $V_\ell I_\ell^*$ has positive real part;
- if there is no generator nor load at ℓ, then I_ℓ is zero and there is zero injected real and reactive power; and
- if there is a load at ℓ, then the injected current will be out of phase with V_ℓ such that $V_\ell I_\ell^*$ has *negative* real part.

The injected reactive power can be positive or negative at a generator or a load bus; however, reactive power injection is typically positive at a generator and negative at a load. We call a bus with no generation nor load a **zero injection bus**. We will return to zero injection buses in Section 12.2.

Because we are interested in voltage magnitudes we are going to re-express (6.9) in terms of voltage magnitude and angle. Because real power is of economic interest, we will also re-express (6.9) in terms of real and reactive power. Let:

- $A_{\ell k} = G_{\ell k} + B_{\ell k}\sqrt{-1}$, $\forall \ell, k$, where we note that by (6.7) and (6.8):
  - we have that $G_{\ell k} < 0$ and $B_{\ell k} > 0$ for $\ell \neq k$, and
  - we have that $G_{\ell \ell} > 0$ and the sign of $B_{\ell \ell}$ is indeterminate but typically less than zero;

- $S_{\ell} = P_{\ell} + Q_{\ell}\sqrt{-1}$, $\forall \ell$, with:
  - for generator buses, $P_{\ell} > 0$ and Q_{ℓ} is typically positive, and
  - for load buses, $P_{\ell} < 0$ and $Q_{\ell} < 0$;

  and
- $V_{\ell} = u_{\ell}\exp(\theta_{\ell}\sqrt{-1})$, $\forall \ell$, with:
  - the voltage magnitude $u_{\ell} \approx 1$ in scaled units to satisfy voltage limits, and
  - the voltage angle θ_{ℓ} typically between $-\pi/4$ and $\pi/4$ radians.

(We have previously used the symbol P to stand for a permutation matrix. For example, see Section 5.3.2.1. We have previously used the symbol Q to stand for the quadratic coefficient matrix of a quadratic function and for a factor of a matrix. In the context of this and several subsequent case studies, however, we will use the symbols P and Q for real and reactive power, respectively.) Using the above definitions, we can separate (6.9) into real and imaginary parts:

$$P_{\ell} = \sum_{k \in J(\ell) \cup \{\ell\}} u_{\ell}u_k[G_{\ell k}\cos(\theta_{\ell} - \theta_k) + B_{\ell k}\sin(\theta_{\ell} - \theta_k)], \qquad (6.10)$$

$$Q_{\ell} = \sum_{k \in J(\ell) \cup \{\ell\}} u_{\ell}u_k[G_{\ell k}\sin(\theta_{\ell} - \theta_k) - B_{\ell k}\cos(\theta_{\ell} - \theta_k)]. \qquad (6.11)$$

The equations (6.10) and (6.11), which are called the **power flow equality constraints**, must be satisfied at each bus ℓ. In some other applications, it may be more useful to write the equations in terms of the real and imaginary parts of the voltage or in terms of the magnitude and angle of the complex power. The choice of representation is driven by our interests.

The relation between P_{ℓ}, Q_{ℓ}, and the voltage magnitudes and angles is *nonlinear*. Although the transmission system is modeled as a linear circuit, our interest in real power (as opposed to voltage and current) has forced us to consider systems of non-linear equations.

6.2.2.6 The power flow problem

Power balance Let us try to satisfy the constraints (6.10) and (6.11) at each bus. Consider a typical generator. It can be controlled so that it injects a specified real and reactive power into the system. The terminal voltage magnitude and angle of the generator will depend on the interaction with the rest of the system. We call a bus with such a generator a PQ bus. Similarly, at a typical load bus the real and reactive power is determined by the loads attached to it. We also call this bus a PQ bus. (The *injected* real power at a load bus is typically negative, while the injected real power at a generator is positive.)

In summary, we specify:

- the real and reactive generations at the generator PQ buses according to the generator control settings, and
- the real and reactive power at the load PQ buses according to supplied data.

For a moment, suppose we specify the real power injection at *all* the generator buses. Recall that the (negative) real power injections have also been specified at all the load buses. Unless the load plus the losses (dissipated in the resistance of the lines) adds up to the sum of the specified generations, we will not be able to satisfy the first law of thermodynamics! In fact, if load plus losses does not add up to generation, the power flow equality constraints cannot be satisfied. A similar analysis applies to the reactive power.

Unless we specify all the injections consistently (that is, unless we already know a solution to the equations), we will find that the equations cannot be satisfied. That is, if we write down the power flow equality constraints (6.10) and (6.11) for all buses and specify the real and reactive generations arbitrarily, we will usually find that we have an inconsistent set of equations. In practice, the control systems of the generators adjust the outputs of generators to match the loads by detecting any imbalance between generation and load through deviation of the electrical frequency from nominal. Our specification of the problem does not capture this aspect of the control of the generators since we assume constant frequency of operation and therefore we must find another approach to solving the equations. (See [8, chapter 11] for discussion of this issue.)

Reference bus A traditional, but *ad hoc* approach to finding a solution to the equations is to single out the reference bus. At this bus, instead of specifying injected real and reactive power, we specify the voltage magnitude. The reference generator is then assumed to produce whatever is needed to balance the real power generation and load and balance the reactive power generation and load for the rest of the system, assuming that such a solution exists.

In other words, following the discussion in Section 3.2.2, we re-arrange the equations involving the real and reactive power injections at the reference bus to eliminate the real and reactive power injections. We solve the reduced set of equations and then substitute for the reference bus real and reactive power injections. We have re-interpreted P_1 and Q_1 to be variables in our formulation and have eliminated these variables by writing them as a function of the rest of the variables.

A physical interpretation of the reference bus is that it supplies whatever real and reactive power is necessary for real and reactive power balance. For this reason, the reference bus is usually chosen to be a generator bus.

The voltage magnitude at the reference bus is typically specified to be $u_1 = 1$ per unit. Recall that the voltage angle at the reference bus is also specified, usually as $\theta_1 = 0°$. (The reference bus is therefore sometimes called a $V\theta$ bus.) We have re-interpreted u_1 and θ_1 to be constant parameters in our problem.

Finding a power flow solution then involves calculating:

(i) the voltage magnitudes and angles at the PQ buses, and
(ii) the real and reactive power that must be injected at the reference bus,

to satisfy the power flow equality constraints. (In actual power systems, many generators are controlled to inject a given amount of real power and also to hold their terminal voltage magnitudes constant. Such a generator is called a PV bus. See [8, chapter 10] for details of the formulation and Exercises 6.5 and 6.6.)

6.2.2.7 Non-linear equations

Suppose that there are n_{PQ} PQ buses, including both the PQ generators and the loads. Let $n = 2n_{PQ}$ and define a vector $x \in \mathbb{R}^n$ consisting of the voltage magnitudes and angles at the PQ buses. Also, for every bus ℓ (that is, including the reference bus as well as the PQ buses) define functions $p_\ell : \mathbb{R}^n \to \mathbb{R}$ and $q_\ell : \mathbb{R}^n \to \mathbb{R}$ by:

$$\forall x \in \mathbb{R}^n, \ p_\ell(x) \ = \ \sum_{k \in \mathbb{J}(\ell) \cup \{\ell\}} u_\ell u_k [G_{\ell k} \cos(\theta_\ell - \theta_k) + B_{\ell k} \sin(\theta_\ell - \theta_k)] - P_\ell,$$

$$(6.12)$$

$$\forall x \in \mathbb{R}^n, \ q_\ell(x) \ = \ \sum_{k \in \mathbb{J}(\ell) \cup \{\ell\}} u_\ell u_k [G_{\ell k} \sin(\theta_\ell - \theta_k) - B_{\ell k} \cos(\theta_\ell - \theta_k)] - Q_\ell.$$

$$(6.13)$$

The right-hand sides of (6.12)–(6.13) re-arrange the terms in power flow equality constraints, (6.10)–(6.11). Compared to (6.10)–(6.11), we have shifted P_ℓ and Q_ℓ, respectively, onto the right-hand sides of (6.10)–(6.11). The functions p_ℓ and q_ℓ represent the net real and reactive power flow, respectively, from bus ℓ into the rest of the system. The power flow equality constraints (6.10)–(6.11) are equivalent to

$p_\ell(x) = 0$ and $q_\ell(x) = 0$. The equations reflect the fact that the net real power flow out of a bus must be zero and that the net reactive power flow out of a bus must be zero. For this reason, (6.10)–(6.11) are also called the real and reactive **power balance equations**.

Finally, define a vector function $g : \mathbb{R}^n \to \mathbb{R}^n$ that includes (6.12) and (6.13) for all the PQ buses, but omits (6.12) and (6.13) for the reference bus. Suppose we can solve:

$$g(x) = \mathbf{0}, \tag{6.14}$$

which consists of n equations in n variables. Then we claim that we can also calculate the real and reactive power injections at the reference bus: we simply substitute into (6.10) and (6.11) for the reference bus. In summary, solving Kirchhoff's equations for the electric power network has been transformed into an equivalent problem:

(i) solve (6.14), which is a system of non-linear simultaneous equations, and
(ii) substitute into (6.10) and (6.11) for the reference bus.

6.2.3 Circuit changes

If a real power injection changes at a bus ℓ then the entries in g corresponding to p_ℓ will change. If a reactive power injection changes at a bus ℓ then the entries in g corresponding to q_ℓ will change. If a transmission line between buses ℓ and k changes, then the entries of g corresponding to p_ℓ, q_ℓ, p_k, and q_k will change. The entries in the admittance matrix A will change in a manner analogous to the changes discussed in Section 4.1.3 for the DC circuit.

6.2.4 Problem characteristics

6.2.4.1 Number of variables and equations

There are the same number of variables as equations in (6.14).

6.2.4.2 Non-existence of direct algorithms

As with the non-linear circuit in Section 6.1, because the equations are non-linear, there is no direct algorithm to solve (6.14) for arbitrary systems. However, it is possible to directly solve some particular systems. See Exercise 6.4 for an example.

6.2.4.3 Number of solutions

In principle, there may be no solutions, one solution, or even multiple solutions to (6.14). As we have discussed, however, power systems are usually designed and operated so that the voltage magnitudes are near to nominal and the voltage angles are relatively close to 0°. If we restrict our attention to solutions such that voltage magnitudes are all close to 1 (and make some other assumptions) then we can find conditions for there to be at most one solution. This is explored in Exercises 6.4 and 6.5. If we cannot be sure that the voltage magnitudes are tightly constrained then the situation is less clear. In general, we may have to combine theoretical understanding with some engineering judgment to assess whether or not we have found *the* solution or just one of several solutions.

6.2.4.4 Admittance matrix

Symmetry The admittance matrix is symmetric. This confers some properties on g that will help us to solve the equations $g(x) = \mathbf{0}$.

Sparsity As in all circuits with a large number of nodes, only a small fraction of the possible lines will actually exist in typical systems. Therefore, the matrix A is sparse and each component of g depends on only a few components of x. We will again exploit sparsity in solving the equations.

Values As mentioned in Section 6.2.2.4, for a typical line, the corresponding line admittance $Y_{\ell k}$ has positive real part and negative imaginary part. If there is a line between buses ℓ and k then the entries $A_{\ell k} = G_{\ell k} + B_{\ell k}\sqrt{-1}$ in the admittance matrix therefore typically satisfy $G_{\ell k} < 0$, $B_{\ell k} > 0$. The diagonal entries $A_{\ell \ell} = G_{\ell \ell} + B_{\ell \ell}\sqrt{-1}$ in the admittance satisfy $G_{\ell \ell} > 0$ and, typically, $B_{\ell \ell} < 0$. The resistance of transmission lines is relatively small compared to the inductance. Furthermore, the shunt elements are often also negligible compared to the inductance. This means that:

$$\forall \ell, \forall k \in \mathbb{J}(\ell) \cup \{\ell\}, |G_{\ell k}| \ll |B_{\ell k}|.$$

This observation will turn out to greatly influence our ability to quickly solve the power flow equations. In Exercise 6.5 we utilize this observation to find conditions for there to be no more than one solution to the equations.

Exercises

Analysis of a non-linear direct current circuit

6.1 Write down the equation for the datum node for the circuit of Figure 6.3. Show that it is redundant. That is, show that the equation expressing Kirchhoff's current law at the datum node is identically equal to some linear combination of the equations expressing Kirchhoff's current law for the other nodes. (See Definition A.54.)

6.2 Consider a circuit with n non-datum nodes consisting of current sources, resistors, and diodes, as illustrated in Figure 6.3. Suppose that there are r resistors and diodes. As in Section 6.1.2.3, suppose that we define $g : \mathbb{R}^n \to \mathbb{R}^n$ to represent the equations in the form $g(x) = \mathbf{0}$, with:

- $x \in \mathbb{R}^n$,
- x_k representing the voltage at node k, and
- $g_\ell : \mathbb{R}^n \to \mathbb{R}$ representing the terms in Kirchhoff's current law at node ℓ.

(i) Consider the Jacobian $J : \mathbb{R}^n \to \mathbb{R}^{n \times n}$ of $g : \mathbb{R}^n \to \mathbb{R}^n$. (See Definition A.36.) Show that the Jacobian is of the form:

$$\forall x \in \mathbb{R}^n, \; J(x) = WG(x)W^\dagger,$$

where:

- $W \in \mathbb{R}^{n \times r}$ has linearly independent rows (see Definition A.55), and
- $G : \mathbb{R}^n \to \mathbb{R}^{r \times r}$ is a matrix-valued function with the following properties:

  – G is diagonal; that is, all off-diagonal entries in G are identically zero, and
  – each diagonal entry of G is strictly positive for all values of x.

(Hint: See Exercise 4.8. The diagonal entries in G are the **incremental conductances** of the resistors and diodes in the circuit. For resistors, this is just the conductance of the resistor. For diodes, this is the derivative of the current to voltage relationship of the diode.)

(ii) Show that for every $x \in \mathbb{R}^n$, $J(x)$ is positive definite. (Hint: Use Exercise 2.27, Part (vi).)

(iii) Show that the equations $g(x) = \mathbf{0}$ have at most one solution. (Hint: Use Theorems 2.2 and 2.3.)

Analysis of an electric power system

6.3 Consider three complex quantities with the same magnitude but with phases $120°$ apart. In particular, consider $u \exp(\theta\sqrt{-1})$, $u \exp([\theta+2\pi/3]\sqrt{-1})$, $u \exp([\theta-2\pi/3]\sqrt{-1})$, with $u, \theta \in \mathbb{R}$. Show that they sum to zero.

6.4 Consider a power system consisting of just two buses and one transmission line:

- bus 1 (the reference bus), where there is a generator, and
- bus 2, where there is load.

Suppose that the reference voltage is specified to be $V_1 = 1\angle 0°$ and that net injection at bus 2, as defined in (6.12), is given by:

$$\forall u_2 \in \mathbb{R}_+, \forall \theta_2 \in \mathbb{R}, \; p_2(u_2, \theta_2) = u_2 \sin\theta_2 + (-P_2).$$

(That is, we assume that $G_{22} = G_{12} = B_{22} = 0$ and $B_{12} = 1$.) Suppose that we also know that $u_2 = 1.0$. (The voltage magnitude can be controlled by **sources of reactive power** such as capacitors.)

(i) What is the largest value of demand $(-P_2)$ for which there is a solution to the equation $p_2(1.0, \theta_2) = 0$? What is the corresponding value of θ_2? We will write $\underline{\theta_2}$ for this value of θ_2.

(ii) What happens if θ_2 is smaller than $\underline{\theta_2}$?

(iii) Show that there are two solutions to the equation $p_2(1.0, \theta_2) = 0$ with $0 \geq \theta_2 > -2\pi$ if $(-P_2) = 0.5$. What are the corresponding values of θ_2?

6.5 Consider a power system with n PV buses, r transmission lines, and no PQ buses. Suppose that:

- the real part of the admittance matrix is the zero matrix, and
- the voltage magnitude at each bus ℓ, including the reference bus, is controlled to be $u_\ell = 1$.

We consider conditions for the real power equations (6.10) for real power balance to have no more than one solution.

Define $\theta \in \mathbb{R}^n$ to be the vector of voltage angles at the PV buses. The voltage angle at the reference bus, θ_1, is not included in the vector θ, but $\theta_1 = 0$ is known.

Since the voltage magnitudes are assumed constant and all equal to 1, we can re-express the real power equation (6.10) for each bus ℓ in the form $g_\ell(\theta) = 0$ where $g_\ell : \mathbb{R}^n \to \mathbb{R}$ is defined by:

$$\forall \theta \in \mathbb{R}^n, g_\ell(\theta) = \sum_{k \in \mathbb{J}(\ell) \cup \{\ell\}} B_{\ell k} \sin(\theta_\ell - \theta_k) - P_\ell,$$

where: P_ℓ is the net generation at bus ℓ; $B_{\ell k} > 0$ for $\ell \neq k$; and $B_{\ell \ell} < 0$. Collect the functions g_ℓ for the PV buses together into the vector function $g : \mathbb{R}^n \to \mathbb{R}^n$. The real power balance equations are then equivalent to $g(\theta) = \mathbf{0}$. (The value of n and the function g as defined is different to the value of n and the function defined in (6.14).)

(i) Show that the Jacobian $J : \mathbb{R}^n \to \mathbb{R}^{n \times n}$ of g is of the form:

$$\forall \theta \in \mathbb{R}^n, J(\theta) = W \mathcal{B}(\theta) W^\dagger,$$

where:

- $W \in \mathbb{R}^{n \times r}$ has linearly independent rows (see Definition A.55), and
- $\mathcal{B} : \mathbb{R}^n \to \mathbb{R}^{r \times r}$ is a matrix-valued function with the following properties:

  - $\mathcal{B}$ is diagonal; that is, all off-diagonal entries in $\mathcal{B}$ are identically zero, and
  - each diagonal entry $\mathcal{B}_{ss}$ of $\mathcal{B}$ corresponds to a line joining buses ℓ and k and is of the form:

    $$\forall \theta \in \mathbb{R}^n, \mathcal{B}_{ss}(\theta) = B_{\ell k} \cos(\theta_\ell - \theta_k).$$

  (Hint: See Exercise 6.2. The diagonal entries in $\mathcal{B}$ are different to the entries in the admittance matrix.)
(ii) Show that if $\theta \in \mathbb{S} = [-\frac{\pi}{4}, \frac{\pi}{4}]^n$ then $J(\theta)$ is positive definite. (Hint: Use Exercise 2.27, Part (vi).)
(iii) Show that the equations $g(\theta) = \mathbf{0}$ have at most one solution for $\theta \in \mathbb{S}$. (Hint: Use Theorems 2.2 and 2.3.)

6.6 As an approximation to the power flow equations, the **DC power flow** is often used. (The term "DC" stands for "direct current" and refers to an electrical analogy between the power system and a related, direct current resistive circuit. See [123, section 4.1.4].) As in Exercise 6.5, we consider a power system with n PV buses and no PQ buses and again suppose that the voltage magnitude at each bus ℓ, including the reference bus, is controlled to be $u_\ell = 1$.

Again define $\theta \in \mathbb{R}^n$ to be the vector of voltage angles at the PV buses. The voltage angle at the reference bus, θ_1, is not included in the vector θ, but $\theta_1 = 0$ is known.

Since the voltage magnitudes are assumed constant and all equal to 1, we can again re-express the real power equation (6.10) for each bus ℓ in the form $g_\ell(\theta) = 0$ where $g_\ell : \mathbb{R}^n \to \mathbb{R}$ is defined by:

$$\forall \theta \in \mathbb{R}^n, g_\ell(\theta) = \sum_{k \in \mathbb{J}(\ell) \cup \{\ell\}} G_{\ell k} \cos(\theta_\ell - \theta_k) + B_{\ell k} \sin(\theta_\ell - \theta_k) - P_\ell.$$

(This function includes terms involving the real part of the admittance matrix since here we do not assume that the real part of the admittance matrix is zero.) Again collect the functions g_ℓ for the PV buses together into the vector function $g : \mathbb{R}^n \to \mathbb{R}^n$. The real power balance equations are again equivalent to $g(\theta) = \mathbf{0}$.

Instead of seeking a solution to $g(\theta) = \mathbf{0}$, however, the DC power flow solves the linear simultaneous equations $J\theta = -g(\mathbf{0})$, where $J \in \mathbb{R}^{n \times n}$ is defined by:

$$\forall \ell \neq 1, \forall k \neq 1, J_{\ell k} = \frac{\partial g_\ell}{\partial \theta_k}(\mathbf{0}).$$

That is, the DC power flow uses a first-order Taylor approximation to $g(\theta)$, linearized about $\theta = \mathbf{0}$, and sets the Taylor approximation equal to zero. (See Section 2.6.3.5.)

   (i) Evaluate the entries in J.
  (ii) Use the DC power flow approximation to estimate the relationship between voltage angles and injections for the two-bus, one-line system specified in Exercise 6.4.
 (iii) Based on the solution to Exercise 6.4, when can you expect the DC power flow approximation to be a poor approximation to the exact solution? (See [3].)

7

Algorithms for non-linear simultaneous equations

In Chapter 5, we introduced triangular factorization and forwards and backwards substitution as a direct algorithm for solving linear equations. In a finite number of steps (assuming infinite precision arithmetic) we could solve the linear equations exactly. We have already remarked in Section 2.4.1.2 that no direct algorithm exists for general non-linear simultaneous equations.

Our approach to solving non-linear simultaneous equations will be iterative. We will start with an initial guess of the solution and try to successively improve on the guess. We will continue until the current iterate becomes sufficiently close to the solution according to a suitable criterion. In general, we will not be able to provide an exact solution, even if we could calculate in infinite precision. The theme of iterative progress towards a solution will recur throughout the rest of the book.

In Section 7.1, we discuss an iterative algorithm called the **Newton–Raphson method**. Then, in Section 7.2 we will introduce some variations on the basic Newton–Raphson method. In Section 7.3 we discuss **local convergence** of the iterates to a solution, while in Section 7.4 we discuss **global convergence**. Finally, we discuss sensitivity and large change analysis in Section 7.5.

The key issues discussed in this chapter are:

- approximating non-linear functions by a **linear approximation** about a point,
- using the linear approximation to improve our estimate of the solution of the non-linear equations and then re-linearizing about the improved solution in an **iterative algorithm**,
- **convergence** of the sequence of iterates produced by repeated re-linearization,
- **variations** that reduce computational effort, and
- **sensitivity** and **large change analysis**.

Most of the material is based on [11, 45, 58, 84] and much more detailed discussions are available in those references.

7.1 Newton–Raphson method

Consider a function $g : \mathbb{R}^n \to \mathbb{R}^n$ and suppose that we want to solve the simultaneous non-linear equations:

$$g(x) = \mathbf{0}. \tag{7.1}$$

Analogously to the case of linear equations with the same number of equations as variables described in Section 4.1.4.1, we call this a **square system of equations**. We will develop an iterative algorithm called the Newton–Raphson method to solve (7.1). As we proceed, we will see that we must restrict ourselves to particular types of functions g to successfully apply the Newton–Raphson method. In Section 7.1.1, we will consider an initial guess and then in Section 7.1.2 use a Taylor approximation about the initial guess to set the stage in Section 7.1.3 for describing the update to the initial guess. In Section 7.1.4 we will present the Newton–Raphson update for the general iterate and then discuss several further issues in Section 7.1.5.

7.1.1 Initial guess

Let $x^{(0)}$ be the initial guess of a solution to (7.1). It should be chosen based on our understanding of the particular problem. The superscript (0) indicates the iteration count of our iterative process. If we are fortunate, then $x^{(0)}$ will satisfy (7.1); that is, $g(x^{(0)}) = \mathbf{0}$. Usually, however, we will not be so fortunate, so that $g(x^{(0)}) \neq \mathbf{0}$ and we seek an updated value of the vector $x^{(1)} = x^{(0)} + \Delta x^{(0)}$ such that:

$$g(x^{(1)}) = g(x^{(0)} + \Delta x^{(0)}) = \mathbf{0}, \tag{7.2}$$

or at least a step direction $\Delta x^{(0)}$ such that $g(x^{(1)})$ is "closer" than $g(x^{(0)})$ to the zero vector.

7.1.2 Taylor approximation

To find an appropriate update $\Delta x^{(0)}$, we will approximate the left-hand side of (7.2) using a **first-order Taylor approximation about** $x^{(0)}$ (Section 2.6.3.5) [72, section 4.1]. That is, we make a linear approximation to g. We will assume that the function $g : \mathbb{R}^n \to \mathbb{R}^n$ is partially differentiable with continuous partial derivatives. (See Definition A.36.) We will first consider the case of a scalar function, in particular the scalar function g_1 given by the first component of the vector function g, and we will then generalize to a vector function. We first discussed first-order Taylor approximations in connection with convex functions in Section 2.6.3.5; however, here we will not be making any assumption about the convexity of the function.

7.1.2.1 Scalar function

Consider the first component of g, namely $g_1 : \mathbb{R}^n \to \mathbb{R}$. If g_1 is partially differentiable (see Definition A.36) with continuous partial derivatives then:

$$
\begin{aligned}
g_1(x^{(1)}) \\
&= g_1(x^{(0)} + \Delta x^{(0)}), \text{ since } x^{(1)} = x^{(0)} + \Delta x^{(0)}, \\
&\approx g_1(x^{(0)}) + \frac{\partial g_1}{\partial x_1}(x^{(0)})\Delta x_1^{(0)} + \frac{\partial g_1}{\partial x_2}(x^{(0)})\Delta x_2^{(0)} + \cdots + \frac{\partial g_1}{\partial x_n}(x^{(0)})\Delta x_n^{(0)}, \\
&= g_1(x^{(0)}) + \sum_{k=1}^{n} \frac{\partial g_1}{\partial x_k}(x^{(0)})\Delta x_k^{(0)}, \\
&= g_1(x^{(0)}) + \frac{\partial g_1}{\partial x}(x^{(0)})\Delta x^{(0)}, \qquad (7.3)
\end{aligned}
$$

where the last equality follows by definition of the partial derivative. (See Definition A.36.)

In (7.3), the symbol "$\approx$" should be interpreted to mean that the difference between the expressions to the left and to the right of the $\approx$ is small compared to $\|\Delta x^{(0)}\|$. Formally, if we define the **remainder at the point** $x^{(0)}$, $e : \mathbb{R}^n \to \mathbb{R}$, by:

$$
\begin{aligned}
\forall \Delta x \in \mathbb{R}^n, e(\Delta x) &= g_1(x^{(0)} + \Delta x) - g_1(x^{(0)}) \\
&\quad - \frac{\partial g_1}{\partial x_1}(x^{(0)})\Delta x_1 - \frac{\partial g_1}{\partial x_2}(x^{(0)})\Delta x_2 - \cdots - \frac{\partial g_1}{\partial x_n}(x^{(0)})\Delta x_n,
\end{aligned}
$$

then, by **Taylor's theorem with remainder** [72, section 4.1], if g_1 is partially differentiable with continuous partial derivatives then:

$$
\lim_{\|\Delta x\| \to 0} \frac{e(\Delta x)}{\|\Delta x\|} = 0.
$$

As first mentioned in Section 2.6.3.5, the expression to the right of the $\approx$ in (7.3) is called a **first-order Taylor approximation**. By Taylor's theorem with remainder, for a partially differentiable function g_1 with continuous partial derivatives, the first-order Taylor approximation about $x = x^{(0)}$ approximates the behavior of g_1 in the vicinity of $x = x^{(0)}$. This is the fundamental reason why linearization is effective in approximating functions. (See Exercise 7.1.)

For points remote from $x^{(0)}$, the first-order Taylor approximation can be poor. Typically, the approximation becomes worse as $\|\Delta x\|$ increases. We will discuss the issue of the region of validity of the Taylor approximation in Section 7.4.2.1.

The first-order Taylor approximation represents a plane that is **tangential** to the graph of the function at the point $x^{(0)}$. As an illustration, suppose that $g_1 : \mathbb{R}^2 \to \mathbb{R}$

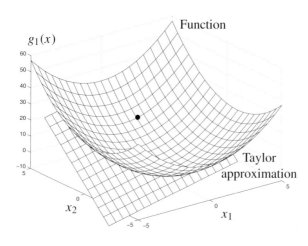

Fig. 7.1. Graph of function repeated from Figure 2.5 and its Taylor approximation about $x^{(0)} = \begin{bmatrix} 1 \\ 3 \end{bmatrix}$.

is the same function as shown in Figure 2.5, so that $n = 2$ and

$$\forall x \in \mathbb{R}^2,\ g_1(x) = (x_1)^2 + (x_2)^2 + 2x_2 - 3.$$

Let $x^{(0)} = \begin{bmatrix} 1 \\ 3 \end{bmatrix}$. Figure 7.1 shows the function g_1 along with the plane representing

its first-order Taylor approximation about $x^{(0)}$. The point $\begin{bmatrix} 1 \\ 3 \\ g_1\begin{bmatrix} 1 \\ 3 \end{bmatrix} \end{bmatrix} \in \mathbb{R}^2 \times$

$\mathbb{R}^1$ is shown as a bullet •. The graph of the function and its first-order Taylor approximation are tangential at $x^{(0)} = \begin{bmatrix} 1 \\ 3 \end{bmatrix}$ as illustrated in Figure 7.1. (We will return to the issue of tangents in Section 14.1.2.)

7.1.2.2 Vector function

We now consider the vector function $g : \mathbb{R}^n \to \mathbb{R}^n$. Since g is a vector function and x is a vector, the Taylor approximation of g involves the $n \times n$ matrix of partial derivatives $\dfrac{\partial g}{\partial x}$ evaluated at $x^{(0)}$ [72, section 2.3]. (See Definition A.36.) The ℓk-th entry of $\dfrac{\partial g}{\partial x}$ is $\dfrac{\partial g_\ell}{\partial x_k}$. A first-order Taylor approximation of g about $x^{(0)}$ yields:

$$g(x^{(0)} + \Delta x^{(0)}) \approx g(x^{(0)}) + \frac{\partial g}{\partial x}(x^{(0)})\Delta x^{(0)},$$

where by the $\approx$ we mean that the norm of the difference between the expressions to the left and to the right of $\approx$ is small compared to $\left\| \Delta x^{(0)} \right\|$. Formally, if we now define the remainder at the point $x^{(0)}$, $e : \mathbb{R}^n \to \mathbb{R}^n$, by:

$$\forall \Delta x \in \mathbb{R}^n, e(\Delta x) \;=\; g(x^{(0)} + \Delta x) - g(x^{(0)})$$

$$- \frac{\partial g}{\partial x_1}(x^{(0)})\Delta x_1 - \frac{\partial g}{\partial x_2}(x^{(0)})\Delta x_2 - \cdots - \frac{\partial g}{\partial x_n}(x^{(0)})\Delta x_n,$$

then, by **Taylor's theorem with remainder** [72, section 4.1], if g is partially differentiable with continuous partial derivatives then:

$$\lim_{\|\Delta x\| \to 0} \frac{\|e(\Delta x)\|}{\|\Delta x\|} = 0.$$

(See Exercise 7.1.) The first-order Taylor approximation again represents a "plane" that is tangential to the graph of the function; however, the situation is much more difficult to visualize for a vector function. The remainder is now a vector function with norm that rapidly approaches zero as the norm of its argument approaches zero.

7.1.2.3 Jacobian

Recall from Section 2.5.3.2 that the matrix of partial derivatives is called the **Jacobian** and we will denote it by $J(\bullet)$. Using this notation, we have:

$$\begin{aligned} g(x^{(1)}) \;&=\; g(x^{(0)} + \Delta x^{(0)}), \text{ by definition of } \Delta x^{(0)}, \\ &\approx\; g(x^{(0)}) + J(x^{(0)})\Delta x^{(0)}, \end{aligned} \qquad (7.4)$$

using the first-order Taylor approximation. The Jacobian is the transpose of the gradient ∇g of g.

In some of our development, we will approximate the Jacobian when we evaluate the right-hand side of (7.4). In this case, the linear approximating function is no longer tangential to f; however, this approximation may still be useful as we will see in Section 7.2.

7.1.3 Initial update

We seek $\Delta x^{(0)}$ that will make the left-hand side of (7.4) equal to zero. However, as we have argued, there is no direct algorithm to find this update for arbitrary g. Instead, let us seek $\Delta x^{(0)}$ that makes the right-hand side of (7.4) equal to zero, which will yield an update that makes the left-hand side of (7.4) *approximately* equal to zero. Setting the right-hand side of (7.4) to zero to solve for $\Delta x^{(0)}$ yields a set of linear simultaneous equations:

$$J(x^{(0)})\Delta x^{(0)} = -g(x^{(0)}), \qquad (7.5)$$

where:

- $J(x^{(0)}) \in \mathbb{R}^{n \times n}$ is the coefficient matrix and is constant given $x^{(0)}$,
- $-g(x^{(0)}) \in \mathbb{R}^n$ is the right-hand side vector and is constant given $x^{(0)}$, and
- $\Delta x^{(0)} \in \mathbb{R}^n$ is the vector of unknowns.

Notice that (7.5) is a linear equation in $\Delta x^{(0)}$ with the same number of variables as equations. We have already discussed the solution of such linear equations in Chapter 5 and know that it may in general have none, one, or many solutions. However, we will initially assume that $J(x^{(0)})$ is non-singular so that a unique solution exists. (We will discuss the case of singular J in Section 7.4.) Having calculated $\Delta x^{(0)}$, we can then calculate $x^{(1)}$, which is the value of the iterate after the first iteration.

7.1.4 General update

Unfortunately, we cannot expect that $x^{(1)}$ will exactly satisfy the equation $g(x^{(1)}) = 0$. However, we hope that $x^{(1)}$ will satisfy the equations "more closely" than the initial estimate $x^{(0)}$. We must repeat the process, updating each successive iterate according to:

$$
\begin{align}
J(x^{(\nu)})\Delta x^{(\nu)} &= -g(x^{(\nu)}), \tag{7.6} \\
x^{(\nu+1)} &= x^{(\nu)} + \Delta x^{(\nu)}, \tag{7.7}
\end{align}
$$

where $x^{(\nu)}$ is the value of x after the ν-th iteration. Equations (7.6)–(7.7) are called the **Newton–Raphson update** and $\Delta x^{(\nu)}$ is the **Newton–Raphson step direction**. We continue until the iterate $x^{(\nu)}$ satisfies (7.1) to sufficient accuracy, where "sufficient accuracy" will be defined more precisely in Section 7.3.

7.1.5 Discussion

The basic Newton–Raphson method has some very desirable properties that we will analyze in Section 7.3. Unfortunately, it has three drawbacks.

(i) The need to calculate the matrix of partial derivatives and solve a system of linear simultaneous equations at each iteration. Even with sparse matrix techniques, this can require considerable effort.

(ii) At some iteration we may find that the linear equation (7.6) does not have a solution, so that the update is not well-defined.

(iii) Even if (7.6) does have a solution at every iteration, the sequence of iterates generated may not converge to the solution of (7.1).

We will discuss some aspects of the first drawback in Section 7.2, indicating the great variety of approaches we can follow to minimize the computational effort. We will select just three of the many approaches:

- the "chord method,"
- the basic Newton–Raphson method, and
- the "quasi-Newton method,"

for further analysis in Section 7.3. The analysis will be "local" in that the initial point must be in the vicinity of the solution for the analysis to apply.

The second and third drawbacks are particularly problematic for iterates that are far from the solution. In Section 7.4, we will discuss convergence from initial points that are remote from the solution and so treat the second and third drawback.

7.2 Variations on the Newton–Raphson method

In this section we will discuss various ways to reduce the effort involved in the basic Newton–Raphson method by *approximating* the Newton–Raphson update. The goal will be to reduce the effort per iteration of the update without significantly disrupting the progress towards the solution. Typically, for a given accuracy of solution, the methods that approximate the update will take more iterations than required by the exact Newton–Raphson update. However, the increase in the number of iterations is often more than compensated by a decrease in the average effort per iteration.

Because the solution of non-linear equations forms the basis of many of the algorithms in the rest of the book, the basic trade-off between the number of iterations and the effort per iteration also applies in the other algorithms. We will discuss the issues in some detail here. We will then refer briefly to the analogous considerations as they appear in later chapters.

We could imagine approximating either or both of g and J in the Newton–Raphson update to reduce the effort. First, consider using approximate values of g in the Newton–Raphson update. Of course, the numerical values of g will always be approximate in that we calculate in finite precision arithmetic, and this poses a fundamental limit to how accurately we can judge whether the equations $g(x) = 0$ are satisfied.

Even if we were able to calculate to infinite precision, we could still consider *deliberately* using approximate values of g. However, if we only evaluate g approximately, then we can only ever hope to solve the equations approximately. In this book, we will not consider in detail issues involved with approximations of g. We will generally assume that g is evaluated accurately. (See, for example, [58, section 5.4] for details in the case of approximate evaluation of g. Moreover, if the calculation of g is subject to **random error** with zero mean and such that the errors on successive evaluations of g are independent then it is possible to "average out" the error by evaluating g multiple times to obtain higher accuracy. See [11, section 2.1].)

On the other hand, we will see that approximating J in the Newton–Raphson update does not inherently prevent us from converging to an accurate solution of the non-linear simultaneous equations. Moreover, under some circumstances, using approximate values for J (or only approximately solving the update equation) will only slightly slow the progress of the iterates towards the solution if the approximation to J is "close enough" to the exact Jacobian in a sense to be made more precise as we proceed through this chapter. In Section 7.2.1, we will consider various ways to approximate the Jacobian in the Newton–Raphson update. In Section 7.2.2, we will mention an iterative approach to approximating the solution to the Newton–Raphson update. In Section 7.2.3, we will describe pre-conditioning as a transformation of the update equations that can facilitate their solution. Finally, in Section 7.2.4, we will mention an approach to reducing the burden of writing software to calculate J.

7.2.1 Approximation of the Jacobian

Suppose that at each iteration we approximate the Jacobian in such a way as to make the update equations easier to solve. That is, at each iteration ν we replace $J(x^{(\nu)})$ by a matrix $\tilde{J}^{(\nu)}$ such that, compared to using $J(x^{(\nu)})$ directly:

   (i) LU factorization of $\tilde{J}^{(\nu)}$ requires less effort (or has already been performed),
   (ii) an inverse of $\tilde{J}^{(\nu)}$ is easier to calculate, or
   (iii) evaluation of $\tilde{J}^{(\nu)}$ is more convenient.

In summary, the use of the approximation $\tilde{J}^{(\nu)}$ reduces the average effort per iteration for performing the update. If the resulting approximation to the Newton–Raphson update satisfies suitable conditions, then it turns out that we will still iterate towards the solution. We will discuss some of these approximations in Sections 7.2.1.1–7.2.1.5.

7.2.1.1 The chord method

Because of the computational cost of forming and factorizing the Jacobian, we seek a method that will reduce the effort. Instead of updating the Jacobian at each iteration, we can think of updating it only occasionally. In the extreme, we form and factorize the Jacobian only once, for the initial update. That is, we define our sequence of iterates to be:

$$J(x^{(0)})\Delta x^{(\nu)} = -g(x^{(\nu)}), \tag{7.8}$$
$$x^{(\nu+1)} = x^{(\nu)} + \Delta x^{(\nu)}, \tag{7.9}$$

so that $\tilde{J}^{(v)} = J(x^{(0)}), \forall v \in \mathbb{Z}_+$. Once we have calculated $J(x^{(0)})$ and LU factorized it, each iteration (7.8)–(7.9) only requires:

- evaluation of $g(x^{(v)})$,
- a single forwards and backwards update to calculate $\Delta x^{(v)}$, and
- a vector addition to calculate $x^{(v+1)}$.

This approximation, which uses $J(x^{(0)})$ at every iteration, is called the **chord method** [58, section 5.4.1] and (7.8)–(7.9) is called the **chord update**. No factorization is required for the chord method after the factorization of the Jacobian evaluated at the initial guess. Therefore, each subsequent iteration requires much less effort than is required for the Newton–Raphson update (7.6)–(7.7). Sometimes, a judicious choice of initial guess provides a Jacobian $J(x^{(0)})$ that is especially easy to form and factorize. (See Section 8.2.4.1 for an example in the case of the electric power system case study from Section 6.2.)

7.2.1.2 The Shamanskii method

Updating the Jacobian every, say, N iterations (instead of only at the initial iteration) is called the **Shamanskii method** [58, section 5.4.3]. In the Shamanskii method, the iterations are:

$$J(x^{(N\lfloor v/N\rfloor)})\Delta x^{(v)} = -g(x^{(v)}), \tag{7.10}$$
$$x^{(v+1)} = x^{(v)} + \Delta x^{(v)}, \tag{7.11}$$

where:

- $\lfloor \bullet \rfloor$ returns the largest integer that is no larger than its argument, and
- we evaluate and factorize the Jacobian only if we have not already factorized it at an earlier iteration.

That is, $\tilde{J}^{(v)} = J(x^{(N\lfloor v/N\rfloor)})$. Again, the average computational effort per iteration for the **Shamanskii update** (7.10)–(7.11) is less than for (7.6)–(7.7).

7.2.1.3 Approximating particular terms

Suppose that we replace small terms in the Jacobian by zero. Sparse factorization techniques then require less effort to factorize the approximate matrix. In the extreme, we might be able to approximate the Jacobian by a matrix that can be inverted with little effort. For example, if the Jacobian is **strongly diagonally dominant**; that is, its diagonal terms are much larger than its off-diagonal, then an approximate inverse is the diagonal matrix having diagonal entries consisting of the inverse of the diagonal entries of the Jacobian.

Exercise 7.3 shows that such approximations should be used with some caution and that the effort involved in the approximations and the performance of the approximation should be compared with the performance of more exact techniques.

7.2.1.4 Analytic approximation to Jacobian

If we are using a numerical model, it can be very difficult to obtain a good numerical estimate of J. We may, however, have an *approximate* analytical model. Then we can combine a numerical evaluation of g with an approximate analytical model of J to use in the Newton–Raphson update. (See Exercise 7.4.)

7.2.1.5 Finite difference approximation to Jacobian

If g is the result of a numerical calculation and there is no analytical model for g, then analytical differentiation to obtain J is not possible. In this case, we have no choice but to approximate J by a **finite difference approximation**. Even if there is an analytical model available, it may be inconvenient to evaluate the Jacobian using the analytical model and we may again choose to use a finite difference approximation. The use of finite differences instead of evaluating the Jacobian is sometimes called a **discrete-Newton method**.

A basic implementation involves approximating the derivative of g in the direction Δx by the average rate of change in the value of g between, for example:

- the point $x^{(\nu)}$ and the point $x^{(\nu)} + \Delta x$, which is called the **forward difference approximation**:
$$J(x^{(\nu)})\Delta x \approx g(x^{(\nu)} + \Delta x) - g(x^{(\nu)});$$

- the point $x^{(\nu)} - \Delta x$ and the point $x^{(\nu)} + \Delta x$, which is called the **central difference approximation**:
$$2J(x^{(\nu)})\Delta x \approx g(x^{(\nu)} + \Delta x) - g(x^{(\nu)} - \Delta x);$$

or

- if $x \in \mathbb{R}$, the point $x^{(\nu)}$ and the point $x^{(\nu-1)}$, which is called the **secant approximation**:
$$\frac{\partial g}{\partial x}(x^{(\nu)}) \approx \frac{g(x^{(\nu)}) - g(x^{(\nu-1)})}{x^{(\nu)} - x^{(\nu-1)}}.$$

(There are other variations on secant approximation, particularly if g is the sum of linear and non-linear terms or is the sum of terms that each depend on only one variable. This is the case in non-linear DC circuit analysis. See [58, section 5.4.5][101] and the discussion below.)

The forward, central, and secant approximations are illustrated in Figure 7.2. Typically, the forward difference will require n evaluations of the vector function g, in addition to the calculation of $g(x^{(\nu)})$, to approximate the n columns of $J(x^{(\nu)})$,

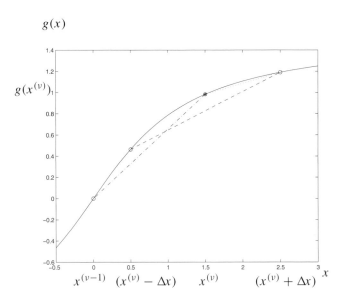

$g(x)$

Fig. 7.2. The finite difference approximations to the
derivative of a function g :
$\mathbb{R} \to \mathbb{R}$ at a point $x^{(\nu)}$. The
function g is illustrated as
a solid curve. The point
$$\begin{bmatrix} x^{(\nu)} \\ g(x^{(\nu)}) \end{bmatrix} = \begin{bmatrix} 1.5 \\ 1 \end{bmatrix} \text{ is}$$
indicated by the ●. The
forward difference approximation with $\Delta x = 1$
is given by the slope of
the dotted line. The cen-
tral difference approximation with $\Delta x = 1$ is given
by the slope of the dashed
line. The secant approxi-
mation for $x^{(\nu-1)} = 0$ is
given by the slope of the
dash-dotted line.

while the central difference will require $2n$ evaluations of g. (See Exercise 7.5.)
The central difference, however, provides higher accuracy under suitable conditions [11, section 1.8]. In special cases, particularly if the Jacobian has a regular
sparsity structure, the Jacobian can be calculated with relatively few evaluations of
g. (See [45, section 4.8.1] for details.)

In principle, the smaller the length of Δx, the more accurately the average rate
of change of g approximates the derivative for the forward and central difference methods, assuming that all calculations are performed to infinite precision.
However, for very small values of $\| \Delta x \|$, the calculation of the difference between
$g(x^{(\nu)} + \Delta x)$ and $g(x^{(\nu)})$ or between $g(x^{(\nu)} + \Delta x)$ and $g(x^{(\nu)} - \Delta x)$ will be subject to significant round-off errors. The choice of the size of Δx is therefore a
compromise. A standard prescription is to choose $\| \Delta x \|$ to be on the order of the
square root of the round-off error in calculating the finite differences. (See [11,
section 1.8][45, sections 4.6.1 and 8.6.1][58, section 5.4.4][84, section 11.4.1] for
a more careful discussion of this issue.)

The secant approximation can be used to evaluate particular entries of J if they
depend on only one variable. This is useful, for example, for the voltage to current
relations of two-terminal circuit components. (See [101].) A variant of the secant
method for functions for which $g(0) \approx 0$ makes the following approximation:

$$\frac{\partial g}{\partial x}(x^{(\nu)}) \approx \frac{g(x^{(\nu)})}{x^{(\nu)}}.$$

It is typically the case that two terminal circuit elements satisfy $g(0) = 0$.

7.2.1.6 Quasi-Newton methods

A drawback of the finite difference method is that it requires many evaluations of the function g to approximate the Jacobian. In this section, we will discuss a technique that uses the successive values of $g(x^{(v)})$ at each iteration to approximate the Jacobian, so avoiding any additional evaluations of g.

Consider a first-order Taylor approximation of g about $x^{(v-1)}$:

$$g(x^{(v-1)} + \Delta x^{(v-1)}) \approx g(x^{(v-1)}) + J(x^{(v-1)})\Delta x^{(v-1)}.$$

Substituting from the Newton–Raphson update equations (7.6)–(7.7) applied to calculate $x^{(v)}$, we obtain:

$$g(x^{(v)}) \approx g(x^{(v-1)}) + J(x^{(v-1)})(x^{(v)} - x^{(v-1)}).$$

Re-arranging, we have:

$$J(x^{(v-1)})(x^{(v)} - x^{(v-1)}) \approx g(x^{(v)}) - g(x^{(v-1)}). \tag{7.12}$$

Quasi-Newton methods [58, chapter 7] involve successively updating each approximation $\tilde{J}^{(v-1)}$ so that the updated approximation $\tilde{J}^{(v)}$ used for calculating $x^{(v+1)}$ satisfies the **quasi-Newton condition**:

$$\forall v > 0, \tilde{J}^{(v)}(x^{(v)} - x^{(v-1)}) = g(x^{(v)}) - g(x^{(v-1)}). \tag{7.13}$$

In (7.13), $x^{(v)} \in \mathbb{R}^n$ and $x^{(v-1)} \in \mathbb{R}^n$ are known and we seek $\tilde{J}^{(v)} \in \mathbb{R}^{n \times n}$ to satisfy the condition. The quasi-Newton method entirely avoids the need to calculate the Jacobian, since it uses the change in g to approximate J. In particular, $\tilde{J}^{(v)}$, (which is used in the calculation of $x^{(v+1)}$) is chosen to mimic the behavior of the change in g that resulted from the choice of $x^{(v)}$ as specified in (7.12). Since there are $(n)^2$ entries in $\tilde{J}^{(v)}$, while the quasi-Newton condition (7.13) specifies n equations, there are typically many matrices that satisfy the quasi-Newton condition (7.13).

Suppose that we have already factorized $\tilde{J}^{(v-1)}$ and that we want to update it to a new approximation $\tilde{J}^{(v)}$ that satisfies the quasi-Newton condition. Surprisingly, if $\tilde{J}^{(v-1)}$ is symmetric then, under mild assumptions, **symmetric rank two updates** (see Section 5.6.2.2) can be found such that by updating $\tilde{J}^{(v-1)}$, the new approximation $\tilde{J}^{(v)}$ satisfies the quasi-Newton condition. (See Exercise 7.6.) The updates are known generally as the **Broyden family** [58, chapter 7], the most popular of which is the **Broyden, Fletcher, Goldfarb, Shanno (BFGS) update** [11, section 1.7].

As discussed in Section 5.6.2.2, if we arrange that $\tilde{J}^{(v-1)}$ and the updated matrix $\tilde{J}^{(v)}$ differ by a symmetric rank two update and if $\tilde{J}^{(v-1)}$ has already been factorized then $\tilde{J}^{(v)}$ can be factorized with additional computational effort that is proportional to $(n)^2$, which is considerably less than the $(n)^3$ effort required for factorization of the Jacobian from scratch. The details can be found in [45, section 4.5.2][58, chapter 7][70, chapter 9][84, section 11.3]. Factorizing $\tilde{J}^{(\nu)}$ requires much less effort than factorizing the exact Jacobian. The resulting update direction is typically a very effective approximation to the Newton–Raphson step direction; however, the approximations and their factors are typically non-sparse [65, section 4]. The convergence rate to a solution is often super-linear: as we will see, this is better than the chord update but not as good as Newton–Raphson.

A typical choice for initialization is $\tilde{J}^{(0)} = \mathbf{I}$; however, $J(x^{(0)})$ or an approximation to it can also be used. Occasionally, it can be worthwhile to "restart" by setting $\tilde{J}^{(\nu)} = \mathbf{I}$.

7.2.2 Iterative algorithms

If the Jacobian is large and non-sparse, then the factorization- or inversion-based techniques that we have discussed so far may not be effective. If the Jacobian is extremely large then it may not even be possible to store the Jacobian conveniently. Nevertheless, it may be possible to calculate the product of the Jacobian and a vector.

Iterative algorithms such as the **conjugate gradient method** mentioned in Section 5.9 only require evaluations of the product of the coefficient matrix and a vector to iterate towards a solution of the linear system. If the Jacobian cannot be easily factorized, then we can try to solve (7.6) approximately using an iterative algorithm. In this case, each iteration of the Newton–Raphson update itself requires several iterations of the iterative algorithm to obtain a suitably accurate approximation to the Newton–Raphson step direction. Details can be found in [45, section 4.8.3][58, chapters 1–3][84, chapter 12].

7.2.3 Pre-conditioning

Pre-conditioning can be used to help with the solution of the update equation if an approximate inverse to the Jacobian is known. A simple "pre-conditioner" is the diagonal matrix consisting of the inverse of the diagonal elements of the Jacobian. As mentioned in Section 5.9, pre-conditioning is often used in combination with iterative methods for the solution of linear equations. Pre-conditioning can also be used with any of the approximations discussed in Section 7.2.1. (See [45, section 4.8.5][58, section 2.5][84, section 12.6] for details and see Section 8.2.4.2 for an example.)

7.2.4 Automatic differentiation

Besides the computational effort involved in calculating the Jacobian, there is also considerable effort involved in *writing* the software to calculate it. If the calculation of g is performed by software that implements a direct algorithm, however, it is possible to systematically transform the software for calculating g into software that calculates the Jacobian. For further details, see [84, section 11.4.2].

7.3 Local convergence of iterative methods

Recall that our goal is to solve the equations $g(x) = \mathbf{0}$. We have introduced the Newton–Raphson method and variations. The Newton–Raphson update generates a sequence of iterates that, in principle, approaches a solution of the equations. We will investigate theoretical conditions for convergence of the sequence of iterates.

Moreover, we have specified how to generate each successive iterate, but not how to stop. We cannot iterate forever, but instead must stop when the current iterate $x^{(\nu)}$ is close enough to the exact solution for our needs, according to some **stopping criterion**. Stopping criteria are extremely important since we must find an answer in a timely manner that we know to be accurate enough for our needs. We discuss empirical characterizations of closeness to an exact solution in Section 7.3.1 that, in conjunction with theoretical conditions, will lead to practical stopping criteria.

In Section 7.3.2, we generalize one of these characterizations of closeness into a powerful theoretical result called the **contraction mapping theorem**, which we then use as part of a convergence proof for the chord method in Section 7.3.3. We also state a result for the Newton–Raphson method in Section 7.3.3. We then discuss the computational effort required to achieve a given solution accuracy in Section 7.3.4.

7.3.1 Closeness to a solution

In this section, we discuss three measures of closeness to a solution that are candidates for use as a stopping criterion. We then discuss using the iteration count and the combination of several stopping criteria.

7.3.1.1 Function value

A natural measure of closeness to a solution is to calculate $\|g(x)\|$ for some norm $\|\bullet\|$. This measures the closeness of satisfaction in terms of the function itself. The infinity norm, $\|\bullet\|_\infty$ (see Section A.3.1), is a typical choice for this criterion. A criterion of the form:

$$\left\|g(x^{(\nu)})\right\|_\infty \leq \epsilon_g, \tag{7.14}$$

for a specified tolerance $\epsilon_g \in \mathbb{R}_{++}$, then requires that *each* equation be satisfied to within a tolerance ϵ_g. This criterion can be interpreted as testing elements of the sequence $\{g(x^{(\nu)})\}_{\nu=0}^{\infty}$ for closeness to $\mathbf{0}$. In practice, since we can only test a finite number of elements, a test of the form (7.14) cannot by itself guarantee that $\{g(x^{(\nu)})\}_{\nu=0}^{\infty}$ converges to $\mathbf{0}$.

A variation on (7.14) is to require that:

$$\left\|g(x^{(\nu)})\right\|_{\infty} \leq \epsilon_g \left\|g(x^{(0)})\right\|_{\infty}. \tag{7.15}$$

For $\epsilon_g < 1$, this condition requires that the satisfaction of the equations be *improved* relative to the satisfaction of the equations by the initial guess.

7.3.1.2 Iteration space

Another measure of closeness of the solution is to suppose that we have a solution, say $x^{\star}$, at hand and measure the distance of our current iterate to the solution. That is, we use $\|x - x^{\star}\|$ as our measure, for some norm $\|\bullet\|$. A typical choice of norm for this criterion is the Euclidean norm, $\|\bullet\|_2$. (See Section A.3.1.) A criterion of the form:

$$\left\|x^{(\nu)} - x^{\star}\right\|_2 \leq \epsilon_x, \tag{7.16}$$

for a specified tolerance $\epsilon_x \in \mathbb{R}_{++}$, then requires that the iterate $x^{(\nu)}$ is close to the actual solution in the sense of Euclidean distance. We are testing elements of the sequence $\{x^{(\nu)}\}_{\nu=0}^{\infty}$ for closeness to $x^{\star}$. A significant drawback of the criterion (7.16) is that we do not know the solution $x^{\star}$. If we did know the solution, then we wouldn't have to perform any iterations!

A variation on (7.16) is to require that:

$$\left\|x^{(\nu)} - x^{\star}\right\|_2 \leq \epsilon_x \left\|x^{(0)} - x^{\star}\right\|_2.$$

For $\epsilon_x < 1$, this condition requires that the distance to the solution be *reduced* relative to the distance of the initial guess from the solution. As with (7.16), this criterion cannot be tested empirically; however, in Section 7.3.1.5 we will see that we can sometimes obtain theoretical results that guarantee a reduction in error of this form.

7.3.1.3 Change in iterate

Another possible way to measure our progress in the decision vector space is to consider the change in $x^{(\nu)}$ from iteration to iteration. That is, we consider $\Delta x^{(\nu)} = x^{(\nu+1)} - x^{(\nu)}$ and a criterion of the form:

$$\left\|\Delta x^{(\nu)}\right\| \leq \epsilon_{\Delta x}, \tag{7.17}$$

300 Algorithms for non-linear simultaneous equations

for a specified tolerance $\epsilon_{\Delta x} \in \mathbb{R}_{++}$. This criterion can be tested without knowledge of the solution. We are testing elements of the sequence $\{\Delta x^{(v)}\}_{v=0}^{\infty}$ for closeness to $\mathbf{0}$. However, even if $\{\Delta x^{(v)}\}_{v=0}^{\infty}$ is known to converge to $\mathbf{0}$, the condition (7.17) is insufficient to guarantee that $\{x^{(v)}\}_{v=0}^{\infty}$ converges, nor *a fortiori* that a particular iterate $x^{(v)}$ is close to a limit. This is explored in Exercise 7.7.

7.3.1.4 Iteration count

Finally, it is common to also limit the total number of iterations. That is, we iterate for no more than, say, N iterations. This criterion, by itself, provides no guarantee that $x^{(N)}$ is close to the solution; however, with suitable additional assumptions about the problem it is sometimes possible to estimate an upper bound on the number of iterations that are necessary to achieve a given accuracy. In Section 7.3.4, we will discuss the qualitative dependence of the number of iterations on the accuracy. Limiting the total number of iterations to this bound (or, perhaps, limiting the total number of iterations to a slightly larger number) then provides a safeguard against iterating forever due to a software error or if the theoretical conditions for the upper bound are not exactly satisfied.

7.3.1.5 Combined stopping criteria

Exercise 7.7 shows that a criterion such as (7.17) that is based on the change in iterates is insufficient to guarantee that the iterates are becoming close to a solution. Since we cannot in practice use a criterion such as (7.16) to evaluate closeness to a solution, it is usual to base stopping criteria on a combination of criteria, such as a combination of criteria of the form (7.14), (7.15), and (7.17). The first criterion, (7.14), ensures that our equations are close to being satisfied, in the sense specified by the norm, the second, (7.15), ensures that the satisfaction has improved compared to the satisfaction by the initial guess, while the third criterion, (7.17), ensures that the update $\Delta x^{(v)}$ has become small. A fourth stopping criterion is to have an explicit limit on the total number of iterations.

Various logical combinations of these three criteria are used in practice to balance the desire to:

- get close to a solution, but
- not perform an excessive number of iterations.

It is also common to require that the combined criteria be satisfied over several successive iterates. Typically, the norms used will be either the L_2 or L_∞ norms. As discussed in Section 3.2.1, it is usually advisable to scale the variables so that a "significant" error in say, x_1, as measured by the norm is roughly the same size numerically in the scaled variables as a significant error in x_2. Similarly, as discussed in Section 3.3.1, it is usually advisable to scale the equations so that a "significant" error in say, g_1, as measured by the norm is roughly the same size numerically in the scaled equations as a significant error in g_2.

A more detailed discussion of stopping criteria, including more examples of criteria, is presented in [45, section 8.2]. Of course, even if a "tight" criterion is satisfied over several iterations, this is no guarantee, by itself, that the infinite sequence of iterates is convergent, nor is it a guarantee that the current iterate is close to the limit. Further conditions must be satisfied before a stopping criteria can be used to reliably judge whether or not a particular iterate is close to the limit of a sequence and close to a solution of the equations. In Section 7.3.2 we will discuss theoretical conditions that can help us to ensure that the stopping criteria, when satisfied, will result in useful answers.

7.3.2 The Cauchy criterion and contraction mappings

In Section 7.3.2.1 we discuss a theoretical criterion, called the Cauchy criterion, that can be used to determine if a sequence is convergent based on information about the sequence of iterates and without reference to the limit itself. In Sections 7.3.2.2 to 7.3.2.5 we define some concepts and then apply the Cauchy criterion to the sequence of iterates generated by a particular class of iterative methods having certain desirable properties. Unfortunately, to express the ideas precisely, we will have to introduce a number of technical definitions. You may want to read these sections through quickly at first to get the overall picture and then re-read them more carefully to follow the details. We will then go on in Section 7.3.3 to apply these ideas to the sequence of iterates generated by the chord method and state an analogous result for the Newton–Raphson method. Most of the material is based on [58, chapter 4].

7.3.2.1 Cauchy sequences

We make the following definition.

Definition 7.1 A sequence $\{x^{(\nu)}\}_{\nu=0}^{\infty}$ is said to be a **Cauchy sequence** or **Cauchy** or satisfy the **Cauchy criterion** if:

$$\forall \epsilon \in \mathbb{R}_{++}, \exists N \in \mathbb{Z}_{+} \text{ such that } (\nu, \nu' \in \mathbb{Z}_{+} \text{ and } \nu, \nu' \geq N) \Rightarrow \left(\left\| x^{(\nu)} - x^{(\nu')} \right\| \leq \epsilon \right).$$

☐

Definition 7.1 says that for any tolerance ϵ, we can find an iteration N such that any two iterates subsequent to iterate N will be apart by no further than a distance ϵ. Being a Cauchy sequence is a stronger condition than requiring that the norm of the successive differences approaches zero. That is, the condition:

$$\forall \epsilon \in \mathbb{R}_{++}, \exists N \in \mathbb{Z}_+ \text{ such that } (\nu \in \mathbb{Z}_+ \text{ and } \nu \geq N) \Rightarrow \left(\left\| x^{(\nu+1)} - x^{(\nu)} \right\| \leq \epsilon \right),$$
$$(7.18)$$

which states that the norm of the difference between successive iterates converges to zero (see Definition 2.9), is insufficient to guarantee that the sequence $\{x^{(\nu)}\}_{\nu=0}^{\infty}$ is Cauchy. As shown in Exercise 7.7, condition (7.18) is insufficient to guarantee convergence of the sequence.

The condition for being a Cauchy sequence is stronger than (7.18), but is *apparently* weaker than the condition for being a convergent sequence. (See Exercise 7.8.) It is somewhat surprising then that for a Cauchy sequence we have:

Lemma 7.1 *A sequence $\{x^{(\nu)}\}_{\nu=0}^{\infty}$ of real vectors converges to a limit in $\mathbb{R}^n$ if and only if it is Cauchy.*

Proof   See [82, chapter 7][111, theorem 3 of chapter 21]. □

In Exercise 7.7, we will see that a sequence can satisfy (7.18) and yet not be Cauchy and also not converge. Lemma 7.1 says that if a sequence *is* Cauchy then it does converge. The advantage of Lemma 7.1 over Definition 2.9 of convergence is that we do not need to know the limit of the sequence to apply Lemma 7.1. Recall that the reason we sought a criterion that can be applied to the difference between iterates is that the direct criterion (7.16) cannot be implemented in practice because we do not know the limit of the sequence. The Cauchy criterion is the key to proving convergence of a sequence of iterates when we do not know its limit.

7.3.2.2 *Lipschitz continuity*

A continuous function does not have any jumps; however, continuous functions can have undesirable properties from the perspective of the Newton–Raphson update and its variants. To rule out these undesirable properties, we define a special type of continuity as follows.

Definition 7.2 A function $\Phi : \mathbb{R}^n \to \mathbb{R}^m$ (or $\Phi : \mathbb{R}^n \to \mathbb{R}^{m \times n}$) is **Lipschitz continuous**:

- on a set $\mathbb{S} \subseteq \mathbb{R}^n$,
- with respect to a norm $\|\bullet\|$ on the domain $\mathbb{R}^n$,
- with respect to a norm $\|\bullet\|$ on the range $\mathbb{R}^m$ (or to a norm on $\mathbb{R}^{m \times n}$), and
- with constant $L \geq 0$,

if:

$$\forall x, x' \in \mathbb{S}, \left\| \Phi(x) - \Phi(x') \right\| \leq L \left\| x - x' \right\|. \qquad (7.19)$$

If $\mathbb{S} = \mathbb{R}^n$ then we say that $\Phi : \mathbb{R}^n \to \mathbb{R}^m$ is Lipschitz continuous with constant L. □

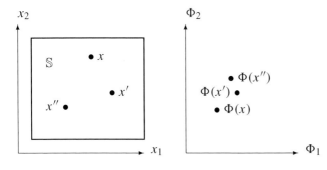

Fig. 7.3. Points x, x', and x'' in a set $\mathbb{S} \subseteq \mathbb{R}^2$ (left panel) and their images $\Phi(x)$, $\Phi(x')$, and $\Phi(x'')$ (right panel) under a Lipschitz continuous function $\Phi : \mathbb{R}^2 \to \mathbb{R}^2$.

Lipschitz continuity is illustrated in Figure 7.3 for a function $\Phi : \mathbb{R}^2 \to \mathbb{R}^2$ and the $\|\bullet\|_2$ norm on both the domain and range. In the figure, three elements x, x', and x'' of a set $\mathbb{S}$ are shown together with their images under the map Φ. In this case, the Euclidean distance between the images of pairs of the points is less than the Euclidean distance between the points. If this is true for *every* pair of points in $\mathbb{S}$, then the function Φ is Lipschitz continuous on $\mathbb{S}$ with respect to the Euclidean norm on its domain and range and with a constant L that is less than one. In general, the Lipschitz constant can be larger than or smaller than one.

If a function is Lipschitz continuous with Lipschitz constant L then it is also Lipschitz continuous with Lipschitz constant equal to any value that is greater than L. For reasons that will become apparent in Section 7.3.2.5, we will usually want to find the smallest value of the Lipschitz constant or find a bound on the Lipschitz constant that is as tight as possible.

Lipschitz continuity is a stronger condition than ordinary continuity as Exercise 7.9 demonstrates. Moreover, the Lipschitz constant L depends on the choice of norm on $\mathbb{R}^n$ and $\mathbb{R}^m$ (or $\mathbb{R}^{m \times n}$.) (See Exercise 7.11.) Condition (7.19) is called a **Lipschitz condition**. For a partially differentiable function Φ with continuous partial derivatives, if the norm of the derivative is bounded by L on a convex set $\mathbb{S}$ then L is a Lipschitz constant for Φ on $\mathbb{S}$. (See Exercise 7.10.)

7.3.2.3 Contraction mapping

In the following we consider the special case of a Lipschitz continuous function with the same domain and range and for which the Lipschitz constant is less than one.

Definition 7.3 A map $\Phi : \mathbb{R}^n \to \mathbb{R}^n$ is called a **contraction mapping** or a **contraction map**:

- on a set $\mathbb{S} \subseteq \mathbb{R}^n$, and
- with respect to a norm $\|\bullet\|$ on $\mathbb{R}^n$,

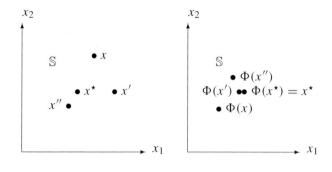

Fig. 7.4. Points x, x', x'', and $x^\star$ in $\mathbb{R}^2$ (left panel) and their images $\Phi(x), \Phi(x'), \Phi(x'')$, and $\Phi(x^\star)$ (right panel) under a function $\Phi : \mathbb{R}^2 \to \mathbb{R}^2$. The point $x^\star$ is a fixed point of Φ because $\Phi(x^\star) = x^\star$.

if $\exists 0 \leq L < 1$ such that:

$$\forall x, x' \in \mathbb{S}, \left\| \Phi(x) - \Phi(x') \right\| \leq L \left\| x - x' \right\|.$$

If $\mathbb{S} = \mathbb{R}^n$ then we say that $\Phi : \mathbb{R}^n \to \mathbb{R}^n$ is a contraction map. □

A map from $\mathbb{R}^n$ to $\mathbb{R}^n$ is a contraction map on $\mathbb{S} \subseteq \mathbb{R}^n$ if it is:

• Lipschitz continuous on $\mathbb{S}$ for one particular norm applied to both its domain and range, and
• the Lipschitz constant is less than one.

That is, under a contraction map the images of two points are closer together than the original two points. A map may be a contraction with respect to one norm and not a contraction with respect to another norm. (See Exercise 7.12.) The map Φ illustrated in Figure 7.3 is a contraction mapping with respect to the Euclidean norm.

7.3.2.4 *General iterative methods and fixed points*

We will discuss generally whether the Newton–Raphson and other iterative methods converge. For convenience, we express a general iterative method in the form:

$$\forall \nu \in \mathbb{Z}_+, \ x^{(\nu+1)} = \Phi(x^{(\nu)}), \tag{7.20}$$

where $\Phi : \mathbb{R}^n \to \mathbb{R}^n$ represents the calculations during a single iteration. (See Exercise 7.13.)

We are interested in whether the sequence of iterates generated by an iterative method in the form (7.20) converges to a solution. To consider the limit of a sequence of iterates generated by such a method, we make the following.

Definition 7.4 A point $x^\star$ is called a **fixed point** of a map $\Phi : \mathbb{R}^n \to \mathbb{R}^n$ if $x^\star = \Phi(x^\star)$. □

Figure 7.4 repeats the points from Figure 7.3 but also includes a point $x^\star$ and its image $\Phi(x^\star)$. Because $\Phi(x^\star) = x^\star$, we observe that $x^\star$ is a fixed point of the map Φ.

Suppose that $x^\star$ is the solution to a particular system of equations. Consider an iterative method of the form (7.20) that was designed to solve the equations. We would hope that if we applied Φ to the solution $x^\star$ then it would return $x^\star$. That is, we would hope that $x^\star = \Phi(x^\star)$, so that $x^\star$ would be a fixed point of Φ.

7.3.2.5 Contraction mapping theorem

We are now ready to present our main result on contraction maps.

Theorem 7.2 *Suppose that $\Phi : \mathbb{R}^n \to \mathbb{R}^n$ is a contraction mapping with Lipschitz constant $0 \leq L < 1$ with respect to some norm $\|\bullet\|$ on a closed set $\mathbb{S} \subseteq \mathbb{R}^n$. Also suppose that $\forall x \in \mathbb{S}, \Phi(x) \in \mathbb{S}$. Then, there exists a unique $x^\star \in \mathbb{S}$ that is a fixed point of Φ. Moreover, for any $x^{(0)} \in \mathbb{S}$, the sequence of iterates generated by the iterative method (7.20) converges to $x^\star$ and satisfies the bound:*

$$\forall \nu \in \mathbb{Z}_+, \left\| x^{(\nu)} - x^\star \right\| \leq (L)^\nu \left\| x^{(0)} - x^\star \right\|. \tag{7.21}$$

Proof The long proof is divided into four parts:

  (i) proving that $\{x^{(\nu)}\}_{\nu=0}^\infty$ is Cauchy and has a limit that is contained in $\mathbb{S}$;
  (ii) proving that the limit is a fixed point of Φ;
  (iii) proving that the fixed point is unique; and
  (iv) proving that the sequence converges to the fixed point according to (7.21).

See Appendix B for details. □

The rate of convergence in the theorem is linear according to Definition 2.10 with a rate constant $C = L$ that is less than one.

7.3.3 The chord and Newton–Raphson methods

In this section we will outline the proof of a convergence result for the chord method and state analogous results for the Newton–Raphson method, called the Kantorovich theorem. In the proof for the chord method, we apply the contraction mapping theorem to the sequence of iterates generated by the chord update. For brevity, our statements do not draw out the full sharpness of the theorems; however, we will see the nature of the restrictions that must be placed on $g : \mathbb{R}^n \to \mathbb{R}^n$ to ensure convergence of the sequence of iterates generated by the chord and Newton–Raphson methods. Sharper theorem statements can be found in [58].

7.3.3.1 The chord method

We have the following convergence result for the chord method. The theorem demonstrates a linear rate of convergence of the error bound to zero. (We quote the result and the proof from [58, section 5.5].)

Theorem 7.3 *Consider a function* $g : \mathbb{R}^n \to \mathbb{R}^n$. *Let* $\|\bullet\|$ *be a norm on* $\mathbb{R}^n$ *and let* $\|\bullet\|$ *also stand for the corresponding induced matrix norm. (See Definition A.30.) Suppose that there exist* $a, b, c,$ *and* $\overline{\rho} \in \mathbb{R}_+$ *such that:*

(i) *g is partially differentiable with continuous partial derivatives at* $x^{(0)}$, *having Jacobian* $J(x^{(0)})$ *satisfying:*

$$\left\| [J(x^{(0)})]^{-1} \right\| \leq a,$$

$$\left\| [J(x^{(0)})]^{-1} g(x^{(0)}) \right\| \leq b,$$

(ii) *g is partially differentiable in a closed ball of radius* $\overline{\rho}$ *about* $x^{(0)}$, *with Jacobian J that is Lipschitz continuous with Lipschitz constant c; that is,*

$$\forall x, x' \in \left\{ x \in \mathbb{R}^n \,\middle|\, \left\| x - x^{(0)} \right\| \leq \overline{\rho} \right\}, \, \| J(x) - J(x') \| \leq c \left\| x - x' \right\|,$$

(iii) $abc < \frac{1}{2}$, *and*

(iv) $\rho_- \leq \overline{\rho}$ *where* $\rho_- = (1 - \sqrt{1 - 2abc})/(ac)$.

Then we have the following.

(i) *In the open ball of radius* $\rho_+ = \min\{\overline{\rho}, (1 + \sqrt{1 - 2abc})/(ac)\}$ *about* $x^{(0)}$ *there is a unique solution* $x^\star$ *of* $g(x) = \mathbf{0}$. *(There may be other solutions outside this ball.)*

(ii) *Consider the chord update (7.8)–(7.9) with* $x^{(0)}$ *as initial guess. The sequence of iterates converges to* $x^\star$ *and each iterate* $x^{(\nu)}$ *is contained in the closed ball of radius* ρ_- *about* $x^{(0)}$. *Furthermore,*

$$\forall \nu \in \mathbb{Z}_+, \, \left\| x^{(\nu)} - x^\star \right\| \leq (ac\rho_-)^\nu \rho_-. \tag{7.22}$$

Proof We define $\Phi : \mathbb{R}^n \to \mathbb{R}^n$ to be the map that represents the update in the chord method. That is:

$$\forall x \in \mathbb{R}^n, \, \Phi(x) = x - [J(x^{(0)})]^{-1} g(x).$$

The proof is divided into four parts:

(i) proving that the iterates stay in $\mathbb{S} = \left\{ x \in \mathbb{R}^n \,\middle|\, \left\| x - x^{(0)} \right\| \leq \rho_- \right\}$;

(ii) proving that Φ is a contraction map with Lipschitz constant $L = ac\rho_- < 1$ so that, by the contraction mapping Theorem 7.2, there exists a unique $x^\star \in \mathbb{S}$ that is a fixed point of Φ;

(iii) proving that the fixed point $x^\star$ of Φ satisfies (7.1) and (7.22); and

(iv) proving that $x^\star$ is the only solution within a distance ρ_+ of $x^{(0)}$.

See Appendix B for details. $\square$

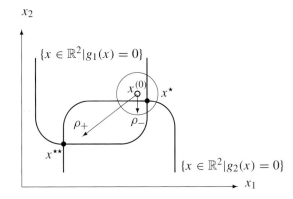

Fig. 7.5. Illustration of chord and Kantorovich theorems.

The first part of the conclusion of the theorem is illustrated in Figure 7.5 for a function $g : \mathbb{R}^2 \to \mathbb{R}^2$. This figure repeats and extends Figure 2.1. The set of points satisfying $g_1(x) = 0$ and the set of points satisfying $g_2(x) = 0$ are shown. The intersections of these curves, $x^\star$ and $x^{\star\star}$, are solutions to the simultaneous equations $g(x) = \mathbf{0}$ and are shown as •. The initial guess $x^{(0)}$ is shown as a ○ at the center of a ball of radius ρ_-. A solution $x^\star$ of the simultaneous equations $g(x) = \mathbf{0}$ lies inside the ball of radius ρ_-. There are no other solutions within a larger ball of radius ρ_+ centered at $x^{(0)}$. However, another solution $x^{\star\star}$ to $g(x) = \mathbf{0}$ lies outside of the ball of radius ρ_+ centered at $x^{(0)}$. (If g were strictly monotone then Theorem 2.2 can be used to show that there is no such other solution $x^{\star\star}$.)

The second part of the conclusion says that the iterates converge to the solution and that the rate of convergence is linear. Because the map Φ defined in the theorem is Lipschitz continuous with Lipschitz constant $L = ac\rho_-$, the error reduces by the factor $ac\rho_-$ at each iteration. Multiplying this factor together for each iteration yields (7.22). Figure 7.6 illustrates a sequence having linear rate of convergence to the solution $x^\star$. The arrows joining $x^\star$ to, respectively, the points $x^{(0)}$, $x^{(1)}$, and $x^{(2)}$, which are illustrated with ○, have lengths that are less than ρ_-, $(ac\rho_-)\rho_-$, and $(ac\rho_-)^2\rho$, respectively.

7.3.3.2 Kantorovich theorem

We quote the Kantorovich theorem essentially as stated in [58, section 5.5].

Theorem 7.4 *(Kantorovich) Consider a function $g : \mathbb{R}^n \to \mathbb{R}^n$. Let $\|\bullet\|$ be a norm on $\mathbb{R}^n$ and let $\|\bullet\|$ also stand for the corresponding induced matrix norm. (See Definition A.30.) Suppose that there exists $a, b, c,$ and $\overline{\rho} \in \mathbb{R}_+$ such that:*

(i) *g is partially differentiable with continuous partial derivatives at $x^{(0)}$, having*

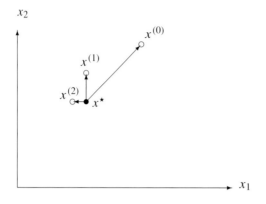

Fig. 7.6. Illustration of linear rate of convergence in chord theorem.

Jacobian $J(x^{(0)})$ satisfying:

$$\left\| [J(x^{(0)})]^{-1} \right\| \leq a,$$

$$\left\| [J(x^{(0)})]^{-1} g(x^{(0)}) \right\| \leq b,$$

(ii) *g is partially differentiable, with Jacobian J that is Lipschitz continuous with Lipschitz constant c in a closed ball of radius $\overline{\rho}$ about $x^{(0)}$; that is,*

$$\forall x, x' \in \left\{ x \in \mathbb{R}^n \,\middle|\, \left\| x - x^{(0)} \right\| \leq \overline{\rho} \right\}, \, \| J(x) - J(x') \| \leq c \| x - x' \|,$$

(iii) *$abc < \frac{1}{2}$, and*

(iv) *$\rho_- \leq \overline{\rho}$ where $\rho_- = (1 - \sqrt{1 - 2abc})/(ac)$.*

(These are the same as conditions (i)–(iv) of the chord theorem, Theorem 7.3.) Then we have the following.

(i) *In the open ball of radius $\rho_+ = \min\{\overline{\rho}, (1 + \sqrt{1 - 2abc})/(ac)\}$ about $x^{(0)}$, there is only one solution $x^\star$ of $g(x) = 0$. (There may be other solutions outside this ball.)*

(ii) *Consider the Newton–Raphson update (7.6)–(7.7) with $x^{(0)}$ as initial guess. The sequence of iterates converges to $x^\star$ and each iterate $x^{(\nu)}$ is contained in the closed ball of radius ρ_- about $x^{(0)}$. Furthermore,*

$$\forall \nu \in \mathbb{Z}_+, \, \left\| x^{(\nu)} - x^\star \right\| \leq \frac{(2abc)^{((2)^\nu)}}{(2)^\nu ac}. \tag{7.23}$$

Proof   See [58] and the references therein. □

The first part of the conclusion is the same as for the chord theorem. The expression $(2abc)^{((2)^\nu)}$ in the second part of the conclusion means $2abc$ raised to the power $(2)^\nu$. At each iteration, the error bound on the norm of the difference between $x^{(\nu)}$ and $x^\star$ is proportional to $(2abc)^{((2)^\nu)}/(2)^\nu$. This factor decreases rapidly from iteration to iteration. According to Definition 2.10, the rate of convergence of the error bound for the Newton–Raphson method is quadratic.

The error bound (7.23) does not imply that the difference between $x^{(\nu)}$ and $x^\star$ decreases at each iteration, only that the bound on the difference decreases. This means that the error may not always be decreasing, which is different to the case for the proof of the chord theorem, which shows (see Appendix B) that the error *itself* decreases by the factor $ac\rho_- < 1$ at each iteration using the chord update. However, even for values of abc close to $\frac{1}{2}$, the iterates calculated according to the Newton–Raphson update will become very close to the solution within just a few iterations.

7.3.3.3 Discussion

Both the chord theorem and the Kantorovich theorem are "local" in that the initial guess must have sufficiently "good" properties (see Items (i) and (iii) of the hypotheses of the theorems) for the iterates to converge. We will discuss approaches to ensuring more "global" convergence properties in Section 7.4.

The parameter c is a measure of the non-linearity of g. If g is linear then J is constant and so arbitrarily small values of c satisfy the conditions of the theorem. In this case, it takes one step to solve linear equations exactly with the chord or Newton–Raphson update: we just solve the linear equations directly.

For non-linear functions, the theorems say the following:

- if the Jacobian is non-singular at the initial guess (so that a is well-defined),
- if the initial guess satisfies the equations sufficiently well (so that the norm b of the initial update:

$$
\begin{aligned}
b &= \left\| [J(x^{(0)})]^{-1} g(x^{(0)}) \right\|, \\
&= \left\| \Delta x^{(0)} \right\|,
\end{aligned}
$$

is small), and
- if the Jacobian does not vary too much over the closed ball of radius $\bar{\rho}$ about $x^{(0)}$ (so that c is small),

then the chord and the Newton–Raphson updates converge to the solution. Moreover:

- the smaller the norm of the inverse of the Jacobian at the initial guess (and therefore the smaller the value of a),
- the closer the satisfaction of the equations by the initial guess (and therefore the smaller the norm of initial update b), and
- the closer that g is to being linear and (therefore the smaller the value of c),

the faster is the approach to the solution. An implication is that if we can transform linear equations to make them "more nearly" linear then this will aid in solving the equations, other things being equal. This is explored in Exercise 7.15, where the function arctan is considered together with two transformed versions. One of the transformed versions removes all the non-linearity: the transformation is the inverse of the arctan function. Generally, we cannot expect to find the exact inverse of the function we are trying to equate to zero.

However, another transformation applied to arctan shows a more typical situation: the transformed function is closer to being linear in the sense that the coefficients in the chord and Kantorovich theorems are smaller. We will return to this issue in the next chapter and in Section 11.1.4.

7.3.4 Computational effort

For direct algorithms, we were able to characterize the computational effort in terms of parameters such as n, the number of entries in the decision vector. In an iterative algorithm, the computational effort also depends on the total number of iterations until the stopping criterion is satisfied.

As suggested in Section 7.3.1.4, we will estimate the number of iterations required to reduce, by a factor of ϵ_x, the bound on the error between the iterate and the solution. To be concrete, we will assume that $\overline{\rho}$ is the best bound we have on the initial error; that is:

$$\left\| x^{(0)} - x^\star \right\| \leq \overline{\rho}, \tag{7.24}$$

and that we want to estimate the number of iterations N such that the error bound is reduced by a factor $\epsilon_x < 1$ so that:

$$\left\| x^{(N)} - x^\star \right\| \leq \epsilon_x \overline{\rho}. \tag{7.25}$$

Let us first assume that each scalar function evaluation requires constant computational effort to evaluate and that matrices and vectors are dense. That is, each of the n entries of g and each of the $(n)^2$ entries of J requires a constant amount of computational effort. We analyze the computational effort for the chord method in Section 7.3.4.1, for the Newton–Raphson method in Section 7.3.4.2, and for the quasi-Newton method in Section 7.3.4.3. We briefly discuss the situation for the other variations in Section 7.3.4.4 and then summarize the performance of the chord, Newton–Raphson, and quasi-Newton methods in Section 7.3.4.5. Finally, in Section 7.3.4.6, we discuss how a more detailed characterization of the computational effort to calculate g and J and to factorize J can be incorporated into the analysis for the other variations on the Newton–Raphson method.

7.3.4.1 Chord method

In the case of the chord method the computations required for N iterations are:

- one evaluation and one factorization of the Jacobian, requiring effort on the order of $(n)^3$, and
- one evaluation of g per iteration, one forwards and backwards substitution per iteration, and one vector addition per iteration, requiring effort on the order of $N(n)^2$.

The overall effort is on the order of $(n)^3 + N(n)^2$ and the average effort per iteration is on the order of $(n)^3/N + (n)^2$. We must find a bound on the size of N that is necessary to satisfy (7.25).

If the hypothesis of Theorem 7.3 holds for the value of $\overline{\rho}$ specified in the bound on the initial error, (7.24), then using (7.22) we have:

$$
\begin{aligned}
\left\| x^{(N)} - x^\star \right\| &\leq (ac\rho_-)^N \rho_-, \\
&= (ac\rho_-)^N \left(\frac{\rho_-}{\overline{\rho}} \right) \overline{\rho}, \\
&\leq (ac\rho_-)^N \left(\frac{\rho_-}{\rho_+} \right) \overline{\rho},
\end{aligned}
$$

since $\rho_+ \leq \overline{\rho}$ by definition. Then (7.25) will be satisfied if:

$$
(ac\rho_-)^N \left(\frac{\rho_-}{\rho_+} \right) \leq \epsilon_x.
$$

Re-arranging this condition we obtain that:

$$
(ac\rho_-)^N \leq \frac{\epsilon_x \rho_+}{\rho_-}.
$$

Taking natural logarithms and re-arranging, we obtain:

$$
N \geq \frac{\ln(\epsilon_x) + \ln(\rho_+) - \ln(\rho_-)}{\ln(ac\rho_-)},
$$

noting that $\ln(ac\rho_-) < 0$, so that the overall effort is on the order of:

$$
(n)^3 + \frac{\ln(\epsilon_x) + \ln(\rho_+) - \ln(\rho_-)}{\ln(ac\rho_-)} (n)^2.
$$

Assuming that a, c, ρ_+, and ρ_- can be bounded approximately independently of $x^{(0)}$, then this means that the computational effort grows with $(n)^3$ and $(n)^2 |\ln(\epsilon_x)|$.

7.3.4.2 Newton–Raphson method

In the case of the Newton–Raphson method, the computations required for N iterations are:

- one evaluation and factorization of the Jacobian per iteration, requiring effort on the order of $N(n)^3$, and
- one evaluation of g per iteration and one forwards and backwards substitution per iteration, requiring effort on the order of $N(n)^2$.

The overall effort is on the order of $N(n)^3$. Again, we must find a bound on the size of N that is necessary to satisfy (7.25).

If the hypothesis of Theorem 7.4 holds for the value of $\bar{\rho}$ specified in the bound on the initial error, (7.24), then using (7.23) we have:

$$
\begin{aligned}
\left\| x^{(N)} - x^\star \right\| &\leq \frac{(2abc)^{((2)^N)}}{(2)^N ac}, \\
&= \frac{(2abc)^{((2)^N)}}{(2)^N ac\bar{\rho}}\bar{\rho}, \\
&\leq \frac{(2abc)^{((2)^N)}}{(2)^N ac\rho_+}\bar{\rho}, \\
&\leq \frac{(2abc)^{((2)^N)}}{ac\rho_+}\bar{\rho},
\end{aligned}
$$

since $\rho_+ \leq \bar{\rho}$ by definition and $(2)^N \geq 1$. Then (7.25) will be satisfied if:

$$
\frac{(2abc)^{((2)^N)}}{ac\rho_+} \leq \epsilon_x.
$$

Re-arranging this condition we obtain that:

$$
(2abc)^{((2)^N)} \leq ac\rho_+\epsilon_x.
$$

Taking natural logarithms, we obtain:

$$
(2)^N \ln(2abc) \leq \ln(ac\rho_+\epsilon_x).
$$

Now $2abc < 1$ by hypothesis, so $\ln(2abc) < 0$ and dividing both sides by the negative number $\ln(2abc)$ yields:

$$
(2)^N \geq \frac{\ln(ac\rho_+\epsilon_x)}{\ln(2abc)}.
$$

Taking natural logarithms again and re-arranging yields:

$$N \geq \frac{\ln\left(\frac{\ln(ac\rho_+\epsilon_x)}{\ln(2abc)}\right)}{\ln(2)},$$

$$= \frac{\ln(|\ln(ac\rho_+\epsilon_x)|) - \ln(|\ln(2abc)|)}{\ln(2)},$$

so that the overall effort is:

$$(n)^3 \frac{\ln(|\ln(ac\rho_+\epsilon_x)|) - \ln(|\ln(2abc)|)}{\ln(2)}.$$

Again assuming that a, c, ρ_+, and ρ_- can be bounded approximately independently of $x^{(0)}$, then for small ϵ_x this means that the computational effort grows with $(n)^3 \ln(|\ln(\epsilon_x)|)$.

For small values of ϵ_x, the required value of N will be smaller for the Newton–Raphson method than for the chord method; however, for large n and relatively large ϵ_x the effort for factorization at each iteration will be unattractive compared to the chord iteration.

7.3.4.3 Quasi-Newton methods

Although we will not discuss the results in detail, quasi-Newton methods can have super-linear convergence to the solution [58, chapter 7]. For the quasi-Newton method, the computations required for N iterations are:

- a symmetric rank two update to a factorization per iteration, requiring effort on the order of $N(n)^2$, and
- one evaluation of g, one forwards and backwards substitution, and one vector addition per iteration, requiring effort on the order of $N(n)^2$.

Assuming super-linear convergence we again find that the number of iterations N grows with $\ln(|\ln(\epsilon_x)|)$ and consequently the computational effort grows with $(n)^2 \ln(|\ln(\epsilon_x)|)$. (See Exercise 2.12.) This effort grows much more slowly with n than for the Newton–Raphson method.

When the iterates converge super-linearly, quasi-Newton methods such as the BFGS method (see Section 7.2.1.6) can have the smallest overall effort for a given reduction in error. For this reason, many software implementations for general purpose use involve a quasi-Newton update to approximate the Newton–Raphson update.

7.3.4.4 Other variations

Besides the chord method and quasi-Newton methods, we discussed several other variations on the Newton–Raphson method in Section 7.2 that involve computational effort per iteration on the order of $(n)^2$. In general, they do not perform as

Number of iterations to satisfy stopping criterion

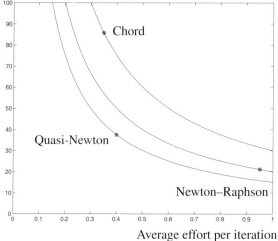

Fig. 7.7. The qualitative trade-off between effort per iteration and number of iterations.

well as the Newton–Raphson method, but they typically perform better than the chord method. In general, the convergence rates will be between the extremes of the rates for the chord and Newton–Raphson methods. Consequently the required value of N to reduce the error by a given factor ϵ_x will also lie between the extremes for the chord and Newton–Raphson methods. For problems with large values of n, it is often the case that $N \ll n$ for each of the variations, so that the variations that avoid a complete factorization at every iteration will be more attractive than the basic Newton–Raphson method.

7.3.4.5 Summary of performance of methods

Figure 7.7 illustrates the typical performance of the chord, Newton–Raphson, and quasi-Newton methods qualitatively. The graph shows both the average effort per iteration and the total number of iterations required to satisfy the criteria. The hyperbolas show curves of equal total effort, where the total effort is equal to the product of the average effort per iteration and the total number of iterations. That is, any point on a given hyperbola represents the same overall effort, with effort increasing from the bottom left hyperbola to the top right hyperbola.

The Newton–Raphson method requires relatively few iterations but the effort per iteration is high. The point illustrated at the lower right of the figure represents the effort per iteration and the number of iterates required to reach a desired accuracy with the Newton–Raphson method. The point lies on the middle of the three hyperbolas.

The chord method requires less effort per iteration on average than the Newton–

Raphson method but may sometimes require more effort overall than the Newton–Raphson method because of the larger number of iterations required to achieve a desired accuracy. As illustrated in Figure 7.7, the point representing the chord iteration has relatively low effort per iteration on average, but the number of iterations required is large. The point representing the chord method lies on the highest effort hyperbola.

Quasi-Newton methods often have the best overall performance, as illustrated in Figure 7.7, because of the reduced effort per iteration compared to the Newton–Raphson method. It is to be emphasized, however, that Figure 7.7 is only a qualitative illustration: it should not be taken literally and particular problems may behave differently to the typical case illustrated in the figure.

7.3.4.6 Calculation of Jacobian

It is sometimes more difficult to calculate entries of J than it is to calculate entries of g. For example, if analytical expressions for the entries of J are unavailable and the central difference approximation is used instead then each entry of J will require at least two evaluations of a scalar function. In this case, we may choose to use a method that uses less information about J but also has a slower rate of convergence, and consequently a larger required number of iterations, because of the savings in the computational effort per iteration. This again points in the direction of using one of the variations instead of the exact Newton–Raphson method. With sparse systems, the trade-offs may change somewhat depending on the effort for factorization versus the effort for forwards and backwards substitution.

The calculations in Sections 7.3.4.1 and 7.3.4.2 about computational effort are based on the assumption that the first iterate is close enough to the solution so that the local analysis can be applied. If the starting point is far from the solution, then the local analysis will not be valid and the rates of convergence will not be as good or the sequence of iterates may not even converge. Again, it is unattractive to form and factorize J exactly because the computational effort to calculate and factorize J is not rewarded with significantly improved performance when we are not close to the solution. Convergence from starting points that are far from the solution will be discussed in Section 7.4.

7.3.5 Discussion

The theoretical results in this section show the great local performance of the chord method and of the Newton–Raphson method. As we will see in Chapter 8, Exercises 8.1 and 8.3, however, the parameters in these theorems are often very difficult to evaluate since they involve properties that hold across regions. It is often easier just to apply the algorithms than to verify the conditions of the convergence theorems. Even when we can calculate the parameters explicitly, their values may not satisfy the requirements of the theorems.

Even if we cannot explicitly use the theorems to prove convergence of the iterates, the theorems can nevertheless provide qualitative insights into convergence. For example, in Section 11.2.3.1, we will use insights from the theorems to understand how a particular transformation helps to improve the solvability of the power system state estimation case study.

Moreover, we can consider iterating from an initial guess until the difference between successive iterates is small and the value of the function is close to zero. We can then try to apply the theorems to the current iterate, $x^{(\nu)}$ say, re-interpreted as a new initial guess. The values of the parameters at this point may enable us to guarantee that the sequence of iterates will converge to the solution.

For initial guesses that are far from the solution we may find, however, that the iterations do not bring us closer to the solution. This necessitates globalization procedures, to be discussed in Section 7.4.

7.4 Globalization procedures

The theorems in the last section are local in nature in that they only apply for initial guesses that are close enough to the solution to satisfy the hypotheses of the theorems. In general, we must consider the possibility that an initial guess is far from the solution and consider the convergence to the solution from arbitrary initial guesses. Such convergence is called **global**.

A characterization of the global convergence of the Newton–Raphson method that requires information at only the initial point is detailed in [110]. However, the information required in that approach involves bounds on higher order derivatives, which are also somewhat difficult to calculate except in special cases.

Instead of characterizing global convergence conditions, we will consider the implications of an initial guess that is far from the solution. For such an initial guess, we must safeguard our algorithm from two related issues:

   (i) singular Jacobian, and

  (ii) excessively large steps.

In Section 7.4.1, we will discuss the first issue. In Section 7.4.2 we will discuss the Armijo step-size rule [6, section 8.3][58, chapter 8][70, section 7.5] as an approach to the second issue. We then make some brief comments about computational effort in Section 7.4.3.

$g(x)$

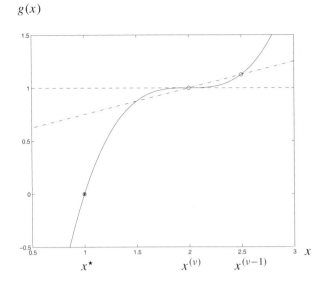

Fig. 7.8. A function with a singular Jacobian at the point $x^{(\nu)} = 2$. The first-order Taylor approximation about $x^{(\nu)}$ is shown dashed. The approximation implied by the secant approximation through $x^{(\nu)}$ and $x^{(\nu-1)}$ is shown as the dash-dotted line.

7.4.1 Singular Jacobian

7.4.1.1 Example

Consider the function of one variable $g : \mathbb{R} \to \mathbb{R}$ defined in (2.6) in Section 2.2.2.2, which we repeat here for reference:

$$\forall x \in \mathbb{R}, g(x) = (x - 2)^3 + 1.$$

This function is illustrated in Figure 7.8 as a solid line. Notice that we have $g(1) = (1 - 2)^3 + 1 = 0$ and that $x^\star = 1$ is the unique solution to $g(x) = 0$. The point $\begin{bmatrix} x^\star \\ g(x^\star) \end{bmatrix} = \begin{bmatrix} 1 \\ 0 \end{bmatrix}$ is illustrated with a • in Figure 7.8.

Suppose that we apply the Newton–Raphson update to solve $g(x) = 0$ at some iteration ν for which $x^{(\nu)} = 2$. The point $\begin{bmatrix} x^{(\nu)} \\ g(x^{(\nu)}) \end{bmatrix} = \begin{bmatrix} 2 \\ 1 \end{bmatrix}$ is illustrated with a ○ in Figure 7.8. However, $J(x) = 3(x - 2)^2$, so that $J(x^{(\nu)}) = J(2) = 0$. Therefore, the Newton–Raphson update equation (7.6) does not have a solution since the first-order Taylor approximation is a horizontal line as illustrated by the dashed line in Figure 7.8. The conditions of Theorems 7.3 and 7.4 cannot be satisfied for this point re-interpreted as an initial guess.

7.4.1.2 Modified factorization

It is clear from the example in the last section that if J is singular at any iterate then the basic Newton–Raphson update will fail. An *ad hoc* approach to this issue is to modify terms in $J(x)$ if it is singular and then solve the resulting equation. In

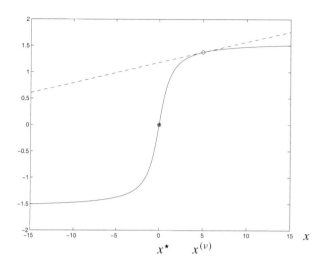

Fig. 7.9. The   inverse tan   function   (shown solid) and its first-order Taylor   approximation about $x^{(\nu)} = 5$ (shown dashed.)   The   point $\begin{bmatrix} x^{(\nu)} \\ g(x^{(\nu)}) \end{bmatrix} = \begin{bmatrix} 5 \\ 1.3734 \end{bmatrix}$ is illustrated with a ∘, while the solution to the equation $g(x) = \mathbf{0}$ is shown with a •.

this example, $g : \mathbb{R} \to \mathbb{R}$ and, if $|J(x^{(\nu)})| < E$ for some threshold $E \in \mathbb{R}_{++}$, then we might replace $J(x^{(\nu)})$ by the secant approximation:

$$\tilde{J}^{(\nu)} = \frac{g(x^{(\nu)}) - g(x^{(\nu-1)})}{x^{(\nu)} - x^{(\nu-1)}}$$

or replace $J(x^{(\nu)})$ by the value E. The dash-dotted line in Figure 7.8 illustrates the secant approximation for this function, given that the previous value of the iterate was $x^{(\nu-1)} = 2.5$. The update can be based on setting this secant approximation equal to zero. (See Exercise 7.17.)

This idea can be generalized to the multi-dimensional case: during factorization of J, if we encounter a small or zero pivot, we simply replace the pivot by a small non-zero number. This modification of the factorization should be performed at any iteration for which $J(x^{(\nu)})$ is singular. We will discuss this **modified factorization** further in Section 10.2.3 when we consider unconstrained minimization.

7.4.2 Step-size selection

7.4.2.1 Region of validity of approximation of function

Even if the Jacobian is non-singular at each iteration, our update is problematic if it suggests a step direction $\Delta x^{(\nu)}$ that is so large that it takes us outside the region of validity of the first-order Taylor approximation of g about $x^{(\nu)}$. A simple example of this is shown by the inverse tangent function, which is illustrated by the solid line in Figure 7.9 [58, section 8.1]. (This function is also considered in Exercise 7.15.)

For $g = \arctan$, the solution of $g(x) = 0$ is $x^\star = 0$. The point $\begin{bmatrix} x^\star \\ g(x^\star) \end{bmatrix} = \mathbf{0}$ is illustrated with a • in Figure 7.9. The point $\begin{bmatrix} x^{(\nu)} \\ g(x^{(\nu)}) \end{bmatrix} = \begin{bmatrix} 5 \\ 1.3734 \end{bmatrix}$ is illustrated with a ○. We have that $\left\| x^{(\nu)} - x^\star \right\| = \left\| 5 - 0 \right\| = 5$, using absolute value as the norm on $\mathbb{R}$. Figure 7.9 also shows the first-order Taylor approximation to the inverse tangent function about $x^{(\nu)} = 5$ as a dashed line.

For $x^{(\nu)} = 5$, using the Newton–Raphson update yields $x^{(\nu)} + \Delta x^{(\nu)} < -15$, so that $\left\| x^{(\nu+1)} - x^\star \right\| > 15$, again using absolute value as the norm on $\mathbb{R}$. That is, the next iterate $x^{(\nu+1)}$ is *further* than $x^{(\nu)}$ from the solution $x^\star = 0$. Furthermore, the value of g is worse. That is:

$$\left\| x^{(\nu+1)} - x^\star \right\| > \left\| x^{(\nu)} - x^\star \right\|,$$
$$\left\| g(x^{(\nu+1)}) \right\| > \left\| g(x^{(\nu)}) \right\|.$$

In this case, the first-order Taylor approximation does not predict the behavior of the inverse tangent function for points that are even a relatively small distance from from $x^{(\nu)}$. If the Newton–Raphson step direction $\Delta x^{(\nu)}$ is so large that it would take the next iterate outside the region of validity of the linear approximation, then we should not move as far as $\Delta x^{(\nu)}$ suggests. Instead, we will consider moving a smaller step in the direction of $\Delta x^{(\nu)}$.

7.4.2.2 Step-size rules

The simplest approach to avoiding updates that take the next iterate outside the region of validity of the linear approximation is to pick a fixed $0 < \alpha < 1$ and use it at each iteration. That is, we modify (7.7) to be:

$$x^{(\nu+1)} = x^{(\nu)} + \alpha \Delta x^{(\nu)},$$

where $0 < \alpha \leq 1$ is fixed for all iterations. This is called the **damped Newton method**. In general, α must be "tuned" for best performance on each problem.

An approach that requires less tuning when J varies significantly is to choose the length of the step at each iteration to bring us closer to solving (7.1). That is, we modify (7.7) to be:

$$x^{(\nu+1)} = x^{(\nu)} + \alpha^{(\nu)} \Delta x^{(\nu)}, \tag{7.26}$$

where $0 < \alpha^{(\nu)} \leq 1$ is chosen at each iteration so that:

$$\left\| g(x^{(\nu)} + \alpha^{(\nu)} \Delta x^{(\nu)}) \right\| < \left\| g(x^{(\nu)}) \right\|. \tag{7.27}$$

If the L_2 norm is chosen in (7.27) then it is possible to choose a suitable $\alpha^{(\nu)}$ if:

• g is partially differentiable with continuous partial derivatives, and

- the step direction $\Delta x^{(\nu)}$ satisfies:

$$[\Delta x^{(\nu)}]^\dagger J(x^{(\nu)})^\dagger g(x^{(\nu)}) < 0. \tag{7.28}$$

For the Newton–Raphson step direction, $\Delta x^{(\nu)} = -[J(x^{(\nu)})]^{-1} g(x^{(\nu)})$. If $J(x^{(\nu)})$ is non-singular, then (7.28) will be satisfied by the Newton–Raphson step direction except at a solution of the simultaneous equations. (See Exercise 7.16.) The variations on the Newton–Raphson method do not automatically ensure that reduction is possible according to (7.27).

7.4.2.3 Armijo step-size rule

A simple way to seek a suitable $\alpha^{(\nu)}$ that satisfies (7.27) is to first try $\alpha^{(\nu)} = 1$, calculate a trial value of $x^{(\nu+1)}$, and check if (7.27) is satisfied. If so, we adopt the trial value as the iterate. If not, we reduce $\alpha^{(\nu)}$ by, say, halving it, and recalculate a new trial value of $x^{(\nu)} + \alpha^{(\nu)} \Delta x^{(\nu)}$ and check (7.27). We proceed until (7.27) is satisfied. The drawback in practice of this approach is that (7.27) does not specify by *how much* the norm of g should decrease to ensure that we obtain a satisfactory improvement in the satisfaction of the equations.

A variation on (7.27) that does specify a "sufficient" decrease requires that:

$$\left\| g(x^{(\nu)} + \alpha^{(\nu)} \Delta x^{(\nu)}) \right\| \le (1 - \delta \alpha^{(\nu)}) \left\| g(x^{(\nu)}) \right\|, \tag{7.29}$$

where $0 < \delta < 1$ is a positive constant. Instead of the strict inequality in (7.27), in (7.29) we specify a non-strict inequality involving an explicit requirement for reduction of the norm of the function.

To understand (7.29), suppose that $\alpha^{(\nu)}$ is small enough so that the first-order Taylor approximation is accurate and also assume that the Newton–Raphson step direction was used. Then:

$$
\begin{aligned}
g(x^{(\nu)} + \alpha^{(\nu)} \Delta x^{(\nu)}) &\approx g(x^{(\nu)}) + J(x^{(\nu)}) \alpha^{(\nu)} \Delta x^{(\nu)}, \text{ since } \alpha^{(\nu)} \text{ is assumed to be} \\
&\qquad \text{small enough so that the first-order Taylor approximation is accurate,} \\
&= g(x^{(\nu)}) - \alpha^{(\nu)} g(x^{(\nu)}), \text{ by definition of } \Delta x^{(\nu)}, \\
&= (1 - \alpha^{(\nu)}) g(x^{(\nu)}).
\end{aligned}
$$

Therefore, taking norms:

$$\left\| g(x^{(\nu)} + \alpha^{(\nu)} \Delta x^{(\nu)}) \right\| \approx (1 - \alpha^{(\nu)}) \left\| g(x^{(\nu)}) \right\|.$$

So, with a step-size of $\alpha^{(\nu)}$, the best we could expect is for $\left\| g(x^{(\nu)} + \alpha^{(\nu)} \Delta x^{(\nu)}) \right\|$ to be reduced by a factor of $(1 - \alpha^{(\nu)})$ compared to $\left\| g(x^{(\nu)}) \right\|$. In practice, we will not achieve a reduction as great as this because:

- the first-order Taylor approximation is not exact, and
- the step direction might not be the exact Newton–Raphson step direction.

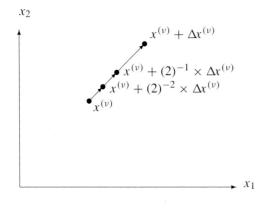

Fig. 7.10. Illustration of back-tracking in Armijo step-size rule.

Consequently, we accept the step-size if the norm of the function is reduced by a factor of only $(1 - \delta\alpha^{(v)})$ instead of requiring a reduction by the smaller factor of $(1 - \alpha^{(v)})$.

Condition (7.29) together with a reduction rule for choosing $\alpha^{(v)}$ is called the **Armijo step-size rule** [6, section 8.3][58, chapter 8][70, section 7.5]. For example, the rule could be to find the largest step length of the form:

$$\alpha^{(v)} = (2)^{-k}, k \geq 0, \tag{7.30}$$

that satisfies (7.29). The condition (7.29) avoids a step-size that is so large that the norm of the equations increases. We seek the largest step-size of the form $(2)^{-k}$ that leads to a fractional reduction in the norm of the equations that satisfies (7.29). To find a step-size of the form (7.30) satisfying (7.29), we can think of first *tentatively* trying the full step of $\Delta x^{(v)}$, which would yield a tentative updated iterate of $x^{(v)} + \Delta x^{(v)}$. If the tentative updated iterate does not satisfy the sufficient decrease criterion (7.29) then we try $x^{(v)} + (2)^{-1} \times \Delta x^{(v)}$. If this tentative updated iterate does not satisfy (7.29) then we try $x^{(v)} + (2)^{-2} \times \Delta x^{(v)}$ and so on. The process of reducing the step-size can be thought of as "back-tracking" from $x^{(v)} + \Delta x^{(v)}$ towards $x^{(v)}$. The first few tentative updates are illustrated in Figure 7.10 for a decision vector $x \in \mathbb{R}^2$.

7.4.2.4 Example

Again consider the inverse tangent function discussed in Section 7.4.2.1. Figure 7.11 repeats the function $g = \arctan$ from Figure 7.9 and its first-order Taylor approximation about $x^{(v)} = 5$, but also illustrates the Armijo rule. As in Figure 7.9, the solution to $g(x) = 0$ is illustrated with a • in Figure 7.11. Figure 7.11 also shows, as a dashed line, the first-order Taylor approximation to the inverse tangent function about $x^{(v)} = 5$, which is illustrated with a ∘.

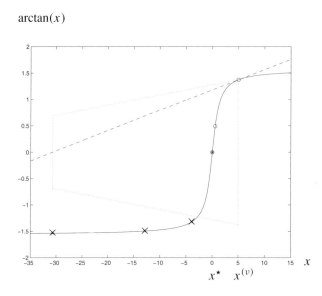

Fig. 7.11. Armijo step-size rule applied to solving equation with arctan function (shown solid). The first-order Taylor approximation about $x^{(\nu)} = 5$ is shown dashed. The point $\begin{bmatrix} x^{(\nu)} \\ g(x^{(\nu)}) \end{bmatrix}$ is illustrated by the right-most o, while the solution $x^{\star} = 0$ to the equation $g(x) = \mathbf{0}$ is shown with a •. The dotted lines bound the region of acceptance for the Armijo rule with $\delta = 0.5$. The three × do not satisfy the Armijo rule. The updated iterate is illustrated by the left-most o.

The Newton–Raphson step direction to solve $g(x) = 0$ is:

$$
\begin{aligned}
\Delta x^{(\nu)} &= -[J(x^{(\nu)})]^{-1} g(x^{(\nu)}), \\
&\approx -35.7,
\end{aligned}
$$

where J is the Jacobian of g. Using this update would yield $x^{(\nu)} + \Delta x^{(\nu)} \approx -30.7$. The left-most × in Figure 7.11 shows the point $\begin{bmatrix} x^{(\nu)} + \Delta x^{(\nu)} \\ g(x^{(\nu)} + \Delta x^{(\nu)}) \end{bmatrix}$. As mentioned in Section 7.4.2.1, using a step-size equal to one does not result in an improved iterate in this case.

The Armijo rule seeks a value of step-size $0 < \alpha^{(\nu)} < 1$ such that (7.29) holds. The dotted lines in Figure 7.11 bound the region satisfying these requirements for $\delta = 0.5$. That is, the dotted lines bound the set of points $\begin{bmatrix} x^{(\nu)} + \alpha^{(\nu)} \Delta x^{(\nu)} \\ \gamma \end{bmatrix}$ that satisfy:

$$
0 \le \alpha^{(\nu)} \le 1,
$$
$$
-(1 - \delta\alpha^{(\nu)}) \left\| g(x^{(\nu)}) \right\| \le \gamma \le (1 - \delta\alpha^{(\nu)}) \left\| g(x^{(\nu)}) \right\|.
$$

Using step-sizes of the form (7.30) results in tentative updated iterates and corresponding function values of:

$$x^{(\nu)} + \Delta x^{(\nu)} \approx -30.7, \qquad g(x^{(\nu)} + \Delta x^{(\nu)}) \approx -1.54,$$
$$x^{(\nu)} + (2)^{-1} \times \Delta x^{(\nu)} \approx -12.9, \qquad g(x^{(\nu)} + (2)^{-1} \times \Delta x^{(\nu)}) \approx -1.49,$$
$$x^{(\nu)} + (2)^{-2} \times \Delta x^{(\nu)} \approx -3.93, \qquad g(x^{(\nu)} + (2)^{-2} \times \Delta x^{(\nu)}) \approx -1.32,$$
$$x^{(\nu)} + (2)^{-3} \times \Delta x^{(\nu)} \approx 0.54, \qquad g(x^{(\nu)} + (2)^{-3} \times \Delta x^{(\nu)}) \approx 0.49.$$

The first three tentative updated iterates fail to satisfy (7.29). They are illustrated by the three × in Figure 7.11. They fall outside the region bounded by the dotted lines. The last tentative updated iterate satisfies (7.29) and is illustrated by the left-most o in Figure 7.11. The updated iterate is therefore $x^{(\nu+1)} = x^{(\nu)} + (2)^{-3} \times \Delta x^{(\nu)} \approx$ 0.54.

7.4.2.5 Choice of δ

If the parameter δ is close to one then it may take many reductions of $\alpha^{(\nu)}$ to satisfy (7.29). Since each tentative update requires a function evaluation, we would prefer to avoid excessive "back-tracking." Consequently, in practice, δ is often chosen to be considerably less than one.

7.4.2.6 Variations

There are other variations on (7.29)–(7.30) that seek to avoid unnecessary "back-tracking." Moreover, instead of repeatedly reducing the step-size until the condition (7.29) is satisfied, we might choose to fit a polynomial equation to the dependence of $\|g(x^{(\nu)} + \alpha \Delta x^{(\nu)})\|$ on the step-size α and then seek the value of α that minimizes the polynomial. This approach is discussed in [58, section 8.3.1]. There are other variants and we will return to this issue when we discuss related issues in unconstrained optimization in Section 10.2.4.

7.4.2.7 Discussion

Step-size rules can significantly aid in convergence from initial guesss that are far from the solution. Their advantages include that:

(i) they can be easily incorporated into any of the methods discussed in this chapter,

(ii) they make an algorithm much more robust in that it can handle initial points that do not satisfy the conditions of our theorems, and

(iii) local to the solution, the step-size rule will typically be satisfied for $\alpha^{(\nu)} = 1$ so that the local convergence properties will hold for the latter iterations.

Nevertheless, we cannot expect the Newton–Raphson method to perform satisfactorily if the Jacobian varies greatly with its argument because the Jacobian will then have a large Lipschitz constant. Fortunately, many problems of practical interest possess Jacobians that are Lipschitz continuous or even approximately constant. Furthermore, as suggested in Exercise 7.15, we will see by example in Sections 11.1 and 11.2 that we can sometimes apply a transformation to a problem that makes the Jacobian more nearly constant and therefore provides better performance than applying the Newton–Raphson update to the untransformed equations.

7.4.3 Computational effort

Computational effort is more difficult to characterize when the initial guess is far from the solution. We will make a qualitative observation to aid in choosing amongst the Newton–Raphson method and its variations. In particular, if we start far from the solution, it is unlikely that the effort to exactly calculate and factorize the Jacobian will be rewarded with fast convergence. That is, the variations on the Newton–Raphson method that require less effort per iteration will tend to perform better overall than the exact Newton–Raphson method.

7.5 Sensitivity and large change analysis

We now suppose that the equations are *parameterized* by a parameter $\chi \in \mathbb{R}^s$. That is, $g : \mathbb{R}^n \times \mathbb{R}^s \to \mathbb{R}^n$. We imagine that we have solved the equations for a base-case value of the parameters, say $\chi = \mathbf{0}$, to find the base-case solution $x^\star$ and that now we are considering the sensitivity of the base-case solution to variation of the parameters around $\chi = \mathbf{0}$.

7.5.1 Sensitivity

7.5.1.1 Implicit function theorem

The **implicit function theorem** (Theorem A.9 in Section A.7.3 of Appendix A) allows us to solve *implicitly* for the solution in terms of the parameters. The following corollary to the implicit function theorem provides us with the sensitivity of the solution to the parameters.

Corollary 7.5 *Let $g : \mathbb{R}^n \times \mathbb{R}^s \to \mathbb{R}^n$ be partially differentiable with continuous partial derivatives. Consider solutions of the equations $g(x; \chi) = \mathbf{0}$, where χ is a parameter. Suppose that $x^\star$ satisfies:*

$$g(x^\star; \mathbf{0}) = \mathbf{0}.$$

We call $x = x^\star$ the base-case solution and $\chi = \mathbf{0}$ the base-case parameters. Define the (parameterized) Jacobian $J : \mathbb{R}^n \times \mathbb{R}^s \to \mathbb{R}^{n \times n}$ by:

$$\forall x \in \mathbb{R}^n, \forall \chi \in \mathbb{R}^s, J(x; \chi) = \frac{\partial g}{\partial x}(x; \chi).$$

Suppose that $J(x^\star; \mathbf{0})$ is non-singular. Then, there is a solution to $g(x; \chi) = \mathbf{0}$ for χ in a neighborhood of the base-case values of the parameters $\chi = \mathbf{0}$. The sensitivity of the solution $x^\star$ to variation of the parameters χ, evaluated at the base-case $\chi = \mathbf{0}$, is given by:

$$\frac{\partial x^\star}{\partial \chi}(\mathbf{0}) = -[J(x^\star; \mathbf{0})]^{-1} K(x^\star; \mathbf{0}),$$

where $K : \mathbb{R}^n \times \mathbb{R}^s \to \mathbb{R}^{n \times s}$ is defined by:

$$\forall x \in \mathbb{R}^n, \forall \chi \in \mathbb{R}^s, K(x; \chi) = \frac{\partial g}{\partial \chi}(x; \chi).$$

☐

If $J(x^\star; \mathbf{0})$ has already been factorized then the calculation of the sensitivity requires one forwards and backwards substitution for each entry of χ.

7.5.1.2 Example

Consider a parameterized version of the function defined in (2.6). In particular, suppose that $g : \mathbb{R} \times \mathbb{R} \to \mathbb{R}$ is defined by:

$$\forall x \in \mathbb{R}, \forall \chi \in \mathbb{R}, g(x; \chi) = (x - 2 - \sin \chi)^3 + 1.$$

We first met this parameterized example in Section 2.7.5.2. We know from Figure 7.8 that the base-case solution is $x^\star = 1$. We consider the sensitivity of the solution to the parameter χ, evaluated at $\chi = 0$. Using Corollary 7.5, we have that the sensitivity is given by:

$$\frac{\partial x^\star}{\partial \chi}(\mathbf{0}) = -[J(x^\star; \mathbf{0})]^{-1} K(x^\star; \mathbf{0}),$$

where $J : \mathbb{R}^n \times \mathbb{R}^s \to \mathbb{R}^{n \times n}$ and $K : \mathbb{R}^n \times \mathbb{R}^s \to \mathbb{R}^{n \times s}$ are defined by:

$$
\begin{aligned}
\forall x \in \mathbb{R}^n, \forall \chi \in \mathbb{R}^s, J(x; \chi) &= \frac{\partial g}{\partial x}(x; \chi), \\
&= 3(x - 2 - \sin \chi)^2, \\
J(x^\star; \mathbf{0}) &= 3, \\
\forall x \in \mathbb{R}^n, \forall \chi \in \mathbb{R}^s, K(x; \chi) &= \frac{\partial g}{\partial \chi}(x; \chi), \\
&= 3(x - 2 - \sin \chi)^2(-\cos \chi), \\
K(x^\star; \mathbf{0}) &= -3.
\end{aligned}
$$

Substituting, the sensitivity is 1.

7.5.2 Large changes

If a parameter in the problem changes very significantly then sensitivity analysis using the base-case may be inadequate. In this case, we may want to solve the problem explicitly for the changed parameter explicitly. We can use the iterative techniques we have developed, using as initial guess the solution to the base-case. (See Exercise 7.18.)

7.6 Summary

In this chapter we have introduced the Newton–Raphson method and its variants as an effective means to iterate towards the solution of non-linear simultaneous equations. We have presented local convergence results and also discussed convergence from initial guesses that are far from the solution. We also discussed sensitivity analysis. In the next chapter we will apply these ideas to our case studies from Chapter 6.

Exercises

Newton–Raphson method

7.1 This exercise concerns Taylor's theorem. Let $g : \mathbb{R}^2 \to \mathbb{R}^2$ be defined by:

$$\forall x \in \mathbb{R}^2, \; g(x) = \begin{bmatrix} \exp(x_1) - x_2 \\ x_1 + \exp(x_2) \end{bmatrix}.$$

  (i) Use Taylor's theorem to linearly approximate $g(x + \Delta x)$ in terms of:

- $g(x)$,
- the Jacobian $J(x)$, and
- Δx.

Write out the linear approximation explicitly for the given g. That is, you must explicitly differentiate g to find the entries in J.

  (ii) Calculate the difference between the exact expression for $g(x + \Delta x)$ and the linear approximation to it. Let us call this difference $e : \mathbb{R}^2 \times \mathbb{R}^2 \to \mathbb{R}$ defined by:

$$\forall x \in \mathbb{R}^2, \forall \Delta x \in \mathbb{R}^2, \; e(x, \Delta x) = g(x + \Delta x) - (\text{the linear approximation}).$$

  (iii) Show that:

$$\frac{\|e(x, \Delta x)\|^2}{\|\Delta x\|^2} \leq \frac{\exp(2x_1)(\exp(\Delta x_1) - 1 - \Delta x_1)^2}{(\Delta x_1)^2}$$

$$+ \frac{\exp(2x_2)(\exp(\Delta x_2) - 1 - \Delta x_2)^2}{(\Delta x_2)^2}.$$

Use the norm given by: $\forall x \in \mathbb{R}^2, \|x\| = \sqrt{(x_1)^2 + (x_2)^2}$.

(iv) Show that $\|e(x, \Delta x)\| / \|\Delta x\| \to 0$ as $\|\Delta x\| \to 0$. Use the norm given by: $\forall x \in \mathbb{R}^2, \|x\| = \sqrt{(x_1)^2 + (x_2)^2}$. Be careful when proving this limit. (Hint: Consider $\|e(x, \Delta x)\|^2 / \|\Delta x\|^2$ and use the previous part together with l'Hôpital's rule to evaluate the limit of the ratio. (See Theorem A.8 in Section A.7.2 of Appendix A.))

7.2 In this exercise we will apply the Newton–Raphson update to solve $g(x) = \mathbf{0}$ where $g : \mathbb{R}^n \to \mathbb{R}^n$ was specified in Section 2.2.2 by (2.5):

$$\forall x \in \mathbb{R}^2, g(x) = \left[\begin{array}{c} (x_1)^2 + (x_2)^2 + 2x_2 - 3 \\ x_1 - x_2 \end{array} \right].$$

(i) Calculate the Jacobian explicitly.
(ii) Calculate the update $\Delta x^{(\nu)}$ according to (7.6) in terms of the current iterate $x^{(\nu)}$.
(iii) Starting with the initial guess $x^{(0)} = \mathbf{0}$, calculate $x^{(1)}$ according to (7.6)–(7.7).
(iv) Calculate $x^{(2)}$ according to (7.6)–(7.7).

Variations on the Newton–Raphson method

7.3 Consider the system of non-linear simultaneous equations $g(x) = \mathbf{0}$ where $g : \mathbb{R}^2 \to \mathbb{R}^2$ is defined by:

$$\forall x \in \mathbb{R}^2, g(x) = \left[\begin{array}{c} (x_1)^2 - 0.1x_1 x_2 \\ (x_2)^2 - 0.1x_1 x_2 \end{array} \right].$$

(i) Write out explicitly the Newton–Raphson update to solve $g(x) = \mathbf{0}$. Invert the Jacobian matrix explicitly using the formula for the inverse of a 2×2 matrix.
(ii) Is there a solution to $g(x) = \mathbf{0}$?
(iii) Starting at $x^{(0)} = \left[\begin{array}{c} 1 \\ 2 \end{array} \right]$, calculate the first three iterates.
(iv) Now replace the exact Jacobian by the approximation $\tilde{J}(x)$ obtained by neglecting the off-diagonal terms of the Jacobian. Write down the new update equations.
(v) Starting at $x^{(0)} = \left[\begin{array}{c} 1 \\ 2 \end{array} \right]$, calculate the first five iterates based on the approximate Jacobian $\tilde{J}(x)$.
(vi) Which method takes less effort overall to achieve a solution satisfying the condition $\|g(x^{(\nu)})\|_\infty < 0.3$: using the exact Jacobian or using the approximate Jacobian? Compare the total number of multiplications, divisions, additions, and subtractions. (Hint: The answer is: "it all depends.")

7.4 Consider a plate capacitor, illustrated in Figure 7.12. (We will consider a similar arrangement in the sizing of interconnects in integrated circuits case study in Section 15.5.) The capacitor consists of two conductors separated by a non-conducting **dielectric**. The upper conductor is of length L and width w and has thickness T. The upper conductor is separated from the lower conductor by a dielectric of thickness d and dielectric constant ε.

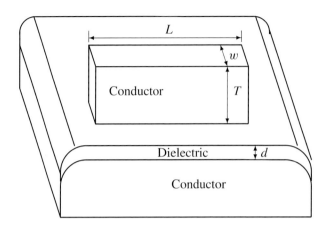

Fig. 7.12. Plate capacitor.

The capacitance, C, of this capacitor appears as a term in a set of non-linear simultaneous equations, $g(x) = \mathbf{0}$ that we are trying to solve. The variables L, w, T, d are entries in the decision vector $x \in \mathbb{R}^n$. In particular,

$$x = \begin{bmatrix} L \\ w \\ T \\ d \\ \xi \end{bmatrix},$$

$$\forall x \in \mathbb{R}^n, g_1(x) = C(x) + \omega(x),$$

where $C, \omega : \mathbb{R}^n \to \mathbb{R}$ are functions and $\xi \in \mathbb{R}^{n-4}$. To apply the Newton–Raphson method to solve the equations, we must evaluate C and its partial derivatives $\dfrac{\partial C}{\partial x}$ at the values of the iterates. (We will ignore the calculation of ω and its derivative.)

If we consider only the **sheet capacitance** and ignore the **fringing capacitance** then the capacitance is approximately given by:

$$\forall x \in \mathbb{R}^n, C(x) \approx \frac{Lw\varepsilon}{d}.$$

In fact, because of fringing capacitance and other issues, the functional dependence of the capacitance is more complicated. We have access to some software that can accurately calculate the capacitance; however, the software does not provide information about the partial derivatives of the capacitance.

(i) Suggest how to combine the results of the software with the approximate capacitor model for use in the Newton–Raphson method to solve $g(x) = \mathbf{0}$.

(ii) Provide an approximation to $\dfrac{\partial C}{\partial x}$. (Make sure you provide approximations to all the entries in the partial derivatives, including with respect to ξ.)

7.5 Show how to approximate the Jacobian using the forward difference approximation with n finite difference calculations. Each finite difference calculation requires the difference between two n-vectors. (Hint: Note that the ℓ-th column of J is equal to the directional derivative of g in the direction I_ℓ.)

7.6 In this exercise we consider updates that satisfy the quasi-Newton condition (7.13). Let $g : \mathbb{R}^n \to \mathbb{R}^n$ and suppose that $\tilde{J}^{(\nu-1)} \in \mathbb{R}^{n \times n}$ is symmetric. Also assume that:

- $(x^{(\nu)} - x^{(\nu-1)})^\dagger [\tilde{J}^{(\nu-1)}]^\dagger (x^{(\nu)} - x^{(\nu-1)}) \neq 0$,
- $(g(x^{(\nu)}) - g(x^{(\nu-1)}))^\dagger (x^{(\nu)} - x^{(\nu-1)}) \neq 0$, and
- $\tilde{J}^{(\nu-1)}(x^{(\nu)} - x^{(\nu-1)}) \neq g(x^{(\nu)}) - g(x^{(\nu-1)})$.

Define the updated approximation $\tilde{J}^{(\nu)} \in \mathbb{R}^{n \times n}$ by $\tilde{J}^{(\nu)} = \tilde{J}^{(\nu-1)} + \Delta J$, where $\Delta J \in \mathbb{R}^{n \times n}$ is defined by:

$$
\Delta J = \frac{-\tilde{J}^{(\nu-1)}(x^{(\nu)} - x^{(\nu-1)})(x^{(\nu)} - x^{(\nu-1)})^\dagger [\tilde{J}^{(\nu-1)}]^\dagger}{(x^{(\nu)} - x^{(\nu-1)})^\dagger [\tilde{J}^{(\nu-1)}]^\dagger (x^{(\nu)} - x^{(\nu-1)})}
$$
$$
+ \frac{(g(x^{(\nu)}) - g(x^{(\nu-1)}))(g(x^{(\nu)}) - g(x^{(\nu-1)}))^\dagger}{(g(x^{(\nu)}) - g(x^{(\nu-1)}))^\dagger (x^{(\nu)} - x^{(\nu-1)})},
$$

where we note that the terms in the denominators are non-zero by assumption.

(i) Show that $\tilde{J}^{(\nu)}$ satisfies the quasi-Newton condition (7.13), which we repeat here:

$$
\tilde{J}^{(\nu)}(x^{(\nu)} - x^{(\nu-1)}) = g(x^{(\nu)}) - g(x^{(\nu-1)}).
$$

(ii) Show that ΔJ is a symmetric rank two update.

Local convergence of iterative methods

7.7 This exercise investigates whether the norm of the change in the iterates of a sequence is, by itself, a good measure of closeness of the iterates to a limit. Suppose that $x \in \mathbb{R}$ is a scalar and that it turned out that:

$$
\forall \nu \in \mathbb{Z}_+, \Delta x^{(\nu)} = \frac{1}{\nu + 1}.
$$

Then $\Delta x^{(\nu)} \to 0$ as $\nu \to \infty$ (see Exercise 2.11), so that for any $\epsilon_{\Delta x} > 0$ the criterion $\left\| \Delta x^{(\nu)} \right\| \leq \epsilon_{\Delta x}$ will eventually be satisfied.

(i) Is $x^{(\nu)}$ ever "close" to a limit? (Hint: Bound $x^{(\nu)}$ from below by an appropriate definite integral of $\frac{1}{t+1}$.)

(ii) Can we use the criterion $\left\| \Delta x^{(\nu)} \right\| \leq \epsilon_{\Delta x}$ alone to measure closeness to a solution?

7.8 Show that any convergent sequence is Cauchy. That is, suppose that $\{x^{(\nu)}\}_{\nu=0}^{\infty}$ converges to $x^\star$ and therefore by definition satisfies:

$$
\forall \epsilon \in \mathbb{R}_{++}, \exists N \in \mathbb{Z}_+ \text{ such that } (\nu \in \mathbb{Z}_+ \text{ and } \nu \geq N) \Rightarrow \left(\left\| x^{(\nu)} - x^\star \right\| \leq \epsilon \right).
$$

Show that the sequence satisfies:

$$\forall \epsilon \in \mathbb{R}_{++}, \exists N \in \mathbb{Z}_+ \text{ such that } (\nu, \nu' \in \mathbb{Z}_+ \text{ and } \nu, \nu' \geq N) \Rightarrow \left(\left\| x^{(\nu)} - x^{(\nu')} \right\| \leq \epsilon \right).$$

(Hint: Use the triangle inequality to bound $\left\| x^{(\nu)} - x^{(\nu')} \right\|$ in terms of $\left\| x^{(\nu)} - x^\star \right\|$ and $\left\| x^{(\nu')} - x^\star \right\|$.)

7.9 Consider the function $\Phi : \mathbb{R} \to \mathbb{R}$ defined by:

$$\forall x \in \mathbb{R}, \Phi(x) = \begin{cases} 1/x, & \text{if } x \neq 0, \\ 0, & \text{otherwise,} \end{cases}$$

let $\mathbb{S} = \{x \in \mathbb{R} | x > 0\}$, and use absolute value as the norm.

(i) Show that Φ is continuous on $\mathbb{S}$.
(ii) Show that Φ is not Lipschitz continuous on $\mathbb{S}$.
(iii) Show that Φ is Lipschitz continuous on $\mathbb{P} = \{x \in \mathbb{R} | x \geq 10^{-47}\}$. Find a Lipschitz constant.

7.10 Suppose that $\Phi : \mathbb{R}^n \to \mathbb{R}^m$ is partially differentiable with continuous partial derivatives on a convex set $\mathbb{S}$ and that, for some induced matrix norm $\|\bullet\|$ (see Definition A.30):

$$\forall x \in \mathbb{S}, \left\| \frac{\partial \Phi}{\partial x}(x) \right\| \leq L.$$

(i) Show that Φ is Lipschitz continuous with constant L for some norms on $\mathbb{R}^n$ and $\mathbb{R}^m$. (Hint: Let $x', x'' \in \mathbb{S}$ and define the function $\phi : [0, 1] \to \mathbb{R}^m$ by:

$$\forall t \in [0, 1], \phi(t) = \Phi(x'' + t(x' - x'')).$$

Then,

$$\Phi(x') - \Phi(x'') = \phi(1) - \phi(0) = \int_0^1 \frac{d\phi}{dt}(t) \, dt.$$

Use the chain rule to evaluate the derivative and then use the definition of the induced matrix norm to bound the derivative in terms of L and the norm of appropriate vectors.)
(ii) Is the result still true if $\mathbb{S}$ is not convex? (Prove or give a counterexample.)

7.11 Let $\Phi : \mathbb{R}^2 \to \mathbb{R}^2$ be defined by:

$$\forall x \in \mathbb{R}^2, \Phi(x) = \begin{bmatrix} 0.9(x_1 - x_2)/\sqrt{2} \\ 0.9(x_1 + x_2)/\sqrt{2} \end{bmatrix}.$$

Calculate the smallest values of L that satisfy (7.19) on $\mathbb{S} = \mathbb{R}^2$ for the following choices of norms on both the domain and range of Φ:

(i) $\|\bullet\| = \|\bullet\|_1$,
(ii) $\|\bullet\| = \|\bullet\|_2$,

(iii) $\|\bullet\| = \|\bullet\|_\infty$.

Notice that there are really two sub-parts to each question:

(a) Find an L that satisfies (7.19). That is, find an L that satisfies:

$$\forall x, x' \in \mathbb{S}, \|\Phi(x) - \Phi(x')\| \leq L \|x - x'\|.$$

(Hint: You can use Exercise 7.10; however, it may be easier to apply the definition of Lipschitz constant directly.)

(b) Prove that the L you have found is actually the smallest value that satisfies the definition.

7.12 Is the map $\Phi : \mathbb{R}^2 \to \mathbb{R}^2$ defined in Exercise 7.11 a contraction map on $\mathbb{R}^2$ with respect to the:

(i) $\|\bullet\|_1$ norm?
(ii) $\|\bullet\|_2$ norm?
(iii) $\|\bullet\|_\infty$ norm?

7.13 Define Φ in such a way that (7.20) represents the Newton–Raphson update (7.6)–(7.7). You can assume that any matrix inversion that you need to perform is well-defined.

7.14 In this exercise we consider applying the chord and Kantorovich theorems to a function. Consider the function $g : \mathbb{R} \to \mathbb{R}$ defined by:

$$\forall x \in \mathbb{R}, g(x) = \exp(x) - 1 - 0.2(x - 1)^2.$$

(i) Calculate the parameters a, b, c for the chord and Kantorovich theorems for applying the Newton–Raphson update to solve $g(x) = 0$ with an initial guess of $x^{(0)} = 0$. Use absolute value as the norm. For calculating c, use the closed ball of radius 0.5 about $x^{(0)}$. To calculate a Lipschitz constant for the Jacobian J, you can use the following result from Exercise 7.10: if $\left| \dfrac{\partial J}{\partial x}(x) \right| \leq L$ for all x in the closed ball of radius 0.5 about $x^{(0)}$ then L is a Lipschitz constant for J in this ball.

(ii) Use the MATLAB function fsolve to solve the equation $g(x) = 0$ for the initial guess of $x^{(0)} = 0$. Report the number of iterations required to satisfy the default stopping criterion and the number of function evaluations.

7.15 In this exercise we consider applying the chord and Kantorovich theorems to a function and transformed versions of it. Consider the function $g : \mathbb{R} \to \mathbb{R}$ defined by:

$$\forall x \in \mathbb{R}, g(x) = \arctan(x).$$

This function is illustrated in Figure 7.9.

(i) Calculate the parameters a, b, c for the chord and Kantorovich theorems for applying the Newton–Raphson update to solve $g(x) = 0$ with an initial guess of $x^{(0)} = 5$. (Use $\overline{\rho} = 5$.)

(ii) Comment on the value of abc for g.

(iii) Use the MATLAB function fsolve to solve the equation $g(x) = 0$ for the initial guess of $x^{(0)} = 5$. Report the number of iterations required to satisfy the default stopping criterion.

(iv) Consider the transformed function $\gamma : \mathbb{R} \to \mathbb{R}$ defined by:

$$\forall x \in \mathbb{R},\ \gamma(x) = \exp(g(x)) - 1.$$

Sketch the graph of γ.

(v) Calculate the parameters a, b, c for γ as defined in Part (iv) for the chord and Kantorovich theorems for applying the Newton–Raphson update to solve $\gamma(x) = 0$ with an initial guess of of $x^{(0)} = 5$. (Use $\bar{\rho} = 5$.)

(vi) Comment on the value of abc for γ as defined in Part (iv).

(vii) Use the MATLAB function fsolve to solve the equation $\gamma(x) = 0$ with γ as defined in Part (iv) for the initial guess of $x^{(0)} = 5$. Report the number of iterations required to satisfy the default stopping criterion.

(viii) Consider the transformed function $\Gamma : \mathbb{R} \to \mathbb{R}$ defined by:

$$\forall x \in \mathbb{R},\ \Gamma(x) = \tan(g(x)).$$

Sketch the graph of Γ.

(ix) Calculate the parameters a, b, c for Γ as defined in Part (viii) for the chord and Kantorovich theorems for applying the Newton–Raphson update to solve $\Gamma(x) = 0$ with an initial guess of $x^{(0)} = 5$.

(x) Comment on the value of abc for Γ as defined in Part (viii).

(xi) Use the MATLAB function fsolve to solve the equation $\Gamma(x) = 0$ with Γ as defined in Part (viii) for the initial guess of $x^{(0)} = 5$. Report the number of iterations required to satisfy the default stopping criterion.

Globalization procedures

7.16 This exercise considers properties of the Newton–Raphson step direction.

(i) Show that the Newton–Raphson step direction satisfies (7.28) at any point $x^{(\nu)}$ such that $J(x^{(\nu)})$ is well-defined and non-singular and such that $x^{(\nu)}$ is not a solution to $g(x) = \mathbf{0}$.

(ii) Show that if (7.28) is satisfied then there exists an $\alpha^{(\nu)}$ that satisfies (7.27). (Hint: Consider a first-order Taylor approximation about $x = x^{(\nu)}$. Consider the remainder.)

7.17 Consider the function $g : \mathbb{R} \to \mathbb{R}$ defined in (2.6) in Section 2.2.2.2, which we repeat here for reference:

$$\forall x \in \mathbb{R},\ g(x) = (x - 2)^3 + 1.$$

The function is illustrated in Figure 7.8. Suppose that the current iterate is $x^{(\nu)} = 2$, as discussed in Section 7.4.1.1. Since the Jacobian is singular at this iterate, we will use the secant approximation instead of the Jacobian to calculate a step direction.

(i) Calculate the step direction based on the secant approximation, given $x^{(\nu-1)} = 2.5$ as illustrated in Figure 7.8.

(ii) Apply the Armijo step-size rule, with $\delta = 0.5$, to find the largest step-size of the form (7.30) satisfying (7.29). Calculate the updated iterate $x^{(\nu+1)}$.

Sensitivity and large change analysis

7.18 Consider the function $g : \mathbb{R} \times \mathbb{R} \to \mathbb{R}$ defined by:

$$\forall x \in \mathbb{R}, \forall \chi \in \mathbb{R}, g(x; \chi) = \arctan(x - \chi) - \chi.$$

(i) Find the base-case solution $x^\star$ to $g(x; 0) = 0$ in the range $-\frac{\pi}{2} < x^\star < \frac{\pi}{2}$.

(ii) Calculate the sensitivity of the base-case solution to variation of χ, evaluated at $\chi = 0$.

(iii) Use the solution from the previous part to estimate the solution of $g(x; 0.1) = 0$.

(iv) Use the MATLAB function `fsolve` to solve the equation $g(x; 0.1) = 0$. Use as initial guess the solution from Part (iii). Report the number of iterations required to satisfy the default stopping criterion.

7.19 Recall the discussion of **eigenvalues** in Section 2.2.2.3. Suppose that $A : \mathbb{R}^s \to \mathbb{R}^{n \times n}$ and consider the parameterized characteristic equation for A:

$$g(\lambda; \chi) = 0,$$

where $g : \mathbb{K} \times \mathbb{R}^s \to \mathbb{K}$ is the parameterized characteristic polynomial defined by:

$$\forall \lambda \in \mathbb{K}, \forall \chi \in \mathbb{R}^s, g(\lambda; \chi) = \det(A(\chi) - \lambda\mathbf{I}),$$

where det is the determinant of the matrix. (See Definition A.10 of the determinant of a matrix.)

For a given value of χ, the characteristic equation $g(\lambda; \chi) = 0$ is one non-linear equation in one variable λ having n (possibly not all distinct) solutions.

(i) Suppose an eigenvalue $\lambda^\star$ is known for the base-case value of $\chi = 0$. Calculate the sensitivity of $\lambda^\star$ to χ, evaluated at $\chi = 0$.

(ii) What if the eigenvalue is repeated?

8

Solution of the non-linear simultaneous equations case studies

In this chapter we will apply the techniques described in the last chapter to the non-linear direct current circuit in Section 8.1 and to the power flow problem in Section 8.2.

8.1 Analysis of a non-linear direct current circuit

In Section 6.1 we developed equations for the non-linear DC circuit illustrated in Figure 6.3, which is repeated in Figure 8.1. The equations describing this circuit were expressed in the form $g(x) = \mathbf{0}$. In Section 8.1.1 we will discuss the Jacobian of g for this problem, while in Sections 8.1.2 and 8.1.3 we discuss the initial guess and the calculation of successive Newton–Raphson iterates. In Section 8.1.4 we investigate the application of the chord and Kantorovich theorems. In Sections 8.1.5 and 8.1.6 we discuss application of the Armijo rule and stopping criteria. In Section 8.1.7 we discuss changes to the circuit.

8.1.1 Jacobian

We must calculate the Jacobian $J = \dfrac{\partial g}{\partial x}$. The function $g : \mathbb{R}^4 \to \mathbb{R}^4$ is defined in (6.6), which we repeat:

$$
\forall x \in \mathbb{R}^4, g(x) = \begin{bmatrix} \left(\frac{1}{R_a}\right) x_1 + i_b(x_1 - x_2) - I_1 \\ -i_b(x_1 - x_2) + \left(\frac{1}{R_c} + \frac{1}{R_d}\right) x_2 + \left(-\frac{1}{R_d}\right) x_3 \\ \left(-\frac{1}{R_d}\right) x_2 + \left(\frac{1}{R_d} + \frac{1}{R_e} + \frac{1}{R_f}\right) x_3 + \left(-\frac{1}{R_f}\right) x_4 \\ \left(-\frac{1}{R_f}\right) x_3 + \left(\frac{1}{R_f}\right) x_4 + i_g(x_4) \end{bmatrix}.
$$

334

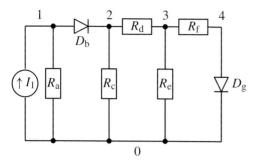

Fig. 8.1. The non-linear DC circuit from Figure 6.3.

From this expression we can obtain the Jacobian $J : \mathbb{R}^4 \to \mathbb{R}^{4\times4}$ defined by:

$$\forall x \in \mathbb{R}^4, \ J(x) =$$

$$
\begin{bmatrix}
\left(\frac{1}{R_a}\right) + \frac{di_b}{dV_b}(x_1 - x_2) & -\frac{di_b}{dV_b}(x_1 - x_2) & 0 & 0 \\
-\frac{di_b}{dV_b}(x_1 - x_2) & \frac{di_b}{dV_b}(x_1 - x_2) + \left(\frac{1}{R_c} + \frac{1}{R_d}\right) & \left(-\frac{1}{R_d}\right) & 0 \\
0 & \left(-\frac{1}{R_d}\right) & \left(\frac{1}{R_d} + \frac{1}{R_e} + \frac{1}{R_f}\right) & \left(-\frac{1}{R_f}\right) \\
0 & 0 & \left(-\frac{1}{R_f}\right) & \left(\frac{1}{R_f}\right) + \frac{di_g}{dV_g}(x_4)
\end{bmatrix}.
$$

$$(8.1)$$

The general case for an arbitrary circuit consisting of resistors, diodes, and current sources was considered in Exercise 6.2. The Jacobian is similar in appearance to the admittance matrix for a linear circuit, with the same sort of sparsity structure:

- non-zeros on the diagonals, and
- non-zeros on the off-diagonals corresponding to branches.

The only qualitative difference is that for the diodes, we have **incremental admittances** evaluated at x instead of admittances. In other words, we have linearized the behavior of the diode about our guess of the operating point. As remarked in Section 6.1.4.3, we can store the information specifying the functional representation of the Jacobian to take advantage of sparsity. Moreover, we can store the value of the Jacobian evaluated at a particular value of x as a sparse matrix and use sparsity techniques when factorizing and performing forwards and backwards substitution.

8.1.2 Initial guess

To start the iterative method, we must provide an initial guess $x^{(0)}$ for the value of the solution $x^\star$. If we know the solution of a similar circuit from previous calculations, then we can use it as $x^{(0)}$. We may know such a previous solution, for example, if we are solving a sequence of circuits that represent the **transient** behavior of the circuit over time at successive time-steps, $t = 0, 1, 2, \ldots$. To calculate the solution for each time-step t except $t = 0$, we can use the solution from the previous time-step $(t - 1)$ as the initial guess for the calculation for time-step t [95, section 10.7]. Of course, we first have to solve the non-linear circuit for $t = 0$ without the benefit of a solution from the previous time-step.

  In the absence of a better guess, $x^{(0)} = \mathbf{0}$ may be a reasonable initial guess for our circuit; however, better guesses will save on computation time and occasionally make the difference between successful and unsuccessful application of the algorithm. In the case of the diode circuit, we may anticipate which diodes are forward conducting and which are not and use this to calculate an initial guess. (See Exercise 8.2 for an example of bounding the solution to the diode circuit.)

8.1.3 Calculation of iterates

The basic Newton–Raphson update (7.6)–(7.7) can now be applied. To calculate $x^{(1)}$, we solve:

$$
\begin{aligned}
J(x^{(0)})\Delta x^{(0)} &= -g(x^{(0)}), \\
x^{(1)} &= x^{(0)} + \Delta x^{(0)}.
\end{aligned}
$$

Successive iterates have an analogous form. Exercise 8.1 applies the Newton–Raphson and chord updates to the non-linear DC circuit from Exercise 6.1.

8.1.4 Application of chord and Kantorovich theorems

Exercise 8.1 suggests that the Newton–Raphson and chord updates generate a sequence of iterates that converge to the solution. The bounds calculated in Exercise 8.2 are consistent with this observation. To guarantee convergence, however, we must apply our theorems. Exercise 8.3 is an extended example of applying the chord and Kantorovich theorems to the circuit.

  As Exercise 8.3 makes clear, applying the chord and Kantorovich theorems can require considerable effort even for simple problems. It is usually easier just to calculate the iterates and empirically judge if they are providing useful answers as Exercise 8.1 shows. However, the theorems:

● confirm that the sequence of iterates converges, and

• estimate the rate of convergence to the solution.

Notice that the theorems will run into difficulty if the entries in the Jacobian vary greatly with their argument because this will cause a large value for the Lipschitz constant c. (See Part (vii) of Exercise 8.3 and compare to Exercise 8.7.) Large variation of the entries in the Jacobian occurs in the diode model and other models with cut-off/cut-on characteristics where the slope of the current versus voltage characteristic varies from near zero to very large.

In Exercise 8.1 we deliberately chose the current source to be small enough so that the diodes are never strongly conducting and therefore the slope of the current versus voltage characteristic varies only over a relatively small range. In practice, we need to solve systems that include diodes that are in the forward conducting region as in Exercises 8.4–8.6. In this case, the theorems will not provide useful results as shown in Exercise 8.7.

Exercise 8.4 indicates that the iterates seem to be converging for the Newton–Raphson update for the 1 amp current source but not converging for the chord update for the 1 amp current source. However, Exercise 8.7 shows that the hypotheses of the chord and Kantorovich theorems do not hold so that we are unsure as to whether the iterates will converge. In fact, in Exercise 8.4,

$$\left\| g(x^{(1)}) \right\|_2 > \left\| g(x^{(0)}) \right\|_2,$$

so that for the first iteration at least, the Newton–Raphson step direction with step-size one is not reducing $\|g(x)\|_2$, although it does bring us towards the solution in the sense of reducing $\|x - x^\star\|_2$. (Notice that $\left\| g(x^{(1)}) \right\|_2 > \left\| g(x^{(0)}) \right\|_2$ is not inconsistent with $\left\| x^{(1)} - x^\star \right\|_2 < \left\| x^{(0)} - x^\star \right\|_2$.) Another problematic issue is that we must calculate to very high precision to get reasonable answers. This is indicative that the Newton–Raphson method is not working well because the "problem" of solving for the Newton–Raphson update is ill-conditioned.

Of course, if we had a starting point closer to the solution, then we would be able to satisfy the hypotheses of the theorems. We could do this, for example, by guessing the voltages across the forward biased diodes. Another good starting point would be the bounds on the solution calculated in Exercise 8.6. However, in general we will not be able to calculate such tight bounds *a priori*.

The basic issue is that the chord and Kantorovich convergence theorems we have presented are *local* in nature. Their conclusions do not help us if we are solving a circuit for the first time and do not know which diodes will be conducting and which will be off. We will discuss step-size rules in Section 8.1.5 to enable convergence from initial guesses that are far from the solution.

8.1.5 Step-size rules

A step-size rule can significantly aid in convergence even when the Jacobian varies greatly. For our circuit, the Armijo rule will guarantee that $\left\| g(x^{(1)}) \right\|_2 < \left\| g(x^{(0)}) \right\|_2$ and improve convergence. This is shown in Exercise 8.9.

The value of δ in the Armijo rule controls the stringency of the acceptance criterion. A small value of δ, such as used in Exercise 8.9, requires only a small reduction in the value of the norm of the function to satisfy the criterion. In contrast, for a large value of δ it may take many calculations of trial values, and many reductions of $\alpha^{(\nu)}$, to find a value that satisfies the Armijo criterion. That is, there will be considerable "back-tracking" from the full step-size. If evaluation of g requires considerable computational effort, then a small value of δ will decrease the amount of "back-tracking."

The Armijo rule guards against the iterations going outside the region of accuracy of the first-order Taylor approximation to g. As suggested in Exercise 8.9 for the Newton–Raphson update, the Armijo rule can consequently improve the convergence. As suggested in Exercise 8.9 for the chord method, the Armijo rule can also mean the difference between convergence and non-convergence. As discussed in Section 7.4.2.6, there are also other methods for aiding in global convergence. The general idea with such approaches is to avoid steps that worsen the value of the iterate and also to avoid taking unnecessarily small steps.

8.1.6 Stopping criteria

We continue with Newton–Raphson or one of the other techniques such as the chord method, using the Armijo rule or some other step-size rule or other approach to guard against excessively large steps, until we obtain a solution that satisfies an appropriate criterion based on "engineering judgment." To determine an appropriate criterion, we should consider the accuracy of the data. For example, the current source in Exercise 8.4 has value 1, but we have not specified the accuracy of the measurement. If the measurement is accurate to, say, 0.1%, then it is superfluous to try to solve the equations to *far* better than this accuracy. On the other hand, we should choose a criterion that is somewhat more stringent than 0.1% to avoid stopping at a point that satisfies the stopping criterion, but which is actually far from a solution [45, section 8.2.3]. A reasonable criterion for this problem, assuming that the measurements were all accurate to around $0.1\% = 10^{-3}$, would be to stop when:

- $\left\| g(x^{(\nu)}) \right\|_\infty \leq 10^{-4}$, and
- $\left\| \Delta x^{(\nu-1)} \right\|_2 \leq 10^{-4}$ or $\left\| \Delta x^{(\nu-1)} \right\|_\infty \leq 10^{-4}$.

We might require that this condition be satisfied over several successive iterations. We might also impose a maximum iteration limit based on our previous experience with the problem or based on analysis using the chord or Kantorovich theorems. As mentioned in Section 7.3.5, we can also try to apply the chord and Kantorovich theorems to the current iterate, $x^{(\nu)}$ say, re-interpreted as a new initial guess in order to obtain an explicit bound on the proximity to the solution.

8.1.7 Circuit changes

Suppose that we have solved the circuit equations for the base-case to a desired accuracy and found an acceptable base-case solution $x^\star$. Now we suppose that the equations are *parameterized* by a parameter $\chi \in \mathbb{R}^s$. That is, $g : \mathbb{R}^n \times \mathbb{R}^s \to \mathbb{R}^n$, with the base-case solution corresponding to $\chi = \mathbf{0}$, and we want to consider how the solution changes with variation in χ. In the next two sections, we consider sensitivity analysis and large change analysis.

8.1.7.1 Sensitivity

Using Corollary 7.5, we can evaluate the sensitivity of the base-case solution to changes in χ, by:

$$\frac{\partial x^\star}{\partial \chi}(\mathbf{0}) = -[J(x^\star; \mathbf{0})]^{-1} K(x^\star; \mathbf{0}),$$

where $J : \mathbb{R}^4 \times \mathbb{R}^s \to \mathbb{R}^{4\times4}$ and $K : \mathbb{R}^4 \times \mathbb{R}^s \to \mathbb{R}^{4\times s}$ are defined by:

$$\forall x \in \mathbb{R}^4, \forall \chi \in \mathbb{R}^s, \ J(x; \chi) = \frac{\partial g}{\partial x}(x; \chi), \ K(x; \chi) = \frac{\partial g}{\partial \chi}(x; \chi).$$

See Exercise 8.10.

8.1.7.2 Large change analysis

If a component or current source changes significantly, we can apply the Newton–Raphson method (or one of the variants) to the changed system using an initial guess for the changed system that is given by the base-case solution $x^\star$ or an estimate of the solution of the changed system obtained by sensitivity analysis. (See Exercise 8.10.)

For a change in a resistor or diode, we can update the Jacobian using a rank one update. In the case of a change in the resistance, we update the Jacobian based on the change in the conductance of the resistor. In the case of a change in a diode, we can update the Jacobian based on the calculated change in the incremental conductance of the diode evaluated at the previous solution $x^\star$. The original factorization can then be updated by the effect of the rank one update. (See Exercise 8.10.)

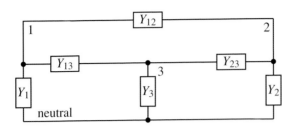

Fig. 8.2. Per-phase equivalent circuit model repeated from Figure 6.10.

If several successive changes are made to the circuit then the updated Jacobian should eventually be factorized again from scratch. Otherwise, accumulated numerical errors from the updates can become large.

8.2 Analysis of an electric power system

In Section 6.2, we developed equations for the power flow problem in the form $g(x) = \mathbf{0}$ for a per-phase equivalent circuit such as shown in Figure 6.10, which is repeated in Figure 8.2.

As with the non-linear DC circuit analyzed in Section 8.1, to apply the Newton–Raphson method we must evaluate the terms in the Jacobian. This will be discussed in Section 8.2.1. We will then go on in Sections 8.2.2 and 8.2.3 to discuss the initial guess and the calculation of iterates using the Newton–Raphson update. We will then discuss approximations to the Jacobian and to the update in Section 8.2.4, discuss step-size rules in Section 8.2.5, and discuss stopping criteria in Section 8.2.6. Finally, in Section 8.2.7 we will discuss circuit changes.

8.2.1 Jacobian

Exercise 6.5 considered the calculation of the Jacobian in the special case where:

- the real part of the admittance matrix is the zero matrix, and
- the voltage magnitude at each bus ℓ is $u_\ell = 1$.

Here we consider the general case.

8.2.1.1 Terms

The entries in $g : \mathbb{R}^n \to \mathbb{R}^n$ are either of the form $p_\ell : \mathbb{R}^n \to \mathbb{R}$, which we repeat from (6.12):

$$\forall x \in \mathbb{R}^n, \; p_\ell(x) = \sum_{k \in J(\ell) \cup \{\ell\}} u_\ell u_k [G_{\ell k} \cos(\theta_\ell - \theta_k) + B_{\ell k} \sin(\theta_\ell - \theta_k)] - P_\ell,$$

or of the form $q_\ell : \mathbb{R}^n \to \mathbb{R}$, which we repeat from (6.13):

$$\forall x \in \mathbb{R}^n, q_\ell(x) = \sum_{k \in J(\ell) \cup \{\ell\}} u_\ell u_k [G_{\ell k} \sin(\theta_\ell - \theta_k) - B_{\ell k} \cos(\theta_\ell - \theta_k)] - Q_\ell.$$

The components in the vector x are either of the form θ_k or of the form u_k, so we have four qualitative types of partial derivative terms corresponding to each combination:

$$\forall x \in \mathbb{R}^n, \frac{\partial p_\ell}{\partial \theta_k}(x) =$$

$$\begin{cases} \sum_{j \in J(\ell)} u_\ell u_j [-G_{\ell j} \sin(\theta_\ell - \theta_j) + B_{\ell j} \cos(\theta_\ell - \theta_j)], & \text{if } k = \ell, \\ u_\ell u_k [G_{\ell k} \sin(\theta_\ell - \theta_k) - B_{\ell k} \cos(\theta_\ell - \theta_k)], & \text{if } k \in J(\ell), \\ 0, & \text{otherwise,} \end{cases}$$

$$\forall x \in \mathbb{R}^n, \frac{\partial p_\ell}{\partial u_k}(x) =$$

$$\begin{cases} 2 u_\ell G_{\ell \ell} + \sum_{j \in J(\ell)} u_j [G_{\ell j} \cos(\theta_\ell - \theta_j) + B_{\ell j} \sin(\theta_\ell - \theta_j)], & \text{if } k = \ell, \\ u_\ell [G_{\ell k} \cos(\theta_\ell - \theta_k) + B_{\ell k} \sin(\theta_\ell - \theta_k)], & \text{if } k \in J(\ell), \\ 0, & \text{otherwise,} \end{cases}$$

$$\forall x \in \mathbb{R}^n, \frac{\partial q_\ell}{\partial \theta_k}(x) =$$

$$\begin{cases} \sum_{j \in J(\ell)} u_\ell u_j [G_{\ell j} \cos(\theta_\ell - \theta_j) + B_{\ell j} \sin(\theta_\ell - \theta_j)], & \text{if } k = \ell, \\ u_\ell u_k [-G_{\ell k} \cos(\theta_\ell - \theta_k) - B_{\ell k} \sin(\theta_\ell - \theta_k)], & \text{if } k \in J(\ell), \\ 0, & \text{otherwise,} \end{cases}$$

$$\forall x \in \mathbb{R}^n, \frac{\partial q_\ell}{\partial u_k}(x) =$$

$$\begin{cases} -2 u_\ell B_{\ell \ell} + \sum_{j \in J(\ell)} u_j [G_{\ell j} \sin(\theta_\ell - \theta_j) - B_{\ell j} \cos(\theta_\ell - \theta_j)], & \text{if } k = \ell, \\ u_\ell [G_{\ell k} \sin(\theta_\ell - \theta_k) - B_{\ell k} \cos(\theta_\ell - \theta_k)], & \text{if } k \in J(\ell), \\ 0, & \text{otherwise.} \end{cases}$$

8.2.1.2 Partitioning by types of terms

To emphasize the four types of terms in the Jacobian, we can partition it into four blocks. To see this, order the entries in g so that all the equations for real power appear first in a sub-vector p followed by all the equations for reactive power in a sub-vector q. Then partition x so that all the voltage angles appear first in a sub-vector θ followed by all the voltage magnitudes in a sub-vector u. Partitioning into the four blocks corresponding to these sub-vectors, we obtain:

$$\forall x \in \mathbb{R}^n, \ J(x) = \begin{bmatrix} J_{p\theta}(x) & J_{pu}(x) \\ J_{q\theta}(x) & J_{qu}(x) \end{bmatrix}, \tag{8.2}$$

where:

$$\forall x \in \mathbb{R}^n, \ J_{p\theta}(x) \ = \ \frac{\partial p}{\partial \theta}(x),$$

$$\forall x \in \mathbb{R}^n, \ J_{pu}(x) \ = \ \frac{\partial p}{\partial u}(x),$$

$$\forall x \in \mathbb{R}^n, \ J_{q\theta}(x) \ = \ \frac{\partial q}{\partial \theta}(x),$$

$$\forall x \in \mathbb{R}^n, \ J_{qu}(x) \ = \ \frac{\partial q}{\partial u}(x).$$

8.2.1.3 Sparsity

Each of the four blocks in (8.2) has the same sparsity structure as the bus admittance matrix. In particular, in each block there are non-zero functions on the diagonal and at the ℓk and $k\ell$ entries for each line ℓk. The zero function appears in each of the other entries in each block.

8.2.1.4 Symmetry

Notice that the blocks $J_{p\theta}$, J_{pu}, $J_{q\theta}$, and J_{qu} are not symmetric and $[J_{pu}]^\dagger \neq J_{q\theta}$. That is, the Jacobian as a whole is not symmetric. In fact, it is also not possible to re-arrange the order of the variables and equations to make it symmetric. In Section 8.2.4.2, however, we will discuss approximations that yield update equations with coefficient matrices that are symmetric.

8.2.1.5 Partitioning by bus number

An alternative to partitioning by the types of terms is to partition the Jacobian into blocks based on the bus number. That is, we re-order the equations and variables so that corresponding entries of $J_{p\theta}(x)$, $J_{pu}(x)$, $J_{q\theta}(x)$, and $J_{qu}(x)$ are grouped together in 2×2 blocks. As discussed in Section 5.5.4.2, we can treat each 2×2 block as a single "entry" in our sparse matrix. This saves on pointers and arrays, therefore reducing overheads of sparse storage.

With the entries arranged as blocks, we can use block pivoting as discussed in Section 5.5.4.2. We can treat each 2×2 block as a single entity in factorization by explicitly inverting the block using the formula for the inverse of a 2×2 matrix. We will not use this approach for solving the power flow problem; however, in some extensions of this problem block pivoting can be exploited to speed up calculations considerably.

8.2.2 *Initial guess*

To apply the Newton–Raphson method to the power flow equations, we must start with an initial guess. If we do not have the solution to a similar problem at hand, then we must make a suitable guess based on our knowledge of the problem. We discuss this guess in this section.

As we described in Section 6.2.2.1, power systems are designed and operated so that the magnitude of the voltages at the buses are all approximately equal to the nominal rated voltage. In the formulation of the power flow problem we normalized all the voltage magnitudes by their nominal values. Therefore, a sensible choice for the initial guess for the voltage magnitude is $u^{(0)} = \mathbf{1}$, where $\mathbf{1}$ is the vector of all ones.

We have indicated that the voltage angles vary somewhat widely, for example, in the range $-45°$ to $45°$. A possible guess for the voltage angle is $\theta^{(0)} = \mathbf{0}$. These choices of initial guess for voltage angle and magnitude are called a **flat start**.

8.2.3 *Calculation of iterates*

The basic Newton–Raphson update (7.6)–(7.7) can now be applied. In terms of the blocks we have defined, the update equations are:

$$\begin{bmatrix} J_{p\theta}(x^{(v)}) & J_{pu}(x^{(v)}) \\ J_{q\theta}(x^{(v)}) & J_{qu}(x^{(v)}) \end{bmatrix} \begin{bmatrix} \Delta\theta^{(v)} \\ \Delta u^{(v)} \end{bmatrix} = -\begin{bmatrix} p(x^{(v)}) \\ q(x^{(v)}) \end{bmatrix}, \tag{8.3}$$

$$\theta^{(v+1)} = \theta^{(v)} + \Delta\theta^{(v)}, \tag{8.4}$$

$$u^{(v+1)} = u^{(v)} + \Delta u^{(v)}. \tag{8.5}$$

8.2.4 *Approximation of the Jacobian and update*

In this section, we describe various approximations that reduce the computational effort required in the update equations.

8.2.4.1 *Chord and Shamanskii updates*

Using a flat start, $x^{(0)} = \begin{bmatrix} \theta^{(0)} \\ u^{(0)} \end{bmatrix} = \begin{bmatrix} \mathbf{0} \\ \mathbf{1} \end{bmatrix}$, as our initial guess, the entries for the Jacobian become:

$$\frac{\partial p_\ell}{\partial \theta_k}(x^{(0)}) = \begin{cases} \sum_{j \in \mathbb{J}(\ell)} B_{\ell j}, & \text{if } k = \ell, \\ -B_{\ell k}, & \text{if } k \in \mathbb{J}(\ell), \\ 0, & \text{otherwise,} \end{cases}$$

$$\frac{\partial p_\ell}{\partial u_k}(x^{(0)}) = \begin{cases} 2G_{\ell\ell} + \displaystyle\sum_{j\in\mathbb{J}(\ell)} G_{\ell j}, & \text{if } k = \ell, \\[2mm] G_{\ell k}, & \text{if } k \in \mathbb{J}(\ell), \\ 0, & \text{otherwise,} \end{cases}$$

$$\frac{\partial q_\ell}{\partial \theta_k}(x^{(0)}) = \begin{cases} \displaystyle\sum_{j\in\mathbb{J}(\ell)} G_{\ell j}, & \text{if } k = \ell, \\[2mm] -G_{\ell k}, & \text{if } k \in \mathbb{J}(\ell), \\ 0, & \text{otherwise,} \end{cases}$$

$$\frac{\partial q_\ell}{\partial u_k}(x^{(0)}) = \begin{cases} -2B_{\ell\ell} - \displaystyle\sum_{j\in\mathbb{J}(\ell)} B_{\ell j}, & \text{if } k = \ell, \\[2mm] -B_{\ell k}, & \text{if } k \in \mathbb{J}(\ell), \\ 0, & \text{otherwise.} \end{cases}$$

Note that $\dfrac{\partial p_\ell}{\partial \theta_k}(x^{(0)})$ and $\dfrac{\partial q_\ell}{\partial u_k}(x^{(0)})$ are symmetric although $\dfrac{\partial p_\ell}{\partial \theta_k}$ and $\dfrac{\partial q_\ell}{\partial u_k}$ are not in general symmetric. In the chord method we use $J(x^{(0)})$ as the approximation to the Jacobian throughout the iterations. In the Shamanskii method, we occasionally update the Jacobian.

8.2.4.2 Approximating particular terms

We have already observed that the Jacobian in the Newton–Raphson update for the power flow equations is sparse and that by taking advantage of sparsity, we can decrease the work in factorization. The necessary effort to solve for the Newton–Raphson update increases with the number of non-zero entries in the Jacobian. We will first approximate the Jacobian by:

(i) neglecting all the terms in the blocks J_{pu} and $J_{q\theta}$, and
(ii) approximating some of the terms in the blocks $J_{p\theta}$ and J_{qu}.

Neglecting terms in the blocks increases the sparsity of the equations. Approximations to the terms in $J_{p\theta}$ and J_{qu} then yield a linear system that is similar to the Jacobian used in the chord update with a flat start. (We will see the relationship between the approximations described here and the chord method in Exercise 8.19.)

Neglecting terms As noted in Section 6.2.4.4, in typical power systems:

(i) the inductive reactance of the series elements is much larger than the resistance, and
(ii) the capacitive susceptance of the shunt elements is much larger than the shunt conductance.

Therefore, for a line between buses ℓ and k, $k \in \mathbb{J}(\ell)$:

$$|G_{\ell k}| \ll |B_{\ell k}|. \tag{8.6}$$

(See Exercise 8.15.)

Furthermore, as mentioned in Section 6.2.2.1 and as discussed in Exercise 6.4, because of flow limits on lines, the real power flow must be limited so that the difference between the voltage angles at each end of a line is not too great. A typical limit is $|\theta_\ell - \theta_k| \leq \frac{\pi}{4}$ and the angle difference is often much smaller. Therefore,

$$
\begin{aligned}
|\sin(\theta_\ell - \theta_k)| &\approx |\theta_\ell - \theta_k|, \quad \text{for small angle differences in radians,} \\
&\ll 1, \quad \text{for small angle differences,} \tag{8.7} \\
\cos(\theta_\ell - \theta_k) &\approx 1, \quad \text{for small angle differences.} \tag{8.8}
\end{aligned}
$$

Since the system is operated so that voltage magnitudes are near to one per unit:

$$u_\ell \approx 1. \tag{8.9}$$

Recall the partition of the Jacobian defined in (8.2). Consider typical terms in the four blocks $J_{p\theta}(x)$, $J_{pu}(x)$, $J_{q\theta}(x)$, and $J_{qu}(x)$. For $k \in \mathbb{J}(\ell)$, assuming that (8.6)–(8.9) hold, we have that:

$$
\begin{aligned}
\frac{\partial p_\ell}{\partial \theta_k}(x) &= u_\ell u_k [G_{\ell k} \sin(\theta_\ell - \theta_k) - B_{\ell k} \cos(\theta_\ell - \theta_k)], \\
&\approx G_{\ell k}(\theta_\ell - \theta_k) - B_{\ell k}, \quad \text{since } u_\ell \approx 1, u_k \approx 1, \text{ and } \cos(\theta_\ell - \theta_k) \approx 1, \\
&\approx -B_{\ell k}, \quad \text{since } |\theta_\ell - \theta_k| \ll 1 \text{ and } |G_{\ell k}| \ll |B_{\ell k}|, \tag{8.10}
\end{aligned}
$$

$$
\begin{aligned}
\left| \frac{\partial p_\ell}{\partial u_k}(x) \right| &= \left| u_\ell [G_{\ell k} \cos(\theta_\ell - \theta_k) + B_{\ell k} \sin(\theta_\ell - \theta_k)] \right|, \\
&\approx |G_{\ell k} + B_{\ell k}(\theta_\ell - \theta_k)|, \quad \text{since } u_\ell \approx 1, \cos(\theta_\ell - \theta_k) \approx 1, \\
&\qquad \text{and } \sin(\theta_\ell - \theta_k) \approx (\theta_\ell - \theta_k), \\
&\ll |B_{\ell k}|, \quad \text{since } |\theta_\ell - \theta_k| \ll 1 \text{ and } |G_{\ell k}| \ll |B_{\ell k}|, \tag{8.11}
\end{aligned}
$$

$$
\begin{aligned}
\left| \frac{\partial q_\ell}{\partial \theta_k}(x) \right| &= \left| u_\ell u_k [-G_{\ell k} \cos(\theta_\ell - \theta_k) - B_{\ell k} \sin(\theta_\ell - \theta_k)] \right|, \\
&\approx |-G_{\ell k} - B_{\ell k}(\theta_\ell - \theta_k)|, \quad \text{since } u_\ell \approx 1, u_k \approx 1, \cos(\theta_\ell - \theta_k) \approx 1, \\
&\qquad \text{and } \sin(\theta_\ell - \theta_k) \approx \theta_\ell - \theta_k), \\
&\ll |B_{\ell k}|, \quad \text{since } |\theta_\ell - \theta_k| \ll 1 \text{ and } |G_{\ell k}| \ll |B_{\ell k}|, \tag{8.12}
\end{aligned}
$$

$$\frac{\partial q_\ell}{\partial u_k}(x) \;=\; u_\ell[G_{\ell k}\sin(\theta_\ell - \theta_k) - B_{\ell k}\cos(\theta_\ell - \theta_k)],$$

$$\approx\; G_{\ell k}(\theta_\ell - \theta_k) - B_{\ell k},\;\; \text{since } u_\ell \approx 1,\, \cos(\theta_\ell - \theta_k) \approx 1,$$

$$\text{and } \sin(\theta_\ell - \theta_k) \approx (\theta_\ell - \theta_k),$$

$$\approx\; -B_{\ell k},\;\; \text{since } |\theta_\ell - \theta_k| \ll 1 \text{ and } |G_{\ell k}| \ll |B_{\ell k}|. \tag{8.13}$$

Therefore, for $k \in \mathbb{J}(\ell)$, if (8.6)–(8.9) are true then the terms $\dfrac{\partial p_\ell}{\partial u_k}(x)$ and $\dfrac{\partial q_\ell}{\partial \theta_k}(x)$

are much smaller in magnitude than the terms $\dfrac{\partial p_\ell}{\partial \theta_k}(x)$ and $\dfrac{\partial q_\ell}{\partial u_k}(x)$, so that we

can often neglect $\dfrac{\partial p_\ell}{\partial u_k}(x)$ and $\dfrac{\partial q_\ell}{\partial \theta_k}(x)$. Similar approximations hold for $k =$

ℓ. (See Exercise 8.16.) These approximations reflect the qualitative observation that real power flow is mostly determined by differences in voltage angles across lines, while reactive power flow is mostly determined by voltage magnitude differences [8, section 10.7] so that the partial derivatives $\dfrac{\partial p_\ell}{\partial u_k}(x)$ and $\dfrac{\partial q_\ell}{\partial \theta_k}(x)$ are relatively small.

If we neglect all the terms in J_{pu} and $J_{q\theta}$, then we can then approximate the Jacobian by:

$$J(x) \approx \begin{bmatrix} J_{p\theta}(x) & \mathbf{0} \\ \mathbf{0} & J_{qu}(x) \end{bmatrix},$$

where $\mathbf{0}$ represents matrices of all zeros of appropriate dimensions. The Newton–Raphson update (8.3)–(8.5) at iteration ν is then approximated by:

$$\begin{bmatrix} J_{p\theta}(x^{(\nu)}) & \mathbf{0} \\ \mathbf{0} & J_{qu}(x^{(\nu)}) \end{bmatrix} \begin{bmatrix} \Delta\theta^{(\nu)} \\ \Delta u^{(\nu)} \end{bmatrix} = - \begin{bmatrix} p(x^{(\nu)}) \\ q(x^{(\nu)}) \end{bmatrix}, \tag{8.14}$$

$$\theta^{(\nu+1)} = \theta^{(\nu)} + \Delta\theta^{(\nu)},$$

$$u^{(\nu+1)} = u^{(\nu)} + \Delta u^{(\nu)}.$$

These are called the **decoupled** Newton–Raphson update equations, since (8.14) separates or decouples into two smaller sets of equations:

$$J_{p\theta}(x^{(\nu)})\Delta\theta^{(\nu)} = -p(x^{(\nu)}), \tag{8.15}$$

$$J_{qu}(x^{(\nu)})\Delta u^{(\nu)} = -q(x^{(\nu)}), \tag{8.16}$$

which are easier to solve than the original larger system. If we use the chord update with a flat start then each system is symmetric.

Approximating terms We can simplify the decoupled equations by using some of the above approximations and some further approximations. In particular, in addition to assuming that $|G_{\ell k}| \ll |B_{\ell k}|$ and that $\cos(\theta_\ell - \theta_k) \approx 1$, we will assume that:

(i) for any bus ℓ, the magnitude of the voltages u_j at buses $j \in \mathbb{J}(\ell)$ is approximately the same as the magnitude of the voltage u_ℓ at ℓ, and

(ii) $B_{\ell\ell} \approx -\sum_{j \in \mathbb{J}(\ell)} B_{\ell j}$.

The first approximation states that the voltage drop along any line is relatively small. This is reasonable so long as lines are not very heavily loaded. The second approximation neglects the contribution of the shunt susceptances to $B_{\ell\ell}$. This is reasonable for short and medium length lines with nominal voltages below about 500 kV.

We will apply these approximations to each entry ℓk of $J_{p\theta}$ and J_{qu}. There are three cases:

(i) diagonal terms, with $k = \ell$;

(ii) off-diagonal terms with a line joining buses ℓ and k; and

(iii) off-diagonal terms with no line joining buses ℓ and k.

We will assume that (8.6)–(8.9) hold in addition to the above assumptions. First, for $k = \ell$:

$$
\begin{aligned}
\frac{\partial p_\ell}{\partial \theta_\ell}(x) &= \sum_{j \in \mathbb{J}(\ell)} u_\ell u_j [-G_{\ell j} \sin(\theta_\ell - \theta_j) + B_{\ell j} \cos(\theta_\ell - \theta_j)], \\
&\approx \sum_{j \in \mathbb{J}(\ell)} (u_\ell)^2 [-G_{\ell j} \sin(\theta_\ell - \theta_j) + B_{\ell j} \cos(\theta_\ell - \theta_j)], \\
&\qquad\qquad \text{assuming } u_j \approx u_\ell \text{ for } j \in \mathbb{J}(\ell), \\
&\approx \sum_{j \in \mathbb{J}(\ell)} (u_\ell)^2 B_{\ell j}, \text{ since } |G_{\ell k}| \ll |B_{\ell k}| \text{ and } \cos(\theta_\ell - \theta_k) \approx 1, \\
&\approx -(u_\ell)^2 B_{\ell\ell}, \text{ since } B_{\ell\ell} \approx -\sum_{j \in \mathbb{J}(\ell)} B_{\ell j}.
\end{aligned}
$$

$$
\begin{aligned}
\frac{\partial q_\ell}{\partial u_\ell}(x) &= -2u_\ell B_{\ell\ell} + \sum_{j \in \mathbb{J}(\ell)} u_j [G_{\ell j} \sin(\theta_\ell - \theta_j) - B_{\ell j} \cos(\theta_\ell - \theta_j)], \\
&\approx -2u_\ell B_{\ell\ell} + \sum_{j \in \mathbb{J}(\ell)} u_\ell [G_{\ell j} \sin(\theta_\ell - \theta_j) - B_{\ell j} \cos(\theta_\ell - \theta_j)], \\
&\qquad\qquad \text{assuming } u_j \approx u_\ell \text{ for } j \in \mathbb{J}(\ell), \\
&\approx -2u_\ell B_{\ell\ell} - \sum_{j \in \mathbb{J}(\ell)} u_\ell B_{\ell j}, \text{ since } |G_{\ell k}| \ll |B_{\ell k}| \text{ and } \cos(\theta_\ell - \theta_k) \approx 1, \\
&\approx -u_\ell B_{\ell\ell}, \text{ assuming } B_{\ell\ell} \approx -\sum_{j \in \mathbb{J}(\ell)} B_{\ell j}.
\end{aligned}
$$

Second, for $k \in \mathbb{J}(\ell)$, we need only use the approximations $|G_{\ell k}| \ll |B_{\ell k}|$ and $\cos(\theta_\ell - \theta_k) \approx 1$ to obtain:

$$\frac{\partial p_\ell}{\partial \theta_k}(x) \approx -u_\ell B_{\ell k} u_k,$$

$$\frac{\partial q_\ell}{\partial u_k}(x) \approx -u_\ell B_{\ell k}.$$

Finally, consider buses k and ℓ such that $k \notin \mathbb{J}(\ell)$ and $k \neq \ell$. For these values of ℓ and k:

$$\frac{\partial p_\ell}{\partial \theta_k}(x) = 0,$$

$$= -u_\ell B_{\ell k} u_k,$$

$$\frac{\partial q_\ell}{\partial u_k}(x) = 0,$$

$$= -u_\ell B_{\ell k}.$$

In summary, the approximations $\dfrac{\partial p_\ell}{\partial \theta_k}(x) \approx -u_\ell B_{\ell k} u_k$ and $\dfrac{\partial q_\ell}{\partial u_k}(x) \approx -u_\ell B_{\ell k}$ apply for all ℓ and k. We will express these approximation values compactly in terms of the imaginary part of the admittance matrix and the voltage vector.

Compact representation Define the matrix B to be the imaginary part of the bus admittance matrix A, except that we delete the row and column of A corresponding to the reference bus. Define U to be the diagonal matrix having diagonal entries equal to the corresponding entries of u. (Recall that the vector u does not include the voltage magnitude for the reference bus.) Similarly, at iteration ν, $U^{(\nu)}$ is defined to be the diagonal matrix having diagonal entries equal to the corresponding entries of $u^{(\nu)}$.

For any ℓ, k, consider the ℓk-th entry of the matrix UBU. It is equal to $u_\ell B_{\ell k} u_k$, which is minus the ℓk-th entry of the approximation to $J_{p\theta}(x)$. Also, consider the ℓk-th entry of the matrix UB. It is equal to $u_\ell B_{\ell k}$, which is minus the ℓk-th entry of the approximation to to $J_{qu}(x)$. We have:

$$J_{p\theta}(x) \approx -UBU,$$
$$J_{qu}(x) \approx -UB.$$

At iteration ν, the decoupled equations (8.15)–(8.16) can therefore be approximated by:

$$-U^{(\nu)}BU^{(\nu)}\Delta\theta^{(\nu)} = -p(x^{(\nu)}), \qquad (8.17)$$
$$-U^{(\nu)}B\Delta u^{(\nu)} = -q(x^{(\nu)}). \qquad (8.18)$$

(See Exercise 8.18.) Exercise 8.19 shows the close connection between the chord method using a flat start and the approximations we have described.

Pre-conditioning and scaling variables Unless $u = 1$ (for example, if we are using a flat start), the matrix $-UB$ is not symmetric. However, since the matrix $U^{(\nu)}$ is diagonal, $U^{(\nu)}$ can be inverted easily so that it can be used to pre-condition (8.17)–(8.18). By moving $U^{(\nu)}$ to the right-hand sides of (8.17) and (8.18) and scaling the variables by defining $\Delta\phi^{(\nu)} = U^{(\nu)}\Delta\theta^{(\nu)}$, we obtain the equivalent system:

$$-B\,\Delta\phi^{(\nu)} \;=\; -[U^{(\nu)}]^{-1}p(x^{(\nu)}), \qquad\qquad (8.19)$$

$$-B\,\Delta u^{(\nu)} \;=\; -[U^{(\nu)}]^{-1}q(x^{(\nu)}). \qquad\qquad (8.20)$$

The coefficient matrix $(-B)$ on the left-hand sides of both (8.19) and (8.20) is constant and symmetric. We have pre-conditioned the system (8.17)–(8.18) to form (8.19)–(8.20).

To solve (8.19) and (8.20), we need only factorize $(-B)$ once, not once per iteration. The factors of $(-B)$ can be used at every iteration ν. This saves considerable time compared to a complete factorization at every iteration, since only forwards and backwards substitutions are then required to calculate $\Delta\phi^{(\nu)}$ and $\Delta u^{(\nu)}$. Once $\Delta\phi^{(\nu)}$ is known, $\Delta\theta^{(\nu)}$ can be calculated using:

$$\Delta\theta^{(\nu)} = [U^{(\nu)}]^{-1}\Delta\phi^{(\nu)}, \qquad\qquad (8.21)$$

which can then be used to evaluate $\theta^{(\nu+1)}$ and substituted into the functions p and q. Slight modifications of these equations are referred to as the **fast decoupled** Newton–Raphson updates. (See [8, section 10.7].)

Discussion The advantage of using a constant coefficient matrix in (8.19) and (8.20) is that it significantly reduces the computational effort per iteration, so that overall the time to converge is usually faster.

However, the approximations we have described are not always very good. (For example, compare Exercises 8.12 and 8.18.) The effect is to slow the rate of convergence and sometimes, in extreme cases, to disrupt convergence if the approximation to the Jacobian is too poor. This is particularly the case for heavily loaded systems where the voltage angles and magnitudes across lines are relatively large. In Section 8.2.5, we briefly discuss the Armijo step-size rule to aid in convergence.

An alternative to using the very approximate Jacobian we have described is to update the Jacobian only every few iterations. A hybrid approach is to begin with these approximations for the first few iterations and then switch to the exact Jacobian. In the next section, we will describe another alternative of using a quasi-Newton method.

8.2.4.3 Quasi-Newton methods

Quasi-Newton methods can also be applied to solve the equations. Equations (8.19) and (8.20) specify a suitable initialization for the approximation to the Jacobian.

8.2.4.4 Iterative methods

Instead of directly solving the linear equations for the Newton–Raphson update, it is also possible to use an iterative algorithm, such as the conjugate gradient method. A pre-conditioned conjugate gradient algorithm for solving the power flow equations is discussed in [39].

8.2.5 Step-size rules

As discussed in Section 8.1.5 for the non-linear DC circuit, a step-size rule can aid in convergence. Exercise 8.13 illustrates the use of the Armijo step-size rule for solving the power flow equations.

8.2.6 Stopping criteria

Typical stopping criteria involve requiring a sufficiently small value of the norm of the:

- change between successive iterates, and
- deviation of the entries of g from zero.

Typically, the infinity norm is used. Since the voltage magnitudes are all normalized by the associated nominal voltages, a given tolerance on all the voltage magnitude changes will result in a comparable bound on the relative change in each of them. Similarly, a single tolerance can be used for all the voltage angles. The deviation of g from $\mathbf{0}$ is called the **mismatch**. Since the values of real and reactive power are all normalized by the base power, a given tolerance on the norm of the mismatch will result in a comparably small absolute bound on all the real and reactive power mismatches.

8.2.7 Circuit changes

In this section, we consider sensitivity and large change analysis. We imagine that the equations are parameterized by $\chi \in \mathbb{R}^s$, with $\chi = \mathbf{0}$ corresponding to the base-case solution, and that we have found a base-case solution $x^\star$.

8.2.7.1 *Sensitivity*

As in Section 8.1.7.1, we can use Corollary 7.5 to evaluate the sensitivity of the base-case solution to changes in χ, by:

$$\frac{\partial x^\star}{\partial \chi}(0) = -[J(x^\star; 0)]^{-1} K(x^\star; 0),$$

where $J : \mathbb{R}^n \times \mathbb{R}^s \to \mathbb{R}^{n \times n}$ and $K : \mathbb{R}^n \times \mathbb{R}^s \to \mathbb{R}^{n \times s}$ are defined by:

$$\forall x \in \mathbb{R}^n, \forall \chi \in \mathbb{R}^s, \ J(x; \chi) = \frac{\partial g}{\partial x}(x; \chi), \ K(x; \chi) = \frac{\partial g}{\partial \chi}(x; \chi).$$

See Exercise 8.21.

8.2.7.2 *Large change analysis*

Large changes to the real and reactive injections into the system can be analyzed by restarting the Newton–Raphson updates based on the solution to the base-case system. If the fast decoupled update equations are used, no changes are necessary to the Jacobian.

Changes to the transmission lines require an update to the Jacobian even if the approximate Jacobian is used. Addition of a transmission line or change in a transmission line parameter involves rank one updates to each of the blocks of the Jacobian. (See Exercise 8.22.)

Exercises

Non-linear DC circuit

8.1 Assume that the current source in the circuit of Figure 8.1 has value 0.05 and that all resistors have value 1. Assume that $q/(\eta \mathcal{K} T) = 40 \text{V}^{-1}$ and $I_{\text{sat}} = 10^{-6} \text{A}$. Let $x^{(0)} = 0$. Perform the following calculations, making sure you keep at least 10 significant figures in all calculations. Report answers to at least 5 significant figures.

   (i) Evaluate $g(x^{(0)})$.
  (ii) Evaluate $J(x^{(0)})$.
 (iii) Factorize $J(x^{(0)})$.
  (iv) Calculate $\Delta x^{(0)}$ and $x^{(1)}$ using the Newton–Raphson update.
   (v) Evaluate $g(x^{(1)})$.
  (vi) Calculate $\Delta x^{(1)}$ and $x^{(2)}$ using the chord update.
 (vii) Evaluate $J(x^{(1)})$.
(viii) Factorize $J(x^{(1)})$.
  (ix) Calculate $\Delta x^{(1)}$ and $x^{(2)}$ using the Newton–Raphson update.
   (x) Compare the effort required to calculate $x^{(2)}$ using the chord and the Newton–Raphson updates and comment on the accuracy.

8.2 In this exercise, we will calculate an upper and a lower bound for the solution $x^\star$ to the circuit of Figure 8.1 specified in Exercise 8.1 based on our specific knowledge of the circuit. That is, we seek $\underline{x}$ and $\bar{x}$ such that:

$$\underline{x} \le x^\star \le \bar{x}.$$

These bounds will help us to evaluate the performance of the iterative techniques used in Exercise 8.1.

(i) Lower bound $\underline{x} \le x^\star$:

  (a) Find a lower bound $\underline{x}$ on $x^\star$ by short-circuiting diode D_g and solving the resulting circuit. You should first combine resistors using the formulas for resistors in series and resistors in parallel to obtain a non-linear problem with just one unknown.

  (b) Explain why short-circuiting the diode provides a lower bound on $x^\star$.

(ii) Upper bound $\bar{x} \ge x^\star$:

  (a) Find an upper bound $\bar{x}$ on $x^\star$ by open-circuiting diode D_g and solving the resulting circuit.

  (b) Explain why open-circuiting the diode provides an upper bound on $x^\star$.

(iii) Comment on the results of Exercise 8.1.

8.3 In this exercise we apply the chord Theorem 7.3 and the Kantorovich Theorem 7.4 to the non-linear DC circuit whose Jacobian J was shown in (8.1) and is repeated here for reference:

$$\forall x \in \mathbb{R}^4,\ J(x) =$$

$$\begin{bmatrix} \left(\frac{1}{R_a}\right) + \frac{di_b}{dV_b}(x_1 - x_2) & -\frac{di_b}{dV_b}(x_1 - x_2) & 0 & 0 \\ -\frac{di_b}{dV_b}(x_1 - x_2) & \frac{di_b}{dV_b}(x_1 - x_2) + \left(\frac{1}{R_c} + \frac{1}{R_d}\right) & \left(-\frac{1}{R_d}\right) & 0 \\ 0 & \left(-\frac{1}{R_d}\right) & \left(\frac{1}{R_d} + \frac{1}{R_e} + \frac{1}{R_f}\right) & \left(-\frac{1}{R_f}\right) \\ 0 & 0 & \left(-\frac{1}{R_f}\right) & \left(\frac{1}{R_f}\right) + \frac{di_g}{dV_g}(x_4) \end{bmatrix}.$$

We use $x^{(0)} = 0$ as the initial guess and we will use the L_2 norm.

(i) Let $P, Q \in \mathbb{R}$ and $A = \begin{bmatrix} P & -P & 0 & 0 \\ -P & P & 0 & 0 \\ 0 & 0 & 0 & 0 \\ 0 & 0 & 0 & Q \end{bmatrix} \in \mathbb{R}^{4\times4}$, which is a symmetric

  matrix. In this case, $\|A\|_2$ is equal to the largest of the absolute values of the eigenvalues of A. (See Section A.3.2.) Show that $\|A\|_2 = \max\{2|P|, |Q|\}$.

(ii) Show that:

$$\forall x, x' \in \mathbb{R}^4,\ J(x) - J(x') = \begin{bmatrix} P(x, x') & -P(x, x') & 0 & 0 \\ -P(x, x') & P(x, x') & 0 & 0 \\ 0 & 0 & 0 & 0 \\ 0 & 0 & 0 & Q(x, x') \end{bmatrix},$$

where $P(x, x') = \dfrac{di_b}{dV_b}(x_1 - x_2) - \dfrac{di_b}{dV_b}(x'_1 - x'_2)$ and $Q(x, x') = \dfrac{di_g}{dV_g}(x_4) - \dfrac{di_g}{dV_g}(x'_4)$.

(iii) Show that:

$$\forall V_b \in \mathbb{R}, \frac{d^2 i_b}{dV_b{}^2}(V_b) \;=\; \left(\frac{q}{\eta \mathcal{K} T}\right)^2 I_{sat} \exp\left(\frac{q V_b}{\eta \mathcal{K} T}\right),$$

$$\forall V_g \in \mathbb{R}, \frac{d^2 i_g}{dV_g{}^2}(V_g) \;=\; \left(\frac{q}{\eta \mathcal{K} T}\right)^2 I_{sat} \exp\left(\frac{q V_g}{\eta \mathcal{K} T}\right).$$

(iv) Show that:

$$|x_4 - x'_4| \;\leq\; \|x - x'\|_2,$$
$$|(x_1 - x'_1) - (x_2 - x'_2)| \;\leq\; \sqrt{2}\,\|x - x'\|_2.$$

(Hint: for the second part, note that:

$$(x_1 - x'_1)^2 + (x_2 - x'_2)^2 \leq [(x_1 - x'_1) + (x_2 - x'_2)]^2 + (x_1 - x'_1)^2 + (x_2 - x'_2)^2,$$

expand, re-arrange, and take positive square roots.)

(v) Let $\overline{\rho} \in \mathbb{R}_{++}$ and show that:

$$\max_{\substack{\|x\|_2,\,\|x'\|_2 \leq \overline{\rho},\\ 0 \leq t \leq 1}} |x'_4 + t(x_4 - x'_4)| \;=\; \overline{\rho},$$

$$\max_{\substack{\|x\|_2,\,\|x'\|_2 \leq \overline{\rho},\\ 0 \leq t \leq 1}} |(x'_1 - x'_2) + t[(x_1 - x_2) - (x'_1 - x'_2)]| \;=\; \sqrt{2}\overline{\rho}.$$

(Hint: Use Theorem 3.11 to decompose the problem hierarchically. First consider the optimizing values of x and x' for given fixed t. Then optimize over values of t. Exercise 3.42 solves a similar problem.)

(vi) Show that:

$$\left(\|x\|_2,\,\|x'\|_2 \leq \overline{\rho}\right) \;\Rightarrow\; \left(|Q(x, x')| \leq \left(\frac{q}{\eta \mathcal{K} T}\right)^2 I_{sat} \exp\left(\frac{q\overline{\rho}}{\eta \mathcal{K} T}\right) \|x - x'\|_2\right),$$

$$\left(\|x\|_2,\,\|x'\|_2 \leq \overline{\rho}\right) \;\Rightarrow$$

$$\left(|P(x, x')| \leq \left(\frac{q}{\eta \mathcal{K} T}\right)^2 I_{sat} \exp\left(\frac{q\sqrt{2}\overline{\rho}}{\eta \mathcal{K} T}\right) \sqrt{2}\,\|x - x'\|_2\right).$$

Note carefully that the coefficients in the exp and of the $\|x - x'\|_2$ are different for the two results. (Hint: For the second result, define the function $\phi : [0, 1] \to \mathbb{R}$ by $\forall t \in [0, 1], \phi(t) = \dfrac{di_b}{dV_b}\left((x'_1 - x'_2) + t[(x_1 - x_2) - (x'_1 - x'_2)]\right)$. Note that

$$|P(x, x')| = |\phi(1) - \phi(0)| = \left| \int_{t=0}^{t=1} \frac{d\phi}{dt}(t)\,dt \right|.$$

Use the chain rule to evaluate the derivative of ϕ.)

(vii) Show that $\forall x, x' \in \mathbb{R}^4$ such that $\|x\|_2, \|x'\|_2 \leq \bar{\rho}$, we have $\|J(x) - J(x')\|_2 \leq c\|x - x'\|_2$, where:

$$c = 2\sqrt{2}\left(\frac{q}{\eta \mathcal{K}T}\right)^2 I_{\text{sat}} \exp\left(\frac{q\sqrt{2}\bar{\rho}}{\eta \mathcal{K}T}\right).$$

That is, show that J is Lipschitz continuous with constant c. (Notice, however, that the Lipschitz "constant" c is parameterized in terms of $\bar{\rho}$ and increases with $\bar{\rho}$.)

(viii) Use MATLAB to evaluate:

$$a = \left\|[J(x^{(0)})]^{-1}\right\|_2 = \left\|[J(0)]^{-1}\right\|_2,$$
$$b = \left\|[J(v^{(0)})]^{-1}g(x^{(0)})\right\|_2 = \left\|[J(0)]^{-1}g(0)\right\|_2.$$

(ix) We know that the current source is of value 0.05 and that the voltage across R_a must therefore be no more than 0.05. We expect the voltages across the other components to be even smaller. Therefore, we know that any solution of the circuit equations must also satisfy $\|x\|_\infty \leq 0.05$. We might guess that the solution will satisfy $\|x\|_2 \leq 0.06$. If we can meet the conditions of the chord and Kantorovich theorems, then these theorems guarantee that there is no solution in the region $\rho_- \leq \|x\|_2 \leq \rho_+ \leq \bar{\rho}$. If we choose a value for $\bar{\rho}$ that is larger than 0.06 and we are able to find values of the coefficients so that we can apply the theorem, we will not obtain any new information about the number of solutions in the region $\|x\|_2 \geq 0.06$ because we already believe that there are no solutions in this region. Since c increases with $\bar{\rho}$ and since abc must be less than $\frac{1}{2}$ to satisfy Item (iii) of the hypotheses of the chord and Kantorovich theorems, we should choose $\bar{\rho}$ to be approximately equal to 0.06 to yield the smallest possible value of abc. Calculate c for $\bar{\rho} = 0.06$.

(x) Check that Items (iii) and (iv) of the hypotheses of the chord and Kantorovich theorems are satisfied for $\bar{\rho} = 0.06$.

(xi) Evaluate ρ_+.

(xii) Explicitly evaluate the error estimates (7.22) and (7.23) for $\nu = 0, 1, 2$.

8.4 Repeat Exercise 8.1 but change the value of the current source to 1. That is, assume that the current source in the circuit of Figure 8.1 has value 1 and that all resistors have value 1. Assume that $q/(\eta \mathcal{K}T) = 40\text{V}^{-1}$ and $I_{\text{sat}} = 10^{-6}\text{A}$. Let $x^{(0)} = 0$. Perform the following calculations, making sure you keep at least 15 significant figures in all calculations. Report answers to at least 5 significant figures.

(i) Evaluate $g(x^{(0)})$.
(ii) Evaluate $J(x^{(0)})$.
(iii) Factorize $J(x^{(0)})$.
(iv) Calculate $\Delta x^{(0)}$ and $x^{(1)}$ using the Newton–Raphson update.
(v) Evaluate $g(x^{(1)})$.
(vi) Calculate $\Delta x^{(1)}$ and $x^{(2)}$ using the chord update.
(vii) Evaluate $J(x^{(1)})$.
(viii) Factorize $J(x^{(1)})$.
(ix) Calculate $\Delta x^{(1)}$ and $x^{(2)}$ using the Newton–Raphson update.

(x) Compare the effort required to calculate $x^{(2)}$ using the chord and the Newton–Raphson updates and comment on the accuracy.

8.5 Use the MATLAB function `fsolve` to solve the circuit problem specified in Exercise 8.4. Solve it two ways:

(i) without specifying the Jacobian to MATLAB so that it uses finite differences to approximate the Jacobian, and
(ii) specifying a MATLAB M-file to evaluate both g and the Jacobian J and specifying that you are supplying the Jacobian by setting the `Jacobian` option to `on` using the `optimset` function.

For each method, report the number of iterations required to satisfy the default stopping criterion and the number of function evaluations.

8.6 Repeat Exercise 8.2 to calculate upper and lower bounds $\underline{x}$ and $\overline{x}$ for the solution $x^\star$ to the circuit problem specified in Exercise 8.4. (That is, assume that the current source in the circuit of Figure 8.1 has value 1 and that all resistors have value 1. Assume that $q/(\eta\mathcal{K}T) = 40\mathrm{V}^{-1}$ and $I_{\mathrm{sat}} = 10^{-6}\mathrm{A}$.)

(i) Find a lower bound $\underline{x}$ on $x^\star$ by short-circuiting diode D_{g} and solving the resulting circuit.
(ii) Find an upper bound $\overline{x}$ on $x^\star$ by open-circuiting diode D_{g} and solving the resulting circuit.
(iii) Comment on the results of Exercise 8.4.

8.7 In this exercise we attempt to apply the chord and Kantorovich theorems.

(i) Calculate a, b, and c for use in the chord and Kantorovich theorems if the current source has value 1. (Hint: You should be able to use the answers to Exercise 8.3 to minimize the calculations. Just make the appropriate modifications. Use, say, $\overline{\rho} = 1.05$.)
(ii) Can you apply the chord and Kantorovich theorems for these values?

8.8 Assume that the current source in the circuit of Figure 8.1 has value 1 and that all resistors have value 1. Assume that $q/(\eta\mathcal{K}T) = 40\mathrm{V}^{-1}$ and $I_{\mathrm{sat}} = 10^{-6}\mathrm{A}$. For each of the following initial guesses, perform one Newton–Raphson update, making sure you keep at least 10 significant figures in all calculations. Report answers to at least 5 significant figures.

(i) As initial guess, use the average of the upper and lower bounds calculated in Exercise 8.6.
(ii) As initial guess, assume that the voltage across each diode is 0.6 volts, but that nodes 2, 3, and 4 are at the same voltage.

8.9 Repeat Exercise 8.4. (That is, assume that the current source in the circuit of Figure 8.1 has value 1 and that all resistors have value 1. Assume that $q/(\eta\mathcal{K}T) = 40\mathrm{V}^{-1}$

and $I_{sat} = 10^{-6}$A. Let $x^{(0)} = \mathbf{0}$.) However, instead of the basic update equation, use the Armijo step-size rule (7.29):

$$\left\| g(x^{(\nu+1)}) \right\| \leq (1 - \delta \alpha^{(\nu)}) \left\| g(x^{(\nu)}) \right\|,$$

at each iteration and find the largest value of $\alpha^{(\nu)}$ of the form of (7.30):

$$\alpha^{(\nu)} = (2)^{-k}, k \geq 0,$$

that satisfies (7.29) to select $\alpha^{(\nu)}$. Use $\delta = 0.01$ and use the Euclidean norm to determine whether a step is accepted. Make sure you keep at least 10 significant figures in all calculations. Report answers to at least 5 significant figures.

(i) Calculate $x^{(1)}$ using the Newton–Raphson step direction and Armijo step-size rule.
(ii) Evaluate $g(x^{(1)})$.
(iii) Calculate $\Delta x^{(1)}$ and $x^{(2)}$ using the chord step direction and Armijo step-size rule.
(iv) Evaluate $J(x^{(1)})$.
(v) Factorize $J(x^{(1)})$.
(vi) Calculate $\Delta x^{(1)}$ and $x^{(2)}$ using the Newton–Raphson step direction and Armijo step-size rule.
(vii) Compare the effort required to calculate $x^{(2)}$ using the chord and the Newton–Raphson step directions and comment on the accuracy.

8.10 Suppose that the current source in the circuit of Figure 8.1 has value 1 and that all resistors have value 1. Assume that the temperature of diode D_b changes so that $q/(\eta \mathcal{K} T) = 38$V^{-1} for diode D_b. Assume that $q/(\eta \mathcal{K} T)$ stays at 40V^{-1} for diode D_g and that the saturation current remains the same at $I_{sat} = 10^{-6}$A for both diodes D_b and D_g. Suppose that we have a solution $x^\star$ of the base-case circuit and the corresponding value of the Jacobian $J(x^\star)$ for the original value of $q/(\eta \mathcal{K} T) = 40$V^{-1}.

(i) Use sensitivity analysis to evaluate the change in the solution. (Use the result of Exercise 8.5 for the solution to the base-case.)
(ii) Show that the change in the Jacobian evaluated at $x^\star$ but with the changed temperature for diode D_b is a symmetric rank 1 update.

Power flow equations

8.11 Consider the four-bus, four-line system illustrated in Figure 8.3. There are no shunt elements and the series admittances $Y_a = B_a\sqrt{-1}$, $Y_b = B_b\sqrt{-1}$, $Y_c = B_c\sqrt{-1}$, and $Y_d = B_d\sqrt{-1}$ have zero conductance.

(i) Write down the bus admittance matrix for the system.
(ii) With bus 1 as the reference bus and the rest of the buses PQ buses, write down the entries of x, the vector of variables in the power flow equations. Make sure you only include the variables in the formulation.
(iii) Explicitly write down the entries of $g(x)$, the vector of functions in the power flow formulation, as defined in (6.12)–(6.13). Make sure you omit the reference bus equations from g.

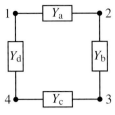

Fig. 8.3. Four-bus, four-line system for Exercise 8.11.

(iv) Write out the elements in the Jacobian (8.2) for the Newton–Raphson update. Assume a flat start; that is, with $x^{(0)}$ having all voltage angles equal to zero radians and all voltage magnitudes equal to one per unit.

8.12 Write a MATLAB program that takes two vectors containing θ and u, and two matrices containing G and B, (the real and imaginary parts of A,) and calculates the terms in the Jacobian evaluated at the given values of θ and u. The program should ask the user for the bus numbers ℓ and k and then print out the four terms in the Jacobian associated with the ℓk-th entry. Test the program on the three-bus system illustrated in Figure 6.10 and repeated in Figure 8.2, assuming that the bus voltages are:

$$
\begin{aligned}
V_1 &= 1\angle 0\text{rad}, \\
V_2 &= 1.05\angle 0.2\text{rad}, \\
V_3 &= 0.95\angle 0.5\text{rad},
\end{aligned}
$$

and that the lines have π-equivalent models with:

- shunt elements purely capacitive with admittance $0.01\sqrt{-1}$ so that the combined shunt elements are:

$$Y_1 = Y_2 = Y_3 = 0.02\sqrt{-1},$$

and
- series elements having admittances:

$$
\begin{aligned}
Y_{12} &= (0.01 + 0.1\sqrt{-1})^{-1}, \\
Y_{23} &= (0.015 + 0.15\sqrt{-1})^{-1}, \\
Y_{31} &= (0.02 + 0.2\sqrt{-1})^{-1}.
\end{aligned}
$$

Provide values of the Jacobian for $\ell = 1$ and $k = 1, 2$, and 3.

8.13 Combine the MATLAB program from Exercise 8.12 with calls to the MATLAB function `lu` to build a program that performs the Newton–Raphson update for the power flow equations. Use the system and data for the three-bus system shown in Figure 8.2 and described in Exercise 8.12. Assume that bus one is the reference bus and that buses two and three are PQ buses. Let $S_2 = 1 + \sqrt{-1}$ and $S_3 = -2 - \sqrt{-1}$, so that bus 2 is a PQ generator bus, while bus 3 is a PQ load bus.

Use a flat start, that is, with $x^{(0)}$ having all voltage angles equal to zero radians and all voltage magnitudes equal to one per unit. Use the Armijo criterion with $\delta = 0.1$. Iterate until:

- the change in successive iterates is smaller than 10^{-4} using the infinity norm, and
- the mismatch in the power flow equations is less than 10^{-3} using the infinity norm.

Specify the voltage angles and voltage magnitudes in your answer as well as the number of iterations necessary to satisfy the stopping criterion.

8.14 Use the MATLAB function `fsolve` to solve the power flow problem specified in Exercise 8.13. Use a flat start as the initial guess. Solve it two ways:

(i) without specifying the Jacobian to MATLAB so that it uses finite differences to approximate the Jacobian,

(ii) specifying a MATLAB M-file to evaluate both g and the Jacobian J and specifying that you are supplying the Jacobian by setting the `Jacobian` option to `on` using the `optimset` function.

For each method, report the number of iterations required to satisfy the default stopping criterion and the number of function evaluations.

8.15 Justify (8.6) for a line between buses ℓ and k so that $k \in \mathbb{J}(\ell)$. Assume that the series resistance $R_{\ell k}$ of the line is much smaller than the series inductive reactance, $X_{\ell k}$.

8.16 Show approximations analogous to (8.10)–(8.13) in the case $k = \ell$. Assume that $|G_{\ell j}| \ll |B_{\ell j'}|$ for all $j, j' \in \mathbb{J}(\ell)$ and neglect shunt elements.

8.17 Modify the MATLAB program in Exercise 8.13 to perform the decoupled Newton–Raphson update (8.15)–(8.16). Use the same stopping criterion as in Exercise 8.13. How many iterations does it take to satisfy the stopping criterion? Also, report the voltage angles and voltage magnitudes in your answer.

8.18 Repeat Exercise 8.12 with the approximations (8.17) and (8.18) instead of the exact Jacobian.

8.19 Show that the approximate Jacobian $\begin{bmatrix} -UBU & 0 \\ 0 & -UB \end{bmatrix}$ is exact in the particular case of no shunt elements, zero series resistance, and a flat start; that is, all voltage angles equal to zero radians and all voltage magnitudes equal to one per unit.

8.20 Modify the MATLAB program in Exercise 8.17 to perform the fast decoupled Newton–Raphson update (8.19)–(8.21) instead of the decoupled Newton–Raphson update (8.15)–(8.16). Use the same stopping criterion as in Exercise 8.13. How many iterations does it take to satisfy the stopping criterion? Also, report the voltage angles and voltage magnitudes in your answer.

8.21 In this exercise, we consider a sensitivity study for the base-case problem specified in Exercise 8.13. In particular, calculate the sensitivity of the solution to changes ΔY_{23} in the admittance Y_{23} in Figure 8.2 from its base-case value of $Y_{23} = (0.015 + 0.15\sqrt{-1})^{-1}$. Calculate the sensitivity with respect to the real and with respect to the imaginary part of ΔY_{23}.

8.22 In this exercise, we consider a contingency study for the base-case problem specified in Exercise 8.13. In particular, suppose that the admittance Y_{23} in Figure 8.2 becomes open circuit due to a line outage. (Neglect any change in the shunt elements.)

(i) Solve the contingency study using the MATLAB function `fsolve` with the solution of the base-case as the initial guess. Report the number of iterations required to satisfy the default stopping criterion.

(ii) Re-solve the contingency study using as initial guess an estimate of the solution of the contingency case based on the sensitivity calculated in Exercise 8.21. (That is, by setting $\Delta Y_{23} = -Y_{23}$.) Report the number of iterations required to satisfy the default stopping criterion.

Part III

Unconstrained optimization

9

Case studies of unconstrained optimization

In this chapter we will introduce two case studies:

   (i) multi-variate linear regression (Section 9.1), and
  (ii) state estimation in an electric power system (Section 9.2).

Both problems will turn out to be unconstrained optimization problems of the special class of **least-squares data fitting problems** [84, chapter 13].

9.1 Multi-variate linear regression

Some of this section is based on [103] and further details can be found there. The development assumes a background in probability. See, for example, [31, 103].

9.1.1 Motivation

In many applications, we have a hypothesized functional relationship between variables. That is, we believe that there are some **dependent variables** that vary according to some function of some **independent variables**. The simplest relationship that we can imagine is a **linear** or **affine** relationship between the variables.

For example, we may be trying to estimate the circuit parameters of a black-box circuit by measuring the relationship between currents and voltages at the terminals of the circuit. We will have to try several values of current and voltage to characterize the circuit parameters. As in the circuit case study of Section 4.1, we could either:

- apply vectors of current injections and measure voltages, interpreting the currents as the independent variables and the voltages as the dependent variables, or
- apply vectors of voltages and measure currents, interpreting the voltages as the independent variables and the currents as the dependent variables.

As another example, we may be testing the efficacy of a combination of drugs or treatments for a particular disease. The independent variables include such variables as:

- the pre-treatment severity of the symptoms for each patient,
- the dosages of the various administered drugs, and
- the particulars of the various treatments.

The dependent variables could include some measure of the recovery of the patient. To characterize the efficacy of the drugs and treatment we must test several, maybe a large number of, patients. Unlike in the circuit case study of Section 4.1, in this example the independent and dependent variables are determined by the underlying cause and effect in the system.

In general, we do not have a complete specification of the function relating the variables (and we often do not know the underlying cause and effect in the system.) For example, if the hypothesized function is linear, the entries in coefficient matrix will typically be unknown to us. These unknown entries are called the **parameters** of the function.

To obtain enough information to estimate the parameters, we can imagine varying the independent variables over successive trials and measuring the resulting variation of the dependent variables. A suitably large number of trials should be examined to fully probe the functional relationship. We assume that the independent variables either:

(i) can be controlled to any desired level, or
(ii) cannot be controlled directly but exhibit a natural variability from trial to trial.

Independent variables that, in principle, can be controlled to any desired level include:

- the injected currents in the circuit estimation example, (given currents as the independent variables), and
- the dosages of drugs and the particulars of various treatments in the drug efficacy example.

Independent variables that naturally vary from trial to trial include the pre-treatment severity of the symptoms in the drug efficacy example.

In this case study, we will examine a general approach to characterizing a linear or affine relationship between the dependent and independent variables. This is called a **multi-variate linear regression** problem. It is also possible to characterize more complex relationships between dependent and independent variables. (See Exercise 9.1.)

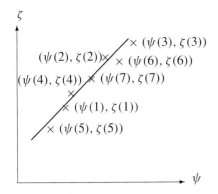

Fig. 9.1. The values of $(\psi(\ell), \zeta(\ell))$ (shown as $\times$) and affine fit.

9.1.2 Formulation

9.1.2.1 Measurement variables

Let us suppose that there is one dependent variable in our problem and call it ζ. The case of multiple dependent variables is a straightforward generalization. Let us suppose that there are $(n-1)$ independent variables. We collect the independent variables together into a vector $\psi \in \mathbb{R}^{n-1}$.

9.1.2.2 Functional relationship

Suppose that we believe that there is an affine relationship between ζ and ψ. That is, we believe that there are unknown constants $\beta \in \mathbb{R}^{n-1}$ and $\gamma \in \mathbb{R}$ such that:

$$\forall \psi \in \mathbb{R}^{n-1}, \zeta = \beta^{\dagger}\psi + \gamma. \tag{9.1}$$

For later convenience, let us collect β and γ together into a vector $x = \begin{bmatrix} \beta \\ \gamma \end{bmatrix} \in \mathbb{R}^{n}$.

9.1.2.3 Trials

We can perform a number of "trials" with varying values for the independent variables ψ. The goal is to discover a value of the vector x that best "satisfies" the measured data. We would like to estimate the "best" values of the parameters $\beta \in \mathbb{R}^{n-1}, \gamma \in \mathbb{R}$ in the relationship (9.1). Let the trials be numbered $\ell = 1, \ldots, m$. We use $\psi(\ell)$ and $\zeta(\ell)$, respectively, to denote the value of the independent variables ψ and the corresponding measured value of the dependent variable ζ for the ℓ-th trial. The situation is illustrated in Figure 9.1 for $n = 2$ and $m = 7$. The independent and dependent variables have been plotted as ordered pairs $(\psi(\ell), \zeta(\ell))$. An affine function of the form (9.1) has also been drawn.

9.1.2.4 Measurement error

For various reasons, the measured value $\zeta(\ell)$ for the ℓ-th trial may not equal $\beta^\dagger \psi(\ell) + \gamma$, but instead will be:

$$\zeta(\ell) = \beta^\dagger \psi(\ell) + \gamma + e_\ell, \tag{9.2}$$

where e_ℓ is the **error** introduced in the ℓ-th trial. The quantity e_ℓ is also called the **residual** implying that it is the remaining discrepancy between the measured value $\zeta(\ell)$ and the value predicted by (9.1) [118, section 4.1.3]. In Figure 9.1, for example, the values of $(\psi(\ell), \zeta(\ell))$ do not exactly fall on the affine function illustrated and, furthermore, there is no affine function that passes through all the points.

In (9.2), we think of the error as occurring in our measurement of $\zeta(\ell)$. However, it is also possible to consider errors in the values of $\psi(\ell)$. For our case study, however, we will assume that $\psi(\ell)$ can be specified or measured to great accuracy. The general case, where there are errors in both $\psi(\ell)$ and $\zeta(\ell)$, and which is called the **total linear regression problem**, is considered in [84, section 13.4].

In the following sections, we consider three possible types of error in the measurement $\zeta(\ell)$. One of them, random error, will be discussed in more detail in Section 9.1.2.5.

Calibration error The error e_ℓ may be due to **calibration error** in the measurement. That is, there exists a function $c : \mathbb{R} \to \mathbb{R}$, called the **calibration function**, such that on each trial ℓ the error is given by $e_\ell = c(\zeta(\ell))$. If c is known then we can calculate:

$$\beta^\dagger \psi(\ell) + \gamma = \zeta(\ell) - c(\zeta(\ell)),$$

and estimate the parameters β and γ that relate ψ to the **calibrated measurement** $\zeta - c(\zeta)$. We will not consider calibration functions in this case study, but the analysis can easily be extended to include this situation if c is known.

Functional error The error e_ℓ may arise because (9.1) does not reflect the actual relationship between ψ and ζ. For example, the actual relationship may be of the form

$$\zeta = \beta^\dagger \psi + \gamma + \psi^\dagger \Gamma \psi, \tag{9.3}$$

where $\Gamma \in \mathbb{R}^{(n-1) \times (n-1)}$ is a matrix of unknown parameters that represent interaction *between* entries of ψ. This could occur in our drug efficacy example if there are important interactions between the drugs. We will not consider functional error in this case study, but in general we must verify whether or not our functional form captures the relationship between the variables. This can be done, in principle, by:

- testing whether or not the functional form captures most of the variation in the data [118, section 4.1.2],
- testing whether or not the estimated parameters specify a function that yield small residuals on the original trial data [118, section 4.1.3], and
- testing whether or not the function can predict the results of other trials not included in the original trial data.

In many cases, we can easily generalize the formulation we describe to represent a more complicated function [118, section 4.2]. For example, Exercise 9.1 illustrates the case for the relationship defined in (9.3).

Random error The error e_ℓ may be **random** with expected value, say, 0. Such errors are present because our model is a simplification of the real world. That is, ζ also depends on other variables besides ψ that we can neither control nor measure easily. Under some circumstances, it may be reasonable to model these errors as random variables that vary independently of the trials.

To illustrate, consider the two examples introduced earlier.

Black-box circuit In our black-box circuit we may not be able to control, for example, ambient temperature, but this will affect the resistances in the circuit and therefore affect the measurements. That is, the ambient temperature also affects ζ. If we perform trials such that each trial is performed at a random temperature, then the measurement will reflect this randomness. It may be reasonable to assume that the temperature is independent of the injected currents. Furthermore, the voltmeter used to measure the circuit response may also introduce further error that is random and independent of the injected currents.

Drug efficacy The recovery of patient ℓ may depend on other variables besides the severity of the symptoms and the drugs and treatment. These variables might include specific properties of each patient's immune system. If we perform trials on a large number of patients, then we would expect to see patients with a variety of immune system properties. It may be reasonable to assume that immune system properties vary randomly from patient to patient and are independent of the symptoms, drugs, and treatment.

Discussion Caution is in order here. For example, the injected currents in the circuit will heat up the resistors, so that the temperature also depends on the injected currents. Similarly, the severity of the patient's symptoms may actually be strongly correlated with the properties of the patient's immune system. We should, in general, be very cautious about asserting independence between the independent variables $\psi(\ell)$ and the error e_ℓ. Nevertheless, for the rest of this case study we will assume that the error is random and uncorrelated with ψ.

9.1.2.5 Random error distribution

In the following, we discuss the use of the central limit theorem to model the distribution of the random error. We then develop a model of the joint probability of errors in several measurements.

Central limit theorem We must model the probability density function [31, chapter 1] of the random variable e_ℓ. Suppose that there are a number of factors that sum to produce the error e_ℓ in trial ℓ. For example, in the circuit example, suppose that the voltmeter consists of a number of parts or components, each of which independently adds a random error into the measurement. The total resulting error e_ℓ is due to the sum of these individual introduced errors.

The **central limit theorem** [31, chapter 2] says that the sum of a large number of independent random variables has a distribution that is approximately **Gaussian**. That is, the probability density function of e_ℓ is:

$$\frac{1}{\sqrt{2\pi}\sigma_\ell} \exp\left(-\frac{(e_\ell - \mu_\ell)^2}{2(\sigma_\ell)^2}\right), \tag{9.4}$$

where μ_ℓ is the expected value of e_ℓ, in our case 0, and σ_ℓ is its standard deviation.

In other cases, a different model may be more appropriate. For example, the error in the measurement might be due to the sum of:

- a relatively small random error with distribution that is well approximated by a Gaussian distribution, and
- a large error due to gross meter failure that occurs with a small probability.

In this case, the resulting error distribution would be more complicated than the basic Gaussian distribution we have discussed. (See [4] for an example of the specification of such a distribution.)

Furthermore, if the independent variables ψ are also subject to random error, then this should also be modeled explicitly. Details for this **total linear regression problem** can be found in [84, section 13.4].

Error correlation The model (9.4) applies to a specific trial ℓ, but we must consider the correlation between the errors on various trials. For example:

(i) the errors on two trials ℓ and ℓ' could be correlated, so that e_ℓ is likely to be positive when $e_{\ell'}$ is positive and e_ℓ is likely to be negative when $e_{\ell'}$ is negative, or

(ii) the errors on separate trials could be uncorrelated.

Let us suppose that only zero mean random error is present and that, furthermore, the errors on different trials are uncorrelated. That is, we model the measurement as (9.2), with the probability density function of e_ℓ given by (9.4) with $\mu_\ell = 0$ and where e_ℓ is uncorrelated with $e_{\ell'}$ for $\ell \neq \ell'$.

The assumption of uncorrelated errors is not always true in practice. For example, in the circuit estimation example, if two trials are chosen to have the same values of the independent variables then it may be the case that the measurement is the same for the two trials and the measurement error is the same for the two trials. That is, the errors may be correlated; however, we will ignore this possibility.

Distribution of dependent variables Since e_ℓ is a random variable then, *prior* to performing trial ℓ, $\zeta(\ell) = \beta^\dagger \psi(\ell) + \gamma + e_\ell$ is also a random variable. The **prior probability density function** of e_ℓ is given by (9.4) with $\mu_\ell = 0$ and so the prior probability density function of the dependent variable $\zeta(\ell)$ is the function $\phi_\ell : \mathbb{R} \to \mathbb{R}$ that is parameterized by $\psi(\ell)$ and x and defined by:

$$\forall \zeta(\ell) \in \mathbb{R}, \, \phi_\ell(\zeta(\ell); \psi(\ell), x) = \frac{1}{\sqrt{2\pi}\sigma_\ell} \exp\left(-\frac{(\zeta(\ell) - \beta^\dagger \psi(\ell) - \gamma)^2}{2(\sigma_\ell)^2} \right).$$

We use a semi-colon to separate the arguments of the function from the parameters $\psi(\ell)$ and x.

Joint measurement distribution We have assumed that e_ℓ and $e_{\ell'}$ are uncorrelated for $\ell \neq \ell'$. If we assume that their distributions are **jointly Gaussian** then they are independent. Therefore, the joint probability density function, $\phi : \mathbb{R}^m \to \mathbb{R}$, is the product of the individual probability densities ([31]):

$$\forall \zeta(1) \in \mathbb{R}, \dots, \forall \zeta(m) \in \mathbb{R}, \, \phi(\zeta(1), \dots, \zeta(m); \psi(1), \dots, \psi(m), x)$$

$$= \prod_{\ell=1}^{m} \frac{1}{\sqrt{2\pi}\sigma_\ell} \exp\left(-\frac{(\zeta(\ell) - \beta^\dagger \psi(\ell) - \gamma)^2}{2(\sigma_\ell)^2} \right). \quad (9.5)$$

This distribution is parameterized by $\psi(1), \dots, \psi(m)$, and x.

9.1.2.6 Problem variables

After performing the trials, the values of $\psi(\ell)$ and $\zeta(\ell)$ are known and we will re-interpret them as *constants*. The unknowns are the parameters β and γ in the relationship (9.1). We have collected together these parameters into the vector x and they will be re-interpreted as the *variables* in our problem formulation since they are the values that are to be determined to solve our regression problem.

9.1.2.7 *Maximum likelihood estimation*

Suppose that we perform the trials and want to find β and γ. We must decide on a criterion for choosing the "best value." Suppose that we are given:

- a collection of measurements $\zeta(1) \in \mathbb{R}, \ldots, \zeta(m) \in \mathbb{R}$,
- values of the parameters $x \in \mathbb{R}^n$, and
- a distance $\delta \in \mathbb{R}_+$.

Now suppose that we consider taking *new* measurements, $\tilde{\zeta}(1), \ldots, \tilde{\zeta}(m)$ using the *same* values of the independent variables. Consider the probability that for all ℓ the value of the ℓ-th new measurement, $\tilde{\zeta}(\ell)$, lies within a distance δ of the value $\zeta(\ell)$. That is, consider the probability that the new measurements $\tilde{\zeta}(1), \ldots, \tilde{\zeta}(m)$ lie in the set:

$$\mathbb{S}(x) = \{\tilde{\zeta}(1) \in \mathbb{R}, \ldots, \tilde{\zeta}(m) \in \mathbb{R} | \zeta(\ell) - \delta \leq \tilde{\zeta}(\ell) \leq \zeta(\ell) + \delta, \forall \ell = 1, \ldots, m\}.$$

This probability is equal to the integral of the joint probability distribution function (9.5) over the set $\mathbb{S}(x)$. For δ small enough, this integral is approximately equal to:

$$\phi(\zeta(1), \ldots, \zeta(m); \psi(1), \ldots, \psi(m), x)(2\delta)^m.$$

We pick $x \in \mathbb{R}^n$ to maximize the probability that the new measurements are in the set $\mathbb{S}(x)$. That is, we maximize the probability that we would observe similar results if we were to repeat the experiment. This means finding x that maximizes:

$$\phi(\zeta(1), \ldots, \zeta(m); \psi(1), \ldots, \psi(m), x)(2\delta)^m,$$

which is equivalent to maximizing:

$$\phi(\zeta(1), \ldots, \zeta(m); \psi(1), \ldots, \psi(m), x)$$

over $x \in \mathbb{R}^n$. That is, we maximize ϕ, *re-interpreted* to be the function $\phi : \mathbb{R}^n \to \mathbb{R}$ defined by:

$$\forall x \in \mathbb{R}^n, \phi(\zeta(1), \ldots, \zeta(m); \psi(1), \ldots, \psi(m), x)$$

$$= \prod_{\ell=1}^{m} \frac{1}{\sqrt{2\pi}\sigma_\ell} \exp\left(-\frac{(\zeta(\ell) - \beta^\dagger \psi(\ell) - \gamma)^2}{2(\sigma_\ell)^2}\right),$$

$$= \prod_{\ell=1}^{m} \frac{1}{\sqrt{2\pi}\sigma_\ell} \exp\left(-\frac{(\psi(\ell)^\dagger \beta + \gamma - \zeta(\ell))^2}{2(\sigma_\ell)^2}\right), \quad (9.6)$$

where $x = \begin{bmatrix} \beta \\ \gamma \end{bmatrix}$.

This criterion for "best" is called **maximum likelihood estimation** and can be justified more formally by Bayes' rule [31, chapter 4][103, section 5.5].

9.1.2.8 Problem

We write the maximum likelihood estimation problem in the form of an uncon-strained optimization problem as:

$$\max_{x \in \mathbb{R}^n} \phi(\zeta(1), \ldots, \zeta(m); \psi(1), \ldots, \psi(m), x). \qquad (9.7)$$

9.1.3 Change of number of trials or correction of data

We may find that after solving the maximum likelihood estimation using trials $1, \ldots, m$ we conduct further trials or find that some of the data might be in error and need to be corrected. We would like to be able obtain an updated estimation without starting from scratch.

9.1.4 Problem characteristics

9.1.4.1 Parameters re-interpreted as variables

We have re-interpreted the *parameters* β and γ of the probability density in (9.5) to be the *variables* in our optimization problem. We interpret $\psi(\ell)$ and $\zeta(\ell)$ to be *known* values once the trials have been completed. So far in this book, functions have been defined very explicitly by first indicating the domain and range and then specifying the functional relation. You may have found this approach rather pedantic. However, in this case study where we re-interpret the parameters and variables, it is very important to be explicit about the variables and domain of functions.

9.1.4.2 Objective

Because of the assumption that the measurements errors for different trials are uncorrelated and jointly Gaussian, the objective:

$$\phi(\zeta(1), \ldots, \zeta(m); \psi(1), \ldots, \psi(m), x),$$

is the product of terms. Each term in the product depends on x.

9.1.4.3 Number of parameters and trials

In principle, the number of parameters, n, could be smaller than, equal to, or greater than the number m of trials. Intuitively, if the number of trials is greater than the number of parameters and the trials "explore" the space of independent variables then there is redundancy in the measurements that can be used to reduce the effects of the random measurements errors. If $m \leq n$ then there is no redundancy and we will not be able to reduce the effects of measurement errors. In practice, we should always have many more trials than the number of parameters we want to estimate.

We will see also that we must have sufficient "variety" in our choice of trials $\psi(\ell)$, as well as sufficient number, to be able to uniquely estimate all the parameters and reduce the effects of measurement error.

If there is not sufficient variety, then we may still want to obtain parameters that maximize Problem (9.7) even though the parameters will not be unique. We will also consider approaches to this case.

9.1.4.4 Generalizations

In some cases, we may have a non-linear relationship between the dependent and independent variables, as in (9.3). The formulation of this problem is explored in Exercise 9.1. A number of other generalizations are explored in [15, chapter 6]. Other aspects of regression analysis are contained in various references, including [84, section 13.3][118].

9.2 Power system state estimation

This case study builds on the multi-variate linear regression case study of Section 9.1 and the electric power system case study introduced in Section 6.2. Again, some familiarity with electric power systems is assumed [8]. The problem we formulate will be a **non-linear regression** problem. A general discussion of non-linear regression is in [84, chapter 13]. Some of this section is based on [123, chapter 12]. Another formulation of this problem is contained in [22, section 3.4.1]. General discussions of state estimation are presented in [1, 80].

9.2.1 Motivation

9.2.1.1 Non-linear regression

Suppose that we hypothesize a non-linear relationship, such as $\zeta = \gamma(\psi)^\beta$, between scalars ψ and ζ with unknown parameters β and γ. A standard approach for this particular non-linear relationship is to take logarithms of both sides to form the equation:

$$\ln(\zeta) = \beta \ln(\psi) + \ln(\gamma).$$

We consider a transformation of the problem. Define variables Ψ and Z such that:

$$\Psi = \ln(\psi),$$
$$Z = \ln(\zeta),$$

That is, we have implicitly defined an onto function $\tau : \mathbb{R}^2_{++} \to \mathbb{R}^2$ specified by:

$$\forall \begin{bmatrix} \psi \\ \zeta \end{bmatrix} \in \mathbb{R}^2_{++}, \tau \left(\begin{bmatrix} \psi \\ \zeta \end{bmatrix} \right) = \begin{bmatrix} \ln(\psi) \\ \ln(\zeta) \end{bmatrix}.$$

Then:

$$Z = \beta \Psi + \Gamma,$$

where $\Gamma = \ln(\gamma)$. We now have a linear regression problem as discussed in Section 9.1, involving the measurement variables Ψ and Z and the unknown parameters β and Γ.

We could apply the same analysis as in Section 9.1 to this linear regression problem, if we assume that measurement of Z is subject to random additive Gaussian error, so that ζ is subject to random *multiplicative* error. This approach can be generalized to any case where the function relating the dependent and independent variables is linear in the unknown parameters, or where the function can be transformed to a function that is linear in the unknown parameters. (See Exercise 9.1.)

The assumption of multiplicative error is an appropriate model for some applications; however, if the error actually adds to ζ then a multiplicative error model is inappropriate. In practice, computational convenience will often suggest a linear regression model even if it does not capture the underlying error probability distribution accurately. The resulting estimator will not in general be equal to the maximum likelihood estimator for the given probability distribution; however, the difference may be small if the measurement errors are small. The resulting estimate may be satisfactory, despite not having a rigorous formulation. In this case, we might refer to the estimate as **satisficing** [109].

In some problems, however, the unknown parameters appear non-linearly in the relationship between dependent and independent variables and, furthermore, there is no transformation that will make them appear linearly. For example, consider a functional relationship between scalars ψ and ζ of the form:

$$\zeta = \gamma (\psi)^\beta + \delta \psi,$$

with unknown parameters β, γ, and δ. We cannot transform this equation in a way such that all the unknown parameters β, γ, and δ (or their transformed versions) appear linearly. Such a problem is called a **non-linear regression** problem. A concrete example of a non-linear regression problem is the state estimation problem in power systems, to be discussed in Section 9.2.1.2.

9.2.1.2 *Power system measurements*

In Section 6.2, we discussed the power flow model where the problem is to solve the power flow equations to find voltage angles and magnitudes in the system, given known values of real and reactive demand and generation. This is most appropriate when we want to, say, *predict* worst-case loadings on the system based on forecasts of peak demand conditions.

In on-line applications, however, we may want to observe the *actual* state of the

system to check if the system is operating within limits. In this case, we will typically have telemetered values of some (but not necessarily all) of the generations
and loads in the system and of some (but not all) of the line flows and voltage
magnitudes. (It is also possible to measure the voltage angles, using accurate time
reference information from the global positioning system [32].)

Telemetered values are subject to both:

(i) meter error, where the meter is working but delivers somewhat inaccurate
data, and
(ii) meter failure, where the meter delivers grossly inaccurate data.

We will consider the first issue here, but note that the second issue of **bad data
detection** presents several additional complications. (See [53] for a general discussion on rejection of bad data from measurements; see [1], [80, chapters 8, 9,
and 13], and [123, chapter 12] for specific discussions in relation to state estimation in power systems; and see [4] for a discussion of the maximum likelihood
estimator in the presence of meter failure.)

The state estimation problem involves finding the voltage angles and magnitudes
in the system that best match the measured values. As we have seen in Section 6.2,
the relationship amongst the measured values and the voltage angles and magnitudes is non-linear and this leads to a non-linear regression problem.

9.2.2 Formulation

9.2.2.1 Measurements

Various quantities are measured at various places in the electric power system and
telemetered for processing. In the following sections, we will consider measurements of:

• real and reactive power injection at a bus,
• real and reactive power flow along a line, and
• voltage magnitude.

Figure 9.2 illustrates a three-bus power system with meters. This is the same system shown in Figure 6.10, except that we have explicitly shown generators at buses
1 and 2 and a load at bus 3 and we have explicitly shown meters. Each meter is
denoted by a filled box: ■. The measured quantities are shown adjacent to the box.

Real and reactive power injection Let $\mathbb{B}$ be the set of buses where there are measurements of the real and reactive power injections into the system. In Figure 9.2,
$\mathbb{B} = \{1\}$. (It is common to measure both real and reactive power together since, as
a practical matter, the equipment needed to measure real power also furnishes data

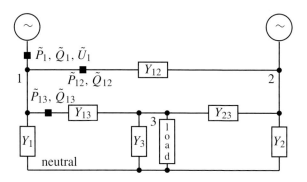

Fig. 9.2. Three-bus power system state estimation problem.

for reactive power.) For each $\ell \in \mathbb{B}$, let $\tilde{P}_\ell$ and $\tilde{Q}_\ell$ be the injection measurement at bus ℓ. The tilde ˜ over the symbol for the measurement signifies that the measured values are approximately, but not exactly, equal to the actual injections P_ℓ and Q_ℓ. At a load bus, these "injection" measurement values will typically be negative.

Real and reactive line flow Let $\mathbb{F}$ be the set of lines where we have line flow measurements. Each element of $\mathbb{F}$ is an ordered pair (ℓ, k) of "from-bus" and "to-bus" and $\tilde{P}_{\ell k}$ and $\tilde{Q}_{\ell k}$ is the measurement of real and reactive power flow on the line from ℓ to k, measured at ℓ in the direction k. In Figure 9.2, $\mathbb{F} = \{(1, 2), (1, 3)\}$. If there were no line losses and no meter errors then a measurement $\tilde{P}_{\ell k}$ taken at one end of a line would be equal and opposite to a measurement $\tilde{P}_{k\ell}$ taken at the other end of the line. Because of line losses, however, these quantities will not sum exactly to zero, even if there are no meter errors. We may have $(\ell, k) \in \mathbb{F}$, yet $(k, \ell) \notin \mathbb{F}$, indicating that there is a meter at the bus ℓ end of the line between ℓ and k but not at the k end of the line. For example, in Figure 9.2, $(1, 2) \in \mathbb{F}$, yet $(2, 1) \notin \mathbb{F}$.

Voltage magnitude Finally, let $\mathbb{U}$ be the set of buses where there are voltage magnitude measurements. In Figure 9.2, $\mathbb{U} = \{1\}$. For each $\ell \in \mathbb{U}$, let $\tilde{U}_\ell$ be the voltage magnitude measurement at bus ℓ.

9.2.2.2 Variables

In Section 6.2, we defined x to be the set of voltage angles and magnitudes in the system, excluding the reference bus. We arbitrarily assigned the voltage magnitude and angle at the reference bus. That is, we specified these as constant parameters in the problem of solving the power flow equations in Section 6.2.

In the state estimation problem, if we do not have any angle measurements, then we can still assign the reference bus to have zero angle since the angle at the reference bus depends on an arbitrary time reference. That is, we can set $\theta_1 = 0$. However, the voltage magnitude at the reference bus and the voltage angles and voltage magnitudes in the rest of the system must be estimated. Let us therefore modify the definition of x in Section 6.2 to include:

- the voltage angles at all buses except the reference bus, and
- the voltage magnitudes at all buses in the system, including the reference bus.

Now $x \in \mathbb{R}^n$, where n is equal to one less than twice the number of buses, so that the vector x has been re-defined compared to Section 6.2.

9.2.2.3 Measurement functions

Recall the definitions of the functions $p_\ell, q_\ell : \mathbb{R}^n \to \mathbb{R}$ in (6.12) and (6.13) that were used in the power flow case study:

$$\forall x \in \mathbb{R}^n, \, p_\ell(x) \quad = \quad \sum_{k \in \mathbb{J}_\ell \cup \{\ell\}} u_\ell u_k [G_{\ell k} \cos(\theta_\ell - \theta_k) + B_{\ell k} \sin(\theta_\ell - \theta_k)] - P_\ell,$$

$$\forall x \in \mathbb{R}^n, \, q_\ell(x) \quad = \quad \sum_{k \in \mathbb{J}_\ell \cup \{\ell\}} u_\ell u_k [G_{\ell k} \sin(\theta_\ell - \theta_k) - B_{\ell k} \cos(\theta_\ell - \theta_k)] - Q_\ell.$$

Let us define new functions by omitting the values of the real and reactive injections, P_ℓ and Q_ℓ. That is, let us define $\tilde{p}_\ell : \mathbb{R}^n \to \mathbb{R}$ and $\tilde{q}_\ell : \mathbb{R}^n \to \mathbb{R}$ to be:

$$\forall x \in \mathbb{R}^n, \, \tilde{p}_\ell(x) \quad = \quad \sum_{k \in \mathbb{J}_\ell \cup \{\ell\}} u_\ell u_k [G_{\ell k} \cos(\theta_\ell - \theta_k) + B_{\ell k} \sin(\theta_\ell - \theta_k)],$$

$$\forall x \in \mathbb{R}^n, \, \tilde{q}_\ell(x) \quad = \quad \sum_{k \in \mathbb{J}_\ell \cup \{\ell\}} u_\ell u_k [G_{\ell k} \sin(\theta_\ell - \theta_k) - B_{\ell k} \cos(\theta_\ell - \theta_k)].$$

With this definition, $\tilde{p}_\ell$ and $\tilde{q}_\ell$ represent the real and reactive power injection at the bus into the series and shunt elements connected to the bus expressed as a function of x. Similarly, each real and reactive line flow and each voltage can be expressed as a function of x. (See Exercise 9.2.) In particular, we define the functions $\tilde{p}_{\ell k}, \tilde{q}_{\ell k} : \mathbb{R}^n \to \mathbb{R}$ to represent the real and reactive power flows on the line joining bus ℓ to bus k, measured at ℓ in the direction of k. We define the function $\tilde{u}_\ell : \mathbb{R}^n \to \mathbb{R}$ to represent the voltage magnitude at node ℓ. These functions that express the exact values of the measured quantities in terms of x are called the **measurement functions**. In summary, we denote the measurement functions by:

$$\tilde{p}_\ell, \tilde{q}_\ell, \qquad \text{for the real and reactive power injection measurements, } \ell \in \mathbb{B},$$
$$\tilde{p}_{\ell k}, \tilde{q}_{\ell k}, \qquad \text{for the real and reactive line flow measurements, } (\ell, k) \in \mathbb{F},$$
$$\tilde{u}_\ell, \qquad \text{for the voltage magnitude measurements, } \ell \in \mathbb{U}.$$

Let us now collect the measurement functions into a vector function $\tilde{g}$ and collect the measurements together into a corresponding vector $\tilde{G}$. (We deviate from our typographical convention by using an upper case letter G to stand for a vector, since we are using the corresponding lower case letter to stand for a vector function.) That is,

$$\forall x \in \mathbb{R}^n, \tilde{g}(x) = \left(\begin{bmatrix} \tilde{p}_\ell(x) \\ \tilde{q}_\ell(x) \end{bmatrix}_{\ell \in \mathbb{B}} \\ \begin{bmatrix} \tilde{p}_{\ell k}(x) \\ \tilde{q}_{\ell k}(x) \end{bmatrix}_{(\ell,k) \in \mathbb{F}} \\ \begin{bmatrix} \tilde{u}_\ell(x) \end{bmatrix}_{\ell \in \mathbb{U}} \right), \tilde{G} = \left(\begin{bmatrix} \tilde{P}_\ell \\ \tilde{Q}_\ell \end{bmatrix}_{\ell \in \mathbb{B}} \\ \begin{bmatrix} \tilde{P}_{\ell k} \\ \tilde{Q}_{\ell k} \end{bmatrix}_{(\ell,k) \in \mathbb{F}} \\ \begin{bmatrix} \tilde{U}_\ell \end{bmatrix}_{\ell \in \mathbb{U}} \right),$$

where we have ordered the entries in $\tilde{g}$ and $\tilde{G}$ to correspond. Let us define a new index set $\mathbb{M}$ that specifies all the measurements. We re-index the entries of $\tilde{g}$ and $\tilde{G}$ using the set $\mathbb{M}$, so that $\tilde{g} = (\tilde{g}_\ell)_{\ell \in \mathbb{M}}$ and $\tilde{G} \in \mathbb{R}^{\mathbb{M}}$, where by $\mathbb{R}^{\mathbb{M}}$ we mean the set of all vectors having entries indexed by the elements in the set $\mathbb{M}$. (See Definition A.5.) That is, $\mathbb{R}^{\mathbb{M}}$ has as many dimensions as there are elements in the set $\mathbb{M}$. In summary, $\tilde{g} : \mathbb{R}^n \to \mathbb{R}^{\mathbb{M}}$ and $\tilde{G} \in \mathbb{R}^{\mathbb{M}}$.

9.2.2.4 *Error distribution*

As in Section 9.1, we expect the measurements to differ from the measurement functions due to meter error. In general, we would expect the error in one measurement to be independent of the measurement in another meter. (One exception to this is where two or more measurements are actually based on a third measurement. This occurs for pairs of real and reactive power measurements. Another exception is where the "measurement error" includes the effects of data transmission errors and several measurements are transmitted over a single error-prone communication link.) Assuming independent Gaussian measurement errors then, as in the multivariate linear regression case study in Section 9.1, we can write the probability density, $\phi : \mathbb{R}^{\mathbb{M}} \to \mathbb{R}$, of the measurement vector $\tilde{G}$ as the product of probability densities:

$$\forall \tilde{G} \in \mathbb{R}^{\mathbb{M}}, \phi(\tilde{G}; x) =$$
$$\prod_{\ell \in \mathbb{B}} \phi_{\tilde{p}_\ell}(\tilde{P}_\ell; x) \prod_{\ell \in \mathbb{B}} \phi_{\tilde{q}_\ell}(\tilde{Q}_\ell; x) \prod_{(\ell,k) \in \mathbb{F}} \phi_{\tilde{p}_{\ell k}}(\tilde{P}_{\ell k}; x) \prod_{(\ell,k) \in \mathbb{F}} \phi_{\tilde{q}_{\ell k}}(\tilde{Q}_{\ell k}; x) \prod_{\ell \in \mathbb{U}} \phi_{\tilde{u}_\ell}(\tilde{U}_\ell; x),$$

where each function $\phi_{\tilde{p}_\ell}(\tilde{P}_\ell; x), \phi_{\tilde{q}_\ell}(\tilde{Q}_\ell; x), \phi_{\tilde{p}_{\ell k}}(\tilde{P}_{\ell k}; x), \phi_{\tilde{q}_{\ell k}}(\tilde{Q}_{\ell k}; x), \phi_{\tilde{u}_\ell}(\tilde{U}_\ell; x)$ represents the probability density function of the corresponding error distribution and is parameterized by x. For example,

$$\forall \tilde{P}_\ell \in \mathbb{R}, \phi_{\tilde{p}_\ell}(\tilde{P}_\ell; x) = \frac{1}{\sqrt{2\pi}\sigma_{\tilde{p}_\ell}} \exp\left(-\frac{(\tilde{p}_\ell(x) - \tilde{P}_\ell)^2}{2(\sigma_{\tilde{p}_\ell})^2}\right),$$

where $\sigma_{\tilde{p}_\ell}$ is the standard deviation of the measurement error of real power at bus ℓ and where we have assumed that the expected error is zero.

After the measurements are made, we can re-interpret ϕ to be a function ϕ : $\mathbb{R}^n \to \mathbb{R}$. That is, as in the multi-variate linear regression case study, we re-interpret ϕ as:

$$\forall x \in \mathbb{R}^n, \phi(\tilde{G}; x) =$$

$$\prod_{\ell \in \mathbb{B}} \phi_{\tilde{p}_\ell}(\tilde{P}_\ell; x) \prod_{\ell \in \mathbb{B}} \phi_{\tilde{q}_\ell}(\tilde{Q}_\ell; x) \prod_{(\ell,k) \in \mathbb{F}} \phi_{\tilde{p}_{\ell k}}(\tilde{P}_{\ell k}; x) \prod_{(\ell,k) \in \mathbb{F}} \phi_{\tilde{q}_{\ell k}}(\tilde{Q}_{\ell k}; x) \prod_{\ell \in \mathbb{U}} \phi_{\tilde{u}_\ell}(\tilde{U}_\ell; x).$$

Our maximum likelihood estimation problem is then:

$$\max_{x \in \mathbb{R}^n} \phi(\tilde{G}; x). \tag{9.8}$$

9.2.3 Change in measurement data

Over time, the state of the power system changes as demand and supply situations change. Consequently, the measured data will change. We will consider how a change in measurement data affects the result.

9.2.4 Problem characteristics

9.2.4.1 Objective

The objective of this problem is very similar to the objective of Problem (9.7) defined in (9.6), except that each term in the product has one of the non-linear functions $\tilde{p}_\ell$, $\tilde{q}_\ell$, $\tilde{p}_{\ell k}$, $\tilde{q}_{\ell k}$, or $\tilde{u}_\ell$ in the exponent instead of the linear function $\psi(\ell)^\dagger \beta + \gamma$ that appears in (9.6).

9.2.4.2 Solvability

Consider the three-bus system in Figure 9.2. There are two PQ buses, at buses 2 and 3, so $n_{PQ} = 2$. Recall that the angle at the reference bus, bus 1, is assigned to be zero. There are $n = 2n_{PQ} + 1 = 5$ variables that must be estimated, namely:

- u_1, u_2, u_3, the voltage magnitudes at buses 1, 2, and 3, and
- θ_2 and θ_3, the voltage angles at buses 2 and 3.

From u_1, u_2, u_3, θ_2, and θ_3, all the real and reactive power injections and line flows in the system can be calculated.

Consider the placement of meters in the system. If all the measurements are concentrated in one part of the system, say near bus 1 as illustrated in Figure 9.2, then it may not be possible to estimate the voltage magnitudes and angles at the other buses. This is because the information from bus 1 does not uniquely identify

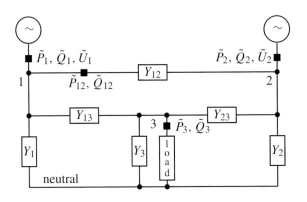

Fig. 9.3. Three-bus power system state estimation problem with spread out measurements.

the generation at bus 2 and the load at bus 3. The measurements shown in the system illustrated in Figure 9.2 do not have enough information to determine all the values of the entries in x.

It is important to have enough measurements in the system and to "spread out" the measurements across the system as illustrated in Figure 9.3. We will see that this requirement is similar to the need for sufficient variety in the trial vectors $\psi(\ell)$ in the multi-variate linear regression problem.

Exercises

Multi-variate linear regression

9.1 Consider a *non-linear* relationship between independent variables ψ and a dependent variable ζ of the form defined in (9.1):

$$\forall \psi \in \mathbb{R}^{n-1}, \zeta = \beta^{\dagger} \psi + \psi^{\dagger} \Gamma \psi,$$

where $\beta \in \mathbb{R}^{n-1}$ and $\Gamma \in \mathbb{R}^{(n-1) \times (n-1)}$. For simplicity, assume that Γ is known to be diagonal. (The generalization to arbitrary Γ is straightforward, but notationally clumsy.) Assume that:

- the measurement $\zeta(\ell)$ for trial ℓ is subject to additive Gaussian error e_{ℓ} with distribution given by (9.4) with $\mu_{\ell} = 0$, and
- the error on trial ℓ is independent of the error on trial ℓ' for $\ell \neq \ell'$ and the joint distribution is jointly Gaussian.

Let $\gamma \in \mathbb{R}^{n-1}$ be the vector with elements that are equal to the corresponding diagonal elements of Γ. Define $x = \begin{bmatrix} \beta \\ \gamma \end{bmatrix} \in \mathbb{R}^{2n-2}$. Formulate the maximum likelihood problem analogous to Problem (9.7) by defining ϕ appropriately.

Power system state estimation

9.2 Consider the functional form of the measurement functions for line flows and voltage magnitudes.

- (i) Explicitly define the form of the real and reactive line flow measurement functions $\tilde{p}_{\ell k}, \tilde{q}_{\ell k} : \mathbb{R}^n \to \mathbb{R}$ for an arbitrary $(\ell, k) \in \mathbb{F}$. (Hint: consider the terms in (6.12) and (6.13) that represent the flow along the line from ℓ to k.)
- (ii) Explicitly define the form of the voltage measurement function $\tilde{u}_\ell : \mathbb{R}^n \to \mathbb{R}$ for an arbitrary $\ell \in \mathbb{U}$.

10

Algorithms for unconstrained minimization

In this chapter we will discuss algorithms for unconstrained minimization problems of the form:

$$\min_{x \in \mathbb{R}^n} f(x),$$

where $x \in \mathbb{R}^n$ and $f : \mathbb{R}^n \to \mathbb{R}$. In Section 10.1, we describe necessary conditions that are satisfied by optimizers of unconstrained problems and describe sufficient conditions to guarantee that a point is an optimizer. Some of the sufficient conditions can be weakened; however, we present them in their simplest versions for expositional clarity. Algorithms for finding optimizers are presented in Section 10.2. As in our analysis of the solution of non-linear simultaneous equations, we will initially not make any assumptions about f, besides partial differentiability of f and continuity of its partial derivatives, but will gradually restrict our attention to those f for which we can construct effective algorithms. We will discuss sensitivity analysis in Section 10.3.

The key issues discussed in this chapter are:

- the notion of **descent directions** to update the iterate to reduce the value of the objective,
- optimality conditions based on **descent directions**,
- optimality conditions for **convex objectives**,
- the development of **iterative algorithms** that seek to successively improve the value of the function, and
- **sensitivity analysis**.

10.1 Optimality conditions

We characterize local optimizers of a problem having an objective that is partially differentiable with continuous partial derivatives. This characterization will in-

381

volve properties of the objective at a candidate solution. Much of the presentation is based on [70, chapter 6].

In Section 10.1.1, we discuss how to move from a point that is not a minimum in a direction that decreases the objective. This will lead to **necessary** conditions for optimality in Section 10.1.2; that is, conditions that must be true at a local optimum. The simplest conditions, the **first-order necessary conditions**, based on the first derivative of the objective, will lead to an algorithm for seeking a minimum. (The conditions are sometimes referred to by their acronym **FONC**.) In the proof of these conditions, we will make the links to the algorithm. We will also discuss **second-order necessary conditions** (or **SONC**) in Section 10.1.3.1 that are based on both the first and second derivatives.

Unfortunately, the first-order conditions are not **sufficient** to guarantee a local minimum. In particular, the conditions also hold at a maximum or other **critical point**. We will therefore also present in Section 10.1.3.2 **second-order sufficient conditions** (or **SOSC**) that are based on both the first and the second derivative of the objective and sufficient to guarantee a minimum. However, the conditions are not necessary; that is, some minimizers will not satisfy the conditions.

In practice, we may find candidate solutions that satisfy the first-order and even the second-order necessary conditions (to within the accuracy allowed by the precision of our calculations or to within a specified tolerance,) but do not "quite" satisfy the second-order sufficient conditions. In this case, we must apply other tests to the problem or apply "engineering judgment" to gauge whether we are at a local optimum or at least have found a useful solution.

Fortunately, the notion of convexity provides conditions for global optimality. Section 10.1.4 discusses necessary and sufficient conditions for global optimality for convex objectives. In contrast to the local optimality conditions in Sections 10.1.2–10.1.3, however, the conditions involving convexity require us to know properties of the objective f throughout $\mathbb{R}^n$.

10.1.1 Descent direction

10.1.1.1 Analysis

In this section we consider when we can move from a candidate point in a way that reduces the objective. We make the following definition.

Definition 10.1 Let $\hat{x} \in \mathbb{R}^n$ and $f : \mathbb{R}^n \to \mathbb{R}$. Then the vector $\Delta x \in \mathbb{R}^n$ is called a **descent direction** for f at $\hat{x}$ if:

$$\exists \overline{\alpha} \in \mathbb{R}_{++} \text{ such that } (0 < \alpha \leq \overline{\alpha}) \Rightarrow (f(\hat{x} + \alpha \Delta x) < f(\hat{x})).$$

□

x_2

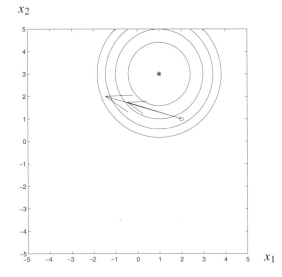

Fig. 10.1. Descent direction (shown as the longer arrow) for a function at a point $\hat{x} = \begin{bmatrix} 2 \\ 1 \end{bmatrix}$, shown as a o. The contours of the function decrease towards $x^\star = \begin{bmatrix} 1 \\ 3 \end{bmatrix}$, which is shown as a •.

That is, Δx is a descent direction for f at $\hat{x}$ if the objective is smaller than $f(\hat{x})$ at points along the line segment $\hat{x} + \alpha \Delta x$ for $\alpha > 0$ and $\alpha \leq \overline{\alpha}$.

10.1.1.2 Example

The situation is illustrated in Figure 10.1, which shows contour sets of the function $f : \mathbb{R}^2 \to \mathbb{R}$ defined by:

$$\forall x \in \mathbb{R}^2, \ f(x) = (x_1 - 1)^2 + (x_2 - 3)^2, \tag{10.1}$$

which we first saw in (2.10) in Section 2.3.1 and was first illustrated in Figure 2.8. The unconstrained minimizer of this function is the point $x^\star = \begin{bmatrix} 1 \\ 3 \end{bmatrix}$, which is illustrated by the • in Figures 2.8 and 10.1.

Superimposed on the contour plot is a point $\hat{x} = \begin{bmatrix} 2 \\ 1 \end{bmatrix}$ shown as a o. At this point are tails of two arrows. The longer arrow illustrates a descent direction Δx at $\hat{x}$, while the shorter arrow shows the vector $\overline{\alpha} \Delta x$, for a value $\overline{\alpha} < 1$. The head of the arrow representing Δx lines on a contour of the function that is higher than the contour associated with $\hat{x}$. The head of the arrow representing $\overline{\alpha} \Delta x$ lies on a contour of the function that is lower than the contour associated with $\hat{x}$. That is, $f(\hat{x} + \overline{\alpha} \Delta x) < f(\hat{x}) < f(\hat{x} + \Delta x)$. Moreover, all points of the form $\hat{x} + \alpha \Delta x$ for $0 < \alpha \leq \overline{\alpha}$ also lie on contours of lower value than that of $\hat{x}$. Note that Δx is a descent direction even though $f(\hat{x} + \Delta x) > f(\hat{x})$. The requirement in the definition is that $f(\hat{x} + \alpha \Delta x) < f(\hat{x})$ for all sufficiently small α.

By definition of descent direction, since there exists Δx that is a descent direction for f at $\hat{x}$ in Figure 10.1, then $\hat{x}$ cannot be a local minimizer of f. As illustrated in

x_2

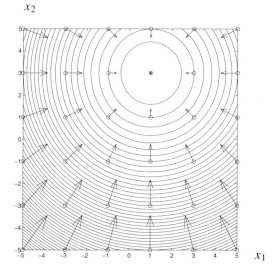

Fig. 10.2. Descent directions for a function at various points. The contours of the function decrease towards $x^\star = \begin{bmatrix} 1 \\ 3 \end{bmatrix}$, which is shown as a •.

Figure 10.1, by moving from $\hat{x}$ in the direction of Δx, we can move to points with lower function values. That is, $\hat{x}$ is not a local minimizer of f.

If we cannot find a descent direction, we might hope that this would indicate that we have found a local minimizer. Unfortunately, this is not true in general as Exercise 10.1 shows. In the special case of a convex function, however, there is a descent direction at a point if and only if the point is not a local minimizer. (See Exercise 10.2.)

Figure 10.2 repeats the contour plot of Figure 10.1 and includes more contours. Superimposed on the contour plot are a collection of points denoted by ◦. At each of these points there is the tail of an arrow. Each arrow is pointed in a descent direction for the function at the point corresponding to the tail of the arrow. In particular, each arrow points in the direction opposite to $\nabla f(x)$ at the point x. Moving from any such point x in the descent direction will reduce the value of the objective. Moreover, the length of each arrow is chosen so that the value $\overline{\alpha} = 1$ will satisfy the definition of the descent direction. At the minimizer of the function, that is, at $x^\star = \begin{bmatrix} 1 \\ 3 \end{bmatrix}$, which is indicated by a • in Figure 10.2, there is no descent direction.

10.1.1.3 Steepest descent step direction

The step direction $\Delta x = -\nabla f(x)$ is called the direction of **steepest descent** because it involves moving in the direction from x in which f is reducing at the greatest rate locally around x. (See Exercise 14.15 in Chapter 14 for details.)

The arrows in Figure 10.2 are proportional to the steepest descent step directions $\Delta x = -\nabla f(x)$ at each point x shown as a $\circ$.

Because we chose a function with circular contour sets, at each point x, the direction $-\nabla f$ points directly to the minimizer and is perpendicular to the contour set at x. We will see that in general the steepest descent step direction does not point directly towards the minimizer.

10.1.1.4 Analysis

We now develop conditions to characterize descent directions. As suggested by Figure 10.2, the conditions involve $\nabla f(x)$.

Lemma 10.1 *Let $f : \mathbb{R}^n \to \mathbb{R}$ be partially differentiable with continuous partial derivatives and let $\hat{x} \in \mathbb{R}^n$, $\Delta x \in \mathbb{R}^n$. Suppose that $\nabla f(\hat{x})^\dagger \Delta x < 0$. Then Δx is a descent direction for f at $\hat{x}$.*

Proof Let $\phi : \mathbb{R} \to \mathbb{R}$ be defined by:

$$\forall t \in \mathbb{R}, \phi(t) = f(\hat{x} + t\Delta x).$$

By the chain rule, $\dfrac{d\phi}{dt}(t) = \dfrac{\partial f}{\partial x}(\hat{x} + t\Delta x)\Delta x$. Evaluating this at $t = 0$ yields:

$$\begin{aligned}
\frac{d\phi}{dt}(0) &= \frac{\partial f}{\partial x}(\hat{x})\Delta x, \\
&= \nabla f(\hat{x})^\dagger \Delta x, \\
&= -2\epsilon,
\end{aligned}$$

say, where $\epsilon > 0$ by assumption.

But, by definition, since f is partially differentiable with continuous partial derivatives,

$$\frac{d\phi}{dt}(0) = \lim_{\alpha \to 0} \frac{f(\hat{x} + \alpha\Delta x) - f(\hat{x})}{\alpha}.$$

(See Definition A.37 and Exercise A.9 in Appendix A.) Let $\bar{\alpha} \in \mathbb{R}_{++}$ be small enough such that

$$(0 < |\alpha| \leq \bar{\alpha}) \Rightarrow \left(\left| \frac{f(\hat{x} + \alpha\Delta x) - f(\hat{x})}{\alpha} - \frac{d\phi}{dt}(0) \right| \leq \epsilon \right).$$

But this means that:

$$(0 < |\alpha| \leq \bar{\alpha}) \Rightarrow \left(\left| \frac{f(\hat{x} + \alpha\Delta x) - f(\hat{x})}{\alpha} - (-2\epsilon) \right| \leq \epsilon \right),$$

which implies that:

$$(0 < |\alpha| \leq \bar{\alpha}) \Rightarrow \left(\frac{f(\hat{x} + \alpha\Delta x) - f(\hat{x})}{\alpha} \leq -\epsilon \right).$$

x_2

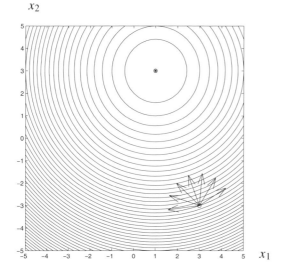

Fig. 10.3. Various descent directions for a function a particular point $\hat{x} \;=\; \begin{bmatrix} 3 \\ -3 \end{bmatrix}$. The contours decrease towards the point $x^\star = \begin{bmatrix} 1 \\ 3 \end{bmatrix}$, which is shown as a •.

(See Exercise 10.3.) So:

$$(0 < \alpha \le \overline{\alpha}) \quad \Rightarrow \quad (f(\hat{x} + \alpha \Delta x) - f(\hat{x}) \le -\alpha \epsilon < 0),$$
$$\Rightarrow \quad (f(\hat{x} + \alpha \Delta x) < f(\hat{x})),$$

and Δx is a descent direction for f at $\hat{x}$. □

Recall that $\nabla f(\hat{x})^\dagger \Delta x$ is called the **directional derivative of f at $\hat{x}$ in the direction** Δx. (See Definition A.37.) Analytically, the condition in Lemma 10.1 that $\nabla f(\hat{x})^\dagger \Delta x < 0$ requires that the directional derivative in the direction Δx be negative. Geometrically, this condition requires that the angle between Δx and $-\nabla f(\hat{x})$ be less than 90° for Δx to be a descent direction. This is illustrated in Figure 10.3 for descent directions from the point $\hat{x} = \begin{bmatrix} 3 \\ -3 \end{bmatrix}$, which is illustrated with a ∘, and for the same function as shown in Figure 10.2. There are seven descent directions illustrated at the point $\hat{x}$.

One of the descent directions illustrated in Figure 10.3, the "middle" one, is the direction $-\nabla f(\hat{x})$. All of the descent directions make an angle of less than 90° with the direction $-\nabla f(\hat{x})$. For example, the direction $-\nabla f(\hat{x})$ makes an angle of 0° with the direction $-\nabla f(\hat{x})$.

Because of the circular contour sets of the function illustrated in Figure 10.3, the steepest descent step direction points directly towards the minimizer and is perpendicular to the contour set at $\hat{x}$. Again, it is to be remembered that this is due to the special case of circular contour sets. We will investigate the more general (and more typical) case in Section 10.2.

We have the following:

Corollary 10.2 *Let $\hat{x} \in \mathbb{R}^n$, let $M \in \mathbb{R}^{n \times n}$ be positive definite, and let $f : \mathbb{R}^n \to \mathbb{R}$ be partially differentiable with continuous partial derivatives and such that $\nabla f(\hat{x}) \neq 0$. Then $\Delta x = -M \nabla f(\hat{x})$ is a descent direction for f at $\hat{x}$.*

Proof Note that $\nabla f(\hat{x})^{\dagger} \Delta x = -\nabla f(\hat{x})^{\dagger} M \nabla f(\hat{x}) < 0$, since M is positive definite and $\nabla f(\hat{x}) \neq 0$. Apply Lemma 10.1. $\square$

To interpret the direction Δx, let us first suppose that $M = \mathbf{I}$, yielding $\Delta x = -\nabla f(\hat{x})$, the steepest descent step direction. Consider a move away from $\hat{x}$ in the direction specified by $\Delta x = -M \nabla f(\hat{x}) = -\nabla f(\hat{x})$. Then if $\dfrac{\partial f}{\partial x_k}(\hat{x}) > 0$ notice that $\Delta x_k = -\dfrac{\partial f}{\partial x_k}(\hat{x}) < 0$, so that we move away from $\hat{x}$ so as to reduce the k-th coordinate of $\hat{x}$ and reduce f, at least for small steps in this direction. Similarly, if $\dfrac{\partial f}{\partial x_k}(\hat{x}) < 0$ then $\Delta x_k > 0$ and we move away from $\hat{x}$ so as to increase the k-th coordinate of $\hat{x}$ and reduce f, at least for small steps in this direction.

We can again use Figure 10.3 to help with interpretation. As discussed above, this figure shows several descent directions at the point $\hat{x} = \begin{bmatrix} 3 \\ -3 \end{bmatrix}$. Also as mentioned above, the "middle" arrow in Figure 10.3 shows the steepest descent step direction at $\hat{x}$, corresponding to the choice $M = \mathbf{I}$. The other directions correspond to other choices of positive definite M and also yield descent directions in that f is also reducing in these directions away from $\hat{x}$. In fact, this allows us to characterize the minimum of an unconstrained problem as the first theorem in Section 10.1.2 shows. Moreover, as we will see in Section 10.2, other choices of positive definite M will turn out to be more desirable than $M = \mathbf{I}$ when the contour sets are not circular.

10.1.2 First-order conditions

In this section, we discuss optimality conditions that are based on the first derivative of the objective function. We call these **first-order conditions**.

10.1.2.1 Necessary conditions

In this section we discuss necessary conditions involving the first derivative of f.

Theorem 10.3 *Let $f : \mathbb{R}^n \to \mathbb{R}$ be partially differentiable with continuous partial derivatives. If $x^{\star}$ is a local minimizer of f then $\nabla f(x^{\star}) = 0$.*

Proof We prove the contra-positive. That is, we prove that if $\nabla f(x^\star) \neq \mathbf{0}$ then $x^\star$ is not a local minimizer. Let $M \in \mathbb{R}^{n \times n}$ be positive definite. By Corollary 10.2, $\Delta x = -M \nabla f(x^\star)$ is a descent direction for f at $x^\star$ and so $x^\star$ is not a local minimizer of f. $\square$

The conditions $\nabla f(x^\star) = \mathbf{0}$ are often referred to as *the* **first-order necessary conditions**, abbreviated **FONC**. However, strictly speaking, the first-order necessary conditions include the rest of the hypothesis of Theorem 10.3.

Inspection of Figure 10.2 shows that $x^\star = \begin{bmatrix} 1 \\ 3 \end{bmatrix}$ is the (unique) local minimizer of the function illustrated. From (10.1) we have that $\nabla f(x) = \begin{bmatrix} 2(x_1 - 1) \\ 2(x_2 - 3) \end{bmatrix}$.

Therefore, $\nabla f(x^\star) = \mathbf{0}$, which is exactly as Theorem 10.3 claims.

The statement and proof of Theorem 10.3, respectively, suggest two approaches to finding a minimizer of f:

(i) solve $\nabla f(x) = \mathbf{0}$, or
(ii) from the current point x, move in the direction $\Delta x = -M \nabla f(x)$, where M is positive definite.

We will see that *both* these ideas should be combined to produce an effective algorithm.

10.1.2.2 Example of insufficiency

Unfortunately, as we noted, the condition $\nabla f(x) = \mathbf{0}$ is not sufficient to guarantee a minimum. We call points that satisfy $\nabla f(x) = \mathbf{0}$ **critical points**. Not all critical points are minimizers. Consider the function $f : \mathbb{R} \to \mathbb{R}$ shown in Figure 10.4 and the derivative of this function shown in Figure 10.5. The function shown in Figure 10.4 has three critical points, illustrated with $\circ$, corresponding to the zeros of the derivative function shown in Figure 10.5:

(i) $\hat{x} = -3$, $f(\hat{x}) = 8$, a local maximizer and maximum of f, respectively,
(ii) $\hat{x} = 0$, $f(\hat{x}) = 0$, a **horizontal inflection point** of f, and
(iii) $x^\star = 3$, $f(x^\star) = -8$, a local minimizer and minimum of f, respectively.

Clearly, the first-order necessary conditions $\nabla f(x) = \mathbf{0}$ are not sufficient to guarantee that we are at a minimum.

10.1.3 Second-order conditions

If f is twice partially differentiable with continuous second partial derivatives, then we can consider using the second partial derivatives to help to determine if we are at a minimizer. Recall that we call the matrix of second partial derivatives

$f(x)$

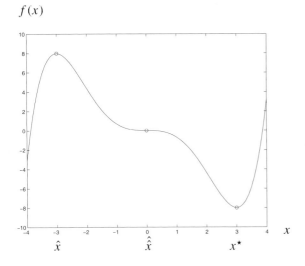

Fig. 10.4. Graph of f and points (illustrated by the ∘) satisfying $\nabla f(x) = \mathbf{0}$ but which may or may not correspond to a minimum.

$\nabla f(x)$

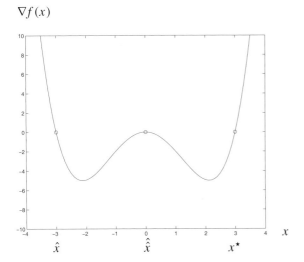

Fig. 10.5. The first derivative ∇f of the function f shown in Figure 10.4.

the **Hessian**. (See Definition A.38.) Optimality conditions based on second derivatives are called **second-order conditions**. We first discuss second-order necessary conditions and then second-order sufficient conditions.

10.1.3.1 Necessary conditions

Analysis The following theorem gives **second-order necessary conditions**, abbreviated **SONC**.

$\nabla^2 f(x)$

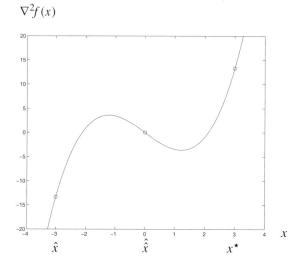

Fig. 10.6. The second derivative $\nabla^2 f$ of the function f shown in Figure 10.4.

Theorem 10.4 *Let $f : \mathbb{R}^n \to \mathbb{R}$ be twice partially differentiable with continuous second partial derivatives and suppose that $x^\star$ is a local minimizer of f. Then:*

$$\nabla f(x^\star) \;=\; \mathbf{0}, \tag{10.2}$$
$$\nabla^2 f(x^\star) \text{ is positive semi-definite.} \tag{10.3}$$

Proof See [70, section 6.1][84, section 10.2]. □

Example To illustrate application of this theorem, again consider the function f shown in Figure 10.4. Its first and second derivatives are shown in Figures 10.5 and 10.6, respectively. Since $f : \mathbb{R} \to \mathbb{R}$ in this case, the Hessian $\nabla^2 f : \mathbb{R} \to \mathbb{R}$ is positive semi-definite if and only if it is non-negative.

The critical points of f are at:

$\hat{x} = -3$. At this point, the Hessian of f, shown in Figure 10.6, is negative and hence not positive semi-definite. Therefore, by Theorem 10.4, $\hat{x} = -3$ cannot be a local minimizer of f. We will discuss how to avoid critical points that are not minimizers of f in Section 10.2.6.

$\hat{\hat{x}} = 0$. At this point, the Hessian of f is zero and hence positive semi-definite. The second-order necessary conditions are satisfied but by inspection of Figure 10.4, $\hat{\hat{x}} = 0$ is clearly not a minimizer. This example shows that in general even the second-order necessary conditions alone are insufficient to guarantee that a candidate point is a minimizer.

To identify whether or not $\hat{\hat{x}} = 0$ is a minimizer using derivative information at

$\hat{x} = 0$ alone requires *higher-order* derivative information at this point. Unfortunately, such information is usually difficult to obtain unless the function is known analytically. Therefore, objectives with horizontal inflection points are difficult to handle because first and second derivatives alone do not characterize minimizers. We will discuss the case of horizontal inflection points further in Section 10.2.6.

$x^\star = 3$. This point is a local minimizer of f. Figure 10.6 and Theorem 10.4 both concur that the Hessian is positive semi-definite.

10.1.3.2 Sufficient conditions

Analysis In the following theorem, we discuss **second-order sufficient conditions** (or **SOSC**) for a point to be a local minimizer of a twice partially differentiable function with continuous second partial derivatives.

Theorem 10.5 *Let $f : \mathbb{R}^n \to \mathbb{R}$ be twice partially differentiable with continuous second partial derivatives and suppose that:*

$$\nabla f(x^\star) = \mathbf{0},$$
$$\nabla^2 f(x^\star) \text{ is positive definite.}$$

Then $x^\star$ is a strict local minimizer of f.

Proof By hypothesis, $\nabla^2 f(x^\star)$ is positive definite and $\nabla^2 f$ is continuous. Therefore, by Exercise 5.19:

$$\exists \epsilon \in \mathbb{R}_{++} \text{ such that } \left(\|x^\star - x\| \leq \epsilon\right) \Rightarrow (\nabla^2 f(x) \text{ is positive definite}). \qquad (10.4)$$

Let Δx be any step direction such that $0 < \|\Delta x\| \leq \epsilon$ and define $\phi : \mathbb{R} \to \mathbb{R}$ by

$$\forall t \in \mathbb{R}, \phi(t) = f(x^\star + t\Delta x).$$

Then:

$$\frac{d\phi}{dt}(t) = \frac{\partial f}{\partial x}(x^\star + t\Delta x)\Delta x,$$

$$\frac{d\phi}{dt}(0) = \frac{\partial f}{\partial x}(x^\star)\Delta x,$$

$$= \nabla f(x^\star)^\dagger \Delta x,$$

$$= \mathbf{0}, \text{ by hypothesis}, \qquad (10.5)$$

$$\frac{d^2\phi}{dt^2}(t) = \Delta x^\dagger \frac{\partial^2 f}{\partial x^2}(x^\star + t\Delta x)\Delta x,$$

$$> 0, \qquad (10.6)$$

where the last inequality follows from (10.4) since $\Delta x \neq 0$ and since:

$$(0 < t \leq 1) \Rightarrow \left(\|x^\star - (x^\star + t\Delta x)\| = t\|\Delta x\| \leq \|\Delta x\| \leq \epsilon\right).$$

We have that $\phi(0) = f(x^\star)$ and:

$$\forall \Delta x \in \mathbb{R}^n, (0 < \|\Delta x\| \le \epsilon) \Rightarrow$$
$$f(x^\star + \Delta x) = \phi(1),$$

$$= \phi(0) + \int_{t=0}^1 \frac{d\phi}{dt}(t)\,dt,$$

by the fundamental theorem of integral calculus applied to ϕ,
(see Theorem A.2 in Section A.4.4.1 of Appendix A),

$$= \phi(0) + \int_{t=0}^1 \left[\frac{d\phi}{dt}(0) + \int_{t'=0}^t \frac{d^2\phi}{dt^2}(t')\,dt' \right]dt,$$

by the fundamental theorem of integral calculus applied to $\dfrac{d\phi}{dt}$,
(see Theorem A.2 in Section A.4.4.1 of Appendix A),

$$= \phi(0) + \frac{d\phi}{dt}(0) + \int_{t=0}^1 \int_{t'=0}^t \frac{d^2\phi}{dt^2}(t')\,dt'\,dt,$$

evaluating the integral of the first term in the integrand,

$$= \phi(0) + \int_{t=0}^1 \int_{t'=0}^t \frac{d^2\phi}{dt^2}(t')\,dt'\,dt, \text{ by (10.5)},$$

$$> f(x^\star), \text{ since the integrand is strictly positive at } t \ne 0 \text{ by (10.6)},$$
(see Theorem A.3 in Section A.4.4.2 of Appendix A).

That is, $x^\star$ is a strict local minimizer. $\square$

Example Continuing with the example from Section 10.1.1.2, note that:

$$\forall x \in \mathbb{R}^2, \nabla^2 f(x) = \begin{bmatrix} 2 & 0 \\ 0 & 2 \end{bmatrix},$$

which is positive definite. Therefore, by Theorem 10.5, the point $x^\star = \begin{bmatrix} 1 \\ 3 \end{bmatrix}$ is a strict local minimizer of f.

Example of insufficiency We emphasize that positive *semi*-definiteness of the second derivative matrix at a critical point $\hat{x}$ is *not* sufficient to guarantee that $\hat{x}$ is a minimizer. For example, consider the function $f : \mathbb{R} \to \mathbb{R}$ defined by:

$$\forall x \in \mathbb{R}, f(x) = -(x)^4,$$

and the point $\hat{x} = [0]$ as illustrated in Figure 10.7. In this case $\nabla f(\hat{x}) = [-4(\hat{x})^3] = [0]$ and $\nabla^2 f(\hat{x}) = [-12(\hat{x})^2] = [0]$ so that:

$$\forall \Delta x \in \mathbb{R}, 0 = \Delta x \nabla^2 f(\hat{x}) \Delta x \ge 0,$$

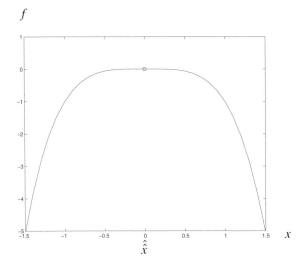

f

Fig. 10.7. A critical point $\hat{\hat{x}} = 0$, illustrated by the ○, where the second derivative matrix is positive semi-definite at $\hat{\hat{x}}$ yet the point is not a minimizer.

and so $\nabla^2 f(\hat{\hat{x}})$ is positive semi-definite. However, $\hat{\hat{x}} = [0]$ is clearly not a minimizer of f. The second derivative $\nabla^2 f$ is not even positive semi-definite for $x \neq [0]$. In the next section we will consider the convex case where f is positive semi-definite everywhere.

10.1.4 Convex objectives

10.1.4.1 First-order sufficient conditions

Analysis If f is twice partially differentiable with continuous partial derivatives and the second derivative matrix of f is positive semi-definite *everywhere* then the objective is convex by Theorem 2.7. In the case of a convex objective, the first-order conditions are necessary and sufficient as the following corollary to Theorem 2.6 shows.

Corollary 10.6 *Let $f : \mathbb{R}^n \to \mathbb{R}$ be convex and partially differentiable with continuous partial derivatives on $\mathbb{R}^n$ and let $x^\star \in \mathbb{R}^n$. If $\nabla f(x^\star) = \mathbf{0}$ then $f(x^\star)$ is the global minimum and $x^\star$ is a global minimizer of f.*

Proof Recall Theorem 2.6. The hypothesis of Theorem 2.6 is satisfied for $\mathbb{S} = \mathbb{R}^n$. Consequently, (2.31) holds, which we repeat:

$$\forall x, x' \in \mathbb{S}, \ f(x) \geq f(x') + \nabla f(x')^\dagger (x - x').$$

Letting $x' = x^\star$ and $\mathbb{S} = \mathbb{R}^n$ in (2.31) and noting that $\nabla f(x^\star) = \mathbf{0}$, we obtain:

$$\forall x \in \mathbb{R}^n, \ f(x) \geq f(x^\star).$$

That is $x^\star$ is a global minimizer of f. □

Summarizing previous results, if f is convex and partially differentiable with continuous partial derivatives then by Theorem 10.3 and Corollary 10.6, $\nabla f(x^\star) = \mathbf{0}$ is a necessary and sufficient condition for global optimality. Moreover, as Exercise 10.4 exemplifies, if f is convex and partially differentiable with continuous partial derivatives but no point satisfies the equations $\nabla f(x) = \mathbf{0}$ then there is no minimizer.

Example Continuing with the example from Sections 10.1.1.2 and 10.1.3.2, note that $\nabla^2 f$ is positive definite so that f is convex. Therefore, by Corollary 10.6, the point $x^\star = \begin{bmatrix} 1 \\ 3 \end{bmatrix}$ is the global minimizer of f.

10.1.4.2 Uniqueness of minimizer

A related result stems from Theorem 2.2. Recall that the Hessian of f is the same as the Jacobian of ∇f. Therefore, by Theorems 2.3 and 2.2, if the Hessian of f is positive definite on $\mathbb{R}^n$ then there can be at most one solution of $\nabla f(x) = \mathbf{0}$. Alternatively, by Theorem 2.7, f is strictly convex and so by Item (iii) of the conclusion of Theorem 2.4, f has a unique minimizer. That is, we have the following.

Theorem 10.7 *Let $f : \mathbb{R}^n \to \mathbb{R}$ be twice partially differentiable with continuous second partial derivatives on $\mathbb{R}^n$. If $\nabla^2 f$ is positive definite throughout $\mathbb{R}^n$ and $\min_{x \in \mathbb{R}^n} f(x)$ possesses a minimum then the associated minimizer is unique.*

Proof Applying Theorems 2.3 and 2.2 to ∇f we find that there is at most one point that satisfies the necessary conditions for minimizing f. Alternatively, Theorem 2.7 and Item (iii) of the conclusion of Theorem 2.4 imply the same result. □

10.2 Approaches to finding minimizers

In Sections 10.2.1 and 10.2.2 we will discuss two approaches to finding minimizers based on, respectively, the proof and the statement of Theorem 10.3, which characterized the first-order necessary conditions. We will discuss the advantages and disadvantages of each and then combine the approaches into an effective algorithm in Section 10.2.3. Initially we will ignore the selection of step-sizes but we will return to this issue in Section 10.2.4. Stopping criteria are discussed in Section 10.2.5. In Section 10.2.6 we discuss how to avoid critical points that are not minimizers. Most of the development is based on [70, chapters 7 and 9].

10.2.1 Steepest descent

Using the steepest descent step direction, we update our iterate according to:

$$x^{(\nu+1)} = x^{(\nu)} - \alpha^{(\nu)} \nabla f(x^{(\nu)}). \tag{10.7}$$

Geometrically, the steepest descent step direction at $x^{(\nu)}$ is perpendicular to the surface of the contour set $\mathbb{C}_f(f(x^{(\nu)}))$ as illustrated in Figure 10.2, where the descent directions illustrated are the steepest descent step directions at each point. (See Exercise 10.6.)

In addition to the step direction, we must also choose the step-size $\alpha^{(\nu)}$. We will discuss some of the issues in selecting a step-size in Section 10.2.4; however, in this section we assume that we know how to pick $\alpha^{(\nu)}$ to reduce the objective. That is, we assume we can find $\bar{\alpha}$ satisfying Definition 10.1.

10.2.1.1 Advantages

The main advantage of the steepest descent step direction $\Delta x^{(\nu)} = -\nabla f(x^{(\nu)})$ is that, unless $\nabla f(x^{(\nu)}) = 0$, it is always possible to find a step-size $\alpha^{(\nu)}$ such that the objective will be reduced from $f(x^{(\nu)})$ by updating the iterate to $x^{(\nu)} - \alpha^{(\nu)} \nabla f(x^{(\nu)})$. Recall that in the absence of further assumptions, a point x satisfying $\nabla f(x) = 0$ may be a minimum, a maximum, or a point of inflection. If f is convex, then Corollary 10.6 shows that if $\nabla f(x^{(\nu)}) = 0$ then $x^{(\nu)}$ is a minimizer of f.

10.2.1.2 Example

Consider the quadratic function illustrated in Figure 10.2 and defined in (10.1). We have: $\nabla f(x) = \begin{bmatrix} 2(x_1 - 1) \\ 2(x_2 - 3) \end{bmatrix}$. Suppose that we use $x^{(0)} = \begin{bmatrix} 3 \\ -5 \end{bmatrix}$ as the initial guess. Then $\nabla f(x^{(0)}) = \begin{bmatrix} 2(3 - 1) \\ 2(-5 - 3) \end{bmatrix} = \begin{bmatrix} 4 \\ -16 \end{bmatrix}$ and the steepest descent step direction at $x^{(0)}$ is $\Delta x^{(0)} = \begin{bmatrix} -4 \\ 16 \end{bmatrix}$. We update according to:

$$x^{(1)} = x^{(0)} + \alpha^{(0)} \Delta x^{(0)} = \begin{bmatrix} 3 \\ -5 \end{bmatrix} + \alpha^{(0)} \begin{bmatrix} -4 \\ 16 \end{bmatrix}.$$

If we were fortunate enough to pick the step-size of $\alpha^{(0)} = 0.5$ then we would find that $x^{(2)} = x^\star = \begin{bmatrix} 1 \\ 3 \end{bmatrix}$ so that we would have reached the minimizer in one iteration.

10.2.1.3 Disadvantages

Besides the difficulty of choosing the step-size in the update, the main disadvantage of the steepest descent step direction is that progress towards the solution may be

Algorithms for unconstrained minimization

x_2

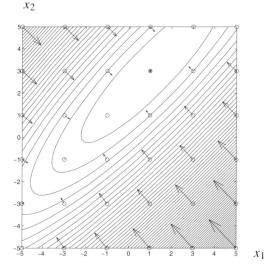

Fig. 10.8. The steepest descent step directions for an objective, defined in (10.8), with contour sets that are highly eccentric ellipses. The contours of the function decrease towards $x^\star = \begin{bmatrix} 1 \\ 3 \end{bmatrix}$, which is shown as a ●.

very slow if the contour sets of the function are very "eccentric." In Figure 10.2, the level sets are circular and so the steepest descent step direction points directly to the global minimizer, $x^\star$. However, as remarked, this is not the typical case in practice.

10.2.1.4 Example

Figure 10.8 shows the steepest descent step directions for a quadratic function $f : \mathbb{R}^2 \to \mathbb{R}$ defined by:

$$\forall x \in \mathbb{R}^2, \ f(x) = (x_1 - 1)^2 + (x_2 - 3)^2 - 1.8(x_1 - 1)(x_2 - 3), \qquad (10.8)$$

which we first met in Section 2.6.3.2. This function is of the form:

$$\forall x \in \mathbb{R}^2, \ f(x) = \frac{1}{2}x^\dagger Q x + c^\dagger x,$$

where:

$$
\begin{aligned}
Q &= \nabla^2 f(x), \\
&= \begin{bmatrix} 2 & -1.8 \\ -1.8 & 2 \end{bmatrix}, \\
c &= \begin{bmatrix} 3.4 \\ -4.2 \end{bmatrix}.
\end{aligned}
$$

This function has the same minimizer, $x^\star = \begin{bmatrix} 1 \\ 3 \end{bmatrix}$, as the function in Figure 10.2, but has eccentric contour sets. This function is more typical of functions encountered in practice. In this case, Figure 10.8 shows that the steepest descent step directions do not point towards $x^\star$.

For a step-size of $\alpha^{(\nu)}$, the next iterate has objective value given by $f\left(x^{(\nu+1)}\right) = f\left(x^{(\nu)} - \alpha^{(\nu)} \nabla f(x^{(\nu)})\right)$. Even if we were able to choose $\alpha^{(\nu)}$ at each iteration to minimize $f\left(x^{(\nu)} - \alpha^{(\nu)} \nabla f(x^{(\nu)})\right)$ *exactly* with respect to $\alpha^{(\nu)}$, it can take many iterations to find the minimum of a quadratic function having eccentric contour sets. The iterates will "zig-zag" back and forth across the axes of the eccentric contour sets, making slow progress towards $x^\star$. We will see in Section 10.2.1.5 that non-quadratic functions with eccentric contour sets will exhibit similarly poor behavior using the steepest descent step direction.

Using the function defined in (10.8), we obtain:

$$\forall x \in \mathbb{R}^2, \nabla f(x) = \begin{bmatrix} 2(x_1 - 1) - 1.8(x_2 - 3) \\ 2(x_2 - 3) - 1.8(x_1 - 1) \end{bmatrix}.$$

Again, suppose that we use $x^{(0)} = \begin{bmatrix} 3 \\ -5 \end{bmatrix}$ as the initial guess. Then:

$$\nabla f(x^{(0)}) = \begin{bmatrix} 2(3 - 1) - 1.8(-5 - 3) \\ 2(-5 - 3) - 1.8(3 - 1) \end{bmatrix} = \begin{bmatrix} 18.4 \\ -19.6 \end{bmatrix},$$

and the steepest descent step direction at $x^{(0)}$ is $\Delta x^{(0)} = \begin{bmatrix} -18.4 \\ 19.6 \end{bmatrix}$. This is consistent with the direction of the arrow at the point $x^{(0)} = \begin{bmatrix} 3 \\ -5 \end{bmatrix}$ in Figure 10.8. We update according to:

$$x^{(1)} = x^{(0)} + \alpha^{(0)} \Delta x^{(0)} = \begin{bmatrix} 3 \\ -5 \end{bmatrix} + \alpha^{(0)} \begin{bmatrix} -18.4 \\ 19.6 \end{bmatrix}.$$

If we find the value of $\alpha^{(0)}$ that minimizes $f(x^{(0)} + \alpha^{(0)} \Delta x^{(0)})$ over choices of $\alpha^{(0)}$, we would find that $x^{(1)} \approx \begin{bmatrix} -1.8467 \\ 0.1628 \end{bmatrix}$, which is relatively far from the minimizer of f.

Figure 10.9 illustrates the progress of iterations using steepest descent step direction, starting at $x^{(0)} = \begin{bmatrix} 3 \\ -5 \end{bmatrix}$, and assuming that at the ν-th iteration the step-size $\alpha^{(\nu)}$ were chosen to minimize $f(x^{(\nu)} + \alpha \Delta x^{(\nu)})$. Figure 10.9 shows that after two iterations of steepest descent we are close to the minimizer of this function. However, starting at $x^{(0)} = \begin{bmatrix} -2 \\ -5 \end{bmatrix}$, the progress is much slower, as illustrated in Figure 10.10, requiring six steepest descent step directions to get close to the minimizer.

In higher dimensions, with n larger than 2, the steepest descent algorithm will

x_2

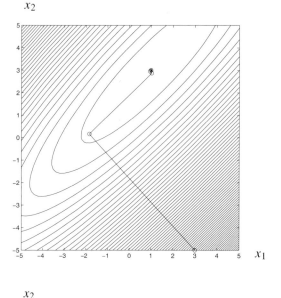

Fig. 10.9. Progress of iterations, shown as ○, using steepest descent step directions for an objective, defined in (10.8), with contour sets that are highly eccentric ellipses. The contours of the function decrease towards $x^\star = \begin{bmatrix} 1 \\ 3 \end{bmatrix}$, which is shown as a ●. The initial guess was $x^{(0)} = \begin{bmatrix} 3 \\ -5 \end{bmatrix}$.

x_2

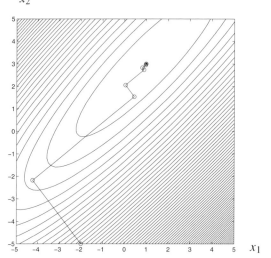

Fig. 10.10. Progress of iterations, shown as ○, using steepest descent step directions for an objective, defined in (10.8), with contour sets that are highly eccentric ellipses. The contours of the function decrease towards $x^\star = \begin{bmatrix} 1 \\ 3 \end{bmatrix}$, which is shown as a ●. The initial guess was $x^{(0)} = \begin{bmatrix} -2 \\ -5 \end{bmatrix}$.

repeatedly take us in directions that do not point directly towards the minimizer. (See Exercise 10.9.) The steepest descent step direction can be arbitrarily close to being at *right angles* to the direction that points towards the minimizer. Moreover, we cannot expect to exactly minimize $f(x^{(\nu)} + \alpha^{(\nu)} \Delta x^{(\nu)})$ over choices of $\alpha^{(\nu)}$ as assumed in Figures 10.9 and 10.10. This typically increases further the number of iterations required to find a useful answer. Finally, if the objective is not quadratic then the performance can be even worse as shown in the following example.

x_2

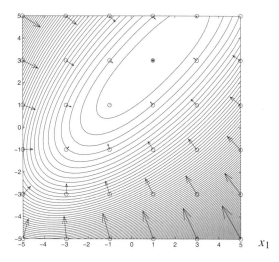

Fig. 10.11. The steepest descent step directions for an objective, defined in (10.9), with contour sets that are perturbed eccentric ellipses. The contours of the function decrease towards $x^\star = \begin{bmatrix} 1 \\ 3 \end{bmatrix}$, which is shown as a •.

10.2.1.5 Example with non-quadratic objective

Consider the function $f : \mathbb{R}^2 \to \mathbb{R}$ defined by:

$$\forall x \in \mathbb{R}^2, \; f(x) = 0.01(x_1 - 1)^4 + 0.01(x_2 - 3)^4 + (x_1 - 1)^2 + (x_2 - 3)^2$$
$$- 1.8(x_1 - 1)(x_2 - 3). \qquad (10.9)$$

This a perturbation of the quadratic function defined in (10.8). Figure 10.11 shows the steepest descent step directions for this function. The contour sets of this function are perturbed ellipses.

The gradient of this function is:

$$\forall x \in \mathbb{R}^2, \; \nabla f(x) = \begin{bmatrix} 0.04(x_1 - 1)^3 + 2(x_1 - 1) - 1.8(x_2 - 3) \\ 0.04(x_2 - 3)^3 - 1.8(x_1 - 1) + 2(x_2 - 3) \end{bmatrix}.$$

Again, suppose that we use $x^{(0)} = \begin{bmatrix} 3 \\ -5 \end{bmatrix}$ as the initial guess. Then, $\nabla f(x^{(0)}) = \begin{bmatrix} 18.72 \\ -40.08 \end{bmatrix}$ and the steepest descent step direction at $x^{(0)}$ is $\Delta x^{(0)} = \begin{bmatrix} -18.72 \\ 40.08 \end{bmatrix}$.

This is consistent with the direction of the arrow at the point $x^{(0)} = \begin{bmatrix} 3 \\ -5 \end{bmatrix}$ in Figure 10.11. We update according to:

$$x^{(1)} = x^{(0)} + \alpha^{(0)} \Delta x^{(0)} = \begin{bmatrix} 3 \\ -5 \end{bmatrix} + \alpha^{(0)} \begin{bmatrix} -18.72 \\ 40.08 \end{bmatrix}.$$

Figure 10.12 shows the progress of a steepest descent algorithm starting at $x^{(0)} =$

x_2

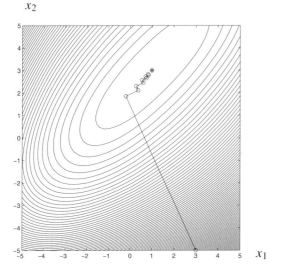

Fig. 10.12. Progress of iterations, shown as o, using steepest descent step directions for an objective, defined in (10.9), with contour sets that are perturbed eccentric ellipses. The contours of the function decrease towards $x^\star = \begin{bmatrix} 1 \\ 3 \end{bmatrix}$, which is shown as a ●. The initial guess was $x^{(0)} = \begin{bmatrix} 3 \\ -5 \end{bmatrix}$.

x_2

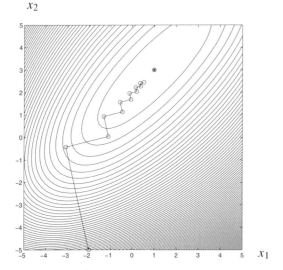

Fig. 10.13. Progress of iterations, shown as o, using steepest descent step directions for an objective, defined in (10.9), with contour sets that are perturbed eccentric ellipses. The contours of the function decrease towards $x^\star = \begin{bmatrix} 1 \\ 3 \end{bmatrix}$, which is shown as a ●. The initial guess was $x^{(0)} = \begin{bmatrix} -2 \\ -5 \end{bmatrix}$.

$\begin{bmatrix} 3 \\ -5 \end{bmatrix}$ and assuming that at the ν-th iteration the step-size $\alpha^{(\nu)}$ were chosen to minimize $f(x^{(\nu)} + \alpha \Delta x^{(\nu)})$. Figure 10.12 shows that many iterations are required to approach the minimizer.

Figure 10.13 shows the progress of a steepest descent algorithm starting at $x^{(0)} = \begin{bmatrix} -2 \\ -5 \end{bmatrix}$, again with the step-size chosen to minimize $f(x^{(\nu)} + \alpha \Delta x^{(\nu)})$ at each iteration. Figure 10.13 shows that the iterates again zig-zag back and forth across the axis of the contour sets and many iterations are required to approach the minimizer.

10.2.2 Solving $\nabla f(x) = 0$

Another approach to minimizing f is based on the observation that $\nabla f(x) = 0$ is a system of either linear or non-linear equations having the same number of equations as variables. We discussed the solution of linear equations in Chapter 5 and the iterative solution of non-linear equations in Chapter 7. We will apply these ideas to the solution of $\nabla f(x) = 0$ in Sections 10.2.2.1 and 10.2.2.2.

10.2.2.1 Linear first-order necessary conditions

Analysis Suppose that $f : \mathbb{R}^n \to \mathbb{R}$ is quadratic of the form:

$$\forall x \in \mathbb{R}^n, \ f(x) = \frac{1}{2}x^\dagger Qx + c^\dagger x,$$

where $Q \in \mathbb{R}^{n \times n}$ and $c \in \mathbb{R}^n$. In this case, the equations $\nabla f(x) = 0$ are linear and of the form $Qx + c = 0$. (See Exercise 10.10.) If the coefficient matrix $Q = \nabla^2 f$ of this linear system is non-singular then we can solve the equations:

$$Qx^\star = -c,$$

by, for example, factorization of Q and forwards and backwards substitution, to find a critical point of f in one step. If the coefficient matrix is singular or ill-conditioned then we can use the QR factorization. (Note that the "Q" factor of the quadratic coefficient matrix Q will, of course, be different to the coefficient matrix Q.) If the coefficient matrix is singular then there will be multiple solutions. (We will explore this case further in Section 11.1.4.)

Example Consider the function $f : \mathbb{R}^2 \to \mathbb{R}$ defined in (10.8):

$$\begin{aligned}
\forall x \in \mathbb{R}^2, \ f(x) &= (x_1 - 1)^2 + (x_2 - 3)^2 - 1.8(x_1 - 1)(x_2 - 3), \\
&= \frac{1}{2}x^\dagger Qx + c^\dagger x,
\end{aligned}$$

where:

$$\begin{aligned}
Q &= \nabla^2 f(x), \\
&= \begin{bmatrix} 2 & -1.8 \\ -1.8 & 2 \end{bmatrix}, \\
c &= \begin{bmatrix} 3.4 \\ -4.2 \end{bmatrix}.
\end{aligned}$$

Solving $Qx^\star = -c$ we obtain the minimizer $x^\star = \begin{bmatrix} 1 \\ 3 \end{bmatrix}$.

10.2.2.2 Non-linear first-order necessary conditions

Analysis For non-linear equations, the basic idea is to apply the Newton–Raphson update to solve $\nabla f(x) = \mathbf{0}$. At the ν-th iteration the Newton–Raphson update to solve $\nabla f(x) = \mathbf{0}$ is given by:

$$\begin{aligned}
\nabla^2 f(x^{(\nu)}) \Delta x^{(\nu)} &= -\nabla f(x^{(\nu)}), \\
x^{(\nu+1)} &= x^{(\nu)} + \alpha^{(\nu)} \Delta x^{(\nu)},
\end{aligned}$$

where $\alpha^{(\nu)}$ is the step-size. The choice of step direction is called the **Newton–Raphson step direction** to minimize f. Initially, we will consider $\alpha^{(\nu)} = 1$ or assume that the step-size has been chosen at the ν-th iteration to minimize $f(x^{(\nu)} + \alpha^{(\nu)} \Delta x^{(\nu)})$. We will discuss the choice of step-size in more detail in Section 10.2.4. The Jacobian of ∇f is the Hessian $\nabla^2 f$ of f, so that the Newton–Raphson step direction involves linearizing ∇f to solve $\nabla f(x) = \mathbf{0}$.

Example with quadratic objective For a quadratic function, the necessary conditions are linear. Nevertheless, we can consider applying the Newton–Raphson update to solve them as though they were non-linear. For a quadratic function $f : \mathbb{R}^n \to \mathbb{R}$ defined by:

$$\forall x \in \mathbb{R}^n, \ f(x) = \frac{1}{2} x^\dagger Q x + c^\dagger x,$$

where $Q \in \mathbb{R}^{n \times n}$ and $c \in \mathbb{R}^n$, the Newton–Raphson step direction is the solution to $Q \Delta x^{(\nu)} = -Q x^{(\nu)} - c$. Using this update with step-size one yields a point satisfying the first-order necessary conditions for minimizing f. (See Exercise 10.10.) That is, at any point, the Newton–Raphson step direction points directly towards the minimizer. Figure 10.14 shows the Newton–Raphson step directions at various points for the function defined in (10.8). They all point towards the minimizer $x^\star = \begin{bmatrix} 1 \\ 3 \end{bmatrix}$.

Example with non-quadratic objective We continue with the function $f : \mathbb{R}^2 \to \mathbb{R}$ from Section 10.2.1.5 defined in (10.9):

$$\forall x \in \mathbb{R}^2, \ f(x) = 0.01(x_1 - 1)^4 + 0.01(x_2 - 3)^4 + (x_1 - 1)^2 + (x_2 - 3)^2$$
$$-1.8(x_1 - 1)(x_2 - 3).$$

The Hessian of this function is:

$$\forall x \in \mathbb{R}^2, \ \nabla^2 f(x) = \begin{bmatrix} 0.12(x_1 - 1)^2 + 2 & -1.8 \\ -1.8 & 0.12(x_2 - 3)^2 + 2 \end{bmatrix}.$$

x_2

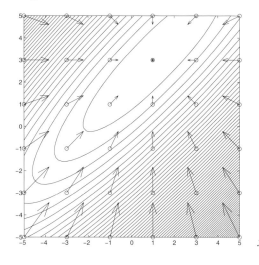

Fig. 10.14. The Newton–Raphson step directions for an objective, defined in (10.8), with contour sets that are highly eccentric ellipses. The contours of the function decrease towards $x^\star = \begin{bmatrix} 1 \\ 3 \end{bmatrix}$, which is shown as a •.

Again, suppose that we use $x^{(0)} = \begin{bmatrix} 3 \\ -5 \end{bmatrix}$ as the initial guess. Then, as in the steepest descent iteration, $\nabla f(x^{(0)}) = \begin{bmatrix} 18.72 \\ -40.08 \end{bmatrix}$. The Newton–Raphson step direction at $x^{(0)}$ is the solution to:

$$\begin{bmatrix} 2.96 & -1.8 \\ -1.8 & 9.68 \end{bmatrix} \Delta x^{(0)} = \begin{bmatrix} -18.72 \\ 40.08 \end{bmatrix}.$$

Solving this, we obtain $\Delta x^{(0)} \approx \begin{bmatrix} -5.250 \\ 3.164 \end{bmatrix}$. We update according to:

$$x^{(1)} = x^{(0)} + \alpha^{(0)} \Delta x^{(0)} = \begin{bmatrix} 3 \\ -5 \end{bmatrix} + \alpha^{(0)} \begin{bmatrix} -5.250 \\ 3.164 \end{bmatrix}.$$

For step-size $\alpha^{(0)} = 1$, we obtain $x^{(1)} = \begin{bmatrix} -2.250 \\ -1.836 \end{bmatrix}$. This value is still relatively far from the minimizer, which is $x^\star = \begin{bmatrix} 1 \\ 3 \end{bmatrix}$.

Figure 10.15 shows the progress of a Newton–Raphson algorithm starting at $x^{(0)} = \begin{bmatrix} 3 \\ -5 \end{bmatrix}$ and assuming that at the ν-th iteration the step-size $\alpha^{(\nu)}$ were chosen to minimize $f(x^{(\nu)} + \alpha \Delta x^{(\nu)})$. Three iterations are required to get close to the minizer. For this function, the Newton–Raphson step direction at the initial guess is not as good as the steepest descent step direction. However, after the first iteration, the progress towards the minimizer is rapid.

Figure 10.16 shows the progress of a Newton–Raphson algorithm starting at

x_2

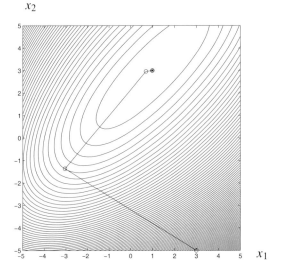

Fig. 10.15. Progress of iterations, shown as ∘, us-
ing Newton–Raphson step
directions for an objective,
defined in (10.9), with contour sets that are perturbed
eccentric ellipses.   The
contours of the function
decrease towards $x^\star$ $=$
$\begin{bmatrix} 1 \\ 3 \end{bmatrix}$, which is shown as a
•. The initial guess was
$x^{(0)} = \begin{bmatrix} 3 \\ -5 \end{bmatrix}$.

x_2

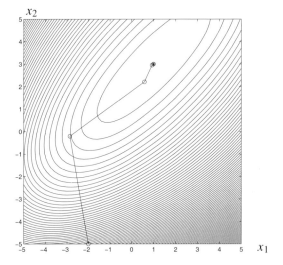

Fig. 10.16. Progress of iterations, shown as ∘, us-
ing Newton–Raphson step
directions for an objective,
defined in (10.9), with contour sets that are perturbed
eccentric ellipses.   The
contours of the function
decrease towards $x^\star$ $=$
$\begin{bmatrix} 1 \\ 3 \end{bmatrix}$, which is shown as a
•. The initial guess was
$x^{(0)} = \begin{bmatrix} -2 \\ -5 \end{bmatrix}$.

$x^{(0)} = \begin{bmatrix} -2 \\ -5 \end{bmatrix}$, again with the step-size chosen to minimize $f(x^{(\nu)} + \alpha \Delta x^{(\nu)})$ at each
iteration. The progress is much faster than for the steepest descent step direction
for the same value of initial guess.

10.2.2.3 Advantages

The advantages of using the Newton–Raphson step direction include the following.

(i) The convergence properties of the Newton–Raphson method discussed in

Section 7.3.3 will be inherited. That is, convergence to the solution of $\nabla f(x) = 0$ will be rapid, at least for initial guesses that are near to a solution of the equations or after the iterate becomes close to a solution of the equations. Ideally, convergence of the sequence of iterates is quadratic. (As we will discuss in Section 10.2.4, we can use a step-size selection rule to aid in convergence from initial guesses that are far from the solution.)

(ii) If f is a quadratic function then, as discussed in Section 10.2.2.2, the Newton–Raphson step direction with step-size $\alpha^{(\nu)} = 1$ takes us to a critical point in just one iteration. (See Exercise 10.10.) Figure 10.14 shows the contour sets of the quadratic objective defined in (10.8) and illustrated in Figure 10.8 but with the Newton–Raphson step directions shown as arrows at various points. The Newton–Raphson step directions point directly to the minimizer of a quadratic function.

(iii) Since $\nabla^2 f(x)$ is symmetric, we can take advantage of symmetry in factorization.

10.2.2.4 Disadvantages

There are some significant disadvantages of using the Newton–Raphson approach.

(i) For non-quadratic objectives and particularly at points that are far from the minimizer, the Newton–Raphson step direction is not necessarily a better direction than the steepest descent step direction. As shown in Figure 10.15, this is the case for the non-quadratic function (10.9) discussed in Section 10.2.2.2 using the Newton–Raphson step direction starting at $x^{(0)} = \begin{bmatrix} 3 \\ -5 \end{bmatrix}$.

(ii) Factorization of the Hessian may require considerable effort if n is large or the Hessian is dense. Moreover, if the function has highly elliptical level sets then the Hessian is an ill-conditioned matrix and we may have to use QR factorization to get useful search directions.

(iii) If $\nabla f(x^{(\nu)})$ is not known analytically then it may be difficult or impossible to directly calculate $\nabla^2 f(x^{(\nu)})$.

(iv) If $\nabla^2 f(x^{(\nu)})$ is not positive definite, then the Newton–Raphson update may take us towards a maximum or a point of inflection. For example, consider the function shown in Figure 10.4. Its derivative and Hessian are illustrated in Figures 10.5 and 10.6. Depending on the proximity of the current iterate $x^{(\nu)}$ to the critical points, the Newton–Raphson update may move towards one or other of the critical points or may cause the update to move far from the critical points. For example, consider the following.

(a) For iterates $x^{(\nu)} < -3$, the Hessian $\nabla^2 f(x^{(\nu)}) < 0$. Therefore, a linear approximation to ∇f yields a line with negative slope. Setting the linear approximation equal to zero will yield an update that brings the iterate closer to $\hat{x} = -3$, a local maximizer of f.

(b) For iterates $x^{(v)} > 3$, the Hessian $\nabla^2 f(x^{(v)}) > 0$. Therefore, a linear approximation to ∇f yields a line with positive slope. Setting the linear approximation equal to zero will yield an update that brings the iterate closer to $x^\star = 3$, a local minimizer of f.

(c) For iterates $-3 \le x^{(v)} \le 3$, the sign of the Hessian $\nabla^2 f(x^{(v)})$ varies with $x^{(v)}$. Therefore, a linear approximation to ∇f yields a line with either positive or negative slope.

1. For $x^{(v)} \approx \pm 2.121$, the Hessian is approximately zero, but $\nabla f(x^{(v)})$ is very different from zero. The linear approximation to ∇f at $x^{(v)}$ is a very poor approximation. Setting the approximation equal to zero yields an update that could be very large in magnitude and a new iterate that is far from any of the critical points.

2. For $x^{(v)} \approx 0$, the Hessian is approximately zero and $\nabla f(x^{(v)})$ is also close to zero. Setting the linear approximation to ∇f equal to zero will yield an update that brings the iterate closer to $\hat{x} = 0$, the horizontal inflection point.

3. For other values of $x^{(v)}$, the update will bring the iterate closer to one of the critical points.

Since we cannot in general predict whether $x^{(0)}$ is close to a minimizer, maximizer, or a point of inflection, we must add safeguards to the basic Newton–Raphson update to seek points satisfying $\nabla f(x) = 0$ that are minimizers of f.

Exercise 10.10 confirms that for a quadratic function with quadratic coefficient matrix $Q \in \mathbb{R}^{n \times n}$, if Q is not positive semi-definite then the "reduction" in objective when using the Newton–Raphson step direction can be negative; that is, the objective can increase if the Newton–Raphson update is applied to a function that has a Hessian that is not positive semi-definite.

Exercise 10.11 shows that the Newton–Raphson step direction is independent of the coordinate system. This is consistent with our observation that the Newton–Raphson step direction always points towards the minimizer of a convex quadratic function.

10.2.3 Generalization of Newton–Raphson and steepest descent

In this section we generalize the Newton–Raphson and steepest descent updates in a way that can combine the advantages of each approach.

10.2.3.1 Uniform treatment of updates

We can treat the Newton–Raphson and steepest descent step directions uniformly by writing:

$$\Delta x^{(\nu)} = -M \, \nabla f(x^{(\nu)}), \qquad (10.10)$$

with $M \in \mathbb{R}^{n \times n}$ positive definite as in Corollary 10.2. The choice $M = \mathbf{I}$ yields the steepest descent step direction, while the choice $M = [\nabla^2 f(x^{(\nu)})]^{-1}$ (if the Hessian $\nabla^2 f$ is non-singular) yields the Newton–Raphson step direction [70, section 7.8 and section 9.1]. We will generalize both of these choices to avoid the disadvantages of both.

10.2.3.2 Modified update

Recall that the Newton–Raphson step direction is not defined if $\nabla^2 f(x^{(\nu)})$ is singular. Furthermore, even if $\nabla^2 f(x^{(\nu)})$ is non-singular, Figure 10.5 and Exercise 10.10 show that if $\nabla^2 f(x^{(\nu)})$ is not positive definite then the Newton–Raphson step direction may take us away from the minimizer. Nevertheless, let us think of calculating $\Delta x^{(\nu)}$ in (10.10) using $M = [\nabla^2 f(x^{(\nu)})]^{-1}$ even if $\nabla^2 f(x^{(\nu)})$ is not positive definite. Of course, we would not explicitly invert $\nabla^2 f(x^{(\nu)})$ except in special cases where an inverse was easy to calculate. Instead, to calculate $\Delta x^{(\nu)}$ satisfying (10.10), we would solve the linear system:

$$\nabla^2 f(x^{(\nu)}) \Delta x^{(\nu)} = -\nabla f(x^{(\nu)}), \qquad (10.11)$$

by, for example, factorizing $\nabla^2 f(x^{(\nu)})$ and performing forwards and backwards substitution.

Consider the factorization of $\nabla^2 f(x^{(\nu)})$. Since $\nabla^2 f(x^{(\nu)})$ is symmetric, we will restrict ourselves to diagonal pivots and factorize it as $LDL^{\dagger}$. Suppose that at the j-th stage of the factorization of $\nabla^2 f(x^{(\nu)})$ there are no positive diagonal pivots available. By Lemma 5.4, this means that $\nabla^2 f(x^{(\nu)})$ is not positive definite, so that the Newton–Raphson step direction, even if it is defined, may not be a descent direction. (See Exercise 10.10.)

Let us modify the factorization by adding a positive quantity E_{jj} to $A_{jj}^{(j)}$, where $A^{(j)}$ is the matrix obtained at the j-th stage of the factorization of $\nabla^2 f(x^{(\nu)})$.

Adding E_{jj} to $A_{jj}^{(j)}$ is equivalent to adding the matrix:

$$
\begin{bmatrix}
0 & & & & & & \\
 & \ddots & & & & & \\
 & & 0 & & & & \\
 & & & E_{jj} & & & \\
 & & & & 0 & & \\
 & & & & & \ddots & \\
 & & & & & & 0
\end{bmatrix}
\tag{10.12}
$$

to $\nabla^2 f(x^{(\nu)})$.

We continue the modified factorization by adding a positive quantity each time we encounter a zero or negative value of $A_{jj}^{(j)}$. We may also add a positive quantity to the pivot if $A_{jj}^{(j)}$ is positive but small in magnitude. The overall effect is to add to $\nabla^2 f(x^{(\nu)})$ a diagonal matrix E, where the j-th diagonal entry of E is E_{jj}. We choose $E_{jj} = 0$ if $A_{jj}^{(j)}$ is positive and large enough in magnitude, while we choose $E_{jj} > 0$ if the unmodified value of $A_{jj}^{(j)}$ is close to zero or is negative. In both cases, we obtain a positive pivot. If $\nabla^2 f(x^{(\nu)})$ is positive definite and so $E = \mathbf{0}$, then (10.10) yields the Newton–Raphson step direction.

By construction, $\nabla^2 f(x^{(\nu)}) + E$ is symmetric and positive definite. By Theorem 5.5, this matrix is invertible and its inverse $M = [\nabla^2 f(x^{(\nu)}) + E]^{-1}$ is also symmetric and positive definite. By Corollary 10.2, the search direction defined by (10.10) using this M is a descent direction.

Recall that in the discussion in Section 7.4.1 of the Newton–Raphson method for solving systems of equations, when we encountered a zero pivot we suggested replacing the pivot by a non-zero number. We did not specify the sign of the replacement. Here, however, we see that if we are trying to solve the first-order necessary conditions to minimize a function, then during factorization we should replace a zero (or negative) pivot by a positive pivot to ensure a descent direction. This is called a **modified factorization** [84, section 10.4]. Furthermore, even if the pivot $A_{jj}^{(j)}$ is positive, if $A_{jj}^{(j)}$ is a small positive number then we should increase it to avoid problems with small non-zero pivots. (If we have a maximization problem, then we should replace an approximately zero or positive pivot by a negative pivot to ensure an **ascent direction**.)

We must choose the size of E_{jj}. Qualitatively, for descent it should be chosen so as to guarantee that the pivot is positive and sufficiently large but not so large as to unnecessarily alter the original matrix. Details on the choice of pivot modification are contained in [45, section 4.4.2.2][84, section 10.4].

10.2.3.3 Further variations

We have seen that the step direction (10.10), with M positive definite, is a descent direction. Furthermore, if $\nabla^2 f(x)$ is positive definite, then $M = [\nabla^2 f(x)]^{-1}$ yields a good choice of step direction at x since for quadratic f a step-size of 1 in this direction will bring us to the minimum of the function. These observations suggest that we have considerable flexibility to either:

(i) construct positive definite approximations to $[\nabla^2 f(x)]^{-1}$, or
(ii) approximately solve the equation:

$$\nabla^2 f(x) \Delta x = -\nabla f(x),$$

in a way that guarantees that for the resulting Δx we have that $\Delta x = -M \nabla f(x)$ for some positive definite M. Many algorithms for unconstrained optimization can be interpreted as one or the other of these alternative approximations. (See [45, chapter 4] and [70, chapters 8 and 9] for examples.) These approximations are introduced for various reasons that are analogous to the reasons discussed for approximating the Jacobian in the Newton–Raphson update for solving systems of non-linear equations, discussed in Section 7.2. Various reasons include the following.

(i) To reduce computational effort, we may only want to factorize $\nabla^2 f(x^{(\nu)})$ at some of the iterations, as in the chord and Shamanskii methods discussed in Sections 7.2.1.1 and 7.2.1.2.
(ii) To reduce computational effort, we may approximate particular terms in the second derivative as discussed in Section 7.2.1.3. For example, we may approximate small off-diagonal terms by zero. If we approximate all the off-diagonal terms by zero then the remaining matrix is diagonal and easily inverted.
(iii) Only approximate analytic models are available for the second derivative as discussed in Section 7.2.1.4. In the extreme, we may approximate the second derivative by $\mathbf{I}$ or by $\lambda \mathbf{I}$ for some $\lambda > 0$. In this case, our iteration becomes simply steepest descent or a scaled version of steepest descent. (A variation of this approach is to approximate some of the terms in the second derivative by $\lambda \mathbf{I}$ for some $\lambda > 0$, which leads to the **Levenberg–Marquardt method**. See [45, section 4.7.3][84, section 13.2] and Section 11.2.3.2.)
(iv) First or second derivative information is not available analytically, even approximately, so finite differences must be used in a **discrete-Newton method** as discussed in Section 7.2.1.5 [45, sections 2.3.5, 4.5.1, 4.6, 4.8, 8.1, 8.6][84, section 11.4].

(v) To reduce computational effort or because we do not have access to the second derivative directly, we may want to build up a positive definite approximation to $\nabla^2 f(x^{(\nu)})$ or its inverse over a series of iterations instead of evaluating it and factorizing it directly, as in the **quasi-Newton method** [45, section 4.5] discussed in Section 7.2.1.6.

(vi) The system (10.11) may be difficult to solve directly because it is large or non-sparse, in which case we can use an iterative algorithm such as the **conjugate gradient method** [45, section 4.8.3] mentioned in Section 7.2.2. **Pre-conditioning** is often used in conjunction with the conjugate gradient method. For example, pre-conditioning by dividing by the diagonal entries was initially discussed in Section 5.7.2 and has the effect of improving the condition number of $\nabla^2 f(x^{(\nu)})$.

Depending on the particular application, one or more of these reasons may be relevant, and this will suggest a particular choice of algorithm. A variety of algorithms are discussed in [45, sections 4.5–4.8]. As with the solution of non-linear equations, there is a compromise between the effort per iteration and the improvement in the objective per iteration.

The variations on the Newton–Raphson method can have better performance than the basic Newton–Raphson method because the reduction in effort required for factorization is large enough to offset the slower rate of convergence. The considerations are similar to the analysis of computational effort for solving systems of non-linear equations that was discussed in Section 7.3.4. The variations on the basic Newton–Raphson method will usually have better performance than steepest descent. The most commonly used method in optimization software appears to be the quasi-Newton method [84, section 11.3].

10.2.4 Step-size

In Sections 10.2.1–10.2.3, we ignored the issue of step-size selection. In Section 10.2.4.1, we will illustrate the need for choosing a step-size. A more detailed discussion of this topic appears in [45, chapter 4][70, sections 7.1–7.5][84, section 10.5]. There are several approaches to step-size selection, including:

- the Armijo step-size rule [58, chapter 8][70, section 7.5][84, section 10.5],
- the Wolfe condition [70, section 7.5][84, section 10.5],
- polynomial approximation [70, section 7.2][84, section 10.5], and
- trust-region methods [84, section 10.6].

We briefly discuss these approaches to selecting step-sizes in Sections 10.2.4.2–10.2.4.6.

10.2.4.1 Need for step-size selection

Let us consider the reduction in f due to updating the iterate at iteration ν. The situation is illustrated in Figure 10.17. A function $f : \mathbb{R} \to \mathbb{R}$ is shown as a solid line together with a quadratic approximation to it, which is shown as a dashed line. The quadratic approximation is taken at the current iterate $x^{(\nu)} = 0.3$. This is the **second-order Taylor approximation** of f about $x^{(\nu)}$. The point $\begin{bmatrix} x^{(\nu)} \\ f(x^{(\nu)}) \end{bmatrix}$ is illustrated by the left-most o in Figure 10.17.

We assume that we have used one of the techniques discussed previously to calculate a step direction at the current iterate $x^{(\nu)}$. For concreteness, suppose that we choose the Newton–Raphson step direction. That is, we solve $\nabla^2 f(x^{(\nu)}) \Delta x^{(\nu)} = -\nabla f(x^{(\nu)})$, obtaining $\Delta x^{(\nu)} = 0.5$, in this case. For this choice, $\check{x} = x^{(\nu)} + \Delta x^{(\nu)} = 0.8$ minimizes the quadratic approximation to f. The value of the quadratic approximation at the point $\check{x} = x^{(\nu)} + \Delta x^{(\nu)}$ is illustrated by the right-most o in Figure 10.17.

In fact, at $\check{x} = x^{(\nu)} + \Delta x^{(\nu)}$, the actual value of the objective $f(\check{x})$ is somewhat higher than the quadratic approximation. The value of the objective $f(\check{x})$ at the point $\check{x}$ is illustrated by the right-most • in Figure 10.17. Note that $f(\check{x}) = f(x^{(\nu)} + \Delta x^{(\nu)})) > f(x^{(\nu)})$. Clearly, choosing a step-size of $\alpha^{(\nu)} = 1$ would lead to an *increase* in the objective, even though we had solved for the Newton–Raphson step direction *exactly*. For this reason, we must consider rules to select a step-size that will guarantee that the objective or its gradient, or both, improve from iteration to iteration.

10.2.4.2 Armijo step-size rule

Analogously to the discussion in Section 7.4.2.3, suppose that we had chosen a step-size $\alpha^{(\nu)}$ that is small enough so that f is accurately represented by a second-order Taylor approximation about $x^{(\nu)}$. Then:

$$f(x^{(\nu)} + \alpha^{(\nu)} \Delta x^{(\nu)})$$

$$\approx f(x^{(\nu)}) + \alpha^{(\nu)} [\nabla f(x^{(\nu)})]^\dagger \Delta x^{(\nu)} + \frac{1}{2} (\alpha^{(\nu)})^2 [\Delta x^{(\nu)}]^\dagger \nabla^2 f(x^{(\nu)}) \Delta x^{(\nu)},$$

by a second-order Taylor approximation,

$$\approx f(x^{(\nu)}) + \alpha^{(\nu)} [\nabla f(x^{(\nu)})]^\dagger \Delta x^{(\nu)} - \frac{1}{2} (\alpha^{(\nu)})^2 [\Delta x^{(\nu)}]^\dagger \nabla f(x^{(\nu)}),$$

assuming that $\Delta x^{(\nu)}$ approximately solves $\nabla^2 f(x^{(\nu)}) \Delta x^{(\nu)} = -\nabla f(x^{(\nu)})$,

$$= f(x^{(\nu)}) + \alpha^{(\nu)} \left(1 - \frac{1}{2}\alpha^{(\nu)}\right) [\nabla f(x^{(\nu)})]^\dagger \Delta x^{(\nu)},$$

$$\leq f(x^{(\nu)}) + \frac{1}{2}\alpha^{(\nu)} [\nabla f(x^{(\nu)})]^\dagger \Delta x^{(\nu)}, \tag{10.13}$$

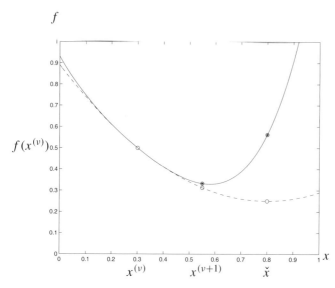

Fig. 10.17. The need for a step-size rule. The function f is illustrated with a solid line together with a quadratic approximation to it, illustrated as a dashed line. The quadratic approximation is a second-order Taylor approximation of f about $x^{(v)} = 0.3$.

where the last inequality is true for $0 \le \alpha^{(v)} \le 1$ (see Exercise 10.13) and assuming that the step direction $\Delta x^{(v)}$ was chosen to satisfy $[\nabla f(x^{(v)})]^\dagger \Delta x^{(v)} < 0$. Recall that, by Lemma 10.1, $[\nabla f(x^{(v)})]^\dagger \Delta x^{(v)} < 0$ is sufficient for $\Delta x^{(v)}$ to be a descent direction for f at $x^{(v)}$ and that all of the updates we have described satisfy this requirement when $\nabla f(x^{(v)}) \ne \mathbf{0}$.

In practice, as in the discussion in Section 7.4.2.3 of step-size rules for updates to solve non-linear equations, we cannot expect that the reduction in the objective will always be as large as predicted by (10.13) because:

(i) the second-order Taylor approximation is only approximate as shown in Figure 10.17, and

(ii) the step direction may only approximately satisfy the Newton–Raphson condition $\nabla^2 f(x^{(v)}) \Delta x^{(v)} = -\nabla f(x^{(v)})$.

This suggests a step-size rule that is analogous to the Armijo rule we discussed in Section 7.4.2.3 for solving non-linear equations. We first choose an acceptance tolerance $0 < \delta < 1$. We start with tentative step-size $\alpha^{(v)} = 1$ and calculate the trial objective $f(x^{(v)} + \alpha^{(v)} \Delta x^{(v)})$. The step-size is accepted if:

$$f(x^{(v)} + \alpha^{(v)} \Delta x^{(v)}) \le f(x^{(v)}) + \frac{\delta}{2} \alpha^{(v)} [\nabla f(x^{(v)})]^\dagger \Delta x^{(v)}. \tag{10.14}$$

Otherwise, reduce the step-size by a factor of, say, one half and repeat the process until an iterate is produced that satisfies (10.14).

For example, consider Figure 10.17 again. The ○ in the middle of the graph illustrates the quadratic approximation to the objective function for the step-size of $\alpha^{(\nu)} = 0.5$. The exact value of the objective for this step-size, $f(x^{(\nu)} + 0.5\Delta x^{(\nu)})$, which is illustrated by the ● just above the middle ○, is only slightly different from the prediction made by the second-order Taylor approximation. Moreover, for values of δ that are not too close to 1, the condition (10.14) would be satisfied and the step-size of $\alpha^{(\nu)} = 0.5$ would be accepted. The updated iterate is $x^{(\nu+1)} = x^{(\nu)} + \alpha^{(\nu)}\Delta x^{(\nu)} = 0.55$. This yields an updated iterate $x^{(\nu+1)}$ that has objective that is significantly improved over $x^{(\nu)}$. This description suggests that a suitable choice of acceptance tolerance δ will be a number that is significantly less than 1.

10.2.4.3 Wolfe condition

The rule for reducing the step-size discussed in the last section does not check for "improvement" in the gradient ∇f. Recall that the first-order necessary conditions are that $\nabla f(x^\star) = \mathbf{0}$ so that we want the update to satisfy $\nabla f(x^{(\nu)} + \alpha^{(\nu)}\Delta x^{(\nu)}) \approx \mathbf{0}$. An alternative to the Armijo step-size rule that makes use of gradient information rather than objective values is provided by the **Wolfe condition**:

$$\left| [\nabla f(x^{(\nu)} + \alpha^{(\nu)}\Delta x^{(\nu)})]^\dagger \Delta x^{(\nu)} \right| \leq \eta \left| [\nabla f(x^{(\nu)})]^\dagger \Delta x^{(\nu)} \right|. \tag{10.15}$$

This condition ensures that the **directional derivative** (see Definition A.37) in the direction Δx evaluated at the next iterate, $[\nabla f(x^{(\nu+1)})]^\dagger \Delta x^{(\nu)}$, is small compared to the directional derivative in the direction Δx at the current iterate, $[\nabla f(x^{(\nu)})]^\dagger \Delta x^{(\nu)}$. Loosely speaking, the Wolfe condition ensures that $\nabla f(x^{(\nu)} + \alpha^{(\nu)}\Delta x^{(\nu)}) \approx \mathbf{0}$.

For values of η not too close to zero, the Wolfe condition (10.15) is satisfied by the updated iterate $x^{(\nu+1)} = f(x^{(\nu)} + 0.5\Delta x^{(\nu)})$ illustrated in Figure 10.17, since $\nabla f(x^{(\nu+1)}) \approx 0$.

10.2.4.4 Combined Armijo and Wolfe conditions

The Wolfe condition (10.15) is often used in conjunction with the Armijo condition (10.14). The Armijo condition (10.14) ensures that the step-size is not so large as to invalidate the quadratic approximation of the objective, while the Wolfe condition (10.15) ensures that the gradient of the objective is reduced sufficiently by the step. Details and convergence results can be found in [45, section 4.3.2.1][84, sections 10.4–10.6].

10.2.4.5 Curve fitting

If f is relatively easy to evaluate, then we can evaluate it at several points along the line $x^{(\nu)} + \alpha\Delta x^{(\nu)}$ for $0 \leq \alpha \leq 1$ and then fit a polynomial curve to it, using, for example, a least-squares fit. (We will see how to do this in Section 11.1.4 and Exercise 11.1.) The fitted curve can then be approximately minimized and the corresponding value of α used as the step-size. If f is itself quadratic then $f(x^{(\nu)} + \alpha \Delta x^{(\nu)})$ is a quadratic function of α. (See Exercise 10.14.)

If either:

- $f(x^{(\nu)} + \alpha \Delta x^{(\nu)})$ is itself a quadratic function, or
- f is not quadratic but we fit a quadratic function to the points,

then we can minimize the quadratic function of α using the following method.

(i) If the coefficient of $(\alpha)^2$ in the quadratic function is positive, then the unconstrained minimum of the quadratic function occurs at the point $x^{(\nu)} + \alpha \Delta x^{(\nu)}$ for α such that the derivative of the quadratic function with respect to α is equal to zero. If this value of α lies outside the range $0 \le \alpha \le 1$ then the closest end-point should be selected.

(ii) If the coefficient of $(\alpha)^2$ in the quadratic function is negative, then the minimizer is one of the end-points $\alpha = 0$ or $\alpha = 1$.

(See Exercise 2.51 for details.)

If a higher-order polynomial curve is fitted, then various special-purpose techniques can be used to find an approximate minimizer of a function of one variable [45, section 4.1.2][70, section 7.2].

10.2.4.6 Trust region

In our approaches so far we have separated the selection of a step direction from the selection of a step-size. In contrast, in a **trust region approach** the selection of an appropriate step direction and step-size both explicitly consider the region over which a second-order Taylor approximation represents the function f accurately. Details and convergence results can be found in [59, section 3.3][84, section 10.6].

10.2.5 Stopping criteria

If f is convex and quadratic, then we can obtain the minimum and a minimizer of f in one Newton–Raphson step. Otherwise, we must iterate until we are satisfied with the value of the objective, the change in iterates, and the gradient of the function. A typical criterion is to require that $\|\nabla f(x^{(\nu)})\|$ and $\|\Delta x^{(\nu)}\|$ be sufficiently small. By Theorem 2.6, if f is convex then the minimizer $x^\star$ of $f(x)$ must satisfy:

$$
\begin{aligned}
f(x^\star) \; &\ge \; f(x^{(\nu)}) + [\nabla f(x^{(\nu)})]^\dagger (x^\star - x^{(\nu)}), \text{ by Theorem 2.6,} \\
&\qquad \text{with } \mathbb{S} = \mathbb{R}^n, x = x^\star, \text{ and } x' = x^{(\nu)}, \\
&\ge \; f(x^{(\nu)}) - \left| [\nabla f(x^{(\nu)})]^\dagger (x^\star - x^{(\nu)}) \right|, \\
&\ge \; f(x^{(\nu)}) - \|\nabla f(x^{(\nu)})\| \, \|x^\star - x^{(\nu)}\|.
\end{aligned}
\tag{10.16}
$$

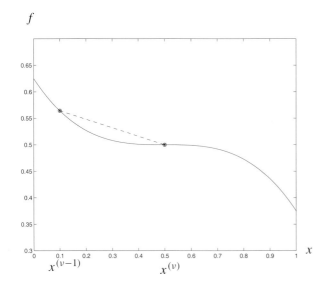

f

Fig. 10.18. Iterate that is a horizontal inflection point of the objective function.

If we know an *a priori* bound on where the optimizer can lie, then we can bound $\left\| x^\star - x^{(\nu)} \right\|$ independently of $x^\star$ by some $\overline{\rho}$, say. Then, we can ensure that $f(x^{(\nu)})$ is within ϵ_f of the value of the global minimum by iterating until $\left\| \nabla f(x^{(\nu)}) \right\| \leq \epsilon_f / \overline{\rho}$. This criterion is often implemented in practice as a slightly different *relative* criterion by testing if:

$$ \left\| \nabla f(x^{(\nu)}) \right\| \leq \frac{\epsilon_f}{\overline{\rho}} \left(1 + |f(x^{(\nu)})| \right). $$

The relative criterion effectively increases the tolerance if $|f(x^{(\nu)})|$ is large. Further details are presented in [45, section 8.2.3.2][84, section 11.5].

As with the solution of non-linear equations, it is usual to combine various criteria to ensure that we are close enough to a solution before stopping and also to safeguard against infinite loops by imposing an upper limit on the number of iterations. Further discussion is contained in [45, section 8.2.3.2][84, section 11.5].

10.2.6 Avoiding critical points that are not minimizers

If, at some iteration ν, we find that $\nabla f(x^{(\nu)}) = \mathbf{0}$ then our basic algorithm terminates unable to make any further progress. (In practice, if $\left\| \nabla f(x^{(\nu)}) \right\| \approx 0$, then we may not be able to make any further progress.) If f is convex and $\nabla f(x^{(\nu)}) = \mathbf{0}$ then, by Corollary 10.6, $x^{(\nu)}$ is a minimizer and $f(x^{(\nu)})$ is a minimum. If f is convex and $\nabla f(x^{(\nu)}) \approx \mathbf{0}$ then $f(x^{(\nu)})$ is close to a minimum in the sense of (10.16).

If f is not convex, then we may be at a point of inflection or a local maximizer. For example, in Figure 10.18, the iterate $x^{(\nu)} = 0.5$ is a horizontal inflection point of the objective. The first-order necessary conditions are satisfied by $x^{(\nu)}$; however, it is clearly not a minimizer.

If the first-order necessary conditions are satisfied, but we can detect that the current iterate is not a minimizer, then one approach is to restart the algorithm by perturbing $x^{(\nu)}$ by a random amount to move it away from the point of inflection or local maximum [45, section 8.1.3.6]. Alternatively, at a horizontal inflection, we can use the previous iterate in a secant approximation as discussed in Section 7.2.1.5, to seek a descent direction. For example, in Figure 10.18, using a secant approximation based on $x^{(\nu-1)}$ and $x^{(\nu)}$ would yield a descent direction. The linear interpolation of f between $x^{(\nu-1)}$ and $x^{(\nu)}$ is shown as a dashed line in Figure 10.18. The secant approximation uses the slope of the interpolating line as the approximation to the directional derivative.

If we are not at a horizontal inflection point then another approach is to look for negative eigenvalues of the Hessian and move in the direction of the corresponding eigenvector [45, section 4.4.2]. (See Exercise 10.16. See Section 2.2.2.3 for discussion of eigenvalues and eigenvectors.)

10.3 Sensitivity

We now suppose that the objective f is *parameterized* by a parameter $\chi \in \mathbb{R}^s$. That is, $f : \mathbb{R}^n \times \mathbb{R}^s \to \mathbb{R}$. We imagine that we have solved the unconstrained minimization problem:

$$\min_{x \in \mathbb{R}^n} f(x; \chi),$$

for a base-case value of the parameters, say $\chi = 0$, to find the base-case solution $x^\star$ and that now we are considering the sensitivity of the base-case solution to variation of the parameters around $\chi = 0$.

10.3.1 Implicit function theorem

As with sensitivity analysis of non-linear equations in Section 7.5, we use the **implicit function theorem** (Theorem A.9 in Section A.7.3 of Appendix A.) The following corollary to the implicit function theorem provides us with the sensitivity of the solution to the parameters.

Corollary 10.8 *Let $f : \mathbb{R}^n \times \mathbb{R}^s \to \mathbb{R}^n$ be twice partially differentiable with continuous second partial derivatives. Consider the minimization problem:*

$$\min_{x \in \mathbb{R}^n} f(x; \chi),$$

where $\chi \in \mathbb{R}^s$ is a parameter. Suppose that $x^\star$ is a local minimizer of this problem for the base-case value of the parameters $\chi = 0$. We call $x = x^\star$ a base-case solution.

Define the (parameterized) Hessian $\nabla^2_{xx} f : \mathbb{R}^n \times \mathbb{R}^s \to \mathbb{R}^{n \times n}$ by:

$$\forall x \in \mathbb{R}^n, \forall \chi \in \mathbb{R}^s, \nabla^2_{xx} f(x; \chi) = \frac{\partial^2 f}{\partial x^2} (x; \chi).$$

Suppose that $\nabla^2_{xx} f(x^\star; \mathbf{0})$ is positive definite, so that $x^\star$ satisfies the second-order suffi-cient conditions for the base-case problem. Then, there is a local minimizer of $f(x; \chi)$ for χ in a neighborhood of the base-case values of the parameters $\chi = \mathbf{0}$ and the local minimizer is a partially differentiable function of χ in this neighborhood. The sensitivity of the local minimizer $x^\star$ with respect to variation of the parameters χ, evaluated at the base-case $\chi = \mathbf{0}$, is given by:

$$\frac{\partial x^\star}{\partial \chi} (\mathbf{0}) = -[\nabla^2_{xx} f(x^\star; \mathbf{0})]^{-1} K(x^\star; \mathbf{0}),$$

where $K : \mathbb{R}^n \times \mathbb{R}^s \to \mathbb{R}^{n \times s}$ is defined by:

$$\forall x \in \mathbb{R}^n, \forall \chi \in \mathbb{R}^s, K(x; \chi) = \frac{\partial^2 f}{\partial x \partial \chi}(x; \chi).$$

The sensitivity of the corresponding local minimum $f^\star$ to variation of the parameters χ, evaluated at the base-case $\chi = \mathbf{0}$, is given by:

$$\frac{\partial f^\star}{\partial \chi} (\mathbf{0}) = \frac{\partial f}{\partial \chi} (x^\star; \mathbf{0}).$$

If $f(\bullet; \chi)$ is convex for χ in a neighborhood of $\mathbf{0}$ then the minimizers and minima are global in this neighborhood.

Proof The sensitivity of the local minimizer follows from Corollary 7.5, noting that by assumption the Hessian is positive definite in a neighborhood of the base-case minimizer and parameters by Exercise 5.19.

The sensitivity of the corresponding local minimum follows by totally differentiating the value of the local minimum $f^\star(\chi) = f(x^\star(\chi); \chi)$ with respect to χ. In particular,

$$\begin{aligned}
\frac{\partial f^\star}{\partial \chi} (\mathbf{0}) &= \frac{d[f(x^\star(\chi); \chi)]}{d\chi} (\mathbf{0}), \\
&= \frac{\partial f}{\partial \chi} (x^\star; \mathbf{0}) + \frac{\partial f}{\partial x} (x^\star; \mathbf{0}) \frac{\partial x^\star}{\partial \chi} (\mathbf{0}),
\end{aligned}$$

on totally differentiating $f(x^\star(\chi); \chi)$ with respect to χ,

$$= \frac{\partial f}{\partial \chi} (x^\star; \mathbf{0}),$$

since the first-order necessary conditions at the base-case are $\frac{\partial f}{\partial x} (x^\star; \mathbf{0}) = 0$.

The global results follow from Corollary 10.6. □

Analogously to the case of linear and non-linear systems of equations, if $\nabla^2_{xx} f(x^\star; \mathbf{0})$ has already been factorized then each sensitivity of $x^\star$ with respect to an entry of

χ requires only a forwards and backwards substitution. The sensitivity of the local minimum is called **the envelope theorem** [119, section 3.3].

10.3.2 Example

Consider the parameterized objective function $f : \mathbb{R}^2 \times \mathbb{R} \to \mathbb{R}$ defined by:

$$\forall x \in \mathbb{R}^2, \forall \chi \in \mathbb{R}, \ f(x; \chi) = (x_1 - \exp(\chi))^2 + (x_2 - 3\exp(\chi))^2 + 5\chi.$$

We first met this example in Section 2.7.5.3. This is a parameterized version of the function defined in (10.1). For $\chi = 0$, the parameterized function is the same as the function defined in (10.1) and from the discussion in Section 10.1.1.2 we know that the base-case solution is $x^\star = \begin{bmatrix} 1 \\ 3 \end{bmatrix}$.

By Corollary 10.8, there is an unconstrained minimizer of $f(\bullet; \chi)$ for χ in a neighborhood of the base-case value of the parameter $\chi = 0$ and the minimizer is a differentiable function of χ in this neighborhood. The sensitivity of the minimizer $x^\star$ with respect to variation of the parameter χ, evaluated at the base-case $\chi = 0$, is given by:

$$\frac{\partial x^\star}{\partial \chi}(0) = -[\nabla_{xx}^2 f(x^\star; 0)]^{-1} K(x^\star; 0),$$

where $\nabla_{xx}^2 f : \mathbb{R}^2 \times \mathbb{R} \to \mathbb{R}^{2\times 2}$ and $K : \mathbb{R}^2 \times \mathbb{R} \to \mathbb{R}^{2\times 1}$ are defined by:

$$\forall x \in \mathbb{R}^2, \forall \chi \in \mathbb{R}, \ \nabla_{xx}^2 f(x; \chi) = \frac{\partial^2 f}{\partial x^2}(x; \chi),$$

$$= \begin{bmatrix} 2 & 0 \\ 0 & 2 \end{bmatrix},$$

$$\nabla_{xx}^2 f(x^\star; 0) = \begin{bmatrix} 2 & 0 \\ 0 & 2 \end{bmatrix},$$

$$\forall x \in \mathbb{R}^2, \forall \chi \in \mathbb{R}, \ K(x; \chi) = \frac{\partial^2 f}{\partial x \partial \chi}(x; \chi),$$

$$= \begin{bmatrix} -2\exp(\chi) \\ -6\exp(\chi) \end{bmatrix},$$

$$K(x^\star; 0) = \begin{bmatrix} -2 \\ -6 \end{bmatrix},$$

where we observe that $\nabla_{xx}^2 f(x^\star; 0)$ is positive definite. That is, the sensitivity is:

$$\frac{\partial x^\star}{\partial \chi}(0) = -[\nabla_{xx}^2 f(x^\star; 0)]^{-1} K(x^\star; 0),$$

$$= -\begin{bmatrix} 2 & 0 \\ 0 & 2 \end{bmatrix}^{-1} \begin{bmatrix} -2 \\ -6 \end{bmatrix},$$

$$= \begin{bmatrix} 1 \\ 3 \end{bmatrix}.$$

The sensitivity of the minimum $f^\star$ to variation of the parameter χ, evaluated at the base-case $\chi = 0$, is given by:

$$\frac{\partial f^\star}{\partial \chi}(0) = \frac{\partial f}{\partial \chi}(x^\star; 0).$$

We have that:

$$\frac{\partial f}{\partial \chi}(x; \chi) = 2(x_1 - \exp(\chi))(-\exp(\chi)) + 2(x_2 - 3\exp(\chi))(-3\exp(\chi)) + 5,$$

and so the sensitivity is:

$$\frac{\partial f^\star}{\partial \chi}(0) = \frac{\partial f}{\partial \chi}(x^\star; 0) = 5.$$

10.4 Summary

In this chapter we have introduced descent directions for reducing the value of an objective function. We described optimality conditions and algorithms based on the conditions. The algorithms made use of step-size rules. Descent will be a continuing theme in the rest of the book. The unconstrained algorithms will be applied to the case studies in the next chapter. We also considered sensitivity analysis.

Exercises

Optimality conditions

10.1 Give an example of an objective $f : \mathbb{R}^2 \to \mathbb{R}$, with f partially differentiable with continuous partial derivatives, and a point $\hat{x} \in \mathbb{R}^2$ such that $\hat{x}$ is *not* a local minimizer of f but there is no descent direction at $\hat{x}$ according to Definition 10.1. (Hint: the function f must be non-convex.)

10.2 Suppose that $f : \mathbb{R}^n \to \mathbb{R}$ is convex and that $x^\star \in \operatorname{argmin}_{x \in \mathbb{R}^n} f(x)$. Let $\hat{x} \in \mathbb{R}^n$. Show that if $\hat{x} \notin \operatorname{argmin}_{x \in \mathbb{R}^n} f(x)$ then $\Delta x = x^\star - \hat{x}$ is a descent direction for f at $\hat{x}$.

10.3 Show that if $|\delta + 2\epsilon| \le \epsilon$ then $-3\epsilon \le \delta \le -\epsilon$.

10.4 Consider the function $f : \mathbb{R} \to \mathbb{R}$ defined by:

$$\forall x \in \mathbb{R}, \ f(x) = \exp(-x).$$

   (i) Calculate ∇f.
   (ii) Calculate $\nabla^2 f$.
   (iii) Show that f is convex.
   (iv) Show that no x exists satisfying $\nabla f(x) = 0$.
   (v) Show that there is no minimizer of $\min_{x \in \mathbb{R}} f(x)$.

Approaches to finding minima

10.5 Suppose that $g : \mathbb{R}^n \to \mathbb{R}^n$ is partially differentiable with continuous partial derivatives and consider solution of the simultaneous equations $g(x) = 0$. Define $f : \mathbb{R}^n \to \mathbb{R}$ by:

$$\forall x \in \mathbb{R}^n, \ f(x) = \frac{1}{2}\|g(x)\|_2^2.$$

Show that the Newton–Raphson step direction at $\hat{x} \in \mathbb{R}^n$ for solving the simultaneous equations $g(x) = 0$ is a descent direction for f at $\hat{x}$ if $g(\hat{x}) \ne 0$.

10.6 Let $f : \mathbb{R}^n \to \mathbb{R}$ be partially differentiable with continuous partial derivatives. Show that, at a point $x^{(\nu)}$, the steepest descent step direction $\Delta x = -\nabla f(x^{(\nu)})$ is perpendicular to the surface of the contour set $C_f(f(x^{(\nu)}))$ at $x^{(\nu)}$. (Hint: Calculate the first-order Taylor approximation to f at $x^{(\nu)}$. Find the set $\mathbb{P}$ of points such that the first-order Taylor approximation is equal to $f(x^{(\nu)})$. That is,

$$\mathbb{P} = \{x \in \mathbb{R} | \nabla f(x^{(\nu)})^\dagger (x - x^{(\nu)}) = 0\}.$$

You can assume that this set is **tangential** to the contour set $C_f(f(x^{(\nu)}))$ at the point $x^{(\nu)}$ (see Definition 13.1) and, moreover, that a direction Δx is perpendicular to the surface of the contour set $C_f(f(x^{(\nu)}))$ at $x^{(\nu)}$ if and only if it is perpendicular to $\mathbb{P}$ at $x^{(\nu)}$. A direction Δx is perpendicular to $\mathbb{P}$ at $x^{(\nu)}$ if for every $x \in \mathbb{P}$, $\Delta x^\dagger (x - x^{(\nu)}) = 0$.

10.7 Consider a quadratic function $f : \mathbb{R}^n \to \mathbb{R}$ defined by:

$$\forall x \in \mathbb{R}^n, \ f(x) = \frac{1}{2}x^\dagger Q x + c^\dagger x,$$

where $Q \in \mathbb{R}^{n \times n}$ is symmetric and $c \in \mathbb{R}^n$. Consider an initial guess $x^{(0)}$ of the unconstrained minimizer of f. Suppose that the steepest descent step direction is used at $x^{(0)}$ with step-size 1 to calculate:

$$x^{(1)} = x^{(0)} - \nabla f(x^{(0)}).$$

Show that $x^{(1)}$ cannot satisfy the first-order necessary conditions for minimizing f unless $(I - Q)(Qx^{(0)} + c) = 0$.

10.8 In this exercise, we use MATLAB to minimize a function.

(i) Use the MATLAB function `fminunc` to minimize $f : \mathbb{R}^2 \to \mathbb{R}$ defined in (10.1):

$$\forall x \in \mathbb{R}^2, \ f(x) = (x_1 - 1)^2 + (x_2 - 3)^2.$$

You should write an MATLAB M-file to evaluate both f and ∇f. Specify that you are supplying the gradient ∇f by setting the `GradObj` option to on using the `optimset` function. Use the steepest descent algorithm by setting the `LargeScale` option to `off` and the `HessUpdate` option to `steepdesc` using the `optimset` function. Use initial guess $x^{(0)} = \begin{bmatrix} 3 \\ -5 \end{bmatrix}$.

(ii) What is the condition number of the quadratic coefficient matrix $Q = \mathbf{I}$?

10.9 In this exercise we use MATLAB to minimize a function.

(i) Use the MATLAB function `fminunc` to minimize $f : \mathbb{R}^2 \to \mathbb{R}$ defined in (10.8):

$$\forall x \in \mathbb{R}^2, \ f(x) = (x_1 - 1)^2 + (x_2 - 3)^2 - 1.8(x_1 - 1)(x_2 - 3).$$

You should write an MATLAB M-file to evaluate both f and ∇f. Specify that you are supplying the gradient ∇f by setting the `GradObj` option to on using the `optimset` function. Use the steepest descent algorithm by setting the `LargeScale` option to `off` and the `HessUpdate` option to `steepdesc` using the `optimset` function. Use initial guess $x^{(0)} = \begin{bmatrix} 3 \\ -5 \end{bmatrix}$. Report the number of iterations required.

(ii) Use the MATLAB function `cond` to evaluate the condition number of the quadratic coefficient matrix:

$$Q = \begin{bmatrix} 2 & -1.8 \\ -1.8 & 2 \end{bmatrix}.$$

(iii) Repeat the first part, but minimize the function $f : \mathbb{R}^4 \to \mathbb{R}$ defined by:

$$\forall x \in \mathbb{R}^4, \ f(x) = (x_1 - 1)^2 + 2(x_2 - 3)^2 + 2(x_3 - 1)^2 + (x_4 - 3)^2$$
$$- 1.8(x_1 - 1)(x_2 - 3) - 1.8(x_2 - 3)(x_3 - 1) - 1.8(x_3 - 1)(x_4 - 3),$$

using initial guess $x^{(0)} = \begin{bmatrix} 3 \\ -5 \\ 3 \\ -5 \end{bmatrix}$. Report the number of iterations required.

(iv) Use the MATLAB function `cond` to evaluate the condition number of the quadratic coefficient matrix:

$$Q = \begin{bmatrix} 2 & -1.8 & 0 & 0 \\ -1.8 & 4 & -1.8 & 0 \\ 0 & -1.8 & 4 & -1.8 \\ 0 & 0 & -1.8 & 2 \end{bmatrix}.$$

10.10 Consider the quadratic function $f : \mathbb{R}^n \to \mathbb{R}$ defined by:

$$\forall x \in \mathbb{R}^n, \ f(x) = \frac{1}{2} x^{\dagger} Q x + c^{\dagger} x,$$

where $Q \in \mathbb{R}^{n \times n}$ is symmetric and non-singular and $c \in \mathbb{R}^n$.

   (i) Prove that the first-order necessary conditions for the problem $\min_{x \in \mathbb{R}^n} f(x)$ are that $Qx^{\star} + c = \mathbf{0}$.
  (ii) Show that from any initial guess $x^{(0)}$ the Newton–Raphson update with step-size $\alpha^{(0)} = 1$ yields a point $x^{(1)}$ that satisfies the first-order necessary conditions.
 (iii) Show that for the Newton–Raphson update the reduction in the objective is given by:

$$f(x^{(0)}) - f(x^{(1)}) = \frac{1}{2} \Delta x^{(0)\dagger} Q \Delta x^{(0)}.$$

This reduction is called the **Newton decrement** [84, section 17.6.2].

  (iv) Can the Newton decrement be negative?

10.11 Let $f : \mathbb{R}^n \to \mathbb{R}$. Consider an onto function $\tau : \mathbb{R}^n \to \mathbb{R}^n$ defined by $\forall \xi \in \mathbb{R}^n, \tau(\xi) = R^{-1}\xi$, with $R \in \mathbb{R}^{n \times n}$ and non-singular. Let the function $\phi : \mathbb{R}^n \to \mathbb{R}$ be defined by $\forall \xi \in \mathbb{R}^n, \phi(\xi) = f(\tau(\xi)) = f(R^{-1}\xi)$. We consider step directions $\Delta \xi$ in ξ coordinates and Δx in x coordinates. We say that the step directions $\Delta \xi$ and Δx **correspond** if $\Delta x = \tau(\Delta \xi) = R^{-1}\Delta \xi$. If an algorithm can be applied in either the x coordinates or the ξ coordinates and its step directions correspond then an algorithm will have similar results when applied in either coordinate system. Otherwise, the choice of coordinate system will be critical in applying the algorithm.

   (i) Show that the Newton–Raphson step direction $\Delta \xi$ for ϕ at a point $\xi^{(\nu)}$ in the ξ coordinates corresponds to the Newton–Raphson step direction Δx for f at the corresponding point $x^{(\nu)} = R^{-1}\xi^{(\nu)}$ in the x coordinates. That is, show that $\Delta x = R^{-1}\Delta \xi$.
  (ii) Show that the steepest descent step directions in ξ and x coordinates do not correspond.

10.12 In this exercise we use MATLAB to minimize two functions.

   (i) Use the MATLAB function `fminunc` to minimize $f : \mathbb{R}^2 \to \mathbb{R}$ defined in (10.8):

$$\forall x \in \mathbb{R}^2, \ f(x) = (x_1 - 1)^2 + (x_2 - 3)^2 - 1.8(x_1 - 1)(x_2 - 3).$$

You should write an MATLAB M-file to evaluate both f and ∇f. Specify that you are supplying the gradient ∇f by setting the `GradObj` option to `on`. Set the `LargeScale` option to `off`. With the other parameter settings set to the default, `fminunc` uses a Broyden, Fletcher, Goldfarb, Shanno (BFGS) quasi-Newton approximation to the Newton–Raphson step direction. (See Section 7.2.1.6.) Use initial guess $x^{(0)} = \begin{bmatrix} 3 \\ -5 \end{bmatrix}$. Report the number of iterations required.

(ii) Repeat the first part, but minimize the function $f : \mathbb{R}^4 \to \mathbb{R}$ defined by:

$$\forall x \in \mathbb{R}^4, \ f(x) = (x_1 - 1)^2 + 2(x_2 - 3)^2 + 2(x_3 - 1)^2 + (x_4 - 3)^2$$
$$- 1.8(x_1 - 1)(x_2 - 3) - 1.8(x_2 - 3)(x_3 - 1) - 1.8(x_3 - 1)(x_4 - 3),$$

using initial guess $x^{(0)} = \begin{bmatrix} 3 \\ -5 \\ 3 \\ -5 \end{bmatrix}$. Report the number of iterations required.

10.13 Show that:

$$(0 \leq \alpha \leq 1) \Rightarrow \alpha \left(1 - \frac{1}{2}\alpha \right) \geq \frac{1}{2}\alpha.$$

10.14 Suppose that $f : \mathbb{R}^n \to \mathbb{R}$ is quadratic. Consider the function $\phi : \mathbb{R} \to \mathbb{R}$ defined by:

$$\forall \alpha \in \mathbb{R}, \phi(\alpha) = f(x^{(\nu)} + \alpha \Delta x^{(\nu)}).$$

Show that ϕ is a quadratic function.

10.15 In this exercise we consider conditions for step-size rules.

(i) Show that $x^{(\nu+1)}$ (for $\nu \geq 0$) cannot be a global maximizer of f if we choose $\alpha^{(\nu)}$ to satisfy:

$$f(x^{(\nu)} + \alpha^{(\nu)} \Delta x^{(\nu)}) < f(x^{(\nu)}).$$

(ii) Is it possible for $x^{(\nu+1)}$ to be a local maximizer?

10.16 Suppose that at some iteration ν we find that $\nabla f(x^{(\nu)}) = 0$ for a quadratic function $f : \mathbb{R}^n \to \mathbb{R}$ defined by:

$$\forall x \in \mathbb{R}^n, \ f(x) = \frac{1}{2}x^\dagger Q x + c^\dagger x,$$

where $Q \in \mathbb{R}^{n \times n}$ is symmetric and $c \in \mathbb{R}^n$. Suppose that $\lambda \in \mathbb{R}$ is an eigenvalue of Q and $\xi \in \mathbb{R}^n$ is the corresponding eigenvector. Show that if $\lambda < 0$ then $\Delta x = \xi$ is a descent direction for f at $x^{(\nu)}$.

Sensitivity

10.17 Show by an example that the conclusion of Corollary 10.8 may fail to hold if the objective is not positive definite at the base-case minimizer and parameters. (Hint: Consider $f : \mathbb{R} \times \mathbb{R} \to \mathbb{R}$ defined by $\forall x \in \mathbb{R}, \forall \chi \in \mathbb{R}, \ f(x; \chi) = x\chi$. Note that $x^\star = [0]$ is a base-case minimizer corresponding to the base-case value of the parameters $\chi = [0]$.)

10.18 Consider the function $f : \mathbb{R} \times \mathbb{R} \to \mathbb{R}$ defined by:

$$\forall x \in \mathbb{R}, \forall \chi \in \mathbb{R}, \; f(x; \chi) = \frac{1}{2}(x - \chi)^2.$$

(i) Find the base-case minimizer of $\min_{x \in \mathbb{R}} f(x; 0)$.
(ii) Calculate the sensitivity of the minimizer to variation of χ, evaluated at $\chi = 0$.
(iii) Calculate the sensitivity of the minimum to variation of χ, evaluated at $\chi = 0$.

10.19 Let $f : \mathbb{R}^n \to \mathbb{R}$, $g : \mathbb{R}^n \to \mathbb{R}^m$, and $h : \mathbb{R}^n \to \mathbb{R}^r$ be twice partially differentiable with continuous second partial derivatives. Recall the definition of the Lagrangian $\mathcal{L} :$ $\mathbb{R}^n \times \mathbb{R}^m \times \mathbb{R}^r \to \mathbb{R}$ and the dual function $\mathcal{D} : \mathbb{R}^m \times \mathbb{R}^r \to \mathbb{R} \cup \{-\infty\}$ from Sections 3.4.1 and 3.4.2, respectively:

$$\begin{aligned}
\forall x \in \mathbb{R}^n, \forall \lambda \in \mathbb{R}^m, \forall \mu \in \mathbb{R}^r, \mathcal{L}(x, \lambda, \mu) &= f(x) + \lambda^\dagger g(x) + \mu^\dagger h(x), \\
\forall \lambda \in \mathbb{R}^m, \forall \mu \in \mathbb{R}^r, \mathcal{D}(\lambda, \mu) &= \inf_{x \in \mathbb{R}^n} \mathcal{L}(x, \lambda, \mu).
\end{aligned}$$

Let $\hat{\lambda} \in \mathbb{R}^m$ and $\hat{\mu} \in \mathbb{R}^r$ and suppose that $\hat{x} \in \mathbb{R}^n$ is a minimizer of $\mathcal{L}(\bullet, \hat{\lambda}, \hat{\mu})$. Further suppose that $\dfrac{\partial^2 \mathcal{L}}{\partial x^2}(\hat{x}, \hat{\lambda}, \hat{\mu})$ is positive definite. Show that $\mathcal{D}$ is partially differentiable in a neighborhood of $\begin{bmatrix} \hat{\lambda} \\ \hat{\mu} \end{bmatrix}$ and that:

$$\nabla \mathcal{D}(\hat{\lambda}, \hat{\mu}) = \begin{bmatrix} g(\hat{x}) \\ h(\hat{x}) \end{bmatrix}.$$

(Hint: Define $\phi : \mathbb{R}^n \times \mathbb{R}^{m+r} \to \mathbb{R}$ by:

$$\forall x \in \mathbb{R}^n, \forall \chi = \begin{bmatrix} \Delta\lambda \\ \Delta\mu \end{bmatrix} \in \mathbb{R}^{m+r}, \; \phi(x; \chi) = f(x) + \left(\hat{\lambda} + \Delta\lambda\right)^\dagger g(x) + \left(\hat{\mu} + \Delta\mu\right)^\dagger h(x),$$

so that $\phi(x; \mathbf{0}) = \mathcal{L}(x, \hat{\lambda}, \hat{\mu})$. Apply Corollary 10.8 and characterize the dual function in a neighborhood of $\begin{bmatrix} \hat{\lambda} \\ \hat{\mu} \end{bmatrix}$.)

11

Solution of the unconstrained minimization case studies

In this chapter, we apply algorithms from Chapter 10 to the two case studies from Chapter 9. We consider the multi-variate linear regression case study in Section 11.1 and the power system state estimation case study in Section 11.2. Both of our case studies will be transformed into **least-squares problems**. Unconstrained optimization algorithms that exploit the special characteristics of least-squares problems are described in [45, section 4.7][84, chapter 13]; however, we will first apply our basic unconstrained optimization algorithm to these problems because in later chapters we will need to solve more general unconstrained problems.

In practice, a special purpose algorithm for least-squares problems can be expected to yield better performance on least-squares problems compared to the performance of a general purpose algorithm for unconstrained problems. That is, as we have discussed previously in Section 2.3.2, we should always in practice try to find the most specifically applicable algorithm for a problem [84, section 13.1]. We will consider such specific algorithms for least-squares problems using further transformations.

11.1 Multi-variate linear regression

In Section 11.1.1 we transform the objective of Problem (9.7) and in Section 11.1.2 we compare the transformed and original problem. In Sections 11.1.3 and 11.1.4 we calculate the derivatives of the transformed objective and present the optimality conditions. In Section 11.1.5, we transform the problem further to avoid numerical ill-conditioning issues. Then, in Section 11.1.6, we relate the optimality conditions to linear regression.

11.1.1 Transformation of objective

Recall Problem (9.7):

$$\max_{x \in \mathbb{R}^n} \phi(\zeta(1), \ldots, \zeta(m); \psi(1), \ldots, \psi(m), x),$$

where $\phi : \mathbb{R}^n \to \mathbb{R}$ was defined in (9.6), which we repeat here:

$$\forall x \in \mathbb{R}^n, \phi(\zeta(1), \ldots, \zeta(m); \psi(1), \ldots, \psi(m), x)$$

$$= \prod_{\ell=1}^{m} \frac{1}{\sqrt{2\pi}\sigma_\ell} \exp\left(-\frac{(\psi(\ell)^\dagger \beta + \gamma - \zeta(\ell))^2}{2(\sigma_\ell)^2}\right).$$

Problem (9.7) is in the form of a maximization problem. In the introduction to Chapter 3, we remarked that we could transform a maximization problem into a minimization problem by using the definition (2.22):

$$\max_{x \in \mathbb{S}} \phi(x) = -\min_{x \in \mathbb{S}}(-\phi(x)).$$

However, by Exercise 3.4, the transformed objective will not in general be convex. Exercise 3.4 does nevertheless suggest a transformation that will produce a convex objective. We will now explore this transformation.

First define $\hat{f} : \mathbb{R}^n \to \mathbb{R}$ by:

$$\forall x \in \mathbb{R}^n, \hat{f}(x) = -\ln(\phi(\zeta(1), \ldots, \zeta(m); \psi(1), \ldots, \psi(m), x)).$$

Then:

$$\forall x \in \mathbb{R}^n, \hat{f}(x) = -\ln\left(\prod_{\ell=1}^{m} \frac{1}{\sqrt{2\pi}\sigma_\ell} \exp\left(-\frac{(\psi(\ell)^\dagger \beta + \gamma - \zeta(\ell))^2}{2(\sigma_\ell)^2}\right)\right),$$

$$= -\sum_{\ell=1}^{m} \left[\ln\left(\frac{1}{\sqrt{2\pi}\sigma_\ell}\right) - \frac{(\psi(\ell)^\dagger \beta + \gamma - \zeta(\ell))^2}{2(\sigma_\ell)^2}\right],$$

$$= \sum_{\ell=1}^{m} \left[\frac{(\psi(\ell)^\dagger \beta + \gamma - \zeta(\ell))^2}{2(\sigma_\ell)^2}\right] - \sum_{\ell=1}^{m} \ln\left(\frac{1}{\sqrt{2\pi}\sigma_\ell}\right),$$

where we recall that:

$$x = \begin{bmatrix} \beta \\ \gamma \end{bmatrix} \in \mathbb{R}^n.$$

The term $-\sum_{\ell=1}^{m} \ln\left(1/(\sqrt{2\pi}\sigma_\ell)\right)$ in the definition of $\hat{f}$ is independent of the variables β and γ and so does not affect the set of minimizers of $\hat{f}$. Furthermore, assuming that $\sigma_\ell = \sigma, \forall \ell = 1, \ldots, m$, we can define $f : \mathbb{R}^n \to \mathbb{R}$ by:

$$
\begin{aligned}
\forall x \in \mathbb{R}^n, f(x) &= \sigma^2 \left[\hat{f}(x) + \sum_{\ell=1}^{m} \ln \left(\frac{1}{\sqrt{2\pi\sigma_\ell}} \right) \right], \\
&= \frac{1}{2} \sum_{\ell=1}^{m} (\psi(\ell)^\dagger \beta + \gamma - \zeta(\ell))^2, \\
&= \frac{1}{2} \sum_{\ell=1}^{m} (A_\ell x - b_\ell)^2, \\
&\qquad \text{where } A_\ell = \begin{bmatrix} \psi(\ell)^\dagger & 1 \end{bmatrix} \in \mathbb{R}^{1 \times n} \text{ and } b_\ell = \zeta(\ell) \in \mathbb{R}, \\
&= \frac{1}{2} (Ax - b)^\dagger (Ax - b), \\
&\qquad \text{where } A = \begin{bmatrix} A_1 \\ \vdots \\ A_m \end{bmatrix} \in \mathbb{R}^{m \times n} \text{ and } b = \begin{bmatrix} b_1 \\ \vdots \\ b_m \end{bmatrix} \in \mathbb{R}^m, \\
&= \frac{1}{2} \| Ax - b \|_2^2.
\end{aligned}
$$

The function f is half of the sum of squares of functions of the form $\psi(\ell)^\dagger \beta + \gamma - \zeta(\ell)$. These functions are the entries in the affine function $Ax - b$. The functions f and ϕ are related by a strictly monotonically decreasing transformation. Therefore, by Theorem 3.1, so long as either:

(i) Problem (9.7) has a maximum, or
(ii) the problem:

$$
\min_{x \in \mathbb{R}^n} f(x), \tag{11.1}
$$

has a minimum,

then they both have the same set of optimizers. Furthermore, their optima are related by the same transformation that relates f and ϕ. Since f is (half of) the sum of squares of terms, we call Problem (11.1) a **least-squares problem**. We refer to the corresponding specification of the affine function defined in (9.1) as a **least-squares fit** to the data.

Since each term $\psi(\ell)^\dagger \beta + \gamma - \zeta(\ell)$ depends linearly on β and γ, Problem (11.1) is further classified as a **linear least-squares problem**. Exercise 11.1 illustrates that *linear* least-squares problems arise whenever the dependent variable $\zeta(\ell)$ can be interpreted as an affine function of the unknown parameters (for a fixed value of the independent variables $\psi(\ell)$) even if the dependent variable depends *non-linearly* on the independent variables.

11.1.2 Comparison of objectives

Since f is a quadratic function, Exercise 10.10 shows that ∇f is linear and that the necessary conditions for a minimum of Problem (11.1) are a set of linear simultaneous equations. This should be contrasted with the necessary conditions for a maximum of Problem (9.7), which are a set of non-linear simultaneous equations since ϕ is non-quadratic. (See Exercise 11.3.)

This means that the transformation from ϕ to f has created a problem for which the optimality conditions in terms of f are simpler than the optimality conditions for the original problem in terms of ϕ. We can solve linear equations with one factorization and one forwards and backwards substitution, whereas solving the non-linear necessary conditions of Problem (9.7) requires an iterative algorithm with *each* iteration typically involving the solution of linear equations. Moreover, we will be able to show that f is convex, so that, by Corollary 10.6, the necessary conditions for optimality of Problem (11.1) are also sufficient.

On the other hand, Exercise 3.4 suggests that $(-\phi)$ is non-convex, so it is not apparent that the necessary conditions of Problem (9.7) are also sufficient for optimality of that problem. In fact, the necessary conditions for Problem (9.7) *are* sufficient for optimality for Problem (9.7). However, the easiest way to prove this is to note that f is convex and then note that the set of points satisfying the necessary conditions for Problem (11.1) is the same as the set of points satisfying the necessary conditions for Problem (9.7). (See Exercise 11.3.)

In Sections 11.1.3 and 11.1.4, we calculate the derivatives of the objective f and then use the results to find optimality conditions for Problem (11.1).

11.1.3 Derivatives of objective

Let us now calculate the first and second derivatives of f. We have (see Exercise 11.2):

$$\nabla f(x) = A^\dagger (Ax - b), \tag{11.2}$$
$$\nabla^2 f(x) = A^\dagger A. \tag{11.3}$$

11.1.4 Optimality conditions

We now use the results from Chapters 2, 3, and 10 to prove a series of results for Problem (11.1).

$\nabla^2 f(x)$ **is positive semi-definite** Since $\nabla^2 f(x) = A^\dagger A$ then, by Part (i) of Exercise 2.27, it is positive semi-definite.

The objective f is convex Since $\nabla^2 f(x)$ is positive semi-definite, then by Theorem 2.7, f is convex.

First-order conditions are sufficient Since f is convex, then by Corollary 10.6, the first-order conditions $\nabla f(x) = \mathbf{0}$ are sufficient for global optimality of Problem (11.1). So if $\nabla f(x) = \mathbf{0}$ has any solutions, then Problem (11.1) has a minimum with minimizers given by the solution or solutions of $\nabla f(x) = \mathbf{0}$.

Solving either problem yields the same set of minimizers By Theorem 3.1, if either Problem (9.7) or Problem (11.1) possesses an optimum, then the set of optimizers of both problems are the same.

In summary, by solving $\nabla f(x) = \mathbf{0}$ for $x^\star = \begin{bmatrix} \beta^\star \\ \gamma^\star \end{bmatrix}$ we will find a maximizer of Problem (9.7).

Setting $\nabla f(x) = \mathbf{0}$ and re-arranging, we obtain $A^\dagger A x = A^\dagger b$, which we can re-write as:

$$\mathcal{A}x = \mathcal{B}, \tag{11.4}$$

where $\mathcal{A} = A^\dagger A$ and $\mathcal{B} = A^\dagger b$. As we have observed, (11.4) is a linear equation in $x = \begin{bmatrix} \beta \\ \gamma \end{bmatrix}$ and so, in principle, can be solved through factorization of $\mathcal{A}$ and a single forwards and backwards update, so long as $\mathcal{A}$ is non-singular. For the sensitivity analysis to be discussed in Section 11.1.7.2, we note that the solution can also be written as $x = [A^\dagger A]^{-1} A^\dagger b$; however, we emphasize that inversion of $A^\dagger A$ is not usually a good way to solve the equations.

11.1.5 Further transformation

There are two basic drawbacks to solving (11.4). First, if $\mathcal{A} = A^\dagger A$ is not positive definite then it is singular and so we will encounter zero pivots in the factorization of $A^\dagger A$. In this case, there will be multiple solutions of (11.4). Even if $A^\dagger A$ is non-singular, however, it is important to realize that the condition number of $A^\dagger A$ can be large. For example, if A is a square non-singular matrix, then the condition number of $A^\dagger A$ is the square of the condition number of A. Consequently, factorizing $A^\dagger A$ into LU factors may lead to numerical difficulties.

This observation suggests that we might try the QR factorization of $A^\dagger A$. However, there is an alternative approach that transforms the problem further and avoids the need to calculate and factorize $A^\dagger A$.

To motivate this alternative approach, note that if we could solve $Ax = b$ then, by (11.2), we would have a solution of $\nabla f(x) = \mathbf{0}$. However, the matrix A typically has more rows than columns, so in general it will not be possible to find a solution of $Ax = b$. Fortunately, by (11.2), all we require is a value of x such that $Ax - b$ is in the null space of $A^\dagger$, since this will be a solution of $\nabla f(x) = \mathbf{0}$. So, instead of calculating and factorizing $A^\dagger A$, we QR factorize A itself to obtain (ignoring any permutations of the rows or columns of A):

$$A = QR,$$

with $Q \in \mathbb{R}^{m \times m}$ unitary, $R = \begin{bmatrix} U \\ \mathbf{0} \end{bmatrix} \in \mathbb{R}^{m \times n}$ upper triangular, with $U \in \mathbb{R}^{n' \times n}$ upper triangular and U having n' linearly independent rows. We have:

$$
\begin{aligned}
\forall x \in \mathbb{R}^n, \ f(x) &= \frac{1}{2}(Ax - b)^\dagger(Ax - b), \\
&= \frac{1}{2}(x^\dagger A^\dagger - b^\dagger)(Ax - b), \\
&= \frac{1}{2}(x^\dagger R^\dagger Q^\dagger - b^\dagger)(QRx - b), \ \text{by definition of } QR, \\
&= \frac{1}{2}(x^\dagger R^\dagger Q^\dagger - b^\dagger QQ^\dagger)(QRx - QQ^\dagger b), \ \text{since } Q \text{ is unitary,} \\
&= \frac{1}{2}(x^\dagger R^\dagger - b^\dagger Q)Q^\dagger Q(Rx - Q^\dagger b), \ \text{on factorizing,} \\
&= \frac{1}{2}(x^\dagger R^\dagger - b^\dagger Q)(Rx - Q^\dagger b), \ \text{because } Q \text{ is unitary,} \\
&= \frac{1}{2}\left(x^\dagger \begin{bmatrix} U \\ \mathbf{0} \end{bmatrix}^\dagger - b^\dagger Q\right)\left(\begin{bmatrix} U \\ \mathbf{0} \end{bmatrix}x - Q^\dagger b\right), \\
&\qquad\quad \text{where } R = \begin{bmatrix} U \\ \mathbf{0} \end{bmatrix}, \\
&= \frac{1}{2}\left\| \begin{bmatrix} U \\ \mathbf{0} \end{bmatrix}x - Q^\dagger b \right\|_2^2, \\
&= \frac{1}{2}\left\| \begin{bmatrix} U \\ \mathbf{0} \end{bmatrix}x - \begin{bmatrix} [Q^\parallel]^\dagger \\ [Q^\perp]^\dagger \end{bmatrix}b \right\|_2^2, \ \text{where } Q = [\, Q^\parallel \quad Q^\perp \,], \\
&\qquad\quad \text{with } Q^\parallel \in \mathbb{R}^{m \times n'}, \ Q^\perp \in \mathbb{R}^{m \times (m-n')}, \\
&= \frac{1}{2}\left\| \begin{bmatrix} Ux - [Q^\parallel]^\dagger b \\ \mathbf{0}x - [Q^\perp]^\dagger b \end{bmatrix} \right\|_2^2, \\
&= \frac{1}{2}\left\| Ux - [Q^\parallel]^\dagger b \right\|_2^2 + \frac{1}{2}\left\| [Q^\perp]^\dagger b \right\|_2^2, \\
&\qquad\quad \text{by definition of the } L_2 \text{ norm.}
\end{aligned}
$$

The last expression is minimized when $Ux = [Q^\parallel]^\dagger b$ since $\frac{1}{2}\left\| [Q^\perp]^\dagger b \right\|_2^2$ is inde-

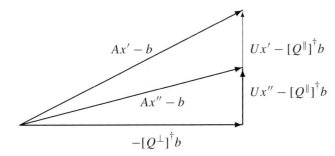

Fig. 11.1. Resolution of the vector $Ax - b$ into two perpendicular vectors for the values $x = x'$ and $x = x''$.

pendent of x. Geometrically, we have resolved the vector $Ax - b$ into the sum of two vectors:

- $Ux - [Q^{\|}]^{\dagger}b$, which depends on x, and
- $0x - [Q^{\perp}]^{\dagger}b = -[Q^{\perp}]^{\dagger}b$, which does not depend on x.

When these two vectors are added together, the length of $Ax - b$ is mostly determined by the length of $-[Q^{\perp}]^{\dagger}b$, so that the dependence of the length of $Ax - b$ on x is "swamped" by the effect of $-[Q^{\perp}]^{\dagger}b$. However, because these two vectors are perpendicular to each other we can write $\|Ax - b\|_2^2$ as the sum of the squares of the lengths of these vectors. The situation is illustrated in Figure 11.1 for two values of x, namely $x = x'$ and $x = x''$. The columns $Q^{\|}$ are such that $[Q^{\|}]^{\dagger}b$ "aligns" with Ux, whereas the columns $Q^{\perp}$ are such that $(-[Q^{\perp}]^{\dagger}b)$ is perpendicular to Ux. This allows us to decompose $\|Ax - b\|_2^2$ as the sum of $\left\|Ux - [Q^{\|}]^{\dagger}b\right\|_2^2$ and $\left\|[Q^{\perp}]^{\dagger}b\right\|_2^2$. The function f is minimized at a value of x that also minimizes $\left\|Ux - [Q^{\|}]^{\dagger}b\right\|_2^2$. If U is non-singular then the first-order necessary conditions for minimizing $\left\|Ux - [Q^{\|}]^{\dagger}b\right\|_2^2$ are $Ux = [Q^{\|}]^{\dagger}b$.

Exercise 11.4 confirms that we can transform the problem of solving the first-order necessary conditions $\nabla f(x) = \mathbf{0}$, which involve the coefficient matrix $A^{\dagger}A$, into the problem of solving a linear equation $Ux = [Q^{\|}]^{\dagger}b$ that does not suffer from the potential ill-conditioning of the matrix $A^{\dagger}A$. Moreover, once we have QR factorized the matrix A, we can obtain the solution to Problem (11.1) by:

- evaluating $y^{\star} = [Q^{\|}]^{\dagger}b$, and
- performing a backwards substitution to solve $Ux^{\star} = y^{\star}$.

(If $n' < n$ then there will be multiple solutions. In this case, any particular solution can be found or the value of x having the smallest L_2 norm can be found by performing the **complete orthogonal factorization** of A instead of QR factorization [45, section 2.2.5.3]. See Section 5.8.2.2.)

The solution $x^* = \begin{bmatrix} \beta^* \\ \gamma^* \end{bmatrix}$ specifies the maximum likelihood estimate of the relationship between the independent and dependent variables:

$$\forall \psi \in \mathbb{R}^{n-1}, \zeta = [\beta^*]^\dagger \psi + \gamma^*.$$

The drawback of this approach for large sparse systems is that Q is usually dense even if A is not dense. For large sparse systems we may use an iterative technique or choose to solve (11.4) (assuming that $A^\dagger A$ is sparse) even despite the potential ill-conditioning of the coefficient matrix.

Exercise 11.6 explores the relationship between finding an approximate solution to an inconsistent linear system $Ax = b$ and least-squares problems that involve the L_2 norm of $Ax - b$. Exercise 11.6 also considers objectives based on L_1 and L_∞ norms of $Ax - b$. See [45, section 4.2.3] for a discussion of how to use a sequence of solutions to least-squares problems to obtain the solution to a problem having the L_1 or L_∞ norm of $Ax - b$ as its objective. We can also treat these norms by transforming the problem into a constrained optimization problem using, for example, the transformation described in Section 3.1.3. We will explore some of these ideas in Part V when we consider least absolute value estimation in Section 15.3.

11.1.6 Relationship of optimality conditions to linear regression

In designing the values of $\psi(\ell)$ for the trials, there are two related issues to be addressed.

(i) Providing enough variety in the trials to ensure that $\nabla^2 f = A^\dagger A$ is positive definite. We discuss this issue in Sections 11.1.6.1 and 11.1.6.2.
(ii) Providing enough redundancy so that the effects of measurement error can be "averaged out." We discuss this briefly in Section 11.1.6.3.

In practice, to provide both variety and redundancy, we must usually have a much larger number of trials m than unknown parameters n so that A will have many more rows than columns.

11.1.6.1 Insufficient variety in the trials

The Hessian of the objective $\nabla^2 f(x) = A^\dagger A$ is independent of x and positive semi-definite, but it is not necessarily positive definite, depending on the nature of the trials, so that it *may* be singular as Exercise 11.7 shows.

If $\nabla^2 f(x)$ is singular then there will be many possible values of the parameters x that satisfy the maximum likelihood criterion in the model (9.1), based on the data from trials $\ell = 1, \ldots, m$. There is insufficient variety in the trials to determine a unique maximum likelihood estimator. In these circumstances it is customary to choose the maximum likelihood estimator that has the smallest L_2 norm. As mentioned above, the complete orthogonal factorization [45, section 2.2.5.3] can be used in this case to obtain the maximum likelihood estimator having the smallest L_2 norm. (See Section 5.8.2.2.)

11.1.6.2 Sufficient variety in the trials

On the other hand, if there is an n element subset $\{\ell_1, \ell_2, \ldots, \ell_n\}$ of the trials $\{1, \ldots, m\}$ such that the n rows of A corresponding to these trials are linearly independent, then $\nabla^2 f(x) = A^\dagger A$ is non-singular. (See Exercise 2.27, Parts (ii) and (iv).) This condition requires that there is enough variety in the trials so that they adequately "cover" the space of all possible trials. Notice that m must be at least as large as n to satisfy the condition.

11.1.6.3 Redundancy and validation of model

We may want to find not only the maximum likelihood estimator but also estimate the variance of the error. This issue is discussed, for example, in [53, chapter 7][103, chapter 7]. A further consideration is that we may also be interested in validating the linear model $\zeta = \beta^\dagger \psi + \gamma$. "Goodness of fit" of the linear regression model is discussed in [84, section 13.3][103, chapter 7]. In general, it requires that m be larger, and typically considerably larger, than n. It is sometimes necessary to increase the number trials in order to obtain enough redundancy [53, chapter 7].

11.1.7 Changes in the problem

In this section, we discuss changes in the problem involving an additional trial and changes to the measured values.

11.1.7.1 Additional trials

If additional trials are added then there will be additional rows added to A and additional entries added to b, necessitating factorization of the augmented A as mentioned in Section 5.6.3.

11.1.7.2 Sensitivity

If a measured value changes then b will change. We consider the sensitivity of the coefficients $\beta^\star$ and $\gamma^\star$ to changes in the measurements. That is, for each $\ell = 1, \ldots, m$, we will imagine that the ℓ-th measurement is actually $\zeta(\ell) + \chi_\ell$, with $\chi \in \mathbb{R}^m$. We calculate the sensitivity of $\beta^\star$ and $\gamma^\star$ to χ, evaluated at $\chi = 0$.

By Corollary 10.8, the sensitivity of the minimizer $x^\star$ is given by:

$$\frac{\partial x^\star}{\partial \chi}(0) = -[\nabla_{xx}^2 f(x^\star; 0)]^{-1} K(x^\star; 0),$$

where $\nabla_{xx}^2 f : \mathbb{R}^n \times \mathbb{R}^m \to \mathbb{R}^{n \times n}$ and $K : \mathbb{R}^n \times \mathbb{R}^m \to \mathbb{R}^{n \times m}$ are defined by:

$$\forall x \in \mathbb{R}^n, \forall \chi \in \mathbb{R}^m, \nabla_{xx}^2 f(x; \chi) = \frac{\partial^2 f}{\partial x^2}(x; \chi),$$
$$= A^\dagger A,$$
$$\forall x \in \mathbb{R}^n, \forall \chi \in \mathbb{R}^m, K(x; \chi) = \frac{\partial^2 f}{\partial x \partial \chi}(x; \chi),$$
$$= -A^\dagger.$$

That is, the sensitivity to χ_ℓ is given by $[A^\dagger A]^{-1} A^\dagger \mathbf{I}_\ell$, where $\mathbf{I}_\ell \in \mathbb{R}^m$ is a vector with zeros in all places except the ℓ-th place, which is a one. Note that this is the same as the solution of a multi-variate linear regression problem that had the same values of independent variables as in the base-case, but where the vector of measurements was changed from b to $\mathbf{I}_\ell$. Using the analysis in Section 11.1.5, we can calculate the sensitivity to χ_ℓ by:

- evaluating $y = [Q^{\parallel}]^\dagger \mathbf{I}_\ell$, and

- performing a backwards substitution to solve $U \dfrac{\partial x^\star}{\partial \chi}(0) = y$.

11.2 Power system state estimation

11.2.1 Transformation of objective

We use a similar transformation to the one in Section 11.1. In contrast to the transformation described in Section 11.1, however, we usually cannot assume that the measurement errors of the meters all have the same standard deviation. We will nevertheless assume that the measurement errors have known standard deviations, $\sigma_\ell, \ell \in \mathbb{M}$. Therefore, we define:

$$\forall x \in \mathbb{R}^n, f(x) = -\ln\phi(\tilde{G}; x) + \sum_{\ell \in \mathbb{M}} \ln\frac{1}{\sqrt{2\pi}\sigma_\ell}, \tag{11.5}$$

where ϕ, $\tilde{G}$, x, σ_ℓ, and $\mathbb{M}$ were defined in Section 9.2.2. Then, by definition of ϕ expressed in terms of the measurement functions $\tilde{g}_\ell$:

$$\forall x \in \mathbb{R}^n, f(x) = \sum_{\ell \in \mathbb{M}} \frac{(\tilde{g}_\ell(x) - \tilde{G}_\ell)^2}{2\sigma_\ell^2},$$

$$= \frac{1}{2}(\tilde{g}(x) - \tilde{G})^\dagger [\Sigma]^{-2}(\tilde{g}(x) - \tilde{G}), \qquad (11.6)$$

where:

- $\Sigma \in \mathbb{R}^{\mathbb{M} \times \mathbb{M}}$ is the diagonal matrix with ℓ-th diagonal entry equal to σ_ℓ, $\ell \in \mathbb{M}$,
- $\tilde{g} : \mathbb{R}^n \to \mathbb{R}^{\mathbb{M}}$ is the vector of all measurement functions, and
- $\tilde{G} \in \mathbb{R}^{\mathbb{M}}$ is the vector of all measurements.

The transformed problem is:

$$\min_{x \in \mathbb{R}^n} f(x), \qquad (11.7)$$

where we recall that $x \in \mathbb{R}^n$ with $n = 2n_{PQ} + 1$. As in Section 11.1, we have a least-squares problem since the objective is (half of) the sum of squares of terms. Each term is of the form $(\tilde{g}_\ell(x) - \tilde{G}_\ell)/\sigma_\ell$. Since $(\tilde{g}(\bullet) - \tilde{G})$ is non-linear, we classify Problem (11.7) as a **non-linear least-squares problem**.

11.2.2 Derivatives of objective

We can differentiate (11.6) to obtain the gradient of f, which we manipulate into a form that makes the calculation of the second derivative more convenient.

$$\forall x \in \mathbb{R}^n, \nabla f(x) = \tilde{J}(x)^\dagger [\Sigma]^{-2}(\tilde{g}(x) - \tilde{G}),$$

$$= \sum_{\ell \in \mathbb{M}} \nabla \tilde{g}_\ell(x)[\Sigma_\ell]^{-2}(\tilde{g}_\ell(x) - \tilde{G}_\ell), \qquad (11.8)$$

$$\forall x \in \mathbb{R}^n, \nabla^2 f(x) = \tilde{J}(x)^\dagger [\Sigma]^{-2}\tilde{J}(x) + \sum_{\ell \in \mathbb{M}} \nabla^2 \tilde{g}_\ell(x)[\Sigma_\ell]^{-2}(\tilde{g}_\ell(x) - \tilde{G}_\ell),$$

$$(11.9)$$

where $\tilde{J}$ is the Jacobian of $\tilde{g}$ and $\nabla \tilde{g}_\ell$ is the transpose of the ℓ-th row of $\tilde{J}$.

11.2.3 Optimality conditions and algorithms

In Section 11.2.3.1 we will compare the optimality conditions of Problems (9.8) and (11.7). Then, in Section 11.2.3.2, we discuss finding points that satisfy the optimality conditions for Problem (11.7).

11.2.3.1 *Qualitative comparison between Problems (9.8) and (11.7)*

The first-order necessary conditions for minimizing Problem (9.8), $\nabla\phi(\tilde{G}; x) = \mathbf{0}$, are non-linear as Exercise 11.9 shows. The first-order necessary conditions for minimizing Problem (11.7), $\nabla f(x) = \mathbf{0}$, are also non-linear since the measurement functions are, in general, non-linear. Let us consider the measurement functions in detail.

    (i) Each voltage magnitude measurement function, $\tilde{u}_k(x) = u_k$, is in fact linear.

    (ii) In Section 8.2.4, we approximated the Jacobian of the power flow equations by a constant matrix. We observed that this approximation was often reasonable in practice. Therefore, the real and reactive injection measurement functions and the real and reactive flow measurement functions are *approximately* linear. This observation and the expression for ∇f, (11.8), mean that the necessary conditions for Problem (11.7), $\nabla f(x) = \mathbf{0}$, are also approximately linear.

Recall the transformation of the objective of the linear regression problem described in Section 11.1. The non-linear objective was transformed into a quadratic objective, so that the necessary conditions for optimality were linear and could be solved, for example, with a single QR factorization, a matrix–vector multiplication, and a backwards update.

In this section, the transformation (11.5) transforms a non-linear objective into an *approximately* quadratic objective, so that, as we have discussed, the necessary conditions for optimality are *approximately* linear. Because the conditions are not exactly linear, however, we must use an algorithm such as the Newton–Raphson method described in 10.2 to iterate towards their solution.

To verify that the Newton–Raphson update will converge, we can try to apply the chord and Kantorovich theorems. Let us use the hypotheses of the theorems to *qualitatively* compare the convergence properties of the Newton–Raphson update when applied to solving the optimality conditions of:

- Problem (9.8); that is, $\nabla\phi(x) = \mathbf{0}$, and
- Problem (11.7); that is, $\nabla f(x) = \mathbf{0}$.

To apply the theorems, we need an estimate of Lipschitz constants for $\nabla^2\phi$ and $\nabla^2 f$. We know that $\nabla\phi$ is non-linear. (See Exercise 11.9.) If ∇f *were* linear, then $\nabla^2 f$ would be constant and $L = 0$ would be a Lipschitz constant for $\nabla^2 f$. Since ∇f is approximately linear, then $\nabla^2 f$ is approximately constant and a Lipschitz constant can be found for $\nabla^2 f$ that is smaller than a Lipschitz constant for $\nabla^2\phi$.

Recall from Exercise 8.3 that the smaller the value of the Lipschitz constant for the Jacobian of the non-linear equations, (that is, the smaller the value of the Lipschitz constant for $\nabla^2 f$ and $\nabla^2 \phi$) the easier it is to satisfy the conditions of the chord and Kantorovich Theorems 7.3 and 7.4, assuming that all else were equal. Therefore, the transformation from ϕ to f has made it easier to satisfy the conditions of the theorems. We expect the radii ρ_-, ρ_+, and $\overline{\rho}$ defined in Theorems 7.3 and 7.4 to be larger for the problem of solving $\nabla f(x) = \mathbf{0}$ than for the problem of solving $\nabla \phi(x) = \mathbf{0}$. That is, we can expect to converge to a solution from a poorer initial guess if we apply the chord or Newton–Raphson update to solve $\nabla f(x) = \mathbf{0}$ instead of applying it to solve $\nabla \phi(x) = \mathbf{0}$. This is an example of using the chord and Kantorovich theorems to qualitatively compare a problem and its transformed version, without explicitly evaluating the parameters required for the theorem.

11.2.3.2 Problem (11.7)

Hessian We now consider the Hessian $\nabla^2 f$ from (11.9) in detail. It consists of the sum of two terms:

(i) $\tilde{J}(x)^\dagger [\Sigma]^{-2} \tilde{J}(x)$, which is of the form $A^\dagger A$ for $A = [\Sigma]^{-1} \tilde{J}(x)$ and so by Exercise 2.27, the matrix $\tilde{J}(x)^\dagger [\Sigma]^{-2} \tilde{J}(x)$, is positive semi-definite, and

(ii) $\sum_{\ell \in \mathbb{M}} \nabla^2 \tilde{g}_\ell(x) [\Sigma_\ell]^{-2} (\tilde{g}_\ell(x) - \tilde{G}_\ell)$, which can turn out to be not positive semi-definite.

Despite the possibility that the second term in the Hessian is not positive semi-definite, we have observed that $\tilde{g}(x)$ is approximately linear. Therefore, each gradient $\nabla \tilde{g}_\ell$, $\ell \in \mathbb{M}$ is an approximately constant vector and $\nabla^2 \tilde{g}_\ell$ is approximately equal to the zero matrix. Moreover, we expect that near to a solution of the problem we will find that $(\tilde{g}_\ell(x) - \tilde{G}_\ell) \approx 0$, unless the measurement $\tilde{G}_\ell$ is grossly in error. Therefore, the second term in the Hessian, being the sum of terms of the form $\nabla^2 \tilde{g}_\ell(x) [\Sigma_\ell]^{-2} (\tilde{g}_\ell(x) - \tilde{G}_\ell)$, will only have a small influence on the Hessian $\nabla^2 f$.

This suggests that f itself may be "approximately" convex, and in practice f often has only one local minimum. However, note well that we have not proved that there is a unique local minimum. For some least-squares problems, the objective can be significantly non-convex and there may be multiple local minima. Fortunately, many least-squares problems, including the power system state estimation problem, are relatively "well-behaved."

Search direction Recall that in defining a search direction, we found that $\Delta x^{(\nu)} = -M \nabla f(x^{(\nu)})$ is a descent direction if M is positive definite. We know that the matrix $\tilde{J}(x)^\dagger [\Sigma]^{-2} \tilde{J}(x)$ is positive semi-definite, even though it may not be positive definite; however, we are not sure about even the positive semi-definiteness of $\nabla^2 f$. Therefore, instead of using the exact Newton–Raphson update, we approximate

$\nabla^2 f$ by its first term:

$$\tilde{J}(x)^{\dagger}[\Sigma]^{-2}\tilde{J}(x), \tag{11.10}$$

which *is* positive semi-definite, and then solve for the approximate update direction:

$$\begin{aligned}
\tilde{J}(x^{(\nu)})^{\dagger}[\Sigma]^{-2}\tilde{J}(x^{(\nu)})\Delta x^{(\nu)} &= -\nabla f(x^{(\nu)}), \\
&= \tilde{J}(x^{(\nu)})[\Sigma]^{-2}(\tilde{G} - \tilde{g}(x^{(\nu)})). \tag{11.11}
\end{aligned}$$

This approximation is called the **Gauss–Newton method** [84, section 13.2].

We must still consider the possibility that $\tilde{J}(x^{(\nu)})^{\dagger}[\Sigma]^{-2}\tilde{J}(x^{(\nu)})$ is not positive definite. We can follow the approach discussed in Section 10.2.3.2 and add terms to the diagonal of the matrix during factorization to ensure that the modified matrix is positive definite. This issue is closely related to the placement of meters in the system, which we will discuss in Section 11.2.4.

Search direction by solving a related linear least-squares problem The use of (11.11) to calculate a search direction suffers from a similar drawback to the solution of (11.4) in the linear case. By defining $A = [\Sigma]^{-1}\tilde{J}(x^{(\nu)})$ and $b = [\Sigma]^{-1}(\tilde{G} - \tilde{g}(x^{(\nu)}))$, note that (11.11) is equivalent to $A^{\dagger}A\Delta x^{(\nu)} = A^{\dagger}b$, which is the same form as the optimality conditions for the multi-variate linear regression problem. We can therefore find $\Delta x^{(\nu)}$ by noting that $\Delta x^{(\nu)}$ is the solution to the *linear* least-squares problem:

$$\min_{\Delta x \in \mathbb{R}^n} \frac{1}{2}\|A\Delta x - b\|_2^2. \tag{11.12}$$

Again, we can use QR factorization of $A = [\Sigma]^{-1}\tilde{J}(x^{(\nu)})$ to avoid the possible ill-conditioning in $A^{\dagger}A$. However, Q will in general be dense and so for large sparse problems we may choose to LU factorize $A^{\dagger}A = \tilde{J}(x^{(\nu)})^{\dagger}[\Sigma]^{-2}\tilde{J}(x^{(\nu)})$ instead.

Levenberg–Marquandt An alternative approach is to approximate the possibly not positive semi-definite term $\sum_{\ell \in \mathbb{M}} \nabla^2 \tilde{g}_\ell(x)[\Sigma_\ell]^{-2}(\tilde{g}_\ell(x) - \tilde{G}_\ell)$ by the positive definite matrix $\lambda \mathbf{I}$, where $\lambda > 0$ is chosen to be large enough to make the resulting approximation of the Hessian positive definite. As mentioned in Section 10.2.3.3, this is called the **Levenberg–Marquandt method** [45, section 4.7.3][84, section 13.2] and is related to the trust region approach that was mentioned in Section 10.2.4 [84, section 10.6].

Further approximation We can further approximate $\tilde{J}$ using the using the fast-decoupled or other approximations to the Jacobian of the power flow equations, as in the discussion of the solution of the power flow equations in Section 8.2.4.2.

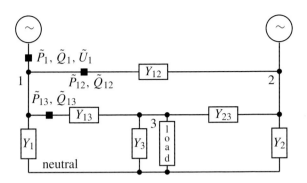

Fig. 11.2. The three-bus power system state estimation problem repeated from Figure 9.2.

11.2.4 Placement of meters in the system

11.2.4.1 Insufficient variety in the measurements

If the measurements are not spread out throughout the system, then $\tilde{J}(x)^{\dagger}[\Sigma]^{-2}\tilde{J}(x)$ can be singular. For example, consider the system in Figure 9.2, which is repeated in Figure 11.2. The are five unknown variables: u_1, θ_2, u_2, θ_3, and u_3. There are seven measurements: $\tilde{P}_1$, $\tilde{Q}_1$, $\tilde{U}_1$, $\tilde{P}_{12}$, $\tilde{Q}_{12}$, $\tilde{P}_{13}$, and $\tilde{Q}_{13}$. However, since:

$$\tilde{p}_1(x) = \tilde{p}_{12}(x) + \tilde{p}_{13}(x),$$
$$\tilde{q}_1(x) = \tilde{q}_{12}(x) + \tilde{q}_{13}(x),$$

there is redundant information concerning bus 1. This would enable us to estimate the voltage magnitude and flows around bus 1, even in the presence of measurement errors. However, there is not enough information to estimate all the voltage and flows in the system.

In fact, similarly to the case discussed in Section 11.1.6.1, the matrix $A^{\dagger}A = \tilde{J}(x)^{\dagger}[\Sigma]^{-2}\tilde{J}(x)$ is singular for every x. We can certainly add diagonal terms to such a singular $\tilde{J}(x)^{\dagger}[\Sigma]^{-2}\tilde{J}(x)$ as suggested in Section 10.2.3 to create a positive definite matrix and obtain a descent direction; however, because $\tilde{J}(x)^{\dagger}[\Sigma]^{-2}\tilde{J}(x)$ is singular for *every* x it turns out that there are many minimizers of f.

If we apply a minimization algorithm and if it converges, it will have converged to some arbitrary minimizer depending on the algorithm details. Since our measurements do not determine the split of power flowing into buses 2 and 3, there will be many sets of voltages and angles θ_2, u_2, θ_3, and u_3 that are consistent with maximizing the likelihood of the observed measurements. We say that the system is **unobservable** [80, chapter 7]. The starting point for the iteration will have a strong influence on the solution. As in the linear least-squares case, we can, in principle, pick one of the multiple minimizers by seeking a solution having smallest norm.

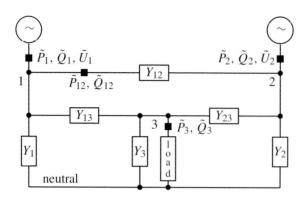

Fig. 11.3. The three-bus power system state estimation problem with spread out measurements repeated from Figure 9.3.

If we are *designing* a measurement system, then singularity of $\tilde{J}(x)^{\dagger}[\Sigma]^{-2}\tilde{J}(x)$ for a candidate meter placement plan suggests that we should add more meters to the plan [80, section 7.3]. If we are *operating* a measurement system and we find that because of, for example, meter failures, the matrix $\tilde{J}(x)^{\dagger}[\Sigma]^{-2}\tilde{J}(x)$ is singular, then we cannot estimate the state completely. In practice, in the latter case, the user of the software usually specifies **pseudo-measurements**; that is, guesses at what the actual measurement would be, based on experience, so that a rough estimate of the complete state can be found [80, section 7.3].

11.2.4.2 Sufficient variety in the measurements

Usually, if there is sufficient variety in the measurements, the positive semi-definite matrix $\tilde{J}(x)^{\dagger}[\Sigma]^{-2}\tilde{J}(x)$ will turn out to be positive definite for almost all, if not all values of x, and hence be non-singular. If it is non-singular then the approximate update equation (11.11) has a unique solution.

For example, for the arrangement in Figure 9.3, which is repeated in Figure 11.3, for almost all values of x there is a five element subset of the rows of $\tilde{J}(x)$ that is linearly independent, so that $\tilde{J}(x)^{\dagger}[\Sigma]^{-2}\tilde{J}(x)$ is non-singular. The Newton–Raphson update and variants will usually bring us towards the unique minimizer of the problem and the system is said to be **observable**. Further details, including methods to determine if the rows of $\tilde{J}$ are linearly independent for almost all values of x, can be found in [62].

Even if the rows of $\tilde{J}$ are linearly independent for almost all values of x, it is possible to encounter singular $\tilde{J}(x^{(\nu)})^{\dagger}[\Sigma]^{-2}\tilde{J}(x^{(\nu)})$ at some iterations ν; however, this is extremely rare in practice for the power system state estimation problem and is also relatively rare for a variety of non-linear least-squares problems. If the matrix does turn out to be singular, then it can be modified during factorization as discussed in Section 11.2.3.2 to approximate it by a positive definite matrix.

11.2.4.3 Sensitivity

We can consider variation of the estimate with variation in the measurement data. This is explored in Exercise 11.11.

Exercises

Multi-variate linear regression

11.1 Consider the maximum likelihood problem formulated in Exercise 9.1. Show that this maximum likelihood problem can be transformed to a problem with objective f that is quadratic in x. That is, we still have a *linear* least-squares problem, even though the measurement function is non-linear in the independent measurement variables.

11.2 Prove (11.2) and (11.3).

11.3 In this exercise we show that the first-order necessary conditions for Problem (9.7) and Problem (11.1) are equivalent.

   (i) Write out explicitly the first-order necessary conditions for a maximum of Problem (9.7).

   (ii) Show that the set of points satisfying the first-order necessary conditions for Problem (9.7) is the same as the set of minimizers of Problem (11.1).

11.4 Let $A \in \mathbb{R}^{m \times n}$, $b \in \mathbb{R}^m$, and let $\nabla f : \mathbb{R}^n \to \mathbb{R}^n$ be as defined in (11.2):

$$\forall x \in \mathbb{R}^n, \nabla f(x) = A^\dagger (Ax - b).$$

Suppose that we have factorized $A = QR$ with $Q \in \mathbb{R}^{m \times m}$ unitary. Moreover, suppose we have partitioned Q into $Q = \begin{bmatrix} Q^\| & Q^\perp \end{bmatrix}$, where $Q^\| \in \mathbb{R}^{m \times n}$, $Q^\perp \in \mathbb{R}^{m \times (m-n)}$, and that $R = \begin{bmatrix} U \\ 0 \end{bmatrix} \in \mathbb{R}^{m \times n}$ is upper triangular with U non-singular. Show that $x^\star = U^{-1}[Q^\|]^\dagger b$ satisfies $\nabla f(x) = 0$.

11.5 In this exercise we solve a multi-variate linear regression problem.

   (i) Use the MATLAB backslash operator to find the affine function:

$$\forall \psi \in \mathbb{R}, \zeta = \beta^\star \psi + \gamma^\star,$$

with $\beta^\star \in \mathbb{R}$ and $\gamma^\star \in \mathbb{R}$ that best fits the following pairs of data $(\psi(\ell), \zeta(\ell))$, for $\ell = 1, \ldots, 7$. Assume that the measurements $\zeta(\ell)$ are subject to independent Gaussian errors of identical standard deviation and zero mean so that the least-squares technique yields the best fit. (Hint: You need to explicitly form the matrix A and vector b discussed in Section 11.1.1.)

ℓ	1	2	3	4	5	6	7
$\psi(\ell)$	0.27	0.2	0.8	0.4	0.2	0.7	0.5
$\zeta(\ell)$	0.3	0.65	0.75	0.4	0.15	0.6	0.5

   (ii) Find the sensitivity of $\beta^\star$ and $\gamma^\star$ to each measurement $\zeta(\ell)$.

11.6 Let $A \in \mathbb{R}^{m \times n}$, $b \in \mathbb{R}^m$, and recall the inconsistent linear system $Ax = b$ discussed in Section 5.8. If there is no x satisfying $Ax = b$ we can instead think of minimizing, for example:

- $f(x) = \|Ax - b\|_1$,
- $f(x) = \|Ax - b\|_1^2$,
- $f(x) = \|Ax - b\|_2$,
- $f(x) = \|Ax - b\|_2^2$,
- $f(x) = \|Ax - b\|_\infty$, or
- $f(x) = \|Ax - b\|_\infty^2$,

with respect to x to find a point $x^\star$ that most nearly satisfies $Ax = b$ in the sense of minimizing f.

Consider the choice $f(x) = \frac{1}{2}\|Ax - b\|_2^2 = \frac{1}{2}(Ax - b)^\dagger(Ax - b)$. Notice that the problem $\min_{x \in \mathbb{R}^n} f(x)$ is a linear least-squares problem and we have shown how to solve it. Can you suggest why it is customary to choose the square of the L_2 norm instead of one of the other norms or their squares when trying to find x that most nearly satisfies $Ax = b$?

11.7 Give an example with $n = 3$ and $m = 4$ of Problem (11.1) where $\nabla^2 f(x)$ is singular.

Power system state estimation

11.8 Suppose that $\hat{x} \in \mathbb{R}^n$ were the true value of the voltage angles and magnitudes in the system. Show that the expected value of $(\tilde{g}_\ell(\hat{x}) - \tilde{G}_\ell)^2/(\sigma_\ell)^2$ is 1. (Hint: Refer to the discussion in Section 9.2.2.4 of the error distribution.)

11.9 In this exercise we investigate the first-order necessary conditions for Problem (9.8).

(i) Write out the first-order necessary conditions for Problem (9.8).
(ii) Show that the first-order necessary conditions for Problem (9.8) would be non-linear even if $\tilde{g}$ were linear.

11.10 Consider the state estimation problem involving the arrangement of meters shown in Figure 9.2, which is repeated in Figure 11.2.

(i) Write down the form of the measurement functions $\tilde{g}$. (Use the results of Exercise 9.2.)
(ii) Calculate the Jacobian $\tilde{J}$ of $\tilde{g}$.
(iii) Show that $\tilde{J}$ has linearly dependent rows.

11.11 Consider the state estimation problem involving the arrangement of meters shown in Figure 9.3, which is repeated in Figure 11.3.

(i) Write down the form of the measurement functions $\tilde{g}$. (Use the results of Exercise 9.2.)

(ii) Calculate f and ∇f.

(iii) Use MATLAB to perform one update of the Newton–Raphson method for solving Problem (11.7) with the Gauss–Newton approximation (11.10) to the Hessian. Use the transmission parameters from Exercise 8.12 and a flat start $\begin{bmatrix} \theta \\ u \end{bmatrix} = \begin{bmatrix} 0 \\ 1 \end{bmatrix}$ as the initial guess for x. Assume that the real power injection measurements at buses 1 and 2 are both equal to 1, the reactive power injection measurements at buses 1 and 2 are both equal to 0.5, and all voltage measurements are equal to 1.0. Also, assume that $\tilde{P}_{12} = 0.5$, $\tilde{Q}_{12} = 0.25$, $\tilde{P}_3 = -2$, and $\tilde{Q}_3 = -1$. Assume measurement error standard deviations of $\sigma_\ell = 0.02$ for all measurements.

(iv) Use the MATLAB function lsqnonlin to solve the problem specified in Part (iii). You should write MATLAB M-files to evaluate both $[\Sigma]^{-1}(\tilde{g} - \tilde{G})$ and $[\Sigma]^{-1}\tilde{J}$. Specify that you are supplying the Jacobian by setting the Jacobian option to on using the optimset function. Use the Gauss–Newton method by specifying both LevenbergMarquadt to off and LargeScale to off using the optimset function and use a flat start $\begin{bmatrix} \theta \\ u \end{bmatrix} = \begin{bmatrix} 0 \\ 1 \end{bmatrix}$ as the initial guess for x.

(v) Repeat the previous part but use the default Levenberg–Marquandt method. Specify that you are supplying the Jacobian by setting the Jacobian option to on using the optimset function.

(vi) Calculate the sensitivity of the solution to changes in each measurement.

(vii) Is the state estimation problem ill-conditioned?

Part IV

Equality-constrained optimization

12

Case studies of equality-constrained optimization

In this chapter we will introduce two case studies:

(i) production, at least-cost, of a commodity from machines, while meeting a total demand (Section 12.1), and

(ii) state estimation in an electric power system where the power injections at some of the buses are known to high accuracy (Section 12.2).

Both problems will turn out to be equality-constrained optimization problems. The first will introduce several new ideas in problem formulation, while the second will build on the state estimation case study from Section 9.2.

12.1 Least-cost production

12.1.1 Motivation

Consider a machine that makes a certain product, requiring some costly input to produce. In many industries it is possible to **stock-pile** the product at low cost from day to day, week to week, or even season to season. In this case, it is natural to try to operate the machine at constant output. Ideally, the constant value of machine output would be matched to either:

- the point of maximum operating efficiency of the machine, or
- some other desirable operating point of the machine.

When demand is lower than production, some of the production goes into the stock-pile. When demand is higher than production, the stocks are used to help to meet demand.

However, if stock-piling is costly or inconvenient or if demand for the product varies rapidly, then to avoid over-supplies and shortages we have to vary production to follow variations in demand. If we have just one machine to make the product, it is easy to decide on the level of the production; however, if there are several

447

machines, and particularly if the operating efficiencies of the machines vary with output and differ from machine to machine, then the problem becomes more interesting.

An extreme example of this problem is in the production of electricity, where energy in a fuel is converted to energy in electricity. Typically the fuel cost is non-zero and it is not practical to stock-pile electrical energy over even very short periods. Moreover, electric generators have efficiencies that vary markedly with output. In electric power, the problem of least-cost production is called **economic dispatch** [22, section 1.8][123, chapter 3].

A variety of scheduling problems arise in manufacturing industries. Many of these problems involve discrete variables; however, we will not treat discrete variables explicitly in this case study, although an extension of the economic dispatch problem, called the **unit commitment** problem, does include discrete variables [79, chapter 12][123, chapter 5]. General references for problems with discrete variables include [22, 67, 85, 92, 122]. (We will treat a rather different kind of scheduling problem in Section 15.2, where we consider optimal routing in data networks. Again, this problem does not involve discrete variables.)

12.1.2 Formulation

12.1.2.1 Variables

Suppose that we own n machines or plants that are producing a commodity or product. We consider the production over a particular period of time. The length T of this period of time should be chosen to be short enough so that the production *per unit time* for the commodity or product by each machine can be well approximated by a constant over the time period T. That is, we are assuming that the plant is in **quasi-steady state**. Consequently, the total amount of commodity produced by a machine over the time period then completely specifies the production per unit time throughout the period by the machine. In practice, this assumption might only be roughly true.

Define $x_k \in \mathbb{R}$ to be the total amount of the commodity produced by machine k over the time period. We collect the production decisions of machines $k = 1, \ldots, n$, into a vector $x \in \mathbb{R}^n$, so that $x = \begin{bmatrix} x_1 \\ \vdots \\ x_n \end{bmatrix}$. The situation for $n = 3$ is illustrated in Figure 12.1, where three machines, represented as circles, are producing, respectively, amounts x_1, x_2, and x_3 of the commodity over the time period.

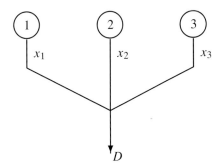

Fig. 12.1. Production
from three machines.

12.1.2.2 Production costs

We suppose that for $k = 1, \ldots, n$ there are functions $f_k : \mathbb{R} \to \mathbb{R}$ such that $f_k(x_k)$ is the cost for machine k to produce x_k over the time period T. We assume that we can choose to take machines out-of-service and assume that if a machine is out-of-service, then it does not cost anything to run. That is, $f_k(0) = 0$. At non-zero production levels, when the machine is in-service, the production cost will typically be non-zero. There may be a discontinuity in f_k at $x_k = 0$ or, as we will discuss in Section 12.1.2.4, it may not be possible to operate the machine at arbitrarily small non-zero levels of production.

12.1.2.3 Objective

A sensible criterion to choose the production levels is to seek the values of x_k, $k = 1, \ldots, n$, that minimize the total cost of production. That is, we would like to minimize the objective $f : \mathbb{R}^n \to \mathbb{R}$ defined by:

$$\forall x \in \mathbb{R}^n, \ f(x) = \sum_{k=1}^{n} f_k(x_k). \tag{12.1}$$

12.1.2.4 Constraints

Machine We assume that machine k has a maximum production capacity, say $\overline{x}_k$, and a minimum production capacity, $\underline{x}_k \geq 0$, reflecting the design capability of the machine. If the machine is in-service then the machine capacity constraints require that:

$$\underline{x}_k \leq x_k \leq \overline{x}_k, \tag{12.2}$$

while if the machine is off, then $x_k = 0$. Typically, $\underline{x}_k \neq 0$ representing the fact that under normal conditions we cannot operate the machine at arbitrarily small non-zero levels of production. (For example, for fossil-fueled electric generators this is due to flame stability in the furnace and due to allowable temperatures in the boiler tubes [123, section 2.1].) The feasible operating set for machine k is therefore:

$$\mathbb{S}_k = \{0\} \cup [\underline{x}_k, \overline{x}_k].$$

The set $\mathbb{S}_k$ is not convex if $\underline{x}_k > 0$. (See Exercise 12.1.) In specifying (12.1) we assumed that each function f_k was defined on the whole of $\mathbb{R}$; however, only the values of f_k on $\mathbb{S}_k$ are relevant to the solution of the problem. In defining f, we have implicitly extrapolated the cost function of each machine from its operating range, as specified by $\mathbb{S}_k$, to the whole of $\mathbb{R}$.

Production Let us assume that during the time period T we face a total demand for the commodity of quantity D, say, as represented in Figure 12.1. To meet demand, we must satisfy the constraint:

$$D = \sum_{k=1}^{n} x_k, \tag{12.3}$$

which we can write in the form $Ax = b$ with either of the following two choices for $A \in \mathbb{R}^{1 \times n}$ and $b \in \mathbb{R}$:

- $A = \mathbf{1}^{\dagger}, b = [D]$, or
- $A = -\mathbf{1}^{\dagger}, b = [-D]$.

For reasons that will be made clear in Section 13.5 when we discuss an economic interpretation of the problem, we prefer to use the second choice for A and b.

Machine and production combined The feasible operating set for all the machines is: $(\prod_{k=1}^{n} \mathbb{S}_k) \subset \mathbb{R}^n$, where the symbol $\prod$ means the **Cartesian product** (see Definition A.4), so that the feasible set for the problem is:

$$\underline{\mathbb{S}} = \left(\prod_{k=1}^{n} \mathbb{S}_k \right) \cap \{x \in \mathbb{R}^n | Ax = b\}.$$

Relaxation For the discussion in this chapter, however, we are going to:

- assume that each machine is in-service and operating, and
- ignore minimum and maximum production capacity constraints.

That is, we are going to relax the set of feasible operating points for machine k from the set $\mathbb{S}_k$ to the whole of $\mathbb{R}$ and correspondingly relax the feasible set for the problem from $\underline{\mathbb{S}}$ to:

$$\mathbb{S} = \{x \in \mathbb{R}^n | Ax = b\}.$$

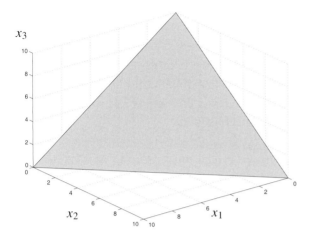

x_3

Fig. 12.2. Part of feasible set $\mathbb{S}$ for least-cost production case study.

Part of the feasible set $\mathbb{S}$ lying in the non-negative orthant is illustrated in Figure 12.2 for $n = 3$ and $D = 10$. In general, $\mathbb{S}$ is a **hyperplane** in $\mathbb{R}^n$. (See Definition A.52.)

12.1.2.5 Problem

Our relaxed optimization problem is:

$$\min_{x \in \mathbb{R}^n} \{ f(x) | Ax = b \}, \tag{12.4}$$

which is a linear equality-constrained optimization problem. As stated above, we have implicitly assumed that each function f_k has been **extrapolated** to being a function defined on the whole of $\mathbb{R}$.

12.1.2.6 Alternative formulation

If the cost function for each machine increases monotonically with production, as is normal for many types of machines, we could also consider solving the inequality-constrained problem:

$$\min_{x \in \mathbb{R}^n} \left\{ f(x) \,\middle|\, D \leq \sum_{k=1}^{n} x_k \right\}, \tag{12.5}$$

which is a further relaxation of our constraints, but which has the same minimum and minimizer as Problem (12.4) if costs are strictly monotonically increasing. (See Exercise 12.2.) The flexibility in the choice of formulation can sometimes be useful in adapting a problem formulation to an algorithm or in proving results about the problem; however, in this chapter we will only consider the equality-constrained version, Problem (12.4). (In Section 15.6.4.1, we use this flexibility to show conditions under which the "optimal power flow" problem is equivalent to a convex problem.)

12.1.2.7 Discussion

As we remarked in Section 3.3.4, if the solution $x^\star$ of the relaxed Problem (12.4) happens to satisfy the omitted minimum and maximum capacity constraints (12.2) so that $x_k^\star \in \mathbb{S}_k$, then the solution of the relaxed Problem (12.4) is optimal for the complete problem including the machine constraints:

$$\min_{x \in \underline{\mathbb{S}}} f(x).$$

If the omitted constraints are not satisfied, then we must consider them explicitly. We will explicitly consider inequality constraints such as the minimum and maximum production capacity constraints (12.2) in Part V, but the feasible set $\mathbb{S}_k$ for machine k is non-convex since it includes the points 0 and $\underline{x}_k$ but not any points between 0 and $\underline{x}_k$. (See Exercise 12.1.) We will not treat such feasible sets in this book. (See [79, chapter 12][123, chapter 5] for several approaches to treating the non-convex feasible set that involve adding discrete variables to the formulation and solving the resulting integer problem. See [22, 67, 85, 92, 122] for more general discussion of problems with discrete variables.)

12.1.3 Change in demand

We can expect that demand will change over time. Consequently, it is important to be able to estimate the change in the costs due to a change in demand from D to $D + \Delta D$, say.

12.1.4 Problem characteristics

12.1.4.1 Objective

Separability Although the objective defined in (12.1) is typically non-linear, it is expressed as the sum of functions, f_k, each of which depends only on a single entry, x_k, of x. That is, the objective is **additively separable**. (See Definition A.23.)

Average production costs Consider the **average cost per unit of production** $f_k(x_k)/x_k$ for machine k producing x_k. This is the ratio of:

- the costs over the time period, divided by
- the production over the time period.

At low levels of production, that is, for relatively small values of x_k, we would expect the average production cost to be relatively high. This is because there are usually costs that must be incurred whenever the plant is in-service and producing non-zero levels of output. These costs include, for example, costs of running auxiliary equipment such as pumps, lighting, computers, and air-conditioning. If we switch the machine off then we can also switch (most of) the auxiliary equipment off and so the auxiliary operating costs would be zero. If the machine is in-service, however, the auxiliary equipment must be operating and, moreover, the operating costs of this auxiliary equipment might not depend on x_k. For low but non-zero values of x_k, the costs of running the auxiliary equipment are averaged over a relatively small quantity of production and so the average costs are high. Figure 12.3 illustrates the average costs for $\underline{x}_k \leq x_k \leq \overline{x}_k$. For $x_k \approx \underline{x}_k$, the average costs are high. (In practice, the costs are typically less smooth than illustrated in Figure 12.3.)

As x_k increases from low levels, the average production costs typically decrease because the costs of operating the auxiliary equipment are averaged over a greater amount of production. We do not expect the average production cost to decrease indefinitely, however, so that for some x_k, the average costs $f_k(x_k)/x_k$ reach a minimum and then begin to increase again for larger values of x_k. The point where $f_k(x_k)/x_k$ is at a minimum is the point of maximum efficiency of the machine. This is illustrated in Figure 12.3.

Production costs If we multiply the values of $f_k(x_k)/x_k$ in Figure 12.3 by x_k, we obtain the production costs $f_k(x_k)$ as illustrated in Figure 12.4. The bullet • at $x_k = 0$, $f_k(x_k) = 0$ indicates that if the machine is out of service then the cost of production is zero. In the presence of auxiliary operating costs, there is a jump in cost between the value $f_k(0) = 0$ at $x_k = 0$ and the values of $f_k(x_k) > 0$ for $x_k > 0$. That is, if we extrapolate the shape of f_k from $\underline{x}_k$ to values $x_k < \underline{x}_k$ then we would find that at $x_k = 0$ the extrapolated value of the production cost function would be greater than zero due to the auxiliary operating costs.

Convexity Let us use the qualitative shape of the average cost curve in Figure 12.4 to try to determine whether or not f_k is convex. If $\underline{x}_k > 0$, then the operating set of the machine, $\mathbb{S}_k = \{0\} \cup [\underline{x}_k, \overline{x}_k]$, is not convex. (See Exercise 12.1.) Recall that the definition of convexity of a function required us to specify a convex test set on which to test the function for convexity.

To consider a convex test set, let us first suppose that $\underline{x}_k = 0$, so that the feasible set $\{0\} \cup [\underline{x}_k, \overline{x}_k] = [\underline{x}_k, \overline{x}_k] = [0, \overline{x}_k]$ *is* convex. Even in this case, however, if there are non-zero auxiliary operating costs then f_k is not a convex function on $[0, \overline{x}_k]$ because of the discontinuity in f_k, as Exercise 12.3 shows.

$f_k(x_k)/x_k$

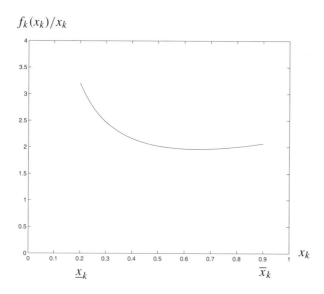

Fig. 12.3. The   average production cost $f_k(x_k)/x_k$ versus production x_k for a typical machine for $\underline{x}_k \leq x_k \leq \underline{x}_k$.

$f_k(x_k)$

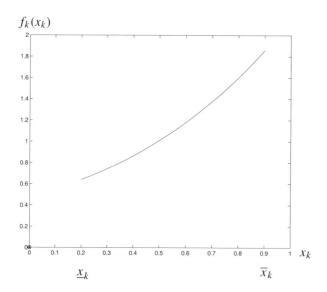

Fig. 12.4. Production cost $f_k(x_k)$ versus production x_k for a typical machine.

To identify a test set on which the objective might be convex, suppose that:

- $\underline{x}_k = 0$ and consider the set $\underline{\mathbb{S}}_k = \{x_k \in \mathbb{R} | 0 < x_k \leq \overline{x}_k\} \subset \mathbb{S}_k$, or
- $\underline{x}_k > 0$ and consider the set $\underline{\mathbb{S}}_k = \{x_k \in \mathbb{R} | \underline{x}_k \leq x_k \leq \overline{x}_k\} \subset \mathbb{S}_k$.

In both cases, $\underline{\mathbb{S}}_k$ is a convex set. Moreover, for both these cases, Figure 12.4 suggests that f_k is convex on $\underline{\mathbb{S}}_k$. The convexity of f_k on $\underline{\mathbb{S}}_k$ arises from a fairly plausible characterization of typical machines. We will assume that the cost function of each machine has been extrapolated to a function that is convex on the *whole* of $\mathbb{R}$.

For example, it is often reasonable to assume that $f_k : \underline{\mathbb{S}}_k \to \mathbb{R}$ is quadratic and of the form:

$$\forall x_k \in \underline{\mathbb{S}}_k, \ f_k(x_k) = \frac{1}{2} Q_{kk}(x_k)^2 + c_k x_k + d_k. \tag{12.6}$$

We can extrapolate f_k to being defined on the whole of $\mathbb{R}$ by simply writing $f_k : \mathbb{R} \to \mathbb{R}$ and extending the definition in (12.6) to all of $\mathbb{R}$, effectively redefining the function value at $x_k = 0$. For convex costs, $Q_{kk} \geq 0$. With non-zero auxiliary costs, $d_k > 0$. We also usually expect that $c_k > 0$.

Adding together the cost functions for all machines, we obtain:

$$\forall x \in \mathbb{R}^n, \ f(x) = \frac{1}{2} x^\dagger Q x + c^\dagger x + d,$$

where:

- $Q \in \mathbb{R}^{n \times n}$ is a diagonal matrix with k-th diagonal entry equal to Q_{kk},
- $c \in \mathbb{R}^n$ has k-th entry equal to c_k, and
- $d = \sum_{k=1}^n d_k \in \mathbb{R}$.

12.1.4.2 Constraint

Eliminating a variable The constraint (12.3) of Problem (12.4) is linear. Because of the linearity, by Corollary 3.7, we can use the equality constraint $Ax = b$ to eliminate one of the variables, say x_1, by writing:

$$x_1 = D - x_2 - \cdots - x_n.$$

Expressing the objective in terms of $x_2, \ldots, x_n$ yields an unconstrained problem with objective $f(\tilde{x})$ where:

$$\begin{aligned}
\tilde{x} &= \begin{bmatrix} D - x_2 - \cdots - x_n \\ x_2 \\ \vdots \\ x_n \end{bmatrix}, \\
&= \begin{bmatrix} D \\ 0 \end{bmatrix} + \begin{bmatrix} -\mathbf{1}^\dagger \\ I \end{bmatrix} \begin{bmatrix} x_2 \\ \vdots \\ x_n \end{bmatrix}, \\
&= \hat{x} + Z\xi, \\
&= \tau(\xi),
\end{aligned}$$

where:

$$\hat{x} = \begin{bmatrix} D \\ 0 \end{bmatrix} \in \mathbb{S},$$

$$Z = \begin{bmatrix} -\mathbf{1}^\dagger \\ I \end{bmatrix} \in \mathbb{R}^{n \times (n-1)},$$

$$\xi = \begin{bmatrix} x_2 \\ \vdots \\ x_n \end{bmatrix} \in \mathbb{R}^{n-1},$$

and where $\tau : \mathbb{R}^{n-1} \to \mathbb{S}$ is defined by:

$$\forall \xi \in \mathbb{R}^{n-1}, \tau(\xi) = \hat{x} + Z\xi,$$

and we note that τ is onto $\mathbb{S}$. The point $\hat{x}$ is a particular solution of the equations $Ax = b$. The matrix Z has columns that form a basis for the null space of A. (See Exercise 12.4.) The objective $f(\tilde{x})$ depends only on $\xi \in \mathbb{R}^{n-1}$. We have transformed the equality-constrained problem into an unconstrained problem with objective $\phi : \mathbb{R}^{n-1} \to \mathbb{R}$ defined by:

$$\forall \xi \in \mathbb{R}^{n-1}, \phi(\xi) = f(\tilde{x}),$$
$$= f\left(\begin{array}{c} D - \mathbf{1}^\dagger \xi \\ \xi \end{array} \right),$$
$$= f(\tau(\xi)).$$

The unconstrained problem:

$$\min_{\xi \in \mathbb{R}^{n-1}} f(\tilde{x}) = \min_{\xi \in \mathbb{R}^{n-1}} \phi(\xi)$$

could then be solved using the techniques developed in Chapter 10.

Elimination of variables is often an effective way to solve a problem with linear constraints. The number of variables will be decreased by the number of linearly independent constraints. In the case of Problem (12.4), the use of one constraint reduces the number of variables to $(n-1)$. If there were, say, m equality constraints, then there would be $(n-m)$ variables in the resulting transformed problem, assuming that the corresponding rows of A were linearly independent.

The objective, ϕ, of the resulting unconstrained problem is not additively separable. Sometimes, the loss of separability can make the problem more difficult to solve so that elimination of variables is not always the best approach.

Treating the constraint directly We will also explore approaches that treat the equality constraints directly. These will also preserve the separability of the objective and provide an intuitive interpretation of the characteristics of the solution.

In fact, these approaches will be derived in Section 13.1 by first eliminating the variables, then writing down the first-order necessary conditions for the resulting unconstrained problem, and then re-interpreting the optimality conditions in terms of the original variables.

12.1.4.3 Solvability

Since:

  (i) we have defined the objective function f on the whole of $\mathbb{R}^n$,
 (ii) the objective increases with increasing values of x_k, for each k, and
(iii) the constraint has a particularly simple form,

there will always be a solution to Problem (12.4). However, the solution might not satisfy the minimum and maximum machine constraints (12.2). That is, the solution to Problem (12.4) might require that a particular machine k operate at a level $x_k < \underline{x}_k$ or at a level $x_k > \overline{x}_k$. This issue will be treated explicitly in Part V when we cover inequality constraints. (As discussed previously, we will not treat non-convex feasible sets such as $\mathbb{S}_k = \{0\} \cup [\underline{x}_k, \overline{x}_k]$ in this book.)

12.2 Power system state estimation with zero injection buses

12.2.1 Motivation

12.2.1.1 Zero injection buses

Recall the power system state estimation problem introduced in Section 9.2. In that problem, we wrote down a measurement equation $\tilde{g}_\ell(x)$ for each measurement $\tilde{G}_\ell$. Consider the situation in Figure 12.5. Bus 2 does not have any load nor generation nor any measurement devices. Such buses are common at intermediate points in electric power systems between generators and load.

  Since there is no generator nor load at bus 2, the "injection" at bus 2 into the rest of the system is known to be exactly zero. In Section 6.2.2.5, we called this bus a **zero injection bus**. Since we know that the injection is zero, we can interpret the zero injection as an *exact* measurement of the real and reactive power, as illustrated in Figure 12.6, where we have shown an injection measurement but have written $P_2 = 0$ and $Q_2 = 0$ to emphasize the fact that we know that the injections are exactly zero. We will try to incorporate this implicit measurement into the formulation of the state estimation problem.

12.2.1.2 Ignoring zero injection buses

Before considering the zero injection buses in detail, we first consider the possibility of ignoring the zero injection buses in the formulation. The drawback of this approach is that if there are many zero injection buses and we ignore them, then

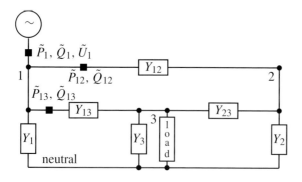

Fig. 12.5. Three-bus electric power system with a bus, bus 2, having neither load nor generation.

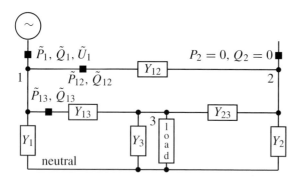

Fig. 12.6. Zero injection bus re-interpreted as an exact measurement.

the remaining system is likely to be unobservable, as defined in Section 11.2.4, because the remaining measurements will be insufficient to completely determine the unknowns. For example, if we use only the measurements shown explicitly in Figure 12.5 in the objective of Problem (9.8), then we do not have enough information to uniquely determine the voltage magnitudes and angles at buses 2 and 3. This is because we have seven measurements that are clustered around bus 1 and these are insufficient to estimate the five unknowns. (See Exercise 12.6.)

12.2.1.3 *Treating zero injection buses as accurate measurements*

Alternatively, we could think of the zero injection at bus 2 as a pair of very accurate real and reactive power measurements having zero value and zero measurement error as illustrated in Figure 12.6. This would give us a total of nine measurements. It turns out that these nine measurements are spread out around the system sufficiently to be able to calculate all five unknowns. These measurements also provide redundancy. (See Exercise 12.6.)

Recall that the optimality conditions developed in Section 11.2 for the state estimation problem involved either the matrix $[\Sigma]^{-1}$ or the matrix $[\Sigma]^{-2}$. If we interpret, say, the power injection at a zero injection bus to be a measurement having value $\tilde{G}_\ell = 0$ and having zero measurement error, then the associated measurement standard deviation would be $\sigma_\ell = 0$. Unfortunately, this makes the corresponding entry of $[\Sigma]^{-1}$ or $[\Sigma]^{-2}$ infinite.

An *ad hoc* approach to this issue would be to pick a small but non-zero value of σ_ℓ for each zero injection bus measurement. We must then compromise between:

(i) making σ_ℓ small enough to approximately represent our certainty that the measurement is zero, and

(ii) making σ_ℓ large enough so that the entries in $[\Sigma]^{-1}$ are not too large.

The entry in $[\Sigma]^{-1}$ corresponding to the zero injection bus measurement $\tilde{G}_\ell = 0$ is $(\sigma_\ell)^{-1}$, which must be "approximately" infinity to enforce satisfaction of the constraint $\tilde{g}_\ell(x) = 0$. We are effectively enforcing the constraint by penalizing the objective for points that do not satisfy the constraint, as discussed in Section 3.1.2.1. Other entries in $[\Sigma]^{-1}$ will be smaller, corresponding to the measurement standard deviations of the actual measurements. Recall that the optimality conditions and algorithms developed in Section 11.2.3 involved factorizing either $\tilde{J}(x)^\dagger [\Sigma]^{-2} \tilde{J}(x)$ or $[\Sigma]^{-1} \tilde{J}(x)$, where $\tilde{J}$ is the Jacobian of $\tilde{g}$. In both cases, the presence of widely differing values in Σ will lead to an ill-conditioned coefficient matrix as discussed in Section 3.1.2.1. For this reason, interpreting zero injection buses as measurements with zero error is somewhat unsatisfactory. However, we will return to this approach in Section 14.5 in discussion of the **merit function**.

12.2.1.4 Treating zero injection buses as equality constraints

The approach we will follow is to explicitly represent the zero injection buses as pairs of equality constraints each of the form $g_\ell(x) = 0$. This approach will be treated in detail in Section 12.2.2.

12.2.2 Formulation

12.2.2.1 Objective

As in Section 11.2, let $\mathbb{M}$ be the set of measurements in the system, not including the injection measurements at the zero injection buses. For each $\ell \in \mathbb{M}$, let the measurements and measurement functions be $\tilde{G}_\ell$ and $\tilde{g}_\ell$, respectively. We can construct a maximum likelihood function and, as in Section 11.2.1, transform it into a function $f : \mathbb{R}^n \to \mathbb{R}$ defined by:

$$\forall x \in \mathbb{R}^n, \ f(x) = \sum_{\ell \in \mathbb{M}} \frac{(\tilde{g}_\ell(x) - \tilde{G}_\ell)^2}{2\sigma_\ell^2}. \tag{12.7}$$

12.2.2.2 Constraints

Let $\mathbb{M}^0$ be the set of real and reactive injections at the zero injection buses. For each $\ell \in \mathbb{M}^0$, let $g_\ell : \mathbb{R}^n \to \mathbb{R}$ be the function representing an injection at a zero injection bus. The power flow equations require that $\forall \ell \in \mathbb{M}^0$, $g_\ell(x) = 0$, so that our estimate of the state x should be consistent with these constraints. We can collect the functions associated with the zero injection buses together into a vector function $g : \mathbb{R}^n \to \mathbb{R}^m$, where m is the number of zero injection bus measurements, which is the number of elements in $\mathbb{M}^0$. That is, $g : \mathbb{R}^n \to \mathbb{R}^m$ is defined by:

$$\forall x \in \mathbb{R}^n, g(x) = (g_\ell(x))_{\ell \in \mathbb{M}^0}, \tag{12.8}$$

so that to satisfy the zero injection bus constraints we require that $g(x) = \mathbf{0}$.

12.2.2.3 Problem

We would still like to obtain the maximum likelihood estimator, but now we must maximize the likelihood over those values of x that are consistent with the zero bus injections. Equivalently, we must minimize the transformed objective over those values of x that are consistent with the zero bus injections. Our problem is therefore:

$$\min_{x \in \mathbb{R}^n} \{f(x)|g(x) = \mathbf{0}\}, \tag{12.9}$$

which is a non-linear equality-constrained optimization problem.

12.2.3 Change in measurement data

As in the unconstrained state estimation case study in Section 9.2, over time, the state of the power system changes as demand and supply situations change. Consequently, the measured data will change. We will consider how a change in measurement data affects the result.

12.2.4 Problem characteristics

12.2.4.1 Objective

As with the basic state estimation problem without zero injection buses introduced in Section 9.2, the objective of Problem (12.9) defined in (12.7) is approximately quadratic. The objective is continuous and differentiable.

12.2.4.2 Constraints

The constraints $g(x) = \mathbf{0}$ are approximately linear; however, since they are not exactly linear we cannot use them to eliminate variables to re-write the problem as an unconstrained optimization in fewer variables. We must develop techniques to deal explicitly with the non-linear equality constraints that appear in Problem (12.9).

12.2.4.3 Solvability

The constraints in the problem are consistent with Kirchhoff's laws and we know from physical principles that there are solutions to Kirchhoff's laws. For example, if the transmission line models have zero values for their shunt elements then we can construct a feasible point: all voltage angles equal to zero and all voltage magnitudes equal to one. (See Exercise 12.7.) In more realistic systems we also expect there to be feasible points and a minimizer.

Exercises

Least-cost production

12.1 Show that if $\underline{x}_k > 0$ then $\mathbb{S}_k = \{0\} \cup [\underline{x}_k, \overline{x}_k]$ is not convex.

12.2 In this exercise we compare solutions of Problems (12.4) and (12.5).

(i) Suppose that Problems (12.4) and (12.5) have minima and minimizers, respectively. Show that if each production cost is monotonically increasing, then Problems (12.4) and (12.5) have the same minima.

(ii) Show that if each production cost is strictly monotonically increasing, then Problems (12.4) and (12.5) have the same set of minimizers. (Hint: Suppose that a minimizer $x^\star$ of Problem (12.5) satisfies $D < \sum_{k=1}^n x_k$.)

12.3 Let $f_k : \mathbb{R} \to \mathbb{R}$ and $\overline{x}_k \in \mathbb{R}_{++}$. Suppose that $f_k(0) = 0$ but that:

$$\lim_{x_k \to 0^+} f_k(x_k) > 0,$$

where $x_k \to 0^+$ means that x_k is approaching 0 from positive values. Show that f_k is not convex on $[0, \overline{x}_k]$.

12.4 Show that the matrix $Z = \begin{bmatrix} -\mathbf{1}^\dagger \\ \mathbf{I} \end{bmatrix} \in \mathbb{R}^{n \times (n-1)}$ has columns that form a basis for the null space of the matrix $A = -\mathbf{1}^\dagger \in \mathbb{R}^{1 \times n}$.

12.5 Recall the discrete-time linear system control problem from Section 4.2. Suppose that we have a linear time-invariant system with state $w(kT) \in \mathbb{R}^m$ at time kT satisfying (4.12):

$$\forall k \in \mathbb{Z}, \, w([k+1]T) = Gw(kT) + hu(kT),$$

where:

- $u(kT) \in \mathbb{R}$ is the input for the time interval from time kT to $(k+1)T$,
- $G \in \mathbb{R}^{m \times m}$ is the state transition matrix, and

- $h \in \mathbb{R}^m$.

Suppose that at time $kT = 0$ the plant is in state $w(0) \in \mathbb{R}^m$ and that we are considering applying inputs for n intervals from $k = 0, \ldots, n - 1$. Moreover, suppose that there is a cost associated with applying the input at each time-step and a cost associated with the state at the end of each time-step. That is, there are functions $f_k : \mathbb{R} \times \mathbb{R}^m \to \mathbb{R}$, $k = 0, \ldots, n - 1$, such that the cost of applying input $u(kT)$ during the time interval from time kT to $(k + 1)T$ and of arriving in state $w([k + 1]T)$ at the end of the interval is $f_k(u(kT), w([k + 1]T))$.

Formulate the problem of minimizing the sum of the costs of controls and states subject to the dynamics of the system being specified by (4.12). (This **optimal control** problem is called a **dynamic programming problem** [7, 10][11, section 1.9]. There are systematic methods to solve such problems involving exploiting the separability of the objective and structure of the constraints.)

Power system state estimation with zero injection buses

12.6 In this exercise we consider observability.

(i) Show that the power system in Figure 12.5 is unobservable. (Hint: Consider Exercise 11.10.)

(ii) Show that the power system in Figure 12.6 is observable if we consider the zero injection bus to provide two highly accurate injection measurements.

12.7 Show that the value of x having all voltage magnitudes equal to one and all voltage angles equal to zero is consistent with the constraints $g(x) = 0$ of Problem (12.9) in the case that all the transmission line models have zero values for their shunt elements.

13

Algorithms for linear equality-constrained minimization

In this chapter we will develop algorithms for constrained optimization problems of the form:

$$\min_{x \in \mathbb{S}} f(x),$$

where $f : \mathbb{R}^n \to \mathbb{R}$ and the feasible set $\mathbb{S}$ is of the form:

$$\mathbb{S} = \{x \in \mathbb{R}^n | g(x) = \mathbf{0}\},$$

with $g : \mathbb{R}^n \to \mathbb{R}^m$ affine. That is, we will consider problems of the form:

$$\min_{x \in \mathbb{R}^n} \{f(x) | Ax = b\}, \tag{13.1}$$

where $A \in \mathbb{R}^{m \times n}$ and $b \in \mathbb{R}^m$ are constants. We call the constraints **linear equality constraints**, although, strictly speaking, it would be more precise to refer to them as **affine equality constraints**. The feasible set defined by the linear equality constraints is convex. (See Exercise 2.36.)

In Section 13.1, we will derive optimality conditions for Problem (13.1). Then, in Section 13.2 we will specialize further to the case where the objective f is convex. That is, we will specialize to convex problems. In Section 13.3, we develop algorithms for solving these problems based on the optimality conditions. In Section 13.4 we will discuss sensitivity analysis. Finally, in Section 13.5, we will apply the algorithms to the least-cost production case study introduced in Section 12.1.

Our approach will involve combining ideas from Chapter 5 on solving systems of linear equations with ideas from Chapter 10 on solving unconstrained optimization problems. The key issues discussed in this chapter are:

- consideration of **descent directions** for the objective that also maintain feasibility for the constraints,
- consideration of the **null space** of the coefficient matrix A to transform the constrained problem into an unconstrained problem,

463

- optimality conditions and the definition and interpretation of the **dual variables** and the **Lagrange multipliers**,
- optimality conditions for **convex problems**, and
- **duality** and **sensitivity analysis**.

13.1 Optimality conditions

We will first discuss descent directions in Section 13.1.1 and then present first-order necessary conditions for local optimality in Section 13.1.2. Following this, we will present second-order sufficient conditions in Section 13.1.3.

13.1.1 Descent directions

13.1.1.1 Conditions for non-minimizer

Analysis Consider a feasible point $\hat{x}$ that is a candidate solution to Problem (13.1). By the discussion in Section 5.8.1.2, every feasible point is of the form $\hat{x} + \Delta x$ where:

$$
\begin{aligned}
\Delta x \in \mathcal{N}(A) &= \{\Delta x \in \mathbb{R}^n | A\Delta x = \mathbf{0}\}, \\
&= \{Z\Delta\xi | \Delta\xi \in \mathbb{R}^{n'}\},
\end{aligned}
$$

where $Z \in \mathbb{R}^{n \times n'}$, with $n' \geq n - m$, is a matrix with columns that form a basis for the null space of A. (See Definition A.50.) We discussed ways to find such a matrix in Section 5.8.1.2.

Now suppose that a vector $\Delta x \in \mathcal{N}(A)$ happened to also satisfy $\nabla f(\hat{x})^\dagger \Delta x < 0$. By Lemma 10.1, such a direction is a descent direction for f at $\hat{x}$. That is:

$$\exists \overline{\alpha} \in \mathbb{R}_{++} \text{ such that } (0 < \alpha \leq \overline{\alpha}) \Rightarrow (f(\hat{x} + \alpha\Delta x) < f(\hat{x})). \tag{13.2}$$

We also have that:

$$
\begin{aligned}
\forall \alpha \in \mathbb{R}, A(\hat{x} + \alpha\Delta x) &= b + \alpha A\Delta x, \\
&= b,
\end{aligned}
$$

so that $\forall \alpha \in \mathbb{R}, \hat{x} + \alpha\Delta x \in \mathbb{S}$. Combining this observation with (13.2) means that we can find feasible points that have lower objective than $f(\hat{x})$, so that $\hat{x}$ is not a minimizer. So, if $\Delta x \in \mathcal{N}(A)$ and $\nabla f(\hat{x})^\dagger \Delta x < 0$ then $\hat{x}$ cannot be a minimizer.

Example The conditions for a point $\hat{x}$ to *not* be a minimizer are illustrated in Figure 13.1, which repeats the contour plot from Figure 10.3 but also includes a

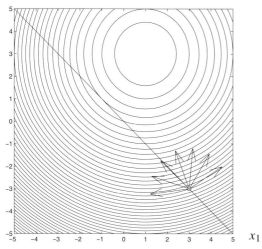

Fig. 13.1. Descent directions for a function at a point $\hat{x} = \begin{bmatrix} 3 \\ -3 \end{bmatrix}$, indicated by the o, and one descent direction that maintains feasibility for the equality constraint corresponding to the feasible set illustrated by the line.

solid line that represents the set of points satisfying the linear equality constraint $Ax = b$, where:

$$A = \begin{bmatrix} 1 & 1 \end{bmatrix} \in \mathbb{R}^{1\times2},$$
$$b = \begin{bmatrix} 0 \end{bmatrix} \in \mathbb{R}^{1}.$$

The circles centered at $x = \begin{bmatrix} 1 \\ 3 \end{bmatrix}$ show the contour sets of the function $f : \mathbb{R}^2 \to \mathbb{R}$ defined in (2.10) and (10.1):

$$\forall x \in \mathbb{R}^2, \, f(x) = (x_1 - 1)^2 + (x_2 - 3)^2.$$

The point $\hat{x} = \begin{bmatrix} 3 \\ -3 \end{bmatrix}$ is indicated by the o.

Several descent directions for the function f at $\hat{x}$ are illustrated by the arrows in Figure 13.1. One of the descent directions Δx is such that $\forall \alpha \in \mathbb{R}, \hat{x} + \alpha \Delta x \in \{x \in \mathbb{R}^2 | Ax = b\}$. The corresponding arrow lies along the line representing the feasible set. In this case, we conclude that $\hat{x}$ cannot be a minimizer for the problem $\min_{x \in \mathbb{R}^2} \{f(x) | Ax = b\}$, since proceeding along the direction Δx from $\hat{x}$ maintains feasibility and, at least for some positive values of α, reduces the objective.

13.1.1.2 Minimizer

Analysis Considering the contrapositive of the statement in Section 13.1.1.1, if $x^\star$ is a minimizer of the linear equality-constrained problem, then for any direction $\Delta x \in \mathcal{N}(A)$, that is, such that $A\Delta x = \mathbf{0}$, we must have that $\nabla f(x^\star)^\dagger \Delta x \not< 0$.

x_2

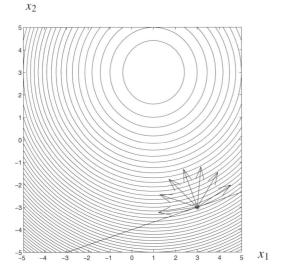

Fig. 13.2. Descent directions for a function at a point $x^\star = \begin{bmatrix} 3 \\ -3 \end{bmatrix}$, indicated by the •, none of which maintains feasibility for the equality constraint corresponding to the feasible set $\mathbb{S}$ illustrated by the line.

Moreover, applying the same argument to the vector $(-\Delta x)$, which is also an element of the null space of A, we must have that $\nabla f(x^\star)^\dagger(-\Delta x) \not< 0$. Combining these two observations, we have that $\nabla f(x^\star)^\dagger \Delta x = 0$.

In summary, if $x^\star$ is a minimizer of the linear equality-constrained problem then for each $\Delta x \in \mathcal{N}(A)$ we must have that $\nabla f(x^\star)^\dagger \Delta x = 0$. That is, $\mathcal{N}(A) \subseteq \{\Delta x \in \mathbb{R}^n | \nabla f(x^\star)^\dagger \Delta x = 0\}$. We can think of these two sets as being "parallel," although in general $\mathcal{N}(A)$ may be a strict subset of $\{\Delta x \in \mathbb{R}^n | \nabla f(x^\star)^\dagger \Delta x = 0\}$.

Example This situation is illustrated in Figure 13.2, which repeats Figure 10.3 but shows the set of points that satisfy a different linear equality constraint to that of Figure 13.1. In particular, the solid line indicates the feasible set $\mathbb{S} = \{x \in \mathbb{R}^2 | x_1 - 3x_2 = 12\}$. The point $x^\star = \begin{bmatrix} 3 \\ -3 \end{bmatrix}$, illustrated with a •, *is* a minimizer of the problem $\min_{x \in \mathbb{R}^2} \{f(x) | x_1 - 3x_2 = 12\}$. Descent directions for the function at the point $x^\star$ are illustrated by arrows.

For the point $x^\star$, none of the descent directions Δx for f at $x^\star$ is such that $x^\star + \alpha \Delta x \in \mathbb{S}$. The point $x^\star$ is a minimizer for the problem $\min_{x \in \mathbb{R}^2} \{f(x) | x_1 - 3x_2 = 12\}$. In particular, there is no direction that simultaneously reduces the objective and maintains feasibility. The set $\{\Delta x \in \mathbb{R}^n | \nabla f(x^\star)^\dagger \Delta x = 0\}$ contains the set $\mathcal{N}(A)$. (In fact, for this example, these two sets are the same.)

13.1.1.3 Geometry of contour set

Tangent plane To develop a geometric interpretation of the conditions for a minimizer, we first make ([72, section 2.5]):

x_2

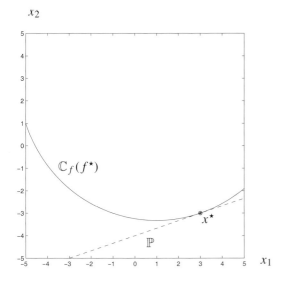

$\mathbb{C}_f(f^\star)$

$\mathbb{P}$

$x^\star$

x_1

Fig. 13.3. Tangent plane $\mathbb{P}$ (shown dashed) to contour set $\mathbb{C}_f(f^\star)$ of f (shown solid) at a point $x^\star = \begin{bmatrix} 3 \\ -3 \end{bmatrix}$, indicated by the bullet ●.

Definition 13.1 Let $f : \mathbb{R}^n \to \mathbb{R}$ be partially differentiable, $x^\star \in \mathbb{R}^n$, and suppose that $\nabla f(x^\star) \neq 0$. Let $f^\star = f(x^\star)$. Then the **tangent plane to the contour set** $\mathbb{C}_f(f^\star) = \{x \in \mathbb{R}^n | f(x) = f^\star\}$ of f at the point $x^\star$ is the set:

$$\mathbb{P} = \{x \in \mathbb{R}^n | \nabla f(x^\star)^\dagger (x - x^\star) = 0\}.$$

For brevity, we will often refer to $\mathbb{P}$ as "the tangent plane to the contour set of f at $x^\star$." If a set $\mathbb{S} \subseteq \mathbb{R}^n$ is contained in $\mathbb{P}$ then we say that "the contour set of f is **tangential** to $\mathbb{S}$ at $x^\star$." □

The tangent plane to the contour set of f at $x^\star$ is obtained by linearizing f at $x^\star$ and then finding the set of points such that the linearized function is equal to $f^\star = f(x^\star)$. (Some authors refer to the related set $\{\Delta x \in \mathbb{R}^n | \nabla f(x^\star)^\dagger \Delta x = 0\}$ as the tangent plane [70, section 10.2].)

Example Consider again the function $f : \mathbb{R}^2 \to \mathbb{R}$ defined in (2.10) and (10.1):

$$\forall x \in \mathbb{R}^2, \ f(x) = (x_1 - 1)^2 + (x_2 - 3)^2,$$

and the point $x^\star = \begin{bmatrix} 3 \\ -3 \end{bmatrix}$. We have that:

$$\forall x \in \mathbb{R}^2, \ \nabla f(x) = \begin{bmatrix} 2(x_1 - 1) \\ 2(x_2 - 3) \end{bmatrix},$$

$$\nabla f(x^\star) = \begin{bmatrix} 4 \\ -12 \end{bmatrix}.$$

Figure 13.3 shows the contour set, $\mathbb{C}_f(f^\star)$, of this function at height $f^\star =$

$f(x^\star) = 40$, as a solid line. The point $x^\star$ lies on the circular contour set and is indicated by the •. The tangent plane to the contour set of f at $x^\star$ is:

$$
\begin{aligned}
\mathbb{P} &= \{x \in \mathbb{R}^2 | \nabla f(x^\star)^\dagger (x - x^\star) = 0\}, \\
&= \left\{ x \in \mathbb{R}^2 \left| \begin{bmatrix} 4 \\ -12 \end{bmatrix}^\dagger \begin{bmatrix} x_1 - 3 \\ x_2 + 3 \end{bmatrix} = 0 \right. \right\}, \\
&= \{x \in \mathbb{R}^2 | x_1 - 3x_2 = 12\}.
\end{aligned}
$$

The tangent plane to the contour set, $\mathbb{P}$, is shown as the dashed line in Figure 13.3.

Example in higher dimension In $\mathbb{R}^n$, the tangent plane is a **hyperplane**; that is, a space of dimension $n - 1$ defined by a single equality constraint. (See Definition A.52.) For example, consider the objective function $f : \mathbb{R}^3 \to \mathbb{R}$ defined by:

$$
\forall x \in \mathbb{R}^3, \ f(x) = (x_1)^2 + (x_2)^2 + (x_3)^2, \tag{13.3}
$$

and the point $x^\star = \begin{bmatrix} 1 \\ 1 \\ 0 \end{bmatrix}$. For this function, we have that:

$$
\forall x \in \mathbb{R}^3, \ \nabla f(x) = \begin{bmatrix} 2x_1 \\ 2x_2 \\ 2x_3 \end{bmatrix},
$$

$$
\nabla f(x^\star) = \begin{bmatrix} 2 \\ 2 \\ 0 \end{bmatrix}.
$$

Figure 13.4 shows the contour set, $\mathbb{C}_f(f^\star)$, of this function of height $f^\star = f(x^\star) = 2$. The point $x^\star$ lies on the spherical contour set and is indicated by the •. The tangent plane to the contour set of f at $x^\star$ is:

$$
\begin{aligned}
\mathbb{P} &= \{x \in \mathbb{R}^3 | \nabla f(x^\star)^\dagger (x - x^\star) = 0\}, \\
&= \left\{ x \in \mathbb{R}^3 \left| \begin{bmatrix} 2 \\ 2 \\ 0 \end{bmatrix}^\dagger \begin{bmatrix} x_1 - 1 \\ x_2 - 1 \\ x_3 - 0 \end{bmatrix} = 0 \right. \right\}, \\
&= \{x \in \mathbb{R}^3 | x_1 + x_2 = 2\}.
\end{aligned}
$$

The tangent plane to the contour set $\mathbb{P}$, is shown as the plane in Figure 13.4. Descent directions for f at $x^\star$ point from $x^\star$ into the sphere.

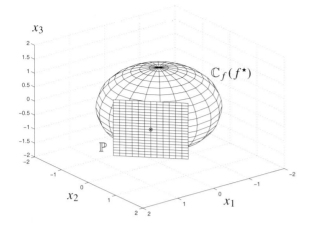

Fig. 13.4. Tangent plane $\mathbb{P}$ to contour set $\mathbb{C}_f(f^\star)$ of f at a point $x^\star = \begin{bmatrix} 1 \\ 1 \\ 0 \end{bmatrix}$, indicated by the •. The contour set is the sphere and the tangent plane is the plane.

13.1.1.4 Geometric interpretation

Analysis In Section 13.1.1.2, we showed that if $x^\star$ is a constrained minimizer of f then:

$$\mathcal{N}(A) \subseteq \{\Delta x \in \mathbb{R}^n | \nabla f(x^\star)^\dagger \Delta x = 0\}.$$

Translating both of these sets by adding $x^\star$ to every element in both sets and noting that $Ax^\star = b$, we have that:

$$
\begin{aligned}
\mathbb{S} &= \{x \in \mathbb{R}^n | Ax = b\}, \\
&= \{x \in \mathbb{R}^n | x = x^\star + \Delta x, \ \Delta x \in \mathcal{N}(A)\}, \ \text{since } Ax^\star = b, \ \text{(see Exercise 13.1)}, \\
&\subseteq \{x \in \mathbb{R}^n | x = x^\star + \Delta x, \ \nabla f(x^\star)^\dagger \Delta x = 0\}, \\
&\qquad \text{since } x^\star \text{ is a constrained minimizer of } f, \\
&= \{x \in \mathbb{R}^n | \nabla f(x^\star)^\dagger (x - x^\star) = 0\}, \\
&= \mathbb{P},
\end{aligned}
$$

which is the tangent plane to the contour set of f at $x^\star$. Geometrically, we can say that the feasible set, $\mathbb{S} = \{x \in \mathbb{R}^n | Ax = b\}$, is contained in the set $\mathbb{P}$, which is the tangent plane to the contour set of f at $x^\star$. More simply, using Definition 13.1, we can also say that the contour set of f is tangential to the feasible set at $x^\star$. This observation is consistent with Figures 13.2 and 13.3.

To see why the inclusion of sets $\mathbb{S} \subseteq \mathbb{P}$ is necessary for $x^\star$ to be a minimizer, we can repeat the argument from Section 13.1.1.2. In particular, suppose that there is some point $\hat{x} \in \mathbb{S}$ that lies *outside* $\mathbb{P}$. That is, $\nabla f(x^\star)^\dagger (\hat{x} - x^\star) \neq 0$. If $\nabla f(x^\star)^\dagger (\hat{x} - x^\star) < 0$ then define $\Delta x = \hat{x} - x^\star$. On the other hand, if $\nabla f(x^\star)^\dagger (\hat{x} - x^\star) > 0$ then define $\Delta x = x^\star - \hat{x}$. In either case, $\nabla f(x^\star)^\dagger \Delta x < 0$ and so Δx is a descent direction for f at $x^\star$ by Lemma 10.1. Moreover, $x^\star + \alpha \Delta x$ is feasible for any α and $x^\star$ is therefore not a minimizer.

x_2

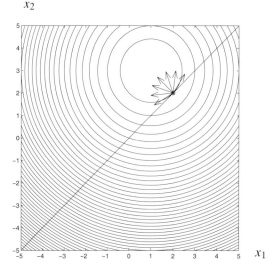

Fig. 13.5. Descent direc-
tions for a function at a
point $x^\star = \begin{bmatrix} 2 \\ 2 \end{bmatrix}$, (indicated by the •), none of
which maintains feasibility
for the equality constraint
illustrated by the line.

Example Recall the example equality-constrained Problem (2.13) first mentioned
in Section 2.3.2.2:

$$\min_{x\in\mathbb{R}^2}\{f(x)|Ax = b\},$$

where $f : \mathbb{R}^2 \to \mathbb{R}$, $A \in \mathbb{R}^{1\times 2}$, and $b \in \mathbb{R}^1$ were defined by:

$$\forall x \in \mathbb{R}^2, \ f(x) \ = \ (x_1 - 1)^2 + (x_2 - 3)^2,$$
$$A \ = \ \begin{bmatrix} 1 & -1 \end{bmatrix},$$
$$b \ = \ \begin{bmatrix} 0 \end{bmatrix}.$$

The (unique) local minimizer is at $x^\star = \begin{bmatrix} 2 \\ 2 \end{bmatrix}$ with minimum $f^\star = 2$.

    The tangent plane to the contour set of f at $x^\star$ is:

$$\mathbb{P} \ = \ \{x \in \mathbb{R}^2 | \nabla f(x^\star)^\dagger (x - x^\star) = \mathbf{0}\},$$
$$= \ \left\{ x \in \mathbb{R}^2 \left| \begin{bmatrix} 2 \\ -2 \end{bmatrix}^\dagger \left(x - \begin{bmatrix} 2 \\ 2 \end{bmatrix} \right) = 0 \right. \right\},$$
$$= \ \{x \in \mathbb{R}^2 | x_1 - x_2 = 0\},$$

which is the same set as the feasible set.

    The situation is illustrated in Figure 13.5, which shows the point $x^\star = \begin{bmatrix} 2 \\ 2 \end{bmatrix}$ as a
•, together with the circular contour sets of the objective, several descent directions
for the objective, and the feasible set. There is no descent direction for f at $x^\star$ that

maintains feasibility of the constraints. The contour set of f is tangential to the feasible set at $x^\star$. (In fact, the contour set of f is coincident with the feasible set.)

Example of strict containment In the previous example, the feasible set was the same as the tangent plane to the contour set. In higher dimensions, it can typically be the case that the feasible set $\mathbb{S} = \{x \in \mathbb{R}^n | Ax = b\}$ is *strictly* contained in $\mathbb{P} = \{x \in \mathbb{R}^n | \nabla f(x^\star)^\dagger (x - x^\star) = 0\}$. For example, consider again the objective function $f : \mathbb{R}^3 \to \mathbb{R}$ defined in (13.3):

$$\forall x \in \mathbb{R}^3, \ f(x) = (x_1)^2 + (x_2)^2 + (x_3)^2.$$

Moreover, suppose that the equality constraints $Ax = b$ are defined by:

$$A = \begin{bmatrix} 1 & 0 & 0 \\ 0 & 1 & 0 \end{bmatrix},$$
$$b = 1.$$

The constraints specify that $x_1 = x_2 = 1$, so that the feasible set is the line in $\mathbb{R}^3$ that is parallel to the x_3-axis and that passes through $x_1 = x_2 = 1$. By inspection, the minimizer of $\min_{x \in \mathbb{R}^3} \{f(x) | Ax = b\}$ is $x^\star = \begin{bmatrix} 1 \\ 1 \\ 0 \end{bmatrix}$.

In this case:

$$\begin{aligned} \mathbb{S} &= \{x \in \mathbb{R}^n | Ax = b\}, \\ &= \{x \in \mathbb{R}^n | x_1 = x_2 = 1\}, \\ \mathbb{P} &= \{x \in \mathbb{R}^n | \nabla f(x^\star)^\dagger (x - x^\star) = 0\}, \\ &= \left\{ x \in \mathbb{R}^n \left| \begin{bmatrix} 2 & 2 & 0 \end{bmatrix} \left(x - \begin{bmatrix} 1 \\ 1 \\ 0 \end{bmatrix} \right) = 0 \right. \right\}, \\ &= \{x \in \mathbb{R}^n | x_1 + x_2 = 2\}. \end{aligned}$$

That is, the tangent plane to the contour set of f at $x^\star$ is a plane, $\mathbb{P}$, in $\mathbb{R}^3$, which strictly contains the feasible set $\mathbb{S}$, which is a line. The situation is illustrated in Figure 13.6, which repeats Figure 13.4 but adds a line that represents $\mathbb{S}$. Recall that descent directions for f at $x^\star$ point into the sphere. That is, no descent directions point along the feasible set $\mathbb{S}$.

13.1.1.5 Summary

In summary, at a minimizer $x^\star$ of Problem (13.1), every descent direction for f at $x^\star$ must lie outside the null space of A. Every point in the null space must not be a descent direction. Translating these sets, we observe that, at a minimizer, the

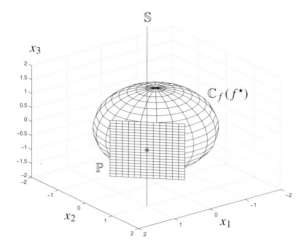

Fig. 13.6. Feasible set $\mathbb{S}$ that is strictly contained in the tangent plane $\mathbb{P}$ to the contour set $\mathbb{C}_f(f^\star)$ of f at a point $x^\star = \begin{bmatrix} 1 \\ 1 \\ 0 \end{bmatrix}$, indicated by the $\bullet$. The contour set is the sphere; the tangent plane is the plane; and the feasible set is the vertical line.

contour set of f is tangential to the feasible set. In the next section we will develop these geometric insights into tractable analytic conditions.

13.1.2 First-order necessary conditions

The discussion in Section 13.1.1 developed necessary conditions in terms of inclusions of *sets* that were defined in terms of the first derivative of f. In this section, we will develop *algebraic* conditions in terms of the first derivative of f. Such algebraic conditions are called **first-order necessary conditions**. As in the unconstrained case, the conditions are sometimes referred to by their acronym **FONC**.

Our approach to developing first-order necessary conditions will first involve a transformation of the problem defined in terms of the null space of A that parallels the discussion in Section 13.1.1. Then, the necessary conditions will be characterized in two other ways in terms of the variables in the original problem.

13.1.2.1 Transformation of problem

Following the discussion in Section 13.1.1, let $Z \in \mathbb{R}^{n \times n'}$, with $n' \geq n - m$, be a matrix with columns that form a basis for the null space of A. As discussed above, the null space of A is specified by:

$$
\begin{aligned}
\mathcal{N}(A) &= \{\Delta x \in \mathbb{R}^n | A\Delta x = \mathbf{0}\}, \\
&= \{Z\Delta\xi | \Delta\xi \in \mathbb{R}^{n'}\}.
\end{aligned}
$$

This allows us to eliminate variables by expressing the vectors in the feasible set in terms of a particular solution to $Ax = b$ and an arbitrary vector in $\mathbb{R}^{n'}$. Suppose that $\hat{x} \in \mathbb{R}^n$ is a particular solution to $Ax = b$. Then (see Exercise 13.1):

$$
\begin{aligned}
\mathbb{S} &= \{x \in \mathbb{R}^n | Ax = b\}, \\
&= \{x \in \mathbb{R}^n | x = \hat{x} + \Delta x, \; A\Delta x = \mathbf{0}, \; \Delta x \in \mathbb{R}^n\}, \\
&= \{\hat{x} + Z\Delta\xi | \Delta\xi \in \mathbb{R}^{n'}\}.
\end{aligned}
$$

That is, we can define an **onto function** $\tau : \mathbb{R}^{n'} \to \mathbb{S}$ by:

$$
\forall \xi \in \mathbb{R}^{n'}, \tau(\xi) = \hat{x} + Z\xi.
$$

(See Definition A.25 of an onto function.) Notice that varying ξ over $\mathbb{R}^{n'}$ allows $\tau(\xi)$ to explore over the feasible set $\mathbb{S}$. Formally, we use Theorem 3.5 to transform the equality-constrained Problem (13.1) into an unconstrained problem. In particular, in the hypothesis of Theorem 3.5, let $\mathbb{P} = \mathbb{R}^{n'}$ and define $\phi : \mathbb{R}^{n'} \to \mathbb{R}$ by:

$$
\forall \xi \in \mathbb{R}^{n'}, \phi(\xi) = f(\tau(\xi)). \tag{13.4}
$$

The function ϕ is called the **reduced function** [84, section 14.2]. By Theorem 3.5:

(i) $\min_{x \in \mathbb{R}^n} \{f(x) | Ax = b\}$ has a minimum if and only if $\min_{\xi \in \mathbb{R}^{n'}} \phi(\xi)$ has a minimum; and

(ii) if either one of the problems in Item (i) possesses a minimum (and consequently, by Item (i), each one possesses a minimum), then:

$$
\min_{\xi \in \mathbb{R}^{n'}} \phi(\xi) = \min_{x \in \mathbb{S}} f(x),
$$

$$
\arg\min_{x \in \mathbb{R}^n} \{f(x) | Ax = b\} = \left\{ \tau(\xi) \,\middle|\, \xi \in \arg\min_{\xi \in \mathbb{R}^{n'}} \phi(\xi) \right\}.
$$

Since $\min_{\xi \in \mathbb{R}^{n'}} \phi(\xi)$ is an *unconstrained* problem, we can use the techniques from Chapter 10 to write down optimality conditions for it. We can solve these optimality conditions in terms of ξ and then find the corresponding value of x in the original problem.

The gradient of ϕ, $\nabla\phi(\bullet) = Z^\dagger \nabla f(\tau(\bullet))$, is called the **reduced gradient** or the **projected gradient** [84, section 14.2]. The transformation into an unconstrained problem leads us to a class of algorithms for solving Problem (13.1) using the reduced gradient that will be presented in Sections 13.3.1.2 and 13.3.2.2.

We now consider the direction corresponding to the reduced gradient in the original decision variables $x \in \mathbb{R}^n$. Referred to the original decision variables x, the reduced gradient $\nabla\phi$ corresponding to a point $\hat{x} \in \mathbb{R}^n$ lies in the direction $ZZ^\dagger \nabla f(\hat{x}) \in \mathbb{R}^n$. The vector $\Delta x = -ZZ^\dagger \nabla f(\hat{x})$, which is opposite to the direction corresponding to the reduced gradient, is a descent direction for f at $\hat{x}$ unless the reduced gradient equals the zero vector. Moreover, if $A\hat{x} = b$ then, for any α, $\hat{x} + \alpha\Delta x$ also satisfies the equality constraints. (See Exercise 13.2.)

13.1.2.2 Necessary conditions in terms of original problem

Analysis Instead of forming the function τ explicitly and introducing the variables ξ each time we want to solve Problem (13.1), we can write down the necessary conditions in terms of the variables x in Problem (13.1). We will develop this approach in the next theorem. The proof essentially repeats the above discussion.

Theorem 13.1 *Suppose that $f : \mathbb{R}^n \to \mathbb{R}$ is partially differentiable with continuous partial derivatives, $A \in \mathbb{R}^{m \times n}$, and $b \in \mathbb{R}^m$. Let $Z \in \mathbb{R}^{n \times n'}$ be a matrix with columns that form a basis for the null space of A. If $x^\star \in \mathbb{R}^n$ is a local minimizer of the problem:*

$$\min_{x \in \mathbb{R}^n} \{f(x) | Ax = b\},$$

then:

$$
\begin{aligned}
Z^\dagger \nabla f(x^\star) &= \mathbf{0}, \\
Ax^\star &= b.
\end{aligned}
\tag{13.5}
$$

Proof See Appendix B for details. □

Example Continuing with the previous equality-constrained Problem (2.13) from Sections 2.3.2.2 and 13.1.1.4, we observe that, by inspection, $Z = \begin{bmatrix} 1 \\ 1 \end{bmatrix} \in \mathbb{R}^{2 \times 1}$ is a matrix with columns that form a basis for the null space:

$$\mathcal{N}(A) = \{\Delta x \in \mathbb{R}^n | A \Delta x = \mathbf{0}\},$$

since:

- $A \Delta x = \mathbf{0}$ if and only if $\Delta x_1 = \Delta x_2$, and
- for $\xi \in \mathbb{R}$, $Z\xi = \begin{bmatrix} \xi \\ \xi \end{bmatrix}$.

Also:

$$
\begin{aligned}
\forall x \in \mathbb{R}^2, \nabla f(x) &= \begin{bmatrix} 2(x_1 - 1) \\ 2(x_2 - 3) \end{bmatrix}, \\
\nabla f(x^\star) &= \begin{bmatrix} 2 \\ -2 \end{bmatrix},
\end{aligned}
$$

so that $x^\star = \begin{bmatrix} 2 \\ 2 \end{bmatrix}$ is not an *unconstrained* minimizer of f. Using these calcula-

tions, we obtain:

$$\begin{aligned} Z^\dagger \nabla f(x^\star) &= \begin{bmatrix} 1 & 1 \end{bmatrix} \begin{bmatrix} 2 \\ -2 \end{bmatrix}, \\ &= [0], \end{aligned}$$

consistent with the conclusion of Theorem 13.1.

13.1.2.3 Lagrange multipliers

Analysis We can characterize the necessary conditions slightly differently in terms of a vector $\lambda^\star \in \mathbb{R}^m$ called the vector of **Lagrange multipliers** with the following.

Theorem 13.2 *Suppose that* $f : \mathbb{R}^n \to \mathbb{R}$ *is partially differentiable with continuous partial derivatives,* $A \in \mathbb{R}^{m \times n}$, *and* $b \in \mathbb{R}^m$. *If* $x^\star \in \mathbb{R}^n$ *is a local minimizer of the problem:*

$$\min_{x \in \mathbb{R}^n} \{f(x) | Ax = b\},$$

then:

$$\exists \lambda^\star \in \mathbb{R}^m \text{ such that } \nabla f(x^\star) + A^\dagger \lambda^\star = 0, \tag{13.6}$$
$$Ax^\star = b. \tag{13.7}$$

Proof ([84, section 14.2].) By Theorem 13.1:

$$Z^\dagger \nabla f(x^\star) = 0, \tag{13.8}$$
$$Ax^\star = b,$$

where $Z \in \mathbb{R}^{n \times n'}$ is a matrix with columns that form a basis for the null space of A. By Theorem A.4 in Appendix A and Exercise 5.47, any vector in $\mathbb{R}^n$ can be written in the form $Z\xi - A^\dagger \lambda$ for some $\xi \in \mathbb{R}^{n'}$ and $\lambda \in \mathbb{R}^m$. In particular, since $\nabla f(x^\star) \in \mathbb{R}^n$, we have:

$$\exists \xi^\star \in \mathbb{R}^{n'}, \exists \lambda^\star \in \mathbb{R}^m \text{ such that } \nabla f(x^\star) = Z\xi^\star - A^\dagger \lambda^\star.$$

Multiplying this expression through by $Z^\dagger$ we obtain:

$$Z^\dagger \nabla f(x^\star) = Z^\dagger Z\xi^\star - Z^\dagger A^\dagger \lambda^\star.$$

But $Z^\dagger \nabla f(x^\star) = 0$ by (13.8), so:

$$Z^\dagger Z\xi^\star - Z^\dagger A^\dagger \lambda^\star = 0.$$

Also $AZ = 0$ by Exercise 5.48, so $Z^\dagger A^\dagger \lambda^\star = 0$ and $Z^\dagger Z\xi^\star = 0$. But this means that $\xi^\star = 0$ since Z has linearly independent columns. (See Exercise 2.27.) That is,

$$\exists \lambda^\star \in \mathbb{R}^m \text{ such that } \nabla f(x^\star) + A^\dagger \lambda^\star = 0,$$

which is (13.6). We already have that $Ax^\star = b$, which is (13.7). $\square$

A vector $\lambda^\star$ satisfying (13.6), given an $x^\star$ that also satisfies (13.7), is called a vector of **Lagrange multipliers** for the problem. The conditions (13.6)–(13.7) are called the **first-order necessary conditions** (or **FONC**) for Problem (13.1), although, strictly speaking, the first-order necessary conditions also include the rest of the hypotheses of Theorem 13.2.

Example Continuing with the previous equality-constrained Problem (2.13) from Sections 2.3.2.2, 13.1.1.4, and 13.1.2.2 we obtain:

$$
\begin{aligned}
\nabla f(x^\star) + A^\dagger[-2] &= \begin{bmatrix} 2 \\ -2 \end{bmatrix} + \begin{bmatrix} 1 \\ -1 \end{bmatrix}[-2], \\
&= 0,
\end{aligned}
$$

which is consistent with Theorem 13.2 for $\lambda^\star = [-2]$.

13.1.2.4 Analytic interpretation

The necessary conditions in Theorems 13.1 and 13.2 are rather abstract. We will offer two analytic interpretations in terms of the Lagrangian to help understand the conditions and then illustrate with an example.

The Lagrangian Recall Definition 3.2 of the **Lagrangian**. For a problem with objective $f : \mathbb{R}^n \to \mathbb{R}$ and equality constraints $Ax = b$, with $A \in \mathbb{R}^{m \times n}$ and $b \in \mathbb{R}^m$, the Lagrangian $\mathcal{L} : \mathbb{R}^n \times \mathbb{R}^m \to \mathbb{R}$ is defined by:

$$
\forall x \in \mathbb{R}^n, \forall \lambda \in \mathbb{R}^m, \mathcal{L}(x, \lambda) = f(x) + \lambda^\dagger(Ax - b), \tag{13.9}
$$

where λ is called the vector of **dual variables** for the problem.

In the context of the Lagrangian and the dual variables, we call $x \in \mathbb{R}^n$ the **primal variables**. The optimal value of the primal variables is the minimizer $x^\star$. The Lagrange multipliers $\lambda^\star$ are the particular value of the dual variables that satisfy the first-order necessary conditions.

We also define the gradients of $\mathcal{L}$ with respect to x and λ by, respectively, $\nabla_x \mathcal{L} = \left[\frac{\partial \mathcal{L}}{\partial x} \right]^\dagger$ and $\nabla_\lambda \mathcal{L} = \left[\frac{\partial \mathcal{L}}{\partial \lambda} \right]^\dagger$. That is:

$$
\begin{aligned}
\nabla_x \mathcal{L}(x, \lambda) &= \nabla f(x) + A^\dagger \lambda, \\
\nabla_\lambda \mathcal{L}(x, \lambda) &= Ax - b.
\end{aligned}
$$

We can interpret the first-order necessary conditions (13.6)–(13.7) in two ways using the Lagrangian $\mathcal{L}$:

Minimization of Lagrangian over primal variables The first-order necessary conditions imply that $x^\star$ is a critical point of the function $\mathcal{L}(\bullet, \lambda^\star)$ that also satisfies the constraints $Ax = b$. If f is convex then, by Corollary 10.6, $x^\star$ is a global minimizer of $\mathcal{L}(\bullet, \lambda^\star)$ minimized as a function of x for fixed $\lambda = \lambda^\star$. We seek a point $x^\star$ that minimizes $\mathcal{L}(\bullet, \lambda^\star)$.

To interpret this observation, first recall that for an *unconstrained* problem, Theorem 10.3 shows that if $x^\star$ is a local minimizer of partially differentiable f having continuous partial derivatives then $\nabla f(x^\star) = 0$. Moving away from $x^\star$ cannot improve the objective, at least for small enough movements.

In the constrained case, however, if $x^\star$ is a local minimizer of Problem (13.1), then by Theorem 13.2, there exists $\lambda^\star \in \mathbb{R}^m$ such that $\nabla f(x^\star) + A^\dagger \lambda^\star = 0$. If $\lambda^\star \neq 0$ then we expect that $\nabla f(x^\star) \neq 0$, as confirmed by the example in Section 13.1.2.2. We could improve the objective by moving away from $x^\star$; however, such a move would violate the constraints. The vector of Lagrange multipliers $\lambda^\star$ "adjusts" the unconstrained optimality conditions by $A^\dagger \lambda^\star$ and this adjustment is just the right amount to "balance" the minimization of the objective against satisfaction of the constraints. For some problems this allows us to solve an *unconstrained* optimization problem to find $x^\star$, so that the equality constraints do not have to be treated explicitly.

Critical point of the Lagrangian The first-order necessary conditions also imply that $\begin{bmatrix} x^\star \\ \lambda^\star \end{bmatrix}$ is a solution of the non-linear simultaneous equations:

$$\nabla_x \mathcal{L}(x, \lambda) = 0, \qquad (13.10)$$
$$\nabla_\lambda \mathcal{L}(x, \lambda) = 0. \qquad (13.11)$$

The second set of equations requires that $x^\star$ be feasible and are linear equations. We seek $\begin{bmatrix} x^\star \\ \lambda^\star \end{bmatrix}$ satisfying $\nabla \mathcal{L}(x^\star, \lambda^\star) = 0$, where $\nabla \mathcal{L} = \begin{bmatrix} \nabla_x \mathcal{L} \\ \nabla_\lambda \mathcal{L} \end{bmatrix}$. That is, $\begin{bmatrix} x^\star \\ \lambda^\star \end{bmatrix}$ is a critical point of $\mathcal{L}$; however, $\begin{bmatrix} x^\star \\ \lambda^\star \end{bmatrix}$ is *not* a minimizer of $\mathcal{L}(\bullet, \bullet)$ over values of $\begin{bmatrix} x \\ \lambda \end{bmatrix}$. (See Exercise 13.6.)

The Lagrangian provides a convenient way to remember the first-order necessary conditions since setting the gradient $\nabla \mathcal{L}$ of the Lagrangian equal to zero reproduces the first-order necessary conditions as in (13.10)–(13.11).

Algorithms As in the unconstrained case, these two interpretations lead us to two (of several) classes of algorithms for solving Problem (13.1):

(i) minimize the Lagrangian over x for a fixed λ and then adjust λ until feasibility is obtained, (Sections 13.3.1.4 and 13.3.2.4), and

(ii) solve the first-order necessary conditions (13.10)–(13.11) for x and λ, (Sections 13.3.1.3 and 13.3.2.3).

Example Continuing with the previous equality-constrained Problem (2.13) from Sections 2.3.2.2, 13.1.1.4,..., 13.1.2.3, the Lagrangian $\mathcal{L} : \mathbb{R}^2 \times \mathbb{R} \to \mathbb{R}$ is defined by:

$$\forall x \in \mathbb{R}^2, \forall \lambda \in \mathbb{R}, \mathcal{L}(x, \lambda) = (x_1 - 1)^2 + (x_2 - 3)^2 + \lambda(x_1 - x_2). \qquad (13.12)$$

Setting the value of the dual variable in the Lagrangian equal to the Lagrange multiplier, $\lambda^\star = [-2]$, we have:

$$\forall x \in \mathbb{R}^2, \mathcal{L}(x, \lambda^\star) = (x_1 - 1)^2 + (x_2 - 3)^2 + (-2)(x_1 - x_2).$$

The first-order necessary conditions for minimizing $\mathcal{L}(\bullet, \lambda^\star)$ are:

$$\begin{aligned}
\nabla_x \mathcal{L}(x, \lambda^\star) &= \begin{bmatrix} 2(x_1 - 1) - 2 \\ 2(x_2 - 3) + 2 \end{bmatrix}, \\
&= \mathbf{0},
\end{aligned}$$

which yields a solution of $x^\star = \begin{bmatrix} 2 \\ 2 \end{bmatrix}$. The contour sets of $\mathcal{L}(\bullet, \lambda^\star)$ are shown in Figure 13.7, confirming that the unconstrained minimizer of $\mathcal{L}(\bullet, \lambda^\star)$ is at $x^\star$, which is illustrated by a $\bullet$ in Figure 13.7. Although $\begin{bmatrix} x^\star \\ \lambda^\star \end{bmatrix}$ is a critical point of $\mathcal{L}$, it *does not* minimize $\mathcal{L}$ as shown in Exercise 13.6.

For other values of the dual variables λ not equal to the Lagrange multipliers $\lambda^\star$, the corresponding minimizer of $\mathcal{L}(\bullet, \lambda)$ will differ from the minimizer of Problem (2.13). For example, for $\tilde{\lambda} = [-5]$, the contour sets of $\mathcal{L}(\bullet, \tilde{\lambda})$ are illustrated in Figure 13.8. The unconstrained minimizer of this function is at $\tilde{x} = \begin{bmatrix} 3.5 \\ 0.5 \end{bmatrix}$, illustrated with a $\circ$ in Figure 13.8, which differs from $x^\star$. The point $\begin{bmatrix} \tilde{x} \\ \tilde{\lambda} \end{bmatrix}$ satisfies (13.10), but does not satisfy (13.11).

13.1.2.5 Relation to geometric interpretation

To see that the first-order necessary conditions imply the geometric observation made in Section 13.1.1, suppose that $\hat{x} \in \mathbb{R}^n$ satisfies:

$$\hat{x} \in \mathbb{S} = \{x \in \mathbb{R}^n | Ax = b\}.$$

x_2

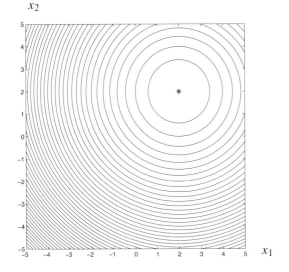

Fig. 13.7. Contour sets for Lagrangian $\mathcal{L}(\bullet, \lambda^\star)$ evaluated at the Lagrange multipliers $\lambda^\star = [-2]$.

x_2

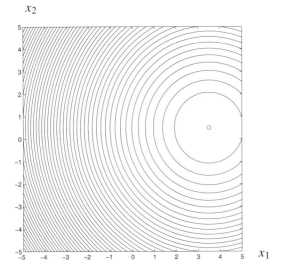

Fig. 13.8. Contour sets for Lagrangian $\mathcal{L}(\bullet, \tilde{\lambda})$ evaluated at value of dual variables $\tilde{\lambda} = [-5]$ not equal to Lagrange multiplers.

Then $A(\hat{x} - x^\star) = \mathbf{0}$ and so $[\lambda^\star]^\dagger A(\hat{x} - x^\star) = \mathbf{0}$. The necessary conditions require that $\nabla f(x^\star)^\dagger + [\lambda^\star]^\dagger A = \mathbf{0}$. Multiplying by $(\hat{x} - x^\star)$ on the right we obtain:

$$
\begin{aligned}
\mathbf{0} &= \left(\nabla f(x^\star)^\dagger + [\lambda^\star]^\dagger A \right) (\hat{x} - x^\star), \\
&= \nabla f(x^\star)^\dagger (\hat{x} - x^\star).
\end{aligned}
$$

Therefore:

$$
\hat{x} \in \mathbb{P} = \{ x \in \mathbb{R}^n | \nabla f(x^\star)^\dagger (x - x^\star) = \mathbf{0} \}.
$$

x_2

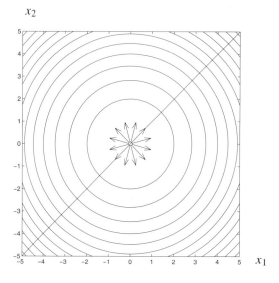

Fig. 13.9. Contour sets for non-convex objective. The objective decreases away from $\hat{x} = \mathbf{0}$.

As we observed in Section 13.1.1, we can say that $\mathbb{S}$ and $\mathbb{P}$ are "parallel," although it may be the case that $\mathbb{S}$ is strictly contained in $\mathbb{P}$. Moreover, the contour set of f is tangential to the feasible set $\mathbb{S}$ at $x^\star$.

13.1.2.6 First-order necessary conditions are not sufficient

Discussion As with unconstrained problems, it is possible for a point $\hat{x}$ to satisfy the first-order necessary conditions (13.6)–(13.7) and yet not be a local minimizer of Problem (13.1).

Example For example, consider the case of Problem (13.1) with $n = 2$ and $m = 1$ and:

$$
\begin{aligned}
\forall x \in \mathbb{R}^2, f(x) &= -\frac{1}{2}(x_1)^2 - \frac{1}{2}(x_2)^2, \\
A &= \begin{bmatrix} 1 & -1 \end{bmatrix}, \\
b &= [0].
\end{aligned}
$$

That is, the problem is $\min_{x \in \mathbb{R}^2} \{ f(x) | x_1 - x_2 = 0 \}$. The contour sets of f are illustrated in Figure 13.9. The contour sets of f are circular, with center at $\hat{x} = \mathbf{0}$. The point $\hat{x} = \mathbf{0}$ is illustrated with a ∘. The objective decreases away from $\hat{x} = \mathbf{0}$ and the arrows indicate various descent directions for f at $\hat{x}$, including two that lie along the feasible set, which is illustrated with a line. That is, $\hat{x}$ cannot be a minimizer of the problem.

However, the point $\hat{x} = \mathbf{0}$ satisfies the first-order necessary conditions (13.6)–

(13.7) with $\hat{\lambda} = [0]$ since:

$$
\begin{aligned}
\nabla f(\hat{x}) + A^{\dagger}\hat{\lambda} &= -\hat{x} + \begin{bmatrix} 1 \\ -1 \end{bmatrix}\hat{\lambda}, \\
&= \mathbf{0} + \begin{bmatrix} 1 \\ -1 \end{bmatrix}0, \\
&= \mathbf{0}, \\
A\hat{x} &= A\mathbf{0}, \\
&= [0], \\
&= b.
\end{aligned}
$$

That is, $\hat{x} = \mathbf{0}$ and $\hat{\lambda} = [0]$ satisfy the first-order necessary conditions for Problem (13.1), but $\hat{x} = \mathbf{0}$ is not a minimizer of this problem. In fact, it is a *maximizer* of f over the feasible set. Similarly, a point of inflection could satisfy the necessary conditions for Problem (13.1) and be neither a maximizer nor a minimizer of the problem. In the next section we discuss sufficient conditions for a minimizer.

13.1.3 Second-order sufficient conditions

We present **second-order sufficient conditions** (or **SOSC**) for local optimality, first in terms of a basis for the null space of the coefficient matrix and then in terms of the Lagrange multipliers.

13.1.3.1 Null space basis

Analysis We have the following conditions in terms of a basis for the null space.

Theorem 13.3 *Suppose that $f : \mathbb{R}^n \to \mathbb{R}$ is twice partially differentiable with continuous second partial derivatives, $A \in \mathbb{R}^{m \times n}$, and $b \in \mathbb{R}^m$. Let $Z \in \mathbb{R}^{n \times n'}$ be a matrix with columns that form a basis for the null space of A. Let $x^{\star} \in \mathbb{R}^n$ and suppose that:*

$$
\begin{aligned}
Z^{\dagger}\nabla f(x^{\star}) &= \mathbf{0}, \\
Ax^{\star} &= b, \\
Z^{\dagger}\nabla^2 f(x^{\star})Z \text{ is positive definite.}
\end{aligned}
$$

Then $x^{\star} \in \mathbb{R}^n$ is a strict local minimizer of the problem $\min_{x \in \mathbb{R}^n}\{f(x)|Ax = b\}$.

Proof See [84, lemma 14.2, section 14.2]. The conditions follow from the second-order sufficient conditions presented in Theorem 10.5 for unconstrained minimization applied to the problem of minimizing the reduced function $\phi : \mathbb{R}^{n'} \to \mathbb{R}$ defined in (13.4):

$$
\forall \xi \in \mathbb{R}^{n'}, \phi(\xi) = f(\tau(\xi)).
$$

□

In addition to the first-order necessary conditions, the second-order sufficient conditions require that $Z^\dagger \nabla^2 f(x^\star) Z$ is positive definite. The condition that the matrix $Z^\dagger \nabla^2 f(x^\star) Z$ is positive definite means that along any direction $\Delta x = Z\xi$ in the null space $\mathcal{N}(A) = \{\Delta x \in \mathbb{R}^n | A\Delta x = \mathbf{0}\}$, if $\Delta x \neq \mathbf{0}$ then $\Delta x^\dagger \nabla^2 f(x^\star) \Delta x > 0$. The function $\nabla^2 \phi$ is called the **reduced Hessian** [84, section 14.2] and the condition requires that the reduced Hessian is positive definite at the minimizer.

Example Continuing with the previous equality-constrained Problem (2.13) from Sections 2.3.2.2, 13.1.1.4,..., 13.1.2.4, we have already verified that $x^\star = \begin{bmatrix} 2 \\ 2 \end{bmatrix}$ and $\lambda^\star = [-2]$ satisfy the first-order necessary conditions. We have:

$$\forall x \in \mathbb{R}^2, \nabla^2 f(x) = \begin{bmatrix} 2 & 0 \\ 0 & 2 \end{bmatrix},$$

$$Z^\dagger \nabla^2 f(x^\star) Z = \begin{bmatrix} 1 & 1 \end{bmatrix} \begin{bmatrix} 2 & 0 \\ 0 & 2 \end{bmatrix} \begin{bmatrix} 1 \\ 1 \end{bmatrix},$$

$$= [4],$$

which is positive definite. Applying Theorem 13.3, we conclude that $x^\star$ is a local minimizer of Problem (2.13).

13.1.3.2 Lagrange multipliers

Analysis As previously, we can also develop second-order sufficient conditions in terms of Lagrange multipliers.

Corollary 13.4 *Suppose that $f : \mathbb{R}^n \to \mathbb{R}$ is twice partially differentiable with continuous second partial derivatives, $A \in \mathbb{R}^{m \times n}$, and $b \in \mathbb{R}^m$. Let $x^\star \in \mathbb{R}^n$ and $\lambda^\star \in \mathbb{R}^m$ satisfy:*

$$\nabla f(x^\star) + A^\dagger \lambda^\star = \mathbf{0},$$
$$Ax^\star = b,$$
$$(A\Delta x = \mathbf{0} \text{ and } \Delta x \neq \mathbf{0}) \Rightarrow (\Delta x^\dagger \nabla^2 f(x^\star) \Delta x > 0).$$

Then $x^\star \in \mathbb{R}^n$ is a local minimizer of the problem $\min_{x \in \mathbb{R}^n} \{f(x)|Ax = b\}$.

Proof The hypotheses of this corollary imply the hypotheses of Theorem 13.3. (See Exercise 13.7.) □

We refer to the conditions in Corollary 13.4 as the **second-order sufficient conditions** (or **SOSC**). In addition to the first-order necessary conditions, the second-order sufficient conditions require that f is twice partially differentiable with continuous second partial derivatives and that:

$$(A\Delta x = \mathbf{0} \text{ and } \Delta x \neq \mathbf{0}) \Rightarrow (\Delta x^\dagger \nabla^2 f(x^\star) \Delta x > 0).$$

This condition on $\nabla^2 f(x^\star)$ is referred to by saying that $\nabla^2 f(x^\star)$ **is positive definite on the null space** $\mathcal{N}(A) = \{\Delta x \in \mathbb{R}^n | A \Delta x = 0\}$. (See Definition A.60.) See [70, section 10.6][79, section 4.B] and Exercise A.18 for tests to determine if a matrix is positive definite on a null space.

Example Continuing with the previous equality-constrained Problem (2.13) from Sections 2.3.2.2, 13.1.1.4,..., 13.1.3.1, we have already verified that $x^\star = \begin{bmatrix} 2 \\ 2 \end{bmatrix}$ and $\lambda^\star = [-2]$ satisfy the first-order necessary conditions. We have that:

$$\nabla^2 f(x^\star) = \begin{bmatrix} 2 & 0 \\ 0 & 2 \end{bmatrix},$$

which is positive definite on $\mathbb{R}^2$ and, in particular, on the null space $\mathcal{N}(A) = \{\Delta x \in \mathbb{R}^n | A \Delta x = 0\}$. Applying Corollary 13.4, we conclude that $x^\star$ is a local minimizer of Problem (2.13). In this example, $\nabla^2 f(x^\star)$ is positive definite and it is therefore positive definite on the null space $\mathcal{N}(A)$. Exercise 13.8 shows an example where at the minimizer $x^\star$ the Hessian $\nabla^2 f(x^\star)$ is not positive definite, even though $\nabla^2 f(x^\star)$ is positive definite on the null space $\mathcal{N}(A)$.

13.2 Convex problems

The convexity of the constraint set in the case of linear constraints allows us to obtain global optimality results for linear equality-constrained problems with convex objectives.

13.2.1 First-order sufficient conditions

13.2.1.1 Analysis

As in the unconstrained case, when we specialize to convex problems, we find that the first-order necessary conditions are also sufficient for optimality.

Theorem 13.5 *Suppose that* $f : \mathbb{R}^n \to \mathbb{R}$ *is partially differentiable with continuous partial derivatives,* $A \in \mathbb{R}^{m \times n}$, *and* $b \in \mathbb{R}^m$. *Consider points* $x^\star \in \mathbb{R}^n$ *and* $\lambda^\star \in \mathbb{R}^m$. *Suppose that:*

   (i) f *is convex on* $\{x \in \mathbb{R}^n | Ax = b\}$,

   (ii) $\nabla f(x^\star) + A^\dagger \lambda^\star = 0$, *and*

   (iii) $Ax^\star = b$.

Then $x^\star$ *is a global minimizer of the problem* $\min_{x \in \mathbb{R}^n} \{f(x) | Ax = b\}$.

Proof Consider any feasible point $x \in \{x \in \mathbb{R}^n | Ax = b\}$. We have:

$$f(x) \geq f(x^\star) + \nabla f(x^\star)^\dagger (x - x^\star), \text{ by Theorem 2.6, noting that:}$$

f is partially differentiable with continuous partial derivatives;

f is convex on the convex set $\{x \in \mathbb{R}^n | Ax = b\}$,

by Item (i) of the hypothesis; and

$x, x^\star \in \{x \in \mathbb{R}^n | Ax = b\}$,

by Item (iii) of the hypothesis and construction,

$$= f(x^\star) - [\lambda^\star]^\dagger A(x - x^\star), \text{ by Item (ii) of the hypothesis,}$$

$$= f(x^\star), \text{ since } Ax = Ax^\star \text{ by Item (iii) of the hypothesis and construction.}$$

Therefore $x^\star$ is a global minimizer of f on $\{x \in \mathbb{R}^n | Ax = b\}$. $\square$

Corollary 13.6 *Suppose that* $f : \mathbb{R}^n \to \mathbb{R}$ *is partially differentiable with continuous partial derivatives,* $A \in \mathbb{R}^{m \times n}$, *and* $b \in \mathbb{R}^m$. *Let* $Z \in \mathbb{R}^{n \times n'}$ *have columns that form a basis for the null space* $\{\Delta x \in \mathbb{R}^n | A\Delta x = 0\}$. *Consider a point* $x^\star \in \mathbb{R}^n$. *Suppose that:*

(i) *f is convex on* $\{x \in \mathbb{R}^n | Ax = b\}$,
(ii) *$Z^\dagger \nabla f(x^\star) = 0$, and*
(iii) *$Ax^\star = b$.*

Then $x^\star$ is a global minimizer of the problem $\min_{x \in \mathbb{R}^n}\{f(x) | Ax = b\}$.

Proof Items (i) and (iii) of the hypothesis of this corollary are the same as the corresponding Items (i) and (iii) of the hypothesis of Theorem 13.5.
Item (ii) of the hypothesis of this corollary says that $Z^\dagger \nabla f(x^\star) = 0$. In the proof of Theorem 13.2, it was proven that:

$$(Z^\dagger \nabla f(x^\star) = 0) \Rightarrow (\exists \lambda^\star \in \mathbb{R}^m \text{ such that } \nabla f(x^\star) + A^\dagger \lambda^\star = 0).$$

That is, Item (ii) of the hypothesis of Theorem 13.5 holds. Therefore, the result then follows from Theorem 13.5. $\square$

13.2.1.2 Example

Continuing with the previous equality-constrained Problem (2.13) from Sections 2.3.2.2, 13.1.1.4,..., 13.1.3.2, we have already verified that $x^\star = \begin{bmatrix} 2 \\ 2 \end{bmatrix}$ and $\lambda^\star = [-2]$ satisfy the first-order necessary conditions. Moreover, f is convex. By Theorem 13.5, this means that $x^\star$ is a global minimizer of Problem (2.13).

13.2.2 Duality

The discussion in Section 13.1.2.4 suggests that if we knew the vector of Lagrange multipliers $\lambda^\star$ we could avoid explicit consideration of the equality constraints if f were convex. Here we discuss one method to find the Lagrange multipliers and indicate some of the issues that arise. In particular, we will see that we generally require strict convexity of f to yield useful results.

13.2.2.1 Dual function

Analysis As we discussed in Section 3.4, we can define a dual problem where the role of variables and constraints is partly or fully swapped [84, chapter 6]. We have observed in Section 13.1.2.4 that if f is convex then $x^\star$ is a global minimizer of $\mathcal{L}(\bullet, \lambda^\star)$ over values of x.

Recall Definition 3.3 of the **dual function** and **effective domain**. For Problem (13.1), the dual function $\mathcal{D} : \mathbb{R}^m \to \mathbb{R} \cup \{-\infty\}$ is defined by:

$$\forall \lambda \in \mathbb{R}^m, \mathcal{D}(\lambda) = \inf_{x \in \mathbb{R}^n} \mathcal{L}(x, \lambda), \tag{13.13}$$

while the effective domain is:

$$\mathbb{E} = \{\lambda \in \mathbb{R}^m | \mathcal{D}(\lambda) > -\infty\},$$

so that the restriction of $\mathcal{D}$ to $\mathbb{E}$ is a function $\mathcal{D} : \mathbb{E} \to \mathbb{R}$.

Example Continuing with the previous equality-constrained Problem (2.13) from Sections 2.3.2.2, 13.1.1.4, ..., 13.2.1.2, the Lagrangian $\mathcal{L} : \mathbb{R}^2 \times \mathbb{R} \to \mathbb{R}$ was defined in (13.12):

$$\forall x \in \mathbb{R}^2, \forall \lambda \in \mathbb{R}, \mathcal{L}(x, \lambda) = (x_1 - 1)^2 + (x_2 - 3)^2 + \lambda(x_1 - x_2).$$

The dual function evaluates the infimum of this function over $x \in \mathbb{R}^2$ for a given λ. That is, by (13.13),

$$
\begin{aligned}
\forall \lambda \in \mathbb{R}, \mathcal{D}(\lambda) &= \inf_{x \in \mathbb{R}^2} \mathcal{L}(x, \lambda), \\
&= \inf_{x \in \mathbb{R}^2} \{(x_1 - 1)^2 + (x_2 - 3)^2 + \lambda(x_1 - x_2)\}.
\end{aligned}
$$

For each $\lambda \in \mathbb{R}$, the objective $\mathcal{L}(\bullet, \lambda)$ in the infimum is partially differentiable with continuous partial derivatives and is strictly convex, so by Corollary 10.6 the first-order necessary conditions for minimizing $\mathcal{L}(\bullet, \lambda)$ are sufficient for global optimality.

The first-order necessary conditions for minimizing $\mathcal{L}(\bullet, \lambda)$ are that:

$$
\begin{aligned}
\nabla_x \mathcal{L}(x, \lambda) &= \begin{bmatrix} 2(x_1 - 1) + \lambda \\ 2(x_2 - 3) - \lambda \end{bmatrix}, \\
&= \mathbf{0},
\end{aligned}
$$

which, for any given $\lambda \in \mathbb{R}$, yields the unique solution of $x^{(\lambda)} = \begin{bmatrix} 1 - \lambda/2 \\ 3 + \lambda/2 \end{bmatrix}$.

Substituting into the Lagrangian $\mathcal{L}$, we obtain:

$$
\begin{aligned}
\forall \lambda \in \mathbb{R}, \mathcal{D}(\lambda) &= \left(1 - \frac{\lambda}{2} - 1\right)^2 + \left(3 + \frac{\lambda}{2} - 3\right)^2 + \lambda\left(1 - \frac{\lambda}{2} - 3 - \frac{\lambda}{2}\right), \\
&= -\frac{(\lambda)^2}{2} - 2\lambda.
\end{aligned}
\tag{13.14}
$$

That is, we have evaluated the dual function explicitly.

13.2.2.2 Dual problem

Analysis In the case that the objective is convex on $\mathbb{R}^n$, if Problem (13.1) has a minimum then the minimum is equal to $\mathcal{D}(\lambda^\star)$, where $\lambda^\star$ is the vector of Lagrange multipliers that satisfies the necessary conditions for Problem (13.1). Moreover, the Lagrange multipliers maximize, over the dual variables λ, the following problem:

$$
\max_{\lambda \in \mathbb{E}} \mathcal{D}(\lambda),
\tag{13.15}
$$

where $\mathcal{D} : \mathbb{E} \to \mathbb{R}$ is the dual function defined in (13.13). Problem (13.15) is called the **dual problem** to Problem (13.1). Problem (13.1) is called the **primal problem** in this context to distinguish it from Problem (13.15).

In particular, we have the following.

Theorem 13.7 *Suppose that $f : \mathbb{R}^n \to \mathbb{R}$ is convex and partially differentiable with continuous partial derivatives, $A \in \mathbb{R}^{m \times n}$, and $b \in \mathbb{R}^m$. Consider the primal problem, Problem (13.1):*

$$
\min_{x \in \mathbb{R}^n} \{f(x) | Ax = b\}.
$$

Also, consider the dual problem, Problem (13.15). We have the following.

 (i) *If the primal problem possesses a minimum then the dual problem possesses a maximum and the optima are equal. That is:*

$$
\min_{x \in \mathbb{R}^n} \{f(x) | Ax = b\} = \max_{\lambda \in \mathbb{E}} \mathcal{D}(\lambda).
\tag{13.16}
$$

(ii) *If:*

  • $\lambda \in \mathbb{E}$,
  • $\min_{x \in \mathbb{R}^n} \mathcal{L}(x, \lambda)$ *exists, and*
  • f *is twice partially differentiable with continuous second partial derivatives and $\nabla^2 f$ is positive definite,*

  then $\mathcal{D}$ is partially differentiable at λ with continuous partial derivatives and

$$
\nabla \mathcal{D}(\lambda) = Ax^{(\lambda)} - b,
\tag{13.17}
$$

 where $x^{(\lambda)}$ is the unique minimizer of $\min_{x \in \mathbb{R}^n} \mathcal{L}(x, \lambda)$.

Proof See [6, theorems 6.2.4 and 6.3.3][11, proposition 3.4.3][70, lemma 1 of chapter 13][84, corollaries 6.1 and 14.2] and (for Item (ii)) Exercise 10.19. This is a special case of Theorem 17.4 to be presented in Chapter 17. □

It is important to note that for some $\lambda \in \mathbb{R}^m$ it is possible for:

- $\inf_{x \in \mathbb{R}^n} \mathcal{L}(x, \lambda)$ to be a real number, so that $\lambda \in \mathbb{E}$, yet for there to be no minimum of $\min_{x \in \mathbb{R}^n} \mathcal{L}(x, \lambda)$, or
- $\nabla^2 f$ to fail to be positive definite so that there are multiple minimizers of the problem $\min_{x \in \mathbb{R}^n} \mathcal{L}(x, \lambda)$.

In either case, the dual function $\mathcal{D}$ may be non-differentiable at $\lambda \in \mathbb{E}$. See Exercises 13.10 and 13.11.

Recall from Theorem 3.12 that the effective domain $\mathbb{E}$ of the dual function is a convex set and that the dual function is concave on $\mathbb{E}$.

Corollary 13.8 *Let* $f : \mathbb{R}^n \to \mathbb{R}$ *be twice partially differentiable with continuous second partial derivatives and with* $\nabla^2 f$ *positive definite,* $A \in \mathbb{R}^{m \times n}$, *and* $b \in \mathbb{R}^m$. *Let* $\mathbb{E}$ *be the effective domain of the dual function.*
If:

- $\mathbb{E} = \mathbb{R}^m$, *and*
- $\forall \lambda \in \mathbb{R}^m$, $\min_{x \in \mathbb{R}^n} \mathcal{L}(x, \lambda)$ *exists,*

then necessary and sufficient conditions for $\lambda^\star \in \mathbb{R}^m$ *to be the maximizer of the dual function are that:*

$$Ax^{(\lambda^\star)} - b = \mathbf{0},$$

where $\{x^{(\lambda^\star)}\} = \operatorname{argmin}_{x \in \mathbb{R}^n} \mathcal{L}(x, \lambda^\star)$. *Moreover, if* $\lambda^\star$ *maximizes the dual function then* $x^{(\lambda^\star)}$ *and* $\lambda^\star$ *satisfy the first-order necessary conditions for Problem (13.1).*

Proof Note that the hypothesis implies that the dual function is finite for all $\lambda \in \mathbb{R}^m$ so that Problem (13.15) is an unconstrained maximization of a real-valued function and, moreover, by Theorem 3.12, $-\mathcal{D}$ is convex and partially differentiable with continuous partial derivatives. By Theorem 10.3 and Corollary 10.6, $\nabla \mathcal{D}(\lambda) = \mathbf{0}$ is necessary and sufficient for λ to be a global maximizer of $\mathcal{D}$. By Theorem 13.7, $\nabla \mathcal{D}(\lambda) = Ax^{(\lambda)} - b$, so the necessary and sufficient conditions for maximizing the dual are that $Ax^{(\lambda)} - b = \mathbf{0}$. Direct substitution shows that $x^{(\lambda^\star)}$ and $\lambda^\star$ satisfy the first-order necessary conditions for Problem (13.1). □

Theorem 13.7 shows that an alternative approach to finding the minimum of Problem (13.1) involves finding the *maximum* of the dual function over $\lambda \in \mathbb{R}^m$. Theorem 3.12 shows that the dual function has at most one local maximum, with necessary and sufficient conditions for the maximizer specified in Corollary 13.8. To seek the maximum of $\mathcal{D}(\lambda)$, we can, for example, utilize the value of the partial derivative of $\mathcal{D}$ from (13.17) in a **steepest ascent** algorithm. Under some circumstances, it is also possible to calculate the second partial derivatives of $\mathcal{D}$ to implement a Newton–Raphson algorithm [70, section 12.3]; however, it is more usual to use steepest ascent or a quasi-Newton method to seek a maximizer of the dual problem.

The results in Theorem 13.7 and Corollary 13.8 can be sharpened in some cases. See [6, theorems 6.2.4 and 6.3.3][11, proposition 3.4.3][84, corollaries 6.1 and 14.2] for details.

Example Continuing with the previous equality-constrained Problem (2.13) from Sections 2.3.2.2, 13.1.1.4,..., 13.2.2.1, we note that $\nabla^2 f$ is positive definite and for each $\lambda \in \mathbb{R}$, $\mathcal{L}(\bullet, \lambda)$ has a unique minimizer, specified by the solution of $\nabla_x \mathcal{L}(x, \lambda) = \mathbf{0}$, so that, by Theorem 13.7, $\mathbb{E} = \mathbb{R}$ and the dual function is partially differentiable with continuous partial derivatives on the whole of $\mathbb{R}$.

Moreover, since the dual function is concave, the first-order necessary conditions to maximize $\mathcal{D}$ are also sufficient. Partially differentiating $\mathcal{D}$ as defined in (13.14) we obtain:

$$\nabla \mathcal{D}(\lambda) = [-\lambda - 2].$$

This is consistent with Theorem 13.7, since:

$$Ax^{(\lambda)} - b = \begin{bmatrix} 1 & -1 \end{bmatrix} \begin{bmatrix} 1 - \lambda/2 \\ 3 + \lambda/2 \end{bmatrix} - [0],$$
$$= [-\lambda - 2].$$

Moreover, $\nabla \mathcal{D}(\lambda^\star) = [0]$ for $\lambda^\star = [-2]$, which is the value of the Lagrange multiplier. Also, $\mathcal{D}(\lambda^\star) = 2$, which is equal to the minimum of Problem (2.13) and $x^{(\lambda^\star)} = \begin{bmatrix} 2 \\ 2 \end{bmatrix}$, which is the minimizer of Problem (2.13).

In this case, it was possible to solve for the dual function explicitly so that we could then evaluate the first-order conditions for maximizing the dual function explicitly. For non-quadratic objectives this is not usually possible and we will typically resort to an iterative algorithm to maximize $\mathcal{D}$.

Wolfe dual In some cases we can write down conditions characterizing the value of the dual function more explicitly than in (13.13) [84, section 14.8.2]. Evaluating the conditions explicitly then facilitates maximization of the dual function.

In particular, consider $\min_{x \in \mathbb{R}^n} \mathcal{L}(\bullet, \lambda)$ for a particular value $\lambda \in \mathbb{R}^m$. Suppose that f is partially differentiable with continuous partial derivatives and that it is convex on $\mathbb{R}^n$. Then by Corollary 10.6, the first-order necessary conditions $\nabla_x \mathcal{L}(x, \lambda) = 0$ are sufficient for minimizing $\mathcal{L}(\bullet, \lambda)$.

Given $\lambda \in \mathbb{R}^m$, if there is a solution to $\nabla_x \mathcal{L}(x, \lambda) = \mathbf{0}$ then we can evaluate the dual function by:

$$\mathcal{D}(\lambda) = \{\mathcal{L}(x, \lambda) | \nabla_x \mathcal{L}(x, \lambda) = \mathbf{0}\},$$

where by the notation on the right-hand side we mean the *value* of $\mathcal{L}(x, \lambda)$ evaluated for a value of x that satisfies $\nabla_x \mathcal{L}(x, \lambda) = \mathbf{0}$, assuming a solution for x exists. That is, for given λ, we solve for the value of x that satisfies $\nabla_x \mathcal{L}(x, \lambda) = \mathbf{0}$ and substitute x and λ into $\mathcal{L}$. If the solution for x is not unique, we pick any solution of $\nabla_x \mathcal{L}(x, \lambda) = \mathbf{0}$, since any solution of the first-order necessary conditions will minimize the convex function $\mathcal{L}(\bullet, \lambda)$.

For some values of λ there may be no solution to $\nabla_x \mathcal{L}(x, \lambda) = \mathbf{0}$. In this case there is no minimum of $\mathcal{L}(\bullet, \lambda)$ and we must explicitly characterize $\inf_{x \in \mathbb{R}^n} \mathcal{L}(x, \lambda)$ to evaluate $\mathcal{D}(\lambda)$.

Using Theorem 13.7, this observation means that under the same assumptions, we can solve for the minimum of Problem (13.1) by using the **Wolfe dual**:

$$\min_{x \in \mathbb{R}^n} \{f(x) | Ax = b\} = \max_{\lambda \in \mathbb{R}^m} \{\mathcal{L}(x, \lambda) | \nabla_x \mathcal{L}(x, \lambda) = \mathbf{0}\}, \qquad (13.18)$$

where we again use the equation $\nabla_x \mathcal{L}(x, \lambda) = \mathbf{0}$ to evaluate x and have tacitly assumed that $\nabla_x \mathcal{L}(x, \lambda) = \mathbf{0}$ has a solution for each λ. We will use the notion of the Wolfe dual in the context of inequality-constrained problems in Section 17.2.2.3.

Discussion It is essential in Theorem 13.7 for f to be convex on the *whole* of $\mathbb{R}^n$, not just on the feasible set. The reason is that the inner minimization of $\mathcal{L}(\bullet, \lambda)$ to evaluate the dual function is taken over the whole of $\mathbb{R}^n$.

Unfortunately, if f is not *strictly* convex then $\mathcal{L}(\bullet, \lambda)$ may have multiple minimizers over x for fixed λ. In this case, it may turn out that some of the minimizers of $\mathcal{L}(\bullet, \lambda^\star)$ do not actually minimize Problem (13.1). Moreover, if there are multiple minimizers of $\mathcal{L}(\bullet, \lambda)$ then $\mathcal{D}(\lambda)$ may not be partially differentiable. Exercise 13.10 explores this situation.

Even when the objective is not strictly convex, (or, indeed, if the feasible set is not convex as we will explore in Section 14.3.2) we can still try to solve the dual problem to obtain $\lambda^\star$ and extract a corresponding value of $x^{(\lambda^\star)}$. In general, we may find that there are multiple minimizers of $\mathcal{L}(\bullet, \lambda)$ for given λ, so that the dual function may not be differentiable. However, it turns out that the step direction $\Delta\lambda = Ax^{(\lambda)} - b$ can still be useful in guiding a search to maximize $\mathcal{D}$. If we find the maximizer $\lambda^\star$ of the dual then we may be able to modify $x^{(\lambda^\star)}$ to obtain a useful solution; however, this is not always easily accomplished. This approach forms the basis of **Lagrangian relaxation** [11, section 6.4.1][35][79, chapter 9], the **subgradient method** [11, section 6.1], and other methods to solve non-differentiable problems that result from "dualizing" a non-convex problem.

If f is strictly convex and the feasible set is $\{x \in \mathbb{R}^n | Ax = b\}$ then, for each

$\lambda \in \mathbb{E}$, if the minimizer of $\mathcal{L}(\bullet, \lambda)$ exists it will be unique and consequently $\mathcal{D}$ is partially differentiable at λ. Unfortunately, a further issue remains. Even for strictly convex f, we may find that the effective domain $\mathbb{E}$ is a strict subset of $\mathbb{R}^m$. This poses some inconvenience in maximizing the dual because the dual problem is not unconstrained. We may not be able to find an initial guess $\lambda^{(0)} \in \mathbb{E}$ nor find updates to increase the value of the dual function. This issue is explored in Exercise 13.11.

In Section 13.2.2.4 we will see that a penalty function approach can be used to to avoid the difficulties that arise when the minimizer of $\mathcal{L}(\bullet, \lambda)$ does not exist for some values of λ (that is, to avoid the case where $\mathbb{E}$ is a strict subset of $\mathbb{R}^m$). Moreover, the penalty function we use can help to ensure that the penalized objective is strictly convex on the whole of $\mathbb{R}^n$, guaranteeing that $\operatorname{argmin}_{x \in \mathbb{R}^n} \mathcal{L}(x, \lambda)$ is a singleton for each λ and so, by Theorem 13.7, that the dual function is partially differentiable. Before discussing this approach in Section 13.2.2.4, however, we will consider the important special case of a separable objective.

13.2.2.3 Separable objective

Analysis If $f : \mathbb{R}^n \to \mathbb{R}$ is separable, so that:

$$\forall x \in \mathbb{R}^n, f(x) = \sum_{k=1}^n f_k(x_k),$$

where $f_k : \mathbb{R} \to \mathbb{R}, k = 1, \dots, n$, then:

$$
\begin{aligned}
\forall \lambda \in \mathbb{E}, \mathcal{D}(\lambda) &= \inf_{x \in \mathbb{R}^n} \mathcal{L}(x, \lambda), \\
&= \min_{x \in \mathbb{R}^n} \mathcal{L}(x, \lambda), \text{ assuming that the minimum exists,} \\
&= \min_{x \in \mathbb{R}^n} f(x) + \lambda^\dagger (Ax - b), \text{ by definition of } \mathcal{L}, \\
&= \min_{x \in \mathbb{R}^n} \left\{ \sum_{k=1}^n f_k(x_k) + \lambda^\dagger \left(\sum_{k=1}^n A_k x_k - b \right) \right\}, \\
&\qquad \text{where } A_k \text{ is the } k\text{-th column of } A, \\
&= \min_{x \in \mathbb{R}^n} \left\{ \sum_{k=1}^n \left(f_k(x_k) + \lambda^\dagger A_k x_k \right) \right\} - \lambda^\dagger b, \\
&\qquad \text{since } \lambda^\dagger b \text{ is independent of } x, \\
&= \sum_{k=1}^n \min_{x_k \in \mathbb{R}} \{ f_k(x_k) + \lambda^\dagger A_k x_k \} - \lambda^\dagger b, \quad (13.19)
\end{aligned}
$$

since each term in the sum depends only on one entry of x. (See Exercise 13.13.) For each fixed $\lambda \in \mathbb{R}^m$, the dual function $\mathcal{D}(\lambda)$ is the sum of:

- a constant $(-\lambda^{\dagger} b)$, and
- n one-dimensional optimization "sub-problems" that can each be evaluated independently.

We have **decomposed** the problem by exploiting the separability of the objective. If there are relatively few constraints but many variables and the objective is separable then maximizing the dual problem involves optimization in a smaller dimension than minimizing the primal problem. Each evaluation of the dual objective and its derivative can be broken up into parallel calculations. In this case, maximizing the dual often turns out to be faster and requires less memory than minimizing the primal problem, since each sub-problem can be treated separately in the evaluation of the dual function.

If the sub-problems $\min_{x_k \in \mathbb{R}} \{ f_k(x_k) + \lambda^{\dagger} A_k x_k \}$ are simple enough then it may be possible to solve them analytically in terms of λ. (See Section 13.3.1.4 for the case of a quadratic objective.) In this case, the dual function $\mathcal{D}$ can be evaluated analytically.

In some applications, the separability does not involve individual entries of x but rather to collections of entries. We can apply an analysis similar to (13.19); however, the sub-problems will then be multi-dimensional problems [100, section 28].

Example Continuing with the previous equality-constrained Problem (2.13) from Sections 2.3.2.2, 13.1.1.4,..., 13.2.2.2, note that the objective is separable. The dual function is:

$$
\begin{aligned}
\forall \lambda \in \mathbb{R}, \mathcal{D}(\lambda) \quad &= \quad \min_{x \in \mathbb{R}^2} \mathcal{L}(x, \lambda), \\
&= \quad \min_{x_1 \in \mathbb{R}} \{ (x_1 - 1)^2 + \lambda x_1 \} + \min_{x_2 \in \mathbb{R}} \{ (x_2 - 3)^2 - \lambda x_2 \}. \quad (13.20)
\end{aligned}
$$

The minimization problem in the definition of the dual has decomposed into two sub-problems. Each of the two convex sub-problems can be solved separately and the result is the same as obtained previously. (The term $\lambda^{\dagger} b$ is zero in this case since $b = [0]$. See Exercise 13.14.)

13.2.2.4 Penalty functions and augmented Lagrangians

In this section, we consider the combined use of penalty functions and duality [70, section 13.4]

Discussion In Section 3.1.2.1, we discussed an approach to approximately solving constrained problems by defining an unconstrained problem with a **penalized objective**. For example, for the problem $\min_{x \in \mathbb{R}^n} \{ f(x) | g(x) = \mathbf{0} \}$, we might try to solve the *unconstrained* problem $\min_{x \in \mathbb{R}^n} f(x) + \Pi \| g(x) \|^2$ for some suitable value of the penalty coefficient $\Pi \in \mathbb{R}_{++}$ and some norm $\|\bullet\|$ such as the L_2 norm. Unfortunately, as mentioned in Section 3.1.2.1, for finite values of Π the solution of the unconstrained problem will usually differ from the solution of the constrained problem. Very large values of Π will make the solutions of the constrained and unconstrained problems nearly coincide; however, the resulting unconstrained problem will then typically be ill-conditioned.

In Section 3.1.2.1 we also observed that we could consider the penalized objective $f + \Pi \|g\|^2$ and retain the constraints. We found that the added term $\Pi \|g\|^2$ did not affect the solution, since it did not alter the objective on the feasible set. However, the additional term could aid in making the objective convex and also reduce the tension between minimizing the objective and satisfying the constraints. In this section, we will explore the addition of the penalty function to the objective in conjunction with an approach involving dual variables. Recall that Theorem 13.7 required the objective to be convex on the whole of $\mathbb{R}^n$, not just on the feasible set. We will use a penalty function to alter the objective for points not in the feasible set so that the penalized objective is strictly convex on the whole of $\mathbb{R}^n$.

Example To see the advantages of a penalized objective in conjunction with a dual approach, consider the problem described in Exercise 3.11, and revisited in Exercise 13.8, having objective function $f : \mathbb{R}^2 \to \mathbb{R}$, equality constraint $Ax = b$, and Lagrangian function $\mathcal{L} : \mathbb{R}^2 \times \mathbb{R} \to \mathbb{R}$ defined, respectively, by:

$$
\begin{aligned}
\forall x \in \mathbb{R}^2, \ f(x) &= -2(x_1 - x_2)^2 + (x_1 + x_2)^2, \\
A &= \begin{bmatrix} 1 & -1 \end{bmatrix}, \\
b &= [0], \\
\forall x \in \mathbb{R}^2, \forall \lambda \in \mathbb{R}, \ \mathcal{L}(x, \lambda) &= f(x) + \lambda^\dagger (Ax - b), \\
&= -2(x_1 - x_2)^2 + (x_1 + x_2)^2 + \lambda(x_1 - x_2).
\end{aligned}
$$

The objective function f is shown in Figure 13.10. The objective is not convex and is not bounded below. Moreover, for any given $\lambda \in \mathbb{R}$, $\mathcal{L}(\bullet, \lambda)$ is not bounded below. (See Exercise 3.46, Part (i).) Therefore:

$$
\forall \lambda \in \mathbb{R}, \ \inf_{x \in \mathbb{R}^n} \mathcal{L}(x, \lambda) = -\infty,
$$

and $\mathbb{E} = \emptyset$. We cannot usefully apply Theorem 13.7.

From Exercise 3.11, we know that the solution to the equality-constrained optimization problem $\min_{x \in \mathbb{R}^n} \{ f(x) | Ax = b \}$ is $x^\star = \mathbf{0}$. Substitution into the necessary conditions shows that corresponding value of the Lagrange multiplier is $\lambda^\star = 0$, so that $\mathcal{L}(\bullet, \lambda^\star) = f(\bullet)$. The primal problem is well-defined, the first-order necessary conditions hold at the minimizer, and $x^\star$ and $\lambda^\star$ satisfy the second-order sufficient conditions.

$f(x)$

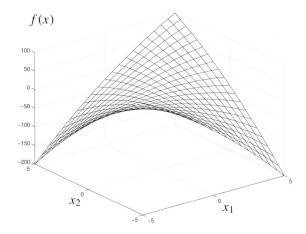

Fig. 13.10. Example non-convex objective function defined in Section 13.2.2.4 and in Exercises 3.11 and 13.8.

$f(x) + \Pi f_{\mathrm{p}}(x)$

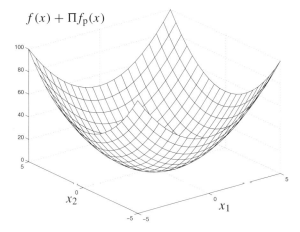

Fig. 13.11. Convex penalized objective function $f + \Pi f_{\mathrm{p}}$ for $\Pi = 3$.

The difficulties in applying Theorem 13.7 arise here because the objective is not strictly convex on $\mathbb{R}^n$. Suppose that instead we consider a penalized objective. That is, we modify the objective to be $f + \Pi f_{\mathrm{p}}$, where $\Pi \in \mathbb{R}_{++}$ and $f_{\mathrm{p}} : \mathbb{R}^n \to \mathbb{R}_+$ is defined by:

$$
\begin{aligned}
\forall x, \; f_{\mathrm{p}}(x) &= \|Ax - b\|_2^2, \\
&= (x_1 - x_2)^2.
\end{aligned}
$$

Theorem 3.2 guarantees that we will not change the minimizer or minimum if we use the penalized objective with a non-negative value of Π. For example, suppose that we choose $\Pi = 3$. The penalized objective is shown in Figure 13.11 and this value of Π is large enough so that the penalized objective is strictly convex.

The Lagrangian of the corresponding problem with penalized objective is called the **augmented Lagrangian**, $\mathcal{L}_p : \mathbb{R}^2 \times \mathbb{R} \to \mathbb{R}$, defined by:

$$
\begin{aligned}
\forall x \in \mathbb{R}^2, \forall \lambda \in \mathbb{R}, \mathcal{L}_p(x, \lambda) &= \mathcal{L}(x, \lambda) + \Pi f_p(x), \\
&= (x_1 + x_2)^2 + \lambda(x_1 - x_2) + (x_1 - x_2)^2,
\end{aligned}
$$

which is strictly convex as a function of x for fixed λ. Moreover, for each $\lambda \in \mathbb{R}$, the minimizer of $\mathcal{L}_p(\bullet, \lambda)$ exists, so that $\mathbb{E} = \mathbb{R}$, and the minimizer is unique, so that the dual function is partially differentiable.

In particular, for the example shown, minimizing $\mathcal{L}_p(\bullet, \lambda^\star)$ over x now yields the minimizer $x^\star$ of the equality-constrained problem, as shown in Exercise 3.46, Part (iii). In general, to achieve this, we must pick Π large enough so that:

- the augmented Lagrangian $\mathcal{L}_p(\bullet, \lambda)$ is strictly convex for each given λ (so that there is at most one minimizer of $\mathcal{L}_p(\bullet, \lambda)$ for each given λ), and
- there is a minimizer of the augmented Lagrangian $\mathcal{L}_p(\bullet, \lambda)$ for each λ.

Analysis We now formalize some of the issues with the augmented Lagrangian in the special case of a quadratic objective.

Consider a quadratic $f : \mathbb{R}^n \to \mathbb{R}$ with quadratic coefficient matrix $Q \in \mathbb{R}^{n \times n}$ and the use of a penalty function $\Pi \|Ax - b\|_2^2$. The Hessian of the augmented Lagrangian $\mathcal{L}_p(\bullet, \lambda)$ for fixed λ is $Q + 2\Pi A^\dagger A$. We have the following.

Theorem 13.9 *Suppose that $Q \in \mathbb{R}^{n \times n}$ is positive definite on the null-space of $A \in \mathbb{R}^{m \times n}$. Then there exists $\Pi > 0$ such that $Q + 2\Pi A^\dagger A$ is positive definite.*

Proof See [11, lemma 3.2.1][70, section 13.4] and Exercise A.18. □

Theorem 13.9 shows that there exists Π such that the augmented Lagrangian is strictly convex as a function of x for fixed λ. In practice we may not know *a priori* how large we need to choose Π. Moreover, if Π is too large, then the minimization of the augmented Lagrangian will be ill-conditioned. An adjustment procedure for finding a value of Π that is large enough to achieve strict convexity and further comments about ill-conditioning appears in [11, section 4.2.1].

It is important to realize that non-strict convexity of the augmented Lagrangian is generally not sufficient to guarantee uniqueness of the minimizer of the augmented Lagrangian. Fortunately, achieving strict convexity is often possible for modest values of Π that do not lead to serious ill-conditioning [70, sections 13.4 and 14.2]. (See Exercises 13.11 and 13.12.) Moreover, the quadratic penalty can often be chosen to make $\mathbb{E} = \mathbb{R}^m$.

The augmented Lagrangian (or the ordinary Lagrangian if the objective is strictly convex) allows us to perform an unconstrained minimization to obtain the minimizer, given that we know the Lagrange multipliers. As previously, we must maximize the dual function over values of the dual variables to seek the Lagrange multipliers.

Separability and the augmented Lagrangian Augmented Lagrangians have a drawback for separable objectives since the penalty function adds "cross-terms" between variables, which prevent decomposition into sub-problems. One approach to preserving separability while maintaining the advantage of convexity of augmented Lagrangians involves linearizing the cross-terms. This approach is discussed in [26] and an application is presented in [61].

13.3 Approaches to finding minimizers

In this section we will briefly describe algorithms for various cases, moving from the most straightforward to the most complex cases. Unless otherwise mentioned, for each case considered, we will tacitly assume that a minimum and minimizer exists. Our basic approach is to solve the first-order necessary conditions for optimality. As the example in Section 13.1.2.6 shows, the necessary conditions are not sufficient for optimality unless the problem satisfies additional properties, such as convexity.

As in our previous discussion of unconstrained optimization, the algorithms will either be:

- direct, typically involving solution of a linear system of equations, or
- iterative, typically requiring at each iteration the solution of a linear equation representing a Newton–Raphson update for solving non-linear equations or an approximation to the Newton–Raphson update.

In principle, evaluation of the coefficient matrices in the linear systems requires calculation of the Hessians of functions. We will proceed as though these Hessians are available and that the resulting linear systems can be conveniently factorized using the basic LU factorization. However, in practice, it may be necessary or desirable to:

- use a variation on the basic Newton–Raphson update along the lines described in Section 10.2.3.3 to avoid the computational effort of evaluation and factorization of the Hessian at each iteration, or
- use a different factorization method such as QR factorization if the equations are ill-conditioned.

We first consider convex quadratic objectives in Section 13.3.1 and then non-quadratic objectives in Section 13.3.2.

13.3.1 Convex quadratic objective

13.3.1.1 Problem

Suppose that the objective function $f : \mathbb{R}^n \to \mathbb{R}$ is of the form:

$$\forall x \in \mathbb{R}^n, \ f(x) = \frac{1}{2}x^\dagger Q x + c^\dagger x,$$

with $c \in \mathbb{R}^n$ and $Q \in \mathbb{R}^{n \times n}$ and symmetric. Exercise 13.15 indicates that for quadratic objectives we should concentrate on the convex case. Therefore, we assume that Q is positive semi-definite, or at least positive semi-definite on the null space $\mathcal{N}(A) = \{\Delta x \in \mathbb{R}^n | A\Delta x = 0\}$. We will consider three approaches to convex quadratic objectives in the next three sections.

13.3.1.2 Null space basis

Optimality conditions Let $Z \in \mathbb{R}^{n \times n'}$ be a matrix with columns that form a basis for the null space $\mathcal{N}(A) = \{\Delta x \in \mathbb{R}^n | A\Delta x = b\}$. By Theorems 13.1 and 13.6, necessary and sufficient conditions for $x^\star \in \mathbb{R}^n$ to be a global minimizer are that:

$$Z^\dagger Q x^\star = -Z^\dagger c, \tag{13.21}$$
$$A x^\star = b. \tag{13.22}$$

Algorithm Equations (13.21) and (13.22) are linear and involve $n' + m$ equations in n variables. By Exercise 13.17, if Q is positive definite on the null space $\mathcal{N}(A) = \{\Delta x \in \mathbb{R}^n | A\Delta x = 0\}$ then the equations will have a unique solution. Otherwise, there will be multiple minimizers.

Example Continuing with the previous equality-constrained Problem (2.13) from Sections 2.3.2.2, 13.1.1.4,..., 13.2.2.3,

$$\min_{x \in \mathbb{R}^2}\{f(x)|Ax = b\},$$

we note that the objective $f : \mathbb{R}^2 \to \mathbb{R}$ is convex and of the form:

$$\forall x \in \mathbb{R}^2, \ f(x) = \frac{1}{2}x^\dagger Q x + c^\dagger x,$$

with:

$$Q = \begin{bmatrix} 2 & 0 \\ 0 & 2 \end{bmatrix}, c = \begin{bmatrix} -2 \\ -6 \end{bmatrix},$$

and the coefficient matrix and right-hand side of the constraints are defined by:

$$A = \begin{bmatrix} 1 & -1 \end{bmatrix}, b = [0].$$

From Section 13.1.2.2:

$$Z = \begin{bmatrix} 1 \\ 1 \end{bmatrix} \in \mathbb{R}^{2 \times 1}$$

is a matrix with columns that form a basis for the null space of A. The matrices and vectors in (13.21)–(13.22) are:

$$
\begin{aligned}
Z^\dagger Q &= \begin{bmatrix} 1 & 1 \end{bmatrix} \begin{bmatrix} 2 & 0 \\ 0 & 2 \end{bmatrix}, \\
&= \begin{bmatrix} 2 & 2 \end{bmatrix}, \\
-Z^\dagger c &= -\begin{bmatrix} 1 & 1 \end{bmatrix} \begin{bmatrix} -2 \\ -6 \end{bmatrix}, \\
&= [8] \\
A &= \begin{bmatrix} 1 & -1 \end{bmatrix}, \\
b &= [0].
\end{aligned}
$$

The equations to be solved are therefore:

$$\begin{bmatrix} 2 & 2 \\ 1 & -1 \end{bmatrix} x^\star = \begin{bmatrix} 8 \\ 0 \end{bmatrix}.$$

The solution to this system is $x^\star = \begin{bmatrix} 2 \\ 2 \end{bmatrix}$.

Discussion The main drawback of this approach is the need to construct the matrix Z and then form and factorize the coefficient matrix $\begin{bmatrix} Z^\dagger Q \\ A \end{bmatrix}$. If the coefficient matrix A is large and sparse, then it may be unattractive to form the matrix Z, which is typically dense. In this case, it can be better to deal with the necessary conditions written in terms of the minimizer $x^\star$ and the Lagrange multipliers $\lambda^\star$ as discussed in the next section.

13.3.1.3 Lagrange multipliers

Optimality conditions By Theorems 13.2 and 13.5, necessary and sufficient conditions for $x^\star \in \mathbb{R}^n$ to be a global minimum are that there exists $\lambda^\star \in \mathbb{R}^m$ such that $x^\star$ and $\lambda^\star$ satisfy:

$$
\begin{aligned}
Qx^\star + A^\dagger \lambda^\star &= -c, & (13.23) \\
Ax^\star &= b. & (13.24)
\end{aligned}
$$

Algorithm Equations (13.23) and (13.24) are linear and involve $n + m$ equations in $n + m$ variables. As shown in Exercise 13.17, if Q is positive definite on the null space $\mathcal{N}(A) = \{\Delta x \in \mathbb{R}^n | A\Delta x = 0\}$ and A has linearly independent rows then the linear simultaneous equations (13.23)–(13.24) will have a unique solution. However, the coefficient matrix of this system:

$$\mathcal{A} = \begin{bmatrix} Q & A^\dagger \\ A & 0 \end{bmatrix}, \tag{13.25}$$

is indefinite, so that a special purpose algorithm for factorization of indefinite matrices should be used, as mentioned in Section 5.4.7. Performing a single forwards and backwards substitution then solves:

$$\begin{bmatrix} Q & A^\dagger \\ A & 0 \end{bmatrix} \begin{bmatrix} x \\ \lambda \end{bmatrix} = \begin{bmatrix} -c \\ b \end{bmatrix}. \tag{13.26}$$

Example Continuing with the previous equality-constrained Problem (2.13) from Sections 2.3.2.2, 13.1.1.4,…, 13.3.1.2, the objective $f : \mathbb{R}^2 \to \mathbb{R}$ is defined by:

$$\forall x \in \mathbb{R}^2, \ f(x) = \frac{1}{2}x^\dagger Q x + c^\dagger x,$$

where:

$$Q = \begin{bmatrix} 2 & 0 \\ 0 & 2 \end{bmatrix}, c = \begin{bmatrix} -2 \\ -6 \end{bmatrix},$$

and the coefficient matrix and right-hand side of the constraints are defined by:

$$A = \begin{bmatrix} 1 & -1 \end{bmatrix}, b = [0].$$

Necessary and sufficient conditions for Problem (2.13) are given by (13.26):

$$\begin{bmatrix} 2 & 0 & 1 \\ 0 & 2 & -1 \\ 1 & -1 & 0 \end{bmatrix} \begin{bmatrix} x \\ \lambda \end{bmatrix} = \begin{bmatrix} 2 \\ 6 \\ 0 \end{bmatrix}.$$

Solving this system we obtain:

$$x^\star = \begin{bmatrix} 2 \\ 2 \end{bmatrix}, \lambda^\star = [-2].$$

Discussion The coefficient matrix $\mathcal{A}$ in (13.25) is sparse if Q and A are sparse. Consequently, solving (13.23)–(13.24) avoids the drawback of the null space basis approach that involved the matrix Z, which is typically dense. Although (13.23)–(13.24) has more equations than (13.21)–(13.22), if (13.23)–(13.24) is sparse then it can be much easier to solve than (13.21)–(13.22) for the reasons discussed in Section 5.5.3.3.

If Q is positive semi-definite but not positive definite, then it may be the case that the minimizer of Problem (13.1) is non-unique. A QR factorization of A specialized to indefinite matrices can be used. (See Section 5.4.7. Note that the "Q" factor of A will, of course, be different to the Hessian Q of f.) If Problem (13.1) has a minimum then it is possible to assign arbitrary values to some of the entries of $x^\star$ and solve for the remaining entries.

If $Q = 0$ so that the problem is actually linear, then it is usually the case that no minimum exists. Furthermore, for some positive semi-definite Q, it is also possible that no minimum exists. (See Exercise 13.18.)

If some of the rows of A are linearly dependent, then the Lagrange multipliers are not unique. (See Exercises 13.4 and 13.19.)

13.3.1.4 Dual maximization

Optimality conditions The dual function $\mathcal{D} : \mathbb{R}^m \to \mathbb{R} \cup \{-\infty\}$ is defined by:

$$\forall \lambda \in \mathbb{R}^m, \mathcal{D}(\lambda) = \inf_{x \in \mathbb{R}^n} \left\{ \frac{1}{2} x^\dagger Q x + c^\dagger x + \lambda^\dagger (Ax - b) \right\}. \tag{13.27}$$

The dual problem is:

$$\max_{\lambda \in \mathbb{E}} \mathcal{D}(\lambda).$$

To evaluate the dual function for a given $\lambda \in \mathbb{E}$ we must minimize the Lagrangian as a function of x. The first-order necessary conditions for the unconstrained minimization problem on the right-hand side of (13.27) are:

$$\nabla_x \mathcal{L}(x, \lambda) = Qx + c + A^\dagger \lambda = 0. \tag{13.28}$$

For the rest of the analysis of dual maximization, we will assume that Q is positive definite so that the unconstrained problem on the right-hand side of (13.27) is strictly convex and (13.28) has a unique solution. We can write down the unique solution to (13.28) as:

$$x^{(\lambda)} = -Q^{-1}(c + A^\dagger \lambda),$$

where $x^{(\lambda)}$ denotes the unique solution for the given value of λ. If f is separable, so that Q is diagonal, and if Q has strictly positive entries on its diagonal, then f is strictly convex and Q^{-1} exists and is easy to calculate. The unique solution $x^{(\lambda)}$ for the given λ is very easy to obtain. Otherwise, if Q^{-1} is not readily available then Q can be factorized and the linear equation $Qx = -(A^\dagger \lambda + c)$ solved.

Substituting the value of $x^{(\lambda)}$ into the objective of the dual function yields the value of the dual function at λ. We seek the maximum of the dual function over all values of λ. Since we have assumed that Q is positive definite then the solution to (13.28) is unique and the dual is partially differentiable by Theorem 13.7. The necessary conditions for maximizing the dual are that:

$$\nabla D(\lambda) = Ax^{(\lambda)} - b = \mathbf{0},$$

where $x^{(\lambda)}$ is the solution to (13.28). Each entry in λ can be increased or decreased depending on whether the corresponding entry of $Ax^{(\lambda)} - b$ is greater than or less than zero.

Algorithm A steepest ascent algorithm based on maximizing the dual function when Q is positive definite involves the following recursion to define the updates:

$$
\begin{array}{rcll}
x^{(\nu)} & = & -Q^{-1}(c + A^\dagger \lambda^{(\nu)}), & (13.29) \\
\Delta\lambda^{(\nu)} & = & Ax^{(\nu)} - b, & (13.30) \\
\lambda^{(\nu+1)} & = & \lambda^{(\nu)} + \alpha^{(\nu)} \Delta\lambda^{(\nu)}, &
\end{array}
$$

where $\alpha^{(\nu)}$ should be chosen to ensure a *sufficient increase* in $D(\lambda^{(\nu+1)})$ compared to $D(\lambda^{(\nu)})$ using, for example, the Armijo criterion described in Section 10.2.4.2. (Alternatively, if an augmented Lagrangian is used then the step-size may be chosen in relation to the penalty coefficient. See [70, sections 13.4 and 13.5] for further details.) If Q is not diagonal nor otherwise easy to invert then (13.29) should be evaluated by solving the linear equation $Qx^{(\nu)} = -(c + A^\dagger \lambda^{(\nu)})$ for $x^{(\nu)}$.

Stopping criterion By Theorem 3.13, $D(\lambda^{(\nu+1)})$ provides a lower bound on the value of the minimum. This lower bound can be incorporated into a stopping criterion if a feasible solution $\hat{x}$ is known. (Calculation of a feasible point from the sequence of dual solutions is described in [6, section 6.5].) We will use this idea more explicitly in the discussion of inequality constraints. (See Exercise 16.25.)

Example Continuing with the previous equality-constrained Problem (2.13) from Sections 2.3.2.2, 13.1.1.4,..., 13.3.1.3, the objective $f : \mathbb{R}^2 \to \mathbb{R}$ is defined by:

$$\forall x \in \mathbb{R}^2,\ f(x) = \frac{1}{2}x^\dagger Qx + c^\dagger x,$$

where:

$$Q = \begin{bmatrix} 2 & 0 \\ 0 & 2 \end{bmatrix}, c = \begin{bmatrix} -2 \\ -6 \end{bmatrix},$$

and the coefficient matrix and right-hand side of the constraints are defined by:

$$A = \begin{bmatrix} 1 & -1 \end{bmatrix}, b = [0].$$

Let $\lambda^{(0)} = [0]$. Then:

$$
\begin{aligned}
x^{(0)} &= -Q^{-1}(c + A^\dagger \lambda^{(0)}), \\
&= -\begin{bmatrix} \frac{1}{2} & 0 \\ 0 & \frac{1}{2} \end{bmatrix}\begin{bmatrix} -2 \\ -6 \end{bmatrix}, \\
&= \begin{bmatrix} 1 \\ 3 \end{bmatrix}, \\
\Delta\lambda^{(0)} &= Ax^{(0)} - b, \\
&= \begin{bmatrix} 1 & -1 \end{bmatrix}\begin{bmatrix} 1 \\ 3 \end{bmatrix} - 0, \\
&= [-2], \\
\lambda^{(1)} &= \lambda^{(0)} + \alpha^{(0)}\Delta\lambda^{(0)}, \\
&= [0] + 1 \times [-2], \text{ picking } \alpha^{(0)} = 1, \\
&= [-2], \\
x^{(1)} &= -Q^{-1}(c + A^\dagger \lambda^{(1)}), \\
&= -\begin{bmatrix} \frac{1}{2} & 0 \\ 0 & \frac{1}{2} \end{bmatrix}\left(\begin{bmatrix} -2 \\ -6 \end{bmatrix} + \begin{bmatrix} 1 \\ -1 \end{bmatrix} \times (-2)\right), \\
&= \begin{bmatrix} 2 \\ 2 \end{bmatrix}, \\
\Delta\lambda^{(1)} &= Ax^{(1)} - b, \\
&= [0],
\end{aligned}
$$

and the dual algorithm has converged in one iteration.

Usually, the dual iteration using steepest ascent requires more than one iteration to obtain a useful estimate of the Lagrange multipliers and minimizer, even if an optimal step-size is chosen. This is because the level sets of the dual function are elliptical and not spherical. In this example, with only one constraint, the steepest ascent algorithm converges rapidly. More generally, this algorithm is most effective when there are only a few constraints but many variables in the primal problem. In principle, the Hessian of the dual function can be calculated or estimated and incorporated into a Newton–Raphson algorithm to facilitate faster convergence to the dual maximizer [70, section 13.1].

Discussion The algorithm adjusts λ until the optimality conditions for the dual are satisfied. Maximizing the dual involves:

- choosing x to satisfy $\nabla f(x) + A^\dagger \lambda^{(\nu)} = 0$ at each iteration, given the current estimate of the Lagrange multiplier, $\lambda^{(\nu)}$, and

- updating the Lagrange multiplier estimate at each iteration so as to more nearly satisfy the constraint (13.7), that is, $Ax = b$.

13.3.2 Non-quadratic objective

13.3.2.1 Problem

Suppose that the objective $f : \mathbb{R}^n \to \mathbb{R}$ is partially differentiable with continuous partial derivatives. We will consider several approaches to this problem.

13.3.2.2 Null space basis

Optimality conditions Let $Z \in \mathbb{R}^{n \times n'}$ be a matrix with columns that form a basis for the null space $\mathcal{N}(A) = \{\Delta x \in \mathbb{R}^n | A\Delta x = b\}$. By Theorem 13.1, necessary conditions for $x^\star \in \mathbb{R}^n$ to be a global minimizer are that $x^\star$ is feasible and that the reduced gradient is zero. That is:

$$Z^\dagger \nabla f(x^\star) = 0, \tag{13.31}$$
$$Ax^\star = b. \tag{13.32}$$

If f is convex then, by Theorem 13.6, the conditions (13.31)–(13.32) are sufficient. If the reduced function ϕ is twice partially differentiable with continuous partial derivatives and its Hessian is positive definite then, by Theorem 13.3, $x^\star$ is a strict local minimizer.

Algorithm Suppose we construct an initial guess $x^{(0)}$ that satisfies the equality constraints. That is, we use the techniques from Section 5.8.1.2 to construct any point that satisfies the equality constraints. At the same time, we can also construct the matrix Z. The set of all solutions to the linear equations is given by:

$$\{x^{(0)} + Z\xi | \xi \in \mathbb{R}^{n'}\}.$$

We can now proceed to minimize the reduced function $\phi : \mathbb{R}^{n'} \to \mathbb{R}$ defined by:

$$\forall \xi \in \mathbb{R}^{n'}, \phi(\xi) = f(x^{(0)} + Z\xi).$$

Any of the unconstrained minimization methods developed in Section 10.2 can be used to minimize this function. For example, a steepest descent algorithm using the reduced gradient $\nabla \phi$ would involve the following recursion to define the iterates:

$$\xi^{(\nu+1)} = \xi^{(\nu)} - \alpha^{(\nu)} \nabla \phi(\xi^{(\nu)}),$$

or equivalently:

$$\begin{aligned}
\xi^{(\nu+1)} &= \xi^{(\nu)} - \alpha^{(\nu)} Z^\dagger \nabla f(x^{(0)} + Z\xi^{(\nu)}), \\
&= \xi^{(\nu)} - \alpha^{(\nu)} Z^\dagger \nabla f(x^{(\nu)}),
\end{aligned}$$

where $x^{(\nu)} = x^{(0)} + Z\xi^{(\nu)}$ and the step-size $\alpha^{(\nu)}$ should be chosen to achieve sufficient decrease in the reduced function $\phi(\xi^{(\nu+1)})$ according to, for example, the Armijo step-size rule as discussed in Section 10.2.4.2. The steepest descent algorithm uses the reduced gradient $\nabla\phi(\xi^{(\nu)}) = Z^\dagger \nabla f(x^{(\nu)})$.

Alternatively, a Newton–Raphson algorithm would involve the update:

$$\begin{aligned}
\nabla^2\phi(\xi^{(\nu)})\Delta\xi^{(\nu)} &= -\nabla\phi(\xi^{(\nu)}), \\
\xi^{(\nu+1)} &= \xi^{(\nu)} + \alpha^{(\nu)}\Delta\xi^{(\nu)},
\end{aligned}$$

or equivalently:

$$\begin{aligned}
Z^\dagger\nabla^2 f(x^{(\nu)})Z\Delta\xi^{(\nu)} &= -Z^\dagger\nabla f(x^{(\nu)}), \\
\xi^{(\nu+1)} &= \xi^{(\nu)} + \alpha^{(\nu)}\Delta\xi^{(\nu)}.
\end{aligned}$$

The Newton–Raphson algorithm uses the reduced Hessian:

$$\nabla^2\phi(\xi^{(\nu)}) = Z^\dagger\nabla^2 f(x^{(\nu)})Z.$$

If zero pivots are encountered in factorizing the reduced Hessian, then the pivots should be modified to be positive to ensure descent of the reduced function. A natural initial guess for ξ is $\xi^{(0)} = \mathbf{0}$, corresponding to an initial guess of $x^{(0)}$ in the x coordinates.

Example Recall the non-quadratic objective $f : \mathbb{R}^2 \to \mathbb{R}$ defined in (10.9) in Section 10.2.1.5:

$$\forall x \in \mathbb{R}^2, \ f(x) = 0.01(x_1 - 1)^4 + 0.01(x_2 - 3)^4 + (x_1 - 1)^2 + (x_2 - 3)^2$$
$$- 1.8(x_1 - 1)(x_2 - 3).$$

Consider the problem $\min_{x\in\mathbb{R}^2}\{f(x)|Ax = b\}$, where $A \in \mathbb{R}^{1\times 2}$ and $b \in \mathbb{R}^1$ are defined by:

$$A = \begin{bmatrix} 1 & -1 \end{bmatrix}, b = [8].$$

By inspection, $Z = \begin{bmatrix} 1 \\ 1 \end{bmatrix}$ is a matrix with columns that form a basis for the null space of A. Consider the initial guess $x^{(0)} = \begin{bmatrix} 3 \\ -5 \end{bmatrix}$, which is feasible for the equality constraint. We perform one iteration of a steepest descent algorithm to minimize the reduced function with initial guess $\xi^{(0)} = [0]$.

We have:

$$
\begin{aligned}
f(x^{(0)}) &= 137.77, \\
\nabla f(x) &= \begin{bmatrix} 0.04(x_1 - 1)^3 + 2(x_1 - 1) - 1.8(x_2 - 3) \\ 0.04(x_2 - 3)^3 - 1.8(x_1 - 1) + 2(x_2 - 3) \end{bmatrix}, \\
\nabla f(x^{(0)}) &= \begin{bmatrix} 18.72 \\ -40.08 \end{bmatrix}, \\
Z^\dagger \nabla f(x^{(0)}) &= [-21.36].
\end{aligned}
$$

Using a step-size of 1, we obtain a tentative update of:

$$
\begin{aligned}
\xi^{(1)} &= \xi^{(0)} - Z^\dagger \nabla f(x^{(0)}), \\
&= [21.36], \\
x^{(0)} + Z\xi^{(1)} &= \begin{bmatrix} 24.36 \\ 16.36 \end{bmatrix}, \\
f(x^{(0)} + Z\xi^{(1)}) &= 3458.8,
\end{aligned}
$$

which is larger than $f(x^{(0)})$, so we must consider a step-size rule. We use the Armijo rule, with the step-size halved until the Armijo condition (10.14) is satisfied. For a step-size of $\alpha^{(0)} = 0.5$, the update still yields an objective that exceeds $f(x^{(0)})$. However, for $\alpha^{(0)} = 0.25$, the Armijo condition is satisfied and we obtain:

$$
\begin{aligned}
\xi^{(1)} &= [5.34], \\
x^{(1)} &= \begin{bmatrix} 8.34 \\ 0.34 \end{bmatrix}, \\
f(x^{(1)}) &= 125.6.
\end{aligned}
$$

(See Exercise 13.25.)

Stopping criterion The algorithm involved unconstrained minimization of the reduced function ϕ. Stopping criteria for unconstrained problems as discussed in Section 10.2.5 can be used for this algorithm.

Discussion Whatever algorithm is used for minimizing ϕ, at each iteration the iterate $x^{(\nu)} = x^{(0)} + Z\xi^{(\nu)}$ is feasible for the equality constraints. That is, we can terminate the calculations at any iteration and obtain a feasible, if not optimal, solution for the problem. In summary, we generate iterates that are:

- feasible at each iteration, satisfying (13.32), and
- in principle, become closer to satisfying the condition (13.31).

Analogously to the discussion of algorithms in Section 10.2, we can also consider the variations that avoid calculation of the reduced Hessian or avoid factorizing the reduced Hessian. We should choose $\alpha^{(\nu)}$ according to a sufficient decrease criterion as discussed in Section 10.2.4.

As with quadratic objectives, the main drawback of this approach is the need to construct the matrix Z. If the coefficient matrix A is large and sparse, then it may be unattractive to form the matrix Z, which is typically dense. In this case, it can be better to deal with the necessary conditions written in terms of the minimizer $x^\star$ and the Lagrange multipliers $\lambda^\star$ as discussed in the next section.

13.3.2.3 Lagrange multipliers

Optimality conditions By Theorem 13.2, first-order necessary conditions for $x^\star \in \mathbb{R}^n$ to be a local minimizer are that there exists $\lambda^\star \in \mathbb{R}^m$ such that $x^\star$ and $\lambda^\star$ satisfy:

$$\nabla f(x^\star) + A^\dagger \lambda^\star = 0, \tag{13.33}$$
$$Ax^\star - b = 0. \tag{13.34}$$

If f is convex, then by Theorem 13.5, the conditions (13.33)–(13.34) are also sufficient. If f is twice partially differentiable with continuous partial derivatives and its Hessian is positive definite on the null space $\mathcal{N}(A) = \{\Delta x \in \mathbb{R}^n | A \Delta x = 0\}$ then, by Theorem 13.4, $x^\star$ is a strict local minimizer.

Algorithm Equations (13.33)–(13.34) are non-linear in $\begin{bmatrix} x \\ \lambda \end{bmatrix}$, involve $n + m$ equations in $n + m$ variables, and can be solved iteratively using the Newton–Raphson method. This algorithm will inherit the convergence properties of the Newton–Raphson method as discussed in Section 7.3.3. In particular, convergence can be quadratic.

Moreover, since the constraints $Ax = b$ are linear, we can, in principle, construct an initial point $x^{(0)}$ that satisfies $Ax^{(0)} = b$ using the techniques discussed in Section 5.8.1. At each subsequent iteration, we try to solve for the Newton–Raphson step direction using:

$$
\mathcal{A} \begin{bmatrix} \Delta x^{(\nu)} \\ \Delta \lambda^{(\nu)} \end{bmatrix} = -\begin{bmatrix} \nabla f(x^{(\nu)}) + A^\dagger \lambda^{(\nu)} \\ Ax^{(\nu)} - b \end{bmatrix},
$$
$$
= -\begin{bmatrix} \nabla f(x^{(\nu)}) + A^\dagger \lambda^{(\nu)} \\ 0 \end{bmatrix}, \tag{13.35}
$$

where $\mathcal{A} \in \mathbb{R}^{(n+m) \times (n+m)}$ is defined by:

$$
\mathcal{A} = \begin{bmatrix} \nabla^2 f(x^{(\nu)}) & A^\dagger \\ A & 0 \end{bmatrix},
$$

and where we have assumed that $Ax^{(\nu)} - b = 0$.

Example Continuing with the non-quadratic objective $f : \mathbb{R}^2 \to \mathbb{R}$ defined in (10.9) in Sections 10.2.1.5 and 13.3.2.2:

$$\forall x \in \mathbb{R}^2, \ f(x) = 0.01(x_1 - 1)^4 + 0.01(x_2 - 3)^4 + (x_1 - 1)^2 + (x_2 - 3)^2$$
$$- 1.8(x_1 - 1)(x_2 - 3),$$

and the problem $\min_{x \in \mathbb{R}^2} \{f(x) | Ax = b\}$, where:

$$A = \begin{bmatrix} 1 & -1 \end{bmatrix}, b = [8],$$

we consider the initial guess $x^{(0)} = \begin{bmatrix} 3 \\ -5 \end{bmatrix}$ and $\lambda^{(0)} = [0]$, and perform one Newton–Raphson update. The coefficient matrix $\mathcal{A}$ in (13.35) is given by:

$$\mathcal{A} = \begin{bmatrix} \nabla^2 f(x^{(0)}) & A^\dagger \\ A & 0 \end{bmatrix},$$

$$= \begin{bmatrix} 2.48 & -1.8 & 1 \\ -1.8 & 9.68 & -1 \\ 1 & -1 & 0 \end{bmatrix},$$

and the right-hand side is:

$$- \begin{bmatrix} \nabla f(x^{(\nu)}) + A^\dagger \lambda^{(\nu)} \\ 0 \end{bmatrix} = \begin{bmatrix} -18.72 \\ 40.08 \\ 0 \end{bmatrix}.$$

Solving (13.35) for these values yields:

$$\begin{bmatrix} \Delta x^{(\nu)} \\ \Delta \lambda^{(\nu)} \end{bmatrix} = \begin{bmatrix} 2.4953 \\ 2.4953 \\ -20.4168 \end{bmatrix}.$$

Using a step-size of one, we obtain:

$$x^{(1)} = \begin{bmatrix} 5.4953 \\ -2.5047 \end{bmatrix},$$

$$\lambda^{(1)} = [-20.4168],$$

with objective value $f(x^{(1)}) = 108.3163$. (See Exercise 13.25.)

Stopping criterion If f is convex and $x^{(\nu)}$ and $\lambda^{(\nu)}$ satisfy $\nabla f(x^{(\nu)}) + A^\dagger \lambda^{(\nu)} = 0$ then $x^{(\nu)}$ minimizes $\mathcal{L}(\bullet, \lambda^{(\nu)})$. Therefore, $\mathcal{D}(\lambda^{(\nu)}) = \mathcal{L}(x^{(\nu)}, \lambda^{(\nu)})$ and so, by Theorem 3.13, $\mathcal{L}(x^{(\nu)}, \lambda^{(\nu)})$ provides a lower bound for $\min_{x \in \mathbb{R}^n} \{f(x) | Ax = b\}$ that can be incorporated into a stopping criterion.

However, $x^{(\nu)}$ will typically only be an approximate minimizer of $\mathcal{L}(\bullet, \lambda^{(\nu)})$. In this case, if f is convex and we want to ensure that $f(x^{(\nu)})$ is within ϵ_f of the minimum, and if there is a known bound on where the minimizer of the problem $\min_{x \in \mathbb{R}^n} \{f(x) | Ax = b\}$ can lie of the form $\|x^\star - x^{(\nu)}\| \leq \bar{\rho}$, then an approach analogous to that in Section 10.2.5 can be taken based on iterating until $\|\nabla f(x^{(\nu)}) + A^\dagger \lambda^{(\nu)}\| \leq \epsilon_f / \bar{\rho}$. (See Exercise 13.23.)

Discussion As in the case of the quadratic objective, even if $\nabla^2 f(x^{(\nu)})$ is positive definite, the coefficient matrix $\mathcal{A}$ in (13.35) is indefinite. To factorize it, we should use a special purpose algorithm as mentioned in Section 5.4.7.

If the coefficient matrix $\mathcal{A}$ in (13.35) is non-singular then we can always solve the equation uniquely. On the other hand, if the coefficient matrix $\mathcal{A}$ in (13.35) is singular then, as discussed in Section 10.2.3, we will have to modify the pivots. However, so long as A has linearly independent rows, we can, in principle, order the pivoting so that we do not have to modify any of the pivots corresponding to the lower right block of the matrix. If none of the pivots in the lower right block of the matrix are modified, then the search direction $\begin{bmatrix} \Delta x^{(\nu)} \\ \Delta \lambda^{(\nu)} \end{bmatrix}$ will satisfy $A \Delta x^{(\nu)} = 0$.

Zero or negative pivots in the top left-hand block $\nabla^2 f(x^{(\nu)})$ should be modified to be positive. This ensures that $\Delta x^{(\nu)}$ corresponds to a descent direction for the reduced function. That is, we modify $\nabla^2 f(x^{(\nu)})$ to be equal to Q, where Q is a positive definite matrix. (See Exercise 13.24.)

In summary, if A has linearly independent rows, we can construct a search direction $\begin{bmatrix} \Delta x^{(\nu)} \\ \Delta \lambda^{(\nu)} \end{bmatrix}$ that satisfies $A \Delta x^{(\nu)} = 0$, modifying pivots in the top left-hand block if necessary to be positive. (If A has linearly dependent rows, it is still possible to find a search direction that satisfies $A \Delta x^{(\nu)} = 0$; however, the factorization scheme is more complicated.)

As in Section 13.3.2.2 for the null space basis approach, we will always stay feasible with respect to the constraints $Ax^{(\nu)} - b = 0$. This means that we can terminate calculations at any iteration and use $x^{(\nu)}$ as a feasible, if not optimal, solution. Again, in principle, the iterates converge to a solution of the first-order necessary conditions. As with the approach in Section 13.3.2.2, the iterates are:

- feasible at each iteration, satisfying (13.34), and
- in principle, become closer to satisfying the condition (13.33).

As in Section 13.3.2.2, we can also:

- use the variations described in Section 10.2.3 to approximate the Hessian or the update to save computational effort,

- use the ideas from Section 10.2.4 to choose the step-size $\alpha^{(\nu)}$ to ensure sufficient descent in f, and
- use the ideas from Section 10.2.5 to develop stopping criteria.

The coefficient matrix $\mathcal{A}$ in (13.35) is sparse if A and $\nabla^2 f$ are sparse. As in Chapter 5, we should take advantage of sparsity in factorization.

13.3.2.4 Dual maximization

Optimality conditions The dual function in this case is $\mathcal{D} : \mathbb{R}^m \to \mathbb{R} \cup \{-\infty\}$ defined by:

$$\forall \lambda \in \mathbb{R}^m, \mathcal{D}(\lambda) = \inf_{x \in \mathbb{R}^n} \{f(x) + \lambda^\dagger (Ax - b)\}.$$

The dual problem is:

$$\max_{\lambda \in \mathbb{E}} \mathcal{D}(\lambda).$$

If we assume that there is a minimum and minimizer of the primal problem and that there is no duality gap, then maximizing the dual function yields the minimum of the primal problem. If the conditions of Corollary 13.8 hold then the optimality conditions for the dual problem are that:

$$\nabla \mathcal{D}(\lambda) = Ax^{(\lambda)} - b,$$
$$= 0,$$

where $x^{(\lambda)}$ is the unique minimizer of $\min_{x \in \mathbb{R}^n} \{f(x) + \lambda^\dagger (Ax - b)\}$.

Algorithm In this case, the algorithm involves the following recursion:

$$x^{(\nu)} \in \arg\min_{x \in \mathbb{R}^n} \{f(x) + [\lambda^{(\nu)}]^\dagger (Ax - b)\}, \qquad (13.36)$$
$$\Delta\lambda^{(\nu)} = Ax^{(\nu)} - b,$$
$$\lambda^{(\nu+1)} = \lambda^{(\nu)} + \alpha^{(\nu)} \Delta\lambda^{(\nu)}.$$

Example Continuing with the objective defined in (10.9),

$$\forall x \in \mathbb{R}^2, f(x) = 0.01(x_1 - 1)^4 + 0.01(x_2 - 3)^4 + (x_1 - 1)^2 + (x_2 - 3)^2$$
$$- 1.8(x_1 - 1)(x_2 - 3),$$

constraints defined by:

$$A = \begin{bmatrix} 1 & -1 \end{bmatrix}, b = [8],$$

and the problem from Sections 10.2.1.5, 13.3.2.2, and 13.3.2.3, we let $\lambda^{(0)} = [0]$, and solve the problem on the right-hand side of (13.36).

We perform one (outer) iteration of a dual maximization algorithm. Since $\lambda^{(0)} =$

[0], Problem (13.36) is equivalent to unconstrained minimization of f. The minimizer is $x^{(0)} = \begin{bmatrix} 1 \\ 3 \end{bmatrix}$, and we have:

$$\begin{aligned}
\Delta\lambda^{(0)} &= Ax^{(0)} - b, \\
&= \begin{bmatrix} 1 & -1 \end{bmatrix} \begin{bmatrix} 1 \\ 3 \end{bmatrix} - [8], \\
&= [-10].
\end{aligned}$$

Using a step-size of $\alpha^{(0)} = 1$, this yields:

$$\begin{aligned}
\lambda^{(1)} &= \lambda^{(0)} + \alpha^{(0)} \Delta\lambda^{(0)}, \\
&= [0] + 1 \times [-10], \\
&= [-10].
\end{aligned}$$

(See Exercise 13.25.)

Stopping criterion Again Theorem 3.13 can be used to show that $\mathcal{D}(\lambda^{(\nu)})$ provides a lower bound on the value of the minimum.

Discussion As in Section 13.3.1.4, maximizing the dual involves:

- satisfying $\nabla f(x^{(\nu)}) + A^{\dagger}\lambda^{(\nu)} = 0$ at each outer iteration, given the current estimate of the Lagrange multiplier, $\lambda^{(\nu)}$, and
- updating the Lagrange multiplier estimate at each outer iteration so as to more nearly satisfy the constraint (13.7).

For each update of λ there are a number of inner iterations to solve Problem (13.36) to sufficient accuracy. Once a minimizer of Problem (13.36) is obtained then, to update λ, the step-size $\alpha^{(\nu)}$ should be chosen to yield a *sufficient increase* in the dual function using, for example, the Armijo criterion as described in Section 10.2.4.2.

If f is separable then the update of x in (13.36) can be performed very easily and in parallel since the dual function separates into n sub-problems. (Notice that the term $-[\lambda^{(\nu)}]^{\dagger}b$ in the objective of the right-hand side of (13.36) is constant with respect to x and so does not enter into the optimization in (13.36) and can be ignored.)

13.4 Sensitivity

We now suppose that the objective f, constraint matrix A, and right-hand side vector b are *parameterized* by a parameter $\chi \in \mathbb{R}^s$. That is, $f : \mathbb{R}^n \times \mathbb{R}^s \to \mathbb{R}$,

$A : \mathbb{R}^s \to \mathbb{R}^{m \times n}$, and $b : \mathbb{R}^s \to \mathbb{R}^m$. We imagine that we have solved the constrained minimization problem:

$$\min_{x \in \mathbb{R}^n} \{ f(x; \chi) | A(\chi)x = b(\chi) \},$$

for a base-case value of the parameters, say $\chi = \mathbf{0}$, to find the base-case minimizer $x^\star$ and base-case Lagrange multipliers $\lambda^\star$ and that now we are considering the sensitivity of the base-case solution to variation of the parameters around $\chi = \mathbf{0}$. In Section 13.4.1, we consider the general case using the implicit function theorem and then in Section 13.4.2 we specialize to the case where only the right-hand side of the equality constraints vary.

13.4.1 General case

As with the sensitivity analysis of non-linear equations in Section 7.5 and of unconstrained minimization in Section 10.3, we use the **implicit function theorem** (Theorem A.9 in Section A.7.3 of Appendix A.) The following corollary to the implicit function theorem provides us with the sensitivity of the solution to the parameters.

Corollary 13.10 *Let* $f : \mathbb{R}^n \times \mathbb{R}^s \to \mathbb{R}^n$ *be twice partially differentiable with continuous second partial derivatives and let* $A : \mathbb{R}^s \to \mathbb{R}^{m \times n}$ *and* $b : \mathbb{R}^s \to \mathbb{R}^m$ *be partially differentiable with continuous partial derivatives. Consider the minimization problem:*

$$\min_{x \in \mathbb{R}^n} \{ f(x; \chi) | A(\chi)x = b(\chi) \}, \tag{13.37}$$

where χ *is a parameter. Suppose that* $x^\star \in \mathbb{R}^n$ *is a local minimizer of Problem (13.37) for the base-case value of the parameters* $\chi = \mathbf{0}$ *with corresponding Lagrange multipliers* $\lambda^\star \in \mathbb{R}^m$. *We call* $x = x^\star$ *a base-case solution and call* $\lambda = \lambda^\star$ *the base-case Lagrange multipliers. Define the (parameterized) Hessian* $\nabla^2_{xx} f : \mathbb{R}^n \times \mathbb{R}^s \to \mathbb{R}^{n \times n}$ *by:*

$$\forall x \in \mathbb{R}^n, \forall \chi \in \mathbb{R}^s, \nabla^2_{xx} f(x; \chi) = \frac{\partial^2 f}{\partial x^2}(x; \chi).$$

Suppose that:

- $\nabla^2_{xx} f(x^\star; \mathbf{0})$ *is positive definite on the null space of* $A(\mathbf{0})$, *so that* $x^\star$ *and* $\lambda^\star$ *satisfy the second-order sufficient conditions for the base-case problem, and*
- $A(\mathbf{0})$ *has linearly independent rows.*

Then, for values of χ *in a neighborhood of the base-case value of the parameters* $\chi = \mathbf{0}$, *there is a local minimum and corresponding local minimizer and Lagrange multipliers for Problem (13.37). Moreover, the local minimum, local minimizer, and Lagrange multipliers are partially differentiable with respect to* χ *and have continuous partial derivatives in this neighborhood.*

We consider the sensitivity with respect to χ_j, the j-th entry of χ. The sensitivity of the local minimizer $x^\star$ and Lagrange multipliers $\lambda^\star$ to χ_j, evaluated at the base-case $\chi = \mathbf{0}$, is given by the solution of:

$$\mathcal{A}\begin{bmatrix} \dfrac{\partial x^\star}{\partial \chi_j}(\mathbf{0}) \\[2mm] \dfrac{\partial \lambda^\star}{\partial \chi_j}(\mathbf{0}) \end{bmatrix} = \begin{bmatrix} -K_j(x^\star;\mathbf{0}) - \left[\dfrac{\partial A}{\partial \chi_j}(\mathbf{0})\right]^\dagger \lambda^\star \\[2mm] -\dfrac{\partial A}{\partial \chi_j}(\mathbf{0})x^\star + \dfrac{\partial b}{\partial \chi_j}(\mathbf{0}) \end{bmatrix}, \tag{13.38}$$

where:

$$\mathcal{A} = \begin{bmatrix} \nabla^2_{xx}f(x^\star;\mathbf{0}) & [A(\mathbf{0})]^\dagger \\ A(\mathbf{0}) & 0 \end{bmatrix},$$

and $K_j : \mathbb{R}^n \times \mathbb{R}^s \to \mathbb{R}^n$ is defined by:

$$\forall x \in \mathbb{R}^n, \forall \chi \in \mathbb{R}^s, \; K_j(x;\chi) = \frac{\partial^2 f}{\partial x \partial \chi_j}(x;\chi).$$

The sensitivity of the local minimum $f^\star$ to χ, evaluated at the base-case $\chi = \mathbf{0}$, is given by:

$$\frac{\partial f^\star}{\partial \chi}(\mathbf{0}) = \frac{\partial \mathcal{L}}{\partial \chi}(x^\star, \lambda^\star; \mathbf{0}),$$

where $\mathcal{L} : \mathbb{R}^n \times \mathbb{R}^m \times \mathbb{R}^s \to \mathbb{R}$ is the parameterized Lagrangian defined by:

$$\forall x \in \mathbb{R}^n, \forall \lambda \in \mathbb{R}^m, \forall \chi \in \mathbb{R}^s, \; \mathcal{L}(x, \lambda; \chi) = f(x;\chi) + \lambda^\dagger(A(\chi)x - b(\chi)).$$

If $f(\bullet; \chi)$ is convex for χ in a neighborhood of $\mathbf{0}$ then the minimizers and minima are global in this neighborhood.

Proof (See [34, theorem 3.2.2] and see Exercise 13.27 for a special case.) The sensitivity of the local minimizer and Lagrange multipliers follows from Corollary 7.5, noting that:

- by assumption and Exercises 5.19 and 5.49, the Hessian $\nabla^2_{xx}f(x;\chi)$ is positive definite on the null space of $A(\chi)$ for x in a neighborhood of the base-case minimizer $x^\star$ and χ in a neighborhood of $\chi = \mathbf{0}$, and
- by assumption and Exercises 5.32 and 13.17, the coefficient matrix $\mathcal{A}$ is non-singular in a neighborhood of the base-case minimizer and parameters,

so that the first-order necessary conditions (13.6)–(13.7) for Problem (13.37) are well-defined and satisfied in a neighborhood of $\chi = \mathbf{0}$ and the sensitivity of the solution of the first-order necessary conditions at $\chi = \mathbf{0}$ is given by the solution of (13.38). Moreover, the second-order sufficient conditions for Problem (13.37) given in Corollary 13.4 are satisfied in this neighborhood.

The sensitivity of the local minimum follows by totally differentiating the value of the local minimum $f^\star(\chi) = f(x^\star(\chi); \chi)$ with respect to χ and noting that the first-order necessary conditions for the minimizer mean that $\dfrac{\partial f}{\partial x}(x^\star; \mathbf{0}) = -[\lambda^\star]^\dagger A(\mathbf{0})$. But:

$$A(\mathbf{0})\frac{\partial x^\star}{\partial \chi}(\mathbf{0}) = -\frac{\partial A}{\partial \chi}(\mathbf{0})x^\star + \frac{\partial b}{\partial \chi}(\mathbf{0}), \tag{13.39}$$

by the second block row of (13.38) evaluated for $j = 1, \ldots, s$ and where, abusing notation, we interpret $\dfrac{\partial A}{\partial \chi}(\mathbf{0})x^\star \in \mathbb{R}^{m \times s}$ as having ℓj-th entry equal to $\sum_{k=1}^{n} \dfrac{\partial A_{\ell k}}{\partial \chi_j}(\mathbf{0})x_k^\star$.

Therefore,

$$
\begin{aligned}
\frac{\partial f^\star}{\partial \chi}(\mathbf{0}) &= \frac{\partial f}{\partial x}(x^\star; \mathbf{0})\frac{\partial x^\star}{\partial \chi}(\mathbf{0}) + \frac{\partial f}{\partial \chi}(x^\star; \mathbf{0}), \quad \text{since } f^\star(\chi) = f(x^\star(\chi); \chi), \\
&= -[\lambda^\star]^\dagger A(\mathbf{0})\frac{\partial x^\star}{\partial \chi}(\mathbf{0}) + \frac{\partial f}{\partial \chi}(x^\star; \mathbf{0}), \\
&= \frac{\partial f}{\partial \chi}(x^\star; \mathbf{0}) - [\lambda^\star]^\dagger \left(-\frac{\partial A}{\partial \chi}(\mathbf{0})x^\star + \frac{\partial b}{\partial \chi}(\mathbf{0}) \right), \quad \text{by (13.39)}, \\
&= \frac{\partial \mathcal{L}}{\partial \chi}(x^\star, \lambda^\star; \mathbf{0}), \quad \text{by definition of } \mathcal{L}.
\end{aligned}
$$

☐

As in the unconstrained case, the sensitivity of the local minimum is sometimes called the **envelope theorem** [119, section 27.4]. The sensitivity result can be sharpened [34, theorem 2.4.4 and chapter 3].

The matrix $\mathcal{A}$ in (13.38) is given by (13.25) in the case of a quadratic objective. If the algorithm in Section 13.3.1.3 was used then $\mathcal{A}$ was factorized to find the minimizer and Lagrange multipliers. Similarly, if the algorithm in Section 13.3.2.3 was used then $\mathcal{A}$ or a close approximation to it was factorized at the last iteration to find the minimizer and Lagrange multipliers. No further factorizations are required to evaluate the sensitivities.

13.4.2 Special case

In this section, we consider a special case of Problem (13.37) where only the right-hand side of the linear equality constraints are varied.

Corollary 13.11 *Consider Problem (13.1), a perturbation vector* $\gamma \in \mathbb{R}^m$, *and a perturbed version of Problem (13.1) defined by:*

$$
\min_{x \in \mathbb{R}^n} \{ f(x) | Ax = b - \gamma \}. \tag{13.40}
$$

Suppose that $f : \mathbb{R}^n \to \mathbb{R}$ *is twice partially differentiable with continuous second partial derivatives,* $A \in \mathbb{R}^{m \times n}$, *and* $b \in \mathbb{R}^m$, *with the rows of* A *linearly independent. Let* $x^\star \in \mathbb{R}^n$ *and* $\lambda^\star \in \mathbb{R}^m$ *satisfy the second-order sufficient conditions in Corollary 13.4 for Problem (13.1):*

$$
\begin{aligned}
\nabla f(x^\star) + A^\dagger \lambda^\star &= \mathbf{0}, \\
Ax^\star &= b, \\
((A\Delta x = \mathbf{0}) \text{ and } (\Delta x \neq \mathbf{0})) &\Rightarrow (\Delta x^\dagger \nabla^2 f(x^\star)\Delta x > 0).
\end{aligned}
$$

Consider Problem (13.40). For values of γ in a neighborhood of the base-case value of the parameters $\gamma = \mathbf{0}$, there is a local minimum and corresponding local minimizer and Lagrange multipliers for Problem (13.40). Moreover, the local minimum, local minimizer, and Lagrange multipliers are partially differentiable with respect to γ and have continuous partial derivatives. The sensitivity of the local minimum to γ, evaluated at the base-case $\gamma = \mathbf{0}$, is equal to $\lambda^\star$. If f is convex then the minimizers and minima are global.

Proof See [11, Proposition 3.2.2] or apply the proof of Corollary 13.10. (See Exercise 13.28.) □

13.4.3 Discussion

A significant part of the effort in proving Corollary 13.10 and Corollary 13.11 is using the implicit function theorem to show that the sensitivity of the minimizer is well-defined. That is, proving that the value of the minimizer is partially differentiable with respect to χ, which then implies that the value of the minimum is partially differentiable. If we assume that the minimizer and minimum are partially differentiable with respect to χ, then the following argument explains why the sensitivity of the minimum is given by the value of the Lagrange multipliers.

Consider Problem (13.40), a perturbation γ, and the corresponding change $\Delta x^\star$ in the minimizer of the perturbed problem. The change in the minimum is:

$$f(x^\star + \Delta x^\star) - f(x^\star)$$
$$\approx \quad \nabla f(x^\star)^\dagger \Delta x^\star, \text{ with equality as } \Delta x^\star \to \mathbf{0},$$
$$= \quad -[\lambda^\star]^\dagger A \Delta x^\star, \text{ by the first-order necessary condition } \nabla f(x^\star) + A^\dagger \lambda^\star = \mathbf{0},$$
$$= \quad [\lambda^\star]^\dagger \gamma,$$

since $A(x^\star + \Delta x^\star) = b - \gamma$, so that $-A \Delta x^\star = \gamma$. But this is true for any such perturbation γ. In the limit as $\gamma \to \mathbf{0}$, the change in the minimum approaches $[\lambda^\star]^\dagger \gamma$. That is, the sensitivity of the minimum is given by $\lambda^\star$.

We can interpret the Lagrange multipliers as the sensitivity of the minimum to changes in γ. In many problems, the specification of constraints represents some judgment about the availability of resources. If there is some flexibility in the availability of resources, perhaps involving the purchase of additional resources at additional cost, then we can use the Lagrange multipliers to help in trading off the change in the optimal objective against the cost of the purchase of additional resources.

The sensitivity result is also extremely useful if we must repeatedly solve problems that differ only in the specification of γ because we can estimate the changes in the solution using the sensitivity.

13.4.4 Example

We continue with Problem (2.13) from Sections 2.3.2.2, 13.1.1.4,..., 13.3.1.4, $\min_{x \in \mathbb{R}^2}\{f(x)|Ax = b\}$, where $f : \mathbb{R}^2 \to \mathbb{R}$, $A \in \mathbb{R}^{1 \times 2}$, and $b \in \mathbb{R}^1$ were defined by:

$$\forall x \in \mathbb{R}^2, \ f(x) = (x_1 - 1)^2 + (x_2 - 3)^2,$$
$$A = \begin{bmatrix} 1 & -1 \end{bmatrix},$$
$$b = \begin{bmatrix} 0 \end{bmatrix}.$$

The minimizer and Lagrange multiplier for this base-case problem are $x^\star = \begin{bmatrix} 2 \\ 2 \end{bmatrix}$ and $\lambda^\star = [-2]$.

However, suppose that the equality constraints changed from $Ax = b$ to $Ax = b - \gamma$. We first met this example, parameterized in a slightly different way, in Section 2.7.5.4. If γ is small enough, the minimum of the perturbed problem differs from the minimum of the original problem by approximately $[\lambda^\star]^\dagger \gamma = (-2)\gamma$. (See Exercise 13.29.)

13.5 Solution of the least-cost production case study

In this section, we solve the least-cost production case study from Section 12.1. We recall the problem in Section 13.5.1, describe algorithms to solve it in Section 13.5.2, discuss the solution, including an economic interpretation, in Section 13.5.3, and sketch sensitivity analysis in Section 13.5.4.

13.5.1 Problem

Recall Problem (12.4): $\min_{x \in \mathbb{R}^n} \{f(x)|Ax = b\}$. Suppose that $n = 3$. Then the coefficient matrix and right-hand side can be specified as:

$$A = \begin{bmatrix} -1 & -1 & -1 \end{bmatrix},$$
$$b = \begin{bmatrix} -D \end{bmatrix}.$$

(The apparently "unnatural" choice of the signs of the entries of A and b will be motivated in the discussion of dual maximization.) As discussed in Section 12.1.4.1, the objective is separable and convex.

In summary, this problem has a convex separable objective and only one equality constraint. Furthermore, the equality constraint is linear. That is, the problem is convex.

13.5.2 Algorithms

We can solve Problem (12.4) by several methods as discussed in the following sections. The calculations are completed in Exercises 13.30 and 13.31.

13.5.2.1 Null space basis

We first construct an initial guess $x^{(0)}$ that is feasible for the equality constraint. For example, by inspection,

$$x^{(0)} = \begin{bmatrix} D \\ 0 \\ 0 \end{bmatrix}$$

is a suitable initial guess. (If the equality constraints are more complicated it may be necessary to use the techniques from Section 5.8.1 to construct an initial guess satisfying the linear equality constraints.)

By Exercise 12.4, a matrix Z with columns that form a basis for the null space of A is $Z = \begin{bmatrix} -1 & -1 \\ 1 & 0 \\ 0 & 1 \end{bmatrix}$. We can form the reduced gradient and update ξ to decrease the reduced objective. Because of the choice of Z, this is equivalent to expressing x_1 in terms of x_2 and x_3 as discussed in Section 12.1.4.2. (See Exercise 13.30, Part (i) and Exercise 13.31, Part (i).)

13.5.2.2 Lagrange multipliers

Writing out each of the entries in the first-order necessary conditions, we obtain:

$$\forall k = 1, \ldots, n, \frac{d f_k}{d x_k}(x_k^\star) - \lambda^\star = 0,$$

$$D - \sum_{k=1}^{n} x_k^\star = 0.$$

We can solve these equations using the Newton–Raphson update, or some approximation to the Newton–Raphson update. If each f_k is quadratic and convex, then the necessary conditions are linear and can be solved directly. If each f_k is strictly convex then there will be a unique minimizer. (See Exercise 13.30, Part (ii) and Exercise 13.31, Part (ii).)

13.5.2.3 Dual maximization

The recursion is:

$$\forall k = 1, \ldots, n, \, x_k^{(\nu)} \in \underset{x_k \in \mathbb{R}}{\operatorname{argmin}} \{ f_k(x_k) - \lambda^{(\nu)} x_k \}, \qquad (13.41)$$

$$\Delta \lambda^{(\nu)} = A x^{(\nu)} - b,$$

$$= D - \sum_{k=1}^{n} x_k^{(\nu)},$$

$$\lambda^{(\nu+1)} = \lambda^{(\nu)} + \alpha^{(\nu)} \Delta \lambda^{(\nu)}.$$

If f_k is quadratic then, at each iteration ν, the k-th sub-problem on the right-hand side of (13.41) can be solved directly in one step by solving the linear necessary conditions. If f_k is not quadratic then (13.41) can be solved by applying the Newton–Raphson update until a value of $x_k^{(\nu)}$ is obtained that satisfies the necessary conditions to within a tolerance. That is, if f_k is non-quadratic, then at each outer iteration ν and for each k we must perform several inner iterations to solve the necessary conditions of (13.41). (See Exercise 13.30, Part (iii) and Exercise 13.31, Part (iii).)

13.5.3 Discussion

Maximizing the dual has a suggestive economic interpretation if we think of λ as the price paid for producing the commodity. The values $\lambda^{(\nu)}$ are tentative prices that are proposed at each iteration by a central purchaser. The goal of the central purchaser is to pick prices such that supply matches demand. The Lagrange multiplier $\lambda^{\star}$ is the final price that matches supply to demand.

Each cost function f_k is associated with a decision-making agent that makes decisions based on:

- its own cost function, and
- the tentative prices.

Each decision-making agent wants to maximize its profits, which is the difference between its revenues and its costs of production. Equivalently, each agent wants to minimize the difference between its costs of production and its revenues. (The difference between costs and revenues will hopefully be negative so that the profits will be positive!) That is, the agent sells a quantity of product x_k based on minimizing the difference between:

- the *cost* of production $f_k(x_k)$ for the quantity x_k, minus
- the *revenues* $x_k \lambda^{(\nu)}$, based on the current value of the dual variable, $\lambda^{(\nu)}$.

The value of $\lambda^{(\nu)}$ is the proposed tentative price per unit of production at the ν-th iteration. For any given price $\lambda^{(\nu)}$, each agent offers for sale an amount given by the solution of (13.41). The solution of (13.41) maximizes the agent's *profit,* that is, revenues minus costs, for the given value of the dual variable. The price is adjusted until the production summed across all agents equals the demand. (If we had made the more "natural" choice of defining A and b to have all positive entries then the dual variable would have been negative and the payment would have been minus the value of the dual variable. We made the "unnatural" definition of negative values in the entries of A and b so that the economic interpretation would not involve the negative of the dual variable.)

At each iteration, the central purchaser adjusts the tentative prices based on comparing the sum of offered productions by the agents to the target value D. If there is not enough production, then the price is raised for the next iteration to encourage more production. Conversely, if production exceeds the target then the price is reduced in the next iteration to discourage production. At the optimum, when the dual variable has converged to the Lagrange multiplier, this price encourages just the right amount of production. This value is sometimes called the **shadow price** [15, section 5.4.4]. Moreover, at the optimum, the "marginal cost of production" for each agent, that is, the derivative of its cost function, is the same for all agents. This idea can be generalized to other contexts. Further discussion of the relation of duality to economic contexts can be found in [119, chapter 6].

13.5.4 Change in demand

Suppose that the demand changes from D to $D + \Delta D$. By Corollary 13.10 or Corollary 13.11, the sensitivity of the minimum of the problem to ΔD, evaluated at $\Delta D = 0$, is $\lambda^\star$. We can also use Corollary 13.10 to find the sensitivity of the minimizer and Lagrange multipliers. (See Exercise 13.32.)

13.6 Summary

We have discussed descent directions for linear equality-constrained optimization problems. Analysis of descent directions yielded optimality conditions, which in turn led to algorithms. We also discussed sensitivity analysis. Finally, we discussed solution of the least-cost production case study.

Exercises

Optimality conditions

13.1 Let $A \in \mathbb{R}^{m \times n}$ and let $Z \in \mathbb{R}^{n \times n'}$, $n' \geq n - m$, be a matrix with columns that form a basis for the null space of A. Let $b \in \mathbb{R}^m$ and suppose that $\hat{x} \in \mathbb{R}^n$ satisfies $Ax = b$. Define $\tau : \mathbb{R}^{n'} \to \mathbb{R}^n$ by:

$$\forall \xi \in \mathbb{R}^{n'}, \tau(\xi) = \hat{x} + Z\xi.$$

(i) Show that $\mathbb{S}' = \mathbb{S}$ where:

$$\begin{aligned} \mathbb{S} &= \{x \in \mathbb{R}^n | Ax = b\}, \\ \mathbb{S}' &= \{\hat{x} + Z\Delta\xi | \Delta\xi \in \mathbb{R}^{n'}\}. \end{aligned}$$

(Hint: First suppose that $x \in \mathbb{S}'$ and show that this implies that $x \in \mathbb{S}$. Then suppose that $x \in \mathbb{S}$ and show that this implies that $x \in \mathbb{S}'$.)

(ii) Show that τ is onto $\mathbb{S}$. (See Definition A.25.)

(iii) Show that $\dfrac{\partial \tau}{\partial \xi} = Z$.

13.2 Let $A \in \mathbb{R}^{m \times n}$, $b \in \mathbb{R}^m$, and let $f : \mathbb{R}^n \to \mathbb{R}$ be partially differentiable with continuous partial derivatives. Let Z be a matrix with columns that form a basis for the null space of A.

(i) Let $\hat{x} \in \mathbb{R}^n$ and suppose that the reduced gradient at $\hat{x}$ is non-zero. That is, suppose that $Z^\dagger \nabla f(\hat{x}) \neq 0$. Show that $\Delta x = -ZZ^\dagger \nabla f(\hat{x})$ is a descent direction for f at $\hat{x}$. (Hint: Use Lemma 10.1 and Exercise 2.27.)

(ii) Suppose that $A\hat{x} = b$. Show that $A(\hat{x} + \alpha \Delta x) = b$ for any value of α, where Δx was specified in Part (i).

13.3 Consider the objective $f : \mathbb{R}^2 \to \mathbb{R}$ defined by:

$$\forall x \in \mathbb{R}^2, f(x) = (x_1 - 1)^2 + (x_2 - 3)^2,$$

and the problem $\min_{x \in \mathbb{R}^2} \{f(x) | Ax = b\}$.

(i) Find a matrix with columns that form a basis for the null space of the matrix $A = \begin{bmatrix} 1 & 1 \end{bmatrix}$. (Hint: See Exercise 12.4.)

(ii) Verify that $x^\star = \begin{bmatrix} 3 \\ -3 \end{bmatrix}$ does not satisfy (13.5) for the problem $\min_{x \in \mathbb{R}^2} \{f(x) | x_1 + x_2 = 0\}$.

(iii) Find a matrix with columns that form a basis for the null space of the matrix $A = \begin{bmatrix} 1 & -3 \end{bmatrix}$.

(iv) Verify that $x^\star = \begin{bmatrix} 3 \\ -3 \end{bmatrix}$ does satisfy (13.5) for the problem $\min_{x \in \mathbb{R}^2} \{f(x) | x_1 - 3x_2 = 12\}$.

13.4 Suppose that $f : \mathbb{R}^n \to \mathbb{R}$ is partially differentiable with continuous partial derivatives, $A \in \mathbb{R}^{m \times n}$, and $b \in \mathbb{R}^m$. Suppose that $x^\star \in \mathbb{R}^n$ is a local minimizer of the problem:

$$\min_{x \in \mathbb{R}^n} \{ f(x) | Ax = b \}.$$

(i) Show that if A has linearly independent rows then there is at most one $\lambda^\star$ that satisfies (13.6) in Theorem 13.2.
(ii) Give an example of a problem where there there are multiple values that satisfy (13.6) in Theorem 13.2.

13.5 Consider the objective $f : \mathbb{R}^2 \to \mathbb{R}$ from Exercise 13.3 defined by:

$$\forall x \in \mathbb{R}^2, \ f(x) = (x_1 - 1)^2 + (x_2 - 3)^2,$$

and the problem $\min_{x \in \mathbb{R}^2} \{ f(x) | Ax = b \}$.

(i) Show that $x^\star = \begin{bmatrix} 3 \\ -3 \end{bmatrix}$ does not satisfy (13.6) for the problem $\min_{x \in \mathbb{R}^2} \{ f(x) | x_1 + x_2 = 0 \}$. (You must show that no $\lambda^\star$ exists that satisfies $\nabla f(x^\star) + A^\dagger \lambda^\star = \mathbf{0}$.)

(ii) Show that $x^\star = \begin{bmatrix} 3 \\ -3 \end{bmatrix}$ does satisfy (13.6) for the problem $\min_{x \in \mathbb{R}^2} \{ f(x) | x_1 - 3x_2 = 12 \}$. (You must find a $\lambda^\star$ that satisfies $\nabla f(x^\star) + A^\dagger \lambda^\star = \mathbf{0}$.)

13.6 Show that $\begin{bmatrix} x^\star \\ \lambda^\star \end{bmatrix} = \begin{bmatrix} 2 \\ 2 \\ -2 \end{bmatrix}$ does not minimize the Lagrangian $\mathcal{L} : \mathbb{R}^2 \times \mathbb{R} \to \mathbb{R}$ defined in (13.12):

$$\forall x \in \mathbb{R}^2, \forall \lambda \in \mathbb{R}, \ \mathcal{L}(x, \lambda) = (x_1 - 1)^2 + (x_2 - 3)^2 + \lambda(x_1 - x_2).$$

13.7 Prove that the hypotheses of Corollary 13.4 imply the hypotheses of Theorem 13.3.

13.8 Recall the problem described in Exercise 3.11, having objective function $f : \mathbb{R}^2 \to \mathbb{R}$ and coefficient matrix and right-hand side defined by:

$$\begin{aligned} \forall x \in \mathbb{R}^2, \ f(x) &= -2(x_1 - x_2)^2 + (x_1 + x_2)^2, \\ A &= \begin{bmatrix} 1 & -1 \end{bmatrix}, \\ b &= [0]. \end{aligned}$$

The objective function is illustrated in Figure 13.10. From Exercise 3.11, we know that the minimizer of the problem $\min_{x \in \mathbb{R}^2} \{ f(x) | x_1 - x_2 = 0 \}$ is $x^\star = \mathbf{0}$.

(i) Show that $\nabla^2 f(x^\star)$ is not positive semi-definite.
(ii) Show that $\nabla^2 f(x^\star)$ is positive definite on the null space $\mathcal{N}(A)$.
(iii) Show that $x^\star = \mathbf{0}$ and $\lambda^\star = [0]$ satisfy the second-order sufficient conditions for this problem.

Convex problems

13.9 Let $\hat{x} \in \mathbb{R}^n$, $A \in \mathbb{R}^{1 \times n}$, $A \neq \mathbf{0}$, $b \in \mathbb{R}$ and consider the hyperplane:

$$\{x \in \mathbb{R}^n | Ax = b\}.$$

Show that the Euclidean distance of $\hat{x}$ to the closest point on the hyperplane is given by:

$$\frac{|A\hat{x} - b|}{\|A\|_2},$$

where $\|A\|_2 = \sqrt{AA^\dagger}$ is the Euclidean norm of the (row vector) A. (Hint: for an arbitrary point $x \in \mathbb{R}^n$, one half of the square of the Euclidean distance from x to $\hat{x}$ is given by the function $f : \mathbb{R}^n \to \mathbb{R}$ defined by:

$$\forall x \in \mathbb{R}^n, \ f(x) = \frac{1}{2} \|x - \hat{x}\|_2^2.$$

Formulate the problem of minimizing f over values of x that lie in the set $\{x \in \mathbb{R}^n | Ax = b\}$. Write down the first-order necessary conditions; solve for the Lagrange multiplier $\lambda^\star \in \mathbb{R}$ explicitly in terms of A, b, and $\hat{x}$; and use this to evaluate $\|x - \hat{x}\|_2^2$.)

13.10 Consider the function $f : \mathbb{R} \to \mathbb{R}$ defined by:

$$\forall x \in \mathbb{R}, \ f(x) = \begin{cases} (x+1)^2, & \text{if } x \leq -1, \\ 0, & \text{if } -1 < x < 1, \\ (x-1)^2, & \text{if } x \geq 1, \end{cases}$$

together with the equality constraint $Ax = b$ specified by $A = [1] \in \mathbb{R}^{1 \times 1}$, $b = [0] \in \mathbb{R}^1$. We consider:

- the problem $\min_{x \in \mathbb{R}} \{f(x) | Ax = b\}$,
- the corresponding dual function, $\mathcal{D} : \mathbb{R} \to \mathbb{R} \cup \{-\infty\}$ defined by:

$$\forall \lambda \in \mathbb{R}, \ \mathcal{D}(\lambda) = \inf_{x \in \mathbb{R}} \mathcal{L}(x, \lambda),$$

and
- the problem of maximizing the dual function:

$$\max_{\lambda \in \mathbb{E}} \mathcal{D}(\lambda).$$

(i) Show that f is differentiable. (Hint: The function is defined in pieces and is continuous. On the interior of each piece, the function is either quadratic or constant and so is differentiable. You only need to check that the derivatives match at the boundaries of the pieces.)

(ii) Show that f is convex. (Hint: f is not twice differentiable, so you cannot use the Hessian to determine convexity. You must either use the definition of a convex function or rely on first derivative information.)

(iii) Find the minimizer of the problem $\min_{x \in \mathbb{R}} \{f(x) | Ax = b\}$ by solving the first-order necessary conditions. (Hint: the problem is convex so that the solution of the first-order necessary conditions are sufficient for optimality.)

(iv) Evaluate the dual function. (Hint: There are three cases that must be considered: $\lambda < 0$, $\lambda = 0$, and $\lambda > 0$. Specify the functional form of the dual function for each case.)
(v) What is the effective domain $\mathbb{E}$ of the dual function?
(vi) For each value of $\lambda \in \mathbb{E}$, specify $\mathrm{argmin}_{x \in \mathbb{R}} \, \mathcal{L}(x, \lambda)$.
(vii) Is $\mathcal{D}$ differentiable at $\lambda = 0$?
(viii) What is the maximizer $\lambda^\star$ of the dual function?
(ix) Does $\mathrm{argmin}_{x \in \mathbb{R}} \, \mathcal{L}(x, \lambda^\star)$ provide the minimizer of $\min_{x \in \mathbb{R}}\{f(x)|Ax = b\}$?

13.11 Consider the function $f : \mathbb{R} \to \mathbb{R}$ defined by:

$$\forall x \in \mathbb{R}, \ f(x) = \exp(-x),$$

together with the equality constraint $Ax = b$ specified by $A = [1] \in \mathbb{R}^{1 \times 1}$, $b = [0] \in \mathbb{R}^1$. We consider:

- the problem $\min_{x \in \mathbb{R}}\{f(x)|Ax = b\}$,
- the corresponding dual function $\mathcal{D} : \mathbb{R} \to \mathbb{R} \cup \{-\infty\}$ defined by:

$$\forall \lambda \in \mathbb{R}, \ \mathcal{D}(\lambda) = \inf_{x \in \mathbb{R}} \mathcal{L}(x, \lambda),$$

and
- the problem of maximizing the dual function:

$$\max_{\lambda \in \mathbb{E}} \mathcal{D}(\lambda).$$

By Exercise 10.4 we know that the function f is convex.

(i) Find the minimizer of the primal problem $\min_{x \in \mathbb{R}}\{f(x)|Ax = b\}$.
(ii) Evaluate the dual function.
(iii) What is the effective domain $\mathbb{E}$ of the dual function?
(iv) Is the dual function differentiable on $\mathbb{E}$? (Hint: Consider $\lambda = 0$ carefully. Does $\min_{x \in \mathbb{R}} \mathcal{L}(x, 0)$ exist?)

13.12 Consider the problem from Exercise 13.11, with objective function $f : \mathbb{R} \to \mathbb{R}$ defined by:

$$\forall x \in \mathbb{R}, \ f(x) = \exp(-x),$$

together with the equality constraint $Ax = b$ specified by $A = [1] \in \mathbb{R}^{1 \times 1}$, $b = [0] \in \mathbb{R}^1$. We consider the problem $\min_{x \in \mathbb{R}}\{f(x)|Ax = b\}$, but form a penalized objective $f + \Pi f_\mathrm{p}$ using the penalty function $f_\mathrm{p} : \mathbb{R} \to \mathbb{R}$ defined by:

$$\forall x \in \mathbb{R}, \ f_\mathrm{p}(x) = \|Ax - b\|_2^2,$$

with $\Pi = 1$.

Consider the minimization in the definition of the corresponding dual function $\mathcal{D}_\mathrm{p} : \mathbb{R} \to \mathbb{R} \cup \{-\infty\}$ defined by:

$$\forall \lambda \in \mathbb{R}, \ \mathcal{D}_\mathrm{p}(\lambda) = \inf_{x \in \mathbb{R}} \mathcal{L}_\mathrm{p}(x, \lambda),$$

where $\mathcal{L}_\mathrm{p} = \mathcal{L} + \Pi f_\mathrm{p}$ and $\mathcal{L}$ is the Lagrangian. Also consider the problem of maximizing the corresponding dual function $\mathcal{D}_\mathrm{p}$.

(i) Show that $f + \Pi f_\mathrm{p}$ is strictly convex.
(ii) Find the minimizer of the primal problem $\min_{x\in\mathbb{R}}\{f(x) + \Pi f_\mathrm{p}(x)|Ax = b\}$.
(iii) What is the effective domain $\mathbb{E}$ of the dual function $\mathcal{D}_\mathrm{p}$?
(iv) Is the dual function $\mathcal{D}_\mathrm{p}$ differentiable on $\mathbb{E}$?

13.13 Prove the equality between the last two lines of (13.19) using the definition of min. That is, prove that for a separable $f : \mathbb{R}^n \to \mathbb{R}$, $A \in \mathbb{R}^{m\times n}$, and $b \in \mathbb{R}^m$ if the problem $\min_{x\in\mathbb{R}^n}\left\{\sum_{k=1}^n \left(f_k(x_k) + \lambda^\dagger A_k x_k\right)\right\}$ has a minimum then each of the sub-problems $\min_{x_k\in\mathbb{R}}\{f_k(x_k) + \lambda^\dagger A_k x_k\}$, $k = 1, \ldots, n$, have minima and:

$$\min_{x\in\mathbb{R}^n}\left\{\sum_{k=1}^n \left(f_k(x_k) + \lambda^\dagger A_k x_k\right)\right\} = \sum_{k=1}^n \min_{x_k\in\mathbb{R}}\{f_k(x_k) + \lambda^\dagger A_k x_k\},$$

where:

$$\forall x \in \mathbb{R}^n,\ f(x) = \sum_{k=1}^n f_k(x_k),$$

with $f_k : \mathbb{R} \to \mathbb{R}$, $k = 1, \ldots, n$.

13.14 Consider the dual function $\mathcal{D} : \mathbb{R} \to \mathbb{R}$ defined in (13.20), which is repeated here for reference:

$$
\begin{aligned}
\forall \lambda \in \mathbb{R},\ \mathcal{D}(\lambda) &= \min_{x\in\mathbb{R}^2} \mathcal{L}(x, \lambda), \\
&= \min_{x_1\in\mathbb{R}}\{(x_1 - 1)^2 + \lambda x_1\} + \min_{x_2\in\mathbb{R}}\{(x_2 - 3)^2 - \lambda x_2\}.
\end{aligned}
$$

(i) Solve each of the two sub-problems explicitly as a function of λ.
(ii) Maximize the dual function using the results from the previous part.
(iii) For the value $\lambda^\star$ that maximizes the dual, calculate the corresponding solutions of the sub-problems.

Approaches to finding minimizers

13.15 ([70, exercise 13, section 10.10]) Let $A \in \mathbb{R}^{m\times n}$, $b \in \mathbb{R}^m$, $c \in \mathbb{R}^n$, and $Q \in \mathbb{R}^{n\times n}$ with Q symmetric.

(i) Suppose that Q is not positive semi-definite on the null space $\mathcal{N}(A) = \{\Delta x \in \mathbb{R}^n | A\Delta x = 0\}$. Also suppose that there is at least one solution to $Ax = b$. Show that the problem $\min_{x\in\mathbb{R}^n}\{\frac{1}{2}x^\dagger Q x + c^\dagger x | Ax = b\}$ is unbounded below. (Hint: By assumption, there exists $\hat{x} \in \mathbb{R}^n$ such that $A\hat{x} = b$ and there exists $\Delta x \in \mathbb{R}^n$ such that $A\Delta x = 0$ and $\Delta x^\dagger Q \Delta x < 0$. Find a descent direction for the objective at $\hat{x}$ and show that moving in this direction maintains feasibility.)
(ii) Prove that $x^\star$ is a local minimizer of the problem $\min_{x\in\mathbb{R}}\{\frac{1}{2}x^\dagger Q x + c^\dagger x + d | Ax = b\}$ if and only if it is a global minimizer.

13.16 Consider the equality-constrained problem $\min_{x \in \mathbb{R}^2}\{f(x)|Ax = b\}$ where $f :$ $\mathbb{R}^2 \to \mathbb{R}$ is defined by:

$$\forall x \in \mathbb{R}^2,\ f(x) = \frac{1}{2}x^\dagger Qx + c^\dagger x,$$

with:

$$Q = \begin{bmatrix} 2 & -1 \\ -1 & 2 \end{bmatrix},\ c = \begin{bmatrix} 4 \\ 3 \end{bmatrix},$$

and the coefficient matrix and right-hand side of the constraints is specified by:

$$A = \begin{bmatrix} 1 & -1 \end{bmatrix},\ b = [0].$$

Eliminate variables to solve the problem by finding a matrix with columns that form a basis for the null space of A.

13.17 In this exercise we consider the coefficient matrix $\mathcal{A}$ defined in (13.25).

(i) Prove that if $Q \in \mathbb{R}^{n \times n}$ is positive definite and $A \in \mathbb{R}^{m \times n}$ has linearly independent rows, then $\mathcal{A}$ defined in (13.25) is non-singular. (Hint: Recall that $\mathcal{A}$ defined in (13.25) is non-singular if $\left(\begin{bmatrix} x \\ \lambda \end{bmatrix} \neq 0 \right) \Rightarrow \left(\begin{bmatrix} Q & A^\dagger \\ A & 0 \end{bmatrix} \begin{bmatrix} x \\ \lambda \end{bmatrix} \neq 0 \right).$)

(ii) Prove that if $Q \in \mathbb{R}^{n \times n}$ is positive definite on the null space $\mathcal{N}(A) = \{\Delta x \in \mathbb{R}^n | A \Delta x = 0\}$ and $A \in \mathbb{R}^{m \times n}$ has linearly independent rows, then $\mathcal{A}$ defined in (13.25) is non-singular.

(iii) Perform block factorization of $\mathcal{A}$ defined in (13.25) into block $LDL^\dagger$ factors, assuming that any inverses you need exist.

(iv) Calculate the inverse of $\mathcal{A}$ defined in (13.25).

13.18 In this exercise we consider the solvability of Problem (13.1).

(i) Give an example, with $n = 2$ and $m = 1$, of Problem (13.1) with linear objective such that there is no minimum.

(ii) Give an example, with $n = 2$ and $m = 1$, of Problem (13.1) with convex quadratic (but non-linear) objective such that there is no minimum.

13.19 Consider Problem (2.13) from Section 2.3.2.2, where $f : \mathbb{R}^2 \to \mathbb{R}$, was defined by:

$$\forall x \in \mathbb{R}^2,\ f(x) = (x_1 - 1)^2 + (x_2 - 3)^2,$$

but change the coefficient matrix and right-hand side to:

$$A = \begin{bmatrix} 1 & -1 \\ 1 & -1 \end{bmatrix},\ b = \begin{bmatrix} 0 \\ 0 \end{bmatrix}.$$

(i) Show that the first-order necessary conditions (13.26) of this problem have many solutions; however, all of them specify the same value of $x^\star$.

(ii) Use the MATLAB function quadprog to solve the problem. Use initial guess $x^{(0)} = \begin{bmatrix} 3 \\ -5 \end{bmatrix}$.

13.20 Consider again the problem $\min_{x \in \mathbb{R}^2}\{f(x)|Ax = b\}$ from Exercise 13.16 where $f : \mathbb{R}^2 \to \mathbb{R}$ is defined by:

$$\forall x \in \mathbb{R}^2, \ f(x) = \frac{1}{2}x^\dagger Q x + c^\dagger x,$$

with:

$$Q = \begin{bmatrix} 2 & -1 \\ -1 & 2 \end{bmatrix}, c = \begin{bmatrix} 4 \\ 3 \end{bmatrix},$$

and the coefficient matrix and right-hand side of the constraints is specified by:

$$A = \begin{bmatrix} 1 & -1 \end{bmatrix}, b = [0].$$

(i) Solve the problem from Exercise 13.16 by solving the first-order necessary conditions (13.26).

(ii) Use the MATLAB function quadprog to solve the problem. Use initial guess
$$x^{(0)} = \begin{bmatrix} 3 \\ -5 \end{bmatrix}.$$

13.21 Consider Problem (13.1) in the case that the objective $f : \mathbb{R}^2 \to \mathbb{R}$ is quadratic of the form:

$$\forall x \in \mathbb{R}^2, \ f(x) = \frac{1}{2}x^\dagger Q x + c^\dagger x,$$

with:

$$Q = \begin{bmatrix} 1 & 0 \\ 0 & 0 \end{bmatrix}, c = \begin{bmatrix} 1 \\ 1 \end{bmatrix},$$

and the coefficient matrix and right-hand side specified by:

$$A = \begin{bmatrix} 0 & 1 \end{bmatrix}, b = \begin{bmatrix} 1 \end{bmatrix}.$$

(i) Calculate the minimum and describe the set of minimizers of this problem by solving the first-order necessary conditions (13.26).

(ii) Use the MATLAB function quadprog to solve the problem. Use initial guess
$$x^{(0)} = \begin{bmatrix} 3 \\ -5 \end{bmatrix}.$$

13.22 Consider again the problem $\min_{x \in \mathbb{R}^2}\{f(x)|Ax = b\}$ from Exercise 13.16 where $f : \mathbb{R}^2 \to \mathbb{R}$ is defined by:

$$\forall x \in \mathbb{R}^2, \ f(x) = \frac{1}{2}x^\dagger Q x + c^\dagger x,$$

with:

$$Q = \begin{bmatrix} 2 & -1 \\ -1 & 2 \end{bmatrix}, c = \begin{bmatrix} 4 \\ 3 \end{bmatrix},$$

and the coefficient matrix and right-hand side of the constraints is specified by:

$$A = \begin{bmatrix} 1 & -1 \end{bmatrix}, b = [0].$$

Solve the problem from Exercise 13.16 by dual maximization. Start with $\lambda^{(0)} = [0]$. Use a step-size of $\alpha^{(\nu)} = 0.5$ at each iteration.

13.23 Let $f : \mathbb{R}^n \to \mathbb{R}$ be convex, $A \in \mathbb{R}^{m \times n}$, and $b \in \mathbb{R}^m$. Consider the algorithm described in Section 13.3.2.3 for solving $\min_{x \in \mathbb{R}^n} \{ f(x) | Ax = b \}$. Suppose that at iteration ν we have that:

$$\left\| \nabla f(x^{(\nu)}) + A^\dagger \lambda^{(\nu)} \right\| = \epsilon_f / \bar{\rho},$$

$$Ax^{(\nu)} = b,$$

and that we also know that the minimizer, $x^\star$, of the problem satisfies $\left\| x^\star - x^{(\nu)} \right\| \leq \bar{\rho}$. Suppose that the problem has a minimum, $f^\star$ say, and show that it satisfies:

$$f^\star \geq f(x^{(\nu)}) - \epsilon_f.$$

(Hint: Use Theorem 2.6 and take a similar approach to that used in Section 10.2.5, noting that:

$$
\begin{aligned}
[\nabla f(x^{(\nu)})]^\dagger (x^\star - x^{(\nu)}) &= [\nabla f(x^{(\nu)}) + A^\dagger \lambda^{(\nu)} - A^\dagger \lambda^{(\nu)}]^\dagger (x^\star - x^{(\nu)}), \\
&= [\nabla f(x^{(\nu)}) + A^\dagger \lambda^{(\nu)}]^\dagger (x^\star - x^{(\nu)}) - [\lambda^{(\nu)}]^\dagger A(x^\star - x^{(\nu)}), \\
&= [\nabla f(x^{(\nu)}) + A^\dagger \lambda^{(\nu)}]^\dagger (x^\star - x^{(\nu)}).)
\end{aligned}
$$

13.24 Let $A \in \mathbb{R}^{m \times n}$, $b \in \mathbb{R}^m$, $f : \mathbb{R}^n \to \mathbb{R}$ be convex and twice partially differentiable and let $x^{(\nu)} \in \mathbb{R}^n$ and $\lambda^{(\nu)} \in \mathbb{R}^m$. Suppose that $Ax^{(\nu)} = b$ and that we use (13.35) to calculate a step direction $\begin{bmatrix} \Delta x^{(\nu)} \\ \Delta \lambda^{(\nu)} \end{bmatrix}$. However, suppose that during the course of factorization of $\mathcal{A}$ we modify pivots in the top left-hand block of $\mathcal{A}$ to be positive so that the system we solve is actually:

$$\begin{bmatrix} Q & A^\dagger \\ A & 0 \end{bmatrix} \begin{bmatrix} \Delta x^{(\nu)} \\ \Delta \lambda^{(\nu)} \end{bmatrix} = - \begin{bmatrix} \nabla f(x^{(\nu)}) + A^\dagger \lambda^{(\nu)} \\ 0 \end{bmatrix},$$

with $Q \in \mathbb{R}^{n \times n}$ positive definite. (We may still need to use a special purpose factorization routine to deal with the indefinite system.) Prove that the resulting direction corresponds to a descent direction for the reduced function, assuming that the reduced gradient, $Z^\dagger \nabla f(x^{(\nu)})$, is non-zero, where Z is a matrix with columns that form a basis for the null space of A. (Hint: By the second block row, $A \Delta x^{(\nu)} = 0$. Moreover, the direction $\Delta x^{(\nu)}$ corresponds to the direction $\Delta \xi^{(\nu)}$ in the variables of the reduced function with $\Delta x^{(\nu)} = Z \Delta \xi^{(\nu)}$. Substitute into the first block row, multiply through by $Z^\dagger$ and note that $Z^\dagger A^\dagger = 0$. Show that the direction $\Delta \xi^{(\nu)}$ is a descent direction for the reduced function.)

13.25 Consider the objective (10.9), defined by:

$$\forall x \in \mathbb{R}^2, \ f(x) = 0.01(x_1-1)^4 + 0.01(x_2-3)^4 + (x_1-1)^2 + (x_2-3)^2 - 1.8(x_1-1)(x_2-3),$$

constraint matrix and vector defined by:

$$A = \begin{bmatrix} 1 & -1 \end{bmatrix}, b = [8],$$

and the problem $\min_{x \in \mathbb{R}^2} \{f(x) | Ax = b\}$.

(i) Using:

$$\xi^{(1)} = [5.34],$$
$$x^{(1)} = \begin{bmatrix} 8.34 \\ 0.34 \end{bmatrix},$$

apply one iteration of the null space basis algorithm in Section 13.3.2.2 to calculate $x^{(2)}$. Use the Armijo rule to calculate the step-size, with the step-size halved until the Armijo condition (10.14) is satisfied.

(ii) Using

$$x^{(1)} = \begin{bmatrix} 5.4953 \\ -2.5047 \end{bmatrix},$$
$$\lambda^{(1)} = [-20.4168],$$

apply one iteration of the Newton–Raphson method described in Section 13.3.2.3 to calculate $x^{(2)}$ and $\lambda^{(2)}$. Use a step-size of $\alpha^{(1)} = 1$.

(iii) Using $\lambda^{(1)} = [-10]$, apply one outer iteration of the dual maximization algorithm in Section 13.3.2.4 to calculate $x^{(1)}$ and $\lambda^{(2)}$.

For the inner iterations, use the MATLAB function `fminunc`, with default options and initial guess of $x^{(0)} = \begin{bmatrix} 1 \\ 3 \end{bmatrix}$ to find the minimizer, $x^{(1)}$, of the problem on the right-hand side of (13.36) for $\nu = 1$. You will have to write a MATLAB M-file to evaluate the Lagrangian.

For the outer iteration, use a step-size of $\alpha^{(1)} = 1$ to calculate $\lambda^{(2)}$.

(iv) Use the MATLAB function `fmincon` with default parameters and initial guess $x^{(0)} = \begin{bmatrix} 1 \\ 3 \end{bmatrix}$ to solve the problem. You will have to write a MATLAB M-file to evaluate f.

Sensitivity

13.26 Show by an example that the conclusion of Corollary 13.10 may fail to hold if the base-case constraint matrix does not have linearly independent rows. (Hint: Consider $A : \mathbb{R} \to \mathbb{R}^{2 \times 1}$ and $b : \mathbb{R} \to \mathbb{R}^2$ defined by:

$$\forall \chi \in \mathbb{R}, A(\chi) = \begin{bmatrix} \chi \\ -\chi \end{bmatrix},$$
$$\forall \chi \in \mathbb{R}, b(\chi) = 0.)$$

13.27 Let $Q : \mathbb{R}^s \rightarrow \mathbb{R}^{n \times n}$, $c : \mathbb{R}^s \rightarrow \mathbb{R}^n$, $A : \mathbb{R}^s \rightarrow \mathbb{R}^{m \times n}$, and $b : \mathbb{R}^s \rightarrow \mathbb{R}^m$ be partially differentiable with continuous partial derivatives. Suppose that $Q(0)$ is symmetric and that it is positive definite on the null space of $A(0)$. Let $f : \mathbb{R}^n \times \mathbb{R}^s \rightarrow \mathbb{R}$ be defined by:

$$\forall x \in \mathbb{R}^n, \forall \chi \in \mathbb{R}^s, f(x) = \frac{1}{2} x^\dagger Q(\chi) x + c(\chi)^\dagger x.$$

Consider the minimization problem:

$$\min_{x \in \mathbb{R}^n} \{f(x; \chi) | A(\chi) x = b(\chi)\},$$

where χ is a parameter.

(i) Write down the first-order necessary conditions for minimizing this problem. (Use the form of the first-order necessary conditions in (13.6)–(13.7).) Use the symbols $x^\star(\chi)$ and $\lambda^\star(\chi)$ for the solution of the first-order necessary conditions at a particular value of χ.

(ii) Show that the first-order necessary conditions have a unique solution for $\chi = 0$.

(iii) Suppose that $A(0)$ has linearly independent rows and calculate the sensitivity of the solution of the first-order necessary conditions to χ, evaluated at $\chi = 0$.

(iv) Show that the sensitivity calculated in Part (iii) is equivalent to (13.38).

13.28 Prove Corollary 13.11 using the proof of Corollary 13.10.

13.29 Consider Problem (2.13) from Section 2.3.2.2:

$$\min_{x \in \mathbb{R}^2} \{f(x) | Ax = b\},$$

where $f : \mathbb{R}^2 \rightarrow \mathbb{R}$, $A \in \mathbb{R}^{1 \times 2}$, and $b \in \mathbb{R}^1$ were defined by:

$$\begin{aligned}
\forall x \in \mathbb{R}^2, f(x) &= (x_1 - 1)^2 + (x_2 - 3)^2, \\
A &= \begin{bmatrix} 1 & -1 \end{bmatrix}, \\
b &= \begin{bmatrix} 0 \end{bmatrix}.
\end{aligned}$$

Suppose that the equality constraints changed from $Ax = b$ to $Ax = b - \gamma$.

(i) Calculate the sensitivity of the minimum to γ, evaluated at $\gamma = [0]$.

(ii) Solve the changed problem explicitly for $\gamma = [0.1]$ and compare to the estimate provided by the sensitivity analysis.

(iii) Repeat the previous part for $\gamma = [1]$.

Solution of the least-cost production case study

13.30 Consider Problem (12.4) in the case that $n = 3$, $D = 5$, and the f_k are of the form:

$$\forall x_1 \in \mathbb{R}, \ f_1(x_1) \ = \ \frac{1}{2}(x_1)^2 + x_1,$$

$$\forall x_2 \in \mathbb{R}, \ f_2(x_2) \ = \ \frac{1}{2} \times 1.1(x_2)^2 + 0.9x_2,$$

$$\forall x_3 \in \mathbb{R}, \ f_3(x_3) \ = \ \frac{1}{2} \times 1.2(x_3)^2 + 0.8x_3.$$

Solve it in four ways.

(i) By eliminating x_1 using a matrix with columns that form a basis for the null space of the coefficient matrix and solving the resulting unconstrained problem.

(ii) By solving the first-order necessary conditions in terms of the minimizer $x^\star$ and the Lagrange multipliers $\lambda^\star$.

(iii) By maximizing the dual function. Use $\lambda^{(0)} = [0]$ as the initial guess and perform steepest ascent with step-size equal to 0.5 at each iteration.

(iv) Using the MATLAB function quadprog, with initial guess $x^{(0)} = \begin{bmatrix} 0 \\ 0 \\ 0 \end{bmatrix}$.

13.31 Repeat Exercise 13.30, but modify each f_k to be:

$$\forall x_1 \in \mathbb{R}, \ f_1(x_1) \ = \ 0.01(x_1)^3 + \frac{1}{2}(x_1)^2 + x_1,$$

$$\forall x_2 \in \mathbb{R}, \ f_2(x_2) \ = \ 0.01(x_2)^3 + \frac{1}{2} \times 1.1(x_2)^2 + 0.9x_2,$$

$$\forall x_3 \in \mathbb{R}, \ f_3(x_3) \ = \ 0.01(x_1)^3 + \frac{1}{2} \times 1.2(x_3)^2 + 0.8x_3.$$

For Part (iii), at each outer iteration, perform two inner iterations. For outer iteration $\nu = 0$, use $x^{(0)} = 0$ as initial guess. For each subsequent outer iteration, use the value of x from the previous outer iteration as initial guess. For Part (iv), use the MATLAB function fmincon with default parameters and initial guess given by the solution of Exercise 13.30.

13.32 Consider the problem specified in Exercise 13.30 but suppose that the demand changes to $D + \Delta D$, where $\Delta D = 0.1$.

(i) Estimate the change in the minimum and minimizer of the problem using the sensitivity analysis results from Corollaries 13.10 and 13.11.

(ii) Calculate the change in the minimum and minimizer of the problem by explicitly re-solving the problem.

(iii) Compare the results.

14

Algorithms for non-linear equality-constrained minimization

In this chapter we will develop algorithms for constrained optimization problems of the form:

$$\min_{x \in \mathbb{S}} f(x),$$

where the feasible set $\mathbb{S}$ is of the form:

$$\mathbb{S} = \{x \in \mathbb{R}^n | g(x) = 0\},$$

and $f : \mathbb{R}^n \to \mathbb{R}$ and $g : \mathbb{R}^n \to \mathbb{R}^m$. That is, we consider problems of the form:

$$\min_{x \in \mathbb{R}^n} \{f(x) | g(x) = 0\}. \tag{14.1}$$

We call the constraints **non-linear equality constraints** although, strictly speaking, it would be more precise to refer to them as **non-affine equality constraints**.

We first investigate properties of non-linear equality constraints in Section 14.1 and then derive optimality conditions in Section 14.2. The optimality conditions we present are not as sharp as possible, but illustrate the general flavor of these results. The optimality conditions will help us to develop algorithms for non-linear equality-constrained minimization problems in Section 14.3. We will discuss the sensitivity of the optimum to changes in the constraints in Section 14.4. Finally, in Section 14.5, we discuss solution of the power system state estimation with zero injection buses case study that was introduced in Section 12.2.

The key issues discussed in this chapter are:

- the notion of a **regular point of constraints** as a characterization of suitable formulations of non-linear equality constraint functions,
- linearization of non-linear constraint functions and consideration of the **null space of the coefficient matrix** of the linearized constraints and the associated **tangent plane**,

529

- optimality conditions and the definition and interpretation of the **Lagrange multipliers,**
- algorithms that seek points that satisfy the optimality conditions,
- use of a **merit function** in the trade-off between satisfaction of constraints and improvement of the objective, and
- **duality** and **sensitivity analysis.**

14.1 Geometry and analysis of constraints

In the case of *linear* equality constraints, the convexity of the feasible set allowed us to consider step directions such that successive iterates were always feasible. That is, for linear equality constraints we can move from a feasible point along a line segment that lies entirely within the feasible set, choosing the direction of the segment to decrease the objective. This motivated the approach of first finding a feasible point and then seeking step directions that kept the iterates feasible and also reduced the value of the objective.

With non-linear constraints, movement from a feasible point along a line segment will usually take us outside the feasible set. Nevertheless, our approach to non-linear equality constraints will be to linearize the equality constraint function about a current iterate. We must explore conditions under which this linearization yields a useful approximation to the original feasible set. The notion of a regular point, to be introduced in Section 14.1.1, provides such a condition. We relate the notion of a regular point of constraints to the notion of the tangent plane to the feasible set in Section 14.1.2 and then discuss the relationship to non-linear equality-constrained optimization in Section 14.1.3.

14.1.1 Regular point

14.1.1.1 Definition

When we use $\{x \in \mathbb{R}^n | g(x) = 0\}$ to represent a feasible set $\mathbb{S}$, we usually have many choices of functions $g : \mathbb{R}^n \to \mathbb{R}^m$ such that $\mathbb{S} = \{x \in \mathbb{R}^n | g(x) = 0\}$. However, some choices of g may be more suitable than others. In this section we characterize suitability of g in terms of the following.

Definition 14.1 Let $g : \mathbb{R}^n \to \mathbb{R}^m$. Then we say that $x^\star$ is a **regular point** of the equality constraints $g(x) = 0$ if:

   (i) $g(x^\star) = 0$,
   (ii) g is partially differentiable with continuous partial derivatives at $x^\star$, and
   (iii) the m rows of the Jacobian $J(x^\star)$ of g evaluated at $x^\star$ are linearly independent.

□

Notice that for $g(x) = \mathbf{0}$ to have *any* regular points, we must have that $m \leq n$, since otherwise the m rows of $J(x^\star)$ cannot be linearly independent. Furthermore, if $x^\star$ is a regular point, then we can find a sub-vector $\omega \in \mathbb{R}^m$ of x such that the $m \times m$ matrix $\dfrac{\partial g}{\partial \omega}(x^\star)$ is non-singular.

14.1.1.2 Example

Consider the function $g : \mathbb{R}^3 \to \mathbb{R}$ defined by:

$$\forall x \in \mathbb{R}^3, \ g(x) = (x_1)^2 + (x_2 + 1)^2 - x_3 - 4,$$

and the point $x^\star = \begin{bmatrix} 1 \\ 3 \\ 13 \end{bmatrix}$. We observe that $x^\star$ is a regular point of the equality constraints $g(x) = 0$ because:

(i) $g(x^\star) = (1)^2 + (3 + 1)^2 - 13 - 4 = 0$,
(ii) g is partially differentiable with Jacobian $J : \mathbb{R}^3 \to \mathbb{R}^{1 \times 3}$ defined by $\forall x \in \mathbb{R}^3$, $J(x) = \begin{bmatrix} 2x_1 & 2(x_2 + 1) & -1 \end{bmatrix}$, which is continuous at $x^\star$, and
(iii) the one row of the Jacobian $J(x^\star)$ of g evaluated at $x^\star$ is given by $J(x^\star) = \begin{bmatrix} 2 & 8 & -1 \end{bmatrix}$, which is a linearly independent row.

14.1.2 Tangent plane

14.1.2.1 Definition

We make the following generalization of Definition 13.1.

Definition 14.2 Let $g : \mathbb{R}^n \to \mathbb{R}^m$ be partially differentiable and $x^\star \in \mathbb{R}^n$. Let $J : \mathbb{R}^n \to \mathbb{R}^{m \times n}$ be the Jacobian of g. Suppose that $x^\star$ is a regular point of the constraints $g(x) = \mathbf{0}$. Then the **tangent plane** to the set $\mathbb{S} = \{x \in \mathbb{R}^n | g(x) = \mathbf{0}\}$ at the point $x^\star$ is the set $\mathbb{T} = \{x \in \mathbb{R}^n | J(x^\star)(x - x^\star) = \mathbf{0}\}$. □

The tangent plane at $x^\star$ is the set of points such that the first-order Taylor approximation to g about $x^\star$ has value $\mathbf{0}$.

14.1.2.2 Example

Consider again the function $g : \mathbb{R}^3 \to \mathbb{R}$ defined in Section 14.1.1.2. Figure 14.1 shows the set $\mathbb{S} = \{x \in \mathbb{R}^3 | g(x) = 0\}$ and the tangent plane $\mathbb{T}$ to the set $\mathbb{S}$ at the regular point $x^\star = \begin{bmatrix} 1 \\ 3 \\ 13 \end{bmatrix}$, which is shown as a bullet $\bullet$. Geometrically, the tangent plane $\mathbb{T}$ to $\mathbb{S}$ at $x^\star$ is **tangential** to the surface of the set $\mathbb{S}$ at the point $x^\star$.

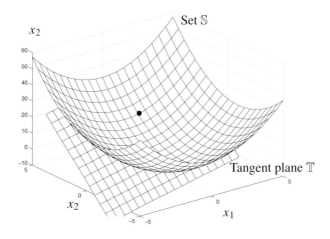

Fig. 14.1. Tangent plane $\mathbb{T}$ to a set $\mathbb{S}$ in $\mathbb{R}^3$ at the point $x^\star = \begin{bmatrix} 1 \\ 3 \\ 13 \end{bmatrix} \in \mathbb{S}$, shown as a •.

14.1.2.3 Affine case

In the case that g is affine of the form:

$$\forall x \in \mathbb{R}^n, \, g(x) = Ax - b,$$

then $J(x) = A$ and the tangent plane $\mathbb{T}$ at a point $x^\star \in \mathbb{S} = \{x \in \mathbb{R}^n | Ax = b\}$ is given by:

$$
\begin{aligned}
\mathbb{T} &= \{x \in \mathbb{R}^n | A(x - x^\star) = \mathbf{0}\}, \\
&= \{x \in \mathbb{R}^n | Ax = b\}, \text{ since } Ax^\star = b \text{ at a feasible point } x^\star, \\
&= \mathbb{S}.
\end{aligned}
$$

That is, in the case that g is affine, the tangent plane $\mathbb{T}$ is the same as the feasible set $\mathbb{S} = \{x \in \mathbb{R}^n | g(x) = \mathbf{0}\}$. In contrast, for non-linear g such as shown in Figure 14.1, the tangent plane $\mathbb{T}$ to $\{x \in \mathbb{R}^n | g(x) = \mathbf{0}\}$ at $x^\star$ is usually different to $\mathbb{S} = \{x \in \mathbb{R}^n | g(x) = \mathbf{0}\}$.

14.1.2.4 Discussion

The concept of a regular point will help us to characterize when the tangent plane $\mathbb{T}$ is a good approximation to the feasible set $\mathbb{S}$, as Exercises 14.1 and 14.3 show. Exercises 14.1 and 14.3 suggest that if $x^\star$ is a regular point of the constraints $g(x) = \mathbf{0}$ then:

- in the vicinity of $x^\star$, the tangent plane $\mathbb{T} = \{x \in \mathbb{R}^n | J(x^\star)(x - x^\star) = \mathbf{0}\}$ is a good approximation to the feasible set $\mathbb{S} = \{x \in \mathbb{R}^n | g(x) = \mathbf{0}\}$,
- a slight perturbation to g will not qualitatively change the feasible set $\mathbb{S} = \{x \in \mathbb{R}^n | g(x) = \mathbf{0}\}$, and

- a slight perturbation to the Jacobian J will not qualitatively change the tangent plane $\mathbb{T} = \{x \in \mathbb{R}^n | J(x^\star)(x - x^\star) = \mathbf{0}\}$.

If $x^\star$ is a regular point of the constraints $g(x) = \mathbf{0}$ then we cannot have situations such as occur in the following.

- Exercise 14.1, Part (iv), where the tangent plane is not qualitatively similar to the feasible set. In Exercise 14.1, Part (iv), $\mathbb{S}$ is a line, while $\mathbb{T}$ is the whole of the x_2–x_3 plane.

- Exercise 14.1, Parts (ii) and (iv), where a slight perturbation of g produces a qualitative change in the feasible set $\mathbb{S} = \{x \in \mathbb{R}^n | g(x) = \mathbf{0}\}$.

- Exercise 14.1, Parts (ii) and (iv), where a slight perturbation of the Jacobian produces a qualitative change in the tangent plane $\mathbb{T} = \{x \in \mathbb{R}^n | J(x^\star)(x - x^\star) = \mathbf{0}\}$ significantly.

The definition of regular point provides one characterization of useful equality constraint functions that is straightforward conceptually and has an intuitive interpretation in terms of the tangent plane. In particular, if $x^\star$ is a regular point of constraints g, then our intuitive *geometric* notion of the tangent plane to the set $\mathbb{S}$ at $x^\star$ coincides with the definition of tangent plane in terms of the analytic representation of the feasible set. With suitable continuity and differentiability assumptions on f and g, the behavior of the objective f of Problem (14.1) for points in the set $\mathbb{T}$ and near to $x^\star$ will be similar to the behavior of the objective for points in the set $\mathbb{S}$ and near to $x^\star$. (See Exercise 14.2.)

If $x^\star$ is *not* a regular point of $g(x) = \mathbf{0}$ then the tangent plane can still be defined *geometrically* as discussed in [70, section 10.2]; however, we cannot conveniently represent it using the Jacobian of g. In particular, the set $\mathbb{T} = \{x \in \mathbb{R}^n | J(x^\star)(x - x^\star) = \mathbf{0}\}$ will not necessarily have the geometric characteristics of the tangent plane if $x^\star$ is not a regular point of the constraints. Since we generally have considerable flexibility in choosing g to represent a feasible set $\mathbb{S}$, we should try to pick g so that feasible points are all regular points.

It is possible to define conditions that are weaker than regularity that also lead to characterizations of useful equality constraint functions; however, we will not discuss these alternative **constraint qualifications** except in the context of nonlinear inequality constraints in Chapter 19. Constraint qualifications are discussed at length in [6, chapter 5].

x_2

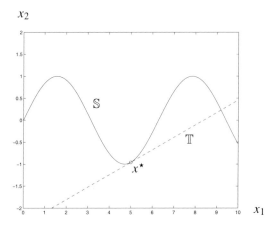

Fig. 14.2. Feasible set $\mathbb{S}$ (shown solid), feasible point $x^\star \in \mathbb{S}$ (shown as o), and tangent plane $\mathbb{T}$ (shown dashed) to $\mathbb{S}$ at $x^\star$.

14.1.3 Relationship of regular points to seeking minimizers

14.1.3.1 Movement from a feasible point

Suppose we are at a point $x^\star$ that we think may be a minimizer for Problem (14.1). As we have remarked, if the equality constraints are non-linear, then moving along a straight line segment away from $x^\star$ will in general take us outside the feasible set $\mathbb{S}$. However, by Exercise 14.2, if $x^\star$ is a regular point of $g(x) = \mathbf{0}$ then we will be close to a point that satisfies the constraints so long as we stay near to $x^\star$ and in the tangent plane $\mathbb{T} = \{x \in \mathbb{R}^n | J(x^\star)(x - x^\star) = \mathbf{0}\}$. Since $\mathbb{T}$ is defined by a linear equality, then by Exercise 2.36 it is convex and we can move along straight line segments within $\mathbb{T}$. Our basic approach to finding a step direction will be to move along straight line segments within $\mathbb{T}$ that are descent directions for the objective f. Of course, for non-convex $\mathbb{S}$ as in Exercise 14.3, by moving along a direction Δx that lies in $\mathbb{T}$ we will actually move at least slightly outside the feasible set $\mathbb{S}$.

For example, consider $n = 2$ and $f : \mathbb{R}^2 \to \mathbb{R}$ defined by:

$$\forall x \in \mathbb{R}^2, f(x) = -x_1. \tag{14.2}$$

Let $g : \mathbb{R}^2 \to \mathbb{R}$ be defined by:

$$\forall x \in \mathbb{R}^2, g(x) = x_2 - \sin(x_1). \tag{14.3}$$

Consider Problem (14.1) for these choices of objective and constraint function. Figure 14.2 shows, as a solid curve, part of the set of points $\mathbb{S}$ satisfying the equality constraint $g(x) = 0$. Also shown is the feasible point $x^\star = \begin{bmatrix} 5 \\ -\sin(5) \end{bmatrix}$, shown as a o, and the tangent plane $\mathbb{T}$ to the feasible set $\mathbb{S}$ at $x^\star$, shown dashed. For this problem, the tangent plane $\mathbb{T}$ is only a good approximation to the feasible set for points that are close to $x^\star$.

x_2

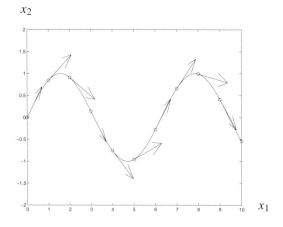

Fig. 14.3. Feasible points (shown as o) and directions along the corresponding tangent planes (shown as arrows).

x_1

We will need to numerically calculate directions that keep us in the set $\mathbb{T}$. Because of finite precision in our calculations, we may not exactly be able to calculate a direction that allows us to stay in $\mathbb{T}$. We will instead move in a direction that lies in a set such as $\tilde{\mathbb{T}} = \{x \in \mathbb{R}^n | \tilde{J}(x^\star)(x - x^\star) = \mathbf{0}\}$, where $\tilde{J} : \mathbb{R}^n \to \mathbb{R}^{m \times n}$ is an approximation to the Jacobian J of g. Regularity ensures that the errors so introduced will not destroy the qualitative resemblance between $\mathbb{T}$, $\tilde{\mathbb{T}}$, and $\mathbb{S}$.

14.1.3.2 Descent

Figure 14.3 shows the same feasible set as illustrated in Figure 14.2. Several feasible points are illustrated by the o. The arrows emanating from the feasible points illustrate directions along the tangent plane at these points. Moving in the direction of the arrows takes us outside the feasible set but reduces the value of the objective defined in (14.2). Evidently, the candidate point $x^\star$ illustrated in Figure 14.2 is not optimal for this problem. (In fact, the problem is unbounded below.)

In summary, each arrow in Figure 14.3 points in a direction such that the objective is decreasing. Paths that stay on the feasible set must follow the curve $g(x) = \mathbf{0}$ and therefore depart from straight line segments. Nevertheless, we will consider paths that, at least initially, follow the tangent plane $\mathbb{T}$.

14.1.3.3 Movement from an infeasible point

If we are at a point $\hat{x}$ that does not satisfy the equality constraints, we will still be interested in moving towards a minimizer of the problem. In this case, we will try to more nearly satisfy the constraints and, at the same time, try to decrease the value of the objective. We will again approximate the feasible points by a set defined in terms of linear equalities. In this case, we define an approximation to

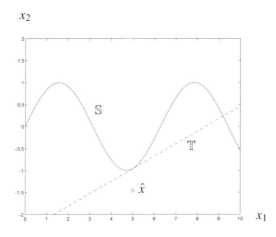

Fig. 14.4. An infeasible point $\hat{x} \notin \mathbb{S}$ and approximation $\mathbb{T}$ (shown dashed) to feasible set $\mathbb{S}$ (shown solid).

the feasible set of the form ([70, section 12.7]):

$$\mathbb{T} = \{x \in \mathbb{R}^n | J(\hat{x})(x - \hat{x}) = -g(\hat{x})\}. \tag{14.4}$$

This definition is consistent with the previous definition of tangent plane in that we are still specifying the set of points such that the first-order Taylor approximation to g about $\hat{x}$ has value $\mathbf{0}$. Linear independence of the rows of J and proximity of $\hat{x}$ to the feasible set will again guarantee that this set closely approximates the feasible set in the vicinity of $\hat{x}$.

Figure 14.4 again shows, as a solid curve, part of the feasible set $\mathbb{S}$ of points satisfying the equality constraint $g(x) = 0$. Also shown is an infeasible point $\hat{x} = \begin{bmatrix} 5 \\ -1.5 \end{bmatrix}$ and the set $\mathbb{T}$ defined according to (14.4), shown dashed. In this particular case, the set $\mathbb{T}$ is tangential to the feasible set $\mathbb{S}$; however, in general this is not the case. (See Exercise 14.5.)

14.1.3.4 *Linear constraints*

If g is affine as in Section 14.1.2.3 and $g(x^\star) = \mathbf{0}$ then $\mathbb{T} = \{x \in \mathbb{R}^n | A(x - x^\star) = \mathbf{0}\}$ is the same as the feasible set, whether or not A has linearly independent rows. However, if A does not have linearly independent rows then the constraints are redundant and a slight perturbation of the coefficient matrix will make the linear approximation to the feasible set empty, as Exercise 14.1, Part (ii) shows. That is, as discussed in Section 2.7.6, a minimization problem with redundant constraints is ill-conditioned. This difficulty can be avoided if calculations are performed in exact arithmetic or if redundant constraints are detected and discarded [70, section 4.7].

14.1.3.5 Formulation of problems

Whether or not g is affine, we should try to formulate the problem to avoid linear dependence of the rows of J since redundant constraints make the problem ill-conditioned. In large-scale problems this is not always easy because the constraints may be compiled from several sources and it may be difficult to identify which constraints are linearly dependent, particularly when there are representation errors in the coefficients used to specify the constraints. This issue is discussed at length in [45, section 6.5.4.1].

Furthermore, we may be able to avoid non-linear equality constraints in some cases. Exercise 14.6 explores this for the problem having objective defined in (14.2) and constraint defined in (14.3). (See also Exercise 14.9.)

14.2 Optimality conditions

In Section 14.2.1, we discuss **first-order necessary conditions** (or **FONC**) for non-linear equality-constrained problems. **Second-order sufficient conditions** (or **SONC**) are discussed in Section 14.2.2.

14.2.1 First-order necessary conditions

In Section 14.2.1.1 we analyze the first-order necessary conditions, showing the relationship to the Lagrangian in Section 14.2.1.2 and to the corresponding linearly constrained problem in Section 14.2.1.3. We present a geometric interpretation of the necessary conditions in Section 14.2.1.4 and an example in Section 14.2.1.5.

14.2.1.1 Analysis

We have the following.

Theorem 14.1 *Consider Problem (14.1) and a point* $x^\star \in \mathbb{R}^n$. *Suppose that:*

   (i) *f is partially differentiable with continuous partial derivatives,*

   (ii) *$x^\star$ is a regular point of the equality constraints $g(x) = 0$. That is:*

      (a) *$g(x^\star) = 0$,*

      (b) *g is partially differentiable with continuous partial derivatives, and*

      (c) *the m rows of the Jacobian $J(x^\star)$ of g evaluated at $x^\star$ are linearly independent.*

Then if $x^\star$ is a local minimizer of Problem (14.1) then:

$$\exists \lambda^\star \in \mathbb{R}^m \text{ such that } \nabla f(x^\star) + J(x^\star)^\dagger \lambda^\star = 0. \tag{14.5}$$

Proof   (See Exercise 14.7 and [11, section 3.1][70, section 10.3][84, section 4.5.1].)
□

As in the linear equality-constrained case, the vector $\lambda^\star$ is again called the vector of **Lagrange multipliers** for the constraints $g(x) = \mathbf{0}$. Sometimes we will refer to:

$$\nabla f(x^\star) + J(x^\star)^\dagger \lambda^\star = \mathbf{0}, \qquad (14.6)$$
$$g(x^\star) = \mathbf{0}, \qquad (14.7)$$

as the **first-order necessary conditions** (or **FONC**) for the solution of a non-linear equality-constrained problem, although, strictly speaking, the first-order necessary conditions also include the other items in the hypothesis of Theorem 14.1.

14.2.1.2 Lagrangian

Recall Definition 3.2 of the **Lagrangian**. For Problem (14.1) the Lagrangian $\mathcal{L}$: $\mathbb{R}^n \times \mathbb{R}^m \to \mathbb{R}$ is defined by:

$$\forall x \in \mathbb{R}^n, \forall \lambda \in \mathbb{R}^m, \mathcal{L}(x, \lambda) = f(x) + \lambda^\dagger g(x).$$

As in the linear case, we can reproduce the first-order necessary conditions (14.6)–(14.7) by setting the gradients of $\mathcal{L}$ with respect to x and λ, respectively, equal to zero.

14.2.1.3 Relationship to linearly constrained problems

The condition (14.5) is the same as the corresponding first-order condition for the *linearly* constrained problem:

$$\min_{x \in \mathbb{R}^n}\{f(x)|J(x^\star)(x - x^\star) = \mathbf{0}\}, \qquad (14.8)$$

which is obtained by replacing the feasible set $\mathbb{S}$ by the tangent plane $\mathbb{T}$ to $\mathbb{S}$ at the minimizer $x^\star$. (See Exercise 14.8.) Regularity of the constraints, in addition to the hypotheses for the linear case, ensures that (14.5) characterizes the necessary conditions in the non-linear equality-constrained case. Unlike in the linear case, the assumption of regularity is important to ensure that there are Lagrange multipliers satisfying (14.5). (See Exercises 14.9 and 14.10.)

14.2.1.4 Geometric interpretation

In the linear equality-constrained case, we interpreted the first-order necessary conditions as requiring that the feasible set be a subset of the tangent plane to the contour set of the objective. We said that the contour set of f was tangential to the feasible set at $x^\star$. In the non-linear equality-constrained case, we can similarly interpret (14.5) as requiring that the feasible set and the contour set be tangential at $x^\star$.

14.2.1.5 Example

As with the linear equality-constrained case, it is possible for $x^\star$ to satisfy the first-order necessary conditions (14.6)–(14.7) and yet not be a local minimizer of Problem (14.1). For example, as in the example in Section 13.1.2.6, if the objective is non-convex then a *maximizer* can satisfy the necessary conditions.

Recall that if the objective is convex and the constraints are *linear* equality constraints, then the first-order conditions are both necessary and sufficient. This is because the problem is then convex, as discussed in Section 13.2.

In the case of non-linear equality constraints, however, we may have an objective $f : \mathbb{R}^n \to \mathbb{R}$ that is convex on $\mathbb{R}^n$, but have a non-convex feasible set. In this case, and unlike the case of linear equality constraints, the necessary conditions are not sufficient in the absence of additional assumptions. For example, consider the case of Problem (14.1) with $n = 2$ and $m = 1$ and:

$$\forall x \in \mathbb{R}^2, f(x) = \frac{1}{2}(x_1)^2 + \frac{1}{2}(x_2)^2, \tag{14.9}$$

$$\forall x \in \mathbb{R}^2, g(x) = \frac{1}{4}(x_1)^2 + (x_2)^2 - 1. \tag{14.10}$$

The circular contour sets of f together with the set of points in $\mathbb{S} = \{x \in \mathbb{R}^n | g(x) = 0\}$ are shown in Figure 14.5. The contour sets of f are circles with center $\mathbf{0}$, while the feasible set $\mathbb{S}$ is illustrated by the heavy ellipse. There are four points that satisfy the first-order necessary conditions.

- Two of the points are $x^\star = \begin{bmatrix} 0 \\ 1 \end{bmatrix}$ and $x^{\star\star} = \begin{bmatrix} 0 \\ -1 \end{bmatrix}$, both with Lagrange multiplier $\lambda^\star = \lambda^{\star\star} = -\frac{1}{2}$, which corresponds to a *minimum* $f^\star = 0.5$ of the objective over the feasible set. The points $x^\star$ and $x^{\star\star}$ are illustrated with $\bullet$ in Figure 14.5.

- The other two points are $\hat{x} = \begin{bmatrix} 2 \\ 0 \end{bmatrix}$ and $\hat{\hat{x}} = \begin{bmatrix} -2 \\ 0 \end{bmatrix}$, both with dual variables $\hat{\lambda} = \hat{\hat{\lambda}} = -2$, which corresponds to a *maximum* $\hat{f} = 2$ of the objective over the feasible set. The points $\hat{x}$ and $\hat{\hat{x}}$ are illustrated with $\circ$ in Figure 14.5.

The feasible set and the contour set are tangential at the points that satisfy the first-order necessary conditions. The tangent planes are shown as dashed lines.

In this example, $m = 1$ and the tangent planes to the objective and to the contour set are coincident. For $m > 1$, at a regular point of the equality constraints satisfying the necessary conditions, the tangent plane to the feasible set will typically be strictly contained in the tangent plane to the contour set as in the example in Section 13.1.1.4.

In Section 14.2.2, we will discuss second-order sufficient conditions to ensure

x_2

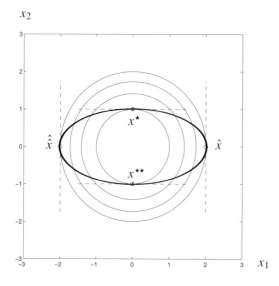

Fig. 14.5. Points $x^\star$, $x^{\star\star}$, $\hat{x}$, and $\hat{\hat{x}}$ that satisfy the first-order necessary conditions but which may or may not be minimizers.

that we are at a minimum and so avoid points such as $\hat{x} = \begin{bmatrix} 2 \\ 0 \end{bmatrix}$ and $\hat{\hat{x}} = \begin{bmatrix} -2 \\ 0 \end{bmatrix}$ that satisfy the necessary conditions but which are not optimal.

14.2.2 Second-order sufficient conditions

14.2.2.1 Analysis

In the following theorem, we present sufficient conditions for a local minimizer of Problem (14.1).

Theorem 14.2 *Suppose that* $f : \mathbb{R}^n \to \mathbb{R}$ *and* $g : \mathbb{R}^n \to \mathbb{R}^m$ *are twice partially differentiable with continuous second partial derivatives. Let* $J : \mathbb{R}^n \to \mathbb{R}^{m \times n}$ *be the Jacobian of* g. *Consider Problem (14.1) and points* $x^\star \in \mathbb{R}^n$ *and* $\lambda^\star \in \mathbb{R}^m$. *Suppose that:*

$$\nabla f(x^\star) + J(x^\star)^\dagger \lambda^\star = \mathbf{0},$$
$$g(x^\star) = \mathbf{0},$$
$$\nabla^2 f(x^\star) + \sum_{\ell=1}^{m} \lambda_\ell^\star \nabla^2 g_\ell(x^\star) \text{ is positive definite on the null space:}$$
$$\mathcal{N} = \{\Delta x \in \mathbb{R}^n | J(x^\star)\Delta x = \mathbf{0}\}. \qquad (14.11)$$

Then $x^\star$ *is a strict local minimizer of Problem (14.1).*

Proof See [70, section 10.5] and Exercise 14.11. □

Compared to the first-order necessary conditions, the **second-order sufficient conditions** (or **SOSC**) in addition require that:

- the objective and constraint functions are twice partially differentiable with continuous second partial derivatives, and
- $x^\star$ and $\lambda^\star$ satisfy (14.11).

In (14.11), the function $\nabla^2_{xx}\mathcal{L} : \mathbb{R}^n \times \mathbb{R}^m \to \mathbb{R}$ defined by:

$$\forall x \in \mathbb{R}^n, \forall \lambda \in \mathbb{R}^m, \nabla^2_{xx}\mathcal{L}(x, \lambda) = \nabla^2 f(x) + \sum_{\ell=1}^{m} \lambda_\ell \nabla^2 g_\ell(x),$$

is called the **Hessian of the Lagrangian**. The condition (14.11) is analogous to the corresponding condition in Corollary 13.4 for linear constraints and can be tested similarly, but involves the Hessians of the equality constraint functions in addition to the Hessian of the objective. (See Exercise 14.12.) It requires that the Hessian of the Lagrangian evaluated at the minimizer and corresponding Lagrange multipliers, $\nabla^2_{xx}\mathcal{L}(x^\star, \lambda^\star)$, is positive definite on the null space $\mathcal{N}$ defined in the theorem. The null space $\mathcal{N}$ is sometimes called the **tangent subspace** [70, section 13.1].

14.2.2.2 Example

Continuing with the example from Section 14.2.1.5, Exercise 14.13 confirms that both of the minimizers of the problem satisfy the second-order sufficient conditions but that both of the other points that satisfy the first-order necessary conditions do not satisfy the second-order sufficient conditions.

14.3 Approaches to finding minimizers

In general, if the constraints are non-linear, we cannot expect to exactly satisfy them. Moreover, even if we could find a feasible point, we cannot descend from a feasible point along a straight line segment and remain feasible. Nevertheless, we can consider algorithms that attempt to satisfy the first-order necessary conditions or use step directions based on the Newton–Raphson update for solving the first-order necessary conditions. In Section 14.3.1 we discuss solution of the first-order necessary conditions and then in Section 14.3.2 we discuss solution by dual maximization.

14.3.1 Solution of first-order necessary conditions

The most straightforward approach to finding a minimizer of Problem (14.1) is to solve the first-order necessary conditions (14.6)–(14.7):

$$\nabla f(x) + J(x)^\dagger \lambda = 0,$$
$$g(x) = 0,$$

which are $n+m$ non-linear equations in $n+m$ variables. In Section 14.3.1.1, we discuss the Newton–Raphson step direction to solve the non-linear equations (14.6)–(14.7). We then then discuss step-size selection in Section 14.3.1.2, feasibility in Section 14.3.1.3, and stopping criteria in Section 14.3.1.4.

14.3.1.1 Newton–Raphson step direction

The Newton–Raphson step direction to solve (14.6)–(14.7) is given by the solution of:

$$\begin{bmatrix} \nabla_{xx}^2 \mathcal{L}(x^{(\nu)}, \lambda^{(\nu)}) & J(x^{(\nu)})^{\dagger} \\ J(x^{(\nu)}) & \mathbf{0} \end{bmatrix} \begin{bmatrix} \Delta x^{(\nu)} \\ \Delta \lambda^{(\nu)} \end{bmatrix} = - \begin{bmatrix} \nabla f(x^{(\nu)}) + J(x^{(\nu)})^{\dagger} \lambda^{(\nu)} \\ g(x^{(\nu)}) \end{bmatrix}.$$

(14.12)

As in Sections 13.3.1.3 and 13.3.2.3, this system is indefinite and an indefinite factorization algorithm should be used. If the coefficient matrix in (14.12) is singular then some of the pivots will have to be modified to make them positive. If $J(x^{(\nu)})$ has linearly independent rows for every iterate, however, we will not need to modify the pivots for the bottom right-hand block of the matrix. This will help us to obtain a direction that brings us closer to satisfying the equality constraints. As discussed for the unconstrained case in Section 10.2.3.2, zero and negative pivots in the top left-hand block should be modified to be positive to ensure that the step direction $\Delta x^{(\nu)}$ is a descent direction for $f + [\lambda^{(\nu)}]^{\dagger} g$ at $x^{(\nu)}$. See [45, section 6.5.4.2] for further comments on the modifications to the Hessian of the Lagrangian in this case.

As in previous approaches to solving systems of non-linear equations, we can approximate the coefficient matrix to reduce the amount of work compared to exact solution of (14.12). For example, we can use a quasi-Newton update to build up an approximation over several iterations or we can factorize an approximation of the coefficient matrix. (See [70, section 14.7] for comments on maintaining a positive definite approximation to the Hessian of the Lagrangian.) These approaches will, in principle, inherit the corresponding convergence rates described for the solution of non-linear equations.

Whether an exact or approximate solution to (14.12) is found, the update is then:

$$\begin{bmatrix} x^{(\nu+1)} \\ \lambda^{(\nu+1)} \end{bmatrix} = \begin{bmatrix} x^{(\nu)} \\ \lambda^{(\nu)} \end{bmatrix} + \alpha^{(\nu)} \begin{bmatrix} \Delta x^{(\nu)} \\ \Delta \lambda^{(\nu)} \end{bmatrix}.$$

14.3.1.2 Selection of step-size

We must choose the step-size. In choosing a step-size $\alpha^{(\nu)}$, we cannot just seek reduction in f because if we are far from satisfying the constraints then we may have to accept an increase in f to obtain a feasible point or to become closer to feasibility. That is, we must trade-off the tension between satisfaction of the constraints and improvement in the objective.

A standard approach to this trade-off is to define a **merit function** [45, section 6.5.3.3] $\phi : \mathbb{R}^n \to \mathbb{R}$ of the form, for example:

$$\forall x \in \mathbb{R}^n, \phi(x) = f(x) + \Pi \|g(x)\|^2, \tag{14.13}$$

for some norm $\|\bullet\|$ and some $\Pi \in \mathbb{R}_{++}$ and use a rule analogous to the Armijo rule or variants to seek a step that leads to sufficient reduction in the merit function ϕ at each iteration. Because the merit function combines the objective with a penalty for constraint violation, there is a trade-off between improving the objective and more closely satisfying the constraints. Unfortunately, the appropriate trade-off, as represented by the value of Π, is often difficult to decide on *a priori*. It may be necessary to update Π during the course of calculation. (See [11, section 4.2.1] for related discussion and see [45, section 6.5.3.3] and [70, section 14.2] for several different merit functions.)

Furthermore, as described in Section 3.3.1, the entries of g should be scaled to ensure that "significant" violation of any constraint involves roughly the same numerical value for each of the entries in the scaled constraint function. We will discuss some of these issues in the context of the power system state estimation with zero injection buses case study in Section 14.5.

A variant on the merit function approach is to replace the objective in Problem (14.1) with the merit function $\phi : \mathbb{R}^n \to \mathbb{R}$ defined in (14.13) [70, section 14.2]. In this case, the Newton–Raphson update will explicitly seek a direction that reduces the merit function. This approach is analogous to the augmented Lagrangian. A drawback of using the merit function as the objective is that the sparsity of the top left-hand block of the coefficient matrix in (14.12) may be spoilt because of the terms due to the second derivative of $\Pi \|g(x)\|^2$.

There are other approaches to trading off the tension between satisfying constraints and improving the objective including:

- a **filter** [36], where the step-size is selected to improve satisfaction of the constraints or the value of the objective or both at each iteration, and
- a **watchdog** [23], where the merit function is allowed to increase for a limited number of iterations.

14.3.1.3 Feasibility

In the discussion so far we have assumed that the iterates will eventually become close enough to being feasible to be useful. In some applications, we might want to be able to terminate at any iteration with an iterate that is close to being feasible. In this case, at each iteration we can first update x to reduce the objective or reduce a merit function and then do a subsidiary search using an iterative technique to return to the feasible set [70, section 11.4]. The hope is that the reduction in objective

is not negated by the movement necessary to restore feasibility. This approach
is used in the **generalized reduced gradient** algorithm, which has traditionally
been the most successful technique for dealing with non-linear constraints [45,
section 6.3][84, section 15.6]. The generalized reduced gradient code of Lasdon is
the most widely used [66].

14.3.1.4 Stopping criteria

We iterate until the first-order necessary conditions are satisfied to sufficient accuracy. Comments on appropriate stopping criteria are presented in [45, sec-
tion 8.2.3]. Notice that we can only expect $x^{(v)}$ to approximately satisfy $g(x^{(v)}) = 0$. Unless the second-order sufficient conditions hold or approximately hold, we
cannot be certain that we are at or close to a local optimum.

14.3.2 Dual maximization

Recall Definition 3.3 of the **dual function**. For Problem (14.1), the dual function
$\mathcal{D} : \mathbb{R}^m \to \mathbb{R} \cup \{-\infty\}$ is defined by:

$$\forall \lambda \in \mathbb{R}^m, \mathcal{D}(\lambda) = \underset{x \in \mathbb{R}^n}{\arg\min} \, \mathcal{L}(x, \lambda),$$

where $\mathcal{L} : \mathbb{R}^n \times \mathbb{R}^m \to \mathbb{R}$ is the Lagrangian. Although the problem is not convex,
we can try to maximize the dual function. The following recursion can be used to
define the iterates:

$$
\begin{aligned}
x^{(v)} &\in \underset{x \in \mathbb{R}^n}{\arg\min} \{ f(x) + [\lambda^{(v)}]^\dagger g(x) \}, \\
\Delta\lambda^{(v)} &= g(x^{(v)}), \\
\lambda^{(v+1)} &= \lambda^{(v)} + \alpha^{(v)} \Delta\lambda^{(v)}.
\end{aligned}
$$

If f or g is non-quadratic then we will have to perform several inner iterations to
approximately minimize the Lagrangian for each outer iteration to update λ.

   In general, we cannot guarantee that the maximum of the dual function equals
the minimum of the primal problem so that there can be a duality gap. Nevertheless,
by Theorem 3.13, the maximum of the dual is a lower bound for the minimum
of the primal problem and the solution of the dual may be a useful guide to the
solution of the primal. Moreover, under second-order sufficient conditions, the
primal value corresponding to the maximum of the dual is at least a *local* minimizer
of Problem (14.1) [70, section 13.1].

   If there are a large number of variables and only a small number of constraints,
then solving the primal problem directly may be prohibitive computationally; however, if the objective is separable it may be feasible to solve the dual problem using

parallel processing. Even if the resulting primal variables x do not satisfy the constraints in the primal problem, it may be possible to apply a heuristic that modifies x to make them feasible. Furthermore, the bound in Theorem 3.13 can be incorporated into a stopping criterion.

14.4 Sensitivity

14.4.1 Analysis

In this section we will analyze a general and a special case of sensitivity analysis for Problem (14.1). For the general case, we suppose that the objective f and equality constraint function g are parameterized by a parameter $\chi \in \mathbb{R}^s$. That is, $f : \mathbb{R}^n \times \mathbb{R}^s \to \mathbb{R}$ and $g : \mathbb{R}^n \times \mathbb{R}^s \to \mathbb{R}^m$. We imagine that we have solved the non-linear equality-constrained minimization problem:

$$\min_{x \in \mathbb{R}^n} \{ f(x; \chi) | g(x; \chi) = \mathbf{0} \}, \tag{14.14}$$

for a base-case value of the parameters, say $\chi = \mathbf{0}$, to find the base-case local minimizer $x^\star$ and the base-case Lagrange multipliers $\lambda^\star$. We now consider the sensitivity of the local minimum of Problem (14.14) to variation of the parameters about $\chi = \mathbf{0}$.

As well as considering the general case of the sensitivity of the local minimum of Problem (14.14) to χ, we also specialize to the case where only the right-hand sides of the equality constraints vary. That is, we return to the special case where $f : \mathbb{R}^n \to \mathbb{R}$ and $g : \mathbb{R}^n \to \mathbb{R}^m$ are not explicitly parameterized. However, we now consider perturbations $\gamma \in \mathbb{R}^m$ and the problem:

$$\min_{x \in \mathbb{R}^n} \{ f(x) | g(x) = -\gamma \}. \tag{14.15}$$

For the parameter values $\gamma = \mathbf{0}$, Problem (14.15) is the same as Problem (14.1). We consider the sensitivity of the local minimum of Problem (14.15) to variation of the parameters about $\gamma = \mathbf{0}$.

We have the following corollary to the implicit function theorem, Theorem A.9 in Section A.7.3 of Appendix A.

Corollary 14.3 *Consider Problem (14.14) and suppose that the functions $f : \mathbb{R}^n \times \mathbb{R}^s \to \mathbb{R}$ and $g : \mathbb{R}^n \times \mathbb{R}^s \to \mathbb{R}^m$ are twice partially differentiable with continuous second partial derivatives. Also consider Problem (14.15) and suppose that the functions $f : \mathbb{R}^n \to \mathbb{R}$ and $g : \mathbb{R}^n \to \mathbb{R}^m$ are twice partially differentiable with continuous second partial derivatives. Let $J = \dfrac{\partial g}{\partial x}$ be the Jacobian of g. Suppose that $x^\star \in \mathbb{R}^n$ and $\lambda^\star \in \mathbb{R}^m$ satisfy:*

- *the second-order sufficient conditions for Problem (14.14) for the base-case value of parameters $\chi = \mathbf{0}$, and*

- *the second-order sufficient conditions for Problem (14.15) for the base-case value of parameters $\gamma = \mathbf{0}$.*

In particular:

- $x^\star$ *is a local minimizer of Problem (14.14) for $\chi = \mathbf{0}$, and*
- $x^\star$ *is a local minimizer of Problem (14.15) for $\gamma = \mathbf{0}$,*

in both cases with associated Lagrange multipliers $\lambda^\star$. Moreover, suppose that the rows of the Jacobians $J(x^\star)$ and $J(x^\star; \mathbf{0})$, respectively, are linearly independent so that $x^\star$ is a regular point of the constraints for the base-case problems.

Then, for values of χ in a neighborhood of the base-case value of the parameters $\chi = \mathbf{0}$, there is a local minimum and corresponding local minimizer and Lagrange multipliers for Problem (14.14). Moreover, the local minimum, local minimizer, and Lagrange multipliers are partially differentiable with respect to χ and have continuous partial derivatives in this neighborhood. The sensitivity of the local minimum $f^\star$ to χ, evaluated at the base-case $\chi = \mathbf{0}$, is given by:

$$\frac{\partial f^\star}{\partial \chi}(\mathbf{0}) = \frac{\partial \mathcal{L}}{\partial \chi}(x^\star, \lambda^\star; \mathbf{0}),$$

where $\mathcal{L} : \mathbb{R}^n \times \mathbb{R}^m \times \mathbb{R}^s \to \mathbb{R}$ is the **parameterized Lagrangian** *defined by:*

$$\forall x \in \mathbb{R}^n, \forall \lambda \in \mathbb{R}^m, \forall \chi \in \mathbb{R}^s, \mathcal{L}(x, \lambda; \chi) = f(x; \chi) + \lambda^\dagger g(x; \chi).$$

Furthermore, for values of γ in a neighborhood of the base-case value of the parameters $\gamma = \mathbf{0}$, there is a local minimum and corresponding local minimizer and Lagrange multipliers for Problem (14.15). Moreover, the local minimum, local minimizer, and Lagrange multipliers are partially differentiable with respect to γ and have continuous partial derivatives. The sensitivity of the local minimum to γ, evaluated at the base-case $\gamma = \mathbf{0}$, is equal to $\lambda^\star$.

Proof See [34, theorem 2.4.4] and [70, section 10.7]. □

14.4.2 *Discussion*

As in the case of linear equality constraints, we can interpret the Lagrange multipliers as the sensitivity of the minimum to changes in γ. Again, this allows us to trade-off the change in the optimal objective against the cost of changing the constraint. The sensitivity result is also extremely useful if we must repeatedly solve problems that differ only in the specification of γ because we can estimate the changes in the minimum using sensitivity analysis. We can also estimate the changes in the minimizer and Lagrange multipliers using sensitivity analysis of the first-order necessary conditions. (See Exercise 14.19.)

14.4.3 Example

Consider Problem (2.14) from Section 2.3.2.2:

$$\min_{x \in \mathbb{R}^2} \{ f(x) | g(x) = 0 \},$$

where $f : \mathbb{R}^2 \to \mathbb{R}$ and $g : \mathbb{R}^2 \to \mathbb{R}$ were defined by:

$$\forall x \in \mathbb{R}^2, \; f(x) = (x_1 - 1)^2 + (x_2 - 3)^2,$$
$$\forall x \in \mathbb{R}^2, \; g(x) = (x_1)^2 + (x_2)^2 + 2x_2 - 3.$$

By Exercise 14.14, the minimizers and Lagrange multipliers of Problem (2.14) satisfy the second-order sufficient conditions and the minimizers are regular points of the constraints. If the equality constraint changes to $g(x) = -\gamma$, where $\gamma = 0.1$, then we can use Corollary 14.3 to approximate the change in the minimum by $0.1\lambda^\star$.

14.5 Solution of power system state estimation with zero injection buses case study

In this section, we solve the power system state estimation with zero injection buses case study from Section 12.2. We recall the problem in Section 14.5.1, describe algorithms to solve it in Section 14.5.2, and sketch sensitivity analysis in Section 14.5.3.

14.5.1 Problem

Recall Problem (12.9):

$$\min_{x \in \mathbb{R}^n} \{ f(x) | g(x) = \mathbf{0} \},$$

where $f : \mathbb{R}^n \to \mathbb{R}$ and $g : \mathbb{R}^n \to \mathbb{R}^m$ were defined in (12.7) and (12.8), respectively:

$$\forall x \in \mathbb{R}^n, \; f(x) = \sum_{\ell \in \mathbb{M}} \frac{(\tilde{g}_\ell(x) - \tilde{G}_\ell)^2}{2\sigma_\ell^2},$$
$$\forall x \in \mathbb{R}^n, \; g(x) = (g_\ell(x))_{\ell \in \mathbb{M}^0}.$$

14.5.2 Algorithms

In this section we discuss the Newton–Raphson step direction, the merit function and step-size, and observability.

14.5.2.1 Newton–Raphson step direction

The most straightforward way to solve this problem is to seek a solution of the necessary conditions (14.6)–(14.7) using the Newton–Raphson step direction given by the solution of (14.12) or some approximation to it that ensures that a descent direction is found for $f + [\lambda^{(\nu)}]^\dagger g$. Possible approximations to the coefficient matrix for the Newton–Raphson step direction include:

- using the fast-decoupled or other approximations to the Jacobian of the power flow equations, as in the discussion of the solution of the power flow equations in Section 8.2.4.2, and
- using the Gauss–Newton or Levenberg–Marquandt approximation to the Hessian of the Lagrangian, as in the discussion of the state estimation problem in Section 11.2.3.2.

14.5.2.2 Merit function and step-size

Recall that f consists of (half of) the sum of squares of terms each of which represent a measurement error for measurement ℓ divided by the standard deviation σ_ℓ of the measurement error. Consequently, each term has expected value of 1 if evaluated at the true value of the voltage angles and magnitudes in the system. (See Exercise 11.8.)

On the other hand, the terms in g represent real and reactive power values that are exactly equal to zero when evaluated at the true value of the voltage angles and magnitudes in the system. In Section 12.2.1.3, we discussed incorporating the zero injection bus measurement equations into the objective by assigning them measurement errors that were smaller than the typical measurement error standard deviation for the measurements. We will take this approach here.

In particular, we can use the merit function ϕ defined in (14.13) with the L_2 norm $\|\bullet\|_2$ and a value of penalty coefficient Π that is somewhat larger than the inverse of the square of a typical real and reactive power measurement error standard deviation. We can interpret the merit function as being a penalized objective, as discussed in Section 12.2.1.3, that uses modest values of the penalty coefficient. The step-size should be selected to ensure sufficient reduction in the merit function defined in (14.13) using a step-size rule such as the Armijo rule. (See Exercise 14.20.)

14.5.2.3 Observability

As in Section 11.2, to ensure that there is a unique maximum likelihood estimator there must be enough measurements and zero bus injections spread around the system to make it observable.

14.5.3 Changes in measurement data

If the measurements change then we can estimate the change in the state using sensitivity analysis. (See Exercise 14.20.)

14.6 Summary

In this chapter we considered the notion of a regular point of constraints as a bridge between equality-constrained problems with linear constraints and equality-constrained problems with non-linear constraints. We developed optimality conditions, algorithms, and sensitivity analysis. We then applied one of the algorithms to the power system state estimation with zero injection buses case study.

Exercises

Geometry and analysis of constraints

14.1 Consider the subset of $\mathbb{R}^3$ defined by the set of points that lie on the x_3 axis. That is, consider the set $\mathbb{S} = \{x \in \mathbb{R}^3 | x_1 = x_2 = 0\}$, which is a line. We are going to consider whether, for various choices of $g : \mathbb{R}^3 \to \mathbb{R}^m$, the set $\{x \in \mathbb{R}^3 | g(x) = \mathbf{0}\}$ is a suitable representation for the set $\mathbb{S} = \{x \in \mathbb{R}^3 | x_1 = x_2 = 0\}$. Consider each of the following definitions of a constraint function, $g : \mathbb{R}^3 \to \mathbb{R}^m$:

(i) $m = 2, \forall x \in \mathbb{R}^3, g(x) = \begin{bmatrix} x_1 \\ x_2 \end{bmatrix}$,

(ii) $m = 3, \forall x \in \mathbb{R}^3, g(x) = \begin{bmatrix} x_1 \\ x_1 \\ x_2 \end{bmatrix}$,

(iii) $m = 2, \forall x \in \mathbb{R}^3, g(x) = \begin{bmatrix} x_1 + (x_2)^2 \\ x_2 \end{bmatrix}$,

(iv) $m = 2, \forall x \in \mathbb{R}^3, g(x) = \begin{bmatrix} (x_1 - 1)^2 + (x_2)^2 - 1 \\ (x_1 + 1)^2 + (x_2)^2 - 1 \end{bmatrix}$.

For each of the functions g defined in (i)–(iv) above, consider its Jacobian $J : \mathbb{R}^3 \to \mathbb{R}^{m \times 3}$, and also consider the functions $\tilde{g} : \mathbb{R}^3 \to \mathbb{R}^m$ and $\tilde{J} : \mathbb{R}^3 \to \mathbb{R}^{m \times 3}$ defined by:

$$\forall x \in \mathbb{R}^3, \tilde{g}(x) = g(x) - 10^{-6} \times \begin{bmatrix} 1 \\ 2 \\ \vdots \\ m \end{bmatrix},$$

$$\forall x \in \mathbb{R}^3, \tilde{J}(x) = J(x) - 10^{-6} \times \begin{bmatrix} 1 & 0 & 0 \\ 2 & 0 & 0 \\ \vdots & \vdots & \vdots \\ m & 0 & 0 \end{bmatrix}.$$

The functions $\tilde{g}$ and $\tilde{J}$ are slight perturbations of g and J, respectively, that simulate the effects of small round-off or representation errors in the computational representations of g and J.

Consider each of the alternative definitions of functions g defined in Parts (i)–(iv) above and their corresponding Jacobians J and the perturbations $\tilde{g}$ and $\tilde{J}$ of g and J, respectively. For each alternative, answer the following.

(a) Determine whether $x^\star = \begin{bmatrix} 0 \\ 0 \\ 0 \end{bmatrix}$ is a regular point of the equality constraints $g(x) = \mathbf{0}$.

(b) Show that the set $\{x \in \mathbb{R}^3 | g(x) = \mathbf{0}\}$ is the same as the set $\mathbb{S} = \{x \in \mathbb{R}^3 | x_1 = x_2 = 0\}$. That is, show that each point contained in $\{x \in \mathbb{R}^3 | g(x) = \mathbf{0}\}$ is an element of $\{x \in \mathbb{R}^3 | x_1 = x_2 = 0\}$ and vice versa.

(c) Describe in words the set $\mathbb{T} = \{x \in \mathbb{R}^3 | J(x^\star)(x - x^\star) = \mathbf{0}\}$.

(d) Describe in words the set $\tilde{\mathbb{S}} = \{x \in \mathbb{R}^3 | \tilde{g}(x) = \mathbf{0}\}$.

(e) Describe in words the set $\tilde{\mathbb{T}} = \{x \in \mathbb{R}^3 | \tilde{J}(x^\star)(x - x^\star) = \mathbf{0}\}$.

(v) We are interested in whether the various sets $\mathbb{T}$, $\tilde{\mathbb{S}}$, and $\tilde{\mathbb{T}}$ are "similar" to $\mathbb{S} = \{x \in \mathbb{R}^3 | x_1 = x_2 = 0\}$ in the sense of whether their qualitative appearance is only slightly different from $\{x \in \mathbb{R}^3 | x_1 = x_2 = 0\}$. Using this notion of "similar" and your answers to the previous parts of the question, answer the following questions for each of the functions g defined in (i)–(iv) above:

(a) Is $x^\star$ a regular point of the equality constraints $g(x) = \mathbf{0}$?

(b) Is $\mathbb{T}$ similar to $\mathbb{S} = \{x \in \mathbb{R}^3 | x_1 = x_2 = 0\}$?

(c) Is $\tilde{\mathbb{S}}$ similar to $\mathbb{S} = \{x \in \mathbb{R}^3 | x_1 = x_2 = 0\}$?

(d) Is $\tilde{\mathbb{T}}$ similar to $\mathbb{S} = \{x \in \mathbb{R}^3 | x_1 = x_2 = 0\}$?

Arrange your answers in a table with:

• rows corresponding to the functions g defined in Parts (i)–(iv), and
• columns corresponding to the answers to Parts (v)(a)–(v)(d).

14.2 Let $g : \mathbb{R}^n \to \mathbb{R}^m$ be partially differentiable with continuous partial derivatives having Jacobian $J : \mathbb{R}^n \to \mathbb{R}^{m \times n}$. Suppose that $x^\star$ is a regular point of the constraints $g(x) = \mathbf{0}$. For convenience, suppose that the $m \times m$ matrix $\hat{J}$ consisting of the first m columns of $J(x^\star)$ is non-singular. Partition $x^\star$ into $x^\star = \begin{bmatrix} \omega^\star \\ \xi^\star \end{bmatrix}$, where $\omega^\star \in \mathbb{R}^m, \xi^\star \in \mathbb{R}^{n-m}$.

(i) Use the implicit function Theorem A.9 to show that there is a neighborhood $\mathbb{P}$ of $\mathbf{0} \in \mathbb{R}^{n-m}$ and a partially differentiable function $\omega : \mathbb{R}^{n-m} \to \mathbb{R}^m$ with continuous partial derivatives such that:

• $\omega(\mathbf{0}) = \omega^\star$,
• ω satisfies:

$$\forall \chi \in \mathbb{P}, \ g\left(\begin{bmatrix} \omega(\chi) \\ \xi^\star + \chi \end{bmatrix}\right) = \mathbf{0},$$

and

- the partial derivative of ω satisfies:

$$\frac{\partial \omega}{\partial \chi}(0) = -\left[\hat{J}\right]^{-1} K,$$

where $K = \frac{\partial g}{\partial \xi}(x^\star)$.

(ii) Consider the tangent plane to $S = \{x \in \mathbb{R}^n | g(x) = 0\}$ at $x^\star$. That is, consider $\mathbb{T} = \{x \in \mathbb{R}^n | J(x^\star)(x - x^\star) = 0\}$. Let $\chi \in \mathbb{R}^{n-m}$, $t \in [0, 1]$, and let $x' = \begin{bmatrix} \omega^\star - t\left[\hat{J}\right]^{-1} K\chi \\ \xi^\star + t\chi \end{bmatrix}$. Show that $x' \in \mathbb{T}$.

(iii) Let $\chi \in \mathbb{P}$, $t = [0, 1]$, and consider the points $x' = \begin{bmatrix} \omega^\star - t\left[\hat{J}\right]^{-1} K\chi \\ \xi^\star + t\chi \end{bmatrix} \in \mathbb{T}$ and $x'' = \begin{bmatrix} \omega(t\chi) \\ \xi^\star + t\chi \end{bmatrix} \in S$. Show that as $t \to 0$, $\frac{\|x' - x''\|}{t} \to 0$.

14.3 Consider the set $S \in \mathbb{R}^2$ defined by $S = \{x \in \mathbb{R}^n | g(x) = 0\}$, where $g : \mathbb{R}^2 \to \mathbb{R}$ was defined in (14.3). That is:

$$\forall x \in \mathbb{R}^2, g(x) = x_2 - \sin(x_1).$$

(i) Is $x^\star = 0$ a regular point of the equality constraint $g(x) = 0$?

(ii) Describe the set $\mathbb{T} = \{x \in \mathbb{R} | J(x^\star)(x - x^\star) = 0\}$, where $J : \mathbb{R}^n \to \mathbb{R}^{m \times n}$ is the Jacobian of g.

(iii) Consider the points in $\mathbb{R}^2$ in the vicinity of $x^\star = 0$. That is, consider $x \in \mathbb{R}^2$ such that $\|x\|_2 \approx \|x^\star\|_2 = 0$. For these points, is the set $\mathbb{T} = \{x \in \mathbb{R}^2 | J(x^\star)(x - x^\star) = 0\}$ qualitatively a good approximation to the set $S = \{x \in \mathbb{R}^2 | g(x) = 0\}$? A sketch and qualitative assessment will suffice.

14.4 Consider $g : \mathbb{R}^2 \to \mathbb{R}$ defined in (14.3). That is:

$$\forall x \in \mathbb{R}^2, g(x) = x_2 - \sin(x_1).$$

For each of the following values of $x^\star$, write out explicitly the set of points in the set:

$$\mathbb{T} = \{x \in \mathbb{R}^2 | J(x^\star)(x - x^\star) = -g(x^\star)\},$$

defined in (14.4).

(i) $x^\star = \begin{bmatrix} 5 \\ -\sin(5) \end{bmatrix}$.

(ii) $x^\star = \begin{bmatrix} 5 \\ -1.5 \end{bmatrix}$.

14.5 Consider $g : \mathbb{R}^2 \to \mathbb{R}$ defined by:

$$\forall x \in \mathbb{R}^2, g(x) = (x_2)^3 - (\sin(x_1))^3.$$

For each of the following values of $x^\star$, write out explicitly the set of points in the set:

$$\mathbb{T} = \{x \in \mathbb{R}^2 | J(x^\star)(x - x^\star) = -g(x^\star)\},$$

defined in (14.4).

(i) $x^\star = \begin{bmatrix} 5 \\ -\sin(5) \end{bmatrix}$.

(ii) $x^\star = \begin{bmatrix} 5 \\ -1.5 \end{bmatrix}$.

14.6 Consider the optimization problem $\min_{x \in \mathbb{R}^2}\{f(x) | g(x) = 0\}$, where $f : \mathbb{R}^2 \to \mathbb{R}$ and $g : \mathbb{R}^2 \to \mathbb{R}$ were defined in (14.2) and (14.3), respectively, which we repeat here:

$$\begin{aligned} \forall x \in \mathbb{R}^2, \ f(x) &= -x_1, \\ \forall x \in \mathbb{R}^2, \ g(x) &= x_2 - \sin(x_1). \end{aligned}$$

Consider the function $\tau : \mathbb{R} \to \mathbb{R}$ defined by:

$$\forall \xi \in \mathbb{R}, \ \tau(\xi) = \begin{bmatrix} \xi \\ \sin(\xi) \end{bmatrix}.$$

(i) Show that τ is onto $\mathbb{S} = \{x \in \mathbb{R}^2 | g(x) = 0\}$.
(ii) Show that the problem $\min_{\xi \in \mathbb{R}} f(\tau(\xi))$, has a linear objective.
(iii) What can you say about the solution of $\min_{x \in \mathbb{R}^2}\{f(x) | g(x) = 0\}$?

Optimality conditions

14.7 Use Exercise 14.2, Part (i), Theorem 10.3, and Theorem A.4 in Section A.6.1.1 of Appendix A to prove Theorem 14.1.

14.8 Show that the first-order necessary conditions for Problem (14.8) yield (14.5).

14.9 ([11, example 1.2 of chapter 3]) Consider the problem $\min_{x \in \mathbb{R}^2}\{f(x) | g(x) = 0\}$ where $f : \mathbb{R}^2 \to \mathbb{R}$ and $g : \mathbb{R}^2 \to \mathbb{R}^2$ are defined by:

$$\begin{aligned} \forall x \in \mathbb{R}^2, \ f(x) &= x_1 + x_2, \\ \forall x \in \mathbb{R}^2, \ g(x) &= \begin{bmatrix} (x_1 - 1)^2 + (x_2)^2 - 1 \\ (x_1 - 2)^2 + (x_2)^2 - 4 \end{bmatrix}. \end{aligned}$$

(i) Show that $x^\star = 0$ is the unique feasible point and (therefore) the unique minimizer for the problem $\min_{x \in \mathbb{R}^2}\{f(x) | g(x) = 0\}$. (Hint: Describe the set of points satisfying each equality constraint $g_1(x) = 0$ and $g_2(x) = 0$ geometrically.)
(ii) Show that $x^\star$ is not a regular point of the constraints $g(x) = 0$.
(iii) Show that no $\lambda^\star$ exists satisfying (14.5).
(iv) Find another specification of the equality constraint functions that specifies the same feasible set and such that $x^\star$ is a regular point of the constraints $g(x) = 0$.

14.10 Consider the problem $\min_{x \in \mathbb{R}^2} \{f(x) | g(x) = \mathbf{0}\}$ where $f : \mathbb{R}^2 \to \mathbb{R}$ and $g : \mathbb{R}^2 \to \mathbb{R}^3$ are defined by:

$$\forall x \in \mathbb{R}^2, \ f(x) \ = \ x_1 + x_2,$$
$$\forall x \in \mathbb{R}^2, \ g(x) \ = \ Ax - b,$$

where $A \in \mathbb{R}^{3 \times 2}$ and $b \in \mathbb{R}^3$ are defined by:

$$A = \begin{bmatrix} 1 & 0 \\ 1 & 0 \\ 0 & 1 \end{bmatrix}, b = \mathbf{0}.$$

(i) Show that $x^\star = \mathbf{0}$ is the unique feasible point and (therefore) the unique minimizer for the problem $\min_{x \in \mathbb{R}^2} \{f(x) | g(x) = \mathbf{0}\}$.
(ii) Show that $x^\star$ is not a regular point of the constraints $g(x) = \mathbf{0}$.
(iii) Show that there exists $\lambda^\star \in \mathbb{R}^3$ satisfying (14.5).
(iv) Specify the set of all $\lambda^\star \in \mathbb{R}^3$ satisfying (14.5).
(v) The constraints are redundant. Suppose we perturb the first row of A to be $\begin{bmatrix} 1 & \chi \end{bmatrix}$, with $\chi \neq 0$. How does the solution change?

14.11 Prove Theorem 14.2. (Hint: Suppose that $x^\star$ is not a strict local minimizer, define an appropriate sequence $\{\Delta x^{(\nu)}\}_{\nu=1}^\infty$, apply Theorem A.7 in Section A.7.1 of Appendix A, and consider an accumulation point $\Delta x^\star$ of the sequence.)

14.12 Consider Problem (14.1) in the case that $g : \mathbb{R}^n \to \mathbb{R}^m$ is affine. That is, there are $A \in \mathbb{R}^{m \times n}$ and $b \in \mathbb{R}^m$ such that:

$$\forall x \in \mathbb{R}^n, \ g(x) = Ax - b.$$

Assume that A has linearly independent rows. Show that Theorem 14.2 specializes to Corollary 13.4.

14.13 In this exercise we consider the points that satisfy the first-order necessary conditions for the problem from Sections 14.2.1.5 and 14.2.2.2.

(i) Show that all four points in Figure 14.5 satisfying the first-order necessary conditions:

$$\nabla f(x) + J(x)^\dagger \lambda \ = \ \mathbf{0},$$
$$g(x) \ = \ \mathbf{0},$$

are regular points of the constraints $g(x) = \mathbf{0}$.
(ii) Show that second-order sufficient conditions are satisfied by the two minimizers and corresponding Lagrange multipliers of the example problem in Section 14.2.1.5.
(iii) Show that second-order sufficient conditions are not satisfied by the other points that satisfy the necessary conditions of the example problem in Section 14.2.1.5.

14.14 Consider Problem (2.14), $\min_{x \in \mathbb{R}^2} \{f(x)|g(x) = 0\}$, from Section 2.3.2.2, where $f : \mathbb{R}^2 \to \mathbb{R}$ and $g : \mathbb{R}^2 \to \mathbb{R}$ were defined by:

$$
\begin{aligned}
\forall x \in \mathbb{R}^2, \, f(x) &= (x_1 - 1)^2 + (x_2 - 3)^2, \\
\forall x \in \mathbb{R}^2, \, g(x) &= (x_1)^2 + (x_2)^2 + 2x_2 - 3.
\end{aligned}
$$

(i) Find all the points satisfying the first-order necessary conditions of Problem (2.14):

$$
\begin{aligned}
\nabla f(x^\star) + J(x^\star)^\dagger \lambda^\star &= 0, \\
g(x^\star) &= 0.
\end{aligned}
$$

(ii) Show that all the points satisfying the first-order necessary conditions are regular points of the constraints $g(x) = 0$.

(iii) Show that second-order sufficient conditions are satisfied by the minimizer and corresponding Lagrange multipliers of Problem (2.14).

(iv) Show that second-order sufficient conditions are not satisfied by the other point that satisfies the first-order necessary conditions of Problem (2.14).

14.15 Let $f : \mathbb{R}^n \to \mathbb{R}$ be partially differentiable with continuous partial derivatives. Show that the steepest descent step direction $-\nabla f(\hat{x})$ for f at a point $\hat{x} \in \mathbb{R}^n$ satisfies:

$$
\frac{-\nabla f(\hat{x})}{\|\nabla f(\hat{x})\|_2} \in \arg \min_{\Delta x \in \mathbb{R}^n} \{\nabla f(\hat{x})^\dagger \Delta x | \|\Delta x\|_2 = 1\}.
$$

(Hint: Solve the first-order necessary conditions for $\min_{\Delta x \in \mathbb{R}^n} \{\nabla f(\hat{x})^\dagger \Delta x | \|\Delta x\|_2 = 1\}$. Show that the solution is unique, that it satisfies the second-order sufficient conditions so that it is the only local minimizer, and use Theorem 2.1 to show that it must be the global minimizer.)

Approaches to finding minimizers

14.16 Consider Problem (2.14), $\min_{x \in \mathbb{R}^2} \{f(x)|g(x) = 0\}$, from Section 2.3.2.2, where $f : \mathbb{R}^2 \to \mathbb{R}$ and $g : \mathbb{R}^2 \to \mathbb{R}$ were defined by:

$$
\begin{aligned}
\forall x \in \mathbb{R}^2, \, f(x) &= (x_1 - 1)^2 + (x_2 - 3)^2, \\
\forall x \in \mathbb{R}^2, \, g(x) &= (x_1)^2 + (x_2)^2 + 2x_2 - 3.
\end{aligned}
$$

Use the MATLAB function fmincon to solve this problem. You should write MATLAB M-files to evaluate $f, g, \nabla f$, and J.

14.17 ([11, figure 5.1.4].) Consider the problem $\min_{x \in \mathbb{R}} \{f(x)|g(x) = 0\}$, where $f : \mathbb{R} \to \mathbb{R}$ and $g : \mathbb{R} \to \mathbb{R}$ are defined by:

$$
\begin{aligned}
\forall x \in \mathbb{R}, \, f(x) &= x, \\
\forall x \in \mathbb{R}, \, g(x) &= (x)^2.
\end{aligned}
$$

(i) Evaluate the dual function.

(ii) Does the dual function have a maximum?

Sensitivity

14.18 Consider the example problem $\min_{x \in \mathbb{R}^2}\{f(x)|g(x) = 0\}$ from Section 14.2.1.5 with objective $f : \mathbb{R}^2 \to \mathbb{R}$ and constraint function $g : \mathbb{R}^2 \to \mathbb{R}$ defined by:

$$\forall x \in \mathbb{R}^2, f(x) = \frac{1}{2}(x_1)^2 + \frac{1}{2}(x_2)^2,$$

$$\forall x \in \mathbb{R}^2, g(x) = \frac{1}{4}(x_1)^2 + \frac{1}{4}(x_2)^2 - 1,$$

but suppose that the constraint changed to $g(x) = -\gamma$, where $\gamma = [-0.1]$. The base-case minimum for $\gamma = [0]$ was $f^\star = 0.5$. By Exercise 14.13, the minimizer $x^\star \in \mathbb{R}$ and corresponding Lagrange multiplier $\lambda^\star \in \mathbb{R}$ satisfy the second-order sufficient conditions and $x^\star$ is a regular point of the constraints. Calculate the change in the minimum of the problem in three ways:

(i) Using the sensitivity analysis in Corollary 14.3.
(ii) Explicitly re-solving the necessary conditions of problem.
(iii) Using the MATLAB function fmincon to solve the problem. You should write MATLAB M-files to evaluate $f, g, \nabla f$, and J.

14.19 Let $f : \mathbb{R}^n \times \mathbb{R}^s \to \mathbb{R}$ and $g : \mathbb{R}^n \times \mathbb{R}^s \to \mathbb{R}^m$ be twice partially differentiable with continuous partial derivatives and consider the problem $\min_{x \in \mathbb{R}^n}\{f(x; \chi)|g(x; \chi) = 0\}$, where χ is a parameter. Let $J : \mathbb{R}^n \times \mathbb{R}^s \to \mathbb{R}^{m \times n}$ be the Jacobian of g. That is, $J = \frac{\partial g}{\partial x}$.

Suppose that $x^\star \in \mathbb{R}^n$ and $\lambda^\star \in \mathbb{R}^m$ satisfy the second-order sufficient conditions for this problem for the base-case value of parameters $\chi = 0$. In particular $x^\star$ is a local minimizer of the problem for $\chi = 0$ with associated Lagrange multipliers $\lambda^\star$. Moreover, suppose that the rows of the Jacobian $J(x^\star; 0)$ are linearly independent so that $x^\star$ is a regular point of the constraints for the base-case value of the parameters $\chi = 0$.

(i) Write down the first-order necessary conditions for minimizing this problem. (Use the form of the first-order necessary conditions in (14.6)–(14.7).) Use the symbols $x^\star(\chi)$ and $\lambda^\star(\chi)$ for the solution of the first-order necessary conditions at a particular value of χ.
(ii) The first-order necessary conditions are non-linear simultaneous equations. Write down the Jacobian corresponding to these equations.
(iii) Find the sensitivity of the solution of the first-order necessary conditions to χ_j, evaluated at $\chi = 0$.

Solution of power system state estimation with zero injection buses case study

14.20 Consider the system shown in Figure 12.6, which is repeated for convenience in Figure 14.6. Using:

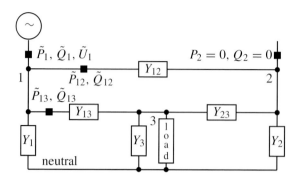

Fig. 14.6. State estimation problem with zero injection buses, repeated from Figure 12.6.

- the parameters from Exercise 8.12, that is, assuming that the lines have π-equivalent models with:
  - shunt elements purely capacitive with admittance $0.01\sqrt{-1}$ so that the combined shunt elements are:

$$Y_1 = Y_2 = Y_3 = 0.02\sqrt{-1},$$

  and
  - series elements having admittances:

$$Y_{12} = (0.01 + 0.1\sqrt{-1})^{-1},$$
$$Y_{23} = (0.015 + 0.15\sqrt{-1})^{-1},$$
$$Y_{31} = (0.02 + 0.2\sqrt{-1})^{-1},$$

- measurements of $\tilde{P}_1 = 1.0$, $\tilde{Q}_1 = 0.5$, and $\tilde{U}_1 = 1.0$, and
- measurement error standard deviation of $\sigma_\ell = 0.02$ for all measurements,

perform the following.

(i) Solve (14.12) for the Newton–Raphson step direction $\begin{bmatrix} \Delta x^{(0)} \\ \Delta \lambda^{(0)} \end{bmatrix}$ using a flat start as the initial guess for $x^{(0)}$ and $\lambda^{(0)} = 0$.

(ii) Calculate the next iterate $\begin{bmatrix} x^{(1)} \\ \lambda^{(1)} \end{bmatrix}$. To select a step-size, $\alpha^{(0)}$, apply the Armijo criterion to the merit function defined in (14.13), using the L_2 norm and $\Pi = (10/\sigma_\ell)^2$. Initially set the step-size to one and halve it until the Armijo criterion (10.14) is satisfied.

(iii) Use the MATLAB function fmincon to solve the problem. You should write MATLAB M-files to evaluate f, g, ∇f, and J. Use a flat start as the initial guess.

(iv) Suppose that the measurements change to $\tilde{P}_1 = 1.05$, $\tilde{Q}_1 = 0.55$, and $\tilde{U}_1 = 0.95$. Use sensitivity analysis of the first-order necessary conditions for this problem to estimate the change in the estimated state from the solution to Part (iii).

(v) Use the MATLAB function fmincon to solve the change-case problem specified in Part (iv) using as initial guess the solution from Part (iii).

(vi) Compare the results of the previous two parts.

Part V

Inequality-constrained optimization

15

Case studies of inequality-constrained optimization

In this chapter we will introduce six case studies:

(i) production, at least-cost, of a commodity from machines that have minimum and maximum machine capacity constraints (Section 15.1),

(ii) optimal routing in a data communications network (Section 15.2),

(iii) least absolute value estimation (Section 15.3),

(iv) optimal margin pattern classification (Section 15.4),

(v) choosing the widths of interconnects between latches and gates in integrated circuits (Section 15.5), and

(vi) the optimal power flow problem in electric power systems (Section 15.6).

The first and third case studies will draw from the previous formulations in Sections 12.1 and 9.1, respectively. The sixth case study *combines* the formulations from Sections 15.1 and 6.2. These three case studies will be introduced briefly, concentrating on the *extensions* from the previous formulations. They further illustrate the idea of incremental model development. The second, fourth, and fifth case studies introduce new material and will be developed in more detail. All six of these case studies will turn out to be optimization problems with both equality and inequality constraints. The first three have linear constraints, while the last three have non-linear constraints. Transformations will be applied to the fourth and fifth to deal with the non-linear constraints.

15.1 Least-cost production with capacity constraints

This case study generalizes the least-cost production case study from Section 12.1.

559

15.1.1 Motivation

Recall the least-cost production case study discussed in Section 12.1. For that problem we ignored the minimum and maximum machine capacity constraints in order to formulate it as equality-constrained Problem (12.4), which we repeat here:

$$\min_{x \in \mathbb{R}^n} \{ f(x) | Ax = b \}.$$

That is, we ignored the inequality constraints. We noted that *if* we found a solution of Problem (12.4) that happened to also satisfy the minimum and maximum machine capacity constraints, then the solution was optimal for the formulation that includes these inequality constraints. In this section, we will consider the case where the solution of Problem (12.4) does not satisfy all the minimum and maximum machine capacity constraints so that these inequality constraints must be considered explicitly [8, section 11.9].

15.1.2 Formulation

15.1.2.1 Objective

As in Section 12.1, we assume that there are n machines with total cost of production $f : \mathbb{R}^n \to \mathbb{R}$ defined by:

$$\forall x \in \mathbb{R}^n, \ f(x) = \sum_{k=1}^{n} f_k(x_k).$$

That is, f is additively separable. (See Definition A.23.)

15.1.2.2 Equality constraints

As in Section 12.1, the n machines face the production constraint (12.3), which we repeat here:

$$D = \sum_{k=1}^{n} x_k.$$

We represented these constraints in the form $Ax = b$ with $A = -\mathbf{1}^\dagger \in \mathbb{R}^{1 \times n}$ and $b = [-D] \in \mathbb{R}^1$.

15.1.2.3 Inequality constraints

In this case study we also explicitly include the minimum and maximum machine capacity constraints of the form (12.2) for each machine:

$$\forall \ell = 1, \ldots, n, \underline{x}_\ell \leq x_\ell \leq \overline{x}_\ell.$$

We summarize these constraints by writing $\underline{x} \leq x \leq \overline{x}$, where $\underline{x} \in \mathbb{R}^n$ and $\overline{x} \in \mathbb{R}^n$ are constant vectors with ℓ-th entries $\underline{x}_\ell$ and $\overline{x}_\ell$, respectively.

15.1.2.4 Problem

We write the complete problem as:

$$\min_{x \in \mathbb{R}^n}\{f(x)|Ax = b, \underline{x} \le x \le \overline{x}\}. \tag{15.1}$$

We are seeking the minimum value of f over x in $\mathbb{R}^n$ such that the sum $\sum_{k=1}^{n} x_k$ equals D and such that $\underline{x} \le x \le \overline{x}$. (As discussed in Section 12.1, we will not consider the extension of this problem to the case where we also allow machines to be switched off.)

15.1.3 Changes in demand and capacity

As in the least-cost production case study in Section 12.1, demand will change over time. Consequently, it is important to be able to estimate the change in the costs due to a change in demand from D to $D + \Delta D$, say.

Moreover, if the capacity of a machine k changes or it fails then the corresponding entries $\overline{x}_k$ and $\underline{x}_k$ of, respectively, $\overline{x}$ and $\underline{x}$ will change.

15.1.4 Problem characteristics

15.1.4.1 Objective

We noted in Section 12.1.4.1 that for machine k if $\underline{x}_k > 0$ then, for typical cost functions, f_k is convex on $[\underline{x}_k, \overline{x}_k]$. Exercise 15.1 explores convexity further for Problem (15.1).

As in Section 12.1.2.5, we will assume that each function f_k has been extrapolated to a function that is convex on the whole of $\mathbb{R}$. Again, this is natural if f_k is specified on $[\underline{x}_k, \overline{x}_k]$ as a quadratic polynomial with a positive quadratic coefficient.

15.1.4.2 Equality constraints

We have already discussed the equality constraint $D = \sum_{k=1}^{n} x_k$ in Section 12.1.2.4.

15.1.4.3 Inequality constraints and the feasible region

As discussed in Section 12.1.2.4 and illustrated in Figure 12.2, the production constraint $D = \sum_{k=1}^{n} x_k$ is a hyperplane in $\mathbb{R}^n$. The lower and upper bound constraints on x define a "box" in $\mathbb{R}^n$. The intersection of the box with the equality constraint restricts the feasible region to being a planar slice through the box. This is illustrated in Figure 15.1 for $n = 3$, $D = 10$, and:

$$\underline{x} = \begin{bmatrix} 1 \\ 2 \\ 3 \end{bmatrix}, \overline{x} = \begin{bmatrix} 4 \\ 5 \\ 6 \end{bmatrix}.$$

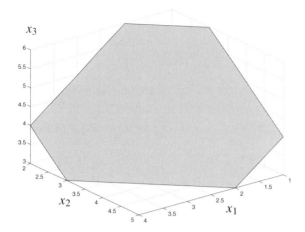

Fig. 15.1. Feasible set for least-cost production case study described in Section 15.1.4.3.

The planar slice is specified by the equality constraint $D = \sum_{k=1}^{n} x_k$ and is the same as illustrated in Figure 12.2.

15.1.4.4 Solvability

By Exercise 15.1, Problem (15.1) is convex. As in our study of equality-constrained optimization, we will find that for convex problems that possess a solution, there are necessary and sufficient conditions for optimality that are based on first-order derivative information only. However, it is certainly possible for there to be no solution to Problem (15.1). For example, the feasible set can be empty as Exercise 15.2 shows.

15.2 Optimal routing in a data communications network

This case study considers one aspect of the operation of a communications network.

15.2.1 Motivation

This case study is based on [9, chapter 5] and assumes some familiarity with data networks. We consider a communications network consisting of communications **links** that join between **nodes**. Users desire to send data from **origin nodes** to **destination nodes** over links between the nodes. Each link has a maximum capacity to transmit data. Usually, several links are incident to each node.

We will assume that data is sent by users to the network in collections called **packets**. For simplicity, each packet is assumed to consist of the same number of bits; that is, the packets are of equal length.

Time elapses between the arrival of each successive packet at an origin node. This **inter-arrival time** between packets is assumed to be random, with **exponential distribution** [9, section 3.3.1]. The parameter of the exponential probability distribution characterizing the inter-arrival times can differ from node to node, but we assume that the underlying probability distributions of the inter-arrival times do not vary over time. This allows us to consider the average traffic on each link due to:

- the distributions of inter-arrival times, and

- a **routing policy**; that is, a decision process for choosing the links on which to send the data.

There are a large number of issues relevant to a data communications network, including: error correcting codes, acknowledgment of message receipt, initialization, synchronization of hardware, reliability, and the expansion of an existing network. (See [9] for details.) We will concentrate on just one of these issues: choosing the links along which to send each message from an origin to a destination pair. We refer to the choice of links, with respect to a given criterion and for given traffic levels between origin–destination pairs, as **optimal routing**. (We consider one extension of optimal routing to include another issue in Exercise 15.4.) We will discuss this problem as though it can be solved centrally; however, in practice the algorithm must be decentralized [9, chapter 5].

We will see that our formulation of the objective of this problem only approximately captures the criterion we discuss. Nevertheless, the objective is useful in avoiding unsatisfactory routing decisions. As in the case of non-linear regression discussed in Section 9.2.1.1, we might better refer to our problem as seeking **satisficing** routing. As mentioned in Section 1.4.7, the word satisficing is a contraction of "satisfying" and "sufficient" [109]. The routing will satisfy the constraints and will (we hope) provide sufficiently good, although perhaps not optimal, routing.

Moreover, traffic levels in a network do in practice change over time, due to patterns of human activity and link failures. We will avoid this issue by assuming that the traffic levels are in **quasi-steady state**. In practice, the implication is that we must occasionally update the routing as conditions change. We will consider how changes in the conditions affect the solution.

Furthermore, we abstract away from the random nature of arrivals by assuming that the variances of the traffic on the links are small enough so that the traffic flow can be well-characterized by just its **expected** or **average** value. We will see that this has implications for specifying the notion of the "capacity" of a link.

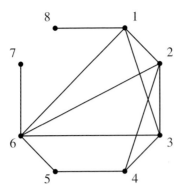

Fig. 15.2. Graphical representation of a data communications network with eight nodes and 12 links.

15.2.2 Formulation

As with several of our previous case studies, we can represent the communications network as a graph. An example network is shown in Figure 15.2. Each of the eight nodes in Figure 15.2 is shown as a bullet •, while each of the 12 links is shown as a line. As in previous case studies involving graphs, the typical number of links is far less than in a complete graph. In the following sections, we will discuss links, nodes, origin–destination pairs, paths, and then the variables, constraints, and objective of the problem.

15.2.2.1 Links

Typical communication links are bi-directional. That is, they can transmit data in both directions. We use a single line to represent the data connections in both directions between the ends of each link, interpreting each line as symbolizing a bi-directional link. We assume that the maximum capacity of a link is the same in each direction, although the generalization to asymmetric links is straightforward. (See Exercise 15.5.)

We write $\mathbb{L}$ for the set of all links in the network, where each link is represented by an ordered pair (i, j) of node numbers, with the first node i being the transmitting node and the second node j being the receiving node. Since links are bi-directional in our formulation, for each link $(i, j) \in \mathbb{L}$, there is also a link $(j, i) \in \mathbb{L}$ that can carry data in the opposite direction. For the data communications network represented in Figure 15.2, we have:

$$\mathbb{L} = \{(1, 8), (8, 1), (1, 2), (2, 1), (1, 3), (3, 1), (1, 6), (6, 1),$$
$$(2, 3), (3, 2), (2, 4), (4, 2), (2, 6), (6, 2), (3, 4), (4, 3),$$
$$(3, 6), (6, 3), (4, 5), (5, 4), (5, 6), (6, 5), (6, 7), (7, 6)\}.$$

The capacity of link (i, j) is denoted by $\overline{y}_{ij} \in \mathbb{R}_{++}$ and represents the maximum number of packets that can be transmitted per second on the link. Because of our assumption of symmetric link capacities, for each link $(i, j) \in \mathbb{L}$, we have that $\overline{y}_{ij} = \overline{y}_{ji}$.

15.2.2.2 Nodes

For our purposes, we can consider the nodes to have three roles, as follows.

- Users put data into the network at nodes. These nodes can be thought of as the **origins** of data.
- A node **switches** arriving data onto one of the links incident to it.
- Users take data out of the network at nodes. These nodes can be thought of as the **destinations** of data.

The equipment that performs these roles is called a **router**.

15.2.2.3 Origin–destination pairs

Some particular pairs of nodes represent **origin–destination pairs**. For example, a user located nearby to node 7 might put data into the network at node 7 and desire to transmit it to someone located near to node 5, so that node 7 is the origin for the data and node 5 is the destination for the data.

Other users might wish to transmit data from node 2 to node 5. We assume that there are m origin–destination pairs and write $\mathbb{W}$ for the set of all origin–destination pairs. In our example, if $(7, 5)$ and $(2, 5)$ are the only origin–destination pairs then:

$$\mathbb{W} = \{(7, 5), (2, 5)\},$$

with $m = 2$. In general, an origin–destination pair $(\ell, \ell') \in \mathbb{W}$ might or might not be joined directly by a link. In our example, neither of the origin–destination pairs $(7, 5)$ $(2, 5)$ is joined directly by a link. If there is no link joining such an origin–destination pair then it is necessary for the data between this pair to traverse several successive links.

15.2.2.4 Paths

A collection of successive links that joins an origin–destination pair is called a **path**. There will typically be several alternative paths joining between each origin–destination pair. Consider again the traffic for the origin–destination pair $(7, 5)$. Since there is only one link incident to node 7, the data that is put into the network at node 7 would be sent via that link, which joins node 7 to node 6. A routing decision must then be made as to which link would receive the data from node 6.

Since the destination for the data is node 5 then a natural choice would be to send the data along the link joining node 6 to node 5. However, if this link was already experiencing considerable traffic then a different choice of links such as, for example, from 6 to 3 to 4 to 5, might be used. Even if an origin–destination pair were joined directly by a link, we still might want to consider several alternate paths for routing between the origin–destination pairs to keep the flow on each link to an acceptable level.

In summary, two paths for the origin–destination pair $(7, 5)$ are:

- links $(7, 6)$ and $(6, 5)$, and
- links $(7, 6)$, $(6, 3)$, $(3, 4)$, $(4, 5)$.

For each origin–destination pair $(\ell, \ell') \in \mathbb{W}$, we write $\mathbb{P}_{(\ell, \ell')}$ for the set of all allowable paths connecting ℓ to ℓ'. We index the paths with consecutive integers. For example, for the origin–destination pair $(7, 5) \in \mathbb{W}$, we will denote:

- the path consisting of links $(7, 6)$ and $(6, 5)$ as path 1, and
- the path consisting of links $(7, 6)$, $(6, 3)$, $(3, 4)$, $(4, 5)$ as path 2.

For the origin–destination pair $(2, 5) \in \mathbb{W}$, we will denote:

- the path consisting of links $(2, 4)$ and $(4, 5)$ as path 3, and
- the path consisting of links $(2, 3)$, $(3, 4)$, $(4, 5)$ as path 4.

We summarize these assignments by $\mathbb{P}_{(7,5)} = \{1, 2\}$, $\mathbb{P}_{(2,5)} = \{3, 4\}$.

We assign a different index k for each allowed path in the network and suppose that there are n paths in all. In our example, if we have described all the allowable paths then $n = 4$.

15.2.2.5 Variables

Since the inter-arrival times are random, then so will be the number of packets in the network, the number of packets on any link, and the number of packets on any path. To characterize the behavior of the network, we consider the **expected** or **average** flow of packets and ignore variance of the distribution of flow.

We define $x_k, k = 1, \ldots, n$, to be the average flow of traffic, in packets per second, on path k. This flow represents the average amount of flow for a particular origin–destination pair that has been assigned to path k. We collect the set of all traffic assignments for all origin–destination pairs together into a vector $x \in \mathbb{R}^n$.

15.2.2.6 Equality constraints

Consider any origin–destination pair $(\ell, \ell') \in \mathbb{W}$. Let the input traffic arrival process for origin–destination pair (ℓ, ℓ') have expected rate of arrival of $b_{(\ell, \ell')}$, in packets per second. In general, we must choose how to share the traffic amongst all the paths that join ℓ to ℓ'. In order that all the arriving traffic for each origin–destination pair (ℓ, ℓ') be transported, we must apportion the flow so that:

$$\forall (\ell, \ell') \in \mathbb{W}, \ \sum_{k \in \mathbb{P}_{(\ell, \ell')}} x_k = b_{(\ell, \ell')}.$$

In our example, the constraints for the origin–destination pairs $(7, 5)$ and $(2, 5)$ are, respectively:

$$x_1 + x_2 = b_{(7,5)},$$
$$x_3 + x_4 = b_{(2,5)}.$$

We collect the entries $b_{(\ell,\ell')}$ for $(\ell, \ell') \in \mathbb{W}$ into a vector $b \in \mathbb{R}^m$. Also, define $A \in \mathbb{R}^{m \times n}$ to be the path to origin–destination pair incidence matrix. That is, define:

$$\forall (\ell, \ell') \in \mathbb{W}, \forall k = 1, \ldots, n, \ A_{(\ell,\ell')k} = \begin{cases} 1, & \text{if } k \in \mathbb{P}_{(\ell,\ell')}, \\ 0, & \text{otherwise.} \end{cases}$$

The entries of A are either 1 or 0. In our example:

$$A = \begin{bmatrix} 1 & 1 & 0 & 0 \\ 0 & 0 & 1 & 1 \end{bmatrix},$$
$$b = \begin{bmatrix} b_{(7,5)} \\ b_{(2,5)} \end{bmatrix},$$

where the rows of A and the entries of b correspond to the origin–destination pairs $(7, 5)$ and $(2, 5)$, respectively. With these definitions, we can write the equality constraints as:

$$Ax = b. \tag{15.2}$$

15.2.2.7 Objective

Discussion There are several criteria that could be used to define an objective function. Unlike the least-cost production case study in Sections 12.1 and 15.1, where the cost of production depended on the amount of production, the operating cost of a data network is generally relatively constant. In particular, the costs of operating the network are relatively independent of the loading on routers and links.

In *delivering service* to customers, however, the quality of service depends on a number of factors, including the delay between sending data and receiving it. In some circumstances, we may be able to distinguish between very valuable service, for which even short delays are intolerable, and less valuable service, for which long delays may not be problematic. For example, some customers may have a contract for high quality service while others may have subscribed to a service that offers only **best-effort** with no guarantees for maximum delay.

In this case study we will assume that there is no such distinction between customers. That is, we consider all customers equally and consider all traffic to have equal priority. (While this may be a poor approximation to reality, it can be readily

Fig. 15.3. A network with an origin–destination pair joined by a path consisting of two links.

generalized to the case of different traffic priorities, so long as the different traffic types can be distinguished.) Our goal will be to minimize the average delay experienced by the traffic in the network.

Delay The delay on a link depends on how much traffic is on the link. When the traffic is nearly as large as the capacity of the link, the delay is longer. We say that the link is **congested**.

It is difficult to obtain an analytic model of the delay in a network. The reason is that the packets interact as they traverse the links, so that the analysis of their statistics is complicated. For example, consider an origin–destination pair (ℓ, ℓ') that is joined by one path, which consists of two successive links (ℓ, j) and (j, ℓ') as shown in Figure 15.3.

We have assumed in our formulation that the inter-arrival time at the origin ℓ is exponentially distributed. The inter-arrival time at node j cannot be exponentially distributed. The reason is that successive packets arriving at j must be separated in time by at least the packet transmission time for the first link and this violates the assumption of exponential distribution [9, section 3.6]. We could, in principle, calculate the distribution of inter-arrival times at node j, but the situation becomes more complicated as we try to analyze larger networks because of the interaction between packets flowing over successive links.

In general, the probability distribution of the inter-arrival times of packets at nodes is difficult to calculate analytically. Consequently the delay experienced by packets is difficult to calculate analytically. (Under somewhat unrealistic assumptions, the traffic on each link can be considered approximately independent of every other link. This, together with several other assumptions, can be used to calculate a delay function that is very similar to the objective we will develop. See [9, chapter 3] for details.)

Congestion model As a *proxy* to calculating the delay experienced by the packets in the network, we define a measure of the congestion on each link. For convenience, we will develop a congestion measure for each link that is convex in the flow on a link and assume that a reasonable proxy to the average delay experienced by all the packets in the network is the sum of the congestion measure across all the links.

Following [9, section 5.4], suppose that a link $(i, j) \in \mathbb{L}$ between nodes i and

j has capacity $\overline{y}_{ij}$ so that the expected flow y_{ij} on this link can never exceed $\overline{y}_{ij}$. In fact, since the instantaneous flow on the link actually varies from its expected value y_{ij}, it must be the case that the expected flow is always strictly less than $\overline{y}_{ij}$. This will affect our formulation of the function representing congestion in this section and also affect the formulation of the inequality constraints to be described in Section 15.2.2.8.

Also suppose that when data is sent on the link, data is queued at the sending end of a link until it is sent on the link. We will posit a congestion measure that depends on the expected flow y_{ij} through the link. In particular, consider the function ϕ_{ij} : $[0, \overline{y}_{ij}) \to \mathbb{R}_+$ defined by:

$$\forall y_{ij} \in [0, \overline{y}_{ij}), \, \phi_{ij}(y_{ij}) = \frac{y_{ij}}{\overline{y}_{ij} - y_{ij}} + \delta_{ij} y_{ij}, \tag{15.3}$$

where:

- δ_{ij} is the sum of the processing delay and the propagation delay through the router and link, and
- the term $\frac{y_{ij}}{\overline{y}_{ij} - y_{ij}}$ is due to queuing at the sending end of the link.

The rapid rise in the congestion function as the flow approaches the capacity models the increase in delay as capacity is reached. This rapid rise can be explained in terms of the random arrival of packets. Even with high expected flow rates, because of random fluctuations, there will occasionally be times when there are no packets in the queue and none on the link. Unfortunately, such "lost opportunities" are lost forever. Therefore, if the expected arrival rate for the link were to equal the capacity of the link then the queue at the sending end of the link would become arbitrarily long, with the result that the average queue length (and the average delay) would be unbounded. For expected flows less than the capacity, the delay increases with increasing expected flow. The first term in the right-hand side of (15.3) qualitatively captures these observations.

Flow The flow y_{ij} on the link is equal to the sum of the flows on all the paths that include link (i, j). We write $\mathbb{F}_{(i,j)}$ for the set of paths that include link (i, j), so that the flow y_{ij} can be expressed as:

$$\forall (i, j) \in \mathbb{L}, \, y_{ij} = \sum_{k \in \mathbb{F}_{(i,j)}} x_k.$$

Define a matrix $C \in \mathbb{R}^{\mathbb{L} \times n}$ by:

$$\forall (i, j) \in \mathbb{L}, \forall k = 1, \ldots, n, \, C_{(i,j)k} = \begin{cases} 1, & \text{if } k \in \mathbb{F}_{(i,j)}, \\ 0, & \text{otherwise.} \end{cases}$$

For each $(i, j) \in \mathbb{L}$, let $C_{(i,j)}$ be the (i, j)-th row of C. Then the flow y_{ij} can be expressed as:

$$\forall (i, j) \in \mathbb{L}, \; y_{ij} = C_{(i,j)} x.$$

Moreover, let $y \in \mathbb{R}^\mathbb{L}$ be a vector with entries y_{ij}, $(i, j) \in \mathbb{L}$. (See Definition A.4.) Then:

$$y = Cx.$$

Additive congestion We have assumed that the congestion measure for each link can be added together to obtain an overall proxy for delay through the network. If we let $\mathbb{P} = \{y \in \mathbb{R}^\mathbb{L} | 0 \le y_{ij} < \overline{y}_{ij}, \forall (i, j) \in \mathbb{L}\}$ then the objective $\phi : \mathbb{P} \to \mathbb{R}$ can be expressed as:

$$\forall y \in \mathbb{P}, \phi(y) = \sum_{(i,j) \in \mathbb{L}} \phi_{ij}(y_{ij}). \tag{15.4}$$

To understand this objective, notice that, for example, both of the paths for origin–destination pair $(7, 5)$ must use the link $(7, 6)$. Flow on either of these paths will contribute to congestion on link $(7, 6)$.

Moreover, paths between various origin–destination pairs will typically have some links in common. For example, recall that:

• path 3 consists of the links $(2, 4)$, $(4, 5)$, and
• path 4 consists of the links $(2, 3)$, $(3, 4)$, $(4, 5)$,

and both of these paths are for the origin–destination pair $(2, 5)$. Traffic on these paths must share the capacity of the link $(4, 5)$ with traffic on path 2, which consists of links $(7, 6)$, $(6, 3)$, $(3, 4)$, $(4, 5)$ for origin–destination pair $(7, 5)$. This means that there will be an interaction between traffic between various origin–destination pairs. The objective we have defined captures the issue that increasing the flow on a path that is incident to a particular link will increase the average delay for *all* paths incident to that link.

It is important to realize that the objective that we have defined does not exactly capture the average delay due to the flows on the paths. It is a proxy to the average delay that is designed to capture the *qualitative* dependence of average delay on the choice of routing. It may be sufficiently accurate to provide guidance to avoid bad routing decisions. That is, it facilitates finding satisficing solutions to the routing problem [109]. An alternative objective is discussed in [9, section 5.4].

15.2.2.8 Inequality constraints and feasible set

All traffic flows must be non-negative. Therefore, we must also include the non-negativity constraints:

$$x \ge \mathbf{0}.$$

Since the capacity of each link $(i, j) \in \mathbb{L}$ is $\overline{y}_{ij}$, the instantaneous flow on link (i, j) can never exceed $\overline{y}_{ij}$. Consequently, the average flow can never exceed $\overline{y}_{ij}$, suggesting constraints of the form:

$$\forall (i, j) \in \mathbb{L}, \; y_{ij} \leq \overline{y}_{ij}.$$

However, as discussed in Section 15.2.2.7, the objective is unbounded if any y_{ij} were to equal $\overline{y}_{ij}$, so we must limit the values of the flows y_{ij} with constraints of the form:

$$\forall (i, j) \in \mathbb{L}, \; y_{ij} < \overline{y}_{ij}.$$

We use the strict inequality because if the assigned flow were to equal the capacity then the congestion function would be unbounded.

To represent these strict inequality constraints explicitly in terms of x, we note that:

$$\forall (i, j) \in \mathbb{L}, \; y_{ij} \; = \; \sum_{k \in \mathbb{F}_{(i,j)}} x_k,$$

$$= \; C_{(i,j)} x.$$

If we define $\overline{y} \in \mathbb{R}^{\mathbb{L}}$ to be a vector with entries $\overline{y}_{ij}$, $(i, j) \in \mathbb{L}$ then we can write the strict inequality constraints as:

$$Cx < \overline{y}. \tag{15.5}$$

The inequality constraints for the problem therefore specify a set of the form:

$$\overline{\mathbb{S}} = \{x \in \mathbb{R}^n | x \geq \mathbf{0}, Cx < \overline{y}\}.$$

15.2.2.9 Problem

The optimization model for optimal routing is:

$$\min_{x \in \mathbb{R}^n} \{f(x) | Ax = b, x \geq \mathbf{0}, Cx < \overline{y}\}, \tag{15.6}$$

where $f : \overline{\mathbb{S}} \to \mathbb{R}$ is defined by:

$$\forall x \in \overline{\mathbb{S}}, \; f(x) \; = \; \phi(Cx),$$

$$= \; \sum_{(i,j) \in \mathbb{L}} \phi_{ij} \left(C_{(i,j)} x \right). \tag{15.7}$$

15.2.3 Changes in links and traffic

We can imagine that a link capacity might change due to, for example, failure of equipment associated with the link or degradation of transmission conditions in a medium. In the first case of equipment failure, we could imagine that the link is completely removed from the network. Alternatively, a partial failure or a communication degradation might reduce the capacity of the link but not remove it from the network. Conversely, a link returning to service or an improvement in transmission conditions would increase capacity. We would like to be able to change the routing to respond to changes in link capacity.

Over time, we also expect that the traffic on the network would change. We would also like to be able to change the routing to respond to changes in traffic.

15.2.4 Problem characteristics

15.2.4.1 Objective

The objective defined in (15.7) of the optimal routing problem is convex and differentiable on $\overline{\mathbb{S}} = \{x \in \mathbb{R}^n | x \geq 0, Cx < \overline{y}\}$, since it is the composition of a linear function with the sum of functions ϕ_{ij}, which are themselves convex. (See Exercise 15.6.) The objective becomes arbitrarily large as the flow on any link approaches its capacity.

15.2.4.2 Equality constraints

The equality constraints are indexed by ordered pairs $(\ell, \ell') \in \mathbb{W}$. This differs from our previous case studies where index sets were subsets of the integers. The equality constraints are affine and the coefficient matrix consists of only zeros and ones.

15.2.4.3 Inequality constraints

There are non-negativity constraints and also strict inequality constraints due to the link capacities. The strict inequality constraints are indexed by the ordered pairs $(i, j) \in \mathbb{L}$. We discussed the potential difficulties with strict inequality constraints in Section 2.3.3. We will see in Section 18.2, however, that because of the form of the objective we can avoid explicit consideration of the strict inequality constraints.

15.2.4.4 Solvability

There may be no feasible solution if there is not enough capacity in the network.

15.3 Least absolute value estimation

This case study generalizes the multi-variate linear regression case study from Section 9.1.

15.3.1 Motivation

Recall the multi-variate linear regression case study from Section 9.1. We transformed that problem into a least-squares problem in Section 11.1.1. In particular, the objective $f : \mathbb{R}^n \to \mathbb{R}$ was defined in Section 11.1.1 to be:

$$\forall x \in \mathbb{R}^n, \ f(x) = \frac{1}{2} \|Ax - b\|_2^2,$$

where:

- $A = \begin{bmatrix} A_1 \\ \vdots \\ A_m \end{bmatrix} \in \mathbb{R}^{m \times n}$,

- $A_\ell = \begin{bmatrix} \psi(\ell)^\dagger & 1 \end{bmatrix} \in \mathbb{R}^{1 \times n}, \ell = 1, \ldots, m$,

- $b = \begin{bmatrix} b_1 \\ \vdots \\ b_m \end{bmatrix} \in \mathbb{R}^m$,

- $b_\ell = \zeta(\ell)$, and
- $(\psi(\ell), \zeta(\ell))$ are the ordered pairs of independent and dependent variables for trial ℓ.

In some contexts, we may find the resulting solution is not **robust** to **outliers** in the data. That is, the quadratic objective allows data from a single trial to significantly affect the resulting estimate of the affine function that best represents the data.

For example, Figure 15.4 repeats the data from Figure 9.1, except that the data for one of the trials, $(\psi(6), \zeta(6))$, is significantly different, perhaps due to a gross failure of a measurement device. This data point is an outlier; that is, it appears to be inconsistent with the data from the other trials. The outlier significantly affects the result of the least-squares problem. The least-squares fit to all of the points in Figure 15.4, including the outlier, is shown by the thick line. This least-squares fit is very different to the least-squares fit shown in Figure 9.1.

If we ignore the point $(\psi(6), \zeta(6))$ then the least-squares fit to the rest of the points is very different from the thick line. A least-squares fit to the rest of the points is shown as the thin line in Figure 15.4. The two least-squares fits are very different. That is, the least-squares fit is very sensitive to gross errors in individual data points. (See Exercise 15.7.)

In these circumstances, we may prefer to use an objective that is less affected by outliers. This provides the motivation for **robust estimation** [53]. One objective that is used to reduce the effect of outliers involves the L_1 norm of $Ax - b$ instead of the Euclidean norm. Instead of squaring the **residuals** $e_\ell = A_\ell x - b_\ell$, as in the least-squares problem, we take the absolute value of them. Outliers, which have

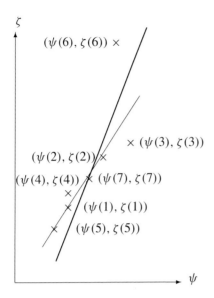

Fig. 15.4. The values of $(\psi(\ell), \zeta(\ell))$, including an outlier, (shown as ×) and least-squares fit (shown as a thick line). The thin line shows the least-squares fit if the data point $(\psi(6), \zeta(6))$ is ignored.

large values of residual, will contribute relatively less to the objective when we use the absolute value rather than the square of the residual.

15.3.2 Formulation

15.3.2.1 Unconstrained problem

Instead of the least-squares objective defined in Section 11.1.1, consider the L_1 norm objective $\phi : \mathbb{R}^n \to \mathbb{R}$ discussed in Exercise 11.6:

$$\forall x \in \mathbb{R}^n, \phi(x) = \|Ax - b\|_1,$$
$$= \sum_{\ell=1}^{m} |A_\ell x - b_\ell|,$$

where $A \in \mathbb{R}^{m \times n}$ and $b \in \mathbb{R}^m$ are as defined in Section 11.1.1 and A_ℓ is the ℓ-th row of A. That is, as in the least-squares problem:

$$A_\ell = \begin{bmatrix} \psi(\ell)^\dagger & 1 \end{bmatrix},$$
$$b = \begin{bmatrix} b_1 \\ \vdots \\ b_m \end{bmatrix}.$$

We define an unconstrained problem:

$$\min_{x \in \mathbb{R}^n} \phi(x). \tag{15.8}$$

As we saw in Section 3.1.4.4, the objective of this problem is non-differentiable because of the absolute values. There are techniques to treat non-differentiable objectives [11, section 6.3] and, moreover, in Section 3.1.4.4, we described how such an objective could be approximated by a smooth objective. However, we will describe a transformation that preserves the problem formulation exactly and removes the non-differentiability of the objective.

15.3.2.2 Transformation

Problem (15.8) can be transformed into an inequality-constrained problem in several steps. As in Section 9.1.2.4, the residual, e_ℓ, for the ℓ-th measurement, is defined by:

$$\forall \ell = 1, \ldots, m, e_\ell = A_\ell x - b_\ell.$$

Each absolute value of a residual can be obtained as:

$$|e_\ell| = \max\{e_\ell, -e_\ell\}, \ell = 1, \ldots, m. \tag{15.9}$$

We then use a similar approach to that used in Theorem 3.4 to evaluate the maximum in (15.9). In particular, by defining variables $z_\ell, \ell = 1, \ldots, m$, and linear constraints:

$$z_\ell \geq e_\ell, \ell = 1, \ldots, m,$$
$$z_\ell \geq -e_\ell, \ell = 1, \ldots, m,$$

we can evaluate $|e_\ell|$ by observing that:

$$|e_\ell| = \min_{z_\ell \in \mathbb{R}}\{z_\ell | z_\ell \geq e_\ell, z_\ell \geq -e_\ell\}.$$

Then note that:

$$\forall x \in \mathbb{R}^n, \phi(x) = \sum_{\ell=1}^{m} |A_\ell x - b_\ell|,$$

$$= \sum_{\ell=1}^{m} |e_\ell|, \text{ where } e_\ell = A_\ell x - b_\ell,$$

$$= \sum_{\ell=1}^{m} \min_{z_\ell \in \mathbb{R}}\{z_\ell | z_\ell \geq e_\ell, z_\ell \geq -e_\ell\}.$$

Combining these observations, we consider the transformed problem:

$$\min_{z \in \mathbb{R}^m, x \in \mathbb{R}^n, e \in \mathbb{R}^m} \{\mathbf{1}^\dagger z | Ax - b - e = \mathbf{0}, z \geq e, z \geq -e\}. \tag{15.10}$$

In (15.10), $z \in \mathbb{R}^m$ is a vector with entries z_ℓ, corresponding to the need to find the maximum of e_ℓ and $-e_\ell$ for each $\ell = 1, \ldots, m$. Exercise 15.8 explores the equivalence between Problems (15.8) and (15.10).

15.3.3 Changes in the number of points and the data

We could imagine adding a new trial and recalculating the estimate of the least-squares fit without starting from scratch. We can also imagine modifying the data for a particular trial.

15.3.4 Problem characteristics

15.3.4.1 Objective

The objective of Problem (15.8) is non-differentiable. Transformation into Problem (15.10) by representing each absolute value using two inequality constraints then yields a differentiable, in fact linear, objective.

15.3.4.2 Constraints

The "cost" of making the objective differentiable is that we have introduced a large number of subsidiary constraints. There are m equality constraints and $2m$ inequality constraints in Problem (15.10), whereas Problem (15.8) was unconstrained.

15.3.4.3 Variables

We have also increased the number of variables, from n to $n + 2m$.

15.3.4.4 Solvability

Problem (15.8) has a minimum and consequently Problem (15.10) also has a minimum.

15.3.4.5 Discussion

If the number of trials m is extremely large then it may be unattractive to solve Problem (15.10). In this case, we may prefer to, for example:

- solve Problem (15.8) using techniques of non-differentiable optimization [11, section 6.3],
- approximate the objective of Problem (15.8) with a smooth function using the approach described in Section 3.1.4.4, or
- use an iterative technique to successively approximate ϕ by smooth functions, such as a sequence of least-squares objectives [45, section 4.2.3].

15.4 Optimal margin pattern classification

This case study considers the classification of, for example, images, into pre-defined classes.

15.4.1 Motivation

There are many problems arising in pattern recognition and classification. In this case study, we will consider the problem of distinguishing between two classes of patterns on the basis of a linear decision function. This case study is based on the discussion in [14] and [15, section 8.6.1]. A number of generalizations of this problem are described in [15, chapter 8].

Geometrically, we seek a hyperplane that separates the two classes of patterns. We assume that the patterns have been **classified** for us. That is, for each pattern we know to which class it belongs. A generalization of this problem involves **clustering** patterns into classes without prior knowledge of the classes.

In practice, we may have more than two classes of patterns. If there are more than two classes then we can, in principle, use several classifiers to distinguish all the classes.

15.4.2 Formulation

15.4.2.1 Classes and training set

For convenience, we label the two classes as class A and class B. We will consider how to find the coefficients that specify a linear decision function in such a way as to provide the best discrimination between classes A and B of patterns.

In particular, we assume that we have r representatives in our **training set**. Potentially, r is very large. The training set is to be used to determine the best linear decision function to separate the classes.

We index the representives in the training set as $\ell = 1, \ldots, r$. The ℓ-th representative consists of two items:

- a **pattern**, namely a vector $\psi(\ell) \in \mathbb{R}^{n-1}$, and
- a value $\zeta(\ell) \in \{-1, 1\}$.

The value $\zeta(\ell)$ shows to which class the pattern belongs. In particular:

$$\forall \ell = 1, \ldots, r, \zeta(\ell) = \begin{cases} 1, & \text{if } \psi(\ell) \text{ is of class A,} \\ -1, & \text{if } \psi(\ell) \text{ is of class B.} \end{cases}$$

That is, for each pattern in the training set we know to which class it belongs. Figure 15.5 shows an example of a training set with $n-1 = 2$. Patterns $\psi(1), \ldots, \psi(4)$ in the bottom half of the figure are of class A, while the patterns $\psi(5), \ldots, \psi(7)$ in the top half of the figure are of class B. That is, $\zeta(1) = \zeta(2) = \zeta(3) = \zeta(4) = 1$ and $\zeta(5) = \zeta(6) = \zeta(7) = -1$. The horizontal line in Figure 15.5 **perfectly discriminates** between classes A and B. That is, the members of class A are all on one side of the line and the members of class B are all on the other side of the line. In general, a line that discriminates between the two classes may not be horizontal

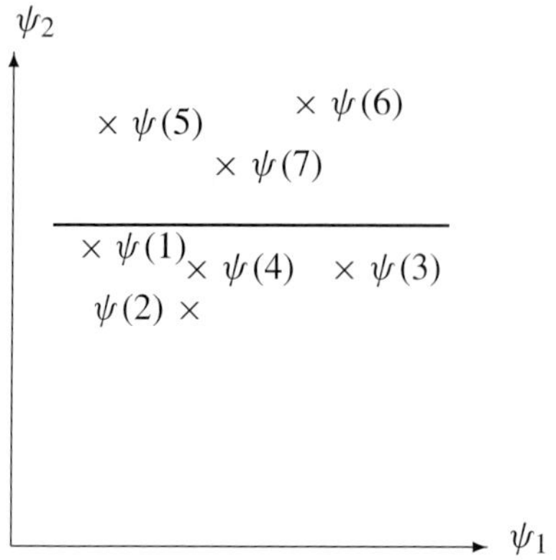

ψ_2

$\times \psi(5)$ $\times \psi(6)$

$\times \psi(7)$

$\times \psi(1)$ $\times \psi(4)$ $\times \psi(3)$

$\psi(2) \times$

ψ_1

Fig. 15.5. Seven example patterns and hyperplane that separates them.

and, moreover, it may not even be possible to find a line that perfectly discriminates between the two classes.

The vectors representing each pattern may have a very large number of entries. That is, $n - 1$ may be very large. For example, a pattern might encode gray-scale levels of pixels in an image and the number of pixels could be millions or more.

15.4.2.2 Feature space

In a variation on this formulation, the patterns $\psi(\ell)$ are *transformed* versions of the ℓ-th original image. That is, the original images are transformed into a **feature space** that extracts features that involve multiple pixels, such as the presence of horizontal lines in the image [14]. For the purposes of our discussion, it does not matter whether we think of the patterns as being "raw" images or transformed images in the feature space; however, the definition of a suitable feature space can be instrumental in discriminating between classes of patterns.

15.4.2.3 Decision function

We consider an affine **decision function** $D : \mathbb{R}^{n-1} \to \mathbb{R}$ defined by:

$$\forall \psi \in \mathbb{R}^{n-1}, D(\psi) = \beta^{\dagger} \psi + \gamma,$$

where the parameters $\beta \in \mathbb{R}^{n-1}$ and $\gamma \in \mathbb{R}$ are to be chosen so that:

$$\forall \ell = 1, \ldots, r, (D(\psi(\ell)) > 0) \Leftrightarrow (\zeta(\ell) = 1). \tag{15.11}$$

That is, the function D can be used to classify the patterns in the training set. Typically, there are many choices of parameters β and γ that will satisfy (15.11). As mentioned above, Figure 15.5 shows a line, which is a hyperplane in $\mathbb{R}^{n-1} = \mathbb{R}^2$, of the form:

$$\{\psi \in \mathbb{R}^{n-1} | D(\psi) = 0\},$$

that divides $\mathbb{R}^{n-1}$ into two half-spaces, one of which contains all the patterns in class A and the other one of which contains all the patterns in class B. That is, the hyperplane separates the patterns and the corresponding decision function D can be used to classify the patterns. There are evidently many choices of β and γ that would satisfy (15.11).

In practice, the parameters that determine the function D are calculated using the training set and the function is then subsequently used to estimate the classes of new, unknown patterns for which we do not know the class. Since we must find the parameters β and γ based on the training set, we must select a suitable criterion for choosing from amongst the values of β and γ that satisfy (15.11).

In principle, if we know, say, the functional form of the probability distribution from which the patterns are drawn then we could estimate the "best" values of the parameters β and γ using a maximum likelihood criterion, as discussed in the multi-variate linear regression case study in Section 9.1. Unfortunately, we usually do not have a lot of information about the patterns that we must subsequently classify and do not know the functional form of the probability distribution from which they are drawn. Consequently, the criterion for choosing the parameters β and γ will be *ad hoc,* aimed at finding a **satisficing** solution [109]. We will seek the β and γ such that the corresponding hyperplane $\{\psi \in \mathbb{R}^{n-1} | D(\psi) = 0\}$ is as far as possible from all the patterns in the training set. That is, we will find the values of β and γ that:

- maximize the minimum distance of any pattern from the hyperplane, and
- allow classification of the two classes of patterns according to (15.11).

We will use the notion of Euclidean distance to define distance. That is, we will use the norm $\|\bullet\|_2$. As Boyd and Vandenberghe describe it, we are trying to find the hyperplane between the two classes that is at the middle of "the thickest slab that separates the two sets of points" [15, section 8.6.1].

15.4.2.4 Variables

The decision vector for this problem consists of $\beta \in \mathbb{R}^{n-1}$ and $\gamma \in \mathbb{R}$. We collect these together into a vector $x = \begin{bmatrix} \beta \\ \gamma \end{bmatrix} \in \mathbb{R}^n$. That is, the parameters that specify the decision function D are the variables for the problem.

15.4.2.5 Objective

We must evaluate the Euclidean distance of a pattern $\psi(\ell)$ from the closest point on the hyperplane:

$$\{\psi \in \mathbb{R}^{n-1} | D(\psi) = 0\}.$$

This distance is given by:

$$\frac{|D(\psi(\ell))|}{\|\beta\|_2},$$

assuming that $\beta \neq \mathbf{0}$. (See Exercise 13.9. The hyperplane is not well-defined if $\beta = \mathbf{0}$.)

Define the set $\mathbb{P} \subset \mathbb{R}^n$ by:

$$\mathbb{P} = \left\{ \begin{bmatrix} \beta \\ \gamma \end{bmatrix} \in \mathbb{R}^n \,\middle|\, \beta \neq \mathbf{0} \right\}.$$

If the decision function D satisfies (15.11) then for each pattern $\psi(\ell)$ and classification $\zeta(\ell)$ we have that:

$$\zeta(\ell)D(\psi(\ell)) = |D(\psi(\ell))|.$$

Therefore, if we choose the coefficients β and γ such that (15.11) holds then, for each pattern $\psi(\ell)$, the distance of $\psi(\ell)$ from the hyperplane is given by the function $\phi_\ell : \mathbb{P} \to \mathbb{R}$ defined by:

$$\begin{aligned}
\forall x \in \mathbb{P}, \, \phi_\ell(x) &= \frac{|D(\psi(\ell))|}{\|\beta\|_2}, \\
&= \frac{\zeta(\ell)D(\psi(\ell))}{\|\beta\|_2}.
\end{aligned}$$

The minimum distance of any point $\psi(\ell)$ to the hyperplane, over all the patterns ℓ, is given by $\phi : \mathbb{R}^n \to \mathbb{R}$ defined by:

$$\forall x \in \mathbb{P}, \, \phi(x) = \min_{\ell=1,\ldots,r} \phi_\ell(x).$$

We call this minimum distance the **margin** between the hyperplane and the patterns. It is a measure of how easy it is to discriminate between the two classes of patterns. In particular, if the margin is "large" then the hyperplane is far from all the patterns and it is easy to distinguish between the two classes in the training set. If the patterns that we must subsequently classify are similar to the training set then we expect that the hyperplane will also be able to classify these new patterns reliably.

15.4.2.6 Constraint

In order for the objective to be well-defined, we must restrict ourselves to choices of $x \in \mathbb{P}$; that is, we must require $\beta \neq \mathbf{0}$. This constraint is not in the form of either an equality or an inequality constraint. Furthermore, $\mathbb{P}$ is neither closed nor convex. (See Exercise 15.10.)

15.4.2.7 Problem

We seek the coefficients $\beta \neq 0$ and γ such that the margin is maximized. Our problem is therefore:

$$\max_{x \in \mathbb{R}^n} \{\phi(x)|\beta \neq 0\}, \tag{15.12}$$

which involves finding the x that maximizes the margin. In the next section, we will transform this problem to remove the minimization embedded in the definition of the objective.

15.4.2.8 Transformation

By Theorem 3.4, we can remove the minimization in the definition of the objective ϕ by defining a subsidiary variable z. (Theorem 3.4 concerns minimizing a function defined as the maximum of several functions. Here we want to maximize a function defined as the minimum of several functions, but the theorem can be applied by recalling that maximizing a function is equivalent to minimizing the negative of the function.) The transformation in Theorem 3.4 allows the maximization with embedded minimization in Problem (15.12) to be transformed in the following manner:

$$\max_{x \in \mathbb{R}^n} \{\phi(x)|\beta \neq 0\}$$

$$= \max_{x \in \mathbb{R}^n} \left\{ \min_{\ell=1,\ldots,r} \phi_\ell(x) \middle| \beta \neq 0 \right\}, \text{ by definition of } \phi,$$

$$= \max_{z \in \mathbb{R}, x \in \mathbb{R}^n} \{z \,|\phi_\ell(x) \geq z, \forall \ell = 1, \ldots, r, \beta \neq 0\}, \text{ by Theorem 3.4,}$$

$$= \max_{z \in \mathbb{R}, x \in \mathbb{R}^n} \left\{ z \middle| \frac{\varsigma(\ell)D(\psi(\ell))}{\|\beta\|_2} \geq z, \forall \ell = 1, \ldots, r, \beta \neq 0 \right\},$$

$$= \max_{z \in \mathbb{R}, x \in \mathbb{R}^n} \{z \,|\varsigma(\ell)(\beta^\dagger \psi(\ell) + \gamma) \geq \|\beta\|_2 \, z, \forall \ell = 1, \ldots, r, \beta \neq 0\},$$

$$\tag{15.13}$$

where we note that $\forall \psi$, $D(\psi) = \beta^\dagger \psi + \gamma$ and we have re-arranged the inequality constraints by multiplying them through by the strictly positive value $\|\beta\|_2$. Instead of solving Problem (15.12), we solve Problem (15.13). If the maximum $z^\star$ of Problem (15.13) is strictly positive then the optimal margin is equal to $z^\star$ and is strictly positive.

15.4.3 Changes

We could consider a change in the problem due to the addition of an extra pattern.

15.4.4 Problem characteristics

15.4.4.1 Objective

The objective z of Problem (15.13) is linear.

15.4.4.2 Constraints

The inequality constraints in Problem (15.13) are non-linear. Each binding inequality constraint at a solution to the problem corresponds to a pattern that is closest to the hyperplane. These are called the **supporting patterns** [14].

As mentioned above, the constraint $\beta \neq \mathbf{0}$ in Problem (15.13) is not in the form of equality or inequality constraints. Moreover, the feasible set of Problem (15.13):

$$\mathbb{S} = \left\{ \begin{bmatrix} z \\ x \end{bmatrix} \in \mathbb{R}^{n+1} \middle| \varsigma(\ell)(\beta^\dagger \psi(\ell) + \gamma) \geq \|\beta\|_2 \, z, \forall \ell = 1, \ldots, r, \beta \neq \mathbf{0} \right\},$$

is not closed and may not be convex. As discussed in Section 2.3.3, feasible sets that are not closed can potentially present difficulties. We will consider further transformation of Problem (15.13) in Sections 18.4 and 20.1.

15.4.4.3 Solvability

If there is no hyperplane that can separate the patterns then Problem (15.13) is infeasible. (See Exercise 15.11.)

15.5 Sizing of interconnects in integrated circuits

This case study considers one aspect of integrated circuit design and assumes some familiarity with digital logic synthesis [95].

15.5.1 Motivation

15.5.1.1 Hierarchical design

The design of digital integrated circuits (ICs) is usually divided into a hierarchy of planning stages. For example, a specification of the functionality of the IC is translated into the logic required to meet the specification. The integrated components to implement the logic must then be laid out on the **floor-plan** of the chip [15, section 8.8.2]. Once the layout is done, there are still various decisions to be made. For example, the widths of the **interconnects** that join one gate to another can be adjusted, within limits, to achieve performance goals.

15.5.1.2 Delay constraints

In high-speed synchronous circuits, one goal is to make sure that the propagation delay on each path from the output of one latch through combinational logic to the input of the next latch is within a limit. Signals propagate between latches and logic on interconnects, which are metal or polysilicon "wires." Each interconnect introduces a delay that adds to the delay through the combinational logic.

Adjusting the width of the interconnects affects the delay. Qualitatively, increasing the width of the interconnect decreases the resistance and increases the capacitance of an interconnect. Decreasing resistance tends to reduce delay because the current from the driving latch or logic is increased, while increasing capacitance tends to increase delay because the increased capacitance requires more current to charge or discharge. If our goal were to *minimize* delay, then the optimal widths would depend on the source impedance of the gates driving the interconnect, the load impedance on the interconnect, the way in which the widths affect the resistance and capacitance, and, potentially, the way in which the load of the interconnect affects the delay through the combinational logic.

ICs are typically designed by dimensioning features to be an *integer* multiple of a length that represents the **minimum feature size** that can be fabricated on a particular fabrication line. It turns out that as the minimum feature size is reduced, the interconnect delay becomes a larger fraction of the overall circuit delay. In this case study we will concentrate on interconnect delay in specifying the functional form of the delay constraints, but recognize that the combinational logic contributes significantly to delay.

15.5.1.3 Area of layout

Another consideration besides delay is that the wider the interconnects, the more area may be required for the circuit. ICs are fabricated on silicon wafers by etching the circuit patterns. The larger the area per circuit, the smaller the number of ICs that can be produced from each wafer and, it turns out, the larger the proportion of those circuits that have defects. Bearing in mind that, for a given fabrication technique, both:

- the cost of production of the IC, and
- the probability of defects,

increase with the area of the chip, we will try to minimize chip area by adjusting the widths of the interconnects, while satisfying the delay constraints. We will assume that the interconnect area is the only issue that affects area; however, many other issues and decisions also affect area and other issues besides area and delay need to be considered, as will be discussed briefly in the next section.

15.5.1.4 Other issues

There are many other goals, such as minimizing power dissipation, and other constraints, such as guaranteeing noise immunity, that must be considered. A practical design must effect a compromise between several criteria that may not be easily comparable and the design must also respect many constraints. Our consideration of area and delay is merely one example of an objective and constraints and our focus is on just one of many phases of IC design. (See Exercise 15.12 for some examples of variations on the formulation.) In seeking a compromise between various goals, we are again seeking a **satisficing** solution [109].

15.5.1.5 Interaction between design levels

Hierarchical decomposition is used in a variety of applications. At each level of the hierarchy, we take as fixed the decisions made at higher levels and seek to optimize the remaining decisions. This approach was described in Section 3.3.5. In principle, the decisions made at lower levels must be "fed back" to the upper levels and we need to iterate between levels.

   In our application, for example, by adjusting the widths of interconnects, we may impact other parts of the design. This may necessitate changes to the layout, which will in turn change the optimal widths. However, we will not consider this issue in detail here and will assume that everything except the widths of the interconnects is fixed. The problem we develop is just a small part of the overall design process.

15.5.2 Formulation

15.5.2.1 Variables

Interconnect widths and lengths Consider Figure 15.6, which shows a schematic diagram of three latches and combinational logic joined by interconnect. The output of latch a at the left of Figure 15.6 drives gate b through a piece of interconnect, labeled 1. Gate b drives a branching interconnect, labeled 2, 3, 4, 5, and 6, which in turn drives two more gates, labeled c and d. These gates drive the pieces of interconnect labeled 7 and 8, which in turn drive latches e and f. The ends of the interconnect at the points where gates and latches are driven are called **sinks**.

Segments The interconnect can be thought of as consisting of **segments**, corresponding to the labeled pieces of interconnect shown in Figure 15.6. We assume that the interconnect can be partitioned into a set of n segments such that each segment has a uniform width along its length. Let the k-th segment have width x_k, thickness T_k, and length L_k, as illustrated in Figure 15.7. (Despite the appearance of Figure 15.7, $L_k \gg T_k > x_k$ in practice. That is, Figure 15.7 is not to scale.) An aluminum or copper segment is shown lying above a silicon dioxide insulating

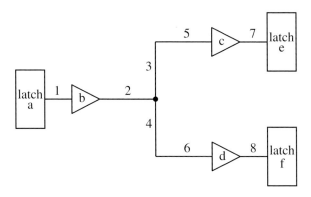

Fig. 15.6. Schematic diagram of gates and latches joined by interconnect.

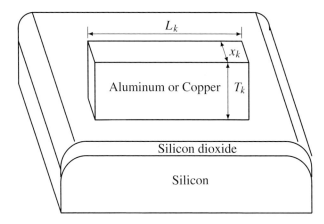

Fig. 15.7. Dimensions of the k-th segment of the interconnect. The figure is not to scale.

layer, which is in turn above a silicon substrate layer. The silicon substrate can be thought of as a "ground plane" at the datum voltage.

There are $n = 8$ segments, $k = 1, \ldots, 8$, shown in Figure 15.6. For a given layout and technology, we will assume that the length and thickness of each segment is a constant; however, we will assume that the width can be adjusted to achieve performance goals. We collect the variables x_k together into the n-vector x.

Discreteness Because we can only dimension features to be an integer multiple of the minimum feature size, x_k can only be chosen from a *discrete* set of alternatives. In a technology that can produce minimum features on the chip of size 65 nm, say, it might be the case that all widths must be an integer multiple of 65 nm. In general, optimizing over a discrete set of alternatives is much more difficult than optimizing over a continuous variable because in the discrete case we:

- cannot use calculus to derive optimality conditions,

- cannot obtain descent directions from purely local first derivative information, and

- cannot make use of convexity to establish global optimality.

If the discretization step of the discrete variable is "small," however, we can often obtain a good approximate answer by assuming that the widths are continuous variables. After solving the continuous problem, we must then convert each continuous width into a discrete value. A practical approach is to "round-off" to the nearest feasible discrete value, but this does not necessarily produce the best discrete alternative. (If the rounded-off solution is feasible we may, at least, be able to bound its sub-optimality.) In this case study, we will neglect discreteness and assume that the widths are continuously variable.

Alternative formulations Another formulation of this problem that avoids the issue of discreteness of widths is described in [25]. In that approach, instead of optimizing over a continuous range of widths x_k for segment k, we consider a finite collection of possible widths, say $\{W_{k1}, \ldots, W_{ks}\}$ for segment k. For example, these widths might correspond to the allowable integer multiples of the minimum feature size. A segment is then specified by a collection of sub-lengths L_{kj}, $j = 1, \ldots, s$ such that $\sum_{j=1}^{s} L_{kj} = L_k$. The value L_{kj} specifies how much of the total length of segment k is of width W_{kj}.

This alternative formulation avoids the discrete variables almost completely: it is still necessary to round off the sub-lengths; however, they are usually much longer than the minimum feature size so that rounding-off these lengths introduces negligible error. Although it seems that there are many more variables in this formulation, it can be the case that far less segments need be used to model the problem. This is an example of a radical transformation of a problem compared to its "natural" formulation. We will not pursue this formulation further; however, details of this formulation are presented in [25]. It is often worthwhile to "step back" from a problem and try to seek an alternative formulation.

15.5.2.2 Objective

We have indicated that our goal is to minimize the area of interconnect. The area $f : \mathbb{R}^n \to \mathbb{R}$ is defined by:

$$\forall x \in \mathbb{R}^n, \ f(x) = \sum_{k=1}^{n} L_k x_k,$$

where L_k is the length of the k-th segment. As remarked, this does not completely characterize the area, but can be a reasonable approximation to the dependence of the area on the interconnect widths. This objective is also related to the dissipation in charging and discharging the interconnect capacitance. (See Exercise 15.12.)

15.5.2.3 Constraints

Upper and lower bounds There are technical limits to how small or how large the interconnects can be. For example, we have indicated that a particular technology will have a minimum feature size. On the other hand, in a particular part of the chip there may only be limited space to increase the width of an interconnect. We can model this with upper and lower limits on the width:

$$\forall k = 1, \ldots, n, \underline{x}_k \leq x_k \leq \overline{x}_k,$$

where $\underline{x}_k$ is the minimum width, for example, the minimum feature size, and $\overline{x}_k$ is the maximum width allowable for the k-th segment.

Bottlenecks There may also be "bottlenecks" in the design; that is, parts of the chip where several segments run side-by-side and the sum of the widths of these segments cannot exceed a limit. This constraint might be expressed in the form:

$$\sum_{k \in \mathbb{B}} x_k \leq \overline{x}_{\mathbb{B}}, \tag{15.14}$$

where $\mathbb{B}$ is the set of segments involved in a particular bottleneck and $\overline{x}_{\mathbb{B}}$ is the maximum total width available for the segments in the set $\mathbb{B}$.

Delay constraints Consider a **path** from a latch through the combinational logic to the input of the next latch. We assume that the paths are labeled $\ell = 1, \ldots, r$. Our performance specification requires that, for each latch-to-latch path ℓ, a signal can propagate from:

- the output of the latch at the beginning of path ℓ,
- through the gates in path ℓ,
- to the input of the latch at the end of path ℓ,

within a maximum allowed time delay that depends on ([56, section 8.5]):

- the clock period,
- the delay from the **clock edge** to when the outputs of latches become valid, and
- the **set-up time** from when the input of latches become valid to the clock edge.

It is reasonable to suppose that the latch-to-latch delay on each path will depend on the widths of the segments (as well as on other fixed parameters.) Therefore, the delay on the ℓ-th path is a function $h_\ell : \mathbb{R}^n \to \mathbb{R}$ depending on the widths and we require that:

$$\forall \ell = 1, \ldots, r, \forall x \in \mathbb{R}^n, h_\ell(x) \leq \overline{h}_\ell, \tag{15.15}$$

where $\overline{h}_\ell$ is the maximum allowed latch-to-latch delay on path ℓ. We collect the delay functions for each path together into a vector function $h : \mathbb{R}^n \to \mathbb{R}^r$. Similarly, we collect the maximum allowed delays into a vector $\overline{h} \in \mathbb{R}^r$.

To evaluate the function h_ℓ we must define "delay" more carefully. Normatively, it is the time difference between:

(i) when the voltage at the output of the latch that is driving path ℓ can be considered to have changed state, and

(ii) when the voltage at the input of the latch that is driven by path ℓ can be considered to have changed state.

In practice, "changing state" is defined on an *ad hoc* basis as when, for example, the voltage waveform has risen to or fallen to within 50%, say, or 90%, say, of its final value.

Calculation of such a delay time is itself somewhat difficult for large circuits since it requires **transient simulation** of the circuit or some approximation to the transient simulation [95, chapters 4 and 5]. Moreover, coupling between nearby pieces of interconnect means that the delay depends partly on the coupling of signals from nearby interconnect. Consequently, the delay is often approximated by a function $\tilde{h}_\ell$ that is easier to calculate than h_ℓ. A typical approximation involves the sum of an approximation to the delays through the gates together with an approximation to the interconnect delays in the path.

We will approximate the gate delays by constants neglecting the effect of the load of the interconnect on the delay through the combinational logic. Because of this assumption, we can re-interpret h_ℓ as being the delay through the interconnect alone, neglecting the gate delays, and reduce the corresponding delay limit $\overline{h}_\ell$ by the sum of the gate delays on path ℓ. That is, we re-define each inequality in (15.15) by reducing the left-hand side and the right-hand side by the sum of the gate delays on path ℓ.

A typical approximation used for the interconnect delay is the **Elmore delay** [33]. The Elmore delay can be calculated from an electrical model of the interconnect. We will initially use the Elmore delay as an approximation to the 50% rise- and fall-times for the pieces of interconnect and assume that our delay constraints (15.15) are expressed in terms of 50% rise- and fall-times. In Section 20.2.2.3, we will also consider how to use more accurate delay models and how to combine analytical information from the Elmore approximation with numerical evaluation of the delay by a more accurate model.

Interconnect electrical model Figure 15.7 shows that the interconnect consists of aluminum or copper conductor separated from the silicon substrate by an insulating dielectric layer of silicon dioxide. As mentioned in Section 15.5.2.1, the substrate

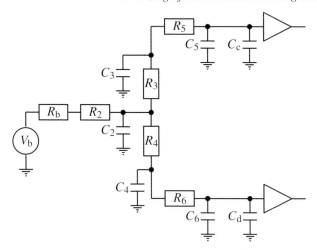

Fig. 15.8. The equivalent circuit of the interconnect between gate b and gates c and d consisting of resistive–capacitive L-segments.

can be considered to be at the voltage of the datum node. The separation of the conductor from the datum voltage means that there is a capacitance between the conductor and substrate. The finite conductivity of the aluminum or copper means that the segment is resistive. This means that each segment of the interconnect is a distributed resistive–capacitive transmission line. (Inductance can usually be ignored except at extremely high frequencies.)

If the segments are chosen to be short enough then their electrical characteristics are well approximated by a lumped resistive–capacitive model as shown in Figure 15.8 for the part of the original circuit shown in Figure 15.6 between gate b and gates c and d. Segment k, for $k = 2, \ldots, 6$, has been represented by a series resistance R_k and shunt capacitance C_k. This representation is called an **L-segment**. The "lower" end of each shunt capacitance is at the datum voltage represented by the "ground" symbol $\equiv$. There are various refinements of this model such as splitting the equivalent shunt capacitance between the two ends of the segment to form a π-equivalent circuit, as we used for modeling electric transmission lines in Section 6.2.2.3.

The resistance of segment k is determined by the **resistivity** ρ_k of the segment and its thickness, length, and width. In particular, resistance is proportional to length but inversely proportional to thickness and width, with the proportionality specified by ρ_k. That is:

$$\forall k = 1, \ldots, n, \ R_k = \rho_k L_k / (T_k x_k),$$
$$= \kappa_{Rk} / x_k, \tag{15.16}$$

where $\kappa_{Rk} = \rho_k L_k / T_k$ is a parameter that depends on the manufacturing process and the length and thickness of the segment.

The capacitance of segment k is determined approximately by the **sheet capaci-**

tance per unit area κ_{Sk}, its **fringing capacitance** per unit length κ_{Fk}, and its length and width. In particular, sheet capacitance increases with the length and width of the segment with constant of proportionality κ_{Sk} and the fringing capacitance increases with the length of the segment with constant of proportionality κ_{Fk}. (See Exercise 7.4.) That is:

$$
\begin{aligned}
\forall k = 1, \ldots, n, \; C_k &= \kappa_{Sk} L_k x_k + \kappa_{Fk} L_k, \\
&= \kappa_{Ck} x_k + C_{Fk},
\end{aligned} \tag{15.17}
$$

where $\kappa_{Ck} = \kappa_{Sk} L_k$ and $C_{Fk} = \kappa_{Fk} L_k$ are parameters that depend on the manufacturing process and the length of the segment.

Gate model We can model the gate driving the interconnect by considering its output transistor. It can be approximately represented by a voltage source driving a resistance. The voltage source can be approximately modeled as a step rise or fall in voltage. The time instant of the step change in voltage coincides with when the gate changes state. (As discussed above, we have incorporated the propagation delay *through* the gate by subtracting it from our delay limit.) The driving gate b is modeled in Figure 15.8 as the voltage source V_b and the driver resistance R_b. (In practice, accurate models of the output of the gate can be much more complicated than just a step voltage in series with constant resistance and models for the delay through the gate can also be more complicated.)

The load presented by **complementary metal-oxide semiconductor** (CMOS) gates at the sinks can be modeled by a capacitance [56, section 3.6][95, section 10.8]. This is shown by C_c and C_d in Figure 15.8 for the inputs to gates c and d, respectively. Other more detailed models may be necessary for other types of gates or for more detailed modeling of delay [95, chapters 10 and 11]. We assume that the capacitive load of a gate that is driven from, say, segment k is lumped into the capacitance C_{Fk}. For example, for segment 5, C_{F5} would include C_c, the input capacitance of gate c.

If we have the opportunity to change the size of the output transistors of the gates, then we can approximately model the gate output resistance as depending inversely on the size of the output transistor. In this case, we include the size of the transistor as a variable and can represent the resistance R_b similarly to (15.16).

Elmore delay Consider a constant voltage source charging a capacitor C through a resistance R. The voltage across the capacitor will exponentially approach the driving voltage. The time-constant of the exponential is RC, so that a reasonable order-of-magnitude estimate for the rise time of the voltage across the capacitor is RC. (In fact, RC is the time for the capacitor voltage to come within $1 - \exp(-1) \approx 0.63$ of the final voltage. See Exercise 15.13.)

For more complicated circuits, such as the multiple resistive–capacitive elements in our interconnect model, the response will be more complicated than a single exponential. However, we can still think of approximating the response by a single exponential. The "Elmore delay" is an estimate of the time constant of a single exponential that approximates the true response. We use this time constant as an estimate of the delay; however, under certain conditions it can be a poor estimate of the delay [50].

Given the lumped L-segment models, the Elmore delay is given by ([33]):

$$\forall \ell = 1, \dots, r, \forall x \in \mathbb{R}^n, \tilde{h}_\ell(x) = \sum_{\mathbb{J} \in \mathbb{P}_\ell} \sum_{j \in \mathbb{J}} \left[R_j \sum_{k \in \mathbb{D}(j)} C_k \right],$$

where:

- $\mathbb{P}_\ell$ is the set of sets of **connected segments** on path ℓ. Two segments are connected if there is a path of segments between them. In a set of connected segments, each pair of segments is connected. For example, for the path ℓ from latch a to latch e in Figure 15.6, we have that $\mathbb{P}_\ell = \{\{1\}, \{2, 3, 5\}, \{7\}\}$, since the path from latch a to latch e consists of three sets of connected segments, namely $\mathbb{J} = \{1\}, \{2, 3, 5\}, \{7\}$. The connected segments are separated by the latches b and c on the path from latch a to latch e.
- $\mathbb{D}(j)$ is the set of **downstream segments** including and between segment j and all sinks that are driven from segment j through connected segments. For example, in Figure 15.6, for $j = 2$, $\mathbb{D}(2) = \{2, 3, 4, 5, 6\}$. For $j = 3$, $\mathbb{D}(3) = \{3, 5\}$.

The Elmore delay is the sum of the resistive–capacitive time-constants of each segment, where:

- the resistive–capacitive time-constant of a segment is equal to the product of the resistance of the segment and all the capacitive load on it, and
- the capacitive load is defined to be the sum of the capacitances of all the connected downstream segments (including the input capacitance of all connected downstream gates and latches.)

Using the lumped resistive–capacitive model (15.16)–(15.17) for each segment, we obtain:

$$\forall \ell = 1, \dots, r, \forall x \in \mathbb{R}^n, \tilde{h}_\ell(x) = \sum_{\mathbb{J} \in \mathbb{P}_\ell} \sum_{j \in \mathbb{J}} \left[\frac{\kappa_{Rj}}{x_j} \sum_{k \in \mathbb{D}(j)} (\kappa_{Ck} x_k + C_{Fk}) \right]. \quad (15.18)$$

We can collect the Elmore delay functions for each path together into a vector function $\tilde{h} : \mathbb{R}^n \to \mathbb{R}^r$, which we use to approximate the actual delay function $h : \mathbb{R}^n \to \mathbb{R}^r$.

15.5.2.4 Problem

The approximate model for minimizing the area subject to the upper and lower constraints on the segment widths and subject to the delay constraints can be written as:

$$\min_{x \in \mathbb{R}^n} \{ f(x) | \tilde{h}(x) \leq \overline{h}, \underline{x} \leq x \leq \overline{x} \}. \tag{15.19}$$

The more accurate delay model is:

$$\min_{x \in \mathbb{R}^n} \{ f(x) | h(x) \leq \overline{h}, \underline{x} \leq x \leq \overline{x} \}, \tag{15.20}$$

where the delays h are calculated, for example, by transient circuit simulation. We will initially consider Problem (15.19), but will then consider the extension to Problem (15.20). Both problems have non-linear inequality constraints. Bottleneck constraints such as (15.14) can also be added to the formulation.

15.5.3 Changes

We could consider changes in parameters such as the sheet or fringe capacitance constants, due to a change in dielectric properties. We could also consider the effect of changing the allowed delays or adding an additional gate in a path. Typically, the addition of a **buffering** gate will reduce the interconnect delay but increase the delay through the combinational logic. The overall effect depends on the relative magnitude of the changes in the interconnect and combinational logic delay.

15.5.4 Problem characteristics

15.5.4.1 Objective

The objective, $f(x)$, of both Problems (15.19) and (15.20) is linear.

15.5.4.2 Constraints

Upper and lower bounds As shown in Exercise 15.1, Part (i), the lower and upper bound constraints $\underline{x} \leq x \leq \overline{x}$ define a convex set.

Delay constraints We focus on Problem (15.19). Consider the constraint function $\tilde{h} : \mathbb{R}^n \rightarrow \mathbb{R}^r$ and the set $\{x \in \mathbb{R}^n | \tilde{h}(x) \leq \overline{h} \}$. Recall that *if* the ℓ-th entry $\tilde{h}_\ell : \mathbb{R}^n \rightarrow \mathbb{R}$ of $\tilde{h}$ were a convex function on $\mathbb{R}^n$ then, by Exercise 2.34, the level sets of $\tilde{h}_\ell$ would be convex. In particular, for each ℓ, the set $\{x \in \mathbb{R}^n | \tilde{h}_\ell(x) \leq \overline{h}_\ell \}$ would be convex and so $\{x \in \mathbb{R}^n | \tilde{h}(x) \leq \overline{h} \}$ would also be a convex set. In summary, if each component function $\tilde{h}_\ell$ *were* a convex function, then $\{x \in \mathbb{R}^n | \tilde{h}(x) \leq \overline{h} \}$ *would be* a convex set and, therefore, $\mathbb{S} = \{x \in \mathbb{R}^n | \tilde{h}(x) \leq \overline{h}, \underline{x} \leq x \leq \overline{x} \}$, the feasible set for Problem (15.19), would be a convex set.

Unfortunately, as Exercise 15.14 shows, the Elmore delay functions $\tilde{h}_\ell$, $\ell = 1, \ldots, r$, are *not* convex functions. We will have to consider the feasible set more carefully to understand whether or not the problem has a unique local minimum.

The constraint functions $\tilde{h}_\ell$ involve the sum of terms each of which is a positive constant times the product of powers of the entries in the decision vector. (See Exercise 15.14.) That is, they are **posynomial functions**. (See Definition 3.1.) This observation will be the key to solving the gate interconnect sizing problem. Posynomial functions occur in many applications.

15.5.4.3 Solvability

If there is no selection of widths that yield delays satisfying the delay constraints, then there may be no feasible solution. This would indicate that some higher level design decisions must be revised before the design procedure can continue. For example, we may find that the delay on a particular path cannot be reduced enough to satisfy the delay constraint. We may need to insert a buffer to break a long segment into two shorter pieces.

15.6 Optimal power flow

In this case study we will consider the optimal power flow problem, which generalizes the least-cost production case study from Sections 12.1 and 15.1 and the power systems analysis case study from Section 6.2. There are many variations and extensions of the basic optimal power flow problem [20][79, chapter 11][123, chapter 13], but we will formulate the most straightforward combination of our previous case studies.

15.6.1 Motivation

15.6.1.1 Generalization of economic dispatch

When applied to electric power systems, the problems described in Sections 12.1 and 15.1 are called **economic dispatch problems**. The equality constraint (12.3) requires that electric generation equal the demand; however, this does not fully characterize the situation in an electricity network. For example, if generators are remote from demand centers then there will be losses incurred in moving power along transmission lines. At the least, (12.3) should be modified to account for losses in this case [8, section 11.10].

15.6.1.2 Constraints on operation

Transmission lines between generation and demand can also limit the feasible choices of generation. In particular, transmission lines have capacity limits on how much real and reactive power can be transmitted along them. If there is limited transmission capacity between a particular generator and the demand, then the feasible choices for generation are limited by the transmission line flow constraints [8, section 11.13]. Additionally, as discussed in the electric power system case study in Section 6.2.2.1, the voltage magnitudes at the buses must be maintained within 5% of nominal in typical systems. This also limits the feasible generation choices.

There may also be other constraints on operation due to emissions limits or fuel availability. See [79, section 11.V].

15.6.1.3 Power flow equations

To check whether or not the line flow and voltage constraints are satisfied, we must expand the detail of representation of the network by explicitly incorporating Kirchhoff's laws, as described in the electric power system case study in Section 6.2.2.4, or by at least incorporating some approximation to Kirchhoff's laws. That is, we should replace the single constraint (12.3) by the power flow equations or by an approximation to them. In this context, (12.3) is sometimes called a **surrogate constraint** because it acts in place of the full set of power flow equations.

15.6.1.4 Other controllable elements

Besides real power generations, we can also consider adjusting any controllable elements in the system so as to minimize costs and meet constraints. Typically, the reactive power generations of the generators and switchable capacitors and other controllable quantities are included in the optimal power flow problem. In fact, elements such as switched capacitors can only be switched in discrete amounts. We will ignore the discrete nature of the reactive power from switched capacitors and assume that they can be represented with continuous variables.

15.6.2 Formulation
15.6.2.1 Variables

In the decision vector, we need to represent:

- real and reactive power generations at the generators, which we will collect together into the vectors P and Q,
- any other controllable quantities in the system, such as the settings of **phase-shifting transformers** [8, section 5.9] and capacitors,
- the voltage magnitudes at every bus in the system, which we collect together into the vector u, and
- the voltage angles at every bus in the system except for the reference bus, which we collect together into the vector θ. (The voltage angle at the reference bus is constant since, as previously, it represents an arbitrary time reference.)

The choice of symbols for the vectors of real and reactive power violates our convention for using lower case letters for vectors; however, we have already used the lower case letters p and q for vector functions in the power flow case study in Section 6.2 and we will need to consider these functions again. Moreover, in that case study, P and Q represented parameters, namely, the specified real and reactive generations at the generator buses (and demands at the demand buses). However, here we have re-interpreted the real and reactive generations to be variables: we keep the same symbol, but re-interpret the meaning.

In this case study we will not treat any controllable quantities besides the real and reactive power generations; however, the formulation can be expanded to encompass other controllable quantities as well. We collect all the variables into the vector:

$$x = \begin{bmatrix} P \\ Q \\ u \\ \theta \end{bmatrix} \in \mathbb{R}^n.$$

In the power flow case study in Section 6.2, the generations at the generators were fixed parameters, except at the reference bus. In this case study, the real and reactive power generations at all generator buses are variables. This is similar to the least-cost production case studies of Sections 12.1 and 15.1, where the real power generations were variables.

In the case studies of Sections 12.1 and 15.1, the decision vector consisted entirely of the real power generations; however, in this case study, the real power generations is just a sub-vector of the decision vector. In the power flow case study in Section 6.2, the decision vector consisted entirely of the voltage magnitudes and angles; however, in this case study the voltage magnitudes and angles is just a sub-vector of the decision vector. This case study generalizes all of these earlier case studies and exemplifies the process of starting with only a few variables and many parameters and gradually re-interpreting the parameters to be variables.

15.6.2.2 Objective

As in the least-cost production case study, a typical objective for optimal power flow is to minimize the total cost of power generation. Let $f : \mathbb{R}^n \to \mathbb{R}$ represent this cost. Typically:

- f depends only on the entries of x corresponding to real power generations; however, in some formulations f also depends somewhat on the entries of x corresponding to reactive power generations, and
- f is separable since the decisions at one generator do not usually affect the costs at any other generators.

15.6.2.3 Equality constraints

The equality constraints for optimal power flow are Kirchhoff's laws expressed in terms of the voltage magnitudes and angles and the real and reactive generations. We considered these equations in detail in Section 6.2 and repeat them here for reference:

$$\forall \ell, \sum_{k \in \mathbb{J}(\ell) \cup \{\ell\}} u_\ell u_k [G_{\ell k} \cos(\theta_\ell - \theta_k) + B_{\ell k} \sin(\theta_\ell - \theta_k)] - P_\ell = 0,$$

$$\forall \ell, \sum_{k \in \mathbb{J}(\ell) \cup \{\ell\}} u_\ell u_k [G_{\ell k} \sin(\theta_\ell - \theta_k) - B_{\ell k} \cos(\theta_\ell - \theta_k)] - Q_\ell = 0,$$

where $\mathbb{J}(\ell)$ is the set of buses joined directly by a line to bus ℓ. These equations represent the fact that, by Kirchhoff's current law, the net real and reactive powers flowing out of a node into the rest of the system must be zero.

In Section 6.2, we expressed these equations in the form:

$$\forall \ell, \, p_\ell(x) = 0,$$
$$\forall \ell, \, q_\ell(x) = 0,$$

where $p_\ell : \mathbb{R}^n \to \mathbb{R}$ and $q_\ell : \mathbb{R}^n \to \mathbb{R}$ were defined in (6.12)–(6.13), which we repeat here for convenience:

$$\forall x \in \mathbb{R}^n, \, p_\ell(x) = \sum_{k \in \mathbb{J}(\ell) \cup \{\ell\}} u_\ell u_k [G_{\ell k} \cos(\theta_\ell - \theta_k) + B_{\ell k} \sin(\theta_\ell - \theta_k)] - P_\ell,$$

$$\forall x \in \mathbb{R}^n, \, q_\ell(x) = \sum_{k \in \mathbb{J}(\ell) \cup \{\ell\}} u_\ell u_k [G_{\ell k} \sin(\theta_\ell - \theta_k) - B_{\ell k} \cos(\theta_\ell - \theta_k)] - Q_\ell.$$

Unlike in Section 6.2.2.6, however, here we must explicitly include a pair of equations for the reference bus, since the objective will depend on generation at the reference bus. Moreover, here the vector x includes not only the voltage magnitudes and angles but also the real and reactive injections at the generator buses. As stated above, the vector x includes the voltage magnitude at the reference bus but does not include the voltage angle at the reference bus. Finally, we collect the equations together into a vector equation similar to the form of (6.14):

$$g(x) = \mathbf{0},$$

where a typical entry of g is of the form of (6.12) or (6.13), but the decision vector x includes the real and reactive generations as well as the voltage magnitudes and angles.

15.6.2.4 Inequality constraints

The inequality constraints include limits on the entries in x, which can be expressed in the form:

$$\underline{x} \leq x \leq \overline{x}.$$

For example, a voltage magnitude limit at bus ℓ could be $0.95 = \underline{u}_\ell \leq u_\ell \leq \overline{u}_\ell = 1.05$. A generator real power limit could be $0.15 = \underline{P}_\ell \leq P_\ell \leq \overline{P}_\ell = 0.7$.

There are also constraints involving functions of x. For example, there are typically angle difference constraints of the form:

$$\forall \ell, \forall k \in \mathbb{J}(\ell), \; -\pi/4 \leq \theta_\ell - \theta_k \leq \pi/4, \tag{15.21}$$

and there might be limits on angle differences between buses that are not joined directly by a line. In addition, transmission line flow constraints can be expressed via the power flow equations in terms of x. That is, we will also have functional constraints of the form:

$$\underline{h} \leq h(x) \leq \overline{h}.$$

A typical constraint might limit the flow on a line that joins bus ℓ to bus k. To understand this constraint, we consider the basic power flow equations. Recall that we derived the power flowing from a given bus into the network by summing the power flowing into the shunt element at node ℓ together with the sum of the powers flowing along each line connected to bus ℓ. This yielded the terms in (6.12)–(6.13).

If we just consider the terms in (6.12)–(6.13) corresponding to the power flowing along a particular line that joins bus ℓ to bus k, say, then we can evaluate the line flow on that line. In particular, if we neglect any shunt elements in the line models, then we can define the real and reactive flow functions for the line joining bus ℓ to bus k by selecting the ℓk terms in the equations (6.12)–(6.13) and noting that $G_{\ell\ell}$ and $B_{\ell\ell}$ are equal to minus the sum of the entries $G_{\ell k}$ and $B_{\ell k}$, respectively. That is, ignoring shunt elements in the line models, the line flow real and reactive power flow functions $p_{\ell k} : \mathbb{R}^n \to \mathbb{R}$ and $q_{\ell k} : \mathbb{R}^n \to \mathbb{R}$ are defined by:

$$\forall x \in \mathbb{R}^n, \; p_{\ell k}(x) = u_\ell u_k [G_{\ell k} \cos(\theta_\ell - \theta_k) + B_{\ell k} \sin(\theta_\ell - \theta_k)] - (u_\ell)^2 G_{\ell k},$$
$$\tag{15.22}$$
$$\forall x \in \mathbb{R}^n, \; q_{\ell k}(x) = u_\ell u_k [G_{\ell k} \sin(\theta_\ell - \theta_k) - B_{\ell k} \cos(\theta_\ell - \theta_k)] + (u_\ell)^2 B_{\ell k}.$$

(Notice that $p_{\ell k}$ and $q_{\ell k}$ in fact only depend on u_ℓ, u_k, θ_ℓ, and θ_k and not on the whole vector x. We will use this observation in Section 15.6.4.1.) If there is a real power flow limit of $\overline{p}_{\ell k}$ on the line joining bus ℓ and k then we represent this limit as an inequality constraint of the form $p_{\ell k}(x) \leq \overline{p}_{\ell k}$ in the inequality constraints $h(x) \leq \overline{h}$. We can also represent constraints on reactive power flow and on the magnitude of the **complex power flow**; that is, the square root of the sum of the squares of the real and the reactive power flows. If there are shunt elements in the line models then the functions $p_{\ell k}$ and $q_{\ell k}$ must be modified accordingly.

15.6.2.5 Problem

The optimal power flow problem is:

$$\min_{x \in \mathbb{R}^n}\{f(x)|g(x) = \mathbf{0}, \underline{x} \le x \le \overline{x}, \underline{h} \le h(x) \le \overline{h}\}. \tag{15.23}$$

The problem has non-linear equality and inequality constraints.

15.6.3 Changes in demand, lines, and generators

We can consider changes in demand at buses and also consider changes in the system. The two basic ways in which the system can change are:

- failure or return to service of a transmission line, and
- failure or return to service of a generator.

When a transmission line fails, power flows redistribute through the remaining lines. An important issue is whether the resulting redistributed flows are within the ratings of the remaining transmission lines. If the new flows exceed the ratings of lines, then **protection equipment** will eventually disconnect them from the system to prevent damage to them. Unfortunately, this will typically lead to yet larger flows on the remaining lines and further disconnections, which can rapidly "cascade" to a complete system **black-out**.

Similarly, if a generator fails, the remaining generators will typically make up the difference in power. However, if there is insufficient capacity to make up the difference then more generators may disconnect from the system. Again, this can cascade to a black-out.

In the case of both types of failures, it is important to be able to assess the effect of the failure. A system that is operated so that it can withstand any single failure is called **secure** and the additional constraints to ensure that any single failure does not result in overloads are called **security constraints** [123]. These constraints can, in principle, be included in the inequality constraints.

15.6.4 Problem characteristics

15.6.4.1 Convexity

Objective As argued in the least-cost production case study in Section 12.1, the objective of this problem is typically convex.

Equality constraints Because the function g is non-linear, the set $\{x \in \mathbb{R}^n | g(x) = 0\}$ is not generally convex. We can argue from two perspectives that this non-convexity does not necessarily create multiple local minima of the optimal power flow problem in practice.

First, following the discussion in Section 8.2.4, we observe that the Jacobian J of g can often be well approximated by a constant; that is, the equations are approximately linear. Since the equations are approximately linear, the feasible set $\{x \in \mathbb{R}^n | g(x) = 0\}$ is not very different from a set defined by a linear equality constraint. Consequently, the convex objective will typically have only a single local minimum on such a set. (In Exercise 17.16 we will consider a version of Problem (15.23) where the constraints are linearized.)

However, we can make a second, stronger argument, based on [24], by considering a relaxation of the feasible set as discussed in Section 3.3.4.1. We observe that if we can "throw away" real and reactive power, then we can replace the power flow equalities with inequalities. That is, we can consider relaxing the power flow equality constraints at each bus ℓ to:

$$p_\ell(x) \leq 0, \tag{15.24}$$
$$q_\ell(x) \leq 0. \tag{15.25}$$

That is, we have relaxed the constraints to requiring that the net power flowing out of a node is at most zero. That is, we allow net power to flow into a node from a line or generator and be "thrown away."

Consider solving the relaxed problem having inequality constraints as specified in (15.24) and (15.25) at each bus ℓ, but with all the other constraints as represented in Problem (15.23). That is, consider the following problem:

$$\min_{x \in \mathbb{R}^n}\{f(x) | g(x) \leq 0, \underline{x} \leq x \leq \overline{x}, \underline{h} \leq h(x) \leq \overline{h}\}. \tag{15.26}$$

In Problem (15.26), the feasible set $\overline{\mathbb{S}} = \{x \in \mathbb{R}^n | g(x) \leq 0, \underline{x} \leq x \leq \overline{x}, \underline{h} \leq h(x) \leq \overline{h}\}$ is a relaxed version of the feasible set of Problem (15.23): $\mathbb{S} = \{x \in \mathbb{R}^n | g(x) = 0, \underline{x} \leq x \leq \overline{x}, \underline{h} \leq h(x) \leq \overline{h}\}$. Suppose we obtain a solution $x^\star \in \overline{\mathbb{S}}$ to Problem (15.26) such that at bus ℓ we have $p_\ell(x^\star) < 0$ or $q_\ell(x^\star) < 0$. In this case, so long as we can dispose of real or reactive power at bus ℓ, then we can consider "throwing away" the difference and re-establishing equality. That is, we can construct a solution $x^{\star\star} \in \mathbb{S}$ to the original equality-constrained problem with the same value of objective and all constraints satisfied.

From a practical perspective, if there is a generator at ℓ then to "throw away" power at bus ℓ we can consider reducing the output of the generator to enable satisfaction of the constraint with equality. This would reduce the objective of the problem since costs typically increase with output. (In fact, electricity is generally not freely disposable and a generator at bus ℓ might be at its lower production limit. That is, we might have $P_\ell^\star = \underline{P}_\ell$. However, we will ignore this issue.) In summary, the inequality-constrained Problem (15.26) that we have described has essentially the same solution as Problem (15.23).

Now we will show that the feasible set defined by the relaxed constraints (15.24) is convex under the assumption that all voltage magnitudes are constant. We will not consider the reactive power constraints (15.25) nor the case where voltage magnitudes can vary, which can introduce multiple local optima. Recall that p_ℓ is defined in (6.12) to be:

$\forall x \in \mathbb{R}^n$,

$$
\begin{aligned}
p_\ell(x) &= \sum_{k \in \mathbb{J}(\ell) \cup \{\ell\}} u_\ell u_k [G_{\ell k} \cos(\theta_\ell - \theta_k) + B_{\ell k} \sin(\theta_\ell - \theta_k)] - P_\ell, \\
&= \sum_{k \in \mathbb{J}(\ell)} u_\ell u_k [G_{\ell k} \cos(\theta_\ell - \theta_k) + B_{\ell k} \sin(\theta_\ell - \theta_k)] + (u_\ell)^2 G_{\ell \ell} - P_\ell, \\
&= \sum_{k \in \mathbb{J}(\ell)} \{ u_\ell u_k [G_{\ell k} \cos(\theta_\ell - \theta_k) + B_{\ell k} \sin(\theta_\ell - \theta_k)] - (u_\ell)^2 G_{\ell k} \} \\
&\quad + (u_\ell)^2 \left(G_{\ell \ell} + \sum_{k \in \mathbb{J}(\ell)} G_{\ell k} \right) - P_\ell,
\end{aligned}
$$

$$\text{on adding and subtracting } (u_\ell)^2 \sum_{k \in \mathbb{J}(\ell)} G_{\ell k},$$

$$
= \sum_{k \in \mathbb{J}(\ell)} p_{\ell k}(x) + (u_\ell)^2 \left(G_{\ell \ell} + \sum_{k \in \mathbb{J}(\ell)} G_{\ell k} \right) - P_\ell,
$$

where, for each $k \in \mathbb{J}(\ell)$, the function $p_{\ell k} : \mathbb{R}^n \to \mathbb{R}$ was defined in (15.22). Moreover, given the assumption that all voltage magnitudes are constant, we can define functions $\hat{p}_{\ell k} : \mathbb{R} \to \mathbb{R}$ by:

$$\forall k \in \mathbb{J}(\ell), \forall \theta_{\ell k} \in \mathbb{R}, \hat{p}_{\ell k}(\theta_{\ell k}) = u_\ell u_k [G_{\ell k} \cos(\theta_{\ell k}) + B_{\ell k} \sin(\theta_{\ell k})] - (u_\ell)^2 G_{\ell k},$$

and we obtain that:

$$\forall \ell, \forall x \in \mathbb{R}^n, p_\ell(x) = \sum_{k \in \mathbb{J}(\ell)} \hat{p}_{\ell k}(\theta_\ell - \theta_k) + (u_\ell)^2 \left(G_{\ell \ell} + \sum_{k \in \mathbb{J}(\ell)} G_{\ell k} \right) - P_\ell.$$

That is, p_ℓ is equal to the sum of $\{(u_\ell)^2 (G_{\ell \ell} + \sum_{k \in \mathbb{J}(\ell)} G_{\ell k}) - P_\ell\}$ plus the sum of terms $\hat{p}_{\ell k}(\theta_\ell - \theta_k)$ each of which depends only on a linear function of two of the entries of x. We will find conditions for $\hat{p}_{\ell k}$ to be convex. By Exercises 2.30 and 3.22 these conditions guarantee that p_ℓ is convex.

We have that the second derivative of $\hat{p}_{\ell k}$ is:

$$\forall \theta_{\ell k} \in \mathbb{R}, \frac{d^2 \hat{p}_{\ell k}}{d\theta_{\ell k}^2}(\theta_{\ell k}) = -u_\ell u_k [G_{\ell k} \cos(\theta_{\ell k}) + B_{\ell k} \sin(\theta_{\ell k})].$$

Recalling that $G_{\ell k} < 0$, $B_{\ell k} > 0$ for $k \in \mathbb{J}(\ell)$, the condition for this term to be positive (and for $\hat{p}_{\ell k}$ to be convex) is that:

$$\forall k \in \mathbb{J}(\ell), |G_{\ell k}| \cos(\theta_{\ell k}) - |B_{\ell k}| \sin(\theta_{\ell k}) \geq 0.$$

This will be true if:

$$-\pi + \arctan\left(\frac{|G_{\ell k}|}{|B_{\ell k}|}\right) \leq \theta_\ell - \theta_k \leq \arctan\left(\frac{|G_{\ell k}|}{|B_{\ell k}|}\right).$$

Additionally incorporating the condition that arises from considering power balance at bus $k \in \mathbb{J}(\ell)$, we obtain that the functions will be convex if for each line joining a bus ℓ to a bus k we have:

$$|\theta_\ell - \theta_k| \leq \min\left\{\arctan\left(\frac{|G_{\ell k}|}{|B_{\ell k}|}\right), \pi - \arctan\left(\frac{|G_{\ell k}|}{|B_{\ell k}|}\right)\right\}. \tag{15.27}$$

(See Exercise 15.15.) Typically, $|G_{\ell k}|/|B_{\ell k}| \approx 0.1$ so this requires that $|\theta_\ell - \theta_k| \leq 0.1$ radian $\approx 6°$, which is somewhat more restrictive than the angle restrictions (15.21) that we previously mentioned for stability limits in Section 15.6.2.4. If (15.27) is satisfied for each ℓ and $k \in \mathbb{J}(\ell)$ (and no lower production limits are binding for generators) then the optimal power flow equality constraints are equivalent to inequality constraints that are specified by a convex inequality constraint function.

Inequality constraints Similarly, if a flow constraint between ℓ and k requires that $p_{\ell k}(x) \leq \overline{P}_{\ell k}$ then the constraint defines a convex set if (15.27) holds.

Discussion We have provided sufficient conditions under which the optimal power flow problem is convex. If these assumptions are violated then there may be multiple local minimizers.

15.6.4.2 Solvability

There are a variety of constraints in the optimal power flow problem and it is easily possible for there to be no solution. In fact, one application of optimal power flow software is to identify when the power system will be operating beyond its ratings. The results are used to inform preventive or remedial action such as switching another generator on to provide more power.

Exercises

Least-cost production with capacity constraints

15.1 In this exercise we consider the convexity of the least-cost production with capacity constraints problem.

(i) Show that the set $\{x \in \mathbb{R}^n | \underline{x} \le x \le \overline{x}\}$ is convex.
(ii) Show that the set $\mathbb{S} = \{x \in \mathbb{R}^n | D = \sum_{k=1}^n x_k, \underline{x} \le x \le \overline{x}\}$ is convex.
(iii) Show that if $\forall k = 1, \ldots, n$, $f_k : \mathbb{R} \to \mathbb{R}$ is convex on $[\underline{x}_k, \overline{x}_k]$ then $f : \mathbb{R}^n \to \mathbb{R}$ defined by $\forall x \in \mathbb{R}^n$, $f(x) = \sum_{k=1}^n f_k(x_k)$ is convex on $\{x \in \mathbb{R}^n | D = \sum_{k=1}^n x_k, \underline{x} \le x \le \overline{x}\}$.

15.2 Give an example with $n = 3$ of Problem (15.1) with no feasible points. (Hint: Modify the problem illustrated in Figure 15.1.)

15.3 Consider the generalization of Problem (15.1) where instead of each machine producing a single commodity, there are two types of commodities, commodity 1 and commodity 2. In particular, suppose that there are three machines, $k = 1, \ldots, 3$, and that the entries x_{2k-1} and x_{2k}, $k = 1, \ldots, 3$, of $x \in \mathbb{R}^6$, represent, respectively, the production by machine k of commodity 1 and commodity 2. Assume that each machine can produce both types of commodities within limits and that there is a convex function that expresses the cost of production in terms of the production of each commodity. That is, the cost of production of machine k is $f_k(x_{2k-1}, x_{2k})$, where $f_k : \mathbb{R}^2 \to \mathbb{R}$ is convex. Moreover, suppose that there is demand for both commodities. Formulate the least-cost production problem for these two commodities. That is, specify the following.

(i) Explicitly define the objective $f : \mathbb{R}^6 \to \mathbb{R}$.
(ii) Explicitly define the equality constraints in the form $Ax = b$.
(iii) Explicitly define the inequality constraints in the form $Cx \le d$.

Optimal routing in a data communications network

15.4 In this exercise we generalize the formulation of Problem (15.6) to that of a **flow control** problem. In Problem (15.6), we are given fixed rates of arrival for traffic between each origin–destination pair. Suppose instead that the rates can be controlled. Moreover, for each origin–destination pair (ℓ, ℓ'), there is a **customer utility** function $u_{(\ell, \ell')} : \mathbb{R} \to \mathbb{R}$ that denominates the value to the customer of receiving a traffic rate $b_{(\ell, \ell')}$ for data. Assume that the customer utilities can be summed across all customers to obtain a total customer utility and that the congestion measure is denominated in the same units. That is we assume that the customer utilities and the congestion measure are all commensurable. (See Section 2.3.1.) Formulate the problem of maximizing the total customer utility minus the congestion while satisfying the link capacity constraints. (See also Exercise 2.8.) That is, specify the following.

(i) Explicitly define the decision vector.
(ii) Explicitly define the objective.
(iii) Explicitly define the feasible set.

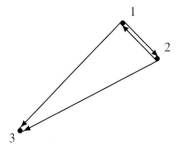

Fig. 15.9. The data com-
munications   network
with   three   nodes   and
four   directed   links   for
Exercise 15.5.

15.5 In this exercise, we generalize the formulation of Problem (15.6) to the case where
the links are not bi-directional. In particular, consider the nodes and *directed* links illustrated in Figure 15.9. There are three nodes and four directed links. (Note that there are
two directed links joining nodes 1 and 2, as indicated by the pair of arrows.) Each link
allows communication in one direction only as specified by the direction of the arrow and
as specified by the set $\mathbb{L}$ defined below. For this network, we are given that:

$$
\begin{aligned}
\text{Links: } \mathbb{L} &= \{(1,2), (2,1), (1,3), (2,3)\}, \\
\text{Link capacities: } \forall (i,j) \in \mathbb{L}, \overline{y}_{ij} &= 2, \\
\text{Origin–destination pairs: } \mathbb{W} &= \{(1,3), (2,3)\}, \\
\text{Paths for origin–destination pair } (1,3): \mathbb{P}_{(1,3)} &= \{1,2\}, \\
\text{Paths for origin–destination pair } (2,3): \mathbb{P}_{(2,3)} &= \{3,4\}, \\
\text{Flow for origin–destination pair } (1,3): b_{(1,3)} &= 1, \\
\text{Flow for origin–destination pair } (2,3): b_{(2,3)} &= 1.
\end{aligned}
$$

Moreover, the allowable paths are:

- path 1, consisting of link $(1,3)$, for origin–destination pair $(1,3)$,
- path 2, consisting of links $(1,2)$, $(2,3)$, for origin–destination pair $(1,3)$,
- path 3, consisting of links $(2,1)$ and $(1,3)$, for origin–destination pair $(2,3)$, and
- path 4, consisting of link $(2,3)$, for origin–destination pair $(2,3)$.

The congestion model is given by functions of the form $\phi_{ij} : [0, \overline{y}_{ij}) \to \mathbb{R}_+$ for each
$(i,j) \in \mathbb{L}$ and defined by:

$$
\forall y_{ij} \in [0, \overline{y}_{ij}), \phi_{ij}(y_{ij}) = \frac{y_{ij}}{\overline{y}_{ij} - y_{ij}}.
$$

Formulate the optimal directed routing problem. That is, specify the following.

   (i) Explicitly define the objective $f : \overline{\mathbb{S}} \to \mathbb{R}$, as defined in (15.7).
   (ii) Explicitly define the equality constraints $Ax = b$, as defined in (15.2).
   (iii) Explicitly define the inequality constraints $Cx < \overline{y}$, as defined in (15.5).

15.6 Show that the objective defined in (15.7) of the optimal routing Problem (15.6) is
convex on $\overline{\mathbb{S}} = \{x \in \mathbb{R}^n | x \geq 0, Cx < \overline{y}\}$. (Hint: See Exercises 2.30 and 3.22.)

Least absolute value estimation

15.7 In this exercise we consider the effect of outliers.

(i) Use the MATLAB function `lsqlin` to find the affine function:

$$\forall \psi \in \mathbb{R}, \zeta = \beta^{\star}\psi + \gamma^{\star},$$

with $\beta^{\star} \in \mathbb{R}$ and $\gamma^{\star} \in \mathbb{R}$ that best fits the following pairs of data $(\psi(\ell), \zeta(\ell))$, for $\ell = 1, \ldots, 7$. Assume that the measurements $\zeta(\ell)$ are subject to independent Gaussian errors of identical standard deviation and zero mean so that a least-squares fit yields the best fit.

ℓ	1	2	3	4	5	6	7
$\psi(\ell)$	0.27	0.2	0.8	0.4	0.2	0.7	0.5
$\zeta(\ell)$	0.3	0.65	0.75	0.4	0.15	1.45	0.5

(ii) Find the sensitivity of $\beta^{\star}$ and $\gamma^{\star}$ to each measurement $\zeta(\ell)$.
(iii) Repeat Part (i), but omit the data point $(\psi(6), \zeta(6))$. That is, find the best least-squares fit to the rest of the data besides $(\psi(6), \zeta(6))$.
(iv) Find the sensitivity of $\beta^{\star}$ and $\gamma^{\star}$ to each measurement $\zeta(\ell)$ when the outlier is omitted.
(v) The data in the table is the same as in Exercise 11.5, except that the data point $(\psi(6), \zeta(6))$ has been altered to equal $(0.7, 1.45)$. Compare the results of the previous parts to the solution of Exercise 11.5.

15.8 Show that Problems (15.8) and (15.10) are equivalent in the sense that the minima are the same and that to each minimizer $x^{\star}$ of Problem (15.8) there is a corresponding minimizer $(z^{\star}, x^{\star}, e^{\star})$ of Problem (15.10), and vice versa.

15.9 In this exercise, we consider the formulation of an estimation problem using the L_∞ norm instead of the L_1 norm. In particular, consider the objective $\phi : \mathbb{R}^n \to \mathbb{R}$ defined by:

$$\forall x \in \mathbb{R}^n, \phi(x) = \max_{\ell=1,\ldots,m} |A_\ell x - b_\ell|,$$

where $A \in \mathbb{R}^{m \times n}$, $b \in \mathbb{R}^m$, and A_ℓ is the ℓ-th row of A. We consider the unconstrained problem:

$$\min_{x \in \mathbb{R}^n} \phi(x).$$

We also consider the transformed problem:

$$\min_{z \in \mathbb{R}, x \in \mathbb{R}^n, e \in \mathbb{R}^m} \{z | Ax - b - e = 0, 1z \geq e, 1z \geq -e\}.$$

Show that these two problems are equivalent in the sense that the minima are the same and that to each minimizer $x^{\star}$ of the first problem there is a corresponding minimizer $(z^{\star}, x^{\star}, e^{\star})$ of the second problem, and vice versa.

Optimal margin pattern classification

15.10 Consider the set $\mathbb{P} \subset \mathbb{R}^n$ by:

$$\mathbb{P} = \left\{ \begin{bmatrix} \beta \\ \gamma \end{bmatrix} \in \mathbb{R}^n \middle| \beta \neq 0 \right\}.$$

(i) Show that $\mathbb{P}$ is not closed.
(ii) Show that $\mathbb{P}$ is not convex.

15.11 Show that if there is no hyperplane that can separate the patterns then Problem (15.13) is infeasible.

Sizing of interconnects in integrated circuits

15.12 Write down the formulations of the following problems.

(i) Find if a feasible solution exists that satisfies the delay constraints and the upper and lower bound constraints on interconnect width.
(ii) Minimize a linear combination of area and delay subject to the upper and lower bound constraints on interconnect width.
(iii) Minimize the power dissipation subject to the delay constraints and the upper and lower bound constraints on interconnect width. (Hint: Assume that the power dissipation is due to the interconnect capacitance being charged and discharged. Assume that each segment of interconnect is, on average, charged up every alternate clock cycle and that the clock frequency is fixed.)

15.13 Consider a voltage source that is initially at zero volts but steps to 1.0 volts at time $t = 0$. Suppose that the voltage source drives a capacitor C through a resistor R and that the capacitor initially has zero volts across it.

(i) Solve for the voltage across the capacitor.
(ii) Show that the time taken for the capacitor voltage to reach $(1 - \exp(-1))$ is equal to RC.

15.14 Consider the Elmore delay function $\tilde{h}_\ell : \mathbb{R}^n \to \mathbb{R}$ defined in (15.18) and repeated here for reference:

$$\forall x \in \mathbb{R}^n, \tilde{h}_\ell(x) = \sum_{\mathsf{J} \in \mathbb{P}_\ell} \sum_{j \in \mathsf{J}} \left[\frac{\kappa R_j}{x_j} \sum_{k \in \mathbb{D}(j)} (\kappa C_k x_k + C_{Fk}) \right].$$

(i) Show that the Elmore delay function is not in general a convex function on $\mathbb{R}^n_{++}$. (Hint: Consider an interconnect that consists of a single segment and take each constant in the Elmore delay function to be of value one. You might find that Exercise 3.32, Part (ii), helps in suggesting an approach.)
(ii) Show that the Elmore delay function is a posynomial function. (See Definition 3.1.)

Optimal power flow

15.15 Consider buses ℓ and k joined directly by a line. Prove that:

$$\left(|\theta_{\ell k}| \leq \min \left\{ \arctan \left(\frac{|G_{\ell k}|}{|B_{\ell k}|} \right), \pi - \arctan \left(\frac{|G_{\ell k}|}{|B_{\ell k}|} \right) \right\} \right) \;\Rightarrow\; \left(\frac{d^2 \hat{p}_{\ell k}}{d\theta_{\ell k}{}^2} (\theta_{\ell k}) > 0 \right),$$

$$\left(|\theta_{k\ell}| \leq \min \left\{ \arctan \left(\frac{|G_{\ell k}|}{|B_{\ell k}|} \right), \pi - \arctan \left(\frac{|G_{\ell k}|}{|B_{\ell k}|} \right) \right\} \right) \;\Rightarrow\; \left(\frac{d^2 \hat{p}_{k\ell}}{d\theta_{k\ell}{}^2} (\theta_{k\ell}) > 0 \right),$$

where $\dfrac{d^2 \hat{p}_{\ell k}}{d\theta_{\ell k}{}^2} : \mathbb{R} \to \mathbb{R}$ and $\dfrac{d^2 \hat{p}_{k\ell}}{d\theta_{k\ell}{}^2} : \mathbb{R} \to \mathbb{R}$ are defined by:

$$\forall \theta_{\ell k} \in \mathbb{R}, \frac{d^2 \hat{p}_{\ell k}}{d\theta_{\ell k}{}^2} (\theta_{\ell k}) = -u_\ell u_k [G_{\ell k} \cos(\theta_{\ell k}) + B_{\ell k} \sin(\theta_{\ell k})],$$

$$\forall \theta_{k\ell} \in \mathbb{R}, \frac{d^2 \hat{p}_{k\ell}}{d\theta_{k\ell}{}^2} (\theta_{k\ell}) = -u_k u_\ell [G_{\ell k} \cos(\theta_{k\ell}) + B_{\ell k} \sin(\theta_{k\ell})].$$

16

Algorithms for non-negatively constrained minimization

In this chapter we will develop algorithms for constrained optimization problems of the form:

$$\min_{x \in \mathbb{S}} f(x),$$

where $f : \mathbb{R}^n \to \mathbb{R}$ and where the feasible set $\mathbb{S}$ is of the form:

$$\mathbb{S} = \{x \in \mathbb{R}^n | g(x) = \mathbf{0}, x \geq \mathbf{0}\},$$

with $g : \mathbb{R}^n \to \mathbb{R}^m$ affine. That is, we will consider problems of the form:

$$\min_{x \in \mathbb{R}^n} \{f(x) | Ax = b, x \geq \mathbf{0}\}, \tag{16.1}$$

where $A \in \mathbb{R}^{m \times n}$ and $b \in \mathbb{R}^m$ are constants. In Problem (16.1), the only inequality constraints are the **non-negativity constraints** $x \geq \mathbf{0}$. These require the decision vector x to lie in the **non-negative orthant** $\mathbb{R}^n_+$. (See Definition A.5.) We refer to Problem (16.1) as a non-negatively constrained problem, where it is understood that it also includes equality constraints in addition to the non-negativity constraints. As mentioned in Section 2.3.2.3, the form of constraints is also referred to as the **standard format**, particularly in the context of linear programming [45, section 5.6.1][84, section 4.2].

The feasible set defined by the linear equality and non-negativity constraints is convex. (See Exercise 2.36.) If f is convex on the feasible set then the problem is convex.

We will develop optimality conditions for Problem (16.1) in Section 16.1, specializing in Section 16.2 to convex problems. The optimality conditions will help us to develop algorithms for non-negatively constrained minimization. In Sections 16.3 and 16.4, we will describe two qualitatively different approaches to finding the minimizer of a non-negatively constrained problem of the form of Problem (16.1).

The key issues in this chapter are:

- optimality conditions for **non-negatively constrained problems** based on the results for equality-constrained problems,
- the **complementary slackness conditions** in the optimality conditions,
- optimality conditions for **convex problems**, and
- **active set** and **interior point** algorithms to seek solutions.

16.1 Optimality conditions

In the following sections we first present first-order necessary conditions and then second-order sufficient conditions for optimality of Problem (16.1).

16.1.1 First-order necessary conditions

In this section we present **first-order necessary conditions** (or **FONC.**)

16.1.1.1 Analysis

We have the following.

Theorem 16.1 *Let* $f : \mathbb{R}^n \to \mathbb{R}$ *be partially differentiable with continuous partial derivatives,* $A \in \mathbb{R}^{m \times n}$, *and* $b \in \mathbb{R}^m$. *Consider Problem (16.1),*

$$\min_{x \in \mathbb{R}^n} \{ f(x) | Ax = b, x \geq 0 \},$$

and a point $x^\star \in \mathbb{R}^n$. *If* $x^\star$ *is a local minimizer of Problem (16.1) then:*

$$
\begin{aligned}
\exists \lambda^\star \in \mathbb{R}^m, \exists \mu^\star \in \mathbb{R}^n \text{ such that: } \nabla f(x^\star) + A^\dagger \lambda^\star - \mu^\star &= 0; \\
M^\star x^\star &= 0; \\
Ax^\star &= b; \\
x^\star &\geq 0; \text{ and} \\
\mu^\star &\geq 0, \qquad (16.2)
\end{aligned}
$$

where $M^\star = \mathrm{diag}\{\mu_\ell^\star\} \in \mathbb{R}^{n \times n}$ *is a diagonal matrix with diagonal entries equal to* $\mu_\ell^\star$, $\ell = 1, \ldots, n$. *The vectors* $\lambda^\star$ *and* $\mu^\star$ *satisfying the conditions (16.2) are called the vectors of Lagrange multipliers for the constraints* $Ax = b$ *and* $x \geq 0$, *respectively. The conditions that* $M^\star x^\star = 0$ *are called the* **complementary slackness conditions.** *The complementary slackness conditions together with the conditions* $x^\star \geq 0$ *and* $\mu^\star \geq 0$ *imply that, for each* ℓ, *either the* ℓ-*th non-negativity constraint* $x_\ell \geq 0$ *is binding or the* ℓ-*th Lagrange multiplier* $\mu_\ell^\star$ *is equal to zero (or both).*

Proof ([84, section 14.4].) This is a special case of Theorem 17.1 to be presented in Chapter 17. We will only sketch the proof of this special case.

Consider the *equality*-constrained problem:

$$\min_{x \in \mathbb{R}^n} \{f(x) | Ax = b, -x_\ell = 0, \forall \ell \in \mathbb{A}(x^\star)\}, \tag{16.3}$$

where $\mathbb{A}(x^\star) = \{\ell \in \{1, \ldots, n\} | x_\ell^\star = 0\}$ is the active set corresponding to the non-negativity constraints $x \geq \mathbf{0}$ for the point $x^\star$. That is, the equality-constrained Problem (16.3) includes as equality constraints the following:

- all of the equality constraints from Problem (16.1), and
- all of the non-negativity constraints of Problem (16.1) that were satisfied with equality by $x^\star$.

That is, the *active* non-negativity constraints from Problem (16.1) at its minimizer $x^\star$ have been included as *equality* constraints in Problem (16.3). The representation of the constraint as $-x_\ell = 0$ rather than as $x_\ell = 0$ is for convenience in interpreting the Lagrange multipliers for equality-constrained Problem (16.3).

The proof involves applying our earlier results for *equality*-constrained problems to Problem (16.3). The proof is divided into three parts:

(i) showing that $x^\star$ is a local minimizer of equality-constrained Problem (16.3),

(ii) using the necessary conditions of the equality-constrained Problem (16.3) to define $\lambda^\star$ and $\mu^\star$ that satisfy the first four lines of (16.2), and

(iii) proving that $\mu^\star \geq \mathbf{0}$ by showing that if a particular Lagrange multiplier were negative, say $\mu_\ell^\star < 0$, then the objective could be reduced by moving in a direction such that x_ℓ increases and so becomes strictly feasible for the constraint $x_\ell \geq 0$. The intuition behind this observation is that if the second-order sufficient conditions held for Problem (16.3) at $x^\star$ then we could apply the sensitivity analysis Corollary 13.11. If we consider changing the constraint from $-x_\ell = 0$ to $-x_\ell = -\gamma$, with $\gamma > 0$, then, if $\mu_\ell^\star < 0$, Corollary 13.11 indicates that the minimum of the changed problem would be lower and x_ℓ would be strictly positive. (See Exercise 16.2.) This means that the constraint $x_\ell \geq 0$ could not have been binding at a minimizer of Problem (16.1) since a strictly positive value of x_ℓ would reduce the objective.

☐

16.1.1.2 Example

Consider Problem (2.15), which we first saw in Section 2.3.2.3:

$$\min_{x \in \mathbb{R}^2} \{x_1 - x_2 | x_1 + x_2 = 1, x_1 \geq 0, x_2 \geq 0\}. \tag{16.4}$$

The feasible set for this problem is shown in Figure 16.1 as a line interval. Consideration of the objective and inspection of Figure 16.1 shows that $x^\star = \begin{bmatrix} 0 \\ 1 \end{bmatrix}$ is the unique minimizer of Problem (16.4), as illustrated in Figure 16.1.

x_2

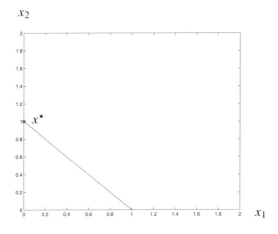

Fig. 16.1. The feasible set (shown as line) and the minimizer $x^\star$ (shown as •) for example problem.

We apply Theorem 16.1 to this non-negatively constrained problem. The objective is linear, and hence partially differentiable with continuous partial derivatives. We have:

$$\forall x \in \mathbb{R}^2,\ f(x) = x_1 - x_2,$$

$$\forall x \in \mathbb{R}^2,\ \nabla f(x) = \begin{bmatrix} 1 \\ -1 \end{bmatrix},$$

$$A = \begin{bmatrix} 1 & 1 \end{bmatrix},$$

$$= \mathbf{1}^\dagger,$$

$$b = [1],$$

where $\mathbf{1}$ is the two-vector of all ones. We claim that $\mu^\star = \begin{bmatrix} 2 \\ 0 \end{bmatrix}$ and $\lambda^\star = [1]$ satisfy (16.2). For:

$$\nabla f(x^\star) + A^\dagger \lambda^\star - \mu^\star = \begin{bmatrix} 1 \\ -1 \end{bmatrix} + \begin{bmatrix} 1 \\ 1 \end{bmatrix} [1] - \begin{bmatrix} 2 \\ 0 \end{bmatrix},$$

$$= \mathbf{0};$$

$$M^\star x^\star = \begin{bmatrix} 2 & 0 \\ 0 & 0 \end{bmatrix} \begin{bmatrix} 0 \\ 1 \end{bmatrix},$$

$$= \mathbf{0};$$

$$Ax^\star = \begin{bmatrix} 1 & 1 \end{bmatrix} \begin{bmatrix} 0 \\ 1 \end{bmatrix},$$

$$= [1],$$

$$= b;$$

$$x^\star = \begin{bmatrix} 0 \\ 1 \end{bmatrix},$$

$$\geq \mathbf{0}; \text{ and}$$

$$\mu^\star = \begin{bmatrix} 2 \\ 0 \end{bmatrix},$$

$$\geq \mathbf{0}.$$

These results concur with Theorem 16.1.

16.1.1.3 Discussion

As in the equality-constrained case, the Lagrange multipliers adjust the unconstrained optimality conditions to balance the constraints against the objective. For the non-negativity constraints the balance is only needed if the objective would encourage the non-negativity constraints to be violated. Consequently, the Lagrange multipliers on the non-negativity constraints are themselves non-negative.

As in the equality-constrained case, we will refer to the equality and inequality constraints specified in (16.2) as *the* first-order necessary conditions, although we recognize that the first-order necessary conditions also include, strictly speaking, the other items in the hypothesis of Theorem 16.1. These conditions are also known as the **Kuhn–Tucker** (KT) or the **Karush–Kuhn–Tucker** (KKT) conditions and a point satisfying the conditions is called a **KKT point**, after the discoverers of these necessary conditions.

As in the linear equality-constrained case, if A does not have linearly independent rows then the Lagrange multipliers may not be unique. Unlike the linear equality-constrained case where the first-order necessary conditions involve only simultaneous equations, we now have inequality constraints on both the minimizer $x^\star$ and on the Lagrange multipliers $\mu^\star$ in the first-order conditions for inequality constraints.

The complementary slackness conditions require that $M^\star x^\star = \mathbf{0}$. For each entry ℓ, this requires that either $\mu_\ell = 0$ or $x_\ell = 0$. That is, for each entry ℓ, the ordered pair $\begin{bmatrix} \mu_\ell \\ x_\ell \end{bmatrix} \in \mathbb{R}^2$ must lie either on the x_ℓ-axis or on the μ_ℓ-axis. (The inequality constraints $x_\ell \geq 0$ and $\mu_\ell \geq 0$ then restrict the ordered pair further to lying either on the non-negative x_ℓ-axis or on the non-negative μ_ℓ-axis.)

It might seem that a straightforward approach to solving for $x \in \mathbb{R}^n$, $\mu \in \mathbb{R}^n$ that satisfy these conditions for each entry ℓ would be to use the Newton–Raphson method. That is, we might think of linearizing the equation $Mx = \mathbf{0}$ about the current values of μ and x at a particular iteration ν and using the linearized equations to construct an update. However this approach is not effective unless we are careful to avoid the boundary of the set defined by $\mu \geq \mathbf{0}$ and $x \geq \mathbf{0}$.

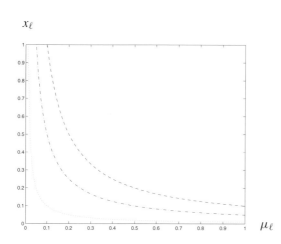

x_ℓ

Fig. 16.2. The complementary slackness condition for the entry ℓ requires that the point $\begin{bmatrix} \mu_\ell \\ x_\ell \end{bmatrix} \in \mathbb{R}^2$ lie either on the μ_ℓ-axis or on the x_ℓ-axis. The hyperbola $\mu_\ell x_\ell = t$ approximates the set of points satisfying the complementary slackness constraints. The dashed curve shows the hyperbola for $t = 0.1$; the dash-dotted curve shows the hyperbola for $t = 0.05$; and the dotted curve shows the hyperbola for $t = 0.01$.

To understand the pitfalls of applying the Newton–Raphson method to seek solutions of the complementary slackness conditions $Mx = \mathbf{0}$, suppose that at iteration ν we had $x_\ell^{(\nu)} = 0$. That is, $x^{(\nu)}$ is on the boundary of $x \geq \mathbf{0}$. In this case, for the particular entry ℓ, linearizing the complementary slackness conditions involves linearizing $\mu_\ell x_\ell$ about $\mu_\ell^{(\nu)}$ and $x_\ell^{(\nu)}$. We obtain:

$$(\mu_\ell^{(\nu)} + \Delta\mu_\ell^{(\nu)})(x_\ell^{(\nu)} + \Delta x_\ell^{(\nu)}) \approx \mu_\ell^{(\nu)} x_\ell^{(\nu)} + x_\ell^{(\nu)} \Delta\mu_\ell^{(\nu)} + \mu_\ell^{(\nu)} \Delta x_\ell^{(\nu)},$$
$$= \mu_\ell^{(\nu)} \Delta x_\ell^{(\nu)},$$

since $x_\ell^{(\nu)} = 0$. Setting this equal to zero yields $\Delta x_\ell^{(\nu)} = 0$. That is, if we ever were at an iterate for which $x_\ell^{(\nu)} = 0$ then the Newton–Raphson update would prevent us from ever moving from this value. Similarly, if $\mu_\ell^{(\nu)} = 0$ at some iteration then linearization of the complementary slackness conditions would prevent any changes in μ_ℓ. Linearizing the complementary slackness constraint does not yield a useful approximation in these cases.

We will see in Section 16.4.3.3 that an effective linearization of this constraint requires us to carefully avoid the possibilities that $\mu_\ell^{(\nu)} = 0$ or $x_\ell^{(\nu)} = 0$. We will see that one way to do this is to *first* approximate the constraint $\mu_\ell x_\ell = 0$ by a hyperbola $\mu_\ell x_\ell = t$, where $t \in \mathbb{R}_{++}$, and then linearize the hyperbolic approximation. Then, we gradually reduce t. Figure 16.2 shows a hyperbolic approximation to the set of points satisfying the complementary slackness constraint for several values of t. As t is reduced, the set of ordered pairs $\begin{bmatrix} \mu_\ell \\ x_\ell \end{bmatrix}$ satisfying $\mu_\ell x_\ell = t$, $\mu_\ell \geq 0$, and $x_\ell \geq 0$ approaches the union of the non-negative μ_ℓ-axis and the non-negative x_ℓ-axis.

16.1.2 Second-order sufficient conditions

In this section we present **second-order sufficient conditions** (or **SOSC**) for Problem (16.1).

16.1.2.1 Analysis

We have:

Theorem 16.2 *Let $f : \mathbb{R}^n \to \mathbb{R}$ be twice partially differentiable with continuous second partial derivatives, $A \in \mathbb{R}^{m \times n}$, and $b \in \mathbb{R}^m$. Consider Problem (16.1),*

$$\min_{x \in \mathbb{R}^n} \{ f(x) | Ax = b, x \geq 0 \},$$

and points $x^\star \in \mathbb{R}^n$, $\lambda^\star \in \mathbb{R}^m$, and $\mu^\star \in \mathbb{R}^n$. Let $M^\star = \mathrm{diag}\{\mu_\ell^\star\}$. Suppose that:

$$
\begin{aligned}
\nabla f(x^\star) + A^\dagger \lambda^\star - \mu^\star &= 0, \\
M^\star x^\star &= 0, \\
Ax^\star &= b, \\
x^\star &\geq 0, \\
\mu^\star &\geq 0, \text{ and}
\end{aligned}
$$

$\nabla^2 f(x^\star)$ *is positive definite on the null space:*

$$\mathcal{N}_+ = \{ \Delta x \in \mathbb{R}^n | A \Delta x = 0; \Delta x_\ell = 0, \forall \ell \in \mathbb{A}_+(x^\star, \mu^\star) \},$$

where $\mathbb{A}_+(x^\star, \mu^\star) = \{ \ell \in \{1, \dots, n\} | x_\ell^\star = 0, \mu_\ell^\star > 0 \}$. Then $x^\star$ is a strict local minimizer of Problem (16.1).

Proof See [45, section 3.3.2]. □

The conditions in the theorem are called the **second-order sufficient conditions** (or **SOSC**.) In addition to the first-order necessary conditions, the second-order sufficient conditions require that:

- f is twice partially differentiable with continuous second partial derivatives, and
- $\nabla^2 f(x^\star)$ is positive definite on the null space $\mathcal{N}_+$ defined in the theorem.

16.1.2.2 Example

Consider the objective $f : \mathbb{R}^2 \to \mathbb{R}$ defined by:

$$\forall x \in \mathbb{R}^2, \; f(x) = (x_1)^2 + (x_2 - 1)^2,$$

and suppose that we do not have any equality constraints. That is, consider the problem:

$$\min_{x \in \mathbb{R}^2} \{ f(x) | x \geq 0 \}.$$

x_2

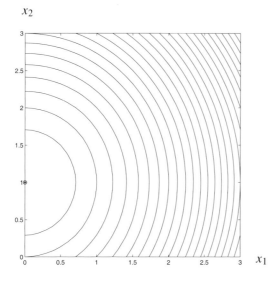

Fig. 16.3. Contour sets of objective function defined in Section 16.1.2.2. The heights of the contours decrease towards the point $x^\star = \begin{bmatrix} 0 \\ 1 \end{bmatrix}$, which is illustrated as a •.

The objective is twice partially differentiable with continuous second partial derivatives. The contour sets of this objective are shown in Figure 16.3.

We claim that the second-order sufficient conditions hold for $x^\star = \begin{bmatrix} 0 \\ 1 \end{bmatrix}$ and $\mu^\star = \mathbf{0}$. The point $x^\star$ is illustrated as a • in Figure 16.3. For this example, the second-order sufficient conditions are that:

$$
\begin{aligned}
\nabla f(x^\star) - \mu^\star &= \begin{bmatrix} 2x_1^\star \\ 2(x_2^\star - 1) \end{bmatrix} - \mu^\star, \\
&= \mathbf{0}, \\
M^\star x^\star &= \mathbf{0}, \\
x^\star &\geq \mathbf{0}, \\
\mu^\star &\geq \mathbf{0},
\end{aligned}
$$

and that $\nabla^2 f(x^\star)$ is positive definite on the null space:

$$
\mathcal{N}_+ = \{\Delta x \in \mathbb{R}^2 | \Delta x_\ell = 0, \forall \ell \in \mathbb{A}_+(x^\star, \mu^\star)\}.
$$

We note that:

$$
\begin{aligned}
\mathbb{A}(x^\star) &= \{1\}, \\
\mathbb{A}_+(x^\star, \mu^\star) &= \{\ell \in \{1, 2\} | x_\ell^\star = 0, \mu_\ell^\star > 0\}, \\
&= \emptyset,
\end{aligned}
$$

$$\mathcal{N}_+ \;=\; \{\Delta x \in \mathbb{R}^2 | \Delta x_\ell = 0, \forall \ell \in \mathbb{A}_+(x^\star, \mu^\star)\},$$
$$=\; \{\Delta x \in \mathbb{R}^2 | \Delta x_\ell = 0, \forall \ell \in \emptyset\},$$
$$=\; \mathbb{R}^2,$$
$$\nabla^2 f(x^\star) \;=\; 2\mathbf{I},$$

which is positive definite on $\mathcal{N}_+ = \mathbb{R}^2$. The second-order sufficient conditions hold at $x^\star = \begin{bmatrix} 0 \\ 1 \end{bmatrix}$ and $\mu^\star = \mathbf{0}$. Note that $\mathbb{A}_+(x^\star, \mu^\star)$ is a strict subset of $\mathbb{A}(x^\star)$ for this example.

16.1.2.3 Discussion

The example in Section 16.1.2.2 shows that the set $\mathbb{A}_+(x^\star, \mu^\star)$ can be a *strict* subset of $\mathbb{A}(x^\star)$, since, compared to $\mathbb{A}(x^\star)$, the set $\mathbb{A}_+(x^\star, \mu^\star)$ omits those constraints ℓ for which $x_\ell^\star = 0$ and $\mu_\ell^\star = 0$. Therefore, the null space specified in Theorem 16.2:

$$\mathcal{N}_+ = \{\Delta x \in \mathbb{R}^n | A \Delta x = \mathbf{0}, \Delta x_\ell = 0, \forall \ell \in \mathbb{A}_+(x^\star, \mu^\star)\},$$

can strictly contain the null space corresponding to the equality constraints and the active inequality constraints. That is, $\mathcal{N}_+$ can strictly contain the null space:

$$\mathcal{N} = \{\Delta x \in \mathbb{R}^n | A \Delta x = \mathbf{0}, \Delta x_\ell = 0, \forall \ell \in \mathbb{A}(x^\star)\},$$

corresponding to the constraints of equality-constrained Problem (16.3), which we repeat here:

$$Ax \;=\; b,$$
$$-x_\ell \;=\; 0, \forall \ell \in \mathbb{A}(x^\star).$$

By Corollary 13.4 of Chapter 13, satisfaction by $x^\star$ of the first-order necessary conditions for equality-constrained Problem (16.3), together with positive definiteness of $\nabla^2 f(x^\star)$ on the null space $\mathcal{N}$, guarantees that $x^\star$ is a strict local minimizer of equality-constrained Problem (16.3). However, this is *insufficient* to guarantee that $x^\star$ is a strict local minimizer of the corresponding inequality-constrained Problem (16.1) if there are any constraints ℓ for which both $x_\ell^\star = 0$ and $\mu_\ell^\star = 0$.

Constraints for which $x_\ell^\star = 0$ and $\mu_\ell^\star = 0$ are called **degenerate constraints**. Intuitively, a degenerate constraint ℓ is only "just" binding. The sensitivity of the objective to changes in x_ℓ is zero. (See Exercise 16.2.) Moreover, there exist feasible movements Δx away from $x^\star$, namely those in which $\Delta x_\ell > 0$, for which the constraint $x_\ell \geq 0$ is no longer binding. Such feasible movements do not satisfy $\Delta x_\ell = 0$. Therefore, to guarantee that we are at a minimizer of Problem (16.1) we must test for positive definiteness of the objective in the larger subspace that allows movement in directions Δx such that $\Delta x_\ell > 0$. If the Hessian is positive definite in these directions then the objective must increase in these directions as

x_2

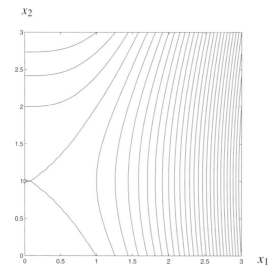

Fig. 16.4. Contour sets of objective function defined in Section 16.1.2.4. The heights of the contours decrease away from the point $\hat{x} = \begin{bmatrix} 0 \\ 1 \end{bmatrix}$, which is illustrated as a ∘, in the direction of increasing values of x_1. The heights of the contours increase away from the point $\hat{x}$ in the direction of increasing or decreasing values of x_2.

we move away from $x^\star$ and consequently we are indeed at a local minimizer of Problem (16.1). That is, if $\nabla^2 f(x^\star)$ is positive definite on $\mathcal{N}_+$ then there can be no feasible descent directions for f at $x^\star$.

In the example in Section 16.1.2.2, the binding constraint was degenerate and we had to check for positive definiteness of the Hessian of the objective on the whole of $\mathbb{R}^2$. In general, we might expect that some constraints might be degenerate and some might not be degenerate so that $\mathcal{N}_+$ might not be the whole of $\mathbb{R}^n$. (See Exercise 16.3.) However, it is very important to check for positive definiteness on the set $\mathcal{N}_+$ and not just on the set $\mathcal{N}$. In the following section, we present another example that illustrates this issue.

16.1.2.4 Example of not satisfying second-order sufficient conditions

For example, suppose that we have the objective $f : \mathbb{R}^2 \to \mathbb{R}$ defined by:

$$\forall x \in \mathbb{R}^2, \ f(x) = -(x_1)^3 + (x_2 - 1)^2.$$

The contour sets of this objective are shown in Figure 16.4. The function decreases with increasing values of x_1.

As in the example in Section 16.1.2.2, suppose that we do not have any equality constraints. That is, consider the problem:

$$\min_{x \in \mathbb{R}^2} \{ f(x) | x \geq \mathbf{0} \}.$$

The problem is unbounded below on the feasible set and therefore has no minimizer.

However, consider the candidate minimizer $\hat{x} = \begin{bmatrix} 0 \\ 1 \end{bmatrix}$ and candidate value of

Lagrange multipliers $\hat{\mu} = \mathbf{0}$. The point $\hat{x}$ is illustrated as a ○ in Figure 16.4. For these values of $\hat{x}$ and $\hat{\mu}$ we have:

$$
\begin{aligned}
\nabla f(\hat{x}) - \hat{\mu} &= \mathbf{0}, \\
\widehat{M}\hat{x} &= \mathbf{0}, \\
\hat{x} &\geq \mathbf{0}, \\
\hat{\mu} &\geq \mathbf{0},
\end{aligned}
$$

where $\widehat{M} = \mathrm{diag}\{\hat{\mu}_\ell\} \in \mathbb{R}^{2 \times 2}$ is a diagonal matrix with diagonal entries equal to $\hat{\mu}_\ell$, $\ell = 1, 2$. That is, $\hat{x}$ and $\hat{\mu}$ satisfy the first-order necessary conditions.

The active set at $\hat{x} = \mathbf{0}$ includes the first non-negativity constraint. That is, $\mathbb{A}(\hat{x}) = \{1\}$. This set is different to $\mathbb{A}_+(\hat{x}, \hat{\mu})$. In fact, $\mathbb{A}_+(\hat{x}, \hat{\mu}) = \{\ell \in \{1, 2\} | x_\ell^\star = 0, \mu_\ell^\star > 0\} = \emptyset$. Therefore, if $\hat{x} = \begin{bmatrix} 0 \\ 1 \end{bmatrix}$ and $\hat{\mu} = \mathbf{0}$ *were* the minimizer and corresponding Lagrange multipliers of this problem, then the constraint $x_1 \geq 0$ would be degenerate.

The Hessian of the objective is:

$$
\nabla^2 f(\hat{x}) = \begin{bmatrix} 0 & 0 \\ 0 & 2 \end{bmatrix}.
$$

The subspace corresponding to the constraints of equality-constrained Problem (16.3) is:

$$
\begin{aligned}
\mathcal{N} &= \{\Delta x \in \mathbb{R}^2 | A \Delta x = \mathbf{0}, \Delta x_\ell = 0, \forall \ell \in \mathbb{A}(\hat{x})\}, \\
&= \{\Delta x \in \mathbb{R}^2 | \Delta x_\ell = 0, \forall \ell \in \{1\}\}, \\
&= \{\Delta x \in \mathbb{R}^2 | \Delta x_1 = \mathbf{0}\}.
\end{aligned}
$$

Note that:

$$
\begin{aligned}
\forall \Delta x \in \mathcal{N}, (\Delta x \neq \mathbf{0}) &\Rightarrow (\Delta x_1 = 0, \Delta x_2 \neq 0), \\
&\Rightarrow (\Delta x^\dagger \nabla^2 f(\hat{x}) \Delta x = 2(\Delta x_2)^2 > 0).
\end{aligned}
$$

That is, the Hessian *is* positive definite on $\mathcal{N}$ and, by Corollary 13.4, $\hat{x}$ is a local minimizer of the equality-constrained problem $\min_{x \in \mathbb{R}^2} \{f(x) | - x_1 = 0\}$. But this is *insufficient* to guarantee local optimality for Problem (16.1). In fact, $\nabla^2 f(\hat{x})$ is not positive definite on the null space $\mathcal{N}_+$ specified in Theorem 16.2:

$$
\begin{aligned}
\mathcal{N}_+ &= \{\Delta x \in \mathbb{R}^2 | A \Delta x = \mathbf{0}, \Delta x_\ell = 0, \forall \ell \in \mathbb{A}_+(\hat{x}, \hat{\mu})\}, \\
&= \{\Delta x \in \mathbb{R}^2 | \Delta x_\ell = 0, \forall \ell \in \emptyset\}, \\
&= \mathbb{R}^2.
\end{aligned}
$$

The second-order sufficient conditions do not hold and $\hat{x}$ is not a minimizer of the problem.

As in the equality-constrained case, this example shows that we may encounter problems where we can find a point that satisfies the first-order necessary conditions but which does not satisfy the second-order sufficient conditions. We must use further judgment in this case to decide if the point is a minimizer. In this example, the contour sets of the objective confirm that $\hat{x}$ is not a minimizer.

16.2 Convex problems
16.2.1 First-order sufficient conditions
16.2.1.1 Analysis

If the constraints consist of linear equality constraints and non-negativity constraints and if f is convex on the feasible set then the problem is convex. In this case, the first-order necessary conditions are also sufficient for optimality.

Theorem 16.3 *Let* $f : \mathbb{R}^n \to \mathbb{R}$ *be partially differentiable with continuous partial derivatives,* $A \in \mathbb{R}^{m \times n}$, *and* $b \in \mathbb{R}^m$. *Consider Problem (16.1),*

$$\min_{x \in \mathbb{R}^n} \{ f(x) | Ax = b, x \geq 0 \},$$

and points $x^\star \in \mathbb{R}^n$, $\lambda^\star \in \mathbb{R}^m$, *and* $\mu^\star \in \mathbb{R}^n$. *Let* $M^\star = \mathrm{diag}\{\mu_\ell^\star\} \in \mathbb{R}^{n \times n}$. *Suppose that:*

(i) *f is convex on $\{x \in \mathbb{R}^n | Ax = b, x \geq 0\}$,*
(ii) *$\nabla f(x^\star) + A^\dagger \lambda^\star - \mu^\star = 0$,*
(iii) *$M^\star x^\star = 0$,*
(iv) *$Ax^\star = b$ and $x^\star \geq 0$, and*
(v) *$\mu^\star \geq 0$.*

Then $x^\star$ is a global minimizer of Problem (16.1).

Proof By Item (iv), $x^\star$ is feasible. Consider any other feasible point $x \in \mathbb{R}^n$. That is, consider x such that:

$$Ax = b, x \geq 0.$$

We have $Ax = Ax^\star = b$, so $A(x - x^\star) = 0$ and:

$$[\lambda^\star]^\dagger A(x - x^\star) = 0. \tag{16.5}$$

We now consider constraints $\ell \in \mathbb{A}(x^\star)$ and constraints $\ell \notin \mathbb{A}(x^\star)$ separately.
For $\ell \notin \mathbb{A}(x^\star)$, we have that $x_\ell^\star > 0$. Consequently, Item (iii) implies that $\mu_\ell^\star = 0$. Therefore,

$$\forall \ell \notin \mathbb{A}(x^\star), \mu_\ell^\star(x_\ell - x_\ell^\star) = 0. \tag{16.6}$$

For $\ell \in \mathbb{A}(x^\star)$, we have that $x_\ell^\star = 0$. Moreover, since $x_\ell \geq 0$ for all ℓ, we have:

$$\forall \ell \in \mathbb{A}(x^\star), \; x_\ell - x_\ell^\star = x_\ell - 0,$$
$$\geq 0.$$

Therefore, since $\mu_\ell^\star \geq 0$, we have:

$$\forall \ell \in \mathbb{A}(x^\star), \mu_\ell^\star(x_\ell - x_\ell^\star) \geq 0. \tag{16.7}$$

Combining (16.6) and (16.7), we have:

$$[\mu^\star]^\dagger(x - x^\star) = \sum_{\ell \in \mathbb{A}(x^\star)} \mu_\ell^\star(x_\ell - x_\ell^\star) + \sum_{\ell \notin \mathbb{A}(x^\star)} \mu_\ell^\star(x_\ell - x_\ell^\star),$$
$$= \sum_{\ell \in \mathbb{A}(x^\star)} \mu_\ell^\star(x_\ell - x_\ell^\star), \text{ by (16.6)},$$
$$\geq 0, \text{ by (16.7)}. \tag{16.8}$$

We have:

$$f(x) \geq f(x^\star) + \nabla f(x^\star)^\dagger(x - x^\star), \text{ by Theorem 2.6, noting that:}$$
$$f \text{ is partially differentiable with continuous partial derivatives;}$$
$$f \text{ is convex on the convex set } \{x \in \mathbb{R}^n | Ax = b, x \geq \mathbf{0}\},$$
$$\text{by Item (i) of the hypothesis; and}$$
$$x, x^\star \in \{x \in \mathbb{R}^n | Ax = b, x \geq \mathbf{0}\},$$
$$\text{by Item (iv) of the hypothesis and construction,}$$
$$= f(x^\star) - [A^\dagger \lambda^\star - \mu^\star]^\dagger(x - x^\star),$$
$$\text{by Item (ii) of the hypothesis,}$$
$$= f(x^\star) - [\lambda^\star]^\dagger A(x - x^\star) + [\mu^\star]^\dagger(x - x^\star),$$
$$= f(x^\star) + [\mu^\star]^\dagger(x - x^\star), \text{ by (16.5)},$$
$$\geq f(x^\star), \text{ by (16.8)}.$$

Therefore, $x^\star$ is a global minimizer of f on $\{x \in \mathbb{R}^n | Ax = b, x \geq \mathbf{0}\}$. □

16.2.1.2 Example

Consider again the problem from Section 16.1.2.2:

$$\min_{x \in \mathbb{R}^2}\{f(x)|x \geq \mathbf{0}\},$$

with objective $f : \mathbb{R}^2 \to \mathbb{R}$ defined by:

$$\forall x \in \mathbb{R}^2, \; f(x) = (x_1)^2 + (x_2 - 1)^2.$$

The objective is partially differentiable with continuous partial derivatives and convex. We have already verified that $x^\star = \begin{bmatrix} 0 \\ 1 \end{bmatrix}$ and $\mu^\star = [0]$ satisfy the first-order necessary conditions. By Theorem 16.3, $x^\star$ is a global minimizer of the problem.

16.3 Approaches to finding minimizers: active set method

In this section, we will discuss the first of two major classes of algorithms for solving inequality-constrained problems. The key to this first approach will be the identification of which constraints are active. If the active inequality constraints could be identified, then they can be treated as equality constraints, much as we did in the proof of Theorem 16.1, and the algorithms we have developed for equality-constrained optimization can be used to solve the problem. However, in general it is difficult to decide *a priori* which constraints are active, and searching over all possible combinations of binding constraints is prohibitive in computation time as shown in Exercise 16.6.

One approach to identifying the active inequality constraints, called the **active set method** is to consider a tentative list of the constraints that we believe are binding at the optimum [84, section 15.4]. This tentative list is called the **working set** and typically consists of the indices of the binding inequalities at the current iterate.

We update the iterates to reduce the objective while temporarily holding as *equalities* those constraints that are in the working set. Since our tentative list may not be the correct list for the minimizer, we must consider how to change this tentative list, either by:

- adding another constraint to the list, which is called **swapping in**, or
- removing a constraint from the list, which is called **swapping out**.

Geometrically, active set algorithms tend to step along the boundary of the region defined by the inequality constraints.

In Section 16.3.1 we discuss the working set, while in Sections 16.3.2, 16.3.3, and 16.3.4, respectively, we discuss swapping in, swapping out, and alternation of swapping in and out. In Section 16.3.5, we discuss finding an initial feasible point. Finally, in Section 16.3.6, we specialize to linear and quadratic objectives. The discussion is drawn from a number of sources, including [45, 70, 84].

16.3.1 Working set

Let us write $\mathbb{W}^{(\nu)}$ for the working set, with the understanding that the working set will change from iteration to iteration and, in some cases, during an iteration, based on values of the iterates. (That is, we will be abusing notation; however, the entries in the working set will always be made clear by the context of the discussion.) The constraints in the working set are treated *temporarily* as equality constraints. A search direction is calculated that seeks the minimizer of an *equality*-constrained problem where the equality constraints consist of:

- all the equality constraints in the original problem, and
- the binding inequality constraints listed in $\mathbb{W}^{(\nu)}$.

If the working set $\mathbb{W}^{(\nu)}$ happens to coincide with the active set for the minimizer $x^\star$ of the inequality-constrained Problem (16.1) then, by the proof of Theorem 16.1, the solution of the equality-constrained problem using $\mathbb{W}^{(\nu)}$ will be $x^\star$.

Inequality constraints are "swapped" in and out of the working set as calculations proceed. After each change to the working set, the updated equality-constrained problem is considered. We will explore this in the following sections.

16.3.2 Swapping in

16.3.2.1 Descent direction

Consider iteration ν, the current value of the iterate $x^{(\nu)}$, and a working set $\mathbb{W}^{(\nu)}$. Suppose that $x^{(\nu)}$ is feasible with respect to all the constraints and strictly feasible with respect to the constraints that are not in the working set $\mathbb{W}^{(\nu)}$. (We will discuss in Section 16.3.5 how to obtain an initial feasible point.)

We consider the equality-constrained problem that is obtained by treating the constraints in the working set as equality constraints and temporarily ignoring the other inequality constraints. That is, we consider the problem:

$$\min_{x \in \mathbb{R}^n} \{ f(x) | Ax = b, -x_\ell = 0, \forall \ell \in \mathbb{W}^{(\nu)} \}. \tag{16.9}$$

We can use the algorithms from Chapter 13 to find a descent direction $\Delta x^{(\nu)}$ at $x^{(\nu)}$ for this equality-constrained problem.

16.3.2.2 Step-size

We now seek movements along the descent direction $\Delta x^{(\nu)}$ that also do not violate any of the inequality constraints that are not in the current working set $\mathbb{W}^{(\nu)}$. That is, we seek a step-size for the update that will maintain feasibility with respect to *all* of the constraints in Problem (16.1).

In particular, consider any inequality constraint ℓ' that is not in the current working set. That is, consider $\ell' \notin \mathbb{W}^{(\nu)}$ so that $x_{\ell'}^{(\nu)} > 0$. For simplicity, first suppose that the objective function decreases along the descent direction for arbitrary step-sizes. Suppose further that an update $\Delta x^{(\nu)}$ based on the current working set and a step-size of 1 would cause inequality constraint ℓ' to be violated because $x_{\ell'}^{(\nu)} + \Delta x_{\ell'}^{(\nu)} < 0$. Then:

- the step-size $\alpha^{(\nu)}$ for the update should be chosen to make constraint ℓ' just binding at the next iterate $x_{\ell'}^{(\nu)} + \alpha^{(\nu)} \Delta x_{\ell'}^{(\nu)}$, and
- the working set should be updated by including constraint ℓ' so that $\mathbb{W}^{(\nu+1)} = \mathbb{W}^{(\nu)} \cup \{\ell'\}$.

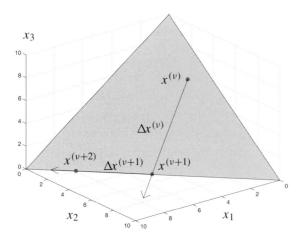

Fig. 16.5. Changes in the working set.

Now we consider the decrease of the objective function along the descent direction. In Section 10.2.4 we discussed the selection of step lengths to ensure sufficient decrease of the objective f. We may find that the function evaluated at $x^{(\nu)}_{\ell'} + \alpha^{(\nu)} \Delta x^{(\nu)}_{\ell'}$ does not satisfy a sufficient decrease criterion. In this case, we should decrease the step size further (and not add the constraint ℓ' to the working set.) Further details about step length selection can be found in [45, section 5.2.2].

16.3.2.3 Example

For example, consider Figure 16.5, which shows the feasible set $\{x \in \mathbb{R}^3 | \mathbf{1}^\dagger x = 10, x \geq \mathbf{0}\}$. This feasible set is an example of a set of the form:

$$\{x \in \mathbb{R}^n | Ax = b, x \geq \mathbf{0}\}, \tag{16.10}$$

where $A = -\mathbf{1}^\dagger \in \mathbb{R}^{m \times n}$ and $b = \begin{bmatrix} -10 \end{bmatrix} \in \mathbb{R}^m$, for $m = 1$ and $n = 3$. This is the same form as the equality constraint in the least-cost production case study of Section 12.1 and we know from Section 12.1.4.2 and Exercise 12.4 that:

$$Z = \begin{bmatrix} -1 & -1 \\ 1 & 0 \\ 0 & 1 \end{bmatrix},$$

is a matrix with columns that form a basis for the null space of A.

Also illustrated in Figure 16.5 is a current iterate $x^{(\nu)} = \begin{bmatrix} 1 \\ 3 \\ 6 \end{bmatrix} \in \mathbb{R}^3_+$ that is feasible for the equality constraint. Since $x^{(\nu)} > \mathbf{0}$, we suppose that the current working set is empty, $\mathbb{W}^{(\nu)} = \emptyset$.

Consider a partially differentiable objective $f : \mathbb{R}^3 \to \mathbb{R}$ such that $\nabla f(x^{(\nu)}) =$

$\begin{bmatrix} 2 \\ -1 \\ 11 \end{bmatrix}$. By the discussion in Section 13.1.2.2, since the vector:

$$\Delta x^{(\nu)} = -ZZ^\dagger \nabla f(x^{(\nu)}),$$
$$= \begin{bmatrix} 6 \\ 3 \\ -9 \end{bmatrix}$$

is non-zero then it is a descent direction for f that lies in the null space of the equality constraint. (There are many other such descent directions and, unless Z is known, it would be inconvenient to evaluate Z in general. However, since we have already found Z, we have used it to find a descent direction in the null space of A.) The vector $\Delta x^{(\nu)}$ is illustrated in Figure 16.5 as the arrow with tail at $x^{(\nu)}$.

Consider movement from $x^{(\nu)}$ along the direction $\Delta x^{(\nu)}$. Summarizing the observations above, moving from $x^{(\nu)}$ in the direction $\Delta x^{(\nu)}$ will simultaneously:

- improve the objective, and
- maintain satisfaction of the equality constraint $\mathbf{1}^\dagger x = 10$.

Let us suppose that the objective decreases along the direction $\Delta x^{(\nu)}$ for step-sizes up to at least 1. That is,

$$\forall \alpha^{(\nu)} \in (0, 1], \ f(x^{(\nu)} + \alpha^{(\nu)} \Delta x^{(\nu)}) < f(x^{(\nu)}).$$

Nevertheless, to maintain feasibility of the next iterate $x^{(\nu+1)}$ with respect to the non-negativity constraints, the update cannot progress past the $x_3 = 0$ plane. We must choose $\alpha^{(\nu)}$ such that:

$$x^{(\nu+1)} = x^{(\nu)} + \alpha^{(\nu)} \Delta x^{(\nu)},$$
$$= \begin{bmatrix} 1 \\ 3 \\ 6 \end{bmatrix} + \alpha^{(\nu)} \begin{bmatrix} 6 \\ 3 \\ -9 \end{bmatrix},$$
$$\geq \mathbf{0}.$$

To satisfy $x^{(\nu+1)} \geq \mathbf{0}$, a step-size of $\alpha^{(\nu)} = \frac{2}{3}$ is chosen so that $x^{(\nu+1)}$ satisfies $x_3^{(\nu+1)} = 0$ and, therefore, $x^{(\nu+1)} = \begin{bmatrix} 5 \\ 5 \\ 0 \end{bmatrix}$. Constraint 3 is added to the working set so that, tentatively, $\mathbb{W}^{(\nu+1)} = \{3\}$.

For the next iteration, we consider movement in a direction $\Delta x^{(\nu+1)}$ such that:

- $\Delta x^{(\nu+1)}$ is a descent direction for the objective f at $x^{(\nu+1)}$,
- movement in the direction $\Delta x^{(\nu+1)}$ maintains feasibility for the equality constraint $\mathbf{1}^\dagger x = 10$, and

- movement in the direction $\Delta x^{(\nu+1)}$ maintains satisfaction of the equality constraint $-x_3 = 0$ implied by the current working set.

For example, suppose that at $x^{(\nu+1)}$ the objective decreases with increasing values of x_1 and decreasing values of x_2. Then a suitable update direction is shown in Figure 16.5 as the arrow labelled $\Delta x^{(\nu+1)}$ having its tail at $x^{(\nu+1)}$ and pointing towards $x = \begin{bmatrix} 10 \\ 0 \\ 0 \end{bmatrix}$.

We would update along the direction $\Delta x^{(\nu+1)}$ until a minimum of the objective was reached or another constraint became binding. In the former case, a point such as $x^{(\nu+2)}$ in Figure 16.5 would be obtained. In the latter case, another constraint would be added to the working set and the procedure would continue. As mentioned above, the iterates typically lie on the boundary of the region defined by the inequality constraints.

16.3.3 Swapping out

16.3.3.1 Descent direction

We can also consider swapping a constraint ℓ'' out of the feasible set. In seeking a descent direction for the equality-constrained Problem (16.9), we can obtain estimates of the Lagrange multipliers for the active constraints. Suppose that for some $\ell'' \in \mathbb{W}^{(\nu)}$ we find that a Lagrange multiplier estimate for the constraint $-x_{\ell''} = 0$ is negative for Problem (16.9). In this case, we can potentially reduce the objective by moving in a direction that makes the constraint non-binding. That is, we should consider removing ℓ'' from the working set.

This approach again follows the proof of Theorem 16.1. In the proof of Theorem 16.1, a negative value of a Lagrange multiplier corresponding to an inequality constraint allowed us to reduce the objective by moving in a direction such that the constraint became strictly feasible. Similarly, a negative Lagrange multiplier estimate for a constraint in the active set signals that the objective can be reduced by removing the constraint from the working set.

In practice, the equality-constrained problems may not be solved to optimality, so that the Lagrange multiplier estimate may be in error. In this case, the working set approach can be prone to "zig-zagging" where constraints repeatedly move in and out of the active set without significant progress. Various strategies have been devised to avoid erroneously swapping a constraint out. Nevertheless, suppose that we choose to swap out constraint ℓ'' to update the working set. Then we revise the working set to be $\mathbb{W}^{(\nu)} \setminus \{\ell''\}$. That is, we remove ℓ'' from the working set. A

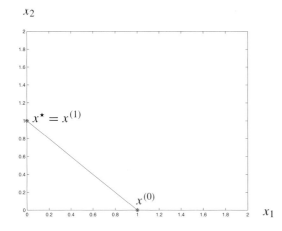

x_2

Fig. 16.6. The trajectory of iterates using the active set algorithm for the example problem. The feasible set is indicated by the solid line.

descent direction is sought for the corresponding equality-constrained problem:

$$\min_{x \in \mathbb{R}^n} \{ f(x) | Ax = b, -x_\ell = 0, \forall \ell \in \mathbb{W}^{(\nu)} \setminus \{\ell''\} \},$$

and the next iterate is calculated based on this descent direction.

16.3.3.2 Example

Consider again Problem (16.4) from Sections 2.3.2.3 and 16.1.1.2:

$$\min_{x_1, x_2 \in \mathbb{R}} \{ x_1 - x_2 | x_1 + x_2 = 1, x_1 \geq 0, x_2 \geq 0 \}.$$

(This is an example of a linear programming problem, which will discuss more specifically in Section 16.3.6.1.) Suppose that we start with the initial guess of $x^{(0)} = \begin{bmatrix} 1 \\ 0 \end{bmatrix}$ for this problem. This initial guess is feasible with respect to all the constraints, is strictly feasible with respect to the inequality constraint $x_1 \geq 0$, and the inequality constraint $x_2 \geq 0$ is active at this initial guess. The situation is illustrated in Figure 16.6.

Working set Since the inequality constraint $x_2 \geq 0$ is active for the initial guess, the initial working set is $\mathbb{W}^{(0)} = \{2\}$.

Descent direction at $x^{(0)}$ We consider the equality-constrained problem:

$$\min_{x_1, x_2 \in \mathbb{R}} \{ x_1 - x_2 | x_1 + x_2 = 1, -x_\ell = 0, \forall \ell \in \mathbb{W}^{(0)} \}$$
$$= \min_{x_1, x_2 \in \mathbb{R}} \{ x_1 - x_2 | x_1 + x_2 = 1, -x_2 = 0 \}, \quad (16.11)$$

and seek a descent direction for it. In fact, however, $x^{(0)}$ is optimal for this problem, but the sign of the Lagrange multiplier for the constraint $-x_2 = 0$ is negative. (See

Exercise 16.7.) That is, we are at the minimizer of the equality-constrained problem but have not found the minimizer of inequality-constrained Problem (16.4).

Update working set We update the working set by removing constraint 2 from it. That is, we now have the revised working set $\mathbb{W}^{(0)} = \emptyset$.

Descent direction at $x^{(0)}$ Since the objective increases with x_1 and decreases with x_2, a descent direction at $x^{(0)}$ for the objective that maintains feasibility for the equality constraints $x_1 + x_2 = 1$ is given by $\Delta x^{(0)} = \begin{bmatrix} -1 \\ 1 \end{bmatrix}$. (See Exercise 16.7.)

16.3.4 Alternation of swapping in and out

In general, we must solve a sequence of problems, alternately swapping in and out. To illustrate, we continue with Problem (16.4) from Sections 2.3.2.3, 16.1.1.2, and 16.3.3.2, starting at $x^{(0)} = \begin{bmatrix} 1 \\ 0 \end{bmatrix}$ and using descent direction $\Delta x^{(0)} = \begin{bmatrix} -1 \\ 1 \end{bmatrix}$.

16.3.4.1 Swapping in

If we move along the descent direction according to $x^{(0)} + \alpha^{(0)} \Delta x^{(0)}$, we find that for $\alpha^{(0)} = 1$, the constraint $x_1 \geq 0$ becomes binding. We obtain the next iterate:

$$
\begin{aligned}
x^{(1)} &= x^{(0)} + \alpha^{(0)} \Delta x^{(0)}, \\
&= \begin{bmatrix} 1 \\ 0 \end{bmatrix} + 1 \begin{bmatrix} -1 \\ 1 \end{bmatrix}, \\
&= \begin{bmatrix} 0 \\ 1 \end{bmatrix},
\end{aligned}
$$

and we update the working set to $\mathbb{W}^{(1)} = \{1\}$.

16.3.4.2 Descent direction

We consider the equality-constrained problem corresponding to the working set $\mathbb{W}^{(1)} = \{1\}$:

$$
\begin{aligned}
&\min_{x_1, x_2 \in \mathbb{R}} \{x_1 - x_2 | x_1 + x_2 = 1, -x_\ell = 0, \forall \ell \in \mathbb{W}^{(1)}\} \\
&= \min_{x_1, x_2 \in \mathbb{R}} \{x_1 - x_2 | x_1 + x_2 = 1, -x_1 = 0\},
\end{aligned} \tag{16.12}
$$

and seek a descent direction for it. In fact, $x^{(1)}$ is the minimizer of this equality-constrained problem and, moreover, the sign of the Lagrange multiplier for the constraint $-x_1 = 0$ is positive. (See Exercise 16.8.) That is, we are at the optimum of the equality-constrained problem and have also found the optimum of inequality-constrained Problem (16.4).

16.3.4.3 Discussion

For this example, since there were only two inequality constraints we took just one swapping out operation and one swapping in operation to find the minimizer. The sequence of iterates is illustrated in Figure 16.6. In general we will find that we will have to successively swap in and out various of the constraints and solve several equality-constrained problems before reaching the minimizer of the original inequality-constrained problem.

16.3.5 Finding an initial feasible guess

The algorithm we have sketched relies on starting with an initial feasible guess. We discussed how to find a point satisfying equality constraints $Ax = b$ in Section 5.8.1; however, in general this approach will not guarantee satisfaction of the non-negativity constraints $x \geq 0$. To find an initial feasible guess for Problem (16.1), we will instead define another optimization problem that is *related* to Problem (16.1) and having the following properties:

- it is easy to find an initial feasible guess for the related problem,
- if Problem (16.1) is feasible, then a minimizer of the related problem yields a feasible initial guess for Problem (16.1), and
- if Problem (16.1) is infeasible, then the minimum of the related problem signals this fact.

The related problem includes the variables $x \in \mathbb{R}^n$ from Problem (16.1) and, additionally, includes **artificial variables** $w \in \mathbb{R}^n$ [6, section 2.7][45, section 5.7][84, section 5.5]. To simplify the discussion of the related problem, let us suppose that $b \geq 0$ (or, swap the sign of any negative entry in b and the signs of the entries in the corresponding row of A.) Consider the following problem, related to Problem (16.1):

$$\min_{x \in \mathbb{R}^n, w \in \mathbb{R}^n} \{\mathbf{1}^\dagger w \,|\, Ax + w = b, x \geq 0, w \geq 0\}. \tag{16.13}$$

Note that $x^{(0)} = \mathbf{0}, w^{(0)} = b \geq \mathbf{0}$ satisfies the equality and inequality constraints of Problem (16.13). We solve Problem (16.13) using the active set method and the feasible initial guess $\begin{bmatrix} x^{(0)} \\ w^{(0)} \end{bmatrix}$. If, at a minimizer $\begin{bmatrix} x^\star \\ w^\star \end{bmatrix}$ of Problem (16.13), we find that $w^\star = \mathbf{0}$ (so that the minimum is $\mathbf{1}^\dagger w^\star = 0$) then $x^\star$ is a a feasible initial guess for Problem (16.1), since:

$$
\begin{aligned}
b &= Ax^\star + w^\star, \text{ since } \begin{bmatrix} x^\star \\ w^\star \end{bmatrix} \text{ is feasible for Problem (16.13),} \\
&= Ax^\star, \text{ since } w^\star = \mathbf{0},
\end{aligned}
$$

$$x^\star \ge \mathbf{0}, \text{ since } \begin{bmatrix} x^\star \\ w^\star \end{bmatrix} \text{ is feasible for Problem (16.13)}.$$

On the other hand, if the minimum is non-zero (so that the minimizer $\begin{bmatrix} x^\star \\ w^\star \end{bmatrix}$ satisfies $w^\star \ne \mathbf{0}$) then Problem (16.1) is infeasible. (See Exercise 16.9.)

The process of finding a feasible initial guess for Problem (16.1) is sometimes
called **phase 1** of optimization. The feasible initial guess is then used as a starting
point by an algorithm to minimize the objective of Problem (16.1) in what is called
phase 2 [70, section 3.5][84, section 5.5].

16.3.6 *Linear and quadratic objectives*

In this section we specialize to linear and to quadratic objectives.

16.3.6.1 *Linear programming*

Analysis Consider a non-negatively constrained linear programming problem:

$$\min_{x \in \mathbb{R}^n} \{c^\dagger x \mid Ax = b, x \ge \mathbf{0}\}. \tag{16.14}$$

As mentioned previously, this is called the **standard format** for linear programs.
For this problem, Theorem 16.1 indicates that, except for:

- the complementary slackness conditions $Mx = \mathbf{0}$, and
- the inequalities $x \ge \mathbf{0}$ and $\mu \ge \mathbf{0}$,

the necessary conditions are *linear* simultaneous equations. The linearity facilitates:

- the calculation of descent directions for the corresponding equality-constrained
  problem,
- avoiding zig-zagging, and
- maintaining feasibility as successive iterates are calculated.

Moreover, since the objective is linear, it is both convex and concave. The linear
minimization Problem (16.14) is equivalent to maximizing the objective $-c^\dagger x$ over
the same feasible set. By Theorem 2.5, there is a maximizer of $-c^\dagger x$ (and therefore
a minimizer of $c^\dagger x$) that is an extreme point of the feasible set. We can restrict
attention to points that are vertices of the feasible set and do not need to consider
points such as $x^{(\nu+2)}$ in Figure 16.5 that are on the boundary but not at a vertex of
the feasible set.

Geometrically, contour sets of the objective are parallel hyperplanes. The minimum of the linear program corresponds to the hyperplane with minimum height

that intersects the feasible set. The intersection will contain a vertex of the feasible set.

Discussion The active set strategy applied to linear programming problems represented in the form of Problem (16.14), together with various techniques to make the constraint swapping and calculation of descent directions more efficient, leads to the **simplex algorithm**, developed in the 1940s by George Dantzig [41, section 3.2]. The simplex algorithm updates the iterates by proceeding from vertex to vertex of the feasible set along edges of the feasible set. (A **simplex** is a set that consists of the convex combinations of a finite number of vertices. The feasible set for the problem is an example of a simplex.) The vertices of the feasible set for Problem (16.14) are points that satisfy equations of the form:

$$Ax = b, -x_\ell = 0, \forall \ell \in \mathbb{W},$$

with $\mathbb{W}$ having $n - m$ members (for $A \in \mathbb{R}^{m \times n}$ having m linearly independent rows.) For example, for the feasible set illustrated in Figure 16.5, the vertices are:

$$\begin{bmatrix} 10 \\ 0 \\ 0 \end{bmatrix}, \begin{bmatrix} 0 \\ 10 \\ 0 \end{bmatrix}, \begin{bmatrix} 0 \\ 0 \\ 10 \end{bmatrix},$$

corresponding, respectively, to the three choices:

$$\mathbb{W} = \{2, 3\}, \mathbb{W} = \{1, 3\}, \mathbb{W} = \{1, 2\}.$$

Each of these choices of working set has $n - m = 3 - 1 = 2$ members.

The form of the feasible set for linear programming leads to important simplifications for updating iterates and swapping in and swapping out. In particular, swapping in and out is performed simultaneously and calculation of a descent direction is facilitated by maintaining and updating factors of an appropriate square sub-matrix of the coefficient matrix of the constraints $Ax = b, -x_\ell = 0, \forall \ell \in \mathbb{W}$. Moreover, the algorithm terminates in a finite number of iterations with the exact minimizer (assuming infinite precision arithmetic). Details can be found in a number of references, including [28][70, part I][84, chapter 5]. The MATLAB function `linprog` uses the simplex algorithm under some circumstances. (See Exercise 16.10.)

For some pathological problems, the simplex algorithm must examine a large proportion of the possible combinations of active inequalities [84, section 9.3]. As shown in Exercise 16.6, the number of combinations is large for n large and would be computationally prohibitive if a large proportion of the combinations had to be examined. However, in practice, the simplex algorithm usually finds a solution of the problem in relatively few iterations. In fact, if we choose linear programming problems "randomly" by choosing the coefficients in the linear and affine functions from particular random distributions, then there are theoretical results that indicate that the simplex algorithm has good *expected* behavior over certain distributions of random problems [84, section 9.5]. Despite the existence of problems where the simplex algorithm is slow, the simplex algorithm and its variants remain the most used and practical optimization algorithms.

There is a vast literature on linear programming. If an optimization problem can be formulated as a linear program (or can be linearized without much loss of accuracy) then it is worthwhile to do so. Many special issues arise in linear programming that allow simplifications of hypotheses and sharpening of conclusions of the theory we have discussed. For example, some linear integer optimization problems have simple solutions in terms of linear programming if all the vertices of the set obtained by relaxing the integrality constraints turn out to have integer coordinates. (See Exercise 16.11.) As another example, a linear programming problem with a feasible solution either has a minimum or is unbounded below [70, section 3.3]. We will not discuss these issues in detail in this book, but refer the interested reader to, for example, [28][67][70, Part I][83][84, Part II].

16.3.6.2 Quadratic programming

As with linear programming, there are also simplifications possible in the case of quadratic objectives [70, section 14.1]. Moreover, there is a large body of active set-based software available to solve quadratic programming problems. The MATLAB function quadprog uses an active set algorithm under some circumstances.

16.3.6.3 Further details

Active set algorithms are covered in detail in, for example, [45, section 5.2][70, section 11.3][84, section 15.4]. We have only introduced them briefly here; however, much software written for optimization problems uses some form of active set algorithm.

16.4 Approaches to finding minimizers: interior point algorithm

A very different approach to solving inequality-constrained problems is not based on identifying the active constraints directly. Conceptually, a "barrier" is erected that prevents violation of all the inequality constraints so that the sequence of iterates remains *strictly* feasible with respect to the inequality constraints. That is, the iterates remain in the interior of the set defined by the inequality constraints. (See Definition 2.5.)

Ideally, the iterates step directly towards the minimizer across the interior of the feasible region, rather than stepping along its boundary as in the active set algorithm. For this reason, the technique is called an **interior point algorithm**. The barrier is a term added to the objective that increases very rapidly as we approach the boundary of the feasible region from its interior. This approach is the topic of significant research, prompted by Karmarkar's presentation of a theoretically and practically fast interior point algorithm for linear programming [57].

In Section 16.4.1, we illustrate the interior point algorithm for the case of a constraint $x \geq 0$, where $x \in \mathbb{R}$. In Section 16.4.2 we define and discuss the approach in more detail, considering the general case of $x \in \mathbb{R}^n$. A computational implementation using the Newton–Raphson update is sketched in Sections 16.4.3 and 16.4.4. We discuss finding an initial feasible guess in Section 16.4.5. We summarize the algorithm in Section 16.4.6 and provide some further discussion in Section 16.4.7. The material in the following sections is based on [70, 73, 84, 102, 126].

16.4.1 Illustration

To illustrate the interior point algorithm, consider the objective $f : \mathbb{R} \to \mathbb{R}$ defined by:

$$\forall x \in \mathbb{R},\ f(x) = x,$$

and a non-negativity constraint $x \geq 0$. We add a **barrier function** for the constraint $x \geq 0$ to the objective $f(x)$ to form the **barrier objective**, $\phi : \mathbb{R}_{++} \to \mathbb{R}$.

The essential characteristic of the barrier function is that it is partially differentiable on the interior of the constraint set but becomes unbounded as the boundary of the constraint set is approached [70, section 12.2]. Two such barrier functions for non-negativity constraints are:

- the **reciprocal** function (see Exercise 16.12), and
- the negative of the logarithm function, which is called the **logarithmic barrier function**.

We will consider the logarithmic barrier function in detail.

In particular, define the logarithmic barrier function $f_b : \mathbb{R}_{++} \to \mathbb{R}$ for the constraints $x \geq 0$ by:

$$\forall x \in \mathbb{R}_{++},\ f_b(x) = -\ln(x).$$

Let $t \in \mathbb{R}_{++}$ be a parameter, called the **barrier parameter** [102, chapter 5]. We add $t f_b$ to the objective f to obtain the barrier objective $\phi : \mathbb{R}_{++} \to \mathbb{R}$ defined by:

$$\begin{aligned}
\forall x \in \mathbb{R}_{++},\ \phi(x) &= f(x) + t f_b(x), \\
&= f(x) - t \ln x.
\end{aligned}$$

$f(x), \phi(x) = f(x) - t \ln(x)$

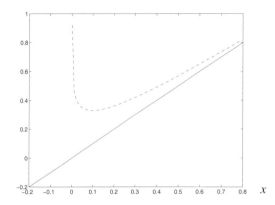

Fig. 16.7. The barrier objective for the constraint $x \geq 0, x \in \mathbb{R}$. The solid curve shows the objective f while the dashed curve shows the barrier objective ϕ for $t = 0.1$ on the interior of the feasible region.

The objective and barrier objective are illustrated in Figure 16.7 for $t = 0.1$. The solid curve shows $f(x)$ for $-0.2 \leq x \leq 0.8$, while the dashed curve shows $\phi(x)$ for $0 < x \leq 0.8$. (We can also imagine that ϕ is an extended real function with $\phi(x)$ defined to be equal to ∞ if x is less than or equal to zero.)

Note that, despite appearances in Figure 16.7, as $x \to \infty$, the barrier objective, as defined, falls below the objective. This is because the barrier function is negative for $x > 1$. This can present problems if the feasible set is unbounded and the objective becomes "flat" as any component of x becomes large. It is possible to define a **modified logarithmic barrier function** that is non-negative; however, we will assume that this issue is not problematic. (See Exercises 16.14 and 16.15 for further discussion of this issue and [84, section 16.2, problem 4][102, section 2.2.3].)

As $x \to 0^+$, $\phi(x) \to \infty$. An algorithm that is trying to minimize ϕ will avoid the vicinity of the boundary of the feasible region. That is, it will produce iterates that are interior to the set defined by the inequality constraint. (See Definition 2.5.) If we use an iterative algorithm starting with an initial guess that is in the interior of the feasible region then we will tend to stay away from the boundary of the feasible region. That is, we will stay in the interior of the feasible region. (We will discuss how to find an initial feasible point that is in the interior of the set defined by the inequality constraints in Section 16.4.5.)

For any fixed $x > 0$, the value of $-t \ln(x)$ approaches 0 as $t \to 0$. That is, as shown in Figure 16.8, the effect of the term $-t \ln(x)$ on the barrier objective becomes negligible for points that are in the interior as we reduce t towards zero. This means that as $t \to 0$, the term $-t \ln(x)$ has the effect of confining the iterates to the feasible region, but within the feasible region it has no effect *asymptotically* as $t \to 0$. As $t \to 0$, the sequence of unconstrained minimizers of ϕ will, under conditions to be established, converge to the constrained minimizer of f.

$-t\ln(x)$

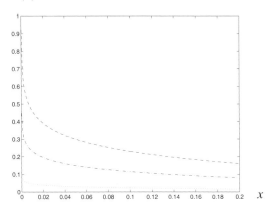

Fig. 16.8. The effect on the barrier function for the constraint $x \geq 0$ as $t \to 0$. The dashed curve shows $-t\ln(x)$ for $t = 0.1$; the dash-dotted curve shows $-t\ln(x)$ for $t = 0.05$; and the dotted curve shows $-t\ln(x)$ for $t = 0.01$.

16.4.2 Outline

In this section we discuss the interior point algorithm for a constraint $x \geq \mathbf{0}$, where $x \in \mathbb{R}^n$. Throughout the rest of this chapter, we will assume that $f : \mathbb{R}^n \to \mathbb{R}$ is (at least) continuous and that Problem (16.1) has a minimum $f^\star$ and at least one minimizer $x^\star$. (There may be more than one minimizer if f is not strictly convex.) We will mention some convexity conditions in Section 16.4.3.

16.4.2.1 Logarithmic barrier function

We define the logarithmic barrier function $f_b : \mathbb{R}^n_{++} \to \mathbb{R}$ for the constraints $x \geq \mathbf{0}$ by:

$$\forall x \in \mathbb{R}^n_{++}, f_b(x) = -\sum_{\ell=1}^{n} \ln(x_\ell). \tag{16.15}$$

The set $\mathbb{R}^n_{++}$ is the **strictly positive orthant**. (See Definition A.5.) The properties of f_b are explored in Exercise 16.13. Part (i) of Exercise 16.13 shows that we can differentiate f_b and use calculus to find descent directions. Part (ii) shows that f_b behaves as a barrier to enforce the constraints $x \geq \mathbf{0}$ and penalize values of x near to the boundary. Part (iii) shows that f_b is convex so that adding a non-negative multiple of it to a convex function results in a convex function.

16.4.2.2 Barrier problem

Given an objective $f : \mathbb{R}^n \to \mathbb{R}$, a barrier function $f_b : \mathbb{R}^n_{++} \to \mathbb{R}$, and a barrier parameter $t \in \mathbb{R}_{++}$, we form the **barrier objective** $\phi : \mathbb{R}^n_{++} \to \mathbb{R}$ defined by:

$$\forall x \in \mathbb{R}^n_{++}, \phi(x) = f(x) + tf_b(x).$$

Instead of solving Problem (16.1), we will consider solving the **barrier problem**:

$$\min_{x \in \mathbb{R}^n} \{\phi(x) | Ax = b, x > \mathbf{0}\}. \tag{16.16}$$

That is, we minimize $\phi(x)$ over values of $x \in \mathbb{R}^n$ that satisfy $Ax = b$ and which are also in the interior of $x \geq \mathbf{0}$.

Problem (16.16) is still inequality-constrained. Furthermore, the constraint is now a *strict* inequality, which makes the problem seem more complicated than Problem (16.1). We discussed the potential disadvantages of an open feasible set such as $\{x \in \mathbb{R}^n | x > \mathbf{0}\}$ in Section 2.3.3. However, in practice, for suitable f, Problem (16.16) can be solved by a technique that considers only the *equality* constraints when seeking a descent direction. For any given value of t, the rapid increase of the barrier as the boundary is approached means that the minimizer of Problem (16.16) is bounded away from the boundary of the set $\{x \in \mathbb{R}^n | x \geq \mathbf{0}\}$.

16.4.2.3 Slater condition

For Problem (16.16) to be useful in finding a solution of Problem (16.1), we need to assume that:

$$\{x \in \mathbb{R}^n | Ax = b, x > \mathbf{0}\} \neq \emptyset,$$

so that Problem (16.16) has a non-empty feasible set. This is called the **Slater condition** [6, chapter 5][11, section 5.3][15, section 5.2.3][84, page 485] and will appear again in the analysis of non-linear inequality constraints in Section 19.3.1. This condition requires the existence of a feasible point that is strictly feasible for the inequality constraints. That is, there must be a feasible interior point. (See Definition 2.5 and [15, section 5.2.3].)

Many constraint systems arising from physical systems satisfy the Slater condition. However, a simple example of constraints that do not satisfy the Slater condition is defined by the following:

$$
\begin{aligned}
A &= \begin{bmatrix} 1 & 1 \end{bmatrix}, \\
b &= \begin{bmatrix} 0 \end{bmatrix}, \\
x &\geq \mathbf{0}.
\end{aligned}
$$

The set $\{x \in \mathbb{R}^2 | Ax = b, x > \mathbf{0}\}$ is empty.

16.4.2.4 Solving the barrier problem

To find the minimizer of Problem (16.16) for any particular value of t, we can start with an initial guess $x^{(0)}$ that satisfies $Ax = b$ and $x > \mathbf{0}$. We then search from $x^{(0)}$ using an iterative algorithm that seeks the value of x that minimizes $\phi(x)$ subject to $Ax = b$.

Since the objective function ϕ of Problem (16.16) becomes arbitrarily large as its argument approaches the boundary of $x \geq \mathbf{0}$, "the search technique (if carefully implemented) will automatically" satisfy $x > \mathbf{0}$ [70, page 370]. The only explicit consideration that needs to be given to the constraints $x > \mathbf{0}$ is to prevent the iterates from going outside the region $x > \mathbf{0}$ by controlling the step-size appropriately. As Luenberger puts it, "although problem [(16.16)] is from a formal viewpoint a[n inequality-]constrained problem, from a computational viewpoint it is" equality-constrained [70, page 370].

16.4.2.5 Sequence of problems

We solve Problem (16.16) not just at one value of t, but for a *sequence* of values of t that approach 0. Allowing $t \to 0$ reduces the effect of the barrier at points in the interior of the feasible set, while still preventing x from violating the constraints $x \geq \mathbf{0}$. We create a sequence of minimizers of Problem (16.16) for values of t that approach 0. The trajectory of minimizers of Problem (16.16) as a function of t is called the **central path** [84, section 17.4][102, section 2.2.4]. Under certain circumstances, the trajectory approaches $x^\star$, a minimizer of Problem (16.1), as t approaches zero.

In principle, we construct a strictly decreasing sequence $\{t^{(\nu)}\}_{\nu=0}^{\infty}$ that converges to zero and for each $t^{(\nu)}$ find a minimizer of Problem (16.16) for the value $t = t^{(\nu)}$. As an initial guess for an iterative algorithm to solve Problem (16.16) for $t = t^{(\nu)}$, $\nu = 1, 2, 3, \ldots$, we can use the minimizer of Problem (16.16) for $t = t^{(\nu-1)}$.

Consider the corresponding sequence $\{x^{(\nu)}\}_{\nu=0}^{\infty}$ of minimizers and Lagrange multipliers for Problem (16.16) for the values $\{t^{(\nu)}\}_{\nu=0}^{\infty}$. Exercise 16.23 shows that if the sequences of minimizers and Lagrange multipliers converge then the limits satisfy the first-order necessary conditions for Problem (16.1). (In Section 16.4.4, we will revisit the approach of solving Problem (16.16). In Section 16.4.5, we will discuss how to obtain an initial feasible guess $x^{(0)}$ for $t = t^{(0)}$.)

16.4.2.6 Example

Consider again Problem (16.4), which we analyzed in Section 16.3.3.2:

$$\min_{x_1,x_2 \in \mathbb{R}} \{x_1 - x_2 | x_1 + x_2 = 1, x_1 \geq 0, x_2 \geq 0\}.$$

The interior point algorithm involves solving the barrier problem, Problem (16.16), for a sequence of values of t that decrease towards zero. For Problem (16.4), the barrier problem is:

$$\min_{x_1,x_2 \in \mathbb{R}} \{x_1 - x_2 - t \ln(x_1) - t \ln(x_2) | x_1 + x_2 = 1, x_1 > 0, x_2 > 0\}. \quad (16.17)$$

Because of its simplicity, we can calculate the minimizer of Problem (16.17) explicitly as a function of t. As discussed in Section 13.1.2.1, we can eliminate x_2 using the equality constraint. That is, we can express the objective as a function of x_1 alone:

$$2x_1 - 1 - t \ln(x_1) - t \ln(1 - x_1). \tag{16.18}$$

We now have an unconstrained problem:

$$\min_{x_1 \in \mathbb{R}} \{2x_1 - 1 - t \ln(x_1) - t \ln(1 - x_1)\}.$$

The reduced objective defined in (16.18) is convex and differentiable. (See Exercise 16.13.) By Theorem 10.3 and Corollary 10.6, first-order necessary and sufficient conditions for minimizing (16.18) are that its derivative be equal to zero. Differentiating (16.18), setting the derivative equal to zero, and re-arranging we find that:

$$(x_1)^2 - x_1(1 + t) + t/2 = 0, \tag{16.19}$$

where we note that both x_1 and $x_2 = 1 - x_1$ must be greater than zero for the objective and derivative to be defined (and for the inequality constraints to be strictly satisfied.) The quadratic equation (16.19) has two solutions, both of which are positive. However, only one of the solutions:

$$x_1 = \frac{1 + t - \sqrt{1 + (t)^2}}{2}, \tag{16.20}$$

yields a value of $x_2 = 1 - x_1$ that satisfies the strict non-negativity constraint for x_2. Substituting, we obtain:

$$x_2 = \frac{1 - t + \sqrt{1 + (t)^2}}{2}. \tag{16.21}$$

In general, we may not be able to conveniently eliminate variables and solve for the minimizer of the barrier problem explicitly as a function of t as we have done for Problem (16.17). Nevertheless, we can think, in principle, of solving the barrier problem for a sequence of decreasing values of t.

Figure 16.9 shows the minimizer given in (16.20) and (16.21) of Problem (16.17) versus t for $t = 1.0, 0.9, \ldots, 0.1$. The minimizers are always in the interior of the set $\{x \in \mathbb{R}^n | x \geq 0\}$. That is, the minimizers are always in $\mathbb{R}^n_{++}$.

As mentioned in Section 16.4.2.5, the trajectory of minimizers of the barrier problem versus the barrier parameter is called the central path. In this example, the central path is contained in a line segment in $\mathbb{R}^2$ because the feasible set is a line segment in $\mathbb{R}^2$; however, in general it will be a curving path through $\mathbb{R}^n$. (See Exercise 16.19.)

For large values of t, the minimizer of Problem (16.17) is far away from the

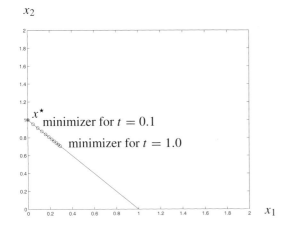

Fig. 16.9. The trajectory of minimizers of Problem (16.17) versus t for $t = 1.0, 0.9, \ldots, 0.1$ shown as o. The minimizer $x^\star$ of Problem (16.4) is shown as a •. The feasible set is indicated by the solid line.

minimizer, $x^\star = \begin{bmatrix} 0 \\ 1 \end{bmatrix}$, of inequality-constrained Problem (16.4); however, as t decreases towards zero, the minimizer of Problem (16.17) approaches $x^\star = \begin{bmatrix} 0 \\ 1 \end{bmatrix}$.

In practice, we find the minimizers of Problem (16.17) for only a finite number of values of t, stopping when t is small enough to guarantee that the minimizer of Problem (16.17) is close enough to a minimizer of Problem (16.4) to satisfy our accuracy requirements. (We will explicitly discuss a stopping criterion in Exercise 16.25 and Section 17.3.1.4.)

Any minimizer of Problem (16.17) is feasible for the constraints $Ax = b, x \geq \mathbf{0}$. Therefore, stopping at any given value of t will yield a feasible, if not optimal, solution of Problem (16.4).

16.4.2.7 Reduction of barrier parameter

For Problem (16.17), we solved for the minimizers explicitly as a function of t. Because we evaluated the minimizer explicitly as a function of t, we could just pick $t = 10^{-10}$, say, and evaluate (16.20)–(16.21) to obtain:

$$x^\star \approx \begin{bmatrix} 5 \times 10^{-11} \\ 1.0000 \end{bmatrix},$$

which is essentially the exact minimizer of Problem (16.4). However, in general, we cannot solve for the minimizer of Problem (16.16) explicitly and we will have to use an iterative algorithm. It turns out that it is very difficult to solve Problem (16.16) from scratch for a small value of t because the initial guess that we can provide for the iterative algorithm leads to a poor update in seeking an unconstrained minimizer. An example of this issue is shown in Exercise 16.20.

In Exercise 16.20, the step direction suggests a step that is far too large, but at least it is in the correct direction. In problems with more than one variable, if the initial guess is far from the minimizer of Problem (16.16) for the current value of the barrier parameter t then the coefficient matrix to determine the step direction can be ill-conditioned. Consequently, with finite precision calculations we may obtain a very poor step direction. Moreover, the step-size to ensure that $x > 0$ will be very small and we will make very slow progress towards the minimizer of the problem.

Instead of trying to minimize the barrier problem from scratch for a small value of t, we start with a large value of t and solve the problem, at least approximately, for this value of t. Then we reduce t and re-solve, using the minimizer (or approximate minimizer) for the previous value of t as the initial point for an iterative algorithm. We continue, successively reducing t and restarting the algorithm with the approximate minimizer from the previous value of t. We will develop this approach using the Newton–Raphson method in the next section.

16.4.3 Newton–Raphson method

In this section, we will discuss how to seek a minimizer of Problem (16.16) for a given value of t using the Newton–Raphson method and how to factorize the resulting Jacobian.

16.4.3.1 Discussion of the barrier problem

To solve Problem (16.16), we will partially ignore the inequality constraints and the domain of the barrier function. That is, computationally, we seek a minimizer of the problem:

$$\min_{x \in \mathbb{R}^n} \{\phi(x) | Ax = b\}. \tag{16.22}$$

We discussed such equality-constrained problems in Part IV. To solve Problem (16.22) we will seek a solution of its first-order necessary conditions and update successive iterates along descent directions for the objective ϕ that maintain feasibility with respect to the equality constraints $Ax = b$. By Theorem 13.2, the first-order necessary conditions of Problem (16.22) are:

$$\nabla \phi(x) + A^{\dagger} \lambda = 0, \tag{16.23}$$
$$Ax - b = 0. \tag{16.24}$$

We must bear in mind, however, that a one-to-one correspondence between minimizers of Problem (16.22) and solutions of the first-order necessary conditions (16.23)–(16.24) can generally only be guaranteed if $\phi = f + tf_b$ satisfies further conditions. For example, sufficient conditions are that f and f_b are convex. Nevertheless, our approach to solving Problem (16.22) is to seek $x^\star$ and $\lambda^\star$ that satisfy the first-order necessary conditions (16.23)–(16.24). These are a set of non-linear simultaneous equations. To solve these equations, we will use the Newton–Raphson method. In Section 16.4.3.2 we introduce a straightforward approach called the primal interior point algorithm, while following that in Section 16.4.3.3 we will develop a second algorithm, called the primal–dual interior point algorithm.

16.4.3.2 Primal interior point algorithm

First, let us investigate a straightforward approach to applying the Newton–Raphson method to solving the first-order necessary conditions (16.23)–(16.24) of Problem (16.22). In particular, consider the first term in (16.23):

$$
\begin{aligned}
\nabla \phi(x) &= \nabla[f(x) + t f_\mathrm{b}(x)], \\
&= \nabla f(x) + t \nabla f_\mathrm{b}(x), \\
&= \nabla f(x) + t \begin{bmatrix} \dfrac{\partial f_\mathrm{b}(x)}{\partial x_1} \\ \vdots \\ \dfrac{\partial f_\mathrm{b}(x)}{\partial x_n} \end{bmatrix}, \\
&= \nabla f(x) + t \begin{bmatrix} -\dfrac{1}{x_1} \\ \vdots \\ -\dfrac{1}{x_n} \end{bmatrix}, \\
&= \nabla f(x) - t[X]^{-1}\mathbf{1}, \\
\nabla^2 \phi(x) &= \nabla^2 f(x) + t[X]^{-2},
\end{aligned}
$$

where $X = \mathrm{diag}\{x_\ell\} \in \mathbb{R}^{n \times n}$ is a diagonal matrix with diagonal entries equal to x_ℓ, $\ell = 1, \ldots, n$. The Newton–Raphson step direction to solve (16.23)–(16.24) is given by:

$$
\begin{bmatrix} \nabla^2 \phi(x^{(\nu)}) & A^\dagger \\ A & \mathbf{0} \end{bmatrix} \begin{bmatrix} \Delta x^{(\nu)} \\ \Delta \lambda^{(\nu)} \end{bmatrix} = \begin{bmatrix} -\nabla \phi(x^{(\nu)}) - A^\dagger \lambda^{(\nu)} \\ b - A x^{(\nu)} \end{bmatrix},
$$

or:

$$
\begin{bmatrix} \nabla^2 f(x^{(\nu)}) + t[X^{(\nu)}]^{-2} & A^\dagger \\ A & \mathbf{0} \end{bmatrix} \begin{bmatrix} \Delta x^{(\nu)} \\ \Delta \lambda^{(\nu)} \end{bmatrix}
$$

$$
= \begin{bmatrix} -\nabla f(x^{(\nu)}) + t[X^{(\nu)}]^{-1}\mathbf{1} - A^\dagger \lambda^{(\nu)} \\ b - A x^{(\nu)} \end{bmatrix}, \tag{16.25}
$$

where $X^{(\nu)} = \mathrm{diag}\{x_\ell^{(\nu)}\} \in \mathbb{R}^{n \times n}$ is a diagonal matrix with diagonal entries equal to $x_\ell^{(\nu)}$, $\ell = 1, \ldots, n$. This update leads to the **primal interior point algorithm**.

We are not going to investigate this algorithm further, except in Section 18.2.1 in the discussion of enforcement of the strict inequality constraints in the case study of optimal routing in a data communication network.

Instead of developing the primal interior point method, we will consider a variant in the next section.

16.4.3.3 Primal–dual interior point algorithm

Instead of the primal interior point algorithm, we will describe an algorithm that incorporates linearization of a hyperbolic approximation to the complementary slackness constraints, as first introduced in Section 16.1.1.3.

New variable and equation We are going to introduce a new variable μ, which will turn out to correspond to the dual variables for the inequality constraints in Problem (16.1). We incorporate the equations:

$$\forall \ell = 1, \ldots, n, \mu_\ell x_\ell = t, \tag{16.26}$$

so that $t/x_\ell = \mu_\ell$. These equations are almost the same as the complementary slackness conditions, except that the right-hand side is equal to the barrier parameter t instead of zero.

As discussed in Section 16.1.1.3, the ordered pairs $\begin{bmatrix} \mu_\ell \\ x_\ell \end{bmatrix}$ that satisfy both the complementary slackness conditions and the non-negativity requirements $\mu_\ell \geq 0$ and $x_\ell \geq 0$ consist of the union of the non-negative x_ℓ-axis and the non-negative μ_ℓ-axis. This set has a severe "kink" at the origin, corresponding to the extreme non-linearity in the complementary slackness condition. Consequently:

- linearization of the complementary slackness conditions at any point on the x_ℓ-axis yields the x_ℓ-axis, while
- linearization of the complementary slackness conditions at any point on the μ_ℓ-axis yields the μ_ℓ-axis.

That is, linearization of the complementary slackness conditions at any point on the x_ℓ- or μ_ℓ-axes yields a linear approximation that does not represent the kink at all. Based on our discussion in Section 7.4.2.7 of the Newton–Raphson method, linearization of the complementary slackness conditions will not yield a useful update.

The approximation in (16.26) instead allows $\begin{bmatrix} \mu_\ell \\ x_\ell \end{bmatrix}$ to lie on a hyperbolic-shaped set as shown in Figure 16.2. Linearization of (16.26), together with an explicit requirement to avoid the x_ℓ- and μ_ℓ-axes, yields a useful update that can approximately represent the kink in the complementary slackness conditions.

We have remarked that we will solve Problem (16.16) for a sequence of decreasing values of t. As $t \to 0$, points that satisfy (16.26) will approach satisfaction of the complementary slackness conditions:

$$Mx = \mathbf{0},$$

for Problem (16.1). That is, as t is reduced, the hyperbolic-shaped sets become closer to the set of points satisfying the complementary slackness conditions and the non-negativity constraints.

We can re-write (16.26) as:

$$X\mu - t\mathbf{1} = \mathbf{0}, \tag{16.27}$$

which we can re-arrange as $\mu = t[X]^{-1}\mathbf{1}$. Recall that the gradient of the barrier objective, $\nabla\phi : \mathbb{R}^n_{++} \to \mathbb{R}^n$, is given by:

$$\forall x \in \mathbb{R}^n_{++}, \nabla\phi(x) = \nabla f(x) - t[X]^{-1}\mathbf{1}.$$

Substituting the expression for $\nabla\phi$ into (16.23) and making the substitution $\mu = t[X]^{-1}\mathbf{1}$, we obtain:

$$\nabla f(x) + A^\dagger\lambda - \mu = \mathbf{0}. \tag{16.28}$$

For convenience, we repeat (16.24):

$$Ax = b. \tag{16.29}$$

Equations (16.27)–(16.29) are equivalent to (16.23)–(16.24) in that:

- a solution of (16.23)–(16.24) satisfies (16.28)–(16.29), given that μ is defined by (16.27), and
- a solution of (16.27)–(16.29) satisfies (16.23)–(16.24).

To summarize, the hyperbolic approximation to the complementary slackness conditions together with (16.28) and (16.29) are equivalent to the first-order necessary conditions for minimizing Problem (16.22).

Moreover, (16.28) and (16.29) are two of the lines of the first-order necessary conditions for Problem (16.1). The condition (16.27) becomes more nearly equivalent to the complementary slackness conditions for Problem (16.1) as $t \to 0$. Instead of seeking $x^\star$ and $\lambda^\star$ that satisfy (16.23)–(16.24), we will seek $x^\star$, $\lambda^\star$, and $\mu^\star$ that satisfy (16.27)–(16.29).

Step direction We can use the Newton–Raphson method to find a step direction to solve (16.27)–(16.29). The iterations start with a guess $\mu^{(0)}$, $x^{(0)}$, and $\lambda^{(0)}$. We linearize (16.27)–(16.29) about $\mu^{(0)}$, $x^{(0)}$, and $\lambda^{(0)}$ and seek an update $\begin{bmatrix} \Delta\mu^{(0)} \\ \Delta x^{(0)} \\ \Delta\lambda^{(0)} \end{bmatrix}$

to make $\begin{bmatrix} \mu^{(1)} \\ x^{(1)} \\ \lambda^{(1)} \end{bmatrix} = \begin{bmatrix} \mu^{(0)} \\ x^{(0)} \\ \lambda^{(0)} \end{bmatrix} + \begin{bmatrix} \Delta\mu^{(0)} \\ \Delta x^{(0)} \\ \Delta\lambda^{(0)} \end{bmatrix}$ more nearly satisfy (16.27)–(16.29). As

mentioned above, we are linearizing (16.27), which is already an approximation to the complementary slackness conditions. The Newton–Raphson step direction to solve (16.27)–(16.29) is given by:

$$\begin{bmatrix} X^{(\nu)} & M^{(\nu)} & 0 \\ -I & \nabla^2 f(x^{(\nu)}) & A^\dagger \\ 0 & A & 0 \end{bmatrix} \begin{bmatrix} \Delta\mu^{(\nu)} \\ \Delta x^{(\nu)} \\ \Delta\lambda^{(\nu)} \end{bmatrix} = \begin{bmatrix} -X^{(\nu)}\mu^{(\nu)} + t\mathbf{1} \\ -\nabla f(x^{(\nu)}) - A^\dagger \lambda^{(\nu)} + \mu^{(\nu)} \\ -Ax^{(\nu)} + b \end{bmatrix},$$

where $M^{(\nu)} = \text{diag}\{\mu_\ell^{(\nu)}\}$ and $X^{(\nu)} = \text{diag}\{x_\ell^{(\nu)}\}$.

Symmetry The Newton–Raphson update equations have a coefficient matrix that is not symmetric. By multiplying the first block row of the equations through by $-[M^{(\nu)}]^{-1}$ on the left, we can create the symmetric system:

$$\begin{bmatrix} -[M^{(\nu)}]^{-1}X^{(\nu)} & -I & 0 \\ -I & \nabla^2 f(x^{(\nu)}) & A^\dagger \\ 0 & A & 0 \end{bmatrix} \begin{bmatrix} \Delta\mu^{(\nu)} \\ \Delta x^{(\nu)} \\ \Delta\lambda^{(\nu)} \end{bmatrix}$$

$$= \begin{bmatrix} x^{(\nu)} - t[M^{(\nu)}]^{-1}\mathbf{1} \\ -\nabla f(x^{(\nu)}) - A^\dagger \lambda^{(\nu)} + \mu^{(\nu)} \\ -Ax^{(\nu)} + b \end{bmatrix}. \tag{16.30}$$

This system is symmetric, but indefinite. In general, to factorize it we must make use of the special factorization algorithms for indefinite matrices as mentioned in Section 5.4.7.

Block pivoting of Jacobian and sparsity issues Unfortunately, the top left-hand block of the coefficient matrix of this system may have entries that are very large and entries that are very small, depending on whether or not the corresponding constraint $x_\ell \geq 0$ is binding. This means that the coefficient matrix can be ill-conditioned. Moreover, the entries in the first sub-vector of the right-hand side involve the differences between numbers that may be large and approximately the same magnitude. This means that the problem of solving for the update direction can be ill-conditioned. An approach to avoiding the ill-conditioning is described in [84, sections 16.3–16.4]; however, it will in general require us to use a QR factorization of the coefficient matrix, which may be unattractive for sparse systems.

Here we will assume that the ill-conditioning does not lead to any significant problems. (This generally requires that $x^{(\nu)}$ and $\mu^{(\nu)}$ are close enough to the exact minimizer and Lagrange multipliers of the barrier problem for the given value of t. A discussion of this issue is contained in [124].) Moreover, we will see that we can deal analytically with the entries in the top left-hand block of the coefficient matrix because of its simple structure. We will do this by block factorizing the Jacobian using the diagonal matrix $-[M^{(v)}]^{-1}X^{(v)}$ as block pivot, noting that we can explicitly invert $-[M^{(v)}]^{-1}X^{(v)}$ to obtain $-[X^{(v)}]^{-1}M^{(v)}$. We obtain:

$$
\begin{bmatrix} \mathbf{I} & \mathbf{0} & \mathbf{0} \\ -[X^{(v)}]^{-1}M^{(v)} & \mathbf{I} & \mathbf{0} \\ \mathbf{0} & \mathbf{0} & \mathbf{I} \end{bmatrix} \begin{bmatrix} -[M^{(v)}]^{-1}X^{(v)} & -\mathbf{I} & \mathbf{0} \\ -\mathbf{I} & \nabla^2 f(x^{(v)}) & A^\dagger \\ \mathbf{0} & A & \mathbf{0} \end{bmatrix}
$$

$$
= \begin{bmatrix} -[M^{(v)}]^{-1}X^{(v)} & -\mathbf{I} & \mathbf{0} \\ \mathbf{0} & \nabla^2 f(x^{(v)}) + [X^{(v)}]^{-1}M^{(v)} & A^\dagger \\ \mathbf{0} & A & \mathbf{0} \end{bmatrix}. \qquad (16.31)
$$

The only terms in the lower right-hand four blocks that have been altered are the diagonal entries in $\nabla^2 f(x^{(v)})$. These terms are increased, effectively "convexifying" the problem [65, section 3]. (See [65, section 3] for discussion regarding adjusting these entries so that they do not become too large.)

Since the diagonal entries in $\nabla^2 f$ are often non-zero, block pivoting on the diagonal matrix $-[M^{(v)}]^{-1}X^{(v)}$ usually introduces no new fill-ins. If we had chosen instead to pivot on any other entries first, before pivoting on the entries of $-[M^{(v)}]^{-1}X^{(v)}$, then we would typically have introduced several fill-ins. Therefore, pivoting on the top left-hand block first is typically consistent with the heuristic for choosing pivots to minimize fill-ins as discussed in Section 5.5.3.2. (In fact, we deliberately ordered the equations and variables to be consistent with the minimum fill-in heuristic.) Another interpretation is that we have used the first block row of (16.30) to eliminate $\Delta\mu^{(v)}$.

After block pivoting, we can directly factorize the remaining bottom right-hand four blocks of the coefficient matrix on the right-hand side of (16.31) using an indefinite factorization algorithm. The system corresponding to these bottom right-hand four blocks is very similar to the system (16.25) in the primal interior point algorithm, except that $[X^{(v)}]^{-1}M^{(v)}$ has replaced $t[X^{(v)}]^{-2}$. These two terms would be equal if (16.27) were satisfied exactly.

Selection of step-size   If we set:

$$
\begin{bmatrix} \mu^{(v+1)} \\ x^{(v+1)} \\ \lambda^{(v+1)} \end{bmatrix} = \begin{bmatrix} \mu^{(v)} \\ x^{(v)} \\ \lambda^{(v)} \end{bmatrix} + \begin{bmatrix} \Delta\mu^{(v)} \\ \Delta x^{(v)} \\ \Delta\lambda^{(v)} \end{bmatrix},
$$

we may find that the new value of the iterate violates the non-negativity constraints on μ or x. To avoid this we may have to take a step that is shorter than the full step

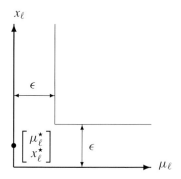

Fig. 16.10. Using a fixed tolerance to enforce non-negativity will prevent convergence to a minimizer.

direction. That is, we modify the update to:

$$
\begin{bmatrix} \mu^{(\nu+1)} \\ x^{(\nu+1)} \\ \lambda^{(\nu+1)} \end{bmatrix}
= \begin{bmatrix} \mu^{(\nu)} \\ x^{(\nu)} \\ \lambda^{(\nu)} \end{bmatrix}
+ \alpha^{(\nu)} \begin{bmatrix} \Delta\mu^{(\nu)} \\ \Delta x^{(\nu)} \\ \Delta\lambda^{(\nu)} \end{bmatrix},
$$

where $\alpha^{(\nu)}$ is the **step-size**.

Similarly to the case of "swapping in" discussed in Section 16.3.2, we should also consider a sufficient decrease criterion for ϕ. That is, in *addition* to constraining $\alpha^{(\nu)}$ so that we do not violate the constraints $x^{(\nu+1)} > \mathbf{0}$ and $\mu^{(\nu+1)} > \mathbf{0}$, we should also include a sufficient decrease criterion for $\phi(x)$.

The strict non-negativity constraints are somewhat problematic. For example, suppose that we implement the requirement of strict non-negativity by choosing a tolerance $\epsilon > 0$ and requiring that the next iterate satisfies $x_\ell^{(\nu+1)} \geq \epsilon, \forall \ell$, and $\mu_\ell^{(\nu+1)} \geq \epsilon, \forall \ell$. A serious drawback of this approach is that *a priori* we do not know how close the minimizer of Problem (16.22) is to the boundary. In particular, our choice of $\alpha^{(\nu)}$ should not prevent the iterates from converging to a minimizer that is very close to the boundary. However, as t is reduced, the minimizers may become close to the boundary. In particular, if a constraint $x_\ell \geq 0$ is binding at the minimizer of Problem (16.1) then we would hope that the $x_\ell^{(\nu)}$ would become close to zero.

Picking a fixed tolerance $\epsilon > 0$ and choosing $\alpha^{(\nu)}$ at every iteration so that $x_\ell^{(\nu+1)} \geq \epsilon, \forall \ell$, and $\mu_\ell^{(\nu+1)} \geq \epsilon, \forall \ell$, will not work. This is because $\mu^{(\nu)}$ and $x^{(\nu)}$ can never in this case get within ϵ of the boundary, as illustrated in Figure 16.10. By complementary slackness, either the entry $x_\ell^\star$ of the minimizer $x^\star$ of Problem (16.22), or the entry $\mu_\ell^\star$ of the vector of Lagrange multipliers $\mu^\star$, or both, will be within ϵ of the boundary. Therefore, our algorithm will be unable to converge to the minimizer if we prevent the iterates from getting closer than ϵ to the boundary.

That is, we must adjust the tolerance so that iterates can, asymptotically, approach the boundary. One scheme is to pick $\alpha^{(\nu)} \leq 1$ at each iteration so that

$\begin{bmatrix} \mu^{(\nu+1)} \\ x^{(\nu+1)} \end{bmatrix}$ is no closer than a fixed *fraction,* say 0.9995, of the distance from the

current iterate $\begin{bmatrix} \mu^{(\nu)} \\ x^{(\nu)} \end{bmatrix}$ to the boundary of $x \geq \mathbf{0}$, $\mu \geq \mathbf{0}$ under the L_∞ norm. With

this choice, $\mu^{(\nu)}$ and $x^{(\nu)}$ *can* approach any point that satisfies the complementary slackness condition. There are many variations on the choice of step-size.

It is also possible to use a different step-size for:

- the *primal* variables x, and
- the *dual* variables μ and λ.

That is, we can update according to:

$$x^{(\nu+1)} = x^{(\nu)} + \alpha_{\text{primal}}^{(\nu)} \Delta x^{(\nu)},$$

$$\begin{bmatrix} \mu^{(\nu+1)} \\ \lambda^{(\nu+1)} \end{bmatrix} = \begin{bmatrix} \mu^{(\nu)} \\ \lambda^{(\nu)} \end{bmatrix} + \alpha_{\text{dual}}^{(\nu)} \begin{bmatrix} \Delta\mu^{(\nu)} \\ \Delta\lambda^{(\nu)} \end{bmatrix},$$

where $\alpha_{\text{primal}}^{(\nu)}$ is chosen to preserve the strict non-negativity of x and $\alpha_{\text{dual}}^{(\nu)}$ is chosen to preserve the strict non-negativity of μ. In fact, the main advantage of the primal–dual algorithm over the primal algorithm is the ability to use a different step-size for x and for μ and λ to control the approach to the boundary of the feasible region separately for these variables. (See, for example, [73, 124].) However, we will not take advantage of this flexibility in our development and examples.

16.4.3.4 Example

Let us apply the primal–dual interior point algorithm to our example Problem (16.4) from Sections 2.3.2.3, 16.1.1.2, 16.3.3.2, and 16.4.2.6.

Terms in update From Section 16.1.1.2, we have the following for Problem (16.4):

$$\forall x \in \mathbb{R}^2, f(x) = x_1 - x_2,$$

$$\forall x \in \mathbb{R}^2, \nabla f(x) = \begin{bmatrix} 1 \\ -1 \end{bmatrix},$$

$$\forall x \in \mathbb{R}^2, \nabla^2 f(x) = \begin{bmatrix} 0 & 0 \\ 0 & 0 \end{bmatrix},$$

$$A = \begin{bmatrix} 1 & 1 \end{bmatrix},$$

$$= \mathbf{1}^\dagger,$$

$$b = [1].$$

Factorization We must factorize the coefficient matrix in (16.30), which we will symbolize by $\mathcal{A}$:

$$\mathcal{A} \;=\; \begin{bmatrix} -[M^{(\nu)}]^{-1}X^{(\nu)} & -\mathbf{I} & \mathbf{0} \\ -\mathbf{I} & \nabla^2 f(x) & A^{\dagger} \\ \mathbf{0} & A & \mathbf{0} \end{bmatrix},$$

$$= \begin{bmatrix} -[M^{(\nu)}]^{-1}X^{(\nu)} & -\mathbf{I} & \mathbf{0} \\ -\mathbf{I} & \mathbf{0} & \mathbf{1} \\ \mathbf{0} & \mathbf{1}^{\dagger} & \mathbf{0} \end{bmatrix}.$$

This matrix is indefinite and, in general, we should use a special purpose factorization algorithm. Here, we will simply apply LU factorization, using the symbols $\mathcal{A}^{(j)}$ and $\mathcal{M}^{(j)}$ for the matrices created at the j-th stage of factorization. (Note that $M^{(\nu)} = \mathrm{diag}\{\mu_\ell^{(\nu)}\}$.) Block pivoting of $\mathcal{A}$ using its top-left block $-[M^{(\nu)}]^{-1}X^{(\nu)}$ as pivot yields $\mathcal{M}^{(1)}$ and $\mathcal{A}^{(1)}$ given by:

$$\mathcal{M}^{(1)} \;=\; \begin{bmatrix} \mathbf{I} & \mathbf{0} & \mathbf{0} \\ -[X^{(\nu)}]^{-1}M^{(\nu)} & \mathbf{I} & \mathbf{0} \\ \mathbf{0} & \mathbf{0} & \mathbf{I} \end{bmatrix},$$

$$\mathcal{A}^{(1)} \;=\; \begin{bmatrix} -[M^{(\nu)}]^{-1}X^{(\nu)} & -\mathbf{I} & \mathbf{0} \\ \mathbf{0} & [X^{(\nu)}]^{-1}M^{(\nu)} & \mathbf{1} \\ \mathbf{0} & \mathbf{1}^{\dagger} & \mathbf{0} \end{bmatrix}.$$

Block pivoting of $\mathcal{A}^{(1)}$ using $[X^{(\nu)}]^{-1}M^{(\nu)}$ as pivot yields:

$$\mathcal{M}^{(2)} = \begin{bmatrix} \mathbf{I} & \mathbf{0} & \mathbf{0} \\ \mathbf{0} & \mathbf{I} & \mathbf{0} \\ \mathbf{0} & -\mathbf{1}^{\dagger}[M^{(\nu)}]^{-1}X^{(\nu)} & \mathbf{I} \end{bmatrix},$$

$$= \begin{bmatrix} \mathbf{I} & \mathbf{0} & \mathbf{0} \\ \mathbf{0} & \mathbf{I} & \mathbf{0} \\ \mathbf{0} & -[x^{(\nu)}]^{\dagger}[M^{(\nu)}]^{-1} & \mathbf{I} \end{bmatrix},$$

$$\mathcal{A}^{(2)} = \mathcal{M}^{(2)}\mathcal{A}^{(1)},$$

$$= \begin{bmatrix} -[M^{(\nu)}]^{-1}X^{(\nu)} & -\mathbf{I} & \mathbf{0} \\ \mathbf{0} & [X^{(\nu)}]^{-1}M^{(\nu)} & \mathbf{1} \\ \mathbf{0} & \mathbf{0} & -\mathbf{1}^{\dagger}[M^{(\nu)}]^{-1}X^{(\nu)}\mathbf{1} \end{bmatrix},$$

$$= \begin{bmatrix} -[M^{(\nu)}]^{-1}X^{(\nu)} & -\mathbf{I} & \mathbf{0} \\ \mathbf{0} & [X^{(\nu)}]^{-1}M^{(\nu)} & \mathbf{1} \\ \mathbf{0} & \mathbf{0} & -[\mu_1^{(\nu)}]^{-1}x_1^{(\nu)} - [\mu_2^{(\nu)}]^{-1}x_2^{(\nu)} \end{bmatrix},$$

so that we can factorize $\mathcal{A}$ into:

$$\mathcal{L} = \begin{bmatrix} I & 0 & 0 \\ [X^{(\nu)}]^{-1}M^{(\nu)} & I & 0 \\ 0 & [x^{(\nu)}]^{\dagger}[M^{(\nu)}]^{-1} & 1^{\dagger} \end{bmatrix},$$

$$\mathcal{U} = \begin{bmatrix} -[M^{(\nu)}]^{-1}X^{(\nu)} & -I & 0 \\ 0 & [X^{(\nu)}]^{-1}M^{(\nu)} & 1 \\ 0 & 0 & -[\mu_1^{(\nu)}]^{-1}x_1^{(\nu)} - [\mu_2^{(\nu)}]^{-1}x_2^{(\nu)} \end{bmatrix}.$$

Initial guess As an initial guess, we pick:

$$\begin{aligned} x_1^{(0)} &= 0.5, \\ x_2^{(0)} &= 0.5, \\ \lambda^{(0)} &= 2, \\ t^{(0)} &= 0.25, \\ \mu_1^{(0)} &= t^{(0)}/x_1^{(0)} = 0.25/0.5 = 0.5, \\ \mu_2^{(0)} &= t^{(0)}/x_2^{(0)} = 0.25/0.5 = 0.5. \end{aligned}$$

The value of $t^{(0)}$ is large enough to yield a useful update direction for the initial guess $x^{(0)}$, $\lambda^{(0)}$, and $\mu^{(0)}$. Exercise 16.22 explores the situation if we instead chose a value of $t^{(0)}$ that is too small.

We chose $x^{(0)}$ to satisfy $Ax^{(0)} = b$. However, $x^{(0)} = \begin{bmatrix} 0.5 \\ 0.5 \end{bmatrix}$ is in the "middle" of the region $Ax = b, x \geq 0$ and is not close to the minimizer of Problem (16.4). Moreover, we will find that $\lambda^{(0)}$ is far from the Lagrange multiplier value. In other words, $x^{(0)}$ has *not* been chosen to make solution especially easy. We chose $\mu^{(0)}$ to satisfy $M^{(0)}x^{(0)} = t^{(0)}1$. In Section 16.4.5, we will discuss in more detail the selection of an initial point satisfying the requirements $x > 0$ and $\mu > 0$.

Step direction The right-hand side of (16.30) is given by:

$$\mathcal{B} = \begin{bmatrix} x^{(0)} - t[M^{(0)}]^{-1}1 \\ -\nabla f(x^{(0)}) - A^{\dagger}\lambda^{(0)} + \mu^{(0)} \\ -Ax^{(0)} + b \end{bmatrix},$$

$$= \begin{bmatrix} x_1^{(0)} - t^{(0)}[\mu_1^{(0)}]^{-1} \\ x_2^{(0)} - t^{(0)}[\mu_2^{(0)}]^{-1} \\ -1 - \lambda^{(0)} + \mu_1^{(0)} \\ 1 - \lambda^{(0)} + \mu_2^{(0)} \\ -x_1^{(0)} - x_2^{(0)} + 1 \end{bmatrix},$$

$$= \begin{bmatrix} 0 \\ 0 \\ -2.5 \\ -0.5 \\ 0 \end{bmatrix}.$$

We must solve the system:

$$\mathcal{A} \begin{bmatrix} \Delta\mu_1^{(0)} \\ \Delta\mu_2^{(0)} \\ \Delta x_1^{(0)} \\ \Delta x_2^{(0)} \\ \Delta\lambda^{(0)} \end{bmatrix} = \mathcal{B},$$

Performing forwards substitution to solve $\mathcal{L}\mathcal{Y} = \mathcal{B}$, we obtain:

$$\mathcal{Y} = \begin{bmatrix} 0 \\ 0 \\ -2.5 \\ -0.5 \\ 3 \end{bmatrix}.$$

Performing backwards substitution to solve $\mathcal{U} \begin{bmatrix} \Delta\mu_1^{(0)} \\ \Delta\mu_2^{(0)} \\ \Delta x_1^{(0)} \\ \Delta x_2^{(0)} \\ \Delta\lambda^{(0)} \end{bmatrix} = \mathcal{Y}$, we obtain:

$$\begin{bmatrix} \Delta\mu_1^{(0)} \\ \Delta\mu_2^{(0)} \\ \Delta x_1^{(0)} \\ \Delta x_2^{(0)} \\ \Delta\lambda^{(0)} \end{bmatrix} = \begin{bmatrix} 1 \\ -1 \\ -1 \\ 1 \\ -1.5 \end{bmatrix}.$$

First iterate If we set:

$$\begin{bmatrix} \mu^{(1)} \\ x^{(1)} \\ \lambda^{(1)} \end{bmatrix} = \begin{bmatrix} \mu^{(0)} \\ x^{(0)} \\ \lambda^{(0)} \end{bmatrix} + \begin{bmatrix} \Delta\mu^{(0)} \\ \Delta x^{(0)} \\ \Delta\lambda^{(0)} \end{bmatrix}, \tag{16.32}$$

we will obtain:

$$
\begin{bmatrix} \mu_1^{(1)} \\ \mu_2^{(1)} \\ x_1^{(1)} \\ x_2^{(1)} \\ \lambda^{(1)} \end{bmatrix} = \begin{bmatrix} 1.5 \\ -0.5 \\ -0.5 \\ 1.5 \\ 0.5 \end{bmatrix},
$$

which will not satisfy the non-negativity constraints on x or μ. Nevertheless, the direction specified by $\begin{bmatrix} \Delta\mu^{(0)} \\ \Delta x^{(0)} \\ \Delta\lambda^{(0)} \end{bmatrix}$ points from $\begin{bmatrix} \mu^{(0)} \\ x^{(0)} \\ \lambda^{(0)} \end{bmatrix}$ in a direction that improves the solution. The full step direction $\begin{bmatrix} \Delta\mu^{(0)} \\ \Delta x^{(0)} \\ \Delta\lambda^{(0)} \end{bmatrix}$, however, can take us too far. Instead, we will update according to:

$$
\begin{bmatrix} \mu^{(1)} \\ x^{(1)} \\ \lambda^{(1)} \end{bmatrix} = \begin{bmatrix} \mu^{(0)} \\ x^{(0)} \\ \lambda^{(0)} \end{bmatrix} + \alpha^{(0)} \begin{bmatrix} \Delta\mu^{(0)} \\ \Delta x^{(0)} \\ \Delta\lambda^{(0)} \end{bmatrix},
$$

where $0 < \alpha^{(0)} < 1$ is chosen to prevent the iterates from going outside $\mu > 0, x > 0$. (As discussed previously, it is also possible to use different step-sizes for the primal variables x and for the dual variables μ and λ.)

For the initial guess $\begin{bmatrix} \mu^{(0)} \\ x^{(0)} \end{bmatrix} = \begin{bmatrix} 0.5 \\ 0.5 \\ 0.5 \\ 0.5 \end{bmatrix}$, the boundary is 0.5 unit away in the L_∞ norm. Using the step-size rule suggested in Section 16.4.3.3, we pick $\alpha^{(0)} \le 1$ so that we come no closer than $(0.9995) \times 0.5$ units of the distance towards the boundary under the L_∞ norm. That is, we choose the largest $\alpha^{(0)} \le 1$ such that:

$$
\alpha^{(0)} \begin{bmatrix} \Delta\mu_1^{(0)} \\ \Delta\mu_2^{(0)} \\ \Delta x_1^{(0)} \\ \Delta x_2^{(0)} \end{bmatrix} \ge -0.9995 \begin{bmatrix} \mu_1^{(0)} \\ \mu_2^{(0)} \\ x_1^{(0)} \\ x_2^{(0)} \end{bmatrix},
$$

which yields $\alpha^{(0)} = 0.49975$ and:

$$
\begin{bmatrix} \mu_1^{(1)} \\ \mu_2^{(1)} \\ x_1^{(1)} \\ x_2^{(1)} \\ \lambda^{(1)} \end{bmatrix} = \begin{bmatrix} 0.99975 \\ 0.00025 \\ 0.00025 \\ 0.99975 \\ 1.250375 \end{bmatrix}.
$$

In general, we could also consider further reduction of $\alpha^{(0)}$ to ensure satisfaction of a sufficient decrease criterion for the objective. However, this particular problem has a linear objective and therefore the objective decreases monotonically with increasing $\alpha^{(0)}$ along the step direction. That is, we should not reduce $\alpha^{(0)}$ further for this problem.

16.4.4 Adjustment of the barrier parameter

In this section, we discuss reduction of the barrier parameter.

16.4.4.1 Sequence of equality-constrained problems

In principle, we could continue iterating with a fixed value $t = t^{(0)}$ until we approach a minimizer $x^{(0)\star}$ of equality-constrained Problem (16.22). We could then use $x^{(0)\star}$ as the starting point for the Newton–Raphson method for Problem (16.22) for a smaller value of t. That is, we would be accurately solving a sequence of equality-constrained problems for points on the central path.

However, we want to reduce t as quickly as possible so that the iterates converge quickly to a minimizer of inequality-constrained Problem (16.1). Taking many iterations to solve equality-constrained Problem (16.22) very close to optimality for a fixed value of t is therefore unattractive, particularly for values of t that are large. This is because we only use the solution to Problem (16.22) as an initial guess for Problem (16.22) for a smaller value of t.

16.4.4.2 Reduction of barrier parameter at every iteration

The minimizer of Problem (16.1) can typically be approached more quickly by reducing t after *every* Newton–Raphson update. For Problem (16.4), we started far from its minimizer with an initial guess of $x^{(0)} = \begin{bmatrix} 0.5 \\ 0.5 \end{bmatrix}$ and used a relatively large value of $t = t^{(0)} = 0.25$. Nevertheless, $x^{(1)}$ is actually very close to the minimizer of inequality-constrained Problem (16.4). That is, $x^{(1)}$ can be thought of as being close to a minimizer of Problem (16.17) for a much smaller value of t than $t^{(0)}$. In this particular case we would like to reduce t significantly. In other cases, we might not be so fortunate and may have to reduce t more slowly. Exercise 16.22 shows that reducing t by too much will yield a very poor update direction because the step-size to maintain non-negativity of the iterates will be very small.

16.4.4.3 Effective value of barrier parameter

We would like a measure of how close the current iterate is to a minimizer of the original inequality-constrained problem and adjust t accordingly. That is, instead of interpreting $x^{(1)}$ as an *approximate* minimizer of Problem (16.22) for $t = t^{(0)}$, we will see if we can interpret $x^{(1)}$ as an *exact* (or nearly exact) minimizer of Problem (16.22) for some other, hopefully smaller, value of t. We think of this value of t as the effective value $t^{(1)}_{\text{effective}}$ for which $x^{(1)}$ is nearly the minimizer of Problem (16.22). We will then pick $t^{(1)} < t^{(1)}_{\text{effective}}$ for the value of t to apply in the next Newton–Raphson update to calculate $x^{(2)}$.

By continuing in this way we will construct a sequence $\{t^{(v)}_{\text{effective}}\}^{\infty}_{v=0}$ and corresponding (approximate) minimizers $x^{(v)}$ of Problem (16.22) for $t = t^{(v)}_{\text{effective}}$. If the sequence $\{t^{(v)}_{\text{effective}}\}^{\infty}_{v=0}$ converges to 0 then we have achieved our goal of a sequence of minimizers of Problem (16.22) with $t \to 0$. We will have avoided the effort of performing many iterations at each value of the barrier parameter t to solve Problem (16.22). Exercise 16.23 explores the case where each $x^{(v)}$ can be interpreted as the minimizer of Problem (16.16) for a value of barrier parameter $t = t^{(v)}_{\text{effective}}$.

To interpret the iterates as in Exercise 16.23, recall that we have been trying to solve (16.27)–(16.29). We are going to interpret $\begin{bmatrix} \mu^{(1)} \\ x^{(1)} \\ \lambda^{(1)} \end{bmatrix}$ together with a value $t^{(1)}_{\text{effective}}$ as nearly satisfying (16.27)–(16.29). We will assume that (16.28) and (16.29) are very nearly satisfied by $\mu^{(1)}$ and $x^{(1)}$. If f is quadratic or linear then (16.28) and (16.29) are linear. If (16.28) and (16.29) are satisfied by $\mu^{(0)}$ and $x^{(0)}$, then they will also be satisfied at each successive iteration. Even if (16.28) and (16.29) are not satisfied exactly by $\mu^{(0)}$ and $x^{(0)}$, the next iterates $\mu^{(1)}$ and $x^{(1)}$ will more nearly satisfy these equations.

Satisfying (16.27) exactly is not possible unless, for all all ℓ, the values of $x^{(1)}_{\ell}\mu^{(1)}_{\ell}$ are the same, and all equal to $t^{(1)}_{\text{effective}}$. However, if we let:

$$t^{(1)}_{\text{effective}} = \frac{[x^{(1)}]^{\dagger}\mu^{(1)}}{n}, \qquad (16.33)$$

where n is the length of x, so that $t^{(1)}_{\text{effective}}$ is the average value of $x^{(1)}_{\ell}\mu^{(1)}_{\ell}$, and, moreover, if the values of $x^{(1)}_{\ell}\mu^{(1)}_{\ell}$ do not vary too much with ℓ, then:

$$X^{(1)}\mu^{(1)} - t^{(1)}_{\text{effective}}\mathbf{1} \approx \mathbf{0}.$$

That is, $x^{(1)}$ and $\mu^{(1)}$ satisfy (16.27) approximately for $t = t^{(1)}_{\text{effective}}$.

16.4.4.4 *Update of barrier parameter*

We now set:

$$t^{(1)} < t^{(1)}_{\text{effective}}.$$

For example, we could choose:

$$t^{(1)} = \frac{t^{(1)}_{\text{effective}}}{n},$$

$$= \frac{[x^{(1)}]^{\dagger}\mu^{(1)}}{(n)^2}.$$

For large n, this reduces t significantly at each step. (See [47, page 212][65, section 1][73][84, chapter 17][102, chapter 7] for alternative interpretations and rules for the adjustment of t.)

We now must solve (or approximately solve) the barrier problem for the updated value $t = t^{(1)}$. As initial guess for the minimizer of the barrier problem for $t = t^{(1)}$ we can use $\mu^{(1)}, x^{(1)}, \lambda^{(1)}$. We calculate the Newton–Raphson step direction $\begin{bmatrix} \Delta\mu^{(1)} \\ \Delta x^{(1)} \\ \Delta\lambda^{(1)} \end{bmatrix}$, and update according to:

$$\begin{bmatrix} \mu^{(2)} \\ x^{(2)} \\ \lambda^{(2)} \end{bmatrix} = \begin{bmatrix} \mu^{(1)} \\ x^{(1)} \\ \lambda^{(1)} \end{bmatrix} + \alpha^{(1)} \begin{bmatrix} \Delta\mu^{(1)} \\ \Delta x^{(1)} \\ \Delta\lambda^{(1)} \end{bmatrix},$$

where $\alpha^{(1)}$ is chosen to ensure that the $x^{(2)}$ and $\mu^{(2)}$ strictly satisfy the non-negativity constraints.

16.4.4.5 Adjustment of barrier parameter in example problem

In Problem (16.4), since $n = 2$ is rather small, we will take an even more aggressive approach and set:

$$t^{(1)} = \frac{1}{10}t^{(1)}_{\text{effective}},$$

$$= 2.499375 \times 10^{-5},$$

for $t^{(1)}_{\text{effective}}$ calculated according to (16.33). The value of $t^{(1)}$ is significantly smaller than $t^{(0)}$. We now calculate the next iterate using forwards and backwards substitution. First, we have $\mathcal{L}\mathcal{Y} = \mathcal{B}$, where:

$$\mathcal{L} = \begin{bmatrix} I & 0 & 0 \\ [X^{(1)}]^{-1}M^{(1)} & I & 0 \\ 0 & [x^{(1)}]^{\dagger}[M^{(1)}]^{-1} & 1 \end{bmatrix},$$

$$= \begin{bmatrix} 1 & 0 & 0 & 0 & 0 \\ 0 & 1 & 0 & 0 & 0 \\ 3999 & 0 & 1 & 0 & 0 \\ 0 & 2.501 \times 10^{-4} & 0 & 1 & 0 \\ 0 & 0 & 2.501 \times 10^{-4} & 3999 & 1 \end{bmatrix},$$

$$
\mathcal{B} = \begin{bmatrix} x_1^{(1)} - t^{(1)}[\mu_1^{(1)}]^{-1} \\ x_2^{(1)} - t^{(1)}[\mu_2^{(1)}]^{-1} \\ -1 - \lambda^{(1)} + \mu_1^{(1)} \\ 1 - \lambda^{(1)} + \mu_2^{(1)} \\ -x_1^{(1)} - x_2^{(1)} + 1 \end{bmatrix},
$$

$$
= \begin{bmatrix} 2.250 \times 10^{-4} \\ 0.899775 \\ -1.251 \\ -0.250 \\ 0 \end{bmatrix},
$$

so that:

$$
\mathcal{Y} = \begin{bmatrix} 2.250 \times 10^{-4} \\ 0.899775 \\ -2.150 \\ -0.250 \\ 1001.050 \end{bmatrix}.
$$

Now we solve $\mathcal{U} \begin{bmatrix} \Delta\mu^{(1)} \\ \Delta x^{(1)} \\ \Delta\lambda^{(1)} \end{bmatrix} = \mathcal{Y}$, where:

$$
\mathcal{U} = \begin{bmatrix} -[M^{(1)}]^{-1}X^{(1)} & -\mathbf{I} & \mathbf{0} \\ \mathbf{0} & [X^{(1)}]^{-1}M^{(1)} & \mathbf{1} \\ \mathbf{0} & \mathbf{0} & -[\mu_1^{(1)}]^{-1}x_1^{(1)} - [\mu_2^{(1)}]^{-1}x_2^{(1)} \end{bmatrix},
$$

$$
= \begin{bmatrix} -2.501 \times 10^{-4} & 0 & -1 & 0 & 0 \\ 0 & -3999 & 0 & -1 & 0 \\ 0 & 0 & 3999 & 0 & 1 \\ 0 & 0 & 0 & 2.501 \times 10^{-4} & 1 \\ 0 & 0 & 0 & 0 & -3999 \end{bmatrix},
$$

so that:

$$
\begin{bmatrix} \Delta\mu_1^{(1)} \\ \Delta\mu_2^{(1)} \\ \Delta x_1^{(1)} \\ \Delta x_2^{(1)} \\ \Delta\lambda^{(1)} \end{bmatrix} = \begin{bmatrix} 1.000 \\ -2.251 \times 10^{-4} \\ -4.751 \times 10^{-4} \\ 4.755 \times 10^{-4} \\ -0.25032 \end{bmatrix}.
$$

x_2

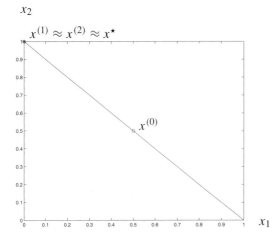

Fig. 16.11. The progress of the primal–dual interior point algorithm in x coordinates for Problem (16.4). The feasible set is indicated by the solid line.

Again, solving for $\alpha^{(1)}$ to bring the next iterate no closer than 0.9995 of the distance to the boundary of $x \geq \mathbf{0}$, $\mu \geq \mathbf{0}$ under the L_∞ norm we find $\alpha^{(1)} = 0.526$ and:

$$
\begin{bmatrix}
\mu_1^{(2)} \\
\mu_2^{(2)} \\
x_1^{(2)} \\
x_2^{(2)} \\
\lambda^{(2)}
\end{bmatrix}
=
\begin{bmatrix}
1.525428 \\
1.317 \times 10^{-4} \\
3.056 \times 10^{-7} \\
0.999999875 \\
1.119
\end{bmatrix}.
$$

After only two iterations, $x^{(2)}$ is extremely close to the minimizer of Problem (16.4), which is $x^\star = \begin{bmatrix} 0 \\ 1 \end{bmatrix}$. (Stopping criteria are discussed in Section 16.4.6.4 and in Exercise 16.25.) The optimal values of the other variables are $\mu^\star = \begin{bmatrix} 2 \\ 0 \end{bmatrix}$ and $\lambda^\star = [1]$. The progress of the algorithm in the x coordinates is shown in Figure 16.11. The set defined by the equality constraints is also shown as a line. The progress of the algorithm in μ and λ coordinates is shown in Figure 16.12.

Asymptotically, $\begin{bmatrix} \mu^{(\nu)} \\ x^{(\nu)} \end{bmatrix}$ can approach the boundary of $\mu \geq \mathbf{0}$ and $x \geq \mathbf{0}$. Even at iteration $\nu = 2$, the change in x is small, the equality constraint $Ax = b$ is satisfied, and $t_{\text{effective}}^{(2)} = 6.61 \times 10^{-5}$, using (16.33). For practical purposes, we could say that the problem has been solved, at least from the perspective of finding an accurate estimate of $x^\star$. More iterations are need to obtain accurate estimates of $\mu^\star$ and $\lambda^\star$. (See Exercises 16.17 and 16.25.)

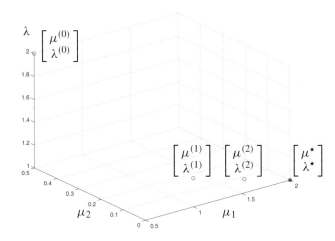

Fig. 16.12. The progress of the primal–dual interior point algorithm in μ and λ coordinates for Problem (16.4).

16.4.4.6 Rate of convergence

For larger and more complex problems, we should expect to take more iterations to approach an accurate answer and we might expect to use a less aggressive reduction of the barrier parameter t at each iteration. Empirically, however, even large problems usually take no more than a few tens of iterations to solve to high accuracy. Variants of this algorithm can be proven to converge super-linearly or quadratically for linear and quadratic programming problems and for some other types of convex objectives. See [84, section 17.7] for the flavor of the results. See [6, section 9.3][86][93][102] for detailed analysis under various assumptions on the objective.

16.4.5 Finding an initial feasible guess

As with the active set algorithm, we must find an initial feasible guess in **phase 1** before proceeding to minimize the objective in **phase 2**. The approach described in Section 16.3.5 used an active set method to solve Problem (16.13) that yielded a point on the boundary of the feasible set of Problem (16.1), assuming that some feasible point existed. That is, an active set method applied to the related Problem (16.13) yielded a point with some or all of the entries of x being zero, which was then suitable as an initial guess for the active set method applied to Problem (16.1) itself.

As discussed in Section 16.4.3.3, however, we require that the initial guess for the primal–dual interior point algorithm satisfies $x > \mathbf{0}$ and $\mu > \mathbf{0}$. We will define another problem related to Problem (16.1) that includes artificial variables and apply the primal–dual interior point algorithm to it. There are a number of possible ways to define the related problem. For example, ([6, section 9.5][12, section 9.4][79, section 6.IX][102, section 20.5]), let $x^{(0)} \in \mathbb{R}^n_{++}$, suppose A has linearly independent rows, define $\tilde{b} = b - Ax^{(0)}$, and consider the problem:

$$\min_{x \in \mathbb{R}^n, w \in \mathbb{R}} \{w \mid Ax + \tilde{b}w = b, x \geq 0, w \geq 0\}. \tag{16.34}$$

Note that $x^{(0)}$ and $w^{(0)} = 1$ satisfies the equality constraints and strictly satisfies the inequality constraints of Problem (16.34) and is therefore a feasible initial guess for this problem that can be used by the primal–dual interior point algorithm. (The values of the dual variables for the non-negativity constraints can be initialized to satisfy $M^{(0)}x^{(0)} = t^{(0)}\mathbf{1}$ and $\sigma^{(0)}w^{(0)} = t^{(0)}\mathbf{1}$, where $\sigma \in \mathbb{R}$ is the dual variable corresponding to the non-negativity constraint $w \geq 0$.) We solve this problem using the primal–dual interior point algorithm and this feasible initial guess. If, at a minimizer $\begin{bmatrix} x^\star \\ w^\star \end{bmatrix}$ of Problem (16.34), we find that $w^\star = 0$ then $x^\star$ satisfies the equality and inequality constraints of Problem (16.1). If the Slater condition holds for Problem (16.1) then there is a minimizer $\begin{bmatrix} x^\star \\ w^\star \end{bmatrix}$ of Problem (16.34) such that $x^\star$ satisfies the equality constraints and strictly satisfies the inequality constraints of Problem (16.1). The primal–dual interior point algorithm can then use $x^\star$ as an initial guess for solving Problem (16.1). If $w^\star > 0$ then Problem (16.1) is infeasible. (See Exercise 16.24. Also see Exercise 3.16 for another example of a related problem that can be used to find an initial guess that strictly satisfies the inequality constraints.)

16.4.6 Summary

The primal–dual interior point algorithm to solve Problem (16.1) has involved many stages and considerable development. We summarize the main steps of the algorithm in this section.

16.4.6.1 Initial guess

The algorithm begins with an initial guess $\begin{bmatrix} \mu^{(0)} \\ x^{(0)} \\ \lambda^{(0)} \end{bmatrix}$ satisfying $Ax^{(0)} = b$, $\mu^{(0)} > 0$, $x^{(0)} > 0$, and with an initial barrier parameter $t^{(0)}$. We may try to arrange that $M^{(0)}x^{(0)} = t^{(0)}\mathbf{1}$.

16.4.6.2 General iteration

Newton–Raphson step direction At the ν-th iteration we solve (16.30) for the

Newton–Raphson step direction $\begin{bmatrix} \Delta\mu^{(\nu)} \\ \Delta x^{(\nu)} \\ \Delta\lambda^{(\nu)} \end{bmatrix}$. The coefficient matrix has been par-

tially block factorized as shown in (16.31). The factorization should be completed
by an algorithm for symmetric indefinite matrices as mentioned in Section 5.4.7
and in conjunction with an optimal ordering algorithm as discussed in Section 5.5
to keep the matrix factors as sparse as possible. Forwards and backwards substitution is then used to evaluate the step direction as discussed in Section 5.2.1. Variants to avoid complete factorization at each iteration are also possible as discussed
in Section 7.2. A further discussion of this is contained in [84, section 16.4].

Step-size The iterate is updated according to:

$$\begin{bmatrix} \mu^{(\nu+1)} \\ x^{(\nu+1)} \\ \lambda^{(\nu+1)} \end{bmatrix} = \begin{bmatrix} \mu^{(\nu)} \\ x^{(\nu)} \\ \lambda^{(\nu)} \end{bmatrix} + \alpha^{(\nu)} \begin{bmatrix} \Delta\mu^{(\nu)} \\ \Delta x^{(\nu)} \\ \Delta\lambda^{(\nu)} \end{bmatrix},$$

where $\alpha^{(\nu)}$ is chosen so that $\mu^{(\nu+1)} > \mathbf{0}$ and $x^{(\nu+1)} > \mathbf{0}$, (and possibly also to satisfy
a sufficient decrease criterion for the barrier objective ϕ.) One rule to guarantee
non-negativity of $\mu^{(\nu+1)}$ and $x^{(\nu+1)}$ is to set:

$$\alpha^{(\nu)} = \min\left\{1.0, 0.9995 \times \left[\min_{\ell\in\{1,\ldots,n\}}\left\{\left.\frac{\mu_\ell^{(\nu)}}{-\Delta\mu_\ell^{(\nu)}}\right| \Delta\mu_\ell^{(\nu)} < 0\right\}\right],\right.$$

$$\left. 0.9995 \times \left[\min_{\ell\in\{1,\ldots,n\}}\left\{\left.\frac{x_\ell^{(\nu)}}{-\Delta x_\ell^{(\nu)}}\right| \Delta x_\ell^{(\nu)} < 0\right\}\right]\right\},$$

but the step-size may have to be reduced further to satisfy the sufficient decrease
criterion for the barrier objective ϕ. (In some implementations, the step-size for x
is chosen separately from the step-size for λ and μ.)

16.4.6.3 Update of barrier parameter

We then update the value of the barrier parameter using a rule such as:

$$t^{(\nu+1)} = \frac{\sum_{\ell=1}^n \mu_\ell^{(\nu+1)} x_\ell^{(\nu+1)}}{(n)^2},$$

where μ and x are of length n. (In some applications, a less aggressive reduction in
the barrier parameter may be necessary. In our example problem, we used a more
aggressive reduction.)

16.4.6.4 Stopping criteria

The iterations continue until $t^{(\nu)}$ is sufficiently reduced, the change in iterates is small, and the first-order necessary conditions of Problem (16.1) are satisfied sufficiently accurately. In the case of linear and quadratic programs, we can use duality to develop a stopping criterion that guarantees closeness of $f(x^{(\nu)})$ to the minimum. In particular, if at each iteration ν we generate iterates $x^{(\nu)} > \mathbf{0}$, $\lambda^{(\nu)}$, and $\mu^{(\nu)} > \mathbf{0}$ that satisfy (16.28)–(16.29) then we can use duality to bound the error in the estimate of the infimum by:

$$f(x^{(\nu)}) - \inf_{x\in\mathbb{R}^n}\{f(x)|Ax = b, x \geq \mathbf{0}\} \leq [\mu^{(\nu)}]^\dagger x^{(\nu)}.$$

If the problem has a minimum and we iterate until:

$$[\mu^{(\nu)}]^\dagger x^{(\nu)} \leq \epsilon_f,$$

then $f(x^{(\nu)})$ will be within ϵ_f of the minimum. (See Exercise 16.25.)

16.4.7 Discussion and variations

The equality constraints (16.29) are linear. Linearization introduces no error in these constraints. If f is quadratic then (16.28) is linear in μ, x, and λ. That is, if f is quadratic then linearizing (16.28) introduces no error so that the Newton–Raphson update can exactly predict the changes necessary to satisfy the conditions (16.28)–(16.29). Similarly, if f is approximately quadratic then the update will approximately predict the changes necessary to satisfy (16.28)–(16.29).

On the other hand, (16.27) is always non-linear and we neglect important terms when we linearize it. A development of the primal–dual algorithm we have described, called the primal–dual **predictor–corrector method**, uses the factorization of (16.30) for two successive updates, one of which is used to bring the iterates closer to being on the central path by reducing the variation of $x_\ell^{(\nu)}\mu_\ell^{(\nu)}$ with ℓ. We will not describe this development in this book, but it is discussed in [84, section 9.6][102, section 20.4.3].

Finally, if the problem formulation requires non-negativity constraints on only some of the entries of x, then the barrier function terms and the corresponding Lagrange multipliers can be omitted for the other, unconstrained, entries. We will make use of this observation in Section 17.3.1.

16.5 Summary

In this chapter, we have described optimality conditions for non-negatively constrained minimization problems, considering also the special case of convex problems. We then considered active set algorithms briefly and interior point algorithms in more detail as algorithms to solve non-negatively constrained problems.

Exercises

Optimality conditions

16.1 Give examples of problems of the form of Problem (16.1) where there is no minimum because of the following.

(i) There are no feasible points.
(ii) The objective f is unbounded below on the feasible set (but f is twice partially differentiable with continuous second partial derivatives on the feasible set and there are feasible points).
(iii) The objective f is not continuous.

16.2 Let $f : \mathbb{R}^n \to \mathbb{R}$ be twice partially differentiable with continuous second partial derivatives, $A \in \mathbb{R}^{m \times n}$, and $b \in \mathbb{R}^m$. Let $\mathbb{A} \subseteq \{1, \dots, n\}$ be any subset of the indices of the decision vector x and suppose that there are r elements in $\mathbb{A}$. Define $C \in \mathbb{R}^{r \times n}$ to be the matrix consisting of those rows of the identity matrix corresponding to the indices in $\mathbb{A}$. For example, if $\mathbb{A} = \{3, 5\}$ then $r = 2$ and:

$$C = \begin{bmatrix} 0 & 0 & 1 & 0 & 0 & \cdots \\ 0 & 0 & 0 & 0 & 1 & \cdots \end{bmatrix} \in \mathbb{R}^{2 \times n}.$$

Suppose that the matrix:

$$\hat{A} = \begin{bmatrix} A \\ C \end{bmatrix},$$

has linearly independent rows. Consider the following generalization of Problem (16.3):

$$\min_{x \in \mathbb{R}^n} \{ f(x) | Ax = b, -x_\ell = 0, \forall \ell \in \mathbb{A} \} = \min_{x \in \mathbb{R}^n} \{ f(x) | Ax = b, -Cx = 0 \}.$$

Suppose that $x^\star \in \mathbb{R}^n$, $\lambda^\star \in \mathbb{R}^m$, and $\mu^\star \in \mathbb{R}^r$ satisfy the second-order sufficient conditions for this problem. That is, $x^\star$ is a local minimizer with Lagrange multipliers $\lambda^\star$ and $\mu^\star$ corresponding to the constraints $Ax = b$ and $-Cx = 0$, respectively.
  Consider the perturbed problem:

$$\min_{x \in \mathbb{R}^n} \{ f(x) | Ax = b, -Cx = -\gamma \},$$

where $\gamma \in \mathbb{R}^r$. Show that the sensitivity of the minimum of the perturbed problem to γ, evaluated at $\gamma = 0$, is $\mu^\star$. (Hint: Use Corollary 13.11.)

16.3 Consider each of the following objectives $f_k : \mathbb{R}^2 \to \mathbb{R}, k = 1, 2, 3$:

(i) $\forall x \in \mathbb{R}^2, f_1(x) = (x_1 - 1)^2 + (x_2 - 1)^2$,
(ii) $\forall x \in \mathbb{R}^2, f_2(x) = (x_1 - 1)^2 + (x_2 + 1)^2$,
(iii) $\forall x \in \mathbb{R}^2, f_3(x) = (x_1)^2 + (x_2 + 1)^2$.
   For each of these objectives, consider the problem $\min_{x \in \mathbb{R}^2} \{ f_k(x) | x \geq 0 \}$ and answer the following.

(a) Evaluate $x^\star$, $\mu^\star$ that satisfy the first-order necessary conditions for the problem.
(b) Describe $\mathbb{A}(x^\star)$.
(c) Describe $\mathcal{N} = \{\Delta x \in \mathbb{R}^n | \Delta x_\ell = 0, \forall \ell \in \mathbb{A}(x^\star)\}$.
(d) Describe $\mathbb{A}_+(x^\star, \mu^\star)$.
(e) Describe $\mathcal{N}_+ = \{\Delta x \in \mathbb{R}^n | \Delta x_\ell = 0, \forall \ell \in \mathbb{A}_+(x^\star, \mu^\star)\}$.
(f) Are any of the binding constraints degenerate?
(g) Is $x^\star$ the minimizer of the problem?

16.4 Consider Problem (16.1) in the case that $f : \mathbb{R}^n \to \mathbb{R}$ is convex and partially differentiable with continuous partial derivatives and that $A = \mathbf{1}^\dagger \in \mathbb{R}^{1 \times n}$, $b \in \mathbb{R}^1$. That is, there is only one equality constraint and it requires that the sum of the entries in x is equal to b. Show that any minimizer $x^\star$ of this problem satisfies the "minimum first derivative length" property [9, section 5.5]:

$$(x_k^\star > 0) \Rightarrow \left(\frac{\partial f}{\partial x_k}(x^\star) \leq \frac{\partial f}{\partial x_\ell}(x^\star), \forall \ell = 1, \ldots, n \right).$$

(Hint: Apply Theorem 16.1.)

Convex problems

16.5 Consider the problem from Section 16.1.2.4:

$$\min_{x \in \mathbb{R}^2} \{f(x) | x \geq \mathbf{0}\},$$

with objective $f : \mathbb{R}^2 \to \mathbb{R}$ defined by:

$$\forall x \in \mathbb{R}^2, \ f(x) = -(x_1)^3 + (x_2 - 1)^2.$$

Also consider the candidate minimizer $\hat{x} = \begin{bmatrix} 0 \\ 1 \end{bmatrix}$ and candidate value of Lagrange multipliers $\hat{\mu} = \mathbf{0}$. Explain which of the hypotheses of Theorem 16.3 are not satisfied by this problem.

Active set method

16.6 Consider Problem (16.1).

(i) Suppose that any of the n non-negativity constraints can be binding or not binding. How many possibilities are there if we want to enumerate all the cases of constraints being binding and not binding?
(ii) A more usual case is that no more than approximately $(n - m)$ of the non-negativity constraints will be binding, where n is the number of variables and m is the number of equality constraints. How many possibilities are there now if we want to enumerate all the cases where there are no more than $(n - m)$ inequality constraints binding out of the total of n inequality constraints?

16.7 In this exercise we consider swapping out.

(i) Consider Problem (16.11):

$$\min_{x_1, x_2 \in \mathbb{R}} \{x_1 - x_2 | x_1 + x_2 = 1, -x_2 = 0\}.$$

Solve the first-order necessary conditions for this problem. Report the value of the minimizer and corresponding Lagrange multipliers.

(ii) Show that the direction $\Delta x^{(0)} = \begin{bmatrix} -1 \\ 1 \end{bmatrix}$ is a descent direction for the objective of

Problem (16.11) at the point $x^{(0)} = \begin{bmatrix} 1 \\ 0 \end{bmatrix}$.

(iii) Show that the direction $\Delta x^{(0)}$ maintains feasibility for $Ax = b$ at $x^{(0)}$.

16.8 Consider Problem (16.12):

$$\min_{x_1, x_2 \in \mathbb{R}} \{x_1 - x_2 | x_1 + x_2 = 1, -x_1 = 0\}.$$

(i) Show that $x^\star = \begin{bmatrix} 0 \\ 1 \end{bmatrix}$ is a minimizer of this problem.

(ii) Show that the corresponding Lagrange multiplier on the constraint $-x_1 = 0$ is positive.

16.9 Let $A \in \mathbb{R}^{m \times n}$, $b \in \mathbb{R}^m$ and consider Problem (16.13):

$$\min_{x \in \mathbb{R}^n, w \in \mathbb{R}^n} \{\mathbf{1}^\dagger w | Ax + w = b, x \geq \mathbf{0}, w \geq \mathbf{0}\}.$$

Suppose that a minimizer, $\begin{bmatrix} x^\star \\ w^\star \end{bmatrix}$, of this problem satisfies $w^\star \neq \mathbf{0}$. Show that $\{x \in \mathbb{R}^n | Ax = b, x \geq \mathbf{0}\} = \emptyset$.

16.10 Solve Problem (16.4),

$$\min_{x \in \mathbb{R}^2} \{x_1 - x_2 | x_1 + x_2 = 1, x_1 \geq 0, x_2 \geq 0\},$$

with the MATLAB function linprog by setting the LargeScale option to off and the Simplex option to on using the optimset function.

16.11 Consider an **integer optimization** problem:

$$\min_{x \in \mathbb{Z}^n} \{c^\dagger x | Ax = b, x \geq \mathbf{0}\},$$

where $\mathbb{Z}$ is the set of integers. Suppose that $b \in \mathbb{Z}^m$ and that $A \in \{0, -1, 1\}^{m \times n}$ is **totally unimodular**; that is, every square sub-matrix of A has determinant equal to either 0, -1, or 1 [67, section 4.12]. Consider the continuous relaxation of this problem:

$$\min_{x \in \mathbb{R}^n} \{c^\dagger x \mid Ax = b, x \geq 0\}.$$

Show that some minimizer of the relaxed problem is also a minimizer of the integer optimization problem. (Hint: You can use the observation made in Section 16.3.6.1 that, for a
linear program, there is a minimizer that is a vertex. Moreover, you can also assume that
([67, theorem 4.12.7]):

- if A is totally unimodular then so is $\begin{bmatrix} A \\ I \end{bmatrix}$, and
- if A is totally unimodular then any sub-matrix of A is also totally unimodular.)

Interior point algorithm

16.12 Consider the **reciprocal barrier function** $f_b : \mathbb{R}^n_{++} \to \mathbb{R}$ for the constraints
$x \geq 0$ defined by:

$$\forall x \in \mathbb{R}^n_{++}, \ f_b(x) = \sum_{\ell=1}^{n} \frac{1}{x_\ell}.$$

(i) Show that the reciprocal barrier function is partially differentiable with continuous
partial derivatives on $\mathbb{R}^n_{++}$.
(ii) Show that for each ℓ, the term $1/x_\ell$ in the definition of the reciprocal barrier function is such that:

$$\forall B \in \mathbb{R}, \exists \underline{x}_\ell \in \mathbb{R}_{++} \text{ such that } \left((0 < x_\ell \leq \underline{x}_\ell) \Rightarrow (1/x_\ell > B) \right).$$

(iii) Show that the reciprocal barrier function is convex on $\mathbb{R}^n_{++}$.

16.13 Consider the logarithmic barrier function $f_b : \mathbb{R}^n_{++} \to \mathbb{R}$ for the constraints $x \geq 0$
defined by:

$$\forall x \in \mathbb{R}^n_{++}, \ f_b(x) = -\sum_{\ell=1}^{n} \ln(x_\ell).$$

(i) Show that the logarithmic barrier function is partially differentiable with continuous partial derivatives on $\mathbb{R}^n_{++}$. (Hint: You can assume that ln is differentiable on
$\mathbb{R}_{++}$ with continuous derivative.)
(ii) Show that for each ℓ, the term $-\ln(x_\ell)$ in the definition of the logarithmic barrier
function is such that:

$$\forall B \in \mathbb{R}, \exists \underline{x}_\ell \in \mathbb{R}_{++} \text{ such that } \left((0 < x_\ell \leq \underline{x}_\ell) \Rightarrow (-\ln(x_\ell) > B) \right).$$

(iii) Show that the logarithmic barrier function is convex on $\mathbb{R}^n_{++}$.

16.14 Consider the **modified logarithmic barrier function** $f_b : \mathbb{R}_{++}^n \to \mathbb{R}$ for the constraints $x \geq 0$ defined by:

$$\forall x \in \mathbb{R}_{++}^n, \, f_b(x) = -\sum_{\ell=1}^n \ln(x_\ell) + \mathbf{1}^\dagger x - n,$$

$$= \sum_{\ell=1}^n (-\ln(x_\ell) + x_\ell - 1),$$

$$= \sum_{\ell=1}^n f_b^\ell(x_\ell),$$

where the functions $f_b^\ell : \mathbb{R}_{++} \to \mathbb{R}, \, \ell = 1, \ldots, n$, are defined by:

$$\forall \ell = 1, \ldots, n, \, \forall x_\ell \in \mathbb{R}_{++}, \, f_b^\ell(x_\ell) = -\ln(x_\ell) + x_\ell - 1.$$

(i) Show that the modified logarithmic barrier function is partially differentiable with continuous partial derivatives on $\mathbb{R}_{++}^n$. (Hint: You can assume that $\ln$ is differentiable on $\mathbb{R}_{++}$ with continuous derivative.)

(ii) Show that for each ℓ, the function f_b^ℓ is such that:

$$\forall B \in \mathbb{R}, \exists \underline{x}_\ell \in \mathbb{R}_{++} \text{ such that } \left((0 < x_\ell \leq \underline{x}_\ell) \Rightarrow (f_b^\ell(x_\ell) > B) \right).$$

(iii) Show that for each ℓ, the function ∇f_b^ℓ is such that:

$$(x_\ell \geq 1) \Rightarrow (\nabla f_b^\ell(x_\ell) \geq 0).$$

(iv) Show that the modified logarithmic barrier function is convex on $\mathbb{R}_{++}^n$.
(v) Show that the modified logarithmic barrier function is non-negative on $\mathbb{R}_{++}^n$.

16.15 Consider the inequality-constrained problem:

$$\min_{x \in \mathbb{R}} \left\{ \frac{-1}{(x+1)^2 + 1} \, \middle| \, x \geq 0 \right\}.$$

(This example appears in [84, section 16.2, exercise 4] and is attributed to Powell.) The objective is shown by the solid curve in Figure 16.13. Although the objective of this problem is not convex, it has convex level sets, is bounded below by -1, and a descent direction for the objective from a feasible point is always in the direction of decreasing x. Problem (16.16) in this case is $\min_{x \in \mathbb{R}} \{\phi(x) \mid x > 0\}$, where the barrier objective $\phi : \mathbb{R}_{++} \to \mathbb{R}$ is defined by:

$$\forall x \in \mathbb{R}_{++}, \, \phi(x) = \frac{-1}{(x+1)^2 + 1} - t \ln(x).$$

The barrier objective is shown as the dashed curve in Figure 16.13 for $t = 0.1$. The barrier objective has a local minimum for $x \approx 0$, but the barrier objective is unbounded below as $x \to \infty$ so that the local minimum is not a global minimum and, indeed, the barrier problem does not have a global minimum. If we restrict ourselves to the vicinity of $x = 0$ then the sequence of local minimizers of the barrier problems as $t \to 0$ will converge to the

$f(x), \phi(x) = \frac{-1}{(x+1)^2+1} - t\ln(x)$

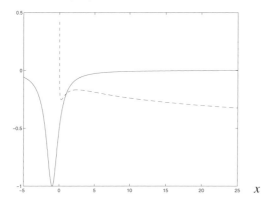

Fig. 16.13. Example of a barrier objective that is un-bounded below. The function f is shown solid while the barrier objective, for $t = 0.1$, is shown dashed.

minimizer $x^\star = 0$ of the inequality-constrained problem. However, if we use an iterative algorithm to solve the barrier problem and start with an initial point that is far from $x = 0$, then the iterative algorithm may generate a sequence of iterates that diverge to ∞.

Instead of the logarithmic barrier, consider using the modified logarithmic barrier function $f_b : \mathbb{R}_{++} \to \mathbb{R}$ for $x \geq 0$ from Exercise 16.14. The modified barrier is defined by:

$$\forall x \in \mathbb{R}_{++}, \ f_b(x) = -\ln(x) + x - 1.$$

Show that the corresponding barrier problem for this barrier function has a global minimum and minimizer for each value of t. (Hint: Use the results from Exercise 16.14 and Theorem 2.1.)

16.16 Prove that if:

- $\{x^{(\nu)}\}_{\nu=0}^{\infty}$ converges to $x^\star$, and
- $\forall \nu, \ Ax^{(\nu)} = b, x^{(\nu)} > 0$,

then $Ax^\star = b$ and $x^\star \geq 0$.

16.17 Consider Problem (16.4), which we repeat here:

$$\min_{x \in \mathbb{R}^2} \{x_1 - x_2 | x_1 + x_2 = 1, x_1 \geq 0, x_2 \geq 0\}.$$

The iterates $x^{(1)}$ and $x^{(2)}$ generated by the primal–dual interior point algorithm were calculated in Section 16.4.4.5. Use the primal–dual interior point algorithm to calculate two more updates, starting at $x^{(2)}$. That is:

(i) Calculate the third iterate $x^{(3)}$.
(ii) Calculate the fourth iterate $x^{(4)}$.
(iii) Comment on the improvement or otherwise in accuracy of each of the components of μ, x, and λ.

For each iteration, use $t^{(\nu)} = \frac{1}{10}t^{(\nu)}_{\text{effective}} = \frac{1}{10n}[x^{(\nu)}]^{\dagger}\mu^{(\nu)}$ and use the step-size rule suggested in Section 16.4.6.2 to allow the next iterate to be no closer to the boundary than a fraction 0.9995 of the distance of the current iterate to the boundary under the L_∞ norm.

16.18 Consider the problem:

$$\min_{x \in \mathbb{R}^3} \{c^\dagger x \mid Ax = b, x \geq 0\},$$

where:

$$c = \begin{bmatrix} 2 \\ -1 \\ 11 \end{bmatrix},$$

$$A = 1^\dagger,$$

$$b = \begin{bmatrix} 10 \end{bmatrix}.$$

(i) Use the primal–dual interior point algorithm to calculate one update to find the minimizer of this problem, starting at $x^{(\nu)} = \begin{bmatrix} 1 \\ 3 \\ 6 \end{bmatrix}$, $\lambda^{(\nu)} = 2$, $t^{(\nu)} = 0.25$, choosing $\mu^{(\nu)}$ to satisfy $M^{(\nu)} x^{(\nu)} = t^{(\nu)} 1$, and using the rule suggested in Section 16.4.6.2 to determine the step-size $\alpha^{(\nu)}$. In particular, calculate $x^{(\nu+1)}$, $\lambda^{(\nu+1)}$, and $\mu^{(\nu+1)}$.

(ii) At the point $x^{(\nu)}$, the gradient of the objective matches that described in Section 16.3.2.3. Compare the update calculated in Part (i) to that calculated in Section 16.3.2.3 and shown in Figure 16.5.

16.19 In this exercise, we sketch the central path for the problem in Exercise 16.18.

(i) Use the MATLAB function `fmincon` to solve Problem (16.16) for the objective and constraints specified in Exercise 16.18 for values of barrier parameter $t = 100, 90, 80, \ldots, 10, 9, 8, \ldots, 1, 0.9, 0.8, \ldots, 0.1$. Enforce the strict non-negativity constraints by requiring $x \geq 0$ and using an initial guess that satisfies the strict non-negativity constraints. That is, solve:

$$\min_{x \in \mathbb{R}^3} \left\{ c^\dagger x - t \sum_{\ell=1}^n \ln(x_\ell) \mid Ax = b, x \geq 0 \right\},$$

where:

$$c = \begin{bmatrix} 2 \\ -1 \\ 11 \end{bmatrix},$$

$$A = 1^\dagger,$$

$$b = \begin{bmatrix} 10 \end{bmatrix}.$$

You should write a MATLAB M-file to evaluate the objective and its gradient and set the `GradObj` option to on using the `optimset` function. For all other options, use default parameters. For $t = 1$, use an initial guess of $x^{(0)} = \begin{bmatrix} 1 \\ 3 \\ 6 \end{bmatrix}$. For each subsequent value of t, use as initial guess the minimizer for the previous value of t.

(ii) Use the results from the previous part to sketch the central path.

16.20 Suppose that we try to minimize the objective defined in (16.18) in Section 16.4.2.6 using the Newton–Raphson step direction to find a zero of the derivative of (16.18).

  (i) Calculate the derivative of (16.18).
  (ii) Calculate the Hessian of (16.18).
  (iii) Calculate the Newton–Raphson step direction to seek a zero of the derivative of (16.18).
  (iv) Evaluate the Newton–Raphson step direction for the first iteration with the values
$$t = 10^{-10} \text{ and } x_1^{(0)} = 0.5.$$
  (v) Why is this step direction problematic?

16.21 Consider the problem:

$$\min_{x \in \mathbb{R}^2} \{ f(x) | x \geq 0 \},$$

where $f : \mathbb{R}^2 \to \mathbb{R}$ is defined by:

$$\forall x \in \mathbb{R}^2, \ f(x) = (x_1 + 1)^2 + (x_2 + 1)^2.$$

Use the primal–dual interior point algorithm to calculate two updates to find the minimizer of this problem, starting at $x^{(0)} = \begin{bmatrix} 5 \\ 5 \end{bmatrix}$ and $t^{(0)} = 1$. That is:

  (i) Calculate $x^{(1)}$.
  (ii) Calculate $x^{(2)}$.

Use the step-size rule suggested in Section 16.4.6.2 and, for $\nu > 0$, use $t^{(\nu)} = \frac{1}{10} t_{\text{effective}}^{(\nu)} = \frac{1}{10n} [x^{(\nu)}]^\dagger \mu^{(\nu)}$.

16.22 In this exercise, we consider a value of the barrier parameter that is too small.

  (i) Calculate the step direction for the example problem in Section 16.4.3.4 for the values $t = 10^{-10}$, $x^{(0)} = \begin{bmatrix} 0.5 \\ 0.5 \end{bmatrix}$, $\lambda^{(0)} = 1.5 + 2 \times 10^{-2}$, and $\mu^{(0)} = \begin{bmatrix} 2 \times 10^{-10} \\ 2 \times 10^{-10} \end{bmatrix}$.
      (Hint: You should be able to re-use some of the calculations from Section 16.4.3.4.)
  (ii) Use the step-size rule suggested in Section 16.4.6.2 to calculate an appropriate step-size so that the updated iterate remains in the interior. Comment on the size of the step-size.

16.23 In this exercise we investigate the relationship between:

  • the limits (or accumulation points) of the sequences $\{\mu^{(\nu)}\}_{\nu=0}^{\infty}$, $\{x^{(\nu)}\}_{\nu=0}^{\infty}$, and $\{\lambda^{(\nu)}\}_{\nu=0}^{\infty}$ generated by the interior point algorithm, and
  • a solution of the first-order necessary conditions for Problem (16.1).

Suppose that the objective function f is partially differentiable with continuous partial derivatives. Consider a sequence of values for the barrier parameter $\{t_{\text{effective}}^{(\nu)}\}_{\nu=0}^{\infty}$ that converge to $t^\star = 0$. Suppose that for each value of ν, the ν-th elements of the sequences $\{\mu^{(\nu)}\}_{\nu=0}^{\infty}$, $\{x^{(\nu)}\}_{\nu=0}^{\infty}$, and $\{\lambda^{(\nu)}\}_{\nu=0}^{\infty}$ satisfy (16.27)–(16.29) for $t = t_{\text{effective}}^{(\nu)}$ and that $x^{(\nu)} > 0$ and $\mu^{(\nu)} > 0$ for every ν. Moreover, assume that $\{\mu^{(\nu)}\}_{\nu=0}^{\infty}$, $\{x^{(\nu)}\}_{\nu=0}^{\infty}$, and $\{\lambda^{(\nu)}\}_{\nu=0}^{\infty}$ converge to $\mu^\star$, $x^\star$, and $\lambda^\star$, respectively. Show that $\mu^\star$, $x^\star$, and $\lambda^\star$ satisfy the first-order necessary conditions for Problem (16.1).

16.24 Let $A \in \mathbb{R}^{m \times n}$, $b \in \mathbb{R}^m$, and $x^{(0)} \in \mathbb{R}^n_{++}$. Suppose that A has linearly independent rows and define $\tilde{b} = b - Ax^{(0)}$ and $w^{(0)} = 1$ and consider Problem (16.34):

$$\min_{x \in \mathbb{R}^n, w \in \mathbb{R}} \{w | Ax + \tilde{b}w = b, x \geq \mathbf{0}, w \geq 0\}.$$

(i) Show that $x^{(0)}$, $w^{(0)}$ satisfies $Ax + \tilde{b}w = b$.
(ii) Show that Problem (16.34) possesses a minimum and minimizer. (Hint: You can use the result that if a linear programming problem has a feasible point then either the problem has a minimum and minimizer or it is unbounded below [70, section 3.3].)
(iii) Consider a minimizer, $\begin{bmatrix} x^\star \\ w^\star \end{bmatrix}$, of Problem (16.34). Show that if $w^\star = 0$ then $x^\star$ satisfies the equality and inequality constraints of Problem (16.1). That is, show that $Ax^\star = b$ and $x^\star \geq \mathbf{0}$.
(iv) Consider a minimizer, $\begin{bmatrix} x^\star \\ w^\star \end{bmatrix}$, of Problem (16.34). Show that if $w^\star > 0$ then Problem (16.1) is infeasible.

16.25 Let $f : \mathbb{R}^n \to \mathbb{R}$ be convex and partially differentiable with continuous partial derivatives, $A \in \mathbb{R}^{m \times n}$, and $b \in \mathbb{R}^m$. Consider the non-negatively constrained problem:

$$\min_{x \in \mathbb{R}^n} \{f(x) | Ax = b, x \geq \mathbf{0}\}.$$

Suppose that we use a primal–dual interior point algorithm such as the one described in Section 16.4.3.3 to solve this problem. Moreover, suppose that, as discussed in Section 16.4.7, at each iteration ν we generate iterates $x^{(\nu)} > \mathbf{0}, \lambda^{(\nu)}$, and $\mu^{(\nu)} > \mathbf{0}$ that exactly satisfy (16.28)–(16.29). That is:

$$\forall \nu, \nabla f(x^{(\nu)}) + A^\dagger \lambda^{(\nu)} - \mu^{(\nu)} = \mathbf{0},$$
$$\forall \nu, Ax^{(\nu)} = b,$$
$$\forall \nu, x^{(\nu)} > \mathbf{0},$$
$$\forall \nu, \mu^{(\nu)} > \mathbf{0}.$$

(In general, it will only be possible to satisfy the first condition exactly at each iteration if f is linear or quadratic.)

(i) What is the Lagrangian for this problem? (Hint: Refer to Section 3.4.1, write the equality constraints as $Ax - b = \mathbf{0}$, and write the inequality constraints as $-Ix \leq \mathbf{0}$.)
(ii) Show that, at each iteration ν, $x^{(\nu)}$ is the global minimizer of the Lagrangian $\mathcal{L}(\bullet, \lambda^{(\nu)}, \mu^{(\nu)})$.
(iii) Evaluate $\mathcal{D}(\lambda^{(\nu)}, \mu^{(\nu)})$.
(iv) Show that we can bound the error in the estimate of the infimum by:

$$f(x^{(\nu)}) - \inf_{x \in \mathbb{R}^n} \{f(x) | Ax = b, x \geq \mathbf{0}\} \leq [\mu^{(\nu)}]^\dagger x^{(\nu)}.$$

(Hint: Note that since $\mu^{(\nu)} \geq \mathbf{0}$, we have by Corollary 3.14:

$$f(x^{(\nu)}) - \inf_{x \in \mathbb{R}^n} \{f(x) | Ax = b, x \geq \mathbf{0}\} \leq f(x^{(\nu)}) - \mathcal{D}(\lambda^{(\nu)}, \mu^{(\nu)}).)$$

16.26 In this exercise, we apply the bound calculated in Exercise 16.25:

$$f(x^{(\nu)}) - \inf_{x \in \mathbb{R}^n} \{f(x) | Ax = b, x \geq \mathbf{0}\} \leq [\mu^{(\nu)}]^\dagger x^{(\nu)},$$

to bound the error in the estimate of the minimum for the initial guess and for the iterates calculated for Problem (16.4) in Sections 16.4.3.4 and 16.4.4.5. Evaluate the bound $[\mu^{(\nu)}]^\dagger x^{(\nu)}$ for $\nu = 0, 1, 2$.

16.27 Solve Problem (16.4),

$$\min_{x \in \mathbb{R}^2} \{x_1 - x_2 | x_1 + x_2 = 1, x_1 \geq 0, x_2 \geq 0\},$$

using the MATLAB function `linprog`. Use the default interior point algorithm.

17

Algorithms for linear inequality-constrained minimization

In this chapter we will develop algorithms for constrained optimization problems of the form:

$$\min_{x \in \mathbb{S}} f(x),$$

where $f : \mathbb{R}^n \to \mathbb{R}$ and where the feasible set $\mathbb{S}$ is of the form:

$$\mathbb{S} = \{x \in \mathbb{R}^n | g(x) = \mathbf{0}, h(x) \leq \mathbf{0}\},$$

with both $g : \mathbb{R}^n \to \mathbb{R}^m$ and $h : \mathbb{R}^n \to \mathbb{R}^r$ affine. That is, we will consider problems of the form:

$$\min_{x \in \mathbb{R}^n} \{f(x) | Ax = b, Cx \leq d\}, \tag{17.1}$$

where $A \in \mathbb{R}^{m \times n}$, $b \in \mathbb{R}^m$, $C \in \mathbb{R}^{r \times n}$, and $d \in \mathbb{R}^r$ are constants. We call the constraints $Cx \leq d$ **linear inequality constraints**, although, strictly speaking, it would be more precise to refer to them as **affine inequality constraints**. The feasible set defined by the linear equality and inequality constraints is convex. (See Exercise 2.36.) If f is convex on the feasible set then the problem is convex. We refer to Problem (17.1) as an inequality-constrained problem, where it is understood that it also includes equality constraints in addition to the inequality constraints.

We will first present the optimality conditions in Section 17.1 for the general case and in Section 17.2 for convex problems. In Section 17.3 we show how to apply the algorithms developed in Chapter 16 for non-negatively constrained optimization to Problem (17.1) through two transformations of Problem (17.1). We discuss sensitivity analysis in Section 17.4.

The key issues in this chapter are:

- optimality conditions for **inequality-constrained problems** based on the results for equality-constrained problems,
- optimality conditions for **convex problems**,

- **transformations** of problems, and
- **duality** and **sensitivity analysis**.

17.1 Optimality conditions

In this section we present first-order necessary and second-order sufficient conditions.

17.1.1 First-order necessary conditions

17.1.1.1 Analysis

We have:

Theorem 17.1 *Suppose that* $f : \mathbb{R}^n \to \mathbb{R}$ *is partially differentiable with continuous partial derivatives,* $A \in \mathbb{R}^{m \times n}, b \in \mathbb{R}^m, C \in \mathbb{R}^{r \times n}, d \in \mathbb{R}^r$. *Consider Problem (17.1):*

$$\min_{x \in \mathbb{R}^n} \{f(x)|Ax = b, Cx \le d\},$$

and a point $x^\star \in \mathbb{R}^n$. *If* $x^\star$ *is a local minimizer of Problem (17.1) then:*

$$\exists \lambda^\star \in \mathbb{R}^m, \exists \mu^\star \in \mathbb{R}^r \text{ such that: } \nabla f(x^\star) + A^\dagger \lambda^\star + C^\dagger \mu^\star = 0;$$
$$M^\star(Cx^\star - d) = 0;$$
$$Ax^\star = b;$$
$$Cx^\star \le d; \text{ and}$$
$$\mu^\star \ge 0, \qquad (17.2)$$

where $M^\star = \text{diag}\{\mu_\ell^\star\} \in \mathbb{R}^{r \times r}$. *The vectors* $\lambda^\star$ *and* $\mu^\star$ *satisfying the conditions (17.2) are called the vectors of Lagrange multipliers for the constraints* $Ax = b$ *and* $Cx \le d$, *respectively. The conditions that* $M^\star(Cx^\star - d) = 0$ *are called the* **complementary slackness conditions**. *They say that, for each* ℓ, *either the* ℓ*-th inequality constraint is binding or the* ℓ*-th Lagrange multiplier is equal to zero (or both).*

Proof ([84, section 14.4].) The proof consists of several steps:

(i) showing that $x^\star$ is a local minimizer of the related equality-constrained problem:

$$\min_{x \in \mathbb{R}^n} \{f(x)|Ax = b, C_\ell x = d_\ell, \forall \ell \in \mathbb{A}(x^\star)\},$$

where the active inequality constraints at $x^\star$ for Problem (17.1) are included as equality constraints,

(ii) using the necessary conditions of the related equality-constrained problem to define $\lambda^\star$ and $\mu^\star$ that satisfy the first four lines of (17.2), and

(iii) proving that $\mu^\star \ge 0$ by showing that if a constraint ℓ, say, had a negative value of its Lagrange multiplier $\mu_\ell^\star < 0$ then the objective could be reduced by moving in a direction such that constraint ℓ becomes strictly feasible.

See Appendix B for details. □

x_2

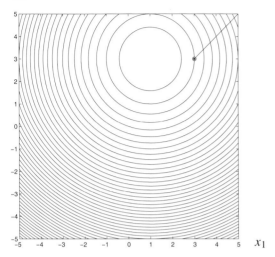

Fig. 17.1. The contour sets of the objective function and feasible set for Problem (2.18). The heights of the contours decrease towards the point $\begin{bmatrix} 1 \\ 3 \end{bmatrix}$. The feasible set is the "half-line" starting at the point $\begin{bmatrix} 3 \\ 3 \end{bmatrix}$, which is also the minimizer and is illustrated with a •.

17.1.1.2 Example

Recall the example quadratic program, Problem (2.18), which we first met in Section 2.3.2.3. The problem is:

$$\min_{x \in \mathbb{R}^2}\{f(x)|Ax = b, Cx \le d\},$$

where:

$$\begin{aligned}
\forall x \in \mathbb{R}^2, f(x) &= (x_1 - 1)^2 + (x_2 - 3)^2, \\
A &= \begin{bmatrix} 1 & -1 \end{bmatrix}, \\
b &= \begin{bmatrix} 0 \end{bmatrix}, \\
C &= \begin{bmatrix} 0 & -1 \end{bmatrix}, \\
d &= \begin{bmatrix} -3 \end{bmatrix}.
\end{aligned}$$

The contours of the objective, the feasible set, and the solution were shown in Figure 2.12, which we repeat in Figure 17.1. In Section 2.3.2, we observed that the solution of this problem was $x^\star = \begin{bmatrix} 3 \\ 3 \end{bmatrix}$.

We claim that $x^\star = \begin{bmatrix} 3 \\ 3 \end{bmatrix}$ together with $\lambda^\star = [-4]$ and $\mu^\star = [4]$ satisfy (17.2) for Problem (2.18). To see this, first observe that:

$$\forall x \in \mathbb{R}^2, \nabla f(x) = \begin{bmatrix} 2 & 0 \\ 0 & 2 \end{bmatrix} x + \begin{bmatrix} -2 \\ -6 \end{bmatrix}.$$

We have:

$$\nabla f(x^\star) + A^\dagger \lambda^\star + C^\dagger \mu^\star$$

$$= \begin{bmatrix} 2 & 0 \\ 0 & 2 \end{bmatrix} \begin{bmatrix} 3 \\ 3 \end{bmatrix} + \begin{bmatrix} -2 \\ -6 \end{bmatrix} + \begin{bmatrix} 1 \\ -1 \end{bmatrix} [-4] + \begin{bmatrix} 0 \\ -1 \end{bmatrix} [4],$$

$$= \mathbf{0};$$

$$\mu^\star(Cx^\star - d) = [4]\left(\begin{bmatrix} 0 & -1 \end{bmatrix} \begin{bmatrix} 3 \\ 3 \end{bmatrix} - [-3] \right),$$

$$= [0];$$

$$Ax^\star = \begin{bmatrix} 1 & -1 \end{bmatrix} \begin{bmatrix} 3 \\ 3 \end{bmatrix},$$

$$= [0],$$

$$= b;$$

$$Cx^\star = \begin{bmatrix} 0 & -1 \end{bmatrix} \begin{bmatrix} 3 \\ 3 \end{bmatrix},$$

$$= [-3],$$

$$\leq [-3],$$

$$= d; \text{ and}$$

$$\mu^\star = [4],$$

$$\geq [0].$$

17.1.1.3 Discussion

As in the non-negatively constrained case, the Lagrange multipliers adjust the unconstrained optimality conditions to balance the constraints against the objective. For the inequality constraints the balance is only needed if the objective would encourage the inequality constraints to be violated. Consequently, the Lagrange multipliers on the inequality constraints are non-negative.

We will again refer to the equality and inequality constraints specified in (17.2) as *the* first-order necessary conditions, although we recognize that the first-order necessary conditions also include, strictly speaking, the other items in the hypothesis of Theorem 17.1. As previously, these conditions are known as the **Kuhn–Tucker** (KT) or the **Karush–Kuhn–Tucker** (KKT) conditions and a point satisfying the conditions is called a **KKT point**.

If the rows of A together with the rows of C corresponding to the binding constraints are not linearly independent then the Lagrange multipliers are not uniquely defined. See [11, section 3.4][45, section 3.3.2] for details and Exercise 17.2 for an example. Linear independence of these rows will play a role in sensitivity analysis in Section 17.4 and in optimality conditions for non-linear inequality constraints to be discussed in Chapter 19. (See Exercise 17.17.)

17.1.1.4 Lagrangian

Recall Definition 3.2 of the **Lagrangian**. For Problem (17.1) the Lagrangian $\mathcal{L}$: $\mathbb{R}^n \times \mathbb{R}^m \times \mathbb{R}^r \to \mathbb{R}$ is defined by:

$$\forall x \in \mathbb{R}^n, \forall \lambda \in \mathbb{R}^m, \forall \mu \in \mathbb{R}^r, \mathcal{L}(x, \lambda, \mu) = f(x) + \lambda^\dagger(Ax - b) + \mu^\dagger(Cx - d).$$

As in the equality-constrained case, define the gradients of $\mathcal{L}$ with respect to x, λ, and μ by, respectively, $\nabla_x \mathcal{L} = \left[\dfrac{\partial \mathcal{L}}{\partial x}\right]^\dagger$, $\nabla_\lambda \mathcal{L} = \left[\dfrac{\partial \mathcal{L}}{\partial \lambda}\right]^\dagger$, and $\nabla_\mu \mathcal{L} = \left[\dfrac{\partial \mathcal{L}}{\partial \mu}\right]^\dagger$.

Evaluating the gradients with respect to x, λ, and μ, we have:

$$
\begin{aligned}
\nabla_x \mathcal{L}(x, \lambda, \mu) &= \nabla f(x) + A^\dagger \lambda + C^\dagger \mu, \\
\nabla_\lambda \mathcal{L}(x, \lambda, \mu) &= Ax - b, \\
\nabla_\mu \mathcal{L}(x, \lambda, \mu) &= Cx - d.
\end{aligned}
$$

Setting the first two of these expressions equal to zero reproduces some of the first-order necessary conditions for the problem. As with equality-constrained problems, the Lagrangian provides a convenient way to remember the optimality conditions. However, unlike the equality-constrained case, in order to recover the first-order necessary conditions for Problem (17.1) we have to:

- add the complementary slackness conditions; that is, $M^\star(Cx^\star - d) = 0$,
- add the non-negativity constraints on μ, that is, $\mu \geq 0$, and
- interpret the third expression as corresponding to inequality constraints; that is, $Cx \leq d$.

If the hypotheses of Theorem 17.1 are satisfied and if, in addition, f is convex then $x^\star$ is a global minimizer of $\mathcal{L}(\bullet, \lambda^\star, \mu^\star)$, where $\lambda^\star$ and $\mu^\star$ are the Lagrange multipliers. Analogously to the equality and non-negatively constrained cases, $\begin{bmatrix} x^\star \\ \lambda^\star \\ \mu^\star \end{bmatrix}$ is not a minimizer of $\mathcal{L}$ but the entries of $\nabla_x \mathcal{L}(x^\star, \lambda^\star, \mu^\star)$ and of $\nabla_\lambda \mathcal{L}(x^\star, \lambda^\star, \mu^\star)$ and some of the entries of $\nabla_\mu \mathcal{L}(x^\star, \lambda^\star, \mu^\star)$ are zero. This observation will play an important role in the discussion of duality in Section 17.2.2.

17.1.2 Second-order sufficient conditions

17.1.2.1 Analysis

In the following, we present second-order sufficient conditions for the minimizer of Problem (17.1).

Theorem 17.2 *Let* $f : \mathbb{R}^n \to \mathbb{R}$ *be twice partially differentiable with continuous second partial derivatives,* $A \in \mathbb{R}^{m \times n}$, $b \in \mathbb{R}^m$, $C \in \mathbb{R}^{r \times n}$, $d \in \mathbb{R}^r$. *Consider Problem (17.1):*

$$\min_{x \in \mathbb{R}^n} \{ f(x) | Ax = b, Cx \leq d \},$$

and points $x^\star \in \mathbb{R}^n$, $\lambda^\star \in \mathbb{R}^m$, *and* $\mu^\star \in \mathbb{R}^r$. *Let* $M^\star = \operatorname{diag}\{\mu_\ell^\star\}$. *Suppose that:*

$$
\begin{aligned}
\nabla f(x^\star) + A^\dagger \lambda^\star + C^\dagger \mu^\star &= \mathbf{0}, \\
M^\star (Cx^\star - d) &= \mathbf{0}, \\
Ax^\star &= b, \\
Cx^\star &\leq d, \\
\mu^\star &\geq \mathbf{0}, \text{ and}
\end{aligned}
$$

$\nabla^2 f(x^\star)$ *is positive definite on the null space:*

$$\mathcal{N}_+ = \{ \Delta x \in \mathbb{R}^n | A\Delta x = \mathbf{0}, C_\ell \Delta x = 0, \forall \ell \in \mathbb{A}_+(x^\star, \mu^\star) \},$$

where C_ℓ *is the* ℓ*-th row of* C *and* $\mathbb{A}_+(x^\star, \mu^\star) = \{ \ell \in \{1, \dots, r\} | C_\ell x^\star = d_\ell, \mu_\ell^\star > 0 \}$. *Then* $x^\star$ *is a strict local minimizer of Problem (17.1).*

Proof See [45, section 3.3.2]. □

The conditions in the theorem are called the **second-order sufficient conditions** (or **SOSC.**) In addition to the first-order necessary conditions, the second-order sufficient conditions require that:

- f is twice partially differentiable with continuous second partial derivatives, and
- $\nabla^2 f(x^\star)$ is positive definite on the null space $\mathcal{N}_+$ defined in the theorem.

17.1.2.2 Example

Recall again the example quadratic program, Problem (2.18) from Sections 2.3.2.3 and 17.1.1.2.

For this problem,

$$
\begin{aligned}
Cx^\star &= d, \\
\mu^\star &= [4], \\
\mathbb{A}_+(x^\star, \mu^\star) &= \{ \ell \in \{1, \dots, r\} | C_\ell x^\star = d_\ell, \mu_\ell^\star > 0 \}, \\
&= \{1\},
\end{aligned}
$$

since the only inequality constraint in this problem is binding and the corresponding Lagrange multiplier is non-zero. Consequently,

$$
\begin{aligned}
\mathcal{N}_+ &= \{ \Delta x \in \mathbb{R}^n | A\Delta x = \mathbf{0}, C_\ell \Delta x = 0, \forall \ell \in \mathbb{A}_+(x^\star, \mu^\star) \}, \\
&= \{ \Delta x \in \mathbb{R}^n | A\Delta x = \mathbf{0}, C\Delta x = 0 \}, \\
&= \{\mathbf{0}\},
\end{aligned}
$$

and $\nabla^2 f(x^\star)$ is positive definite on this null space by definition.

17.1.2.3 Discussion

The sets $\mathcal{N}_+$ and $\mathbb{A}_+(x^\star, \mu^\star)$ have analogous roles to their roles in the case of non-negativity constraints presented in Section 16.1.2. If $\nabla^2 f(x^\star)$ is positive definite on $\mathcal{N}_+$ then there can be no feasible descent directions for f at $x^\star$. As in the non-negatively constrained case, the set $\mathbb{A}_+(x^\star, \mu^\star)$ can be a *strict* subset of $\mathbb{A}(x^\star)$, since it omits those constraints ℓ for which $C_\ell x^\star = d_\ell$ and $\mu_\ell^\star = 0$. Therefore, the null space specified in Theorem 17.2:

$$\mathcal{N}_+ = \{\Delta x \in \mathbb{R}^n | A \Delta x = \mathbf{0}, C_\ell \Delta x = 0, \forall \ell \in \mathbb{A}_+(x^\star, \mu^\star)\},$$

can strictly contain the null space corresponding to the equality constraints and the active inequality constraints; that is $\mathcal{N}_+$ can strictly contain the null space:

$$\mathcal{N} = \{\Delta x \in \mathbb{R}^n | A \Delta x = \mathbf{0}, C_\ell \Delta x = 0, \forall \ell \in \mathbb{A}(x^\star)\}.$$

As in the non-negatively constrained case, constraints for which $C_\ell x^\star = d_\ell$ and $\mu_\ell^\star = 0$ are called **degenerate constraints**.

17.1.2.4 Example of degenerate constraints

Recall again Problem (2.18) from Sections 2.3.2.3, 17.1.1.2 and 17.1.2.2. Consider the following modified version of Problem (2.18):

$$\min_{x \in \mathbb{R}^2}\{f(x) | Ax = b, Cx \le \hat{d}\}, \tag{17.3}$$

where:

$$
\begin{aligned}
\forall x \in \mathbb{R}^2, \ f(x) &= (x_1 - 1)^2 + (x_2 - 3)^2, \\
A &= \begin{bmatrix} 1 & -1 \end{bmatrix}, \\
b &= \begin{bmatrix} 0 \end{bmatrix}, \\
C &= \begin{bmatrix} 0 & -1 \end{bmatrix}, \\
\hat{d} &= \begin{bmatrix} -2 \end{bmatrix}.
\end{aligned}
$$

First consider the relaxation of Problem (17.3) where we neglect the inequality constraint. This relaxation yields Problem (2.13), which we first met in Section 2.3.2.2 and which has minimizer $x^\star = \begin{bmatrix} 2 \\ 2 \end{bmatrix}$. Now notice that:

$$Cx^\star = [-2] \le [-2] = \hat{d},$$

so that $x^\star$ is feasible for Problem (17.3). By Theorem 3.10, $x^\star$ is a also minimizer of Problem (17.3). The contours of the objective, the feasible set, and the minimizer are shown in Figure 17.2.

x_2

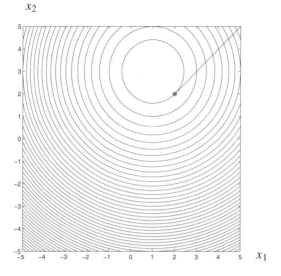

Fig. 17.2. The contour sets of the objective function and feasible set for Problem (17.3). The heights of the contours decrease towards the point $\begin{bmatrix} 1 \\ 3 \end{bmatrix}$. The feasible set is the "half-line" starting at the point $\begin{bmatrix} 2 \\ 2 \end{bmatrix}$, which is also the minimizer and is illustrated with a •.

We claim that $x^\star$ together with $\lambda^\star = [-2]$ and $\mu^\star = [0]$ satisfy the first-order necessary conditions for Problem (17.3). To see this, observe that:

$$\nabla f(x^\star) + A^\dagger \lambda^\star + C^\dagger \mu^\star$$
$$= \begin{bmatrix} 2 & 0 \\ 0 & 2 \end{bmatrix} \begin{bmatrix} 2 \\ 2 \end{bmatrix} + \begin{bmatrix} -2 \\ -6 \end{bmatrix} + \begin{bmatrix} 1 \\ -1 \end{bmatrix} [-2] + \begin{bmatrix} 0 \\ -1 \end{bmatrix} [0],$$
$$= \mathbf{0};$$

$$\mu^\star (Cx^\star - \hat{d}) = [0] \left(\begin{bmatrix} 0 & -1 \end{bmatrix} \begin{bmatrix} 2 \\ 2 \end{bmatrix} - [-2] \right),$$
$$= [0] \times [0],$$
$$= [0];$$

$$Ax^\star = \begin{bmatrix} 1 & -1 \end{bmatrix} \begin{bmatrix} 2 \\ 2 \end{bmatrix},$$
$$= [0],$$
$$= b;$$

$$Cx^\star = \begin{bmatrix} 0 & -1 \end{bmatrix} \begin{bmatrix} 2 \\ 2 \end{bmatrix},$$
$$= [-2],$$
$$\leq [-2],$$
$$= \hat{d}; \text{ and}$$

$$\mu^\star = [0],$$
$$\geq [0].$$

Notice that $Cx^\star = \hat{d}$ and $\mu^\star = [0]$, so that the constraint $Cx \le \hat{d}$ is degenerate. For this problem:

$$
\begin{aligned}
\mathbb{A}_+(x^\star, \mu^\star) &= \{\ell \in \{1, \ldots, r\} | C_\ell x^\star = \hat{d}_\ell, \mu_\ell^\star > 0\}, \\
&= \emptyset, \\
\mathcal{N}_+ &= \{\Delta x \in \mathbb{R}^2 | A\Delta x = \mathbf{0}, C_\ell \Delta x = 0, \forall \ell \in \mathbb{A}_+(x^\star, \mu^\star)\}, \\
&= \{\Delta x \in \mathbb{R}^2 | A\Delta x = \mathbf{0}\}, \\
&= \{\Delta x \in \mathbb{R}^2 | \Delta x_1 = \Delta x_2\}.
\end{aligned}
$$

We have that:

$$
\forall x \in \mathbb{R}^2, \nabla^2 f(x) = \begin{bmatrix} 2 & 0 \\ 0 & 2 \end{bmatrix},
$$

which is positive definite on $\mathbb{R}^2$ and therefore also positive definite on $\mathcal{N}_+$. Therefore, the second-order sufficient conditions hold and, by Theorem 17.2, $x^\star$ is a strict local minimizer of Problem (17.3).

17.1.2.5 Example of second-order sufficient conditions not holding

Consider the following modified version of Problem (17.3) from Section 17.1.2.4:

$$
\min_{x \in \mathbb{R}^2} \{\phi(x) | Ax = b, Cx \le \hat{d}\}, \tag{17.4}
$$

where $\phi : \mathbb{R}^2 \to \mathbb{R}$ is defined by:

$$
\forall x \in \mathbb{R}^2, \phi(x) = -f(x).
$$

That is, we are minimizing $(-f)$ instead of f.

We claim that $\hat{x} = \begin{bmatrix} 2 \\ 2 \end{bmatrix}$ together with $\hat{\lambda} = [2]$ and $\hat{\mu} = [0]$ satisfy the first-order necessary conditions for Problem (17.4). To see this, first observe that:

$$
\forall x \in \mathbb{R}^2, \nabla \phi(x) = \begin{bmatrix} -2 & 0 \\ 0 & -2 \end{bmatrix} x + \begin{bmatrix} 2 \\ 6 \end{bmatrix}.
$$

We have:

$$\nabla \phi(\hat{x}) + A^\dagger \hat{\lambda} + C^\dagger \hat{\mu}$$

$$= \begin{bmatrix} -2 & 0 \\ 0 & -2 \end{bmatrix} \begin{bmatrix} 2 \\ 2 \end{bmatrix} + \begin{bmatrix} 2 \\ 6 \end{bmatrix} + \begin{bmatrix} 1 \\ -1 \end{bmatrix} [2] + \begin{bmatrix} 0 \\ -1 \end{bmatrix} [0],$$

$$= \mathbf{0};$$

$$\hat{\mu}(C\hat{x} - \hat{d}) = [0]\left(\begin{bmatrix} 0 & -1 \end{bmatrix} \begin{bmatrix} 2 \\ 2 \end{bmatrix} - [-2]\right),$$

$$= [0] \times [0],$$

$$= [0];$$

$$A\hat{x} = \begin{bmatrix} 1 & -1 \end{bmatrix} \begin{bmatrix} 2 \\ 2 \end{bmatrix},$$

$$= [0],$$

$$= b;$$

$$C\hat{x} = \begin{bmatrix} 0 & -1 \end{bmatrix} \begin{bmatrix} 2 \\ 2 \end{bmatrix},$$

$$= [-2],$$

$$\leq [-2],$$

$$= \hat{d}; \text{ and}$$

$$\hat{\mu} = [0],$$

$$\geq [0].$$

Notice that again $C\hat{x} = \hat{d}$ and $\hat{\mu} = [0]$. Therefore, if $\hat{x} = \begin{bmatrix} 2 \\ 2 \end{bmatrix}$ and $\hat{\mu} = [0]$ *were* the minimizer and the Lagrange multiplier corresponding to the constraint $Cx \leq \hat{d}$, then this constraint would be degenerate. For this problem:

$$\mathbb{A}_+(\hat{x}, \hat{\mu}) = \{\ell \in \{1, \ldots, r\} | C_\ell \hat{x} = \hat{d}_\ell, \hat{\mu}_\ell > 0\},$$

$$= \emptyset,$$

$$\mathcal{N}_+ = \{\Delta x \in \mathbb{R}^2 | A\Delta x = \mathbf{0}, C_\ell \Delta x = 0, \forall \ell \in \mathbb{A}_+(\hat{x}, \hat{\mu})\},$$

$$= \{\Delta x \in \mathbb{R}^2 | A\Delta x = \mathbf{0}\},$$

$$= \{\Delta x \in \mathbb{R}^2 | \Delta x_1 = \Delta x_2\}.$$

However, we note that $\nabla^2 \phi(\hat{x}) = \begin{bmatrix} -2 & 0 \\ 0 & -2 \end{bmatrix}$ is not positive definite on $\mathcal{N}_+$. (See Exercise 17.5.) Therefore the second-order sufficient conditions do not hold. In fact, $\hat{x}$ is not a minimizer of the problem, since the objective can be reduced by moving away from $\hat{x}$ along the equality constraint so as to make the inequality constraint strictly feasible. This can be seen from Figure 17.2, on noting that the contours of ϕ are the same as those of f, except that the heights of the contours of ϕ decrease *away* from the point $\begin{bmatrix} 1 \\ 3 \end{bmatrix}$. (In fact, $\hat{x}$ maximizes ϕ over the feasible set.)

If we had *erroneously* considered the null space:

$$
\begin{aligned}
\mathcal{N} &= \{\Delta x \in \mathbb{R}^2 | A\Delta x = \mathbf{0}, C_\ell \Delta x = 0, \forall \ell \in \mathbb{A}(\hat{x})\}, \\
&= \{\Delta x \in \mathbb{R}^2 | \Delta x_1 = \Delta x_2, -\Delta x_2 = 0\}, \\
&= \{\mathbf{0}\},
\end{aligned}
$$

then we would not have realized that $\hat{x}$ is not a minimizer.

17.2 Convex problems

As previously, convexity allows us to obtain global optimality results.

17.2.1 First-order sufficient conditions

17.2.1.1 Analysis

If the constraints consist of linear equality and inequality constraints and if f is convex on the feasible set then the problem is convex. Again, in this case, the first-order necessary conditions are also sufficient for optimality.

Theorem 17.3 *Suppose that $f : \mathbb{R}^n \to \mathbb{R}$ is partially differentiable with continuous partial derivatives, $A \in \mathbb{R}^{m \times n}, b \in \mathbb{R}^m, C \in \mathbb{R}^{r \times n}, d \in \mathbb{R}^r$. Consider Problem (17.1):*

$$
\min_{x \in \mathbb{R}^n} \{ f(x) | Ax = b, Cx \le d \},
$$

and points $x^\star \in \mathbb{R}^n, \lambda^\star \in \mathbb{R}^m$, and $\mu^\star \in \mathbb{R}^r$. Let $M^\star = \operatorname{diag}\{\mu_\ell^\star\}$. Suppose that:

(i) *f is convex on $\{x \in \mathbb{R}^n | Ax = b, Cx \le d\}$,*
(ii) *$\nabla f(x^\star) + A^\dagger \lambda^\star + C^\dagger \mu^\star = \mathbf{0}$,*
(iii) *$M^\star (Cx^\star - d) = \mathbf{0}$,*
(iv) *$Ax^\star = b$ and $Cx^\star \le d$, and*
(v) *$\mu^\star \ge \mathbf{0}$.*

Then $x^\star$ is a global minimizer of Problem (17.1).

Proof The proof is very similar to the proof of Theorem 16.3 in Chapter 16. See Appendix B for details. $\square$

In addition to the first-order necessary conditions, the first-order sufficient conditions require that f is convex on the feasible set.

17.2.1.2 Example

Again consider Problem (2.18) from Sections 2.3.2.3, 17.1.1.2, and 17.1.2.2. In Section 17.1.1.2, we observed that $x^\star = \begin{bmatrix} 3 \\ 3 \end{bmatrix}$, $\lambda^\star = [-4]$, and $\mu^\star = [4]$ satisfy the first-order necessary conditions for this problem. Moreover, f is twice continuously differentiable with continuous partial derivatives and the Hessian is positive definite. Therefore, f is convex and $x^\star$ is the global minimizer of the problem.

17.2.2 Duality

As we discussed in Section 3.4 and as in the discussion of linear equality constraints in Section 13.2.2, we can define a dual problem where the role of variables and constraints is partly or fully swapped [84, chapter 6]. We again recall some of the discussion from Section 3.4 in the following sections.

17.2.2.1 Dual function

Analysis We have observed in Section 17.1.1.4 that if f is convex then $x^\star$ is a global minimizer of $\mathcal{L}(\bullet, \lambda^\star, \mu^\star)$. Recall Definition 3.3 of the **dual function** and **effective domain**. For Problem (17.1), the dual function $\mathcal{D} : \mathbb{R}^m \times \mathbb{R}^r \to \mathbb{R} \cup \{-\infty\}$ is defined by:

$$\forall \begin{bmatrix} \lambda \\ \mu \end{bmatrix} \in \mathbb{R}^{m+r}, \mathcal{D}(\lambda, \mu) = \inf_{x \in \mathbb{R}^n} \mathcal{L}(x, \lambda, \mu). \tag{17.5}$$

The effective domain of $\mathcal{D}$ is:

$$\mathbb{E} = \left\{ \begin{bmatrix} \lambda \\ \mu \end{bmatrix} \in \mathbb{R}^{m+r} \middle| \mathcal{D}(\lambda, \mu) > -\infty \right\}.$$

Recall that by Theorem 3.12, $\mathbb{E}$ is convex and $\mathcal{D}$ is convex on $\mathbb{E}$.

Example We continue with Problem (2.18) from Sections 2.3.2.3, 17.1.1.2, ..., 17.2.1.2. The problem is:

$$\min_{x \in \mathbb{R}^2} \{ f(x) | Ax = b, Cx \le d \},$$

where:

$$\forall x \in \mathbb{R}^2, f(x) = (x_1 - 1)^2 + (x_2 - 3)^2,$$
$$A = \begin{bmatrix} 1 & -1 \end{bmatrix},$$
$$b = \begin{bmatrix} 0 \end{bmatrix},$$
$$C = \begin{bmatrix} 0 & -1 \end{bmatrix},$$
$$d = \begin{bmatrix} -3 \end{bmatrix}.$$

The Lagrangian $\mathcal{L} : \mathbb{R}^2 \times \mathbb{R} \times \mathbb{R} \to \mathbb{R}$ for this problem is defined by:

$$\forall x \in \mathbb{R}^2, \forall \lambda \in \mathbb{R}, \forall \mu \in \mathbb{R},$$
$$
\begin{aligned}
\mathcal{L}(x, \lambda, \mu) &= f(x) + \lambda^\dagger(Ax - b) + \mu^\dagger(Cx - d), \\
&= (x_1 - 1)^2 + (x_2 - 3)^2 + \lambda\begin{bmatrix} 1 & -1 \end{bmatrix}x + \mu\left(\begin{bmatrix} 0 & -1 \end{bmatrix}x + 3\right).
\end{aligned}
$$

For any given λ and μ, the Lagrangian $\mathcal{L}(\bullet, \lambda, \mu)$ is strictly convex and therefore, by Corollary 10.6, the first-order necessary conditions $\nabla_x \mathcal{L}(x, \lambda, \mu) = 0$ are sufficient for minimizing $\mathcal{L}(\bullet, \lambda, \mu)$ and, moreover, a minimizer exists, so that the inf in the definition of $\mathcal{D}$ can be replaced by min. Furthermore, there is a unique minimizer $x^{(\lambda,\mu)}$ corresponding to each value of λ and μ. In particular, we have:

$$\forall x \in \mathbb{R}^2, \forall \lambda \in \mathbb{R}, \forall \mu \in \mathbb{R},$$
$$
\begin{aligned}
\nabla_x \mathcal{L}(x, \lambda, \mu) &= \nabla f(x) + A^\dagger \lambda + C^\dagger \mu, \\
&= \begin{bmatrix} 2 & 0 \\ 0 & 2 \end{bmatrix}x + \begin{bmatrix} -2 \\ -6 \end{bmatrix} + \begin{bmatrix} 1 \\ -1 \end{bmatrix}\lambda + \begin{bmatrix} 0 \\ -1 \end{bmatrix}\mu,
\end{aligned}
$$

$$\forall \lambda \in \mathbb{R}, \forall \mu \in \mathbb{R},$$
$$
\begin{aligned}
x^{(\lambda,\mu)} &= -\begin{bmatrix} 2 & 0 \\ 0 & 2 \end{bmatrix}^{-1}\left[\begin{bmatrix} -2 \\ -6 \end{bmatrix} + \begin{bmatrix} 1 \\ -1 \end{bmatrix}\lambda + \begin{bmatrix} 0 \\ -1 \end{bmatrix}\mu\right], \\
&= \begin{bmatrix} 1 \\ 3 \end{bmatrix} + \begin{bmatrix} -0.5 \\ 0.5 \end{bmatrix}\lambda + \begin{bmatrix} 0 \\ 0.5 \end{bmatrix}\mu. \qquad (17.6)
\end{aligned}
$$

Consequently, the effective domain is $\mathbb{E} = \mathbb{R} \times \mathbb{R}$ and the dual function $\mathcal{D} : \mathbb{R} \times \mathbb{R} \to \mathbb{R}$ is given by:

$$\forall \begin{bmatrix} \lambda \\ \mu \end{bmatrix} \in \mathbb{R}^2, \mathcal{D}(\lambda, \mu) = \inf_{x \in \mathbb{R}^n} \mathcal{L}(x, \lambda, \mu),$$
$$
\begin{aligned}
&= \mathcal{L}(x^{(\lambda,\mu)}, \lambda, \mu), \text{ since } x^{(\lambda,\mu)} \text{ minimizes } \mathcal{L}(\bullet, \lambda, \mu), \\
&= (x_1^{(\lambda,\mu)} - 1)^2 + (x_2^{(\lambda,\mu)} - 3)^2 \\
&\quad + \lambda\begin{bmatrix} 1 & -1 \end{bmatrix}x^{(\lambda,\mu)} + \mu\left(\begin{bmatrix} 0 & -1 \end{bmatrix}x^{(\lambda,\mu)} + 3\right), \\
&= -\frac{1}{2}(\lambda)^2 - \frac{1}{4}(\mu)^2 - 2\lambda - \frac{1}{2}\mu\lambda,
\end{aligned}
$$

on substituting from (17.6) for $x^{(\lambda,\mu)}$.

17.2.2.2 Dual problem

Analysis As in the equality-constrained case, if the objective is convex on $\mathbb{R}^n$ then the minimum of Problem (17.1) is equal to $\mathcal{D}(\lambda^\star, \mu^\star)$, where $\lambda^\star$ and $\mu^\star$ are the Lagrange multipliers that satisfy the necessary conditions for Problem (17.1). As in the equality-constrained case, under certain conditions, the Lagrange multipliers can be found as the maximizer of the **dual problem**:

$$\max_{\left[\begin{smallmatrix}\lambda\\\mu\end{smallmatrix}\right]\in\mathbb{E}} \{\mathcal{D}(\lambda,\mu)|\mu \geq \mathbf{0}\}, \tag{17.7}$$

where $\mathcal{D} : \mathbb{E} \rightarrow \mathbb{R}$ is the dual function defined in (17.5). Again, Problem (17.1) is called the **primal problem** in this context to distinguish it from Problem (17.7).

These observations are embodied in the following. (Also see [6, theorems 6.2.4 and 6.3.3][11, proposition 5.2.1] [15, section 5.2.3] [84, corollaries 6.1 and 14.2] for generalizations.)

Theorem 17.4 *Suppose that* $f : \mathbb{R}^n \rightarrow \mathbb{R}$ *is convex and partially differentiable with continuous partial derivatives,* $A \in \mathbb{R}^{m\times n}$, $b \in \mathbb{R}^m$, $C \in \mathbb{R}^{r\times n}$, *and* $d \in \mathbb{R}^r$. *Consider the primal problem, Problem (17.1):*

$$\min_{x\in\mathbb{R}^n}\{f(x)|Ax = b, Cx \leq d\}.$$

Also, consider the dual problem, Problem (17.7). We have the following.

(i) *If the primal problem possesses a minimum then the dual problem possesses a maximum and the optima are equal. That is:*

$$\min_{x\in\mathbb{R}^n}\{f(x)|Ax = b, Cx \leq d\} = \max_{\left[\begin{smallmatrix}\lambda\\\mu\end{smallmatrix}\right]\in\mathbb{E}}\{\mathcal{D}(\lambda,\mu)|\mu \geq \mathbf{0}\}.$$

(ii) *If:*

- $\begin{bmatrix}\lambda\\\mu\end{bmatrix} \in \mathbb{E}$,
- $\min_{x\in\mathbb{R}^n}\mathcal{L}(x,\lambda,\mu)$ *exists, and*
- f *is twice partially differentiable with continuous second partial derivatives and* $\nabla^2 f$ *is positive definite,*

then $\mathcal{D}$ *is partially differentiable at* $\begin{bmatrix}\lambda\\\mu\end{bmatrix}$ *with continuous partial derivatives and:*

$$\begin{bmatrix}\nabla_\lambda\mathcal{D}(\lambda,\mu)\\\nabla_\mu\mathcal{D}(\lambda,\mu)\end{bmatrix} = \nabla\mathcal{D}(\lambda,\mu) = \begin{bmatrix}Ax^{(\lambda,\mu)} - b\\Cx^{(\lambda,\mu)} - d\end{bmatrix}, \tag{17.8}$$

where $x^{(\lambda,\mu)}$ *is the unique minimizer of* $\min_{x\in\mathbb{R}^n}\mathcal{L}(x,\lambda,\mu)$.

Proof ([6, theorems 6.2.4 and 6.3.3][11, proposition 3.4.3][70, lemma 1 of chapter 13][84, corollaries 6.1 and 14.2].)

(i) Suppose that Problem (17.1) possesses a minimum with minimizer $x^\star$. By Theorem 17.1,

$$\exists\lambda^\star \in \mathbb{R}^m, \exists\mu^\star \in \mathbb{R}^r \text{ such that } \mathbf{0} = \begin{aligned}&\nabla f(x^\star) + A^\dagger\lambda^\star + C^\dagger\mu^\star,\\&\nabla_x\mathcal{L}(x^\star,\lambda^\star,\mu^\star),\end{aligned}$$

where we note that $\mathcal{L}(\bullet, \lambda^\star, \mu^\star)$ is convex and partially differentiable, so that, by Corollary 10.6, $x^\star$ is also a minimizer of $\mathcal{L}(\bullet, \lambda^\star, \mu^\star)$. Therefore,

$$
\begin{aligned}
\mathcal{D}(\lambda^\star, \mu^\star) &= \inf_{x \in \mathbb{R}^n} \mathcal{L}(x, \lambda^\star, \mu^\star), \\
&= \mathcal{L}(x^\star, \lambda^\star, \mu^\star), \text{ because } x^\star \text{ minimizes } \mathcal{L}(\bullet, \lambda^\star, \mu^\star), \\
&= f(x^\star) + [\lambda^\star]^\dagger (Ax^\star - b) + [\mu^\star]^\dagger (Cx^\star - d), \text{ by definition,} \\
&= f(x^\star), \text{ since } x^\star \text{ is feasible and, by Theorem 17.1,} \\
&\qquad \mu^\star_\ell (C_\ell x^\star - d_\ell) = 0, \forall \ell = 1, \dots, r, \\
&\geq \mathcal{D}(\lambda, \mu), \forall \lambda \in \mathbb{R}^m, \forall \mu \in \mathbb{R}^r_+, \text{ by Theorem 3.13.}
\end{aligned}
$$

That is, $\begin{bmatrix} \lambda^\star \\ \mu^\star \end{bmatrix}$ maximizes the dual function over $\lambda \in \mathbb{R}^m$ and $\mu \in \mathbb{R}^r_+$:

$$
\begin{aligned}
f(x^\star) &= \max_{\begin{bmatrix} \lambda \\ \mu \end{bmatrix} \in \mathbb{E}} \{\mathcal{D}(\lambda, \mu) | \mu \geq 0\}, \\
&= \mathcal{D}(\lambda^\star, \mu^\star).
\end{aligned}
$$

(ii) See Exercise 10.19.

□

As in the equality-constrained case, it is possible for $\mathcal{D}$ to not be partially differentiable at a point $\begin{bmatrix} \lambda \\ \mu \end{bmatrix} \in \mathbb{E}$ if:

- $\mathcal{L}(\bullet, \lambda, \mu)$ is bounded below (so that $\inf_{x \in \mathbb{R}^n} \mathcal{L}(x, \lambda, \mu) \in \mathbb{R}$) yet the minimum $\min_{x \in \mathbb{R}^n} \mathcal{L}(x, \lambda, \mu)$ does not exist, or
- $\nabla^2 f$ is not positive definite and $\min_{x \in \mathbb{R}^n} \mathcal{L}(x, \lambda, \mu)$ has multiple minimizers.

Corollary 17.5 *Let $f : \mathbb{R}^n \to \mathbb{R}$ be twice partially differentiable with continuous second partial derivatives and with $\nabla^2 f$ positive definite, $A \in \mathbb{R}^{m \times n}, b \in \mathbb{R}^m, C \in \mathbb{R}^{r \times n}, d \in \mathbb{R}^r$. Consider Problem (17.1):*

$$
\min_{x \in \mathbb{R}^n} \{f(x) | Ax = b, Cx \leq d\},
$$

the Lagrangian of this problem, and the effective domain $\mathbb{E}$ of the dual function. If:

- *the effective domain $\mathbb{E}$ contains $\mathbb{R}^m \times \mathbb{R}^r_+$, and*
- *for each $\lambda \in \mathbb{R}^m$ and $\mu \in \mathbb{R}^r_+$, $\min_{x \in \mathbb{R}^n} \mathcal{L}(x, \lambda, \mu)$ exists,*

then necessary and sufficient conditions for $\begin{bmatrix} \lambda^\star \\ \mu^\star \end{bmatrix} \in \mathbb{R}^{m+r}$ to be the maximizer of the dual problem:

$$
\max_{\begin{bmatrix} \lambda \\ \mu \end{bmatrix} \in \mathbb{E}} \{\mathcal{D}(\lambda, \mu) | \mu \geq 0\},
$$

are:

$$M^\star(Cx^{(\lambda^\star,\mu^\star)} - d) = 0;$$
$$Ax^{(\lambda^\star,\mu^\star)} = b;$$
$$Cx^{(\lambda^\star,\mu^\star)} - d \leq 0; \text{ and}$$
$$\mu^\star \geq 0,$$

where $\{x^{(\lambda^\star,\mu^\star)}\} = \operatorname{argmin}_{x \in \mathbb{R}^n} \mathcal{L}(x, \lambda^\star, \mu^\star)$ *and* $M^\star = \operatorname{diag}\{\mu_\ell^\star\}$. *Moreover, if* $\begin{bmatrix} \lambda^\star \\ \mu^\star \end{bmatrix}$
maximizes the dual problem then $x^{(\lambda^\star,\mu^\star)}$, $\lambda^\star$, *and* $\mu^\star$ *satisfy the first-order necessary conditions for Problem (17.1).*

Proof Note that the hypothesis implies that the dual function is finite for all $\lambda \in \mathbb{R}^m$ and all $\mu \in \mathbb{R}_+^r$ so that the dual problem is a non-negatively constrained maximization of a real-valued function and, moreover, by Theorem 3.12, $-\mathcal{D}$ is convex and partially differentiable with continuous partial derivatives on the convex set:

$$\left\{ \begin{bmatrix} \lambda \\ \mu \end{bmatrix} \in \mathbb{R}^{m+r} \,\middle|\, \mu \geq 0 \right\}.$$

Moreover, by Theorem 17.4,

$$\nabla_\lambda \mathcal{D}(\lambda^\star, \mu^\star) = Ax^{(\lambda^\star,\mu^\star)} - b,$$
$$\nabla_\mu \mathcal{D}(\lambda^\star, \mu^\star) = Cx^{(\lambda^\star,\mu^\star)} - d.$$

Applying Theorems 17.1 and 17.3 to the dual problem and some substitution yields the conclusion. (See Exercise 17.7.) □

Theorem 17.4 shows that an alternative approach to finding the minimum of Problem (17.1) involves finding the *maximum* of the dual function $\mathcal{D}(\lambda, \mu)$ over $\lambda \in \mathbb{R}^m$, $\mu \in \mathbb{R}^r$, $\mu \geq 0$. Theorem 3.12 shows that the dual function has at most one local maximum. To seek the maximum of $\mathcal{D}(\lambda, \mu)$ over $\lambda \in \mathbb{R}^m$, $\mu \in \mathbb{R}^r$, $\mu \geq 0$, we can, for example, utilize the value of the gradient of $\mathcal{D}$ from (17.8) as part of an active set or interior point algorithm. As in the equality-constrained case, under some circumstances, it is also possible to calculate the Hessian of $\mathcal{D}$ [70, section 12.3].

Example Continuing with Problem (2.18) from Sections 2.3.2.3, 17.1.1.2, ..., 17.2.2.1, we recall that the effective domain of the dual function is $\mathbb{E} = \mathbb{R} \times \mathbb{R}$ and the dual function $\mathcal{D} : \mathbb{R} \times \mathbb{R} \to \mathbb{R}$ is:

$$\forall \begin{bmatrix} \lambda \\ \mu \end{bmatrix} \in \mathbb{R}^2, \mathcal{D}(\lambda, \mu) = -\frac{1}{2}(\lambda)^2 - \frac{1}{4}(\mu)^2 - 2\lambda - \frac{1}{2}\mu\lambda,$$

with unique minimizer of the Lagrangian specified by (17.6). The dual function is twice partially differentiable with continuous second partial derivatives. In particular,

$$
\forall \begin{bmatrix} \lambda \\ \mu \end{bmatrix} \in \mathbb{R}^2, \nabla D(\lambda, \mu) = \begin{bmatrix} -2 - \lambda - \mu/2 \\ -\lambda/2 - \mu/2 \end{bmatrix},
$$

$$
\forall \begin{bmatrix} \lambda \\ \mu \end{bmatrix} \in \mathbb{R}^2, \nabla^2 D(\lambda, \mu) = \begin{bmatrix} -1 & -0.5 \\ -0.5 & -0.5 \end{bmatrix}.
$$

We claim that $\begin{bmatrix} \lambda^\star \\ \mu^\star \end{bmatrix} = \begin{bmatrix} -4 \\ 4 \end{bmatrix}$ maximizes the dual function over $\mu \geq [0]$. In particular,

$$
\nabla D(\lambda^\star, \mu^\star) = \mathbf{0},
$$

$\mu^\star > [0]$, and $\nabla^2 D$ is negative definite. Consequently, $\begin{bmatrix} \lambda^\star \\ \mu^\star \end{bmatrix}$ is the unique maximizer of Problem (17.7).

We also observe that $\lambda^\star = [-4]$ and $\mu^\star = [4]$ satisfy the conditions specified in Corollary 17.5 for maximizing the dual. To see this, we first use (17.6) to evaluate $x^{(\lambda,\mu)}$ at $\lambda^\star = [-4]$ and $\mu^\star = [4]$. We obtain $x^{(\lambda^\star,\mu^\star)} = \begin{bmatrix} 3 \\ 3 \end{bmatrix}$. The necessary and sufficient conditions in Corollary 17.5 for maximizing the dual are satisfied, since:

$$
\mu^\star(Cx^{(\lambda^\star,\mu^\star)} - d) = [4] \left(\begin{bmatrix} 0 & -1 \end{bmatrix} \begin{bmatrix} 3 \\ 3 \end{bmatrix} - [-3] \right),
$$
$$
= [0];
$$
$$
Ax^{(\lambda^\star,\mu^\star)} - b = \begin{bmatrix} 1 & -1 \end{bmatrix} \begin{bmatrix} 3 \\ 3 \end{bmatrix},
$$
$$
= [0];
$$
$$
Cx^{(\lambda^\star,\mu^\star)} - d = \begin{bmatrix} 0 & -1 \end{bmatrix} \begin{bmatrix} 3 \\ 3 \end{bmatrix} - [-3],
$$
$$
= [0],
$$
$$
\leq [0]; \text{ and}
$$
$$
\mu^\star = [4],
$$
$$
\geq [0].
$$

Moreover, $x^{(\lambda^\star,\mu^\star)}$, $\lambda^\star$, and $\mu^\star$ satisfy the first-order necessary conditions for Problem (2.18).

Discussion As in the equality-constrained case, it is essential in Theorem 17.4 for f to be convex on the *whole* of $\mathbb{R}^n$, not just on the feasible set. The reason is again that the inner minimization of $\mathcal{L}(\bullet, \lambda, \mu)$ is taken over the whole of $\mathbb{R}^n$.

Unfortunately, if f is not strictly convex then $\mathcal{L}(\bullet, \lambda, \mu)$ may have multiple minimizers over x for fixed λ and μ. In this case, it may turn out that some of the minimizers of $\mathcal{L}(\bullet, \lambda^\star, \mu^\star)$ do not actually minimize Problem (17.1). Moreover, if there are multiple minimizers of $\mathcal{L}(\bullet, \lambda, \mu)$ then $\mathcal{D}(\lambda, \mu)$ may be not partially differentiable. The issues are similar to the equality-constrained case. We will see in Section 17.2.2.3, however, that in the particular cases of linear and of strictly convex quadratic programs, we can calculate the dual function and characterize the effective domain explicitly. This allows us to use duality for the not strictly convex case of linear programs.

Problem (17.7) is itself an inequality-constrained optimization problem; however, depending on the structure of Problem (17.1), it may be easier to solve than Problem (17.1). In particular, the dual problem is non-negatively constrained of the form of Problem (16.1), except that the non-negativity constraints apply to only some of the decision vector, and we can apply essentially the same algorithms as we developed for Problem (16.1). We will take this approach in Section 17.3.2.

17.2.2.3 Dual of linear and quadratic programs

In the case of linear and of strictly convex quadratic programs, we can characterize the effective domain and the dual function explicitly by solving the first-order necessary conditions for minimizing the Lagrangian:

$$\nabla_x \mathcal{L}(x, \lambda, \mu) = \mathbf{0}.$$

The approach parallels that of the Wolfe dual, described in Section 13.2.2.2. We first consider the case of linear objective and then strictly convex quadratic objectives. The analysis is based in part on [6, section 6.6][84, section 14.8].

Linear program Suppose that the objective is linear and the constraint functions are all affine. In particular, suppose $f : \mathbb{R}^n \to \mathbb{R}$ is of the form:

$$\forall x \in \mathbb{R}^n, \ f(x) = c^\dagger x,$$

for some $c \in \mathbb{R}^n$. We have:

$$\forall x \in \mathbb{R}^n, \forall \lambda \in \mathbb{R}^m, \forall \mu \in \mathbb{R}^r, \mathcal{L}(x, \lambda, \mu) = c^\dagger x + \lambda^\dagger (Ax - b)$$
$$+ \mu^\dagger (Cx - d),$$
$$\forall x \in \mathbb{R}^n, \forall \lambda \in \mathbb{R}^m, \forall \mu \in \mathbb{R}^r, \nabla_x \mathcal{L}(x, \lambda, \mu) = c + A^\dagger \lambda + C^\dagger \mu.$$

The first-order necessary and sufficient conditions for minimizing the Lagrangian are that $c + A^\dagger \lambda + C^\dagger \mu = \mathbf{0}$. These conditions do not involve x, but also do not necessarily have a solution for all values of λ and μ. In particular, if $c + A^\dagger \lambda +$

$C^\dagger \mu \neq \mathbf{0}$ then $\mathcal{L}(\bullet, \lambda, \mu)$ is unbounded below and $\begin{bmatrix} \lambda \\ \mu \end{bmatrix} \notin \mathbb{E}$. Conversely, if $c + A^\dagger \lambda + C^\dagger \mu = \mathbf{0}$ then, after substituting, we find that:

$$
\begin{aligned}
\mathcal{D}(\lambda, \mu) &= -\lambda^\dagger b - \mu^\dagger d, \\
&> -\infty.
\end{aligned}
$$

That is:

$$
\mathbb{E} = \left\{ \begin{bmatrix} \lambda \\ \mu \end{bmatrix} \in \mathbb{R}^{m+r} \middle| c + A^\dagger \lambda + C^\dagger \mu = \mathbf{0} \right\},
$$

$$
\forall \begin{bmatrix} \lambda \\ \mu \end{bmatrix} \in \mathbb{E}, \mathcal{D}(\lambda, \mu) = -\lambda^\dagger b - \mu^\dagger d.
$$

(See Exercise 17.8.)

We now substitute the characterization of the dual function and effective domain into the definition of the dual problem and apply Theorem 17.4. In particular, if $\min_{x \in \mathbb{R}^n} \{c^\dagger x | Ax = b, Cx \leq d\}$ possesses a minimum then by Theorem 17.4:

$$
\begin{aligned}
&\min_{x \in \mathbb{R}^n} \{c^\dagger x | Ax = b, Cx \leq d\} \\
&= \max_{\begin{bmatrix} \lambda \\ \mu \end{bmatrix} \in \mathbb{E}} \{\mathcal{D}(\lambda, \mu) | \mu \geq \mathbf{0}\}, \\
&= \max_{\begin{bmatrix} \lambda \\ \mu \end{bmatrix} \in \mathbb{R}^{m+r}} \{\mathcal{D}(\lambda, \mu) | c + A^\dagger \lambda + C^\dagger \mu = \mathbf{0}, \mu \geq \mathbf{0}\}, \\
&\quad \text{since } \mathbb{E} = \left\{ \begin{bmatrix} \lambda \\ \mu \end{bmatrix} \in \mathbb{R}^{m+r} \middle| c + A^\dagger \lambda + C^\dagger \mu = \mathbf{0} \right\}, \\
&= \max_{\begin{bmatrix} \lambda \\ \mu \end{bmatrix} \in \mathbb{R}^{m+r}} \{-\lambda^\dagger b - \mu^\dagger d | c + A^\dagger \lambda + C^\dagger \mu = \mathbf{0}, \mu \geq \mathbf{0}\}, \\
&\quad \text{since } \mathcal{D}(\lambda, \mu) = -\lambda^\dagger b - \mu^\dagger d \text{ for } c + A^\dagger \lambda + C^\dagger \mu = \mathbf{0}, \\
&= -\min_{\begin{bmatrix} \lambda \\ \mu \end{bmatrix} \in \mathbb{R}^{m+r}} \{\lambda^\dagger b + \mu^\dagger d | c + A^\dagger \lambda + C^\dagger \mu = \mathbf{0}, \mu \geq \mathbf{0}\}, \\
&= -\min_{\begin{bmatrix} \lambda \\ \mu \end{bmatrix} \in \mathbb{R}^{m+r}} \left\{ \begin{bmatrix} b \\ d \end{bmatrix}^\dagger \begin{bmatrix} \lambda \\ \mu \end{bmatrix} \middle| \begin{bmatrix} A \\ C \end{bmatrix}^\dagger \begin{bmatrix} \lambda \\ \mu \end{bmatrix} = -c, \mu \geq \mathbf{0} \right\}. \quad (17.9)
\end{aligned}
$$

The dual problem in the last line of (17.9) has a linear objective, linear equality constraints, and non-negativity constraints on the variables μ. We observe that there is at least one point in the feasible set of the dual problem,

$$\left\{ \begin{bmatrix} \lambda \\ \mu \end{bmatrix} \in \mathbb{R}^{m+r} \middle| \begin{bmatrix} A \\ C \end{bmatrix}^\dagger \begin{bmatrix} \lambda \\ \mu \end{bmatrix} = -c, \mu \geq \mathbf{0} \right\},$$

namely the Lagrange multipliers $\begin{bmatrix} \lambda^\star \\ \mu^\star \end{bmatrix}$ that correspond to the minimizer $x^\star$ of the primal problem. We say that the problem is **dual feasible**.

We have transformed a primal problem with n variables, m equality constraints, and r inequality constraints into a dual problem with $m + r$ variables, n equality constraints, and r inequality constraints. The dual of a linear program is therefore also a linear program. However, the form of the inequality constraints in the dual is simpler than in the primal problem since in the dual problem they are non-negativity constraints on some of the variables rather than general linear inequalities. Moreover, we have substituted for the solution of the embedded inner problem in the dual. Under an appropriate re-definition of variables, the dual problem is essentially in the same form as Problem (16.1). The only difference is that the non-negativity constraints only apply to the variables μ and not to λ. (See Exercise 17.12.)

Quadratic program Suppose that the objective is quadratic and the constraint functions are all affine. In particular, suppose $f : \mathbb{R}^n \to \mathbb{R}$ is of the form:

$$\forall x \in \mathbb{R}^n, \ f(x) = \tfrac{1}{2}x^\dagger Q x + c^\dagger x,$$

for some $Q \in \mathbb{R}^{n \times n}$ and $c \in \mathbb{R}^n$. We assume that Q is positive definite so that the objective is strictly convex. (See [6, section 6.6] for further discussion in the case of Q positive semi-definite.) We have:

$$\forall x \in \mathbb{R}^n, \forall \lambda \in \mathbb{R}^m, \forall \mu \in \mathbb{R}^r, \ \mathcal{L}(x, \lambda, \mu) = \tfrac{1}{2}x^\dagger Q x + c^\dagger x + \lambda^\dagger (Ax - b)$$
$$+ \mu^\dagger (Cx - d),$$
$$\forall x \in \mathbb{R}^n, \forall \lambda \in \mathbb{R}^m, \forall \mu \in \mathbb{R}^r, \ \nabla_x \mathcal{L}(x, \lambda, \mu) = Qx + c + A^\dagger \lambda + C^\dagger \mu.$$

The first-order necessary conditions for minimizing $\mathcal{L}(\bullet, \lambda, \mu)$ are that $Qx + c + A^\dagger \lambda + C^\dagger \mu = \mathbf{0}$. Since Q is positive definite, this condition has a solution for all values of λ and μ, namely $x = -Q^{-1}[c + A^\dagger \lambda + C^\dagger \mu]$. After substituting, we find that:

$$\forall \begin{bmatrix} \lambda \\ \mu \end{bmatrix} \in \mathbb{R}^{m+r}, \ \mathcal{D}(\lambda, \mu) = -\tfrac{1}{2}[c + A^\dagger \lambda + C^\dagger \mu]^\dagger Q^{-1}[c + A^\dagger \lambda + C^\dagger \mu]$$
$$- \lambda^\dagger b - \mu^\dagger d,$$
$$> -\infty,$$

so that $\mathbb{E} = \mathbb{R}^{m+r}$. (See Exercise 17.9.)

As in the case of a linear program, we now substitute the characterization of the dual function and effective domain into the definition of the dual problem. In particular, if $\min_{x \in \mathbb{R}^n} \{\frac{1}{2} x^\dagger Q x + c^\dagger x | A x = b, C x \leq d\}$ possesses a minimum then by Theorem 17.4:

$$\min_{x \in \mathbb{R}^n} \{\tfrac{1}{2} x^\dagger Q x + c^\dagger x | A x = b, C x \leq d\}$$

$$= \max_{\left[\begin{smallmatrix} \lambda \\ \mu \end{smallmatrix}\right] \in \mathbb{E}} \{\mathcal{D}(\lambda, \mu) | \mu \geq 0\},$$

$$= \max_{\left[\begin{smallmatrix} \lambda \\ \mu \end{smallmatrix}\right] \in \mathbb{R}^{m+r}} \left\{ \left. \begin{array}{c} -\frac{1}{2}[c + A^\dagger \lambda + C^\dagger \mu]^\dagger Q^{-1} [c + A^\dagger \lambda + C^\dagger \mu] \\ - \lambda^\dagger b - \mu^\dagger d \end{array} \right| \mu \geq 0 \right\},$$

$$= - \min_{\left[\begin{smallmatrix} \lambda \\ \mu \end{smallmatrix}\right] \in \mathbb{R}^{m+r}} \left\{ \left. \begin{array}{c} \frac{1}{2}[c + A^\dagger \lambda + C^\dagger \mu]^\dagger Q^{-1} [c + A^\dagger \lambda + C^\dagger \mu] \\ + \lambda^\dagger b + \mu^\dagger d \end{array} \right| \mu \geq 0 \right\}. \quad (17.10)$$

The dual problem in the last line of (17.10) has a quadratic objective and non-negativity constraints. We have transformed a primal problem with n variables, m equality constraints, and r inequality constraints into a dual problem with $m + r$ variables and r inequality constraints. The dual of a quadratic program is therefore also a quadratic program. Again, the form of the inequality constraints in the dual is simpler than in the primal problem since they are non-negativity constraints. If we solve the problem in the last line of (17.10) for optimal $\lambda^\star$ and $\mu^\star$ then the minimizer, $x^\star$, of the primal problem can be recovered as $x^\star = -Q^{-1}[c + A^\dagger \lambda^\star + C^\dagger \mu^\star]$.

Discussion There is considerable literature on the relationship between primal and dual linear programs [84, chapter 6] and on primal and dual quadratic programs. The standard treatment of duality in linear programming differs from the way we have discussed it here, there are a variety of special cases, and we have omitted many details. For example, we have not discussed how to recover a minimizer of the primal problem from the solution of the dual of a linear program.

Furthermore, **primal–dual algorithms** (including the primal–dual interior point algorithm described in Section 16.4.3.3) represent *both* the primal and dual variables and simultaneously solve for both the minimizer and the Lagrange multipliers [70, section 4.6]. The primal–dual interior point algorithm is therefore essentially the same whether it is applied to the primal or dual problem. Nevertheless, in Exercise 16.25 and Section 17.3.1.4 we will use duality to develop stopping criteria for primal–dual algorithms.

17.2.2.4 Partial duals

Analysis A variant of the dual problem is often very useful. We can define the **partial dual** with respect to some of the constraints [70, section 13.1]. For example, we define $\mathcal{D}_= : \mathbb{R}^m \to \mathbb{R} \cup \{-\infty\}$ and $\mathcal{D}_\le : \mathbb{R}^r \to \mathbb{R} \cup \{-\infty\}$ by:

$$\forall \lambda \in \mathbb{R}^m, \mathcal{D}_=(\lambda) \quad = \quad \inf_{x \in \mathbb{R}^n} \{f(x) + \lambda^\dagger (Ax - b) | Cx \le d\},$$

$$\forall \mu \in \mathbb{R}^r, \mathcal{D}_\le(\mu) \quad = \quad \inf_{x \in \mathbb{R}^n} \{f(x) + \mu^\dagger (Cx - d) | Ax = b\}.$$

The function $\mathcal{D}_=$ is called the partial dual with respect to the equality constraints, while $\mathcal{D}_\le$ is called the partial dual with respect to the inequality constraints. As before, under conditions of convexity, we have:

Theorem 17.6 *Suppose that $f : \mathbb{R}^n \to \mathbb{R}$ is convex and partially differentiable with continuous partial derivatives, $A \in \mathbb{R}^{m \times n}$, $b \in \mathbb{R}^m$, $C \in \mathbb{R}^{r \times n}$, and $d \in \mathbb{R}^r$. Suppose that Problem (17.1) possesses a minimum. Then:*

$$\min_{x \in \mathbb{R}^n} \{f(x) | Ax = b, Cx \le d\} = \max_{\lambda \in \mathbb{E}_=} \{\mathcal{D}_=(\lambda)\} = \max_{\mu \in \mathbb{E}_\le} \{\mathcal{D}_\le(\mu) | \mu \ge 0\},$$

where $\mathcal{D}_=$ is the partial dual with respect to the equality constraints and $\mathbb{E}_=$ is its effective domain and $\mathcal{D}_\le$ is the partial dual with respect to the inequality constraints and $\mathbb{E}_\le$ is its effective domain.

Proof See [11, section 3.4][70, section 13.1]. □

It is also possible to take a partial dual with respect to only some of the equality or some of the inequality constraints or some of both of the equality and inequality constraints, leaving the other constraints explicitly in the problem. See [11, section 3.4] for details.

Separable problems To see an example of the usefulness of partial duality, consider the case where:

- f is separable and strictly convex, so that $f(x) = \sum_{k=1}^n f_k(x_k)$, and
- the inequality constraints consist only of upper and lower bound constraints $\underline{x} \le x \le \overline{x}$.

Then:

$$\forall \lambda \in \mathbb{R}^m, \mathcal{D}_=(\lambda) \quad = \quad \min_{x \in \mathbb{R}^n} \{f(x) + \lambda^\dagger (Ax - b) | Cx \le d\},$$

$$= \quad \min_{x \in \mathbb{R}^n} \{f(x) + \lambda^\dagger (Ax - b) | \underline{x} \le x \le \overline{x}\},$$

$$= \min_{x \in \mathbb{R}^n} \left\{ \left. \sum_{k=1}^{n} f_k(x_k) + \lambda^\dagger \sum_{k=1}^{n} A_k x_k - \lambda^\dagger b \; \right| \; \begin{matrix} \underline{x}_k \leq x_k \leq \overline{x}_k, \\ \forall k = 1, \ldots, n \end{matrix} \right\},$$

where A_k is the k-th column of A,

$$= \min_{x \in \mathbb{R}^n} \left\{ \left. \sum_{k=1}^{n} \left(f_k(x_k) + \lambda^\dagger A_k x_k \right) - \lambda^\dagger b \; \right| \; \begin{matrix} \underline{x}_k \leq x_k \leq \overline{x}_k, \\ \forall k = 1, \ldots, n \end{matrix} \right\},$$

on re-arranging,

$$= \sum_{k=1}^{n} \min_{x_k \in \mathbb{R}} \{ f_k(x_k) + \lambda^\dagger A_k x_k | \underline{x}_k \leq x_k \leq \overline{x}_k \} - \lambda^\dagger b, \qquad (17.11)$$

on swapping the minimum and the summation, noting that there is no coupling between the sub-problems because of the form of the upper and lower bound constraints. (See Exercise 17.11.)

This means that, for a given value of λ, the dual with respect to the equality constraints is the sum of:

- a constant $(-\lambda^\dagger b)$, and
- n one-dimensional optimization sub-problems that can each be evaluated independently.

The primal problem has been **decomposed** into a collection of sub-problems using the partial dual. In general, if the problem has constraints that couple between sub-problems, then by dualizing with respect to these **coupling constraints** we can decompose the problem into the sub-problems. If each sub-problem is simple enough, it may be possible to evaluate its minimizer and minimum explicitly without resorting to an iterative technique. This applies to the least-cost production case study from Section 15.1 and will be described in detail in Section 18.1.2.2.

17.3 Approaches to finding minimizers

In this section we will show two basic ways in which inequality-constrained Problem (17.1) can be transformed into the form of Problem (16.1) from Chapter 16. We can then use the algorithmic development from Chapter 16 to solve Problem (17.1).

17.3.1 Primal algorithm

17.3.1.1 Transformation

In this section we use slack variables to transform Problem (17.1) into a non-negatively constrained problem.

Slack variables To handle the inequality constraints of the primal problem, we consider the following problem incorporating slack variables as introduced in Section 3.3.2:

$$\min_{x \in \mathbb{R}^n, w \in \mathbb{R}^r} \{f(x) | Ax = b, Cx + w = d, w \geq 0\}. \qquad (17.12)$$

The variables w are called the slack variables because they account for the "slack" in the constraints $Cx \leq d$. By Theorem 3.8, Problem (17.12) is equivalent to Problem (17.1).

Relation to non-negatively constrained minimization We just showed the equivalence between Problem (17.1) and Problem (17.12). We will now show that Problem (17.12) can be solved using the algorithms developed in Sections 16.3 and 16.4 for non-negatively constrained minimization.

In Problem (17.12), if we consider:

- $\begin{bmatrix} x \\ w \end{bmatrix} \in \mathbb{R}^{n+r}$ to be the decision vector,

- f to be the objective, and

- $\begin{bmatrix} Ax - b \\ Cx + w - d \end{bmatrix} = 0$, or equivalently, $\begin{bmatrix} A & 0 \\ C & I \end{bmatrix} \begin{bmatrix} x \\ w \end{bmatrix} = \begin{bmatrix} b \\ d \end{bmatrix}$ to be the equality constraints,

then Problem (17.12) can be expressed in the form of Problem (16.1) (except that we have non-negativity constraints on just w and not on the whole of the decision vector $\begin{bmatrix} x \\ w \end{bmatrix}$.) The equivalent problem is:

$$\min_{x \in \mathbb{R}^n, w \in \mathbb{R}^r} \left\{ f(x) \left| \begin{bmatrix} A & 0 \\ C & I \end{bmatrix} \begin{bmatrix} x \\ w \end{bmatrix} = \begin{bmatrix} b \\ d \end{bmatrix}, w \geq 0 \right. \right\}. \qquad (17.13)$$

Under an appropriate re-definition of variables, this is essentially in the same form as Problem (16.1). The only difference is that the non-negativity constraints only apply to the variables w and not to x. (See Exercises 17.6 and 17.12.)

In the next section, we will apply the primal–dual interior point algorithm from Section 16.4 to Problem (17.13).

17.3.1.2 Primal–dual interior point algorithm

In this section we consider the barrier objective and problem and associated assumptions for Problem (17.13), which has non-negativity constraints $w \geq 0$.

Barrier objective and problem Analogously to the discussion in Section 16.4.2.2, given a barrier function $f_b : \mathbb{R}^r_{++} \to \mathbb{R}$ for the constraints $w \geq 0$ and a barrier parameter $t \in \mathbb{R}_{++}$, we form the **barrier objective** $\phi : \mathbb{R}^n \times \mathbb{R}^r_{++} \to \mathbb{R}$ defined by:

$$\forall x \in \mathbb{R}^n, \forall w \in \mathbb{R}^r_{++}, \phi(x, w) = f(x) + tf_b(w).$$

(Note that there are no barrier function terms corresponding to the entries in x because there are no constraints in Problem (17.1) of the form $x \geq 0$. See Section 17.3.1.4 for representing non-negativity constraints $x \geq 0$.)

Instead of solving (17.13), we will consider solving the **barrier problem**:

$$\min_{x \in \mathbb{R}^n, w \in \mathbb{R}^r} \left\{ \phi(x, w) \left| \begin{bmatrix} A & 0 \\ C & I \end{bmatrix} \begin{bmatrix} x \\ w \end{bmatrix} = \begin{bmatrix} b \\ d \end{bmatrix}, w > 0 \right. \right\}. \tag{17.14}$$

We seek (approximate) minimizers of Problem (17.14) for a decreasing sequence of values of the barrier parameter.

Slater condition As in the case of non-negativity constraints described in Section 16.4.2.2, in order to apply the interior point algorithm effectively, we must assume that the **Slater condition** holds so that there are feasible points for Problem (17.14). That is, we assume that $\{x \in \mathbb{R}^n | Ax = b, Cx < d\} \neq \emptyset$.

Equality-constrained problem To solve Problem (17.14), we can take a similar approach to the primal–dual interior point algorithm for non-negativity constraints presented in Section 16.4 of Chapter 16. In particular, we can partially ignore the inequality constraints and the domain of the barrier function and seek a solution to the following linear equality-constrained problem:

$$\min_{x \in \mathbb{R}^n, w \in \mathbb{R}^r} \left\{ \phi(x, w) \left| \begin{bmatrix} A & 0 \\ C & I \end{bmatrix} \begin{bmatrix} x \\ w \end{bmatrix} = \begin{bmatrix} b \\ d \end{bmatrix} \right. \right\}, \tag{17.15}$$

which has first-order necessary conditions:

$$\nabla f(x) + A^\dagger \lambda + C^\dagger \mu = 0, \tag{17.16}$$
$$Ax = b, \tag{17.17}$$
$$Cx + w = d, \tag{17.18}$$
$$t \nabla f_b(w) + \mu = 0, \tag{17.19}$$

where λ and μ are the dual variables on the constraints $Ax = b$ and $Cx + w = d$, respectively. (See Exercise 17.13.) We can use the techniques for minimization of linear equality-constrained problems from Section 13.3.2 of Chapter 13 to solve Problem (17.15). In particular, in Section 17.3.1.3, we will consider the

Newton–Raphson method for solving the first-order necessary conditions of Problem (17.15).

Logarithmic barrier function As in the primal–dual interior point algorithm for non-negativity constraints, we will use the logarithmic barrier function. That is:

$$\forall w \in \mathbb{R}^r_{++}, f_b(w) = -\sum_{\ell=1}^{r} \ln(w_\ell),$$

$$\forall w \in \mathbb{R}^r_{++}, \nabla f_b(w) = -[W]^{-1}\mathbf{1},$$

where $W = \text{diag}\{w_\ell\} \in \mathbb{R}^{r \times r}$ is a diagonal matrix with diagonal entries equal to w_ℓ, $\ell = 1, \ldots, r$. Substituting the expression for ∇f_b into (17.19) and re-arranging, we obtain:

$$W\mu - t\mathbf{1} = \mathbf{0}. \tag{17.20}$$

17.3.1.3 Newton–Raphson method

Analysis The Newton–Raphson step direction to solve (17.20) and (17.16)–(17.18) is given by the solution of:

$$\begin{bmatrix} M^{(\nu)} & 0 & 0 & W^{(\nu)} \\ 0 & \nabla^2 f(x^{(\nu)}) & A^\dagger & C^\dagger \\ 0 & A & 0 & 0 \\ I & C & 0 & 0 \end{bmatrix} \begin{bmatrix} \Delta w^{(\nu)} \\ \Delta x^{(\nu)} \\ \Delta \lambda^{(\nu)} \\ \Delta \mu^{(\nu)} \end{bmatrix} = \begin{bmatrix} -W^{(\nu)}\mu^{(\nu)} + t\mathbf{1} \\ -\nabla f(x^{(\nu)}) - A^\dagger \lambda^{(\nu)} - C^\dagger \mu^{(\nu)} \\ b - Ax^{(\nu)} \\ d - Cx^{(\nu)} - w^{(\nu)} \end{bmatrix},$$

where $M^{(\nu)} = \text{diag}\{\mu_\ell^{(\nu)}\}$ and $W^{(\nu)} = \text{diag}\{w_\ell^{(\nu)}\}$. As in the case of the primal–dual interior point algorithm for non-negativity constraints that was discussed in Section 16.4.3.3, we can re-arrange these equations to make them symmetric and use block pivoting on the top left-hand block of the matrix since the top left-hand block is diagonal. This results in a system that is similar to (13.35), except that a diagonal block of the form $[M^{(\nu)}]^{-1}W^{(\nu)}$ is added to the Hessian $\nabla^2 f(x^{(\nu)})$. Issues regarding solving the first-order necessary conditions, such as factorization of the indefinite coefficient matrix, approximate solution of the conditions, sparsity, and step-size selection, are similar to those described in Sections 13.3.2.3 and 16.4.3.3.

Example In this section, we will apply the primal–dual algorithm to the example quadratic program, Problem (2.18), from Sections 2.3.2.3, 17.1.1.2, …, 17.2.2.2. Recall that the problem is:

$$\min_{x \in \mathbb{R}^2}\{f(x)|Ax = b, Cx \le d\},$$

where:

$$\begin{align}
\forall x \in \mathbb{R}^2, f(x) &= (x_1 - 1)^2 + (x_2 - 3)^2, \\
A &= \begin{bmatrix} 1 & -1 \end{bmatrix}, \\
b &= \begin{bmatrix} 0 \end{bmatrix}, \\
C &= \begin{bmatrix} 0 & -1 \end{bmatrix}, \\
d &= \begin{bmatrix} -3 \end{bmatrix}.
\end{align}$$

The Newton–Raphson update for the corresponding barrier problem is:

$$
\begin{bmatrix}
\mu^{(\nu)} & 0 & 0 & 0 & w^{(\nu)} \\
0 & 2 & 0 & 1 & 0 \\
0 & 0 & 2 & -1 & -1 \\
0 & 1 & -1 & 0 & 0 \\
1 & 0 & -1 & 0 & 0
\end{bmatrix}
\begin{bmatrix}
\Delta w^{(\nu)} \\
\Delta x^{(\nu)} \\
\Delta \lambda^{(\nu)} \\
\Delta \mu^{(\nu)}
\end{bmatrix}
=
\begin{bmatrix}
-w^{(\nu)}\mu^{(\nu)} + t \\
-2(x_1^{(\nu)} - 1) - \lambda^{(\nu)} \\
-2(x_2^{(\nu)} - 3) + \lambda^{(\nu)} + \mu^{(\nu)} \\
-x_1^{(\nu)} \\
x_2^{(\nu)} \\
-3 + x_2^{(\nu)} - w^{(\nu)}
\end{bmatrix}.
$$

Exercise 17.14 discusses the calculation of iterates.

17.3.1.4 Other issues

In this section, we discuss adjustment of the barrier parameter, the initial guess, a stopping criterion, and non-negativity and lower and upper bound constraints on x.

Adjustment of barrier parameter To reduce the barrier parameter, we can use the approach described in Section 16.4.4 of Chapter 16. See Exercise 17.14 for an example.

Initial guess We can take an approach analogous to that in Section 16.4.5 to find an initial feasible guess for Problem (17.13) that is strictly feasible for the non-negativity constraints.

Stopping criterion Exercise 16.25 investigated the use of a primal–dual algorithm such as the one described in Section 16.4.3.3 to solve a non-negatively constrained problem:

$$\min_{x \in \mathbb{R}^n} \{ f(x) | Ax = b, x \geq 0 \}.$$

where $f : \mathbb{R}^n \to \mathbb{R}$ is convex and partially differentiable with continuous partial derivatives, $A \in \mathbb{R}^{m \times n}$, and $b \in \mathbb{R}^m$. Exercise 16.25 showed that if we iterate until:

$$[\mu^{(\nu)}]^\dagger x^{(\nu)} \leq \epsilon_f,$$

then $f(x^{(v)})$ will be within ϵ_f of the minimum of the non-negatively constrained problem. The corresponding condition for Problem (17.1) is to iterate until:

$$[\mu^{(v)}]^\dagger w^{(v)} \le \epsilon_f,$$

where μ is now the vector of dual variables corresponding to the constraints $w \ge 0$ (and corresponding to the constraints $Cx \le d$.)

Non-negativity and lower and upper bound constraints on x If we add constraints of the form $x \ge 0$ to Problem (17.1) then we can also include them in the barrier function and Problem (17.14). Similarly, it is also possible to treat constraints of the form $\underline{x}_\ell \le x_\ell \le \overline{x}_\ell$ by using a barrier function of the form:

$$-t \left(\ln(x_\ell - \underline{x}_\ell) + \ln(\overline{x}_\ell - x_\ell) \right).$$

This will be discussed in more detail in Section 18.1.2.1 in the context of the least-cost production with capacity constraints case study.

17.3.2 Dual algorithm

In this section we treat Problem (17.1) through problem transformations involving duality with respect to some or all of the inequality and equality constraints. We then discuss non-quadratic objectives.

17.3.2.1 Inequality constraints

In Section 17.2.2.2, we showed that, by taking the dual with respect to all of the constraints, an inequality-constrained problem of the form of Problem (17.1) could be transformed into a dual problem with the dual function defined in (17.5) as its objective. We observed in Section 17.2.2.4 that we could also take the dual with respect to just the inequality constraints. Under convexity assumptions, the dual and primal problems had the same optima. If the objective is strictly convex then the minimizer of the primal problem can be recovered as the unconstrained minimizer of $\mathcal{L}(\bullet, \lambda^\star, \mu^\star)$, where $\begin{bmatrix} \lambda^\star \\ \mu^\star \end{bmatrix}$ is the maximizer of the dual problem.

Whereas Problem (17.1) has general linear inequality constraints, taking the dual with respect to all the constraints or with respect to just the inequality constraints yields a dual problem where the inequality constraints are non-negativity constraints on variables only. We can apply algorithms developed for Problem (16.1). For example, the algorithms developed in Sections 16.3 and 16.4 can be applied to the dual problem; however, we have to swap the roles of primal and dual variables compared to the discussion in Sections 16.3 and 16.4. (See Exercise 17.12. In the case of the primal–dual interior point algorithm presented in Section 16.4, the dual variables are already explicitly represented in the algorithm so there is no advantage to taking the dual with respect to all of the constraints for use in the primal–dual algorithm.)

17.3.2.2 Equality constraints

Taking the dual with respect to the equality constraints yields a dual problem with no equality nor inequality constraints, but with inner problems having inequality constraints. To maximize the dual function, we can apply the algorithms developed in Section 10.2. Taking the dual with respect to only some of the equality constraints yields a dual problem with equality constraints. We can apply the algorithms developed in Section 13.5.2. As discussed in Section 17.2.2.4, the use of partial duality for problems with separable objectives can yield inner problems with a simple structure.

17.3.2.3 Non-quadratic objectives

Although the dual can be found for general non-quadratic objectives, it is often not as useful because the non-linearity of the optimality conditions in the definition of the dual function prevents us from simplifying the objective of the dual as in the linear and quadratic cases. If the primal problem is non-convex, we can still apply the algorithm to the dual problem; however, we must be more cautious about interpreting the results since the corresponding value of the primal variables may be infeasible or not optimal for the primal problem.

17.4 Sensitivity

17.4.1 Analysis

In this section we will analyze a general and a special case of sensitivity analysis for Problem (17.1). For the general case, we suppose that the objective f, equality constraint matrix A, right-hand side vector b, inequality constraint matrix C, and right-hand side vector d are parameterized by a parameter $\chi \in \mathbb{R}^s$. That is, $f : \mathbb{R}^n \times \mathbb{R}^s \to \mathbb{R}$, $A : \mathbb{R}^s \to \mathbb{R}^{m \times n}$, $b : \mathbb{R}^s \to \mathbb{R}^m$, $C : \mathbb{R}^s \to \mathbb{R}^{r \times n}$, and $d : \mathbb{R}^s \to \mathbb{R}^r$. We imagine that we have solved the inequality-constrained minimization problem:

$$\min_{x \in \mathbb{R}^n} \{ f(x; \chi) | A(\chi)x = b(\chi), C(\chi)x \leq d(\chi) \}, \qquad (17.21)$$

for a base-case value of the parameters, say $\chi = \mathbf{0}$, to find the base-case local minimizer $x^\star$ and the base-case Lagrange multipliers $\lambda^\star$ and $\mu^\star$. We now consider the sensitivity of the local minimum of Problem (17.21) to variation of the parameters about $\chi = \mathbf{0}$.

As well as considering the general case of the sensitivity of the local minimum of Problem (17.21) to χ, we also specialize to the case where only the right-hand sides of the equality and inequality constraints vary. That is, we return to the special case where $f : \mathbb{R}^n \to \mathbb{R}$, $A \in \mathbb{R}^{m \times n}$, $b \in \mathbb{R}^m$, $C \in \mathbb{R}^{r \times n}$, and $d \in \mathbb{R}^r$ are not explicitly parameterized. However, we now consider perturbations $\gamma \in \mathbb{R}^m$ and $\eta \in \mathbb{R}^r$ and the problem:

$$\min_{x \in \mathbb{R}^n} \{ f(x) | Ax = b - \gamma, Cx \le d - \eta \}. \tag{17.22}$$

For the parameter values $\gamma = \mathbf{0}$ and $\eta = \mathbf{0}$, Problem (17.22) is the same as Problem (17.1). We consider the sensitivity of the local minimum of Problem (17.22) to variation of the parameters about $\gamma = \mathbf{0}$ and $\eta = \mathbf{0}$.

We have the following corollary to the implicit function theorem, Theorem A.9 in Section A.7.3 of Appendix A.

Corollary 17.7 *Consider Problem (17.21) and suppose that $f : \mathbb{R}^n \times \mathbb{R}^s \to \mathbb{R}$ is twice partially differentiable with continuous second partial derivatives and that $A : \mathbb{R}^s \to \mathbb{R}^{m \times n}$, $b : \mathbb{R}^s \to \mathbb{R}^m$, $C : \mathbb{R}^s \to \mathbb{R}^{r \times n}$, and $d : \mathbb{R}^s \to \mathbb{R}^r$ are partially differentiable with continuous partial derivatives. Also consider Problem (17.22) and suppose that the function $f : \mathbb{R}^n \to \mathbb{R}$ is twice partially differentiable with continuous second partial derivatives. Suppose that $x^\star \in \mathbb{R}^n$, $\lambda^\star \in \mathbb{R}^m$, and $\mu^\star \in \mathbb{R}^r$ satisfy:*

- *the second-order sufficient conditions for Problem (17.21) for the value of parameters $\chi = \mathbf{0}$, and*
- *the second-order sufficient conditions for Problem (17.22) for the value of parameters $\gamma = \mathbf{0}$ and $\eta = \mathbf{0}$.*

In particular:

- *$x^\star$ is a local minimizer of Problem (17.21) for $\chi = \mathbf{0}$, and*
- *$x^\star$ is a local minimizer of Problem (17.22) for $\gamma = \mathbf{0}$ and $\eta = \mathbf{0}$,*

in both cases with associated Lagrange multipliers $\lambda^\star$ and $\mu^\star$. Moreover, suppose that the matrix $\hat{A}$ has linearly independent rows, where $\hat{A}$ is the matrix with rows consisting of:

- *the m rows of A (or $A(\mathbf{0})$), and*
- *those rows C_ℓ of C (or of $C(\mathbf{0})$) for which $\ell \in \mathbb{A}(x^\star)$.*

Furthermore, suppose that there are no degenerate constraints at the base-case solution. Then, for values of χ in a neighborhood of the base-case value of the parameters $\chi = \mathbf{0}$, there is a local minimum and corresponding local minimizer and Lagrange multipliers for Problem (17.21). Moreover, the local minimum, local minimizer, and Lagrange multipliers are partially differentiable with respect to χ and have continuous partial derivatives in this neighborhood. The sensitivity of the local minimum $f^\star$ to χ, evaluated at the base-case $\chi = \mathbf{0}$, is given by:

$$\frac{\partial \mathcal{L}}{\partial \chi}(x^\star, \lambda^\star, \mu^\star; \mathbf{0}),$$

where $\mathcal{L} : \mathbb{R}^n \times \mathbb{R}^m \times \mathbb{R}^r \times \mathbb{R}^s \to \mathbb{R}$ is the **parameterized Lagrangian** *defined by:*

$$\forall x \in \mathbb{R}^n, \forall \lambda \in \mathbb{R}^m, \forall \mu \in \mathbb{R}^r, \forall \chi \in \mathbb{R}^s,$$
$$\mathcal{L}(x, \lambda, \mu; \chi) = f(x; \chi) + \lambda^\dagger (A(\chi)x - b(\chi)) + \mu^\dagger (C(\chi)x - d(\chi)).$$

Furthermore, for values of γ and η in a neighborhood of the base-case value of the parameters $\gamma = 0$ and $\eta = 0$, there is a local minimum and corresponding local minimizer and Lagrange multipliers for Problem (17.22). Moreover, the local minimum, local minimizer, and Lagrange multipliers are partially differentiable with respect to γ and η and have continuous partial derivatives. The sensitivities of the local minimum to γ and η, evaluated at the base-case $\gamma = 0$ and $\eta = 0$, are equal to $\lambda^\star$ and $\mu^\star$, respectively.

Proof See [34, theorem 3.2.2] and [70, section 10.8] for details. □

17.4.2 Discussion

We can again interpret the Lagrange multipliers as the sensitivity of the minimum to the right-hand sides of the equality constraints and inequality constraints. We can use the Lagrange multipliers to help in trading off the change in the optimal objective against the cost of modifying the constraints. As in the case of non-linear equality constraints described in Section 14.4, we can again use sensitivity analysis of the first-order necessary conditions to estimate the changes in the minimizer and Lagrange multipliers.

Corollary 17.7 does not apply directly to linear programming problems; however, sensitivity analysis can also be applied to linear programming and, as with linear programming in general, the linearity of both objective and constraints leads to various special cases. For example, the range of validity of the sensitivity analysis can be determined as a by-product of the sensitivity analysis. See [84, sections 6.4 6.5][102, section 5.2 and chapter 19].

17.4.3 Example

Consider Problem (2.18) from Sections 2.3.2.3, 17.1.1.2, ..., 17.3.1.3, which has objective $f : \mathbb{R}^2 \to \mathbb{R}$ and constraints $Ax = b$ and $Cx \le d$ defined by:

$$
\begin{aligned}
\forall x \in \mathbb{R}^2, \ f(x) &= (x_1 - 1)^2 + (x_2 - 3)^2, \\
A &= \begin{bmatrix} 1 & -1 \end{bmatrix}, \\
b &= \begin{bmatrix} 0 \end{bmatrix}, \\
C &= \begin{bmatrix} 0 & -1 \end{bmatrix}, \\
d &= \begin{bmatrix} -3 \end{bmatrix}.
\end{aligned}
$$

We have already verified that the second-order sufficient conditions are satisfied at the base-case solution. Moreover, the matrix

$$
\hat{A} = \begin{bmatrix} A \\ C \end{bmatrix} = \begin{bmatrix} 1 & -1 \\ 0 & -1 \end{bmatrix},
$$

has linearly independent rows, and, furthermore, the inequality constraint is not degenerate at the base-case solution. Suppose that the inequality constraint was changed to $Cx \leq d - \eta$. We first met this example, parameterized in a slightly different way, in Section 2.7.5.5. If η is small enough, then by Corollary 17.7 the minimum of the perturbed problem differs from the minimum of the original problem by approximately $[\mu^\star]^\dagger \eta$. (See Exercise 17.21.)

17.5 Summary

In this chapter, we considered linear inequality-constrained problems and showed that they could be solved using the techniques developed for non-negatively constrained problems in two ways:

(i) using slack variables, and
(ii) using duality.

We also considered sensitivity analysis.

Exercises

Optimality conditions

17.1 ([34, example 2.1.1]) Let $f : \mathbb{R}^2 \to \mathbb{R}$, $C \in \mathbb{R}^{2\times2}$, and $d \in \mathbb{R}^2$ be defined by:

$$\forall x \in \mathbb{R}^2, f(x) = \frac{1}{2}(x_1 + 1)^2 + \frac{1}{2}(x_2 + 1)^2,$$

$$C = \begin{bmatrix} 1 & -1 \\ -1 & -1 \end{bmatrix},$$

$$d = 0.$$

Consider the problem:

$$\min_{x\in\mathbb{R}^2}\{f(x)|Cx \leq d\} = \min_{x\in\mathbb{R}^2}\{\frac{1}{2}(x_1 + 1)^2 + \frac{1}{2}(x_2 + 1)^2|x_2 \geq x_1, x_2 \geq -x_1\}.$$

(i) Find the minimizer of this problem. (Hint: Sketch the feasible set and the contour sets of the objective.)
(ii) What are the Lagrange multipliers?
(iii) Are either of the constraints degenerate?

17.2 Suppose that $f : \mathbb{R}^n \to \mathbb{R}$ is partially differentiable with continuous partial derivatives, $A \in \mathbb{R}^{m\times n}$, $b \in \mathbb{R}^m$, $C \in \mathbb{R}^{r\times n}$, and $d \in \mathbb{R}^r$. Suppose that $x^\star \in \mathbb{R}^n$ is a local minimizer of the problem:

$$\min_{x\in\mathbb{R}^n}\{f(x)|Ax = b, Cx \leq d\}.$$

(i) Suppose that the matrix $\hat{A}$ has linearly independent rows, where $\hat{A}$ is the matrix with rows consisting of:

- the m rows of A, and
- those rows C_ℓ of C for which $\ell \in \mathbb{A}(x^\star)$.

Show that the there is at most one value of the vector of Lagrange multipliers.

(ii) Give an example of a problem where the Lagrange multipliers are not unique.

(Hint: See Exercise 13.4.)

17.3 Use Theorem 17.1 to show that the first-order necessary conditions for Problem (16.1) are:

$$
\begin{aligned}
\nabla f(x) + A^\dagger \lambda - \mu &= 0; \\
Mx &= 0; \\
Ax - b &= 0; \\
x &\geq 0; \text{ and} \\
\mu &\geq 0,
\end{aligned}
$$

where $M = \text{diag}\{\mu_\ell\} \in \mathbb{R}^{n \times n}$. (Hint: Define the inequality constraints by specifying $C = -\mathbf{I}$ and $d = \mathbf{0}$ in Problem (17.1) so that $(Cx \leq d) \Leftrightarrow (x \geq \mathbf{0})$ and write down the resulting first-order necessary conditions from Theorem 17.1. The second line of these conditions are the complementary slackness conditions for Problem (16.1).)

17.4 Write down the Lagrangian for Problem (16.1). (Hint: Define the inequality constraints by specifying $C = -\mathbf{I}$ and $d = \mathbf{0}$ in Problem (17.1) so that $(Cx \leq d) \Leftrightarrow (x \geq \mathbf{0})$.)

17.5 Consider the function $\phi : \mathbb{R}^2 \to \mathbb{R}$ defined by:

$$\forall x \in \mathbb{R}^2, \phi(x) = -(x_1 - 1)^2 - (x_2 - 3)^2.$$

(i) Show that the Hessian of ϕ is not positive definite.
(ii) Show that the Hessian of ϕ is not positive definite on $\mathcal{N}_+ = \{\Delta x \in \mathbb{R}^2 | \Delta x_1 = \Delta x_2\}$.

17.6 In this exercise we consider the generalization of Problem (16.1) where we only have non-negativity constraints on some of the entries of x. Let $f : \mathbb{R}^n \to \mathbb{R}$ be partially differentiable with continuous partial derivatives, $A \in \mathbb{R}^{m \times n}$, and $b \in \mathbb{R}^m$. Let $\mathbb{A} \subseteq \{1, \ldots, n\}$ be any subset of the indices of the decision vector x and suppose that there are r elements in $\mathbb{A}$. Write down the first-order necessary conditions for the following problem:

$$\min_{x \in \mathbb{R}^n} \{ f(x) | Ax = b, x_\ell \geq 0, \forall \ell \in \mathbb{A} \}.$$

(Hint: Define $C \in \mathbb{R}^{r \times n}$ to be the matrix consisting of the negative of those rows of the identity matrix corresponding to the indices in $\mathbb{A}$. For example, if $\mathbb{A} = \{3, 5\}$ then $r = 2$ and:

$$
C = \begin{bmatrix} 0 & 0 & -1 & 0 & 0 & \cdots \\ 0 & 0 & 0 & 0 & -1 & \cdots \end{bmatrix} \in \mathbb{R}^{2 \times n}.
$$

Consider the problem $\min_{x \in \mathbb{R}}\{f(x)|Ax = b, Cx \leq \mathbf{0}\}$.)

Convex problems

17.7 Under the hypotheses of Corollary 17.5, apply Theorems 17.1 and 17.3 to the dual problem to derive the necessary and sufficient conditions for maximizing the dual problem. (Hint: Let $w \in \mathbb{R}^r$ be the dual variables corresponding to the non-negativity constraints $\mu \geq 0$ in the dual problem. Find the first-order necessary conditions and show that they are also sufficient. Eliminate w.)

17.8 Let $c \in \mathbb{R}^n$, $A \in \mathbb{R}^{m \times n}$, $b \in \mathbb{R}^m$, $C \in \mathbb{R}^{r \times n}$, and $d \in \mathbb{R}^r$. Consider the problem $\min_{x \in \mathbb{R}^n} \{c^\dagger x \mid Ax = b, Cx \leq d\}$. Show that the corresponding dual function $\mathcal{D} : \mathbb{E} \to \mathbb{R}$ and its effective domain $\mathbb{E}$ satisfy:

$$
\mathbb{E} = \left\{ \begin{bmatrix} \lambda \\ \mu \end{bmatrix} \in \mathbb{R}^{m+r} \,\middle|\, c + A^\dagger \lambda + C^\dagger \mu = 0 \right\},
$$

$$
\forall \begin{bmatrix} \lambda \\ \mu \end{bmatrix} \in \mathbb{E}, \mathcal{D}(\lambda, \mu) = -\lambda^\dagger b - \mu^\dagger d.
$$

17.9 Let $Q \in \mathbb{R}^{n \times n}$ be positive definite, $c \in \mathbb{R}^n$, $A \in \mathbb{R}^{m \times n}$, $b \in \mathbb{R}^m$, $C \in \mathbb{R}^{r \times n}$, and $d \in \mathbb{R}^r$. Consider the problem $\min_{x \in \mathbb{R}^n} \{\frac{1}{2} x^\dagger Q x + c^\dagger x \mid Ax = b, Cx \leq d\}$. Show that the corresponding dual function $\mathcal{D} : \mathbb{E} \to \mathbb{R}$ and its effective domain $\mathbb{E}$ satisfy:

$$
\mathbb{E} = \mathbb{R}^{m+r},
$$

$$
\forall \begin{bmatrix} \lambda \\ \mu \end{bmatrix} \in \mathbb{E}, \mathcal{D}(\lambda, \mu) = -\tfrac{1}{2}[c + A^\dagger \lambda + C^\dagger \mu]^\dagger Q^{-1}[c + A^\dagger \lambda + C^\dagger \mu] - \lambda^\dagger b - \mu^\dagger d.
$$

17.10 Consider the non-negatively constrained problem:

$$
\min_{x \in \mathbb{R}^n} \{f(x) \mid x \geq 0\},
$$

where $f : \mathbb{R}^n \to \mathbb{R}$ is convex and partially differentiable with continuous partial derivatives.

   (i) Write down the definition of the dual function. (Hint: Use Exercise 17.4.)

   (ii) Suppose that $f : \mathbb{R}^n \to \mathbb{R}$ is quadratic and of the form:

$$
\forall x \in \mathbb{R}^n, f(x) = \frac{1}{2} x^\dagger Q x + c^\dagger x,
$$

with $Q \in \mathbb{R}^{n \times n}$ symmetric and positive definite, having inverse Q^{-1}. Use the optimality conditions in the definition of the dual function to explicitly evaluate the dual function.

   (iii) Explain why taking the dual of a non-negatively constrained problem (or taking the partial dual with respect to the non-negativity constraints) is not likely, in itself, to be useful from a computational perspective.

17.11 Prove the equality between the last two lines of (17.11) using the definition of min. That is, prove that for $f_k : \mathbb{R} \to \mathbb{R}$, $k = 1, \ldots, n$, $A \in \mathbb{R}^{m \times n}$, and $b \in \mathbb{R}^m$ if the problem $\min_{x \in \mathbb{R}^n} \left\{ \sum_{k=1}^n \left(f_k(x_k) + \lambda^\dagger A_k x_k \right) \middle| \underline{x}_k \leq x_k \leq \overline{x}_k, \forall k = 1, \ldots, n \right\}$ has a minimum then each of the sub-problems $\min_{x_k \in \mathbb{R}} \{ f_k(x_k) + \lambda^\dagger A_k x_k \,|\, \underline{x}_k \leq x_k \leq \overline{x}_k \}$, $k = 1, \ldots, n$, have minima and:

$$\min_{x \in \mathbb{R}^n} \left\{ \sum_{k=1}^n (f_k(x_k) + \lambda^\dagger A_k x_k \,\middle|\, \underline{x}_k \leq x_k \leq \overline{x}_k, \forall k = 1, \ldots, n \right\}$$

$$= \sum_{k=1}^n \min_{x_k \in \mathbb{R}} \{ f_k(x_k) + \lambda^\dagger A_k x_k \,|\, \underline{x}_k \leq x_k \leq \overline{x}_k \}.$$

(Hint: Exercise 13.13 treats a similar situation in the case of no inequality constraints.)

Approaches to finding minimizers

17.12 In this exercise we consider the relationship between Problem (16.1) and the various problem forms discussed in Chapter 17.

(i) Consider Problem (17.13):

$$\min_{x \in \mathbb{R}^n, w \in \mathbb{R}^r} \left\{ f(x) \,\middle|\, \begin{bmatrix} A & 0 \\ C & I \end{bmatrix} \begin{bmatrix} x \\ w \end{bmatrix} = \begin{bmatrix} b \\ d \end{bmatrix}, w \geq 0 \right\}.$$

Show that this problem can be written in the form:

$$\min_{\mathcal{X} \in \mathbb{R}^N} \{ \phi(\mathcal{X}) \,|\, \mathcal{A}\mathcal{X} = \mathcal{B}, \mathcal{X}_\geq \geq 0 \},$$

where $\mathcal{X}_\geq \in \mathbb{R}^r$ consists of the last r entries of $\mathcal{X}$. Explicitly define $N, \mathcal{X}, \phi, \mathcal{A}$, and $\mathcal{B}$.

(ii) Consider the problem in the last line of (17.9):

$$\min_{\begin{bmatrix} \lambda \\ \mu \end{bmatrix} \in \mathbb{R}^{m+r}} \left\{ \begin{bmatrix} b \\ d \end{bmatrix}^\dagger \begin{bmatrix} \lambda \\ \mu \end{bmatrix} \,\middle|\, \begin{bmatrix} A \\ C \end{bmatrix}^\dagger \begin{bmatrix} \lambda \\ \mu \end{bmatrix} = -c, \mu \geq 0 \right\}.$$

Show that this problem can be written in the form:

$$\min_{\mathcal{X} \in \mathbb{R}^N} \{ \mathcal{C}^\dagger \mathcal{X} \,|\, \mathcal{A}\mathcal{X} = \mathcal{B}, \mathcal{X}_\geq \geq 0 \},$$

where $\mathcal{X}_\geq \in \mathbb{R}^r$ consists of the last r entries of $\mathcal{X}$. Explicitly define $N, \mathcal{X}, \mathcal{C}, \mathcal{A}$, and $\mathcal{B}$.

(iii) Consider the problem in the last line of (17.10):

$$\min_{\begin{bmatrix} \lambda \\ \mu \end{bmatrix} \in \mathbb{R}^{m+r}} \left\{ \frac{1}{2} [c + A^\dagger \lambda + C^\dagger \mu]^\dagger Q^{-1} [c + A^\dagger \lambda + C^\dagger \mu] + \lambda^\dagger b + \mu^\dagger d \,\middle|\, \mu \geq 0 \right\}.$$

Show that this problem can be written in the form:

$$\min_{\mathcal{X} \in \mathbb{R}^N} \left\{ \frac{1}{2} \mathcal{X}^\dagger \mathcal{Q}\mathcal{X} + \mathcal{C}^\dagger \mathcal{X} + \mathcal{D} \,\middle|\, \mathcal{X}_\geq \geq 0 \right\},$$

where $\mathcal{X}_\geq \in \mathbb{R}^r$ consists of the last r entries of $\mathcal{X}$. Explicitly define $N, \mathcal{X}, \mathcal{Q}, \mathcal{C}$, and $\mathcal{D}$.

17.13 Consider Problem (17.15):

$$\min_{x\in\mathbb{R}^n, w\in\mathbb{R}^r} \left\{ f(x) + t f_{\mathrm{b}}(w) \left| \begin{bmatrix} A & 0 \\ C & I \end{bmatrix} \begin{bmatrix} x \\ w \end{bmatrix} = \begin{bmatrix} b \\ d \end{bmatrix} \right. \right\},$$

where $f : \mathbb{R}^n \to \mathbb{R}$, $f_{\mathrm{b}} : \mathbb{R}^r_{++} \to \mathbb{R}$, $A \in \mathbb{R}^{m\times n}$, $b \in \mathbb{R}^m$, $C \in \mathbb{R}^{r\times n}$, and $d \in \mathbb{R}^r$. Ignoring the issue of the domain of f_{b}, show that the problem has first-order necessary conditions:

$$\begin{aligned} \nabla f(x) + A^\dagger \lambda + C^\dagger \mu &= 0, \\ Ax &= b, \\ Cx + w &= d, \\ t\nabla f_{\mathrm{b}}(w) + \mu &= 0, \end{aligned}$$

where λ and μ are the dual variables on the constraints $Ax = b$ and $Cx + w = d$, respectively.

17.14 Consider Problem (2.18):

$$\min_{x\in\mathbb{R}^2}\{f(x)|Ax = b, Cx \le d\},$$

where:

$$\begin{aligned} \forall x \in \mathbb{R}^2, \ f(x) &= (x_1 - 1)^2 + (x_2 - 3)^2, \\ A &= \begin{bmatrix} 1 & -1 \end{bmatrix}, \\ b &= \begin{bmatrix} 0 \end{bmatrix}, \\ C &= \begin{bmatrix} 0 & -1 \end{bmatrix}, \\ d &= \begin{bmatrix} -3 \end{bmatrix}. \end{aligned}$$

(i) Perform three iterations of the primal–dual interior point algorithm described in Section 17.3.1 for this problem. The Newton–Raphson update was presented in Section 17.3.1.3. Use as initial guess:

$$\mu^{(0)} = [0.25], \ x_1^{(0)} = 5, \ x_2^{(0)} = 5, \ w^{(0)} = [2], \ \lambda_1^{(0)} = 0, \ \lambda_2^{(0)} = 0, \ t^{(0)} = 0.5.$$

For $\nu > 0$, use $t^{(\nu)} = \frac{1}{10} t^{(\nu)}_{\text{effective}} = \frac{1}{10r}[w^{(\nu)}]^\dagger \mu^{(\nu)}$ and, at each iteration, allow the next iterate to be no closer to the boundary than a fraction 0.9995 of the distance of the current iterate to the boundary under the L_∞ norm.

(ii) Evaluate $[\mu^{(\nu)}]^\dagger w^{(\nu)}$ for $\nu = 0, 1, 2, 3$.

17.15 Consider Problem (2.18):

$$\min_{x\in\mathbb{R}^2}\{f(x)|Ax = b, Cx \le d\},$$

where

$$\forall x \in \mathbb{R}^2, \, f(x) = (x_1 - 1)^2 + (x_2 - 3)^2,$$
$$A = \begin{bmatrix} 1 & -1 \end{bmatrix},$$
$$b = \begin{bmatrix} 0 \end{bmatrix},$$
$$C = \begin{bmatrix} 0 & -1 \end{bmatrix},$$
$$d = \begin{bmatrix} -3 \end{bmatrix}.$$

(i) Use the MATLAB function quadprog to find the minimizer and minimum of the problem. Use as initial guess:

$$x_1^{(0)} = 5, x_2^{(0)} = 5.$$

(ii) Form the dual of the problem.

(iii) Use the MATLAB function quadprog to find the maximum of the dual problem. Use as initial guess:

$$\mu^{(0)} = [0.25], \lambda^{(0)} = [0].$$

17.16 Using the DC power flow approximation developed in Exercise 6.6 to approximate the real power flows on the lines, use the MATLAB function quadprog to solve a **DC optimal power flow** that minimizes the cost of production of the generators subject to linearized constraints on the line flows. That is, consider a version of Problem (15.23) where the constraints are linearized. Use the line data from Exercise 8.12. That is, the π-equivalent line models have:

- shunt elements purely capacitive with admittance $0.01\sqrt{-1}$ so that the combined shunt elements are:

$$Y_1 = Y_2 = Y_3 = 0.02\sqrt{-1},$$

and
- series elements having admittances:

$$Y_{12} = (0.01 + 0.1\sqrt{-1})^{-1},$$
$$Y_{23} = (0.015 + 0.15\sqrt{-1})^{-1},$$
$$Y_{31} = (0.02 + 0.2\sqrt{-1})^{-1}.$$

Furthermore, assume the following.

- There are generators at bus 1 and bus 2 and a real power load of 1 at bus 3.
- All lines have real power flow limits of 0.75.
- All voltage magnitudes are set to 1.0 per unit so that u can be ignored in the formulation.
- Zero cost for reactive power production and no constraints on reactive power production nor on reactive power flow so that Q can be ignored in the formulation.

- Costs for real power production at the generators:

$$f_1(P_1) \quad = \quad P_1 \times 1\frac{\$}{\text{per unit}} + (P_1)^2 \times 0.1\frac{\$}{(\text{per unit})^2},$$

$$f_2(P_2) \quad = \quad P_2 \times 1.1\frac{\$}{\text{per unit}} + (P_2)^2 \times 0.05\frac{\$}{(\text{per unit})^2},$$

where P_k is the real power production at generator $k = 1, 2$, with $0 \le P_k \le 1$ for each generator.
- No other constraints on production.

Use as initial guess $P^{(0)} = \mathbf{0}$ and $\theta^{(0)} = \mathbf{0}$. Use as stopping criterion that all of the following are satisfied:

- $t_{\text{effective}} < 10^{-5}$, and
- the change in successive iterates is less than 0.0001 per unit.

Sensitivity

17.17 Show by an example that the conclusion of Corollary 17.7 may fail to hold if the matrix $\hat{A}$ defined in Corollary 17.7 does not have linearly independent rows. (Hint: Consider $C : \mathbb{R} \to \mathbb{R}^{2 \times 1}$ and $d : \mathbb{R} \to \mathbb{R}^2$ defined by:

$$\forall \chi \in \mathbb{R}, C(\chi) \quad = \quad \begin{bmatrix} \chi \\ -\chi \end{bmatrix},$$

$$\forall \chi \in \mathbb{R}, d(\chi) \quad = \quad \mathbf{0}.)$$

17.18 Show by an example that the conclusion of Corollary 17.7 may fail to hold if there is a degenerate constraint at the base-case solution. (Hint: ([34, example 2.1.1]) Consider the problem in Exercise 17.1, but change the objective to $f : \mathbb{R}^2 \times \mathbb{R} \to \mathbb{R}$ defined by:

$$\forall x \in \mathbb{R}^2, \forall \chi \in \mathbb{R}, \ f(x; \chi) = \frac{1}{2}(x_1 + 1 - \chi)^2 + \frac{1}{2}(x_2 + 1)^2.$$

Consider the trajectory of the minimizer as χ varies around $\chi = [0]$.

17.19 ([34, example 2.1.2]) Let $f : \mathbb{R} \times \mathbb{R} \to \mathbb{R}$ be defined by:

$$\forall x \in \mathbb{R}, \forall \chi \in \mathbb{R}, \ f(x; \chi) = x\chi.$$

Consider the problem $\min_{x \in \mathbb{R}}\{f(x; \chi) | x \ge -1\}$.
  (i) Find the minimum, set of minimizers, and Lagrange multipliers for the base-case problem where $\chi = 0$.
  (ii) Find the minimum, minimizer, and Lagrange multipliers for $\chi > 0$.
  (iii) Show that there is no minimum for $\chi < 0$.
  (iv) Which of the hypotheses of Corollary 17.7 fail to hold at the base-case for this problem?

17.20 Consider the sensitivity result in Corollary 17.7 for Problem (17.21):

$$\min_{x \in \mathbb{R}^n} \{ f(x; \chi) | A(\chi)x = b(\chi), C(\chi)x \le d(\chi) \},$$

where $\chi \in \mathbb{R}^s$, $f : \mathbb{R}^n \times \mathbb{R}^s \to \mathbb{R}$, $A : \mathbb{R}^s \to \mathbb{R}^{m \times n}$, $b : \mathbb{R}^s \to \mathbb{R}^m$, $C : \mathbb{R}^s \to \mathbb{R}^{r \times n}$, and $d : \mathbb{R}^s \to \mathbb{R}^r$. Use the sensitivity result for Problem (17.21) to prove the sensitivity result in Corollary 17.7 for Problem (17.22):

$$\min_{x \in \mathbb{R}^n} \{ f(x) | Ax = b - \gamma, Cx \le d - \eta \}.$$

17.21 Consider Problem (2.18), which has objective $f : \mathbb{R}^2 \to \mathbb{R}$ and equality constraint $Ax = b$ defined by:

$$\begin{aligned}
\forall x \in \mathbb{R}^2, f(x) &= (x_1 - 1)^2 + (x_2 - 3)^2, \\
A &= \begin{bmatrix} 1 & -1 \end{bmatrix}, \\
b &= \begin{bmatrix} -3 \end{bmatrix}.
\end{aligned}$$

However, suppose that the inequality constraint was changed to $Cx \le d - \eta$, with $C \in \mathbb{R}^{1 \times 2}$ and $d \in \mathbb{R}^1$ defined by:

$$\begin{aligned}
C &= \begin{bmatrix} 0 & -1 \end{bmatrix}, \\
d &= \begin{bmatrix} -3 \end{bmatrix}.
\end{aligned}$$

Let $\eta = [0.1]$.

(i) Use Corollary 17.7 to estimate the change in the minimum due to the change in the inequality constraint.

(ii) Solve the change-case problem explicitly and compare the result to that obtained by sensitivity analysis.

18

Solution of the linear inequality-constrained case studies

In this chapter we solve the case studies that can be formulated as or transformed to linear inequality-constrained minimization problems. These case studies are:

- least-cost production with capacity constraints (Section 18.1),
- optimal routing in a data communications network (Section 18.2),
- least absolute value estimation (Section 18.3), and
- optimal margin pattern classification (Section 18.4).

18.1 Least-cost production with capacity constraints

In this section, we solve the least-cost production with capacity constraints case study from Section 15.1. We recall and analyze the problem in Section 18.1.1, describe algorithms in Section 18.1.2, and sketch sensitivity analysis in Section 18.1.3.

18.1.1 Problem and analysis

Recall Problem (15.1):

$$\min_{x \in \mathbb{R}^n} \{f(x) | Ax = b, \underline{x} \le x \le \overline{x}\},$$

where the equality constraints are represented in the form $A = -\mathbf{1}^{\dagger}$, $b = [-D]$. This problem has:

- a convex separable objective,
- one equality constraint, and
- two inequality constraints for each variable.

The inequality constraints are simple bounds on variables. We can solve this problem using slight modifications of the algorithms developed in Section 17.3.

18.1.2 Algorithms

18.1.2.1 Primal–dual interior point algorithm

We can use a primal–dual interior point algorithm to solve the problem. For each variable x_ℓ, in order to enforce the bounds $\underline{x}_\ell \leq x_\ell \leq \overline{x}_\ell$, the corresponding term in the barrier objective is:

$$-t \left(\ln(x_\ell - \underline{x}_\ell) + \ln(\overline{x}_\ell - x_\ell) \right).$$

(See Exercise 18.1.)

18.1.2.2 Dual algorithm

Alternatively, we can solve the dual problem by taking the partial dual with respect to the equality constraints. This decomposes the problem into a set of sub-problems, one for each machine k, each with two bound constraints $\underline{x}_k \leq x_k \leq \overline{x}_k$ as discussed in Section 17.2.2.4. The objective of each sub-problem is convex and (at least approximately) quadratic consisting of the cost function for the corresponding machine together with a term involving the latest estimate of the Lagrange multiplier.

Suppose that for each k, the cost f_k of machine k is quadratic and of the form defined in (12.6):

$$\forall x_k \in \mathbb{S}_k, \ f_k(x_k) = \frac{1}{2} Q_{kk}(x_k)^2 + c_k x_k + d_k.$$

Then, for a particular value of the dual variable λ, we obtain constrained sub-problems of the form:

$$\forall k = 1, \ldots, n, \ \min_{x_k \in \mathbb{R}} \left\{ \frac{1}{2} Q_{kk}(x_k)^2 + c_k x_k + d_k - \lambda x_k | \underline{x}_k \leq x_k \leq \overline{x}_k \right\}.$$

The *unconstrained* minimizer of the objective of each sub-problem is given by setting the derivative of the objective equal to zero. That is, the unconstrained minimizer of the objective of each sub-problem is:

$$x_k = \frac{1}{Q_{kk}}(\lambda - c_k).$$

If the unconstrained minimizer is within the range allowed by the upper and lower bound constraints then, by Theorem 3.10, the unconstrained minimizer is also the minimizer of the constrained sub-problem. If the unconstrained minimizer lies outside the range allowed by the bound constraints then the minimizer of the sub-problem is the nearest bound. (See Exercise 2.51 for details.)

In summary, for a given value of the dual variable λ, the corresponding minimizer of the inner problem in the definition of the partial dual is $x^{(\lambda)}$, where:

$$\forall k = 1, \ldots, n, \; x_k^{(\lambda)} = \min\left\{\overline{x}_k, \max\left\{\underline{x}_k, \frac{1}{Q_{kk}}(\lambda - c_k)\right\}\right\}.$$

This expression is very easy to evaluate. Substituting the solution $x_k^{(\lambda)}$ into the expression for the dual, we obtain:

$$\forall \lambda \in \mathbb{R}, \; \mathcal{D}(\lambda) = \sum_{k=1}^{n} f_k(x_k^{(\lambda)}) + \lambda\left(D - \sum_{k=1}^{n} x_k^{(\lambda)}\right).$$

The dual variable can be updated using a steepest ascent algorithm based on the satisfaction of the equality constraint according to:

$$
\begin{aligned}
\Delta\lambda &= \nabla\mathcal{D}(\lambda), \\
&= Ax^{(\lambda)} - b, \\
&= D - \sum_{k=1}^{n} x_k^{(\lambda)}.
\end{aligned}
$$

(See Exercise 18.1.) Since each machine cost function is strictly convex, the minimizer of the primal problem can be found from the solution of the dual algorithm. As in Section 13.5.3, we can interpret λ as the tentative price per unit of production.

18.1.3 Changes in demand and capacity

Corollary 17.7 can be used to estimate the changes in costs due to a change in demand or capacity. (See Exercise 18.2.)

18.2 Optimal routing in a data communications network

In this section, we solve the optimal routing in a data communications network case study from Section 15.2. We recall and analyze the problem in Section 18.2.1, sketch algorithms in Section 18.2.2, and sketch sensitivity analysis in Section 18.2.3.

18.2.1 Problem and analysis

Recall Problem (15.6):

$$\min_{x \in \mathbb{R}^n}\; \{f(x)\,|\,Ax = b, x \geq 0, Cx < \overline{y}\},$$

where $f : \overline{\mathbb{S}} \to \mathbb{R}$, with $\overline{\mathbb{S}} = \{x \in \mathbb{R}^n | x \geq 0, Cx < \overline{y}\}$, was defined in (15.7), which we repeat here:

$$\forall x \in \overline{\mathbb{S}}, \ f(x) \quad = \quad \phi(Cx),$$
$$= \quad \sum_{(i,j) \in \mathbb{L}} \phi_{ij} \left(C_{(i,j)} x \right).$$

The delay function ϕ_{ij} in the objective increases without bound as a flow approaches its capacity. Consequently, assigning a flow to be arbitrarily close to the link capacity can never be optimal.

In fact, the delay function has the same form as the **reciprocal barrier function** investigated in Exercise 16.12. As with the logarithmic barrier function, this means that the strict inequality constraints:

$$Cx < \overline{y},$$

can be ignored so long as:

- an initial feasible solution can be found that satisfies these constraints, and
- a step size is chosen at each iteration to avoid going outside the feasible region.

We effectively have a problem with a barrier objective that enforces the strict inequality constraints $Cx < \overline{y}$ and that must be solved for a single fixed value of the barrier parameter. That is, to solve Problem (15.6) we can effectively solve the problem:

$$\min_{x \in \mathbb{R}^n} \{ f(x) | Ax = b, x \geq 0 \}. \tag{18.1}$$

A Newton–Raphson step direction to minimize f subject to equality constraints would be similar to the update for the primal interior point algorithm described in Section 16.4.3.2.

18.2.2 Algorithms

Problem (18.1) is non-negatively constrained and these constraints can be treated using an active set or interior point algorithm, so long as we ensure that the step-size is chosen at each iteration to also satisfy $Cx < \overline{y}$. A step-size rule analogous to that for the primal–dual interior point algorithm from Section 16.4.3.3 can be used to ensure satisfaction of the strict inequality constraints $Cx < \overline{y}$. (See Exercise 18.3.)

18.2.3 Changes in links and traffic

Corollary 17.7 and extensions can be used to estimate the changes in optimal routing to respond to a change in traffic or link capacities. (See Exercise 18.4.)

18.3 Least absolute value estimation

In this section, we solve the least absolute value estimation case study from Section 15.3. We recall the problem in Section 18.3.1, sketch algorithms in Section 18.3.2, and sketch sensitivity analysis in Section 18.3.3.

18.3.1 Problem

Recall Problem (15.10):

$$\min_{z\in\mathbb{R}^m, x\in\mathbb{R}^n, e\in\mathbb{R}^m} \{\mathbf{1}^\dagger z | Ax - b - e = \mathbf{0}, z \geq e, z \geq -e\}.$$

This problem has a linear objective and linear inequality constraints.

18.3.2 Algorithms

We can solve this problem using the primal or the dual algorithms developed in Section 17.3. The solution to the corresponding *least-squares* estimation problem can provide a suitable initial guess for $x^{(0)}$. (See Exercise 18.6.)

18.3.3 Changes in the number of points and data

Corollary 17.7 and extensions can be used to estimate the changes in parameters specifying the affine fit if additional data points are added or if the data changes. (See Exercise 18.6.)

18.4 Optimal margin pattern classification

In this section, we solve the optimal margin pattern classification case study from Section 15.4. We recall and analyze the problem in Section 18.4.1, describe algorithms in Section 18.4.2, and sketch sensitivity analysis in Section 18.4.3.

18.4.1 Problem and analysis

Recall Problem (15.13):

$$\max_{z\in\mathbb{R}, x\in\mathbb{R}^n} \left\{ z \,\middle|\, \varsigma(\ell)(\beta^\dagger \psi(\ell) + \gamma) \geq \|\beta\|_2\, z, \forall \ell = 1, \ldots, r, \beta \neq \mathbf{0} \right\},$$

where $x = \begin{bmatrix} \beta \\ \gamma \end{bmatrix}$. This problem has the drawback that its feasible set is not closed and may not be convex. Furthermore, the inequality constraints are non-linear.

In the following sections, we will discuss two ways to *further* transform the problem. Both transformations rest on the observation that, given a maximizer

$\begin{bmatrix} z^\star \\ x^{\star\star} \end{bmatrix}$ of Problem (15.13) and a constant $\kappa \in \mathbb{R}_{++}$, then $\begin{bmatrix} z^\star \\ x^\star \end{bmatrix} = \begin{bmatrix} z^\star \\ x^{\star\star}/\kappa \end{bmatrix}$ is also

a maximizer of Problem (15.13). This simply reflects the fact that the coefficients in the equation for a hyperplane can be scaled without changing the hyperplane. In particular, suppose that Problem (15.13) has maximizer:

$$\begin{bmatrix} z^\star \\ x^{\star\star} \end{bmatrix} = \begin{bmatrix} z^\star \\ \beta^{\star\star} \\ \gamma^{\star\star} \end{bmatrix},$$

and let $\kappa \in \mathbb{R}_{++}$. Consider the candidate solution $\begin{bmatrix} z^\star \\ x^\star \end{bmatrix}$ defined by:

$$\begin{bmatrix} z^\star \\ x^\star \end{bmatrix} = \begin{bmatrix} z^\star \\ \beta^\star \\ \gamma^\star \end{bmatrix},$$

$$= \begin{bmatrix} z^\star \\ \beta^{\star\star}/\kappa \\ \gamma^{\star\star}/\kappa \end{bmatrix}. \tag{18.2}$$

We observe that $\begin{bmatrix} z^\star \\ x^\star \end{bmatrix}$ is also a maximizer of Problem (15.13) with the same maximum. (See Exercise 18.7.)

18.4.1.1 First approach to transforming constraints

The first way to transform the problem into an inequality-constrained problem is to choose $\kappa = \|\beta^{\star\star}\|_2$ in (18.2). That is, if there is a maximizer to Problem (15.13) then there is a maximizer that satisfies $\beta^\star = \beta^{\star\star}/\|\beta^{\star\star}\|_2$, so that $\|\beta^\star\|_2 = 1$. That is, we can impose the additional constraint $\|\beta\|_2 = 1$ in Problem (15.13) without changing its maximum. Furthermore, since $\|\beta\|_2 = 1$ implies that $\beta \neq \mathbf{0}$, we can ignore the constraint $\beta \neq \mathbf{0}$. We can use Theorem 3.10 to show that if Problem (15.13) has a maximum then maximizing the objective over the "smaller" feasible set:

$$\hat{\mathbb{S}} = \left\{ \begin{bmatrix} z \\ x \end{bmatrix} \in \mathbb{R}^{n+1} \,\middle|\, \zeta(\ell)(\beta^\dagger \psi(\ell) + \gamma) \geq \|\beta\|_2\, z, \, \forall \ell = 1, \dots, r, \, \|\beta\|_2 = 1 \right\},$$

will yield the same maximum as Problem (15.13) and, moreover, the maximizer specifies the same hyperplane.

The smaller feasible set $\hat{\mathbb{S}}$ is closed and bounded, which as we saw in Section 2.3.3 avoids the difficulties that non-closed and unbounded sets present. However, a constraint of the form $\|\beta\|_2 = 1$ is still difficult to handle directly because it defines a non-convex set. One way to deal with this is to convert the representation into polar coordinates. (See Exercise 3.19.) However, a more straightforward further transformation is to note that if Problem (15.13) has a maximum then the value of the maximum is given by:

$$\max_{z \in \mathbb{R}, x \in \mathbb{R}^n} \left\{ z \, \big| \, \zeta(\ell)(\beta^\dagger \psi(\ell) + \gamma) \geq \|\beta\|_2 \, z, \, \forall \ell = 1, \ldots, r, \, \beta \neq \mathbf{0} \right\}$$

$$= \max_{z \in \mathbb{R}, x \in \mathbb{R}^n} \left\{ z \, \big| \, \zeta(\ell)(\beta^\dagger \psi(\ell) + \gamma) \geq \|\beta\|_2 \, z, \, \forall \ell = 1, \ldots, r, \, \|\beta\|_2 = 1 \right\},$$

by the argument above,

$$= \max_{z \in \mathbb{R}, x \in \mathbb{R}^n} \left\{ z \, \big| \, \zeta(\ell)(\beta^\dagger \psi(\ell) + \gamma) \geq z, \, \forall \ell = 1, \ldots, r, \, \|\beta\|_2 = 1 \right\},$$

since $\|\beta\|_2 = 1$,

$$= \max_{z \in \mathbb{R}, x \in \mathbb{R}^n} \left\{ z \, \big| \, \zeta(\ell)(\beta^\dagger \psi(\ell) + \gamma) \geq z, \, \forall \ell = 1, \ldots, r, \, \|\beta\|_2 \leq 1 \right\},$$

where we note that any maximizer $\begin{bmatrix} z^\star \\ x^\star \end{bmatrix} = \begin{bmatrix} z^\star \\ \beta^\star \\ \gamma^\star \end{bmatrix}$ of the last problem will satisfy

$\|\beta^\star\|_2 = 1$, since if $\|\beta^\star\|_2 < 1$ then we could find a feasible solution having a larger objective by dividing both $z^\star$ and $x^\star$ by $\|\beta^\star\|_2$. (See Exercise 18.8.) The *relaxation* of the problem to having the larger feasible set with the constraint $\|\beta\|_2 \leq 1$ yields a convex feasible set with the same maximum as Problem (15.13) and its maximizer specifies the same hyperplane as a maximizer of Problem (15.13). Since $\|\beta\|_2$ is not smooth, we will use the equivalent condition $\|\beta\|_2^2 \leq 1$.

By defining $C \in \mathbb{R}^{r \times n}$ to have ℓ-th row:

$$C_\ell = -\zeta(\ell) \begin{bmatrix} \psi(\ell)^\dagger & 1 \end{bmatrix},$$

and noting that $z - \zeta(\ell)(\beta^\dagger \psi(\ell) + \gamma) = z + C_\ell x$, we can transform the problem to the equivalent problem:

$$\max_{z \in \mathbb{R}, x \in \mathbb{R}^n} \{ z \, | \, \mathbf{1}z + Cx \leq \mathbf{0}, \, \|\beta\|_2^2 \leq 1 \}, \tag{18.3}$$

where we have squared the norm of β to obtain a differentiable function. (See Exercise 18.9.) This problem has a linear objective, r linear inequality constraints, and one quadratic inequality constraint. We will treat the solution of this formulation of the problem in Section 20.1. (See [15, section 8.6.1] for a slightly different transformation of this problem.)

18.4.1.2 Second approach to transforming constraints

We will present a second transformation of Problem (15.13) that yields a problem with quadratic objective and linear constraints. Consider a maximizer $\begin{bmatrix} z^\star \\ x^{\star\star} \end{bmatrix} =$

$\begin{bmatrix} z^\star \\ \beta^{\star\star} \\ \gamma^{\star\star} \end{bmatrix}$ of Problem (15.13). Suppose that $z^\star \in \mathbb{R}_{++}$ so that the margin is strictly positive. Since β^{**} is feasible, we have that $\beta^{**} \neq \mathbf{0}$. We can choose $\kappa = \|\beta^{**}\|_2 \, z^\star$ in (18.2). Consequently, if there is a maximizer to Problem (15.13) with positive margin then there is a maximizer that satisfies $\beta^\star = \beta^{**}/(\|\beta^{**}\|_2 \, z^\star)$, so that $\|\beta^\star\|_2 \, z^\star = 1$.

As in the first approach to transforming the constraints in Section 18.4.1.1, we can impose the additional constraint $\|\beta\|_2 \, z = 1$ in Problem (15.13) without changing its maximum. Furthermore, since $\|\beta\|_2 \, z = 1$ implies that $\beta \neq \mathbf{0}$, we can again ignore the constraint $\beta \neq \mathbf{0}$. We can again use Theorem 3.10 to show that if Problem (15.13) has a maximizer and strictly positive maximum $z^\star$ then $z^\star$ will also be the maximum of a problem having the same objective but with "smaller" feasible set:

$$
\underline{\mathbb{S}} = \left\{ \begin{bmatrix} z \\ x \end{bmatrix} \in \mathbb{R}^{n+1} \; \middle| \; \frac{\varsigma(\ell) D(\psi(\ell))}{\|\beta\|_2} \geq z, \, \forall \ell = 1, \ldots, r, \, \|\beta\|_2 \, z = 1 \right\}.
$$

Moreover, if Problem (15.13) has a maximum and maximizer, then at least one of maximizers of the problem is an element of $\underline{\mathbb{S}}$. That is, if Problem (15.13) has a maximum and the margin is positive then the value of the maximum is given by:

$$
\max_{z \in \mathbb{R}, x \in \mathbb{R}^n} \left\{ z \; \middle| \; \varsigma(\ell)(\beta^\dagger \psi(\ell) + \gamma) \geq \|\beta\|_2 \, z, \, \forall \ell = 1, \ldots, r, \, \beta \neq \mathbf{0} \right\}
$$

$$
= \max_{z \in \mathbb{R}, x \in \mathbb{R}^n} \left\{ z \; \middle| \; \varsigma(\ell)(\beta^\dagger \psi(\ell) + \gamma) \geq \|\beta\|_2 \, z, \, \forall \ell = 1, \ldots, r, \, \|\beta\|_2 \, z = 1 \right\},
$$

by the argument above,

$$
= \max_{z \in \mathbb{R}, x \in \mathbb{R}^n} \left\{ z \; \middle| \; \varsigma(\ell)(\beta^\dagger \psi(\ell) + \gamma) \geq 1, \, \forall \ell = 1, \ldots, r, \, \|\beta\|_2 \, z = 1 \right\},
$$

since $\|\beta\|_2 \, z = 1$,

$$
= \max_{z \in \mathbb{R}, x \in \mathbb{R}^n} \left\{ \frac{1}{\|\beta\|_2} \; \middle| \; \varsigma(\ell)(\beta^\dagger \psi(\ell) + \gamma) \geq 1, \, \forall \ell = 1, \ldots, r, \, \|\beta\|_2 \, z = 1 \right\},
$$

since $z = 1/\|\beta\|_2$,

$$
= \max_{x \in \mathbb{R}^n} \left\{ \frac{1}{\|\beta\|_2} \; \middle| \; \varsigma(\ell)(\beta^\dagger \psi(\ell) + \gamma) \geq 1, \, \forall \ell = 1, \ldots, r \right\}, \quad \text{by Corollary 3.7,}
$$

on eliminating the variable z using the constraint $\|\beta\|_2 \, z = 1$,

$$
= \left[\frac{1}{\min_{x \in \mathbb{R}^n} \left\{ \|\beta\|_2 \; \middle| \; \varsigma(\ell)(\beta^\dagger \psi(\ell) + \gamma) \geq 1, \, \forall \ell = 1, \ldots, r \right\}} \right],
$$

by Theorem 3.1, since the reciprocal function is monotonically decreasing. As in Section 18.4.1.1, by defining $C \in \mathbb{R}^{r \times n}$ to have ℓ-th row:

$$
C_\ell = -\varsigma(\ell) \begin{bmatrix} \psi(\ell)^\dagger & 1 \end{bmatrix},
$$

and defining $d = -\mathbf{1} \in \mathbb{R}^r$, we can transform the problem in the denominator to the equivalent problem:

$$\min_{x \in \mathbb{R}^n} \left\{ \frac{1}{2} \|\beta\|_2^2 \, \Big| \, Cx \leq d \right\}, \tag{18.4}$$

which has a quadratic objective and linear constraints and so is a quadratic program. (The factor $\frac{1}{2}$ in the objective has been included to be consistent with our
conventions for quadratic functions. The norm $\|\beta\|_2$ has been squared to make it
differentiable.) If Problem (18.4) has a minimizer $x^\star = \begin{bmatrix} \beta^\star \\ \gamma^\star \end{bmatrix}$ and $\beta^\star \neq \mathbf{0}$ then the
optimal margin is given by $1/\|\beta^\star\|_2$. (See Exercise 18.9.)

18.4.2 Algorithms

18.4.2.1 Primal algorithm

Problem (18.4) has a convex quadratic objective, linear inequality constraints, and
no equality constraints. A quadratic programming algorithm, such as the algorithm
in Section 17.3.1, can be applied to Problem (18.4). See Exercise 18.11.

If the number, r, of patterns is extremely large then a further relaxation of the
problem may be much easier to solve. In particular, we can first solve the problem
using only some of the patterns to find a tentative separating hyperplane. The
feasible set using only some of the patterns is a relaxed version of the feasible
set of Problem (18.4). Then the rest of the patterns are searched until a pattern is
found that is not correctly identified by the tentative separating hyperplane. The
problem is re-solved with the new pattern incorporated and the process repeated. If
a separating hyperplane is found after only a modest number of patterns are added
then we have avoided the computational effort of solving the problem will all r
constraints explicitly represented.

18.4.2.2 Dual algorithm

The dual of Problem (18.4) has a quadratic objective, non-negativity constraints,
and one linear equality constraint. (See Exercise 18.10.)

18.4.3 Changes

Adding a pattern would add an extra row to the inequality constraints $Cx \leq d$. The
relaxation procedure described in Section 18.4.2.1 can be applied or the dual can
be updated and solved.

Exercises

Least-cost production with capacity constraints

18.1 In this exercise, we add minimum and maximum capacity constraints to the problem
from Exercise 13.30. Consider Problem (15.1) in the case that $n = 3$, $D = 5$, and the f_k
are of the form:

$$\forall x_1 \in \mathbb{R}, \; f_1(x_1) \;=\; \frac{1}{2}(x_1)^2 + x_1,$$

$$\forall x_2 \in \mathbb{R}, \; f_2(x_2) \;=\; \frac{1}{2} \times 1.1 \times (x_2)^2 + 0.9 \times x_2,$$

$$\forall x_3 \in \mathbb{R}, \; f_3(x_3) \;=\; \frac{1}{2} \times 1.2 \times (x_3)^2 + 0.8 \times x_3.$$

Also, suppose that the minimum and maximum capacity constraints are specified by:

$$\underline{x} = \begin{bmatrix} 1 \\ 1 \\ 2 \end{bmatrix}, \; \overline{x} = \begin{bmatrix} 4 \\ 5 \\ 6 \end{bmatrix}.$$

Solve it in three ways.

(i) By performing three iterations of the primal–dual interior point algorithm. Use
initial guess $x^{(0)} = \begin{bmatrix} 2.5 \\ 3 \\ 4 \end{bmatrix}$, $\lambda^{(0)} = [0]$, and initial value of barrier parameter
$t^{(0)} = 1$. Note that there are $r = 6$ constraints corresponding to the six entries
of $w \in \mathbb{R}^6$. Define $\mu^{(0)}$ to satisfy $M^{(0)}w^{(0)} = t^{(0)}\mathbf{1}$. For $\nu > 0$, use $t^{(\nu)} = \frac{1}{10}t^{(\nu)}_{\text{effective}} = \frac{1}{10r}[w^{(\nu)}]^\dagger \mu^{(\nu)}$ and, at each iteration, allow the next iterate to be no
closer to the boundary than a fraction 0.9995 of the distance of the current iterate
to the boundary under the L_∞ norm.

(ii) By maximizing the partial dual with respect to the equality constraints. At each
iteration of the algorithm to maximize the dual, explicitly solve the inner inequality-
constrained problem using the discussion from Section 18.1.2.2. Use $\lambda^{(0)} = [0]$ as
the initial guess and perform steepest ascent of the dual function with step-size
equal to 0.5 at each iteration.

(iii) Using the MATLAB function quadprog. Use initial guess $x^{(0)} = \begin{bmatrix} 0 \\ 0 \\ 0 \end{bmatrix}$.

18.2 Consider the solution of Exercise 18.1.

(i) Using sensitivity analysis, estimate the minimum if demand changes to $D = 5.1$.
(ii) Use the MATLAB function quadprog to calculate the minimum if demand changes
to $D = 5.1$. Compare the result to the previous part.
(iii) Using sensitivity analysis, estimate the minimum if the capacity of machine 3
changes to $\overline{x}_3 = 5$.
(iv) Use the MATLAB function quadprog to calculate the minimum if the capacity of
machine 3 changes to $\overline{x}_3 = 5$. Compare the result to the previous part.

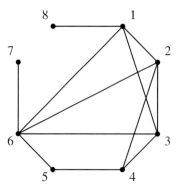

Fig. 18.1. The graph of the data communications network with eight nodes and 12 links from Section 15.2.

Optimal routing in a data communications network

18.3 Consider the optimal routing problem described in Section 15.2 with data communications network illustrated in Figure 18.1. For this network, we are given that:

$$
\begin{aligned}
\mathbb{L} &= \{(1,8), (8,1), (1,2), (2,1), (1,3), (3,1), (1,6), (6,1), \\
&\qquad (2,3), (3,2), (2,4), (4,2), (2,6), (6,2), (3,4), (4,3), \\
&\qquad (3,6), (6,3), (4,5), (5,4), (5,6), (6,5), (6,7), (7,6)\}, \\
\forall (i,j) \in \mathbb{L}, \bar{y}_{ij} &= 2, \\
\mathbb{W} &= \{(7,5), (2,5)\}, \\
\mathbb{P}_{(7,5)} &= \{1,2\}, \\
\mathbb{P}_{(2,5)} &= \{3,4\}, \\
b_{(7,5)} &= 1, \\
b_{(2,5)} &= 1.
\end{aligned}
$$

Moreover, the allowable paths are:

- path 1, consisting of links $(7,6)$, $(6,5)$, for origin–destination pair $(7,5)$,
- path 2, consisting of links $(7,6)$, $(6,3)$, $(3,4)$, $(4,5)$, for origin–destination pair $(7,5)$,
- path 3, consisting of links $(2,4)$, $(4,5)$, for origin–destination pair $(2,5)$, and
- path 4, consisting of links $(2,3)$, $(3,4)$, $(4,5)$, for origin–destination pair $(2,5)$.

The congestion model is given by functions of the form $\phi_{ij} : [0, C_{ij}) \rightarrow \mathbb{R}_+$ for each $(i,j) \in \mathbb{L}$ and defined by (15.3), which we repeat here:

$$
\forall y_{ij} \in [0, \bar{y}_{ij}), \phi_{ij}(y_{ij}) = \frac{y_{ij}}{\bar{y}_{ij} - y_{ij}} + \delta_{ij} y_{ij}.
$$

For simplicity, we assume that the processing delay and propagation delay through each link is negligible so that $\delta_{ij} = 0$ for each $(i,j) \in \mathbb{L}$. Consider optimal routing Problem (18.1) for this network.

(i) Perform three iterations of the primal–dual interior point algorithm to solve the optimal routing problem. For the initial guess $x^{(0)}$, assign half of the flow for each origin–destination pair to each corresponding path. Also set $\lambda^{(0)} = [0]$ and initial value of barrier parameter $t^{(0)} = 1$. Note that there are $n = 4$ non-negativity constraints corresponding to the four entries of $x \in \mathbb{R}^4$. Define $\mu^{(0)}$ to satisfy

$M^{(0)}x^{(0)} = t^{(0)}\mathbf{1}$. For $v > 0$, use $t^{(v)} = \frac{1}{10}t^{(v)}_{\text{effective}} = \frac{1}{10n}[x^{(v)}]^{\dagger}\mu^{(v)}$ and, at each iteration, allow the next iterate to be no closer to the boundary than a fraction 0.9995 of the distance of the current iterate to the boundary under the L_{∞} norm.

(ii) Use the MATLAB function fmincon to solve the problem. For the initial guess $x^{(0)}$, assign half of the flow for each origin–destination pair to each corresponding path. Represent the strict inequality constraints as non-strict inequality constraints.

(iii) Verify that the solution from Part (ii) satisfies the minimum first derivative length property that was proved in Exercise 16.4. (This allows for a more efficient algorithm for solving this problem. See [9, section 5.5] for details.)

18.4 Consider the solution of Exercise 18.3.

(i) Using sensitivity analysis of the first-order necessary conditions, estimate the optimal routing if the expected rate of arrival for both origin–destination pairs changes to 1.1.

(ii) Use the MATLAB function fmincon to calculate the optimal routing if the expected rate of arrival for both origin–destination pairs changes to 1.1. Compare the result to the previous part.

(iii) Using sensitivity analysis of the first-order necessary conditions, estimate the optimal routing if the capacity of each link changes to 1.5.

(iv) Use the MATLAB function fmincon to calculate the optimal routing if the capacity of each link changes to 1.5. Compare the result to the previous part.

18.5 Consider the optimal routing problem described in Exercise 15.5 having directed links. The nodes and the directed links were illustrated in Figure 15.9, which is repeated in Figure 18.2. There are three nodes and four directed links. For this network, we are given that:

$$
\begin{aligned}
\text{Links: } \mathbb{L} &= \{(1, 2), (2, 1), (1, 3), (2, 3)\}, \\
\text{Link capacities: } \forall (i, j) \in \mathbb{L}, \overline{y}_{ij} &= 2, \\
\text{Origin–destination pairs: } \mathbb{W} &= \{(1, 3), (2, 3)\}, \\
\text{Paths for origin–destination pair } (1, 3)\text{: } \mathbb{P}_{(1,3)} &= \{1, 2\}, \\
\text{Paths for origin–destination pair } (2, 3)\text{: } \mathbb{P}_{(2,3)} &= \{3, 4\}, \\
\text{Flow for origin–destination pair } (1, 3)\text{: } b_{(1,3)} &= 1, \\
\text{Flow for origin–destination pair } (2, 3)\text{: } b_{(2,3)} &= 1.
\end{aligned}
$$

Moreover, the allowable paths are:

- path 1, consisting of link $(1, 3)$, for origin–destination pair $(1, 3)$,
- path 2, consisting of links $(1, 2)$, $(2, 3)$, for origin–destination pair $(1, 3)$,
- path 3, consisting of links $(2, 1)$, $(1, 3)$, for origin–destination pair $(2, 3)$, and
- path 4, consisting of link $(2, 3)$, for origin–destination pair $(2, 3)$.

The congestion model is given by functions of the form $\phi_{ij} : [0, \overline{y}_{ij}) \to \mathbb{R}_+$ for each $(i, j) \in \mathbb{L}$ and defined by:

$$
\forall y_{ij} \in [0, \overline{y}_{ij}), \phi_{ij}(y_{ij}) = \frac{y_{ij}}{\overline{y}_{ij} - y_{ij}}.
$$

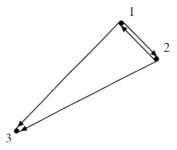

Fig. 18.2. The graph of the data communications network with three nodes and four directed links for Exercise 18.5.

(i) Use the MATLAB function fmincon to solve optimal routing Problem (18.1) for this network. As initial guess $x^{(0)}$, assign half of the flow for each origin–destination pair to each corresponding path. Represent the strict inequality constraints as non-strict inequality constraints.

(ii) Verify that the solution satisfies the minimum first derivative length property that was proved in Exercise 16.4. (As in the case of bi-directional links, this allows for a more efficient algorithm for solving this problem. See [9, section 5.5] for details.)

Least absolute value estimation

18.6 In this exercise we consider best fits to data in the sense of least absolute error.

(i) Use the MATLAB function fmincon to find the affine function:

$$\forall \psi \in \mathbb{R}, \zeta = \alpha \psi + \beta,$$

with $\alpha \in \mathbb{R}$ and $\beta \in \mathbb{R}$, that best fits the following pairs of data $(\psi(\ell), \zeta(\ell))$, for $\ell = 1, \ldots, 7$, in a least absolute error sense. Use as initial guess the solution from Exercise 11.5.

ℓ	1	2	3	4	5	6	7
$\psi(\ell)$	0.27	0.2	0.8	0.4	0.2	0.7	0.5
$\zeta(\ell)$	0.3	0.65	0.75	0.4	0.15	0.6	0.5

(ii) Compare the result of the previous part to the solution to Exercise 11.5.

(iii) Now suppose that the data point $(\psi(6), \zeta(6))$ is altered to equal $(0.7, 1.45)$. Use sensitivity analysis of the first-order necessary conditions for Part (i) to estimate the parameters for the best affine fit to the altered data in the least absolute error sense. You will have to calculate the sensitivity with respect to $\chi = \Delta\zeta(6)$.

(iv) Again suppose that the data point $(\psi(6), \zeta(6))$ is altered to equal $(0.7, 1.45)$. Use the MATLAB function fmincon to find the parameters for the best affine fit to the altered data in the least absolute error sense. Use as initial guess the solution from Part (i).

(v) Compare the result of Parts (iii) and (iv) to the solution to Exercise 15.7.

Optimal margin pattern classification

18.7 Suppose that Problem (15.13) has maximizer:

$$\begin{bmatrix} z^\star \\ x^{\star\star} \end{bmatrix} = \begin{bmatrix} z^\star \\ \beta^{\star\star} \\ \gamma^{\star\star} \end{bmatrix}.$$

(i) Let $\kappa \in \mathbb{R}_{++}$. Show that:

$$\begin{bmatrix} z^\star \\ x^\star \end{bmatrix} = \begin{bmatrix} z^\star \\ x^{\star\star}/\kappa \end{bmatrix}$$

is also a maximizer of Problem (15.13).

(ii) Show that:

$$\begin{bmatrix} z^\star \\ x^\star \end{bmatrix} = \begin{bmatrix} z^\star \\ x^{\star\star}/\|\beta^{\star\star}\|_2 \end{bmatrix}$$

is also a maximizer of Problem (15.13). (Why is $\beta^{\star\star} \neq \mathbf{0}$?)

(iii) Suppose that $z^\star \in \mathbb{R}_{++}$. Show that:

$$\begin{bmatrix} z^\star \\ x^\star \end{bmatrix} = \begin{bmatrix} z^\star \\ x^{\star\star}/(\|\beta^{\star\star}\|_2 \, z^\star) \end{bmatrix}$$

is also a maximizer of Problem (15.13). (Why is $\beta^{\star\star} z^\star \neq \mathbf{0}$?)

18.8 Consider the following two problems:

$$\max{}_{z\in\mathbb{R}, x\in\mathbb{R}^n} \left\{ z \,\middle|\, \zeta(\ell)(\beta^\dagger \psi(\ell) + \gamma) \geq z, \forall \ell = 1, \dots, r, \|\beta\|_2 = 1 \right\},$$
$$\max{}_{z\in\mathbb{R}, x\in\mathbb{R}^n} \left\{ z \,\middle|\, \zeta(\ell)(\beta^\dagger \psi(\ell) + \gamma) \geq z, \forall \ell = 1, \dots, r, \|\beta\|_2 \leq 1 \right\}.$$

Suppose that the first problem has a maximum.

(i) Show that the second problem has a maximum.
(ii) Show that both problems have the same maximum and that their maximizers specify the same hyperplane.

(Hint: Consider a feasible solution $\begin{bmatrix} z^{\star\star} \\ x^{\star\star} \end{bmatrix} = \begin{bmatrix} z^{\star\star} \\ \beta^{\star\star} \\ \gamma^{\star\star} \end{bmatrix}$ of the second problem. Show that if $\|\beta^{\star\star}\|_2 < 1$ then we could find a feasible solution having a larger objective by dividing both $z^{\star\star}$ and $x^{\star\star}$ by $\|\beta^{\star\star}\|_2$.)

18.9 In this exercise we consider the transformed versions of Problem (15.13).

(i) Show that if there is no hyperplane that can separate the patterns then Problem (18.3) is infeasible.
(ii) Show that if there is no hyperplane that can separate the patterns then Problem (18.4) is infeasible.

18.10 Consider Problem (18.4):

$$\min_{x \in \mathbb{R}^n} \left\{ \frac{1}{2} \|\beta\|_2^2 \,\middle|\, Cx \le d \right\}.$$

Show that the dual of Problem (18.4) has a quadratic objective, non-negativity constraints, and one linear equality constraint.

18.11 Consider the following patterns and their classification:

ℓ	1	2	3	4	5	6	7
$\psi(\ell)$	$\begin{bmatrix} 1.25 \\ -0.2 \end{bmatrix}$	$\begin{bmatrix} 0.0 \\ 1.05 \end{bmatrix}$	$\begin{bmatrix} 0.8 \\ 0.45 \end{bmatrix}$	$\begin{bmatrix} 0.4 \\ 0.45 \end{bmatrix}$	$\begin{bmatrix} 0.2 \\ 0.8 \end{bmatrix}$	$\begin{bmatrix} 0.7 \\ 0.85 \end{bmatrix}$	$\begin{bmatrix} 0.5 \\ 0.7 \end{bmatrix}$
$\zeta(\ell)$	1	1	1	1	-1	-1	-1

(i) Use the MATLAB function `quadprog` to find the solution of Problem (18.4) for these data.

(ii) Explicitly write out the dual of Problem (18.4) for these data.

(iii) Use the MATLAB function `quadprog` to find the solution of the dual of Problem (18.4) for these data.

19

Algorithms for non-linear inequality-constrained minimization

In this chapter we will develop algorithms for constrained optimization problems of the form:

$$\min_{x \in \mathbb{S}} f(x),$$

where $f : \mathbb{R}^n \to \mathbb{R}$ and where the feasible set $\mathbb{S}$ is of the form:

$$\mathbb{S} = \{x \in \mathbb{R}^n | g(x) = \mathbf{0}, h(x) \leq \mathbf{0}\},$$

with both $g : \mathbb{R}^n \to \mathbb{R}^m$ and $h : \mathbb{R}^n \to \mathbb{R}^r$ non-linear. That is, we will consider problems of the form:

$$\min_{x \in \mathbb{R}^n} \{f(x) | g(x) = \mathbf{0}, h(x) \leq \mathbf{0}\}. \tag{19.1}$$

We refer to Problem (19.1) as a non-linear inequality-constrained problem, where it is understood that it also includes non-linear equality constraints in addition to the inequality constraints.

We first investigate properties of non-linear equality constraints in Section 19.1 and then derive optimality conditions in Section 19.2. We consider the convex case in Section 19.3. As in previous chapters, the optimality conditions we present are not as sharp as possible, but illustrate the general flavor of the results. The optimality conditions will help us to develop algorithms for non-linear inequality-constrained minimization problems in Section 19.4. We will discuss sensitivity analysis in Section 19.5.

The key issues discussed in this chapter are:

- the notion of a **regular point of constraints** as one characterization of suitable formulations of non-linear equality and inequality constraint functions,
- linearization of non-linear constraint functions and consideration of the **null space of the coefficient matrix** of the linearized constraints and the associated **tangent plane**,

723

- optimality conditions and the definition and interpretation of the **Lagrange multipliers**,
- the **Slater condition** as an alternative characterization of suitable formulation of constraint functions for convex problems,
- algorithms that seek points that satisfy the optimality conditions,
- use of a **merit function** in the trade-off between satisfaction of constraints and improvement of the objective, and
- **sensitivity analysis**.

19.1 Geometry and analysis of constraints

In the case of linear equality and inequality constraints, the convexity of the feasible set allowed us to consider step directions such that successive iterates were always feasible. That is, we could move from a feasible point along a line segment that lies entirely within the feasible set, choosing the direction of the segment to decrease the objective. This motivated the approach of first finding a feasible point and then seeking step directions that kept the iterates feasible and also reduced the value of the objective.

Similarly to our approach to non-linear equality constraints, for non-linear inequality constraints we will again linearize the constraint functions g and h about a current iterate and seek step directions. We must explore conditions under which this linearization yields a useful approximation to the original feasible set. The notion of a regular point, introduced in Section 14.1.1 for non-linear equality-constrained problems and suitably generalized here for non-linear inequality constraints, provides one such **constraint qualification**.

19.1.1 Regular point

As in the non-linear equality-constrained case, when we use the representation $\{x \in \mathbb{R}^n | g(x) = 0, h(x) \le 0\}$ for a feasible set $\mathbb{S}$, we usually have many choices of functions $g : \mathbb{R}^n \to \mathbb{R}^m$ and $h : \mathbb{R}^n \to \mathbb{R}^r$ such that $\mathbb{S} = \{x \in \mathbb{R}^n | g(x) = 0, h(x) \le 0\}$. However, some choices of g and h may be more suitable than others. In this section we characterize suitability of g and h in terms of the following.

Definition 19.1 Let $g : \mathbb{R}^n \to \mathbb{R}^m$ and $h : \mathbb{R}^m \to \mathbb{R}^r$. Then we say that $x^\star$ is a **regular point** of the constraints $g(x) = 0$ and $h(x) \le 0$ if:

   (i) $g(x^\star) = 0$ and $h(x^\star) \le 0$,

   (ii) g and h are both partially differentiable with continuous partial derivatives, and

   (iii) the matrix $\hat{A}$ has linearly independent rows, where $\hat{A}$ is the matrix with rows consisting of:

       - the m rows of the Jacobian $J(x^\star)$ of g evaluated at $x^\star$, and

- those rows $K_\ell(x^\star)$ of the Jacobian K of h evaluated at $x^\star$ for which $\ell \in \mathbb{A}(x^\star)$.

The matrix $\hat{A}$ consists of the rows of $J(x^\star)$ together with those rows of $K(x^\star)$ that correspond to the active constraints. If there are no equality constraints then the matrix $\hat{A}$ consists of the rows of $K(x^\star)$ corresponding to active constraints. If there are no binding inequality constraints then $\hat{A} = J(x^\star)$. If there are no equality constraints and no binding inequality constraints then the matrix $\hat{A}$ has no rows and, by definition, it has linearly independent rows.

☐

Notice that for $x^\star$ to be a regular point of the constraints $g(x) = \mathbf{0}$ and $h(x) \leq \mathbf{0}$, we must have that $m + \hat{r} \leq n$, where $\hat{r}$ is the number of active inequality constraints at $x^\star$, since otherwise the $m + \hat{r}$ rows of $\hat{A}$ cannot be linearly independent. Furthermore, if $x^\star$ is a regular point, then we can find a sub-vector $\omega \in \mathbb{R}^{m+\hat{r}}$ of x such that the $(m + \hat{r}) \times (m + \hat{r})$ matrix consisting of the corresponding $m + \hat{r}$ columns of $\hat{A}$ is non-singular.

At a regular point of inequality constraints, linearization of the equality constraints and of the binding inequality constraints yields a useful approximation to the feasible set or its boundary, at least locally in the vicinity of the regular point. For this reason, and as in the case of non-linear equality constraints, the definition of a regular point provides one characterization of useful equality and inequality constraint functions.

19.1.2 Example

Recall the **dodecahedron** from Section 2.3.2.3 and illustrated in Figure 2.14. Figure 19.1 repeats Figure 2.14.

The dodecahedron can be described as the set of points satisfying the inequality constraints $h(x) \leq \mathbf{0}$, with $h : \mathbb{R}^3 \to \mathbb{R}^{12}$ affine:

$$\forall x \in \mathbb{R}^3, h(x) = Cx - d,$$

where:

- $C \in \mathbb{R}^{12 \times 3}$ with each row of C not equal to the zero vector (see Exercise 19.1), and
- $d \in \mathbb{R}^{12}$.

The Jacobian of h is $K = C$ and the ℓ-th row of K is the ℓ-th row of C, which we will denote by C_ℓ.

We will consider whether or not each point $x^\star \in \mathbb{R}^3$ is a regular point of the constraints $h(x) \leq \mathbf{0}$. First, if $h(x^\star) \not\leq \mathbf{0}$ so that $x^\star$ is not in the dodecahedron then $x^\star$ is not a regular point by definition.

If $h(x^\star) \leq \mathbf{0}$ then we consider the matrix $\hat{A}$ consisting of the rows C_ℓ of C for

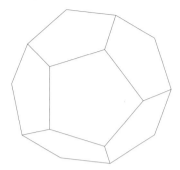

Fig. 19.1. The dodecahedron in $\mathbb{R}^3$ repeated from Figure 2.14.

which $\ell \in \mathbb{A}(x^\star)$. We consider whether or not $\hat{A}$ has linearly independent rows. There are several cases.

(i) $x^\star$ is in the **interior** of the dodecahedron. That is, $h(x^\star) = Cx^\star - d < \mathbf{0}$, $\mathbb{A}(x^\star) = \emptyset$, $\hat{A}$ has no rows, and so $x^\star$ is a regular point by definition.

(ii) $x^\star$ is on a **face** of the dodecahedron but not on an **edge** or **vertex**. That is, exactly one constraint ℓ is binding, $\mathbb{A}(x^\star) = \{\ell\}$, $\hat{A} = C_\ell$, where C_ℓ is the ℓ-th row of C. The single row of $\hat{A}$ is linearly independent, since it is a single row that is not equal to the zero vector.

(iii) $x^\star$ is on an **edge** but not a **vertex** of the dodecahedron. That is, exactly two constraints ℓ, ℓ' are binding, $\mathbb{A}(x^\star) = \{\ell, \ell'\}$, and

$$\hat{A} = \begin{bmatrix} C_\ell \\ C_{\ell'} \end{bmatrix}.$$

Since the corresponding two faces of the dodecahedron are not parallel then the two corresponding rows of C, namely C_ℓ and $C_{\ell'}$, are linearly independent.

(iv) $x^\star$ is on a **vertex** of the dodecahedron. That is, exactly three constraints ℓ, ℓ', and ℓ'' are binding, $\mathbb{A}(x^\star) = \{\ell, \ell', \ell''\}$, and

$$\hat{A} = \begin{bmatrix} C_\ell \\ C_{\ell'} \\ C_{\ell''} \end{bmatrix}.$$

The corresponding three faces are oblique to each other and therefore the three corresponding rows of C are linearly independent.

In summary, every feasible point is a regular point of the constraints $h(x) \le \mathbf{0}$.

If we assume that the dodecahedron in Figure 19.1 is a **regular solid** (in the sense of solid geometry) then each face is a regular pentagon and the *opposite* faces of the dodecahedron are parallel. The corresponding rows of C (and corresponding rows of K) are the same to within a multiplicative constant. That is, the rows of K corresponding to opposite faces of the dodecahedron are not linearly independent. However, the constraints corresponding to two opposite faces are never both binding at the same point. Consequently, they will never both be included in the matrix $\hat{A}$ in Definition 19.1.

Now consider adding an additional inequality constraint corresponding to a plane that just grazes the dodecahedron at one of its vertices, say $x^\star$. This constraint is redundant. To represent this additional inequality constraint, we augment an additional row to C to form $\tilde{C} \in \mathbb{R}^{13 \times 3}$ and augment an additional entry to d to form $\tilde{d} \in \mathbb{R}^{13}$. We define the function $\tilde{h} : \mathbb{R}^3 \rightarrow \mathbb{R}^{13}$ to consist of the entries of h together with a thirteenth entry $\tilde{h}_{13} : \mathbb{R}^3 \rightarrow \mathbb{R}$ defined by:

$$\forall x \in \mathbb{R}^3, \tilde{h}_{13}(x) = \tilde{C}_{13} x - \tilde{d}_{13}.$$

We now have that $\{x \in \mathbb{R}^3 | h(x) \leq \mathbf{0}\} = \{x \in \mathbb{R}^3 | \tilde{h}(x) \leq \mathbf{0}\}$. However, the vertex $x^\star$ is not a regular point of the constraints $\tilde{h}(x) \leq \mathbf{0}$ because there are *four* constraints active at $x^\star$ and the four corresponding rows of $\tilde{C}$ cannot be linearly independent in $\mathbb{R}^3$. Since $\{x \in \mathbb{R}^3 | h(x) \leq \mathbf{0}\}$ and $\{x \in \mathbb{R}^3 | \tilde{h}(x) \leq \mathbf{0}\}$ represent the *same* set, it is important to realize that whether or not a point $x^\star$ is a regular point of the constraints depends on the choice of representation of the constraints.

Although this example involves affine inequality constraint functions for simplicity, similar observations apply to non-linear constraint functions, which are the central topic of this chapter. As with non-linear equality constraints, we should seek g and h such that feasible points are all regular points.

19.2 Optimality conditions

In Section 19.2.1 we present first-order necessary conditions and in Section 19.2.2 we present second-order sufficient conditions.

19.2.1 First-order necessary conditions

19.2.1.1 Analysis

We have:

Theorem 19.1 *Suppose that the functions $f : \mathbb{R}^n \rightarrow \mathbb{R}$, $g : \mathbb{R}^n \rightarrow \mathbb{R}^m$, and $h : \mathbb{R}^n \rightarrow \mathbb{R}^r$ are partially differentiable with continuous partial derivatives. Let $J : \mathbb{R}^n \rightarrow \mathbb{R}^{m \times n}$ and $K : \mathbb{R}^n \rightarrow \mathbb{R}^{r \times n}$ be the Jacobians of g and h, respectively. Consider Problem (19.1):*

$$\min_{x \in \mathbb{R}^n} \{f(x) | g(x) = \mathbf{0}, h(x) \leq \mathbf{0}\}.$$

Suppose that $x^\star \in \mathbb{R}^n$ is a regular point of the constraints $g(x) = \mathbf{0}$ and $h(x) \le \mathbf{0}$. If $x^\star$ is a local minimizer of Problem (19.1) then:

$$\exists \lambda^\star \in \mathbb{R}^m, \exists \mu^\star \in \mathbb{R}^r \text{ such that: } \nabla f(x^\star) + J(x^\star)^\dagger \lambda^\star + K(x^\star)^\dagger \mu^\star = \mathbf{0};$$
$$M^\star h(x^\star) = \mathbf{0};$$
$$g(x^\star) = \mathbf{0};$$
$$h(x^\star) \le \mathbf{0}; \text{ and}$$
$$\mu^\star \ge \mathbf{0}, \quad (19.2)$$

where $M^\star = \mathrm{diag}\{\mu_\ell^\star\} \in \mathbb{R}^{r \times r}$. The vectors $\lambda^\star$ and $\mu^\star$ satisfying the conditions (19.2) are called the vectors of Lagrange multipliers for the constraints $g(x) = \mathbf{0}$ and $h(x) \le \mathbf{0}$, respectively. The conditions that $M^\star h(x^\star) = \mathbf{0}$ are called the **complementary slackness conditions**. *They say that, for each ℓ, either the ℓ-th inequality constraint is binding or the ℓ-th Lagrange multiplier is equal to zero (or both).*

Proof ([70, section 10.8].) □

As previously, we refer to the equality and inequality constraints in (19.2) as the **first-order necessary conditions** (or **FONC**). As in the case of non-linear equality constraints, the condition that $x^\star$ be a regular point of the constraints is again called a **constraint qualification**. In Section 19.3.1, we will see an alternative constraint qualification for the case of convex problems.

19.2.1.2 Lagrangian

Recall Definition 3.2 of the **Lagrangian**. Analogously to the discussion in Section 17.1.1.4, by defining the Lagrangian $\mathcal{L} : \mathbb{R}^n \times \mathbb{R}^m \times \mathbb{R}^r \to \mathbb{R}$ by:

$$\forall x \in \mathbb{R}^n, \forall \lambda \in \mathbb{R}^m, \forall \mu \in \mathbb{R}^r, \mathcal{L}(x, \lambda, \mu) = f(x) + \lambda^\dagger g(x) + \mu^\dagger h(x),$$

we can again reproduce some of the first-order necessary conditions as:

$$\nabla_x \mathcal{L}(x^\star, \lambda^\star, \mu^\star) = \mathbf{0},$$
$$\nabla_\lambda \mathcal{L}(x^\star, \lambda^\star, \mu^\star) = \mathbf{0},$$
$$\nabla_\mu \mathcal{L}(x^\star, \lambda^\star, \mu^\star) \le \mathbf{0}.$$

19.2.1.3 Example

Recall the example non-linear program, Problem (2.19), from Section 2.3.2.3:

$$\min_{x \in \mathbb{R}^3} \{ f(x) | g(x) = \mathbf{0}, h(x) \le 0 \},$$

where $f : \mathbb{R}^3 \to \mathbb{R}$, $g : \mathbb{R}^3 \to \mathbb{R}^2$, and $h : \mathbb{R}^3 \to \mathbb{R}$ are defined by:

$$\forall x \in \mathbb{R}^3, \, f(x) \;=\; (x_1)^2 + 2(x_2)^2,$$
$$\forall x \in \mathbb{R}^3, \, g(x) \;=\; \begin{bmatrix} 2 - x_2 - \sin(x_3) \\ -x_1 + \sin(x_3) \end{bmatrix},$$
$$\forall x \in \mathbb{R}^3, \, h(x) \;=\; [\sin(x_3) - 0.5].$$

We claim that $x^\star = \begin{bmatrix} 0.5 \\ 1.5 \\ \pi/6 \end{bmatrix}$, $\lambda^\star = \begin{bmatrix} 6 \\ 1 \end{bmatrix}$, and $\mu^\star = [5]$ satisfy the first-order necessary conditions in Theorem 19.1. First, $x^\star$ is feasible. Now let $J : \mathbb{R}^3 \to \mathbb{R}^{2\times3}$ and $K : \mathbb{R}^3 \to \mathbb{R}^{1\times3}$ be the Jacobians of g and h, respectively. Then:

$$\forall x \in \mathbb{R}^3, \, \nabla f(x) \;=\; \begin{bmatrix} 2x_1 \\ 4x_2 \\ 0 \end{bmatrix},$$

$$\forall x \in \mathbb{R}^3, \, J(x) \;=\; \begin{bmatrix} 0 & -1 & -\cos(x_3) \\ -1 & 0 & \cos(x_3) \end{bmatrix},$$

$$J(x^\star) \;=\; \begin{bmatrix} 0 & -1 & -\cos(\pi/6) \\ -1 & 0 & \cos(\pi/6) \end{bmatrix},$$

$$\forall x \in \mathbb{R}^3, \, K(x) \;=\; \begin{bmatrix} 0 & 0 & \cos(x_3) \end{bmatrix},$$
$$K(x^\star) \;=\; \begin{bmatrix} 0 & 0 & \cos(\pi/6) \end{bmatrix}.$$

Note that $\hat{A} = \begin{bmatrix} J(x^\star) \\ K(x^\star) \end{bmatrix}$ has linearly independent rows so that $x^\star$ is a regular point of the constraints. Moreover,

$$\nabla f(x^\star) + J(x^\star)^\dagger \lambda^\star + K(x^\star)^\dagger \mu^\star$$

$$= \begin{bmatrix} 1 \\ 6 \\ 0 \end{bmatrix} + \begin{bmatrix} 0 & -1 \\ -1 & 0 \\ -\cos(\pi/6) & \cos(\pi/6) \end{bmatrix} \begin{bmatrix} 6 \\ 1 \end{bmatrix} + \begin{bmatrix} 0 \\ 0 \\ \cos(\pi/6) \end{bmatrix} 5,$$

$$= \mathbf{0};$$

$$\mu^\star h(x^\star) = [5] \times [0],$$
$$= [0];$$
$$g(x^\star) = \mathbf{0};$$
$$h(x^\star) = [0],$$
$$\le [0]; \text{ and}$$
$$\mu^\star = [5],$$
$$\ge [0].$$

That is, $x^\star = \begin{bmatrix} 0.5 \\ 1.5 \\ \pi/6 \end{bmatrix}$, $\lambda^\star = \begin{bmatrix} 6 \\ 1 \end{bmatrix}$, and $\mu^\star = [5]$ satisfy the first-order necessary conditions in Theorem 19.1. (See also Exercise 3.35.)

19.2.2 Second-order sufficient conditions

19.2.2.1 Analysis

Theorem 19.2 *Suppose that the functions* $f : \mathbb{R}^n \to \mathbb{R}$, $g : \mathbb{R}^n \to \mathbb{R}^m$, *and* $h : \mathbb{R}^n \to \mathbb{R}^r$ *are twice partially differentiable with continuous second partial derivatives. Let* $J : \mathbb{R}^n \to \mathbb{R}^{m \times n}$ *and* $K : \mathbb{R}^n \to \mathbb{R}^{r \times n}$ *be the Jacobians of* g *and* h, *respectively. Consider Problem (19.1):*

$$\min_{x \in \mathbb{R}^n} \{f(x) | g(x) = \mathbf{0}, h(x) \le \mathbf{0}\},$$

and points $x^\star \in \mathbb{R}^n$, $\lambda^\star \in \mathbb{R}^m$, *and* $\mu^\star \in \mathbb{R}^r$. *Let* $M^\star = \mathrm{diag}\{\mu_\ell^\star\}$. *Suppose that:*

$$\begin{align}
\nabla f(x^\star) + J(x^\star)^\dagger \lambda^\star + K(x^\star)^\dagger \mu^\star &= \mathbf{0}, \\
M^\star h(x^\star) &= \mathbf{0}, \\
g(x^\star) &= \mathbf{0}, \\
h(x^\star) &\le \mathbf{0}, \\
\mu^\star &\ge \mathbf{0}, \ and
\end{align}$$

$$\nabla^2 f(x^\star) + \sum_{\ell=1}^m \lambda_\ell^\star \nabla^2 g_\ell(x^\star) + \sum_{\ell=1}^r \mu_\ell^\star \nabla^2 h_\ell(x^\star) \ is \ positive \ definite \ on \ the \ null \ space:$$

$$\begin{align}
\mathcal{N}_+ &= \{\Delta x \in \mathbb{R}^n | J(x^\star)\Delta x = \mathbf{0}, K_\ell(x^\star)\Delta x = 0, \forall \ell \in \mathbb{A}_+(x^\star, \mu^\star)\}, \\
where \ \mathbb{A}_+(x^\star, \mu^\star) &= \{\ell \in \{1, \ldots, r\} | h_\ell(x^\star) = 0, \mu_\ell^\star > 0\}.
\end{align}$$

Then $x^\star$ *is a strict local minimizer of Problem (19.1).*

Proof See [11, proposition 3.3.2][70, section 10.8]. $\square$

The conditions in the theorem are called the **second-order sufficient conditions** (or **SOSC**). The function $\nabla^2_{xx}\mathcal{L} : \mathbb{R}^n \times \mathbb{R}^m \times \mathbb{R}^r \to \mathbb{R}$ defined by:

$$\forall x \in \mathbb{R}^n, \forall \lambda \in \mathbb{R}^m, \forall \mu \in \mathbb{R}^r,$$

$$\nabla^2_{xx}\mathcal{L}(x, \lambda, \mu) = \nabla^2 f(x) + \sum_{\ell=1}^m \lambda_\ell \nabla^2 g_\ell(x) + \sum_{\ell=1}^r \mu_\ell \nabla^2 h_\ell(x),$$

is again called the **Hessian of the Lagrangian**. In addition to the first-order necessary conditions, the second-order sufficient conditions require that:

- f, g, and h are twice partially differentiable with continuous second partial derivatives, and

• the Hessian of the Lagrangian evaluated at the minimizer and corresponding Lagrange multipliers, $\nabla^2_{xx}\mathcal{L}(x^\star, \lambda^\star, \mu^\star)$, is positive definite on the null space $\mathcal{N}_+$ defined in the theorem.

The sets $\mathcal{N}_+$ and $\mathbb{A}_+$ have analogous roles to their roles in the non-negatively constrained case presented in Section 16.1.2 and the linear inequality-constrained case presented in Section 17.1.2. Again, constraints ℓ for which $\mu^\star_\ell = 0$ and $h_\ell(x^\star) = 0$ are called **degenerate constraints**.

19.2.2.2 Example

Continuing with Problem (2.19) from Sections 2.3.2.3 and 19.2.1.3, we note that f, g, and h are twice partially differentiable with continuous second partial derivatives. By the discussion in Section 19.2.1.3, the first-order necessary conditions are satisfied by $x^\star = \begin{bmatrix} 0.5 \\ 1.5 \\ \pi/6 \end{bmatrix}$, $\lambda^\star = \begin{bmatrix} 6 \\ 1 \end{bmatrix}$, and $\mu^\star = [5]$. We have:

$$\mathbb{A}(x^\star) = \mathbb{A}_+(x^\star, \mu^\star) = \{1\},$$

since the one inequality constraint is binding and the corresponding Lagrange multiplier is non-zero. That is, the constraint is not degenerate. Finally,

$$
\begin{aligned}
\mathcal{N}_+ &= \{\Delta x \in \mathbb{R}^n | J(x^\star)\Delta x = 0, K_\ell(x^\star)\Delta x = 0, \forall \ell \in \mathbb{A}_+(x^\star, \mu^\star)\}, \\
&= \{\Delta x \in \mathbb{R}^n | J(x^\star)\Delta x = 0, K_1(x^\star)\Delta x = 0\}, \\
&= \{0\},
\end{aligned}
$$

so that the Hessian of the Lagrangian $\nabla^2_{xx}\mathcal{L}(x^\star, \lambda^\star, \mu^\star)$ is positive definite on the null space $\mathcal{N}_+$. That is $x^\star$, $\lambda^\star$, and $\mu^\star$ satisfy the second-order sufficient conditions.

19.3 Convex problems

In this section, we consider the case where $g : \mathbb{R}^n \to \mathbb{R}^m$ is affine and $h : \mathbb{R}^n \to \mathbb{R}^r$ is convex. That is, we consider the following problem:

$$\min_{x \in \mathbb{R}^n}\{f(x) | Ax = b, h(x) \le 0\}, \tag{19.3}$$

where $A \in \mathbb{R}^{m \times n}$ and $b \in \mathbb{R}^m$. If $f : \mathbb{R}^n \to \mathbb{R}$ is convex on the feasible set then Problem (19.3) is convex.

We discuss first-order necessary conditions for Problem (19.3) in Section 19.3.1, then first-order sufficient conditions in Section 19.3.2, and finally discuss duality in Section 19.3.3.

19.3.1 First-order necessary conditions

19.3.1.1 Slater condition

In the case of affine g and convex h, we can obtain first-order necessary conditions with an alternative constraint qualification to the assumption of regular constraints that was presented in Section 19.1.1. In particular, we will assume that:

$$\{x \in \mathbb{R}^n | Ax = b, h(x) < 0\} \neq \emptyset. \tag{19.4}$$

This alternative constraint qualification is called the **Slater condition** [6, chapter 5][11, section 5.3][15, section 5.2.3][84, page 485]. The Slater condition was first introduced in Section 16.4.2.3 in the context of the interior point algorithm for linear inequality-constrained problems. We will see in Section 19.4.1.2 that we also need to make a similar assumption for applying the interior point algorithm to non-linearly constrained problems. As mentioned in Section 16.4.2.3, many constraint systems arising from physical problems satisfy the Slater condition.

19.3.1.2 Analysis

We have the following.

Theorem 19.3 *Suppose that $f : \mathbb{R}^n \rightarrow \mathbb{R}$ and $h : \mathbb{R}^n \rightarrow \mathbb{R}^r$ are partially differentiable with continuous partial derivatives and with h convex, $A \in \mathbb{R}^{m \times n}$, and $b \in \mathbb{R}^m$. Let $K : \mathbb{R}^n \rightarrow \mathbb{R}^{r \times n}$ be the Jacobian of h. Consider Problem (19.3) and suppose that the Slater condition (19.4) holds. If $x^\star \in \mathbb{R}^n$ is a local minimizer of Problem (19.3) then:*

$$\exists \lambda^\star \in \mathbb{R}^m, \exists \mu^\star \in \mathbb{R}^r \text{ such that: } \nabla f(x^\star) + A^\dagger \lambda^\star + K(x^\star)^\dagger \mu^\star = 0;$$
$$M^\star h(x^\star) = 0;$$
$$Ax^\star = b;$$
$$h(x^\star) \leq 0; \text{ and}$$
$$\mu^\star \geq 0,$$

where $M^\star = \text{diag}\{\mu_\ell^\star\} \in \mathbb{R}^{r \times r}$.

Proof ([11, section 5.3].) □

19.3.2 First-order sufficient conditions

19.3.2.1 Analysis

In the convex case, the first-order necessary conditions are also sufficient for optimality.

Theorem 19.4 *Suppose that $f : \mathbb{R}^n \rightarrow \mathbb{R}$ and $h : \mathbb{R}^n \rightarrow \mathbb{R}^r$ are partially differentiable with continuous partial derivatives, $A \in \mathbb{R}^{m \times n}$, and $b \in \mathbb{R}^m$. Let $K : \mathbb{R}^n \rightarrow \mathbb{R}^{r \times n}$*

be the Jacobian of h. Consider Problem (19.3) and points $x^\star \in \mathbb{R}^n$, $\lambda^\star \in \mathbb{R}^m$, and $\mu^\star \in \mathbb{R}^r$. Let $M^\star = \text{diag}\{\mu_\ell^\star\}$. Suppose that:

    (i) h is convex,
    (ii) f is convex on $\{x \in \mathbb{R}^n | Ax = b, h(x) \leq 0\}$,
    (iii) $\nabla f(x^\star) + A^\dagger \lambda^\star + K(x^\star)^\dagger \mu^\star = 0$,
    (iv) $M^\star h(x^\star) = 0$,
    (v) $Ax^\star = b$ and $h(x^\star) \leq 0$, and
    (vi) $\mu^\star \geq 0$.

Then $x^\star$ is a global minimizer of Problem (19.3).

Proof The proof is very similar to the proofs of Theorem 16.3 in Chapter 16 and of Theorem 17.3 in Chapter 17. See Appendix B for details. $\square$

In addition to the first-order necessary conditions in Theorem 19.3, the first-order sufficient conditions require that f is convex on the convex feasible set.

19.3.2.2 Example

Let $f : \mathbb{R}^2 \to \mathbb{R}$ and $h : \mathbb{R}^2 \to \mathbb{R}$ be defined by:

$$\forall x \in \mathbb{R}^2, f(x) = x_1 + x_2,$$
$$\forall x \in \mathbb{R}^2, h(x) = (x_1)^2 + (x_2)^2 - 2.$$

Consider the problem:

$$\min_{x \in \mathbb{R}^2} \{f(x) | h(x) \leq 0\}.$$

Figure 19.2 illustrates the contour sets of the objective and the feasible set. We observe that both f and h are partially differentiable with continuous partial derivatives and convex. We claim that $x^\star = -\mathbf{1}$ is the global minimizer with $\mu^\star = [0.5]$ the corresponding Lagrange multiplier. To see this, note that:

$$\forall x \in \mathbb{R}^2, \nabla f(x) = 1,$$
$$\forall x \in \mathbb{R}^2, K(x) = \begin{bmatrix} 2x_1 & 2x_2 \end{bmatrix},$$
$$K(x^\star) = \begin{bmatrix} -2 & -2 \end{bmatrix},$$
$$\nabla f(x^\star) + K(x^\star)^\dagger \mu^\star = 1 + \begin{bmatrix} -2 & -2 \end{bmatrix} \times [0.5],$$
$$= 0;$$
$$\mu^\star h(x^\star) = 0;$$
$$h(x^\star) = [0],$$
$$\leq [0]; \text{ and}$$
$$\mu^\star = [0.5],$$
$$\geq [0],$$

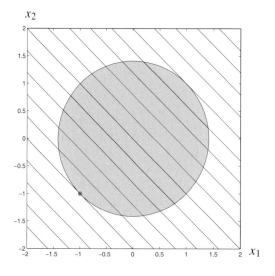

x_2

Fig. 19.2. Contour sets of objective function defined in Section 19.3.2.2 with feasible set shaded. The heights of the contours decrease to the left and down. The minimizer, $x^\star = -1$, is indicated with the ●.

so that $x^\star = -1$ and $\mu^\star = [0.5]$ satisfy the first-order sufficient conditions. The point $x^\star = -1$ is illustrated as a ● in Figure 19.2.

19.3.3 Duality

As we discussed in Section 3.4 and as in the discussion of linear equality constraints in Section 13.2.2 and linear inequality constraints in Section 17.2.2, we can define a dual problem where the role of variables and constraints is partly or fully swapped [84, chapter 6].

19.3.3.1 Dual function

Analysis If f and h are convex and g is affine then $\mathcal{L}(\bullet, \lambda, \mu)$ is convex for $\mu \geq 0$ and so $x^\star$ is a global minimizer of $\mathcal{L}(\bullet, \lambda^\star, \mu^\star)$. Recall Definition 3.3 of the **dual function** and **effective domain**. For Problem (19.3), the dual function $\mathcal{D} : \mathbb{R}^m \times \mathbb{R}^r \to \mathbb{R} \cup \{-\infty\}$ is defined by:

$$\forall \lambda \in \mathbb{R}^m, \mu \in \mathbb{R}^r, \mathcal{D}(\lambda, \mu) = \inf_{x \in \mathbb{R}^n} \mathcal{L}(x, \lambda, \mu). \tag{19.5}$$

The effective domain of $\mathcal{D}$ is:

$$\mathbb{E} = \left\{ \begin{bmatrix} \lambda \\ \mu \end{bmatrix} \in \mathbb{R}^{m+r} \,\middle|\, \mathcal{D}(\lambda, \mu) > -\infty \right\}.$$

Recall that by Theorem 3.12, $\mathbb{E}$ is convex and $\mathcal{D}$ is convex on $\mathbb{E}$.

Example Continuing with the example problem from Section 19.3.2.2, the Lagrangian $\mathcal{L} : \mathbb{R}^2 \times \mathbb{R} \to \mathbb{R}$ for this problem is defined by:

$$
\begin{aligned}
\forall x \in \mathbb{R}^2, \forall \mu \in \mathbb{R}, \mathcal{L}(x, \mu) &= f(x) + \mu^\dagger h(x), \\
&= x_1 + x_2 + \mu((x_1)^2 + (x_2)^2 - 2).
\end{aligned}
$$

For $\mu > 0$, the Lagrangian $\mathcal{L}(\bullet, \mu)$ is strictly convex and therefore, by Corollary 10.6, the first-order necessary conditions $\nabla_x \mathcal{L}(x, \mu) = 0$ are sufficient for minimizing $\mathcal{L}(\bullet, \mu)$ and, moreover, a minimizer exists, so that the inf in the definition of $\mathcal{D}$ can be replaced by min. Furthermore, there is a unique minimizer $x^{(\mu)}$ corresponding to each value of $\mu > 0$. In particular, we have:

$$
\begin{aligned}
\forall x \in \mathbb{R}^2, \forall \mu \in \mathbb{R}, \nabla_x \mathcal{L}(x, \mu) &= \nabla f(x) + K(x)^\dagger \mu, \\
&= \begin{bmatrix} 1 + 2\mu x_1 \\ 1 + 2\mu x_2 \end{bmatrix}, \\
\forall \mu \in \mathbb{R}_{++}, x^{(\mu)} &= \begin{bmatrix} -1/(2\mu) \\ -1/(2\mu) \end{bmatrix}, \\
\forall \mu \in \mathbb{R}_{++}, \mathcal{D}(\mu) &= -\frac{1}{2\mu} - 2\mu.
\end{aligned}
$$

On the other hand, if $\mu \leq 0$ then the objective in the dual function is unbounded below. Consequently, the effective domain is $\mathbb{E} = \mathbb{R}_{++}$.

19.3.3.2 Dual problem

Analysis If the Lagrangian $\mathcal{L}(\bullet, \lambda^\star, \mu^\star)$ is convex on $\mathbb{R}^n$ then the minimum of Problem (19.3) is equal to $\mathcal{D}(\lambda^\star, \mu^\star)$, where $\lambda^\star$ and $\mu^\star$ are the Lagrange multipliers that satisfy the necessary conditions for Problem (19.3). As in the equality-constrained case and the linear inequality-constrained case, under certain conditions, the Lagrange multipliers can be found as the maximizer of the **dual problem**:

$$
\max_{\begin{bmatrix} \lambda \\ \mu \end{bmatrix} \in \mathbb{E}} \{ \mathcal{D}(\lambda, \mu) | \mu \geq \mathbf{0} \}, \tag{19.6}
$$

where $\mathcal{D} : \mathbb{E} \to \mathbb{R}$ is the dual function defined in (19.5). Again, Problem (19.3) is called the **primal problem** in this context to distinguish it from Problem (19.6).

Similarly to Theorem 17.4, these observations are embodied in the following. (Also see [6, theorems 6.2.4 and 6.3.3][11, proposition 5.2.1] [15, section 5.2.3] [84, corollaries 6.1 and 14.2] for generalizations.)

Theorem 19.5 *Suppose that* $f : \mathbb{R}^n \to \mathbb{R}$ *and* $h : \mathbb{R}^n \to \mathbb{R}$ *are convex and partially differentiable with continuous partial derivatives,* $A \in \mathbb{R}^{m \times n}$, *and* $b \in \mathbb{R}^m$. *Consider* *the primal problem, Problem (19.3):*

$$\min_{x \in \mathbb{R}^n} \{ f(x) | Ax = b, h(x) \le 0 \},$$

and suppose that the Slater condition (19.4) holds. Also, consider the dual problem, Problem (19.6). We have that:

(i) *If the primal problem possesses a minimum then the dual problem possesses a maximum and the optima are equal. That is:*

$$\min_{x \in \mathbb{R}^n} \{ f(x) | Ax = b, h(x) \le 0 \} = \max_{\begin{bmatrix} \lambda \\ \mu \end{bmatrix} \in \mathbb{E}} \{ \mathcal{D}(\lambda, \mu) | \mu \ge 0 \}.$$

(ii) *If:*

- $\begin{bmatrix} \lambda \\ \mu \end{bmatrix} \in \mathbb{E}$,
- $\min_{x \in \mathbb{R}^n} \mathcal{L}(x, \lambda, \mu)$ *exists, and*
- *f and h are twice partially differentiable with continuous second partial derivatives, $\nabla^2 f$ is positive definite, and $\nabla^2 h_\ell, \ell = 1, \ldots, r$, are all positive definite,*

then $\mathcal{D}$ is partially differentiable at $\begin{bmatrix} \lambda \\ \mu \end{bmatrix}$ with continuous partial derivatives and:

$$\nabla \mathcal{D}(\lambda, \mu) = \begin{bmatrix} A x^{(\lambda, \mu)} - b \\ h(x^{(\lambda, \mu)}) \end{bmatrix}. \tag{19.7}$$

Proof The proof is very similar to the proof of Theorem 17.4 in Chapter 17. See Appendix B for details. □

As in the equality-constrained and linear inequality-constrained cases, it is possible for $\mathcal{D}$ to not be partially differentiable at a point $\begin{bmatrix} \lambda \\ \mu \end{bmatrix} \in \mathbb{E}$ if:

- $\mathcal{L}(\bullet, \lambda, \mu)$ is bounded below (so that $\inf_{x \in \mathbb{R}^n} \mathcal{L}(x, \lambda, \mu) \in \mathbb{R}$) yet the minimum $\min_{x \in \mathbb{R}^n} \mathcal{L}(x, \lambda, \mu)$ does not exist, or
- there are multiple minimizers of $\min_{x \in \mathbb{R}^n} \mathcal{L}(x, \lambda, \mu)$.

Corollary 19.6 *Let $f : \mathbb{R}^n \to \mathbb{R}$ and $h : \mathbb{R}^n \to \mathbb{R}^r$ be twice partially differentiable with continuous second partial derivatives, $\nabla^2 f$ be positive definite, and $\nabla^2 h_\ell, \ell = 1, \ldots, r$, all be positive definite; $A \in \mathbb{R}^{m \times n}$; and $b \in \mathbb{R}^m$. Consider Problem (19.3):*

$$\min_{x \in \mathbb{R}^n} \{ f(x) | Ax = b, h(x) \le 0 \},$$

the Lagrangian of this problem, and the effective domain $\mathbb{E}$ of the dual function. If:

- *the effective domain $\mathbb{E}$ contains $\mathbb{R}^m \times \mathbb{R}^r_+$, and*
- *for each $\lambda \in \mathbb{R}^m$ and $\mu \in \mathbb{R}^r_+$, $\min_{x \in \mathbb{R}^n} \mathcal{L}(x, \lambda, \mu)$ exists,*

then necessary and sufficient conditions for $\begin{bmatrix} \lambda^\star \\ \mu^\star \end{bmatrix} \in \mathbb{R}^{m+r}$ *to be the maximizer of the dual problem:*

$$\max_{\begin{bmatrix} \lambda \\ \mu \end{bmatrix} \in \mathbb{E}} \{\mathcal{D}(\lambda, \mu) | \mu \geq \mathbf{0}\},$$

are:

$$
\begin{aligned}
M^\star h(x^{(\lambda^\star, \mu^\star)}) &= \mathbf{0}; \\
Ax^{(\lambda^\star, \mu^\star)} &= b; \\
h(x^{(\lambda^\star, \mu^\star)}) &\leq \mathbf{0}; \text{ and} \\
\mu^\star &\geq \mathbf{0},
\end{aligned}
$$

where $\{x^{(\lambda^\star, \mu^\star)}\} = \text{argmin}_{x \in \mathbb{R}^n} \mathcal{L}(x, \lambda^\star, \mu^\star)$ *and* $M^\star = \text{diag}\{\mu_\ell^\star\}$. *Moreover, if* $\lambda^\star$ *and* $\mu^\star$ *maximize the dual problem then* $x^{(\lambda^\star, \mu^\star)}$, $\lambda^\star$, *and* $\mu^\star$ *satisfy the first-order necessary conditions for Problem (19.3).*

Proof The proof is very similar to the proof of Corollary 17.5 in Chapter 17. See Appendix B for details. $\square$

Theorem 19.5 shows that, as in the linear inequality-constrained case, an alternative approach to finding the minimum of Problem (19.3) involves finding the *maximum* of the dual function over $\lambda \in \mathbb{R}^m$ and $\mu \in \mathbb{R}_+^r$. Theorem 3.12 shows that the dual function has at most one local maximum. To seek the maximum of $\mathcal{D}(\lambda, \mu)$ over $\lambda \in \mathbb{R}^m$, $\mu \in \mathbb{R}_+^r$, we can, for example, utilize the value of the gradient of $\mathcal{D}$ from (19.7) as part of an active set or interior point algorithm. As in the equality-constrained and linear inequality-constrained cases, under some circumstances, it is also possible to calculate the Hessian of $\mathcal{D}$ [70, section 12.3].

Example Continuing with the dual of the example problem from Sections 19.3.2.2 and 19.3.3.1, we recall that the effective domain is $\mathbb{E} = \mathbb{R}_{++}$ and the dual function $\mathcal{D} : \mathbb{R}_{++} \to \mathbb{R}$ is:

$$\forall \mu \in \mathbb{R}_{++}, \mathcal{D}(\mu) = -\frac{1}{2\mu} - 2\mu.$$

Moreover, for each $\mu \in \mathbb{R}_{++}$ the dual function is twice partially differentiable. In particular,

$$
\begin{aligned}
\forall \mu \in \mathbb{R}_{++}, \nabla \mathcal{D}(\mu) &= \frac{1}{2(\mu)^2} - 2, \\
\forall \mu \in \mathbb{R}_{++}, \nabla^2 \mathcal{D}(\mu) &= -\frac{1}{4(\mu)^3}, \\
&< 0.
\end{aligned}
$$

Although we cannot apply Corollary 19.6 directly because $\mathbb{E} = \mathbb{R}_{++}$ does not contain $\mathbb{R}_+$, we note that, by inspection of $\mathcal{D}$, $\mu^\star = [0.5]$ maximizes the dual over $\mathbb{E}$. Moreover, the corresponding minimizer of the Lagrangian, $x^{(\mu^\star)}$, together with $\mu^\star$ satisfy the first-order necessary conditions for the primal problem.

Discussion As in the equality and linear inequality-constrained cases, it is essential in Theorem 19.5 for f and h to be convex on the *whole* of $\mathbb{R}^n$, not just on the feasible set. The reason is again that the inner minimization of $\mathcal{L}(\bullet, \lambda, \mu)$ is taken over the whole of $\mathbb{R}^n$. Furthermore, we generally require strict convexity of f and h to ensure that there are not multiple minimizers of the Lagrangian. The issues are similar to the discussion in Section 17.2.2.2.

Problem (19.6) is non-negatively constrained of the form of Problem (16.1) and so we can apply essentially the same algorithms as we developed for Problem (16.1). We will take this approach in Section 19.4.2.

19.3.3.3 Partial duals

As in Section 17.2.2.4, it is also possible to take the partial dual with respect to some of the equality and some of the inequality constraints.

19.4 Approaches to finding minimizers

19.4.1 Primal algorithm

19.4.1.1 Transformation

In Section 17.3.1.1, we transformed linear inequality-constrained Problem (17.1) into non-negatively constrained Problem (17.12) through the use of slack variables. This allowed the use of an algorithm for non-negatively constrained minimization to solve Problem (17.1).

We will take a similar approach here to Problem (19.1). In particular, to handle the inequality constraints involving h we consider the following problem:

$$\min_{x \in \mathbb{R}^n, w \in \mathbb{R}^r} \{ f(x) | g(x) = \mathbf{0}, h(x) + w = \mathbf{0}, w \geq \mathbf{0} \}. \tag{19.8}$$

By Theorem 3.8, Problems (19.1) and (19.8) are equivalent.

19.4.1.2 Primal–dual interior point algorithm

In this section we outline a primal–dual interior point algorithm for Problem (19.8). See [65] for further details.

Barrier objective and problem Analogously to the discussion in Section 16.4.2.2 and as in the discussion in Section 17.3.1.2, given a barrier function $f_b : \mathbb{R}^r_{++} \to \mathbb{R}$ and a barrier parameter $t \in \mathbb{R}_{++}$, we form the **barrier objective** $\phi : \mathbb{R}^n \times \mathbb{R}^r_{++} \to \mathbb{R}$ defined by:

$$\forall x \in \mathbb{R}^n, \forall w \in \mathbb{R}^r_{++}, \phi(x, w) = f(x) + t f_b(w).$$

Instead of solving Problem (19.8), we will consider solving the **barrier problem**:

$$\min_{x \in \mathbb{R}^n, w \in \mathbb{R}^r} \{\phi(x, w) | g(x) = \mathbf{0}, h(x) + w = \mathbf{0}, w > \mathbf{0}\}. \qquad (19.9)$$

That is, we minimize $\phi(x, w)$ over values of $x \in \mathbb{R}^n$ and $w \in \mathbb{R}^r$ that satisfy $g(x) = \mathbf{0}$ and $h(x) + w = \mathbf{0}$ and which are also in the interior of $w \geq \mathbf{0}$. We then decrease the barrier parameter t.

Slater condition Analogously to the discussion in Sections 16.4.2.2 and 17.3.1.2, we must assume that Problem (19.9) is feasible. That is, we assume that $\{x \in \mathbb{R}^n | g(x) = \mathbf{0}, h(x) < \mathbf{0}\} \neq \emptyset$. We again call this the **Slater condition**.

Equality-constrained problem To solve Problem (19.9), we can take a similar approach as in the primal–dual interior point algorithm for non-negativity constraints presented in Section 16.4.3.3 and for linear inequality constraints presented in Section 17.3.1.2. In particular, we can partially ignore the inequality constraints and the domain of the barrier function and seek a solution to the following non-linear equality-constrained problem:

$$\min_{x \in \mathbb{R}^n, w \in \mathbb{R}^r} \{\phi(x, w) | g(x) = \mathbf{0}, h(x) + w = \mathbf{0}\}, \qquad (19.10)$$

which has first-order necessary conditions:

$$\nabla f(x) + J(x)^{\dagger}\lambda + K(x)^{\dagger}\mu = \mathbf{0}, \qquad (19.11)$$
$$g(x) = \mathbf{0}, \qquad (19.12)$$
$$h(x) + w = \mathbf{0}, \qquad (19.13)$$
$$t\nabla f_b(w) + \mu = \mathbf{0}, \qquad (19.14)$$

where J and K are the Jacobians of g and h, respectively, and λ and μ are the dual variables on the constraints $g(x) = \mathbf{0}$ and $h(x) + w = \mathbf{0}$, respectively. We can use the techniques for minimization of non-linear equality-constrained problems from Section 14.3 to solve Problem (19.10). In particular, in Section 19.4.1.3, we will consider the Newton–Raphson method for solving the first-order necessary conditions of Problem (19.10).

Logarithmic barrier function As in the primal–dual interior point algorithm for non-negativity constraints and for linear inequality constraints, we will use the logarithmic barrier function. That is:

$$\forall w \in \mathbb{R}^r_{++}, \; f_b(w) \;=\; -\sum_{\ell=1}^{r} \ln(w_\ell),$$

$$\forall w \in \mathbb{R}^r_{++}, \; \nabla f_b(w) \;=\; -[W]^{-1}\mathbf{1},$$

where $W = \mathrm{diag}\{w_\ell\} \in \mathbb{R}^{r \times r}$. Substituting the expression for ∇f_b into (19.14) and re-arranging, we again obtain:

$$W\mu - t\mathbf{1} = \mathbf{0}. \tag{19.15}$$

19.4.1.3 Newton–Raphson method

The Newton–Raphson step direction to solve (19.15) and (19.11)–(19.13) is:

$$
\begin{bmatrix}
M^{(\nu)} & \mathbf{0} & \mathbf{0} & W^{(\nu)} \\
\mathbf{0} & \nabla^2_{xx}\mathcal{L}(x^{(\nu)}, \lambda^{(\nu)}, \mu^{(\nu)}) & J(x^{(\nu)})^\dagger & K(x^{(\nu)})^\dagger \\
\mathbf{0} & J(x^{(\nu)}) & \mathbf{0} & \mathbf{0} \\
\mathbf{I} & K(x^{(\nu)}) & \mathbf{0} & \mathbf{0}
\end{bmatrix}
\begin{bmatrix}
\Delta w^{(\nu)} \\
\Delta x^{(\nu)} \\
\Delta \lambda^{(\nu)} \\
\Delta \mu^{(\nu)}
\end{bmatrix}
$$

$$
=
\begin{bmatrix}
-W^{(\nu)}\mu^{(\nu)} + t\mathbf{1} \\
-\nabla f(x^{(\nu)}) - J(x^{(\nu)})^\dagger \lambda^{(\nu)} - K(x^{(\nu)})^\dagger \mu^{(\nu)} \\
-g(x^{(\nu)}) \\
-h(x^{(\nu)})
\end{bmatrix},
$$

where $M^{(\nu)} = \mathrm{diag}\{\mu^{(\nu)}_\ell\}$ and $W^{(\nu)} = \mathrm{diag}\{w^{(\nu)}_\ell\}$. As in the case of the primal–dual interior point algorithm for non-negativity constraints that was discussed in Section 16.4.3.3 and for linear inequality constraints that was discussed in Section 17.3.1.3, we can re-arrange these equations to make them symmetric and use block pivoting on the top left-hand block of the matrix since the top left-hand block is diagonal. This results in a system that is similar to (14.12), except that a diagonal block of the form $[M^{(\nu)}]^{-1} W^{(\nu)}$ is added to the Hessian of the Lagrangian. Issues regarding solving the first-order necessary conditions, such as factorization of the indefinite coefficient matrix, approximate solution of the conditions, sparsity, the merit function, step-size selection, and feasibility, are similar to those described in Sections 14.3.1 and 16.4.3.3.

19.4.1.4 Other issues

In this section, we discuss adjustment of the barrier parameter, the initial guess, and a stopping criterion.

Adjustment of barrier parameter To reduce the barrier parameter, we can again use the approach described in Section 16.4.4 of Chapter 16.

Initial guess We can again take an approach analogous to that in Section 16.4.5 to find an initial feasible guess for Problem (19.8). However, the effort to find such a guess in a phase 1 approach may be significant. An alternative is to begin with $w^{(0)} > \mathbf{0}$, $x^{(0)}$, $\lambda^{(0)}$, $\mu^{(0)} > \mathbf{0}$ that do not necessarily satisfy the equality constraints $g(x) = \mathbf{0}$ nor $h(x) + w = \mathbf{0}$. Feasibility is approached during the course of iterations from this **infeasible start** [15, section 11.3.1][65, section 1][126].

Stopping criterion As for the linear inequality-constrained case discussed in Section 17.3.1.4, we can develop a stopping criterion based on duality using Theorem 3.13. If f or h are non-quadratic or g is non-linear, however, we can typically only approximately evaluate the dual function.

19.4.2 Dual algorithm

Problem (19.6):

$$\max_{\left[\begin{smallmatrix}\lambda\\\mu\end{smallmatrix}\right]\in\mathbb{E}} \{\mathcal{D}(\lambda, \mu)|\mu \geq \mathbf{0}\},$$

has non-negativity constraints. If the dual function can be evaluated conveniently, then the algorithms from Section 16.3 and 16.4 for non-negativity constraints can be applied to the dual problem. For example, if the objective and inequality constraint function are quadratic and strictly convex and the equality constraints are linear then the dual function can be evaluated through the solution of a linear equation. (See Exercise 19.6 and [15, appendix B] for further results involving quadratic objectives and inequality constraint functions.) A dual algorithm can be particularly attractive if there are only a few constraints or if a **partial dual** is taken with respect to only some of the constraints [70, section 13.1].

19.5 Sensitivity

19.5.1 Analysis

In this section we will analyze a general and a special case of sensitivity analysis for Problem (19.1). For the general case, we suppose that the objective f, equality constraint function g, and inequality constraint function h are parameterized by a parameter $\chi \in \mathbb{R}^s$. That is, $f : \mathbb{R}^n \times \mathbb{R}^s \to \mathbb{R}$, $g : \mathbb{R}^n \times \mathbb{R}^s \to \mathbb{R}^m$, and

$h : \mathbb{R}^n \times \mathbb{R}^s \to \mathbb{R}^r$. We imagine that we have solved the non-linear inequality-constrained minimization problem:

$$\min_{x \in \mathbb{R}^n}\{f(x; \chi)|g(x; \chi) = 0, h(x; \chi) \le 0\}, \qquad (19.16)$$

for a base-case value of the parameters, say $\chi = 0$, to find the base-case solution $x^\star$ and the base-case Lagrange multipliers $\lambda^\star$ and $\mu^\star$. We now consider the sensitivity of the minimum of Problem (19.16) to variation of the parameters about $\chi = 0$.

As well as considering the general case of the sensitivity of the minimum of Problem (19.16) to χ, we also specialize to the case where only the right-hand sides of the equality and inequality constraints vary. That is, we return to the special case where $f : \mathbb{R}^n \to \mathbb{R}$, $g : \mathbb{R}^n \to \mathbb{R}^m$, and $h : \mathbb{R}^n \to \mathbb{R}^r$ are not explicitly parameterized. However, we now consider perturbations $\gamma \in \mathbb{R}^m$ and $\eta \in \mathbb{R}^r$ and the problem:

$$\min_{x \in \mathbb{R}^n}\{f(x)|g(x) = -\gamma, h(x) \le -\eta\}. \qquad (19.17)$$

For the parameter values $\gamma = 0$ and $\eta = 0$, Problem (19.17) is the same as Problem (19.1). We consider the sensitivity of the minimum of Problem (19.17) to variation of the parameters about $\gamma = 0$ and $\eta = 0$.

We have the following corollary to the implicit function theorem, Theorem A.9 in Section A.7.3 of Appendix A.

Corollary 19.7 *Consider Problem (19.16) and suppose that the functions $f : \mathbb{R}^n \times \mathbb{R}^s \to \mathbb{R}$, $g : \mathbb{R}^n \times \mathbb{R}^s \to \mathbb{R}^m$, and $h : \mathbb{R}^n \times \mathbb{R}^s \to \mathbb{R}^r$ are twice partially differentiable with continuous second partial derivatives. Also consider Problem (19.17) and suppose that the functions $f : \mathbb{R}^n \to \mathbb{R}$, $g : \mathbb{R}^n \to \mathbb{R}^m$, and $h : \mathbb{R}^n \to \mathbb{R}^r$ are twice partially differentiable with continuous second partial derivatives. Suppose that $x^\star \in \mathbb{R}^n$, $\lambda^\star \in \mathbb{R}^m$, and $\mu^\star \in \mathbb{R}^r$ satisfy:*

- *the second-order sufficient conditions for Problem (19.16) for the value of parameters $\chi = 0$, and*
- *the second-order sufficient conditions for Problem (19.17) for the value of parameters $\gamma = 0$ and $\eta = 0$.*

In particular:

- *$x^\star$ is a local minimizer of Problem (19.16) for $\chi = 0$, and*
- *$x^\star$ is a local minimizer of Problem (19.17) for $\gamma = 0$ and $\eta = 0$,*

in both cases with associated Lagrange multipliers $\lambda^\star$ and $\mu^\star$. Moreover, suppose that $x^\star$ is a regular point of the constraints for the base-case problems and that there are no degenerate constraints at the base-case solution.

Then, for values of χ in a neighborhood of the base-case value of the parameters $\chi = 0$, there is a local minimum and corresponding local minimizer and Lagrange multipliers for Problem (19.16). Moreover, the local minimum, local minimizer, and Lagrange multipliers are partially differentiable with respect to χ and have continuous partial derivatives in this neighborhood. The sensitivity of the local minimum $f^\star$ to χ, evaluated at *the base-case $\chi = 0$, is given by:*

$$\frac{\partial \mathcal{L}}{\partial \chi}(x^\star, \lambda^\star, \mu^\star; 0),$$

where $\mathcal{L} : \mathbb{R}^n \times \mathbb{R}^m \times \mathbb{R}^r \times \mathbb{R}^s \to \mathbb{R}$ is the **parameterized Lagrangian** *defined by:*

$$\forall x \in \mathbb{R}^n, \forall \lambda \in \mathbb{R}^m, \forall \mu \in \mathbb{R}^r, \forall \chi \in \mathbb{R}^s,$$
$$\mathcal{L}(x, \lambda, \mu; \chi) = f(x; \chi) + \lambda^\dagger g(x; \chi) + \mu^\dagger h(x; \chi).$$

Furthermore, for values of γ and η in a neighborhood of the base-case value of the parameters $\gamma = 0$ and $\eta = 0$, there is a local minimum and corresponding local minimizer and Lagrange multipliers for Problem (19.17). Moreover, the local minimum, local minimizer, and Lagrange multipliers are partially differentiable with respect to γ and η and have continuous partial derivatives. The sensitivities of the local minimum to γ and η, evaluated at the base-case $\gamma = 0$ and $\eta = 0$, are equal to $\lambda^\star$ and $\mu^\star$, respectively.

Proof See [34, theorem 2.4.4] and [70, section 10.8] for details. □

19.5.2 Discussion

We can again interpret the Lagrange multipliers as the sensitivity of the minimum to the right-hand side of the equality constraints and inequality constraints. We can use the Lagrange multipliers to help in trading off the change in the optimal objective against the cost of modifying the constraints. As in the case of non-linear equality constraints described in Section 14.4 and linear inequality constraints described in Section 17.4, we can again use sensitivity analysis of the first-order necessary conditions to estimate the changes in the minimizer and Lagrange multipliers.

19.5.3 Example

Continuing with Problem (2.19) from Sections 2.3.2.3, 19.2.1.3, and 19.2.2.2, we have already verified that the second-order sufficient conditions are satisfied at the base-case solution, that $x^\star$ is a regular point of the constraints, and that there are no degenerate constraints. Suppose that the first entry in the equality constraint changed to $2 - x_2 - \sin(x_3) = -\gamma_1$ and that the inequality constraint changed to $\sin(x_3) - 0.5 \leq -\eta$. Then, by Corollary 19.7, if γ_1 and η are small enough the change in the minimum is given approximately by $\lambda_1^\star \gamma_1 + \mu^\star \eta = 6\gamma_1 + 5\eta$. (See Exercise 19.8.)

19.6 Summary

In this chapter we have considered problems with non-linear equality and inequality constraints, providing optimality conditions. We considered the convex case and sketched application of the primal–dual interior point method and dual algorithm to these problems. Finally, we provided sensitivity analysis.

Exercises

Geometry and analysis of constraints

19.1 Let $C \in \mathbb{R}^{r \times n}$ and $d \in \mathbb{R}^r$. Suppose that the first row of C is equal to the zero vector. Characterize the set $\{x \in \mathbb{R}^n | Cx \leq d\}$ in the cases that:

(i) $d_1 < 0$,
(ii) $d_1 \geq 0$.

19.2 Consider the set $\mathbb{S} \in \mathbb{R}^2$ defined by $\mathbb{S} = \{x \in \mathbb{R}^n | h(x) \leq 0\}$, where $h : \mathbb{R}^2 \to \mathbb{R}$ is defined by:

$$\forall x \in \mathbb{R}^2, \, g(x) = x_2 - \sin(x_1).$$

This exercise is similar to Exercise 14.3.

(i) Is $x^\star = \mathbf{0}$ a regular point of the inequality constraint $h(x) \leq 0$?
(ii) Describe the set $\mathbb{T} = \{x \in \mathbb{R} | K(x^\star)(x - x^\star) = \mathbf{0}\}$, where K is the Jacobian of h.
(iii) Consider the points in $\mathbb{R}^2$ in the vicinity of $x^\star = \mathbf{0}$. That is, consider $x \in \mathbb{R}^2$ such that $\|x\|_2 \approx \|x^\star\|_2 = 0$. For these points, is the set $\mathbb{T} = \{x \in \mathbb{R}^2 | K(x^\star)(x - x^\star) = \mathbf{0}\}$ qualitatively a good approximation to the set $\mathbb{P} = \{x \in \mathbb{R}^2 | h(x) = 0\}$?

Optimality conditions

19.3 ([84, example 14.12]) Consider the problem $\min_{x \in \mathbb{R}^2} \{f(x) | h(x) \leq 0\}$ where $f : \mathbb{R}^2 \to \mathbb{R}$ and $h : \mathbb{R}^2 \to \mathbb{R}^2$ are defined by:

$$
\begin{aligned}
\forall x \in \mathbb{R}^2, \, f(x) &= x_1 + x_2, \\
\forall x \in \mathbb{R}^2, \, h(x) &= \begin{bmatrix} (x_1 - 1)^2 + (x_2)^2 - 1 \\ (x_1 + 1)^2 + (x_2)^2 - 1 \end{bmatrix}.
\end{aligned}
$$

(i) Show that $x^\star = \mathbf{0}$ is the unique feasible point and (therefore) the unique minimizer for the problem $\min_{x \in \mathbb{R}^2} \{f(x) | h(x) \leq 0\}$. (Hint: Describe the set of points satisfying each inequality constraint geometrically.)
(ii) Show that $x^\star$ is not a regular point of the constraints $h(x) \leq 0$.
(iii) Show that the problem does not satisfy the Slater condition (19.4).
(iv) Show that no $\mu^\star$ exists satisfying (19.2).

(v) Find another specification of the incquality constraint functions (possibly involving more than two inequality constraints) that specifies the same feasible set and such that $x^\star$ is a regular point of the constraints $h(x) \leq \mathbf{0}$.

Convex problems

19.4 Consider the problem $\min_{x \in \mathbb{R}^2} \{ f(x) | h(x) \leq 0 \}$ from Section 19.3.2.2 with objective $f : \mathbb{R}^2 \to \mathbb{R}$ and inequality constraint function $h : \mathbb{R}^2 \to \mathbb{R}$ be defined by:

$$\forall x \in \mathbb{R}^2, \ f(x) = x_1 + x_2,$$
$$\forall x \in \mathbb{R}^2, \ h(x) = (x_1)^2 + (x_2)^2 - 2.$$

Show that the problem satisfies the Slater condition (19.4).

Approaches to finding minimizers

19.5 Consider Problem (19.10):

$$\min_{x \in \mathbb{R}^n, w \in \mathbb{R}^r} \{ \phi(x, w) | g(x) = \mathbf{0}, h(x) + w = \mathbf{0} \},$$

where $\phi : \mathbb{R}^n \times \mathbb{R}^r_{++} \to \mathbb{R}$ is defined by:

$$\forall x \in \mathbb{R}^n, \forall w \in \mathbb{R}^r_{++}, \phi(x, w) = f(x) + t f_b(w),$$

where $f : \mathbb{R}^n \to \mathbb{R}$, $g : \mathbb{R}^n \to \mathbb{R}^m$, $h : \mathbb{R}^n \to \mathbb{R}^r$, and where $f_b : \mathbb{R}^r_{++} \to \mathbb{R}$ is the logarithmic barrier function defined by:

$$\forall w \in \mathbb{R}^r_{++}, f_b(w) = - \sum_{\ell=1}^{r} \ln(w_\ell).$$

Ignoring the issue of the domain of the barrier function, show that Problem (19.10) has first-order necessary conditions given by (19.11)–(19.14):

$$\nabla f(x) + J(x)^\dagger \lambda + K(x)^\dagger \mu = \mathbf{0},$$
$$g(x) = \mathbf{0},$$
$$h(x) + w = \mathbf{0},$$
$$t \nabla f_b(w) + \mu = \mathbf{0},$$

where J and K are the Jacobians of g and h, respectively.

19.6 Let $Q \in \mathbb{R}^{n \times n}$ be positive definite, $Q^{(\ell)} \in \mathbb{R}^{n \times n}$, $\ell = 1, \ldots, r$, all be positive definite, $c \in \mathbb{R}^n$, $c^{(\ell)} \in \mathbb{R}^n$, $\ell = 1, \ldots, r$, $d^{(\ell)} \in \mathbb{R}$, $\ell = 1, \ldots, r$, $A \in \mathbb{R}^{m \times n}$, $b \in \mathbb{R}^m$, and define $f : \mathbb{R}^n \to \mathbb{R}$ and $h : \mathbb{R}^n \to \mathbb{R}^r$ by:

$$\forall x \in \mathbb{R}^n, \ f(x) = \frac{1}{2} x^\dagger Q x + c^\dagger x,$$
$$\forall x \in \mathbb{R}^n, \forall \ell = 1, \ldots, r, \ h_\ell(x) = \frac{1}{2} x^\dagger Q^{(\ell)} x + [c^{(\ell)}]^\dagger x + d^{(\ell)},$$

and consider the problem $\min_{x \in \mathbb{R}^n}\{f(x)|Ax - b, h(x) \leq 0\}$ and its Lagrangian $\mathcal{L} : \mathbb{R}^n \times \mathbb{R}^m \times \mathbb{R}^r \to \mathbb{R}$ and dual function $\mathcal{D} : \mathbb{R}^m \times \mathbb{R}^n \to \mathbb{R} \cup \{-\infty\}$ defined by:

$$\forall x \in \mathbb{R}^n, \forall \lambda \in \mathbb{R}^m, \forall \mu \in \mathbb{R}^r, \mathcal{L}(x, \lambda, \mu) \;=\; f(x) + \lambda^\dagger(Ax - b) + \mu^\dagger h(x),$$
$$\forall \lambda \in \mathbb{R}^m, \mu \in \mathbb{R}^r, \mathcal{D}(\lambda, \mu) \;=\; \inf_{x \in \mathbb{R}^n} \mathcal{L}(x, \lambda, \mu).$$

(i) Let $\lambda \in \mathbb{R}^m$ and $\mu \in \mathbb{R}^r_+$ and find necessary and sufficient conditions to minimize $\mathcal{L}(\bullet, \lambda, \mu)$.

(ii) Evaluate the dual function for $\lambda \in \mathbb{R}^m$, $\mu \in \mathbb{R}^r_+$ and show that the effective domain contains $\mathbb{R}^m \times \mathbb{R}^r_+$.

(iii) For a fixed value of $\mu \in \mathbb{R}^r_+$, find the maximizer of the dual over $\lambda \in \mathbb{R}^m$.

Sensitivity

19.7 Use the general case result in Corollary 19.7 for Problem (19.16) to prove the special case result in Corollary 19.7 for Problem (19.17).

19.8 Consider the example non-linear program, Problem (2.19), from Section 2.3.2.3:

$$\min_{x \in \mathbb{R}^2}\{f(x)|g(x) = \mathbf{0}, h(x) \leq 0\},$$

where $f : \mathbb{R}^2 \to \mathbb{R}$, $g : \mathbb{R}^2 \to \mathbb{R}^2$, and $h : \mathbb{R}^2 \to \mathbb{R}$ are defined by:

$$\forall x \in \mathbb{R}^2, f(x) \;=\; (x_1)^2 + 2(x_2)^2,$$
$$\forall x \in \mathbb{R}^2, g(x) \;=\; \begin{bmatrix} 2 - x_2 - \sin(x_3) \\ -x_1 + \sin(x_3) \end{bmatrix},$$
$$\forall x \in \mathbb{R}^2, h(x) \;=\; \sin(x_3) - 0.5.$$

(i) Use the MATLAB function fmincon to find the minimizer and minimum of this problem. Write MATLAB M-files to evaluate functions and Jacobians. Use an initial guess of $x^{(0)} = \begin{bmatrix} 1 \\ 1 \\ 0 \end{bmatrix}$.

(ii) Now suppose that the first entry in the equality constraint changed to $2 - x_2 - \sin(x_3) = -\gamma_1$, where $\gamma_1 = 0.1$. Use Corollary 19.7 to estimate the change in the minimum.

(iii) Use the MATLAB function fmincon to find the minimizer and minimum of the problem in Part (ii). Compare to the result of Part (ii).

(iv) Now suppose that the inequality constraint changed to $\sin(x_3) - 0.5 \leq -\eta$, where $\eta = 0.1$. Use Corollary 19.7 to estimate the change in the minimum.

(v) Use the MATLAB function fmincon to find the minimizer and minimum of the problem in Part (iv). Compare to the result of Part (iv).

(vi) Now suppose that the first entry in the equality constraint changed to $2 - x_2 - \sin(x_3) = -\gamma_1$, where $\gamma_1 = 0.1$, and the inequality constraint changed to $\sin(x_3) - 0.5 \leq -\eta$, where $\eta = 0.1$. Use Corollary 19.7 to estimate the change in the minimum.

(vii) Use the MATLAB function `fmincon` to find the minimizer and minimum of the problem in Part (vi). Compare to the result of Part (vi).

20

Solution of the non-linear inequality-constrained case studies

In this chapter we will solve the non-linear inequality-constrained case studies:

- optimal margin pattern classification (Section 20.1),
- sizing of interconnects in integrated circuits (Section 20.2), and
- optimal power flow (Section 20.3).

20.1 Optimal margin pattern classification

Recall the first transformation of the optimal margin pattern classification case study in Section 18.4.1.1. This transformation yielded the maximization Problem (18.3), which we recast into a minimization problem as:

$$\min_{z \in \mathbb{R}, x \in \mathbb{R}^n} \{ -z \mid \mathbf{1}z + Cx \leq \mathbf{0}, \|\beta\|_2^2 \leq 1 \}. \tag{20.1}$$

This problem has a linear objective, r linear inequality constraints, and one convex quadratic inequality constraint. This can be solved using the algorithms developed in Section 19.4. Exercise 20.1 shows that the dual of Problem (20.1) is equivalent to a quadratic program.

20.2 Sizing of interconnects in integrated circuits

In this section, we solve the sizing of interconnects in integrated circuits case study from Section 15.5. We recall and analyze the problem in Section 20.2.1, describe algorithms in Section 20.2.2, and sketch sensitivity analysis in Section 20.2.3.

20.2.1 Problem and analysis

Recall Problem (15.19):

$$\min_{x \in \mathbb{R}^n} \{ f(x) \mid \tilde{h}(x) \leq \overline{h}, \underline{x} \leq x \leq \overline{x} \},$$

which used the Elmore delay approximation $\tilde{h}$ to the actual delay h. This problem has a linear objective but has inequality constraints defined in terms of functions that are, in general, non-convex as shown in Exercise 15.14. However, as discussed in Section 15.5.4 and Exercise 15.14, the objective and constraint functions are **posynomial**. (See Definition 3.1.)

As discussed in Exercise 3.33, each posynomial function can be transformed into a convex function through a transformation involving the exponential of each entry of the decision vector and the logarithm of the function. (See Exercise 20.2.) The transformed problem is convex and therefore possesses at most one local minimum. Because the transformation of the decision vector is one-to-one and onto and the transformations of the objective and constraints are monotonically increasing then, by Theorems 3.1, 3.5, and 3.9, the original problem also possesses at most one local minimum.

20.2.2 Algorithms

20.2.2.1 Primal algorithm

In principle, we can apply the optimization techniques developed in Section 19.4 to either the original problem or the transformed problem and be guaranteed that any local minimum is the global minimum. (See Exercise 20.2.) However, since the inequality constraint functions are not convex in the original problem, the Hessian of the Lagrangian for the original problem will typically not be positive definite and so we can expect that pivots will be modified significantly during factorization, potentially retarding the progress towards the minimizer.

20.2.2.2 Dual algorithm

Since the transformed problem is convex, we can also dualize the transformed problem. (See Exercise 20.2.) Further transformation of the dual problem is possible to simplify the dual problem to having linear constraints. (See [6, sections 11.5.1–11.5.2] for details.)

20.2.2.3 Accurate delay model

Recall Problem (15.20):

$$\min_{x\in\mathbb{R}^n}\{f(x)|h(x) \leq \overline{h}, \underline{x} \leq x \leq \overline{x}\},$$

which used the more accurate delay model h instead of the Elmore delay model $\tilde{h}$. In general, we cannot expect that h will have any particular functional form. That is, we cannot expect that h will be posynomial. However, $\tilde{h}$ may be a reasonable approximation of h. The algorithms we have described typically require both function evaluations *and* derivative evaluations. To solve the problem with the more accurate delay model, we can combine accurate delay values calculated according to h with approximate first and second derivatives calculated from the functional form of $\tilde{h}$. Furthermore, we can apply such an algorithm to the original problem or to the transformed problem.

20.2.3 Changes

Corollary 19.7 and extensions can be used to estimate the changes in area and width due to changes in parameters and allowed delays. (See Exercise 20.4.)

20.3 Optimal power flow

Recall Problem (15.23):

$$\min_{x \in \mathbb{R}^n} \{ f(x) | g(x) = \mathbf{0}, \underline{x} \le x \le \overline{x}, \underline{h} \le h(x) \le \overline{h} \}.$$

This problem has non-linear objective and equality and inequality constraint functions. As argued in Section 15.6.4.1, however, under certain assumptions the problem is equivalent to a convex problem. We can use the primal–dual interior point algorithm sketched in Section 19.4.1 to solve it [126]. (See Exercise 20.5.) Corollary 19.7 and extensions can be used to estimate the changes in costs due to changes in demand and changes in line and generator capacities. (See Exercise 20.6.)

Exercises

Optimal margin pattern classification

20.1 Consider Problem (20.1):

$$\min_{z \in \mathbb{R}, x \in \mathbb{R}^n} \{ -z | \mathbf{1}z + Cx \le \mathbf{0}, \|\beta\|_2^2 \le 1 \}.$$

(i) Write down the Lagrangian $\mathcal{L}$ and the dual function $\mathcal{D}$ explicitly for this problem. For convenience in the rest of the exercise, partition x into $x = \begin{bmatrix} \beta \\ \gamma \end{bmatrix}$ and partition C into $C = \begin{bmatrix} \hat{C} & c \end{bmatrix}$, where $\hat{C} \in \mathbb{R}^{r \times (n-1)}$ is the first $n-1$ columns of C and $c \in \mathbb{R}^m$ is its last column. Use $\mu \in \mathbb{R}^r$ for the dual variables for the constraints $\mathbf{1}z + Cx \le \mathbf{0}$ and use $\sigma \in \mathbb{R}$ for the dual variable for the constraint $\|\beta\|_2^2 \le 1$.

(ii) Find the effective domain $\mathbb{E}$ of the dual function.

(iii) Evaluate the dual function $\mathcal{D}$ on the effective domain.

(iv) Use hierarchical decomposition Theorem 3.11 to simplify the dual problem by optimizing over $\sigma \ge 0$.

(v) Show that the resulting problem is equivalent to a quadratic program.

(See [15, section 8.6.1] for the dual of a related problem.)

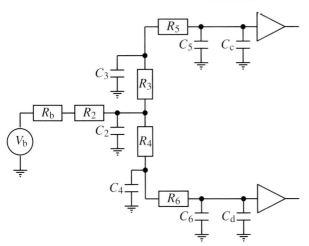

Fig. 20.1. The equivalent circuit of interconnect between gate b and gates c and d with resistive–capacitive segments, repeated from Figure 15.8.

Sizing of interconnects in integrated circuits

20.2 Consider Problem (15.19):

$$\min_{x \in \mathbb{R}^n} \{ f(x) | \tilde{h}(x) \le \overline{h}, \underline{x} \le x \le \overline{x} \}.$$

(i) Apply the transformation described in Exercise 3.33 to Problem (15.19) to form a convex problem.
(ii) Prove that any local minimizer of Problem (15.19) is a global minimizer.
(iii) Find the dual of the transformed problem from Part (i).

20.3 Consider the resistive–capacitive model of interconnect shown in Figure 15.8 and repeated in Figure 20.1. The figure shows part of the circuit between gate b (having driving voltage V_b) and buffers c and d (having input capacitances C_c and C_d, respectively).

Suppose that the buffer input capacitances are $C_c = C_d = 1$ in normalized capacitive units and that $R_b = 1$ in normalized resistive units. Also assume that the length of each segment is equal to 1 in normalized length units. Assume that we can choose the width of any of the segments $k = 2, 3, \ldots, 6$ illustrated in the figure, which will have the effect of changing the segment resistance and capacitance. In normalized width units, the k-th segment can have width x_k in the range:

$$\underline{x}_k = 1 \le x_k \le 5 = \overline{x}_k.$$

We assume that the width can be chosen as a continuous variable. The dependence of resistance on width is specified in (15.16), which we repeat:

$$\forall k = 1, \ldots, n, \ R_k = \kappa_{Rk}/x_k,$$

where $\kappa_{Rk} = 1$ in normalized units for all segments. The dependence of capacitance on width is specified in (15.17), which we repeat:

$$\forall k = 1, \ldots, n, \ C_k = \kappa_{Ck} x_k + C_{Fk},$$

where $\kappa_{Ck} = 1$ and $C_{Fk} = 1$ in normalized units.

(i) Calculate the Elmore delay from the driving voltage V_b to the input of buffer c, assuming that all segment widths are equal to 1.

(ii) Use the MATLAB function `fmincon` to solve the problem of finding the widths that minimize the Elmore delay from the driving voltage V_b to the input of buffer c.

(iii) For the widths chosen in Part (ii), find the Elmore delay from the driving voltage V_b to the input of buffer d.

(iv) Use the MATLAB function `fmincon` to solve the problem of finding the widths that minimize the maximum of the Elmore delays from:

- the driving voltage V_b to the input of buffer c, and
- the driving voltage V_b to the input of buffer d.

(Hint: You should first transform the problem using Theorem 3.4.)

(v) Define $\overline{h} \in \mathbb{R}$ to be the average of the delays calculated in Parts (i) and (iv). Use the MATLAB function `fmincon` to solve Problem (15.19) where we want to minimize the area of interconnect subject to the delay constraint that the delays from voltage V_b to buffer c and to buffer d are both less than or equal to $\overline{h}$.

(vi) Use the MATLAB function `fmincon` to solve the problem in Part (v) using the transformed version of the problem from Exercise 20.2.

20.4 Consider the solution of Exercise 20.3, Part (v).

(i) Using sensitivity analysis, estimate the change in the minimum area if the delay constraints were changed so that the delays were both required to be less than $0.9\overline{h}$.

(ii) Use the MATLAB function `fmincon` to calculate the minimum area if the delay constraints were changed so that the delays were both required to be less than $0.9\overline{h}$.

(iii) Compare the results of the previous parts.

Optimal power flow

20.5 Using the program developed in Exercise 8.13 as a basis, develop an optimal power flow program that minimizes the cost of production of the generators using the primal-dual interior point algorithm. Test the program using the line data from Exercise 8.12. That is, the π-equivalent line models have:

- shunt elements purely capacitive with admittance $0.01\sqrt{-1}$ so that the combined shunt elements are:

$$Y_1 = Y_2 = Y_3 = 0.02\sqrt{-1},$$

and
- series elements having admittances:

$$Y_{12} = (0.01 + 0.1\sqrt{-1})^{-1},$$
$$Y_{23} = (0.015 + 0.15\sqrt{-1})^{-1},$$
$$Y_{31} = (0.02 + 0.2\sqrt{-1})^{-1}.$$

Furthermore, assume the following.

- There are generators at bus 1 and bus 2 and a load of $1 + 0.5\sqrt{-1}$ at bus 3.
- All lines have real power flow limits of 0.75.
- All voltage magnitudes constrained to be between 0.95 and 1.05 per unit.
- Zero cost for reactive power production.
- Costs for real power production at the generators:

$$f_1(P_1) = P_1 \times 1 \frac{\$}{\text{per unit}} + (P_1)^2 \times 0.1 \frac{\$}{(\text{per unit})^2},$$

$$f_2(P_2) = P_2 \times 1.1 \frac{\$}{\text{per unit}} + (P_2)^2 \times 0.05 \frac{\$}{(\text{per unit})^2},$$

where P_k is the real power production at generator $k = 1, 2$, with $0 \leq P_k \leq 1$ for each generator.
- No other constraints on production.

Use as initial guess:

$$x^{(0)} = \begin{bmatrix} P^{(0)} \\ Q^{(0)} \\ u^{(0)} \\ \theta^{(0)} \end{bmatrix} = \begin{bmatrix} 0 \\ 0 \\ 1 \\ 0 \end{bmatrix}.$$

Use as stopping criterion that all of the following are satisfied:

- $t_{\text{effective}} < 10^{-5}$,
- power flow equations are satisfied to within 0.0001 per unit, and
- the change in successive iterates is less than 0.0001 per unit.

Compare your result to that of Exercise 17.16.

20.6 Consider the optimal power flow problem specified in Exercise 20.5.

(i) Use the MATLAB function fmincon to solve the problem. Use as initial guess $x^{(0)}$ as specified in Exercise 20.5.
(ii) Using sensitivity analysis, estimate the change in the minimum if the real and reactive power load at bus 3 increased by 5%.
(iii) Use the MATLAB function fmincon to calculate the minimum if the real and reactive power load at bus 3 increased by 5%. Use the solution from Part (i) as initial guess. Compare the result to the previous part.

References

[1] A. Abur and A. Gomez-Exposito. *Power System State Estimation*. New York: Marcel Dekker, 2004.

[2] F. L. Alvarado, W. F. Tinney, and M. K. Enns. Sparsity in large-scale network computation. In: C. T. Leondes (editor), *Advances in Electric Power and Energy Conversion System Dynamics and Control*. San Diego, CA: Academic Press, Inc., 1991.

[3] R. Baldick. Variation of distribution factors with loading. *IEEE Transactions on Power Systems*, **18**(4):1316–1323, 2003.

[4] R. Baldick, K. A. Clements, Z. Pinjo-Dzigal, and P. W. Davis. Implementing nonquadratic objective functions for state estimation and bad data rejection. *IEEE Transactions on Power Systems*, **12**(1):376–382, 1997.

[5] R. Baldick, A. B. Kahng, A. Kennings, and I. L. Markov. Efficient optimization by modifying the objective function: Applications to timing-driven VLSI layout. *IEEE Transactions on Circuits and Systems I: Fundamental Theory and Applications*, **48**(8):947–956, 2001.

[6] M. S. Bazaraa, H. D. Sherali, and C. M. Shetty. *Nonlinear Programming: Theory and Algorithms*. New York: John Wiley and Sons, Inc., Second Edition, 1993.

[7] R. E. Bellman and S. E. Dreyfus. *Applied Dynamic Programming*. Princeton: Princeton University Press, 1962.

[8] A. R. Bergen and V. Vittal. *Power Systems Analysis*. Upper Saddle River, NJ: Prentice-Hall, Second Edition, 2000.

[9] D. Bertsekas and R. Gallager. *Data Networks*. Upper Saddle River, NJ: Prentice Hall, Second Edition, 1992.

[10] D. P. Bertsekas. *Dynamic Programming and Optimal Control*. Belmont, MA: Athena Scientific, 1995.

[11] D. P. Bertsekas. *Nonlinear Programming*. Belmont, MA: Athena Scientific, 1995.

[12] D. Bertsimas and J. N. Tsitsiklis. *Linear Optimization*. Belmont, MA: Athena Scientific, 1997.

[13] M. Bhattacharya and P. Mazumder. Augmentation of SPICE for simulation of circuits containing resonant tunneling diodes. *IEEE Transactions on Computer-Aided Design of Integrated Circuits and Systems*, **20**(1):39–50, 2001.

[14] B. Boser, I. Guyon, and V. Vapnik. A training algorithm for optimal margin classifiers. In: *Proceedings of the Fifth Annual ACM Workshop on Computational Learning Theory COLT*. ACM Press, July 1992.

[15] S. Boyd and L. Vandenberghe. *Convex Optimization*. Cambridge and New York: Cambridge University Press, 2004.

[16] S. P. Boyd and C. H. Barratt. *Linear Controller Design: Limits of Performance*. Englewood Cliffs, NJ: Prentice Hall, 1991.

[17] M. A. Branch and A. Grace. *The Optimization Toolbox User's Guide*. The MathWorks, Inc., www.mathworks.com, Natick, MA, 1996.

[18] A. Brooke, D. Kendrick, and A. Meeraus. *GAMS User's Guide*. Redwood City, CA: The Scientific Press, 1990.

[19] F. M. Callier and C. A. Desoer. *Multivariable Feedback Systems*. New York: Springer, 1982.

[20] J. Carpentier. Contribution a l'etude du dispatching economique. *Bulletin de la Societe Française Electriciens*, **3**:431–437, 1962.

[21] G. W. Carter and A. Richardson. *Techniques of Circuit Analysis*. Cambridge: Cambridge University Press, 1972.

[22] E. Castillo, A. J. Conejo, P. Pedregal, R. Garcia, and N. Alguacil. *Building and Solving Mathematical Programming Models in Engineering and Science*. New York: John Wiley and Sons, Inc., 2002.

[23] R. M. Chamberlain, M. J. D. Powell, D. Lemarechal, and H. C. Pedersen. The watchdog technique for forcing convergence in algorithms for constrained optimization. *Mathematical Programming Study*, **16**:1–17, 1982.

[24] H. Chao and S. Peck. A market mechanism for electric power transmission. *Journal of Regulatory Economics*, **10**(1):25–59, 1996.

[25] C. C. N. Chu and D. F. Wong. A new approach to simultaneous buffer insertion and wire sizing. In: *IEEE/ACM International Conference on Computer-Aided Design*, pages 614–621. IEEE/ACM, November 1997.

[26] G. Cohen and D. L. Zhu. Decomposition coordination methods in large scale optimization problems: The nondifferentiable case and the use of augmented Lagrangians. In: J. B. Cruz (editor), *Advances in Large Scale Systems, Volume 1*, pages 203–266. Greenwich, CT: JAI Press Inc., 1984.

[27] C. Cuvelier, A. Segal, and A. van Steenhoven. *Finite Element Methods and Navier–Stokes Equations*. Boston, MA: Kluwer Academic, 1986.

[28] G. B. Dantzig and M. N. Thapa. *Linear Programming, 1: Introduction*. New York: Springer, 1997.

[29] J. E. Dennis and R. B. Schnabel. *Numerical Methods for Unconstrained Optimization and Nonlinear Equations*. Englewood Cliffs, NJ: Prentice Hall, 1983.

[30] J. J. Dongarra, G. A. Geist, and C. H. Romine. ALGORITHM 710 FORTRAN subroutines for computing the eigenvalues and eigenvectors of a general matrix by reduction to general tridiagonal form. *ACM Transactions on Mathematical Software*, **18**(4):392–400, 1992.

[31] R. Durrett. *Probability: Theory and Examples*. Pacific Grove, CA: Wadsworth and Brooks/Cole, 1991.

[32] A. El-Rabbany. *Introduction to GPS: The Global Positioning System*. Boston, MA: Artech House, 2002.

[33] W. C. Elmore. The transient response of damped linear networks with particular regard to wide-band amplifiers. *Journal of Applied Physics*, **19**(1):55–63, 1948.

[34] A. V. Fiacco. *Introduction to Sensitivity and Stability Analysis in Nonlinear Programming*. New York: Academic Press, 1983.

[35] M. L. Fisher. The Lagrangian relaxation method for solving integer programming problems. *Management Science*, **27**(1):1–18, 1981.

[36] R. Fletcher and S. Leyffer. Nonlinear Programming without a Penalty Function. Technical Report Numerical Analysis Report NA/171, University of Dundee, September 1997.

[37] F. Fourer and S. Mehrotra. Solving Symmetric Indefinite Systems in an Interior-point Method for Linear Programming. Technical Report 92-01, Department of Industrial Engineering and Management Sciences, Northwestern University, Evanston, IL, 1992.

[38] R. Fourer, D. M. Gay, and B. W. Kernighan. *AMPL: A Mathematical Programming Language*. Murray Hill, NJ: AT&T Bell Laboratories, 1989.

[39] F. Galiana, H. Javidi, and S. McFee. On the application of a pre-conditioned conjugate gradient algorithm to power network analysis. *IEEE Transactions on Power Systems*, **9**(2):629–636, 1994.

[40] M. R. Garey and D. S. Johnson. *Computers and Intractability: A Guide to the Theory of NP-Completeness*. San Francisco, CA: W. H. Freeman and Company, 1979.

[41] A. M. Geoffrion. Elements of large scale mathematical programming part I: Concepts. *Management Science*, **16**(11):652–675, 1970.

[42] A. M. Geoffrion. Elements of large scale mathematical programming part II: Synthesis of algorithms and bibliography. *Management Science*, **16**(11):676–691, 1970.

[43] A. M. Geoffrion. Generalized Benders decomposition. *Journal of Optimization Theory and Applications*, **10**(4):237–260, 1972.

[44] P. E. Gill, W. Murray, M. A. Saunders, and M. H. Wright. SNOPT: an SQP algorithm for large-scale constrained optimization. *SIAM Review*, **47**(1):99–131, 2005.

[45] P. E. Gill, W. Murray, and M. H. Wright. *Practical Optimization*. London: Academic Press, Inc., 1981.

[46] D. E. Goldberg. *Genetic Algorithms in Search, Optimization, and Machine Learning*. Reading, MA: Addison-Wesley Publishing, 1989.

[47] C. C. Gonzaga. Path-following methods in linear programming. *SIAM Review*, **34**(2):167–224, 1992.

[48] P. Grogono. *Programming in Pascal*. Reading, MA: Addison-Wesley, 1978.

[49] C. Guéret, C. Prins, and M. Sevaux. *Applications of optimization with Xpress-MP*. Northants, United Kingdom: Dash Optimization, 2002.

[50] R. Gupta, B. Krauter, B. Tutuianu, J. Willis, and L. T. Pileggi. The Elmore delay as a bound for RC trees with generalized input signals. In: *ACM/IEEE Conference on Design Automation*, pages 364–369. IEEE/ACM, June 1995.

[51] M. Held, P. Wolfe, and H. P. Crowder. Validation of subgradient optimization. *Mathematical Programming*, **6**:62–88, 1974.

[52] D. S. Hochbaum, editor. *Approximation Algorithms for NP-Hard Problems*. Boston, MA: PWS Publishing, 1997.

[53] P. J. Huber. *Robust Statistics*. New York: John Wiley and Sons, 1981.

[54] P. A. Jensen and J. F. Bard. *Operations Research Models and Methods*. Hoboken: John Wiley and Sons, 2003.

[55] T. Kailath. *Linear Systems*. Prentice-Hall Information and System Sciences Series. Englewood Cliffs, NJ: Prentice-Hall, 1980.

[56] S.-M. Kang and Y. Leblebici. *CMOS Digital Integrated Circuits*. Boston, MA: McGraw-Hill, Second Edition, 1999.

[57] N. K. Karmarkar. A new polynomial-time algorithm for linear programming. *Combinatorics*, **4**:273–295, 1984.

[58] C. T. Kelley. *Iterative Methods for Linear and Nonlinear Equations*. Philadelphia, PA: SIAM, 1995.

[59] C. T. Kelley. *Iterative Methods for Optimization*. Philadelphia, PA: SIAM, 1999.

[60] B. W. Kernighan and D. M. Ritchie. *The C Programming Language*. Englewood Cliffs, NJ: Prentice-Hall, Inc., 1978.

[61] B. Kim and R. Baldick. Coarse-grained distributed optimal power flow. *IEEE Transactions on Power Systems*, **12**(2):932–939, 1997.

[62] G. R. Krumpholz, K. A. Clements, and P. W. Davis. Power system observability: A practical algorithm using network topology. *IEEE Transactions on Power Apparatus and Systems*, **99**:1534–1542, 1980.

[63] P. R. Kumar and P. Varaiya. *Stochastic Systems: Estimation, Identification, and Adaptive Control*. Prentice-Hall Information and System Sciences Series. Englewood Cliffs, NJ: Prentice Hall, 1986.

[64] R. E. Larson and J. L. Casti. *Principles of Dynamic Programming, Part I: Basic Analytic and Computational Methods*. New York: Marcel Dekker, Inc., 1978.

[65] L. S. Lasdon, J. Plummer, and G. Yu. Primal–dual and primal interior point algorithms for general nonlinear programs. *ORSA Journal on Computing*, **7**(3):321–332, 1995.

[66] L. S. Lasdon and A. D. Warren. Generalized reduced gradient software for linearly and nonlinearly constrained problems. In: H. J. Greenberg (editor), *Design and Implementation of Optimization Software*, NATO Advanced Study Institutes Series. Series E, Applied Sciences No. 28. Alphen aan den Rijn, Netherlands: Sijthoff and Noordhoff, 1978.

[67] E. L. Lawler. *Combinatorial Optimization: Networks and Matroids*. New York: Holt, Rinehart and Winston, 1976.

[68] Lindo Systems, Inc., Chicago, IL. *The LINDO API User's Manual*, 2005. Available from www.lindo.com.

[69] M. Livio. *The Equation That Couldn't Be Solved*. New York, NY: Simon and Schuster, 2005.

[70] D. G. Luenberger. *Linear and Nonlinear Programming*. Reading, MA: Addison-Wesley Publishing Company, Second Edition, 1984.

[71] H. M. Markowitz. The elimination form of the inverse and its application to linear programming. *Management Science*, **3**:255–269, 1957.

[72] J. E. Marsden and A. B. Tromba. *Vector Calculus*. New York: W. H. Freeman and Company, 2nd Edition, 1981.

[73] R. Marsten, R. Subramanian, M. Saltzman, I. Lustig, and D. Shanno. Interior point methods for linear programming: Just call Newton, Lagrange, and Fiacco and McCormick. *Interfaces*, **20**(4):105–116, 1990.

[74] MathWorks, Inc. *The Student Edition of MATLAB*. Englewood Cliffs, NJ: Prentice-Hall, 1995.

[75] Microsoft Corporation, Washington. *Microsoft Excel*, 1994.

[76] K. M. Miettinen. *Nonlinear Multiobjective Optimization*. Boston, MA: Kluwer Academic Publishers, 1998.

[77] A. R. Mitchell and R. A. Wait. *The Finite Element Method in Partial Differential Equations*. London and New York: Wiley, 1977.

[78] C. B. Moler and G. W. Stewart. An algorithm for generalized matrix eigenvalue problems. *SIAM Journal on Numerical Analysis*, **10**(2):241–256, 1973.

[79] J. Momoh. *Electric Power System Applications of Optimization*. New York and Basel: Marcel Dekker, Inc., 2001.

[80] A. Monticelli. *State Estimation in Electric Power Systems: A Generalized Approach*. Power Electronics and Power Systems. Boston, MA: Kluwer Academic, 1999.

[81] J. J. Moré and S. J. Wright. *Optimization Software Guide*. Philadelphia, PA: Society for Industrial and Applied Mathematics, 1993.

[82] J. R. Munkres. *Topology: A First Course*. Englewood Cliffs, NJ: Prentice-Hall, Inc., 1975.

[83] B. A. Murtagh. *Advanced Linear Programming: Computation and Practice*. New York and London: McGraw-Hill International Book, 1981.

[84] S. G. Nash and A. Sofer. *Linear and Nonlinear Programming*. New York: McGraw-Hill, 1996.

[85] G. L. Nemhauser and L. A. Wolsey. *Integer and Combinatorial Optimization*. New York: John Wiley and Sons, 1988.

[86] Y. Nesterov and A. S. Nemirovskii. *Interior-point Polynomial Algorithms in Convex Programming*. Philadelphia, PA: SIAM, 1993.

[87] A. Neumaier. Complete search in continuous global optimization and constraint satisfaction. *Acta Numerica*, **13**:271–369, 2004.

[88] The Numerical Algorithms Group Ltd, Oxford, UK. *NAG C Library Manual, Mark 7*, 2002. Available from www.nag.co.uk.

[89] K. Ogata. *Discrete-Time Control Systems*. Englewood Cliffs, NJ: Prentice-Hall, Inc., 1987.

[90] V. Y. Pan. Solving a polynomial equation: Some history and recent progress. *SIAM Review*, **39**(2):187–220, 1997.

[91] P. M. Pardalos and J. B. Rosen. *Constrained Global Optimization: Algorithms and Applications*. Lecture Notes in Computer Science: 268. Berlin and New York: Springer-Verlag, 1987.

[92] R. G. Parker and R. L. Rardin. *Discrete Optimization*. San Diego, CA: Academic Press, Inc., 1988.

[93] J. Peng, C. Roos, and T. Terlaky. *Self-Regularity: A New Paradigm for Primal-Dual Interior-Point Algorithms*. Princeton and Oxford: Princeton University Press, 2002.

[94] G. Peters and J. H. Wilkinson. Eigenvectors of real and complex matrices by *LR* and *QR* triangularizations. *Numerische Mathematik*, **16**(3):181–204, 1970.

[95] L. T. Pillage, R. A. Rohrer, and C. Visweswariah. *Electronic Circuit and System Simulation Methods*. New York: McGraw-Hill, Inc., 1995.

[96] E. Polak. *Computational Methods in Optimization*. New York: Academic Press, 1971.

[97] B. T. Polyak. Minimization of unsmooth functionals. *USSR Computational Mathematics and Mathematical Physics*, **9**(3):14–29, 1969.

[98] S. J. Qin and T. A. Badgwell. An overview of industrial model predictive control technology. In: *Proceedings of the Fifth International Conference on Chemical Process Control: Chemical Process Control-V, Tahoe City, CA*. New York: American Institute of Chemical Engineers, 1997.

[99] N. S. Rau. *Optimization Principles: Practical Applications to the Operation and Markets of the Electric Power Industry*. Piscataway, NJ: IEEE Press, 2003.

[100] R. T. Rockafellar. *Convex Analysis*. Princeton, NJ: Princeton University Press, 1970.

[101] R. Rohrer. Successive secants in the solution of nonlinear network equations. In: H. S. Wilf and F. Harary (editors), *Mathematical Aspects of Electrical Network Analysis, Volume III*, pages 103–112. Providence, RI: American Mathematical Society, 1971.

[102] C. Roos, T. Terlaky, and J.-P. Vial. *Theory and Algorithms for Linear Optimization*. Chichester: John Wiley and Sons, 1997.

[103] S. M. Ross. *Introduction to Probability and Statistics for Engineers and Scientists*. New York: John Wiley and Sons, 1987.

[104] H. L. Royden. *Real Analysis*. New York and London: Macmillan, Second Edition, 1968.

[105] S. Schaible. Fractional programming: A recent survey. *Journal of Statistics and Management Systems*, **5**(1–3):63–86, 2002.

[106] K. Shimuzu, Y. Ishizuka, and J. F. Bard. *Nondifferentiable and Two-Level Mathematical Programming*. Boston, MA: Kluwer Academic Publishers, 1997.

[107] N. Z. Shor. *Minimization Methods for Non-Differentiable Functions*. Berlin: Springer-Verlag, 1985.

[108] P. P. Silvester and R. L. Ferrari. *Finite Elements for Electrical Engineers*. Cambridge: Cambridge University Press, 1996.

[109] H. A. Simon. A behavioral model of rational choice. *Quantitative Journal of Economics*, **69**:174–183, 1955.

[110] S. Smale. Newton's method estimates from data at one point. In: R. E. Ewing, K. I. Gross, and C. F. Martin (editors), *The Merging of Disciplines: New Directions in Pure, Applied, and Computational Mathematics*, pages 185–196. New York: Springer Verlag, 1986.

[111] M. Spivak. *Calculus*. London and Menlo Park: W. A. Benjamin, Inc., 1967.

[112] I. Stewart. *Galois Theory*. London: Chapman and Hall, 1973.

[113] M. Tawarmalani and N. V. Sahinidis. *Convexification and Global Optimization in Continuous and Mixed-Integer Nonlinear Programming*. Nonconvex Optimization and Its Applications. Boston, MA: Kluwer Academic, 2002.

[114] G. B. Thomas, Jr. and R. L. Finney. *Calculus and Analytic Geometry*. Reading, MA: Addison-Wesley, Ninth Edition, 1996.

[115] W. M. Thorburn. Occam's razor. *Mind*, **XXIV**(2):287–288, 1915.

[116] R. J. Vanderbei. Symmetric Quasi-definite Matrices. Technical Report SOR-91-10, School of Engineering and Applied Science, Department of Civil Engineering and Operations Research, Princeton University, 1991.

[117] G. N. Vanderplaats. *Numerical Optimization Techniques for Engineering Design*. New York: McGraw-Hill, Inc., 1984.

[118] S. B. Vardeman and J. M. Jobe. *Basic Engineering Data Collection and Analysis*. Pacific Grove, CA: Duxbury Brooks/Cole, 2001.

[119] H. R. Varian. *Microeconomic Analysis*. New York: W. W. Norton and Company, Third Edition, 1992.

[120] Visual Numerics, Inc., San Ramon, CA. *IMSL C Numerical Library: User's Guide*, 2003. Available from www.vni.com.

[121] J. H. Wilkinson. *The Algebraic Eigenvalue Problem*. Oxford and New York: Oxford University Press, 1965.

[122] L. A. Wolsey. *Integer Programming*. New York: John Wiley and Sons, Inc., 1998.

[123] A. J. Wood and B. F. Wollenberg. *Power Generation, Operation, and Control*. New York: Wiley, Second Edition, 1996.

[124] S. J. Wright. *Primal Dual Interior Point Methods*. Philadelphia, PA: SIAM, 1997.

[125] F. F. Wu. Course Notes for EE215, Power Systems. Department of Electrical Engineering and Computer Sciences, University of California, Berkeley, 1988.

[126] Y.-C. Wu, A. S. Debs, and R. E. Marsten. A direct nonlinear predictor-corrector primal-dual interior point algorithm for optimal power flows. *IEEE Transactions on Power Systems*, **9**(2):876–883, 1994.

Index